INDUSTRIAL INSTRUMENTATION FUNDAMENTALS

AUSTIN E. FRIBANCE

Professor, Mechanical Department
Rochester Institute of Technology

McGRAW-HILL BOOK COMPANY

New York St. Louis San Francisco
London Toronto Sydney

DEDICATION

This book is dedicated to the memory of my father and mother who gave so much—and asked so little.

INDUSTRIAL INSTRUME N FUNDAMENTALS

Library of Congress Catalo mber: 61-14684

07-022370-X

17 18 19 KPKP 7 9 8

Preface

The increasing use of industrial process instrumentation has created the need for more people trained in the use of modern measuring instruments. They should be competent to plan, design, install, operate, and maintain complex systems which produce materials of high quality. "Industrial Instrumentation Fundamentals" has been written to assist in the training of such personnel.

In the fields of modern research there is a greater and greater requirement for precise measurements as more and more complex experiments are carried out. Students at all levels are faced with making extensive sets of measurements under widely varying conditions. For the values to be meaningful and accurate, it is absolutely essential that the student or research worker—as well as industry—observe the precautions and limitations which industry has discovered over many years with large expenditures of both money and time. Since the principles and limitations which must be observed by industry are the same as those that must be observed everywhere if the measured values are to be significant, the emphasis of this text is on commercial, or industrial, installations. Students should realize that the principles involved are the important elements and that it is of little significance whether a measurement is made for a large corporation or by a baccalaureate candidate. The principles of industrial measuring techniques are not restricted to commercial measurements; however, their commercial uses and limitations show the precautions that all must employ.

"Industrial Instrumentation Fundamentals" provides its readers with the following:

1. A review of the basic principles governing the selection and employment of materials, the strength of materials, and related subjects.
2. A study of the fundamental laws concerning forces, force balance, and torque balance, and their employment in instrument design and use.
3. A consideration of the laws of electricity and electronics (including transistors and semiconductors), which are necessary if modern equipment is to be understood and serviced.

4. A detailed consideration of the principles employed in each of the various groups of instruments.
5. A summary for each class of instruments and guides to their selection and use.
6. A comprehensive study and listing of the factors which make each type of installation successful.
7. Problems that will illustrate and require the use of the principles presented.
8. A series of questions which will serve to stimulate thought and emphasize the most important parts of the chapter.

The entire emphasis is on understanding, not memorizing, and the effort is to present a logical development and application of principles rather than specific apparatus. The considerations always are:

1. How can the measurement be made?
2. What factors will make the instrument function well?
3. What alternative methods might be used?

The material in this text has been arranged so that the basic principles used in instrumentation are thoroughly understood before their applications in the process industry are encountered. Since some of the principles governing the basic measurements of pressure, temperature, level, and flow are so different, they have been considered separately. The use of each primary component of a complex mechanism is carefully analyzed and then the proper interrelationship of each part is shown. The common parts are brought together in the subject summaries and the application notes.

ACKNOWLEDGMENTS

This volume represents the contributions of many companies and hundreds of individuals; some were involved less directly than others, but their contributions in terms of material, assistance, and encouragement are gratefully acknowledged. No one person could hope to assemble and evaluate the hundreds of man-years of experience which certain sections represent; they are the result of many years of manufacturing experience and represent the considered judgment of some of the most representative instrument manufacturers of this country.

In particular, the assistance given by the various companies and individuals listed below is gratefully acknowledged; it must also be kept in mind that many whose names do not appear here have helped in very significant ways—incomplete records of certain events prevent listing some individuals who certainly should have been recognized.

This book would never have been written without the active support and assistance of such men as George Howard, Chief Application Engineer, Henry Stoll, Manager, Flow Measurements, both of Taylor Instrument Companies, Gerald Akins, Chief Instrument Engineer, Kodak Park Works, Eastman Kodak Company, and W. H. Furry, Director of Training, Foxboro Company, who aided greatly in procurement, evaluation, and preparation of the material. With their encouragement, the task was begun. In addition, the author should acknowledge the encouragement and help of Professor Frank B. Moore, Pennsylvania State University, and Professor George H. LeCain, Head, Mechanical Department, Rochester Institute of Technology. Finally, the author is greatly indebted to Miss Betty Weatherhog for typing the manuscript.

The Bristol Company, Waterbury, Conn., D. C. Sanford, Manager, Applications Engineering Department, and K. P. Neale, C. Q. Sandt, L. E. Smith, G. VanVessen, and W. C. Virbila.

The Eastman Kodak Company, Rochester, N.Y., Gerald Akins, Chief Instrument Engineer, Kodak Park Works.

The Fischer and Porter Company, Hatboro, Pa., Robert Rice, formerly Vice President, Sales, who arranged for consultations with William Buzzard, R. H. Condell, and Frank Houpt.

Foxboro Company, Foxboro, Mass., Norman Anderson, Donald Barta.

The Hagan Chemical and Controls Co., Pittsburgh, Pa., Joseph E. Cotton, Jr., Assistant Sales Manager, Control Division, and P. E. Straight and Franklin Michaels.

The Leeds and Northrup Company, Philadelphia, Pa., Royal A. Dresher, Director of Customer Training.

Manning, Maxwell and Moore Co., Stratford, Conn., Garland Roper, Chief Engineer, Microsen Division, and Edmund R. Lehman, Manager, Application Engineering.

Mine Safety Appliances, Pittsburgh, Pa., J. P. Sherwin, Assistant Sales Manager.

Minneapolis-Honeywell Regulator Company, Philadelphia, Pa., Louis Gess, Consultant.

The Perkin-Elmer Corporation, Norwalk, Conn., Newell Claudy, Advertising Supervisor.

The Taylor Instrument Companies, Rochester, N.Y., Anthony Turner and Norman Gollin.

Tracerlabs, Inc., Waltham, Mass., Robert J. Allen, Assistant Public Relations Manager.

The Shell Oil Company, New York, N.Y., J. G. Kerley, Manufacturing Engineering Department.

Norwood Controls, Norwood, Mass., Michael Altfillisch, Manager, Technical Sales, and Messrs. Baring and Eliot.

AUSTIN E. FRIBANCE

Contents

Part I *Principles Used in Instrumentation*

Part II *Instrumentation in the Process Industry*

Part I — Principles Used in Instrumentation

1 Introduction

The modern world, of which we are a part, exists in its present form largely because we as individuals have made ourselves specialists to some degree and because vast amounts of wealth have been concentrated on the production of the things which we need. Instead of being jacks of all trades, and masters of none, we have, even in routine tasks, become extremely efficient. In this way, each of us can produce far more in a given time than could our fathers or grandfathers. In order to employ vast amounts of capital, yet keep the selling price down, it has been necessary to employ the available wealth in creating productive situations in which the efficient use of modern technologies could create goods of high quality, but at maximum rates and with a minimum of spoilage and waste.

Importance of the subject. In this study, we are very much interested in the measurements and the control of the physical conditions required for the mass production of high-quality products. Unless high output can be attained, the unit cost becomes far too high and the product cannot be sold. Unless the customer can be assured uniform quality at each purchase, the market soon is lost. If large amounts of material are produced in order to get a given amount of salable product, then the price again will soon eliminate the product.

Over the years, the physical sciences have been used to augment our sense of temperature, viscosity, pressure, and color. We are incapable of sensing, within narrow limits, differences in temperature, speed, color, or light intensity. Therefore, we have produced mechanical aides that never get tired, seldom go on strike, and that not only are extremely accurate and sensitive in their response but keep those characteristics for extended periods of time.

As individuals we cannot maintain maximum efficiency for more than a very short time. Even if our senses were sufficiently acute, which they are not, more than three people per 24-hr day would be needed to monitor each station or group of stations that one person could sense and keep under control on any one shift.

For most types of control jobs that have to be done today, people are not adequate. Even if people had adequate sensing abilities, there aren't enough people to do all of the many tasks required. We could also say truly that our industrial civilization and all its automobiles, refrigerators, textiles, shoes, and airplanes could not exist in adequate quantities without the mechanical helpers called *instruments* and *controllers*.

The selection, installation, maintenance, and operation of these systems are therefore of great importance to our entire economy. Adequate, effective instrumentation provides one way to increase quality while maintaining or reducing prices in the face of higher costs of materials, supplies, and labor.

How measurements are made. It may come as quite a surprise to learn that there are very few measurements which are made directly. We are using the word "measurement" to tell us the length, the weight, the temperature, the color, or the change in one of these characteristics of a material. Distances can be measured directly by finding how many times a foot or an inch is contained in the length, or volumes can be measured by finding how many times a quart or gallon jug can be filled. Measurements of this sort are the exception, however. Precise measurements of length, to a thousandth of an inch, are made by counting the number of turns and the fractional part of a turn made by a precision screw which has 40 threads per inch. Thus, the screw advances $\frac{1}{40}$ in. each revolution, or 0.025 in. per revolution. If we divide the circumference into 25 equal parts, each part will represent $\frac{1}{1000}$ in.

We tell time by the angular position of the hands of a watch or clock. We interpret the positions to mean that time has elapsed; a further assumption is that the hands are turning at a constant rate. Weight is often determined on scales that really measure the amount that a spring stretches or the point at which a given weight must be placed to furnish a given turning effort. The speedometer in a car indicates speed by the position of a pointer which actually shows the effect of rotating a magnet near a metallic piece which is restrained by a spring.

We "measure" the light used in taking a photograph; actually, it is the current in the meter circuit which is measured. We judge the speed of a car by the sound it makes, just as we judge the speed of the wind by its whistle.

We could go on through a long list. We have learned by long experience to interpret one result as meaning that something else has happened. We have become so skilled in certain areas that we are not in the least troubled at having to interpret changes in the volume of mercury in a thermometer as meaning that the temperature has changed.

Thus, if we want to measure almost any property or quality, we have to find out what physical changes are taking place at the same time and can be related to the characteristic in which we are interested. We must interpret everything in the field of measurements: an extension of a bellows suggests a force acting on it; changes in the length of a rod suggest variation in temperature or stress.

It is because of this dependence upon interpretation, and a need to measure properties that are related to those in which we are interested, that it becomes so important for us to know how the various common materials used in the industrial instrumentation field will change with pressure, temperature, humidity, twisting, and so forth.

Division of the subject. It is because we measure things or quantities so indirectly that the subject of instrumentation is made somewhat complex. We may use so many different types of responses to measure the same thing that the classification becomes difficult. There are two general classifications of instruments that we shall make:

1. Those employing purely mechanical means
2. Those employing primarily electromechanical techniques

Therefore, the very fundamental measurements of pressure, level, flow, and temperature are studied first for their mechanical construction and their basic principles. Then the same type of measurement will be studied for those situations in which electrical techniques may be employed. It will be necessary to employ the basic principles of physics, strength of materials, and mechanics for both general classifications.

If we are to understand thoroughly what is going on in each instrument under various conditions, we must learn:

1. The basic principles of physics, mechanics, electricity, and electronics
2. How the instruments are built, that is, what basic principles are employed in their construction and not why arm A is 2 in. long when it might have been 3 in. long
3. How the principles are related in a single device to give the wonderful performance that is possible today with modern measuring and controlling means

By the time that the first eight chapters and their related fundamentals have been mastered, almost all of the basic science which is required will have been studied. We will see how electricity and electronics can be used to extend tremendously the range and types of measurements. Chapters 16 to 23, therefore, will not be subdivided into the two general groups of electrical and nonelectrical. Such a division is made only to simplify the teaching of basic principles, with the hope that the principles will be more apparent and therefore more readily understood.

The chapters which contain a Summary and Application Notes will tie together and show in proper perspective each instrument and its relationship to all others of the same or related types. By the time that all the chapters have been mastered, we will also have found that:

1. The number of basic principles is small.
2. Most measurements employ one, two, or more simple relationships.
3. The most complex instrument is composed of a number of simple parts.
4. Emphasis has been placed on how to analyze and find the simple relationships.
5. Methodical analysis is the only way to approach a situation that calls for understanding of the equipment or repair or replacement of some defective part.

Throughout this study, if we are seeking real mastery of the basic principles and their applications in the field of industrial measurements, we must ask ourselves repeatedly: "How does it do this, and why does it do it?" This text is not to teach you what nut or bolt must be tightened; rather the purpose of the text is to show you why there is such a piece and what it probably will be doing in related groups of instruments, not just that of a specific make. Basic principles, what they are and how they are applied, are of the greatest concern in this book, no matter what the chapter heading may be.

Because of the extreme complexity of the field of automatic control, it is possible within reasonable limits of cost and space to consider here only the portion that is termed industrial measurement.

2 Understanding the Instrument Application

Let us assume that you have just reported for work. It is a new job with a company for which you have not previously worked. After the foreman has left you with one of the older employees, your co-worker tells you to go to the stockroom for a thermometer. Soon you are back for more information, because all the thermometers given in Table 2·1 are on hand. You are now told that the temperature to be measured is 600°F. But the stockkeeper now wants additional information; he asks you what the *span* must be. When you look puzzled, he explains that *span* refers to the number of units the instrument indicates; *range* indicates the maximum value. If a thermometer scale extends from 600 to 1000°, the span is 400° and the range is 1000°. After you get the answer to his question about span, he wants to know whether you want an averaging instrument or a hot-spot temperature. If it were a long hot walk each time back to your co-worker in order to get all this additional information, you would soon appreciate that the selection of an instrument may require vastly more background information than the range and span and the difference between average and hot-spot measurements. For example, some instruments have a fast response; others are sluggish.

Table 2·1 Types of Thermometers

Type	Temp range, °F	Temp span, °F
Mercury actuated	800	200 to 400
Hydrocarbon filled	600	200
Gas filled	1000	400
Vapor pressure	300	100
Resistance	−300 to +1000	50
Bimetallic	−100 to +1000	300

The process. As we will use the term, the *process* refers to the operations which must be carried out to create the desired end product. It may be as simple as the mixing of two liquids or the heating of a liquid for a given period of time. Or it may refer to a complex sequence in converting crude oil into gasoline or some related product. Again, it might involve the measurement of film density, or mass spectroscopy, or infrared absorption.

It should be evident that selecting an instrument for a process or application which we do not know intimately may be just as effective as being told to hire a truck without first knowing what the truck must haul. Coal and iron ore certainly will need trucks which are markedly different from those used to carry a load of feathers or empty cardboard boxes. In each case, the selection depends upon what is to be done.

By this time, it should be obvious that the primary requirement for successful application of measuring and control devices is that we know the process in all its details: we must find out just what raw materials are used, when they are mixed, for how long, and at what temperature. We must find out what temperature, pressure, or flow variations have an undesirable effect upon the final or intermediate products.

Generally, the chemical engineer is the one who establishes the original conditions under which the reactions are to take place. He should know the answers to the questions raised in the preceding paragraph, or he should have approximate answers. It may be that he has set up a process that cannot be controlled by existing equipment. It is only when the process engineer and the instrument engineer have fully investigated the process requirements and the availability of instruments that we can tell reasonably well whether or not the process is practicable. This involves a frank, realistic study of all items. It may be that the process can be improved by a different sequence or by the use of less critical materials.

So before we undertake a study of the instruments which will be used, let us list some of the things that we need to know about a process or application before we can recommend the use of one instrument or another.

1. What are the maximum temperatures, pressures, levels, rates of flow, and so on?

2. What are the usual temperatures, pressures, levels, rates of flow, and so on?

3. What spans are desired for each of the measurements in (1) and (2)?

4. What safety precautions must be observed? (For example, mercury cannot be used safely in photographic film manufacture and electricity must be used cautiously in explosive atmospheres of gases or dusts.)

5. Are high speeds of response an important factor? (If the reaction is at the rate of many degrees per second rise or fall, the sensing element may have a large dynamic error.)

6. Is the process continuous or batch?
7. What are the most critical process requirements?

   a. The temperatures must be controlled within what limits?
   b. Within what limits must flow be controlled?
   c. What are the tolerances for ratio mixing?
   d. Is the purity of a product to be determined by the boiling point of the product?

8. Must the process be continuous once it has started? For how long?

9. Can the process be shut down for instrumentation repairs or replacement?

10. What is the probable value of the material in process that would have to be scrapped in case of a severe instrument failure?

11. What are the normal flow variations?

12. Will the process affect the sensing elements? (If films of gas, liquids, or heavy viscous materials cover a thermometer well, or even the bulb itself, the response will be greatly delayed and control may be very difficult. Therefore, thought must be given to keeping thermometer bulbs and wells in proper condition. A thermometer well is a tube used to protect the sensing bulb from corrosion or pressure or to permit change of instruments without shutting down the process. Whenever a well is used, a dynamic error is introduced, because the response time is increased by the poorer heat path between the bulb and the fluid which it is to measure.)

The questions raised here are of the type that must be asked about any instrumentation measurement. In addition, there are several other questions which should be asked:

1. How many ways are there for making the desired measurement? What are the probable costs?

2. What are the most promising ways of making the measurement? Are they compatible with existing equipment?

3. Can the equipment be serviced and maintained?

Number of instruments available. There is such a variety of available instruments in terms of principle of operation, range, and application that it may be well to note what just one manufacturer offers:

Temperature Measurement

1. Liquid-filled thermometers (Class IA)
2. Liquid-filled thermometers (Class IB)
3. Vapor-pressure thermometers (Class II)
4. Gas-filled thermometers (Class IIIB)
5. Electrical-resistance thermometers—five types
6. Thermocouples—four types

Pressure Measurements

7. Mechanical pressure gages—nine quite different types
8. Pressure transmitters, pneumatic
9. Pressure transmitters, electric

Others

10. Flowmeters—three basically different types with six variations of one type
11. Pneumatic differential pressure transmitter
12. Electrical differential pressure transmitter
13. Pneumatic differential vapor-pressure transmitter
14. Primary flow devices—venturi, orifice plate, flow nozzle, pitot tube
15. Automatic batch metering
16. Liquid-level determination—five different ways
17. Liquid-density determination
18. Humidity determination
19. Dew point determination
20. Radiation devices for thickness and weight measurement
21. Moisture control for paper
22. Position and motion devices, electric and pneumatic
23. pH measurements
24. Oxidation-reduction measurements
25. Solution conductivity
26. Electric strain gages
27. Gas analysis

If only one manufacturer offers such an array of instruments, the total number of instruments on the market is tremendous. Therefore, in order to make a wise choice, the suitability of any given device must be determined from a consideration of the operating conditions and the measurements that must be made.

With such a variety to choose from and know about, the best that we can hope to do is understand the basic principles upon which the various devices are based. The number of principles is quite small, so the task should not be difficult.

3 Basic Behavior of Materials

In the field of instrumentation and process control, we find it imperative to know how materials in their many forms act or react when heated or cooled, stretched or compressed, twisted or bent. We also have to know some of the less obvious characteristics which are changed by these external situations; for example, are the electrical characteristics of the materials altered, do the materials act as tiny generators?

It should be recognized at the start that the metallurgists have done a magnificent job in making so many metals whose behavior is so predictable. As more and more is learned of the fundamental nature of materials, even higher degrees of accuracy, greater permanence of calibration, and simpler adjustments can be hoped for. However, at the present time, we know very little, relatively, about the composition of matter. Most of what we know is based on experience rather than theoretical studies.

However, we now have every reason to believe that all matter, which includes all materials without regard for size, shape, or use, is made up of certain basic building blocks. Occurring naturally are 92 *elements* which are found in various percentages in everything that we use, be it food, water, or clothing.

The smallest particle of matter occurring naturally that can be recognized as having physical characteristics and properties different from any other material is the *molecule*. The molecule is made up of still smaller parts called *atoms,* which for any given *element* are always alike. To get a better concept of what happens in the molecules and atoms, we need to be able to picture them. The atom is often compared to the solar system because it has a nucleus, or center, about which some extremely small particles, called *electrons,* appear to revolve. The atoms of the different elements have centers of different mass and electric charge, while the electrons arrange themselves in a more or less orderly fashion in rings (orbits) which are at different distances from the center. If the center, or nucleus, has but one charge and there is but one electron moving around it, then the material is hydrogen, a very active gas capable of diffusing through steel. But if there are two electrons circling about the nucleus, then the material is the inert gas helium. If one more electron is added to this group, it will start a second orbit farther from the center, and the properties of the element will be totally unlike the properties of helium. As more and more electrons are added, all of the elements known to man are formed.

At this point, the question may be raised: "What does this have to do with industrial measurement?" It is at the heart of it, because we must know something of the composition and structural behavior of the materials before we can hope to understand, and interpret correctly, what is happening to them.

Long investigation has shown that when the "electronic moons" move around the nucleus, they will move farther from the center, and thus take up more room, when heat energy is added to them. Conversely, when heat energy is removed (they become colder), they move nearer the center. Picture, if you will, a carousel with long ropes hanging from the top of a rotating mast—this corresponds to the nucleus—and attach cars (electrons) to each rope. Now start to rotate the mast. At low speeds (at which there is little energy of motion) the space taken up is small, but rotate the mast at a higher speed and the cars will move out farther from the center. Energy was needed to get the cars moving fast enough so that they would have a larger orbit. Remove energy from this rotating system by applying a brake and, as the speed of rotation becomes less, the cars come nearer to the center. This is very nearly what happens to the electrons on their paths. With the addition of more heat energy (higher temperatures), the electrons take up more space because each nucleus and its moons need more space to hold them. As the electrons lose energy (heat is lost or removed when the temperature is reduced), the nucleus and its moons occupy smaller amounts of space.

Another analogy might be to think of 100 or 1,000 people grouped quite closely together when the temperature is 70°F. They take up a certain amount of space. Now let the temperature rise to 100°F. Each person now will probably want more room to allow any breeze to cool him. In any event, he will want more space under the new conditions. Or let the temperature drop toward the freezing point; the people will huddle together for warmth. They take up more space when they are warm and less when they are cold. The very important principle to remember is that when heat energy is added to materials, almost every one of them will become longer, wider, and thicker.

Lineal expansion. This is one of the most important principles used in thermometry, the measurement of temperature. There are many

Table 3·1 Average Coefficients of Lineal Expansion

Material	Coefficient of expansion, per °C	Coefficient of expansion, per °F
Aluminum	0.000022	0.000012
Brass	0.000019	0.000011
Copper	0.000017	0.0000094
Glass, ordinary	0.0000095	0.0000053
Glass, pyrex	0.0000036	0.0000020
Invar (nickel-steel alloy)	0.0000009	0.0000005
Iron	0.000012	0.0000067
Steel	0.000011	0.0000061

measurements of temperature which depend entirely upon a change in dimension to provide the clue which will tell how much of a change in temperature must have taken place. The change of dimension with temperature is well known for each of the common materials. Most of the changes are quite small for temperature changes of 5 to 10°. Table 3·1 lists the coefficients for some of the more common materials used in industrial instrumentation. A coefficient of lineal expansion gives the fractional change in length for a temperature change of 1°.

Coefficients of expansion are used to determine what change in dimension should occur for any given temperature change. It should be remembered that the values given in Table 3·1 are "average" for the range specified. (For problem solving, it is assumed that the values given in the table hold.) But a close study of the materials listed in the table would reveal that they do not always behave in exactly the same way for given temperature changes. In some regions they expand at a faster than average rate; in others, they respond more slowly. This has particular significance for the instrument designer and the instrument man; they must know that there are nonlinear responses, that is, that the response does not always vary directly, or exactly, with the temperature change. The measurements taken from the dial of a bimetal thermometer show how the arc distance for each 100° will vary over the scale.

It is quite probable that the dials for two bimetal thermometers of the same style, size, and temperature range would not be interchangeable if the thermometers had been made not at the same time, but, for example, five years apart. The reason probably would lie in the fact that the bimetals used to actuate the thermometers were not from the same batch of material and hence were different in their characteristics. A dial to go with a thermometer produced at still another time probably would be different from either of the first two.

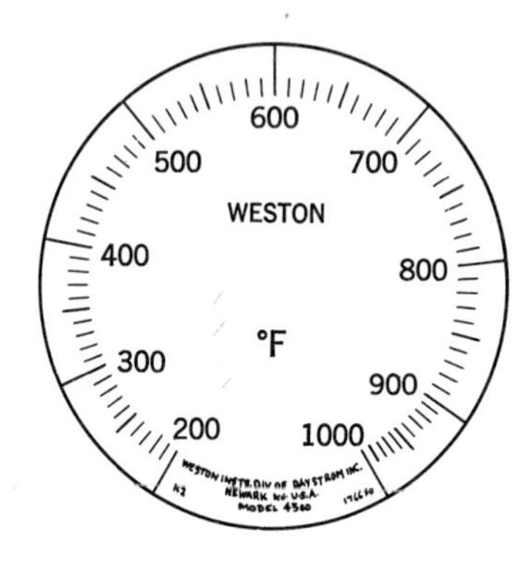

Fig. 3·1 The nonlinear expansion rates of the materials in the sensing element create unequal scale divisions. *(Weston Electrical Instrument Corp.)*

Table 3·1 tells us which materials have the highest rate of expansion and enables us to calculate the change in dimension or volume for a given temperature change. The coefficient for volume change is approximately 3 times that for linear change.

EXAMPLE 3·1 Determine the change in length of a copper rod 10 in. long when it is heated from 100 to 1100°F.

SOLUTION. From the table we find that copper has an average expansion coefficient of 0.0000094 for each degree. Coefficients of expansion are usually expressed in *millionths,* or more precisely by the use of a *standard form* number: a number multiplied by some power of 10. For example, 0.000006 in./(in.)(°F) may be written as 6/1,000,000, or as $6/10^6$ in./(in.)(°F). If numerator and denominator are both multiplied by 10^{-6}, the fraction becomes

$$\frac{6(10^{-6})}{10^6(10^{-6})} = \frac{6(10^{-6})}{10^{\circ}} = \frac{6(10^{-6})}{1} = 6(10^{-6})$$

In this case, if we multiply the original length of 10 in. by the change per degree for each inch of length, we find that 10 in. has been increased by

$$(0.0000094)(10)(1100 - 100) = 0.094 \text{ in.}$$

If the temperature rise had been 500°F, the change in length would have been about half as much. Thus, we can generalize that the change in length (or other dimension) varies directly, but not linearly, with the change in temperature. A change of 0.03 in. or 0.06 in. would be enough to move a pointer, close a contact, or do many other things that should be done.

EXAMPLE 3·2 Each of two pieces of metal, one iron and one brass, has a length of 10 in. at 70°F. What will be the lengths at 1070°F?

SOLUTION. For the iron, the length is

$$10 \text{ in.} + (10 \text{ in.})(1070°\text{F} - 70°\text{F})(0.0000067 \text{ in./(in.)(°F)}) = 10.067 \text{ in.}$$

For the brass, the length is

$$10 \text{ in.} + (10 \text{ in.})(1070°\text{F} - 70°\text{F})(0.000011 \text{ in./(in.)(°F)}) = 10.110 \text{ in.}$$

Table 3·2 Cubical Expansion of Liquids *

Material	Temp span, deg	Coefficient of expansion, per °C
Ethyl alcohol, 99%	27–46	$1.12(10^{-3})$
Benzene	11–81	$1.237(10^{-3})$
Carbon tetrachloride	0–76	$1.236(10^{-3})$
Mercury	0–100	$0.181(10^{-3})$
Petroleum, density 0.846	24–120	$0.955(10^{-3})$
Water	0–33	$0.207(10^{-3})$

* When values in different tables do not agree, it is probably due to taking the average over a different span.

From Example 3·2 we see that brass and iron have quite an appreciable difference in change of length with change in temperature. If we had drilled and securely fastened the two pieces of metal together, one would therefore try to pull the other. The one which had the greater length would be pulled into an arc by the more slowly expanding material. The brass, having the greater rate of expansion, would be on the outside, where the length would be greater. In this simple way, we would have made a bimetal "thermometer," one whose amount of bending depends upon the temperature. Examples 3·1 and 3·2 show how temperature changes can be computed and indicate how small may be the dimensional changes that are used in temperature-sensing devices.

Volumetric changes. Table 3·2 differs from Table 3·1 in that it contains a list of liquids and their expansion coefficients. It will be noted that these coefficients are for changes in volume and not for linear, or one-dimensional, changes. It should be noted that they are approximately 3 times the linear coefficients.

Liquids undergo the same sort of change with increases in temperature that metals undergo; that is, they take up more space when the atoms in their molecules have larger orbits. This change in volume is the basis for the ordinary mercury-glass thermometer. As the mercury increases in volume, it takes up available space in the calibrated portion of the tube.

But if the mercury were in a closed, sealed system, the attempt of the mercury to expand would be resisted by the housing. This would create a high pressure which could be transmitted to other parts of a system to cause pointers to move or bellows to stretch. This is another fundamental property that we shall investigate in more detail later.

Our physical world consists of solids, liquids, and gases. Their basic chemical composition may be the same. But with solids, the holding force between molecules gives the material a fixed, definite shape. Liquids, on the other hand, are less tightly bound, and they shape themselves to their containers. But gases, which have only weak attractive forces between molecules, move about so as to fill completely the enclosed spaces which confine them. Sometimes the addition of heat can cause a solid, such as ice, to become a liquid, and then the addition of more heat converts it into a gas. The heat energy gives the gas molecules the ability to move with speeds that are related to the temperature. As these molecules pursue their random movements, they hit the walls of the container. The hotter they are, the harder they hit. We can easily prove this by connecting a pressure gage to the container. If we heat the container and the gas in it, we shall find a definite relationship between the temperature and the pressure. This relationship is one of the best we have for showing how well one characteristic or property—in this case the pressure created by the activity of the gas molecules—is related to another—that is, a change in temperature. The gas thermometer is one of our standard ways for determining temperature.

In the study of physics, the relationship between pressure and temperature for a constant volume is known as Charles's law, which states that, when the volume of a gas is kept constant, the pressure of a gas changes linearly with the absolute temperature.

At this point we shall have to digress and consider the more common and pertinent temperature scales. To understand what follows, we must learn just how the scales differ and what their values signify.

Temperature scales. The common Fahrenheit temperature scale is based on the freezing and boiling points of water: 32° and 212°. The centigrade system is also based on the freezing and boiling points of

Fig. 3·2 Absolute pressure versus temperature; (*a*) centigrade scale, (*b*) absolute scale. Note that the slopes of the lines are the same but are displaced 273°.

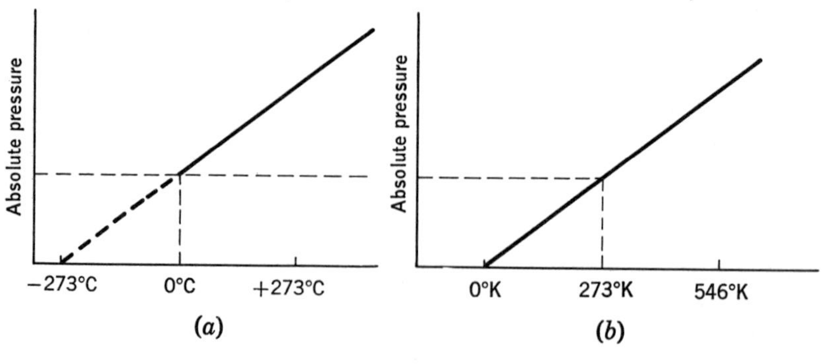

water but has them as 0° and 100°. In either case the measurements are made under standard conditions of pressure. The scales used are purely artificial and arbitrary; any other pair of numbers might have been used, inasmuch as there is no particular significance in either pair.

Experiments with gases showed that when the absolute gas pressures were related to the temperatures, the graphs reproduced in Fig. 3·2 were obtained. The graphs indicate that *if* the gas pressures decreased in a regular manner, the pressure caused by the movement of the molecules should become zero at a temperature of −273°C. In many ways, it is easier to work with scales which do not have negative signs. For this reason, therefore, a new scale was devised; it started at the point at which thermal molecular motion begins. It will be noted that the *absolute temperature* is found by adding 273° to the centigrade value. This particular absolute temperature scale is named for one of the famous British scientists, Lord Kelvin. Temperatures so designated are specified as degrees Kelvin or °K.

Figure 3·3*a* shows the relationship between absolute pressures and temperature when the Fahrenheit scale is used. This indicates that the pressure caused by the thermal activity of the molecules would cease at −459.6° if nothing altered the linear change.

Again, it is easier to use a positive scale which starts at zero (Fig. 3·3*b*). This scale is called the *Rankine* scale in honor of another noted British scientist. The absolute-temperature scales are used because there is a linear relationship between temperature and the total pressure while we are dealing with gases. *All gas laws, and their various relationships, are based upon the use of one or the other of the absolute-temperature scales.*

We shall also develop the relationship between the Fahrenheit and the centigrade systems, the more usual commercial and industrial scales, Fig. 3·4. The Fahrenheit scale has 180 units between the boiling point and the freezing point of water, while the centigrade scale has but 100

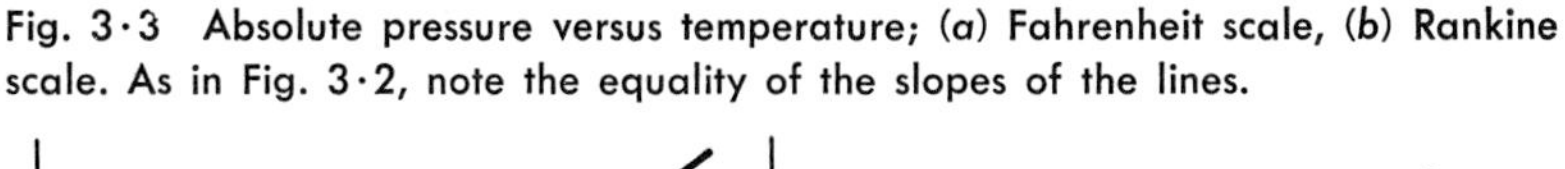
Fig. 3·3 Absolute pressure versus temperature; (*a*) Fahrenheit scale, (*b*) Rankine scale. As in Fig. 3·2, note the equality of the slopes of the lines.

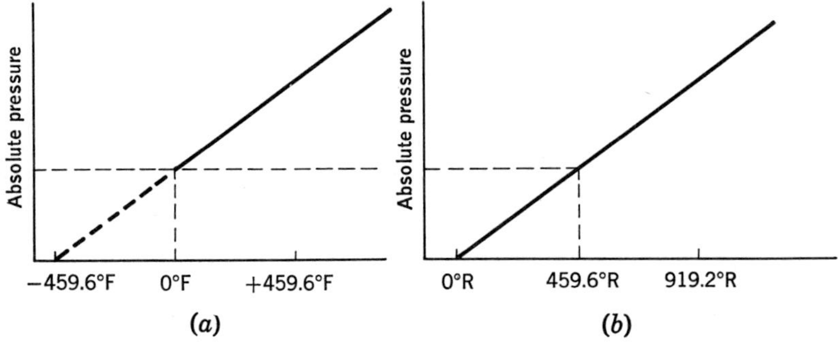

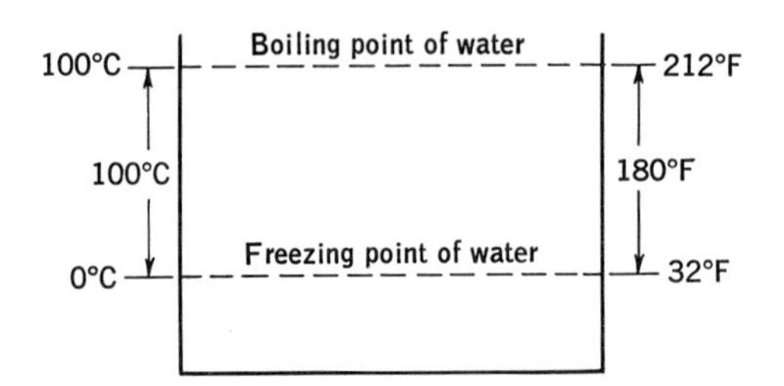

Fig. 3·4 A comparison of the centigrade and Fahrenheit scales shows that 100° of the former equals 180° on the latter.

units to measure the same temperature range. Therefore, 180 Fahrenheit units equal 100 centigrade units, or $1°F = {}^{100}\!/_{180}°C = {}^{5}\!/_{9}°C$. Also, $1°C = {}^{9}\!/_{5}°F$. Because the Fahrenheit scale starts at 32° instead of 0°, we have to subtract 32 from each reading before multiplying by the factor of $^{5}\!/_{9}$.

Thus 150°F is converted to the centigrade scale as follows:

$$150 - 32 = 118°F$$

$$^{5}\!/_{9}(118°) = 65.5°C$$

Alternatively,

$$^{5}\!/_{9}(150 - 32) = 65.5°C$$

When converting from degrees Fahrenheit to degrees centigrade, the number will decrease to slightly more than half of its original value.

To convert from the centigrade to the Fahrenheit scale, the procedure is reversed. For example, to change 65.5°C to degrees Fahrenheit, we write

$$^{9}\!/_{5}(65.5) + 32 = 118 + 32 = 150°F$$

When converting from degrees centigrade to degrees Fahrenheit, the number will almost double in value in most instances. This will provide a quick approximate check.

Gas-pressure measurements. There is another term that we must learn to use correctly: *absolute pressure.* We have previously used the term "absolute temperature" to tell us that the temperature scale starts from the point at which the thermal activity of the atoms is zero. In this case, we want to start our pressure scale from the point of zero pressure.

Because of its familiarity, we generally forget that we live at the bottom of an ocean of air which pushes down on the surface of the earth just as water pushes down on the bottom of the ocean. To get to the point of zero pressure, we have to rise to the top of the ocean, perhaps 200 to 300 miles above the earth. At this point, where there is no air, there is no pressure because there are almost no molecules to hit upon the sensing device. But at the bottom of the sea of air in which we live, the molecules are forced close together by the squeezing effect of the weight of all the molecules above them. There they exist in such numbers that they normally exert a force of 14.7 psi under standard conditions (sea level and 20°C).

If our pressure-measuring gage reads pressure above atmosphere, the gage reads *zero* when air pressure is free to act on it (we are also assuming that it is sensitive enough to show a pressure of 14 psi). To get the true or total pressure, we must correct all gage pressure readings by adding 14.7 psi to the gage pressure. *Gage pressure* is always understood to be pressure above atmospheric.

EXAMPLE 3·3 A tire gage reads 30 psi. Find the absolute pressure in the tire.

SOLUTION

$$\text{Absolute pressure} = \text{gage pressure} + 14.7 \text{ psi}$$
$$= 30 + 14.7 = 44.7 \text{ psia (pounds per square inch absolute)}$$

EXAMPLE 3·4 If the absolute pressure is 100 psi, what will the usual type of pressure gage read?

SOLUTION

$$\text{Absolute pressure} - 14.7 \text{ psi} = \text{gage pressure}$$
$$100 - 14.7 = 85.3 \text{ psig (pounds per square inch gage)}$$

Oftentimes there are processes or conditions such that both gas temperatures and gas pressures change. Here we must consider the effect upon pressure of (1) changing the thermal activity and thus the gas pressure and (2) changing the gas pressure by putting more or fewer molecules into a given space. If instead of changing the volume occupied by a gas, we leave it fixed in value, then the pressure developed will depend upon the force with which the gas molecules strike the walls of the box or tank. They will strike harder and thus create greater pressures if they can be made to travel faster; they will create smaller pressures if they strike the walls at lower speeds.

Now the energy of the molecules, and consequently their speed of travel, depends on the amount of *heat* which they possess. But the amount of heat can be measured by the absolute temperatures of the gases. Consequently,

$$\text{Pressure} \propto \text{heat energy}$$

$$\text{Heat energy} = k \text{ (absolute temperature)}$$

Therefore by substitution,

$$\text{Pressure} = k'T_{\text{ab}}$$

EXAMPLE 3·5 If the gage pressure of air in a truck tire is 100 psi when the temperature is 100°F, what will it be when the temperature is 200°F?

SOLUTION

$$P_{1\,\text{ab}} = k'T_{1\,\text{ab}}$$

$$P_{2\,\text{ab}} = k'T_{2\,\text{ab}}$$

Then dividing $P_{1\,\text{ab}}$ by $P_{2\,\text{ab}}$, we get

$$\frac{P_{1\,\text{ab}}}{P_{2\,\text{ab}}} = \frac{k'T_{1\,\text{ab}}}{k'T_{2\,\text{ab}}}$$

$$\frac{100 + 14.7}{P_{2\,\text{ab}}} = \frac{k'(100 + 459)}{k'(200 + 459)}$$

$$\frac{114.7}{P_{2\,\text{ab}}} = \frac{559}{659} = 0.85$$

$$P_{2\,\text{ab}} = \frac{114.7}{0.85} = 135 \text{ psi}$$

$$P_{2\,\text{gage}} = 135 - 14.7 = 120.3 \text{ psi}$$

EXAMPLE 3·6 An automobile tire had a gage pressure of 30 psi at 0°C, but after two hours of running at high speed while overloaded, it had reached a temperature of 100°C. What was the pressure in the tire when it was hot?

SOLUTION

$$\frac{P_{1\,\text{ab}}}{P_{2\,\text{ab}}} = \frac{k'T_{1\,\text{ab}}}{k'T_{2\,\text{ab}}}$$

$$\frac{30 + 14.7}{P_{2\,\text{ab}}} = \frac{k'(0 + 273)}{k'(100 + 273)}$$

$$\frac{44.7}{P_{2\,\text{ab}}} = \frac{273}{373} = 0.733$$

$$P_{2\,\text{ab}} = \frac{44.7}{0.733} = 61.1 \text{ psia}$$

$$P_{2\,\text{gage}} = P_{2\,\text{ab}} - 14.7 = 61.1 - 14.7 = 46.4 \text{ psig}$$

The relationship between pressure and temperature, *for constant volume conditions,* is known as Charles's law. It is generally written in the form

$$\frac{P_{1\,\text{ab}}}{P_{2\,\text{ab}}} = \frac{T_{1\,\text{ab}}}{T_{2\,\text{ab}}} \tag{3·1}$$

Boyle's law. Boyle's law states that when the temperature is kept constant, the pressure of a gas varies inversely with the volume. Place a given amount of gas in a larger container, or allow it to expand into a larger container, and its pressure will drop. Conversely, crowd the gas into a smaller space and there will be more molecules to hit the walls in any given period of time, so the pressure will rise. Stated another way:

$$\frac{\text{Original pressure}}{\text{New pressure}} = \frac{\text{new volume}}{\text{original volume}}$$

This may also be written:

$$(\text{Original pressure})(\text{original volume}) = (\text{new pressure})(\text{new volume}) \qquad (3\cdot 2)$$

EXAMPLE 3·7 A container of 30 ft^3 is filled with hydrogen gas. A common pressure gage reads 100 psi. What will the pressure be if the gas is compressed to fill a 10-ft^3 container?

SOLUTION. If we use Eq. (3·2), we get

$$\begin{aligned} (100 \text{ psi} + 14.7 \text{ psi})(30 \text{ ft}^3) &= x(10 \text{ ft}^3) \\ 3441 \text{ (psia)(ft}^3) &= 10x \text{ ft}^3 \\ x &= 344.1 \text{ psia} \\ &= 344.1 - 14.7 = 329.4 \text{ psig} \end{aligned}$$

It will be noted that *the temperatures and pressures used in all gas volume-pressure relationships are absolute.*

If we combine the effects of temperature and volume change, we can state them mathematically:

$$\frac{(\text{Original pressure})(\text{original volume})}{\text{Original temperature}} = \frac{(\text{new pressure})(\text{new volume})}{\text{new temperature}} \qquad (3\cdot 3)$$

EXAMPLE 3·8 If a container holding 100 ft^3 of hydrogen with a gage pressure of 85.3 psi at 27°C were allowed to expand to 1,000 ft^3 at 127°C, what would be the new pressure?

SOLUTION. Rewrite Eq. (3·3) as

$$\frac{P_1V_1}{T_1} = \frac{P_2V_2}{T_2}$$

Then substitute:

$$\frac{(85.3 + 14.7)(100)}{27 + 273} = \frac{(P_2)(1000)}{127 + 273}$$

$$P_2 = 13.3 \text{ psia}$$

Gases expand to fill any space. The answer in Example 3·8 tells us that the gas expands so much that the pressure the molecules can exert is less than they would exert under conditions of standard temperature and pressure. If we subtract 14.7 from 13.3 we get −1.4 psi, which tells us that we have a *vacuum,* or a pressure less than atmospheric. Such conditions are found many times.

Resistance changes with temperature. Another characteristic of materials which is used many times in industrial instrumentation is the change in *resistance* of a material because of either a change in temperature or a change in length brought about by some mechanical means. To

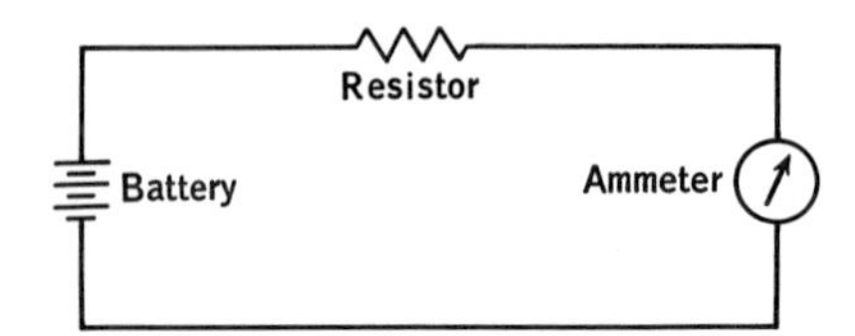

Fig. 3·5 A simple schematic representation of a series circuit, in which the current passes first through one component and then through another.

understand what is happening, we have to make another digression to consider a simple electric circuit.

One of the simplest circuits includes a battery, some wire such as copper wire, some resistance wire made possibly of nickel-silver, and an ammeter, Fig. 3·5. The battery acts as a pump pushing a current, which is measured by the meter, through the wire. The size of the current depends upon the difficulty (called the *resistance*) with which the electrons can be sent along the wire. The relationship of the pressure, current, and resistance was stated by the German scientist, Ohm, who found that

$$\text{Current, or quantity of electrons} = \frac{\text{voltage, or pressure}}{\text{resistance}}$$

These same relationships will be used to explain the behavior of valves, piping, and so on. That is, the expression is often used for relationships which do not involve electrical quantities. Expressed in symbols, the statement of Ohm's law is

$$I\text{, or current} = \frac{E\text{, or voltage}}{R\text{, or resistance}}$$

or

$$I\text{, amp} = \frac{E\text{, volts}}{R\text{, ohms}}$$

Three people are honored in this expression, because all the units are named for scientists: volts for an Italian, Volta; ohms for a German, Ohm; and amperes for a Frenchman, Ampère.

If we know any two of the three quantities I, E, and R, we can find the third. If R can be made to change for any reason, the current indicated by the meter will also change, but in the opposite direction; that is, making the value of R smaller will cause the current to increase.

Example 3·9 If a 6-volt battery and a 36-ohm resistance are connected in series—which means that the current goes first through one and then through the other, or that there is but one path for the current—what current will be indicated by the meter? (See Fig. 3·5.)

SOLUTION

$$I\text{, amp} = \frac{E\text{, volts}}{R\text{, ohms}}$$

$$I = \frac{6 \text{ volts}}{36 \text{ ohms}} = \tfrac{1}{6}\text{, or 0.16, amp}$$

If the resistance is reduced to 18 ohms,

$$I = \frac{6 \text{ volts}}{18 \text{ ohms}} = \tfrac{1}{3}\text{, or 0.33, amp}$$

We are particularly interested in the fact that changes in resistance are brought about by changes in temperature. Almost every pure metal will have its resistance increase as its temperature rises, and because this change takes place in a very definite pattern, we can calculate or determine what the value of the resistance should be at any given temperature.

Certain other materials, often called semiconductors because their values of resistance greatly exceed those for copper, silver, and aluminum, have their resistance decrease with temperature. Not only do the resistances decrease, but they change at rates which are many times those for the more common metals. Table 3·3 and Figs. 3·6 and 3·7 show how widely these values range.

Table 3·3 Thermistor Thermometers

Material:	A	B *	C	D
Temp, °C	Nominal resistance, ohms			
−25	.....	......	87,500	610,000
0	5,000	325,000	37,500	153,000
25	2,000	100,000	18,000	48,500
50	900	33,000	9,700	17,300
75	460	13,000	5,500	7,100
100	250	6,000	3,700	3,400
150	95	1,600	.....	870
200	.....	500		
300	.....	80		

* These data are used in Fig. 3·7.

SOURCE: From Nomograph B 1443, Bell System Technical Publication, Becker, Green, and Pearson.

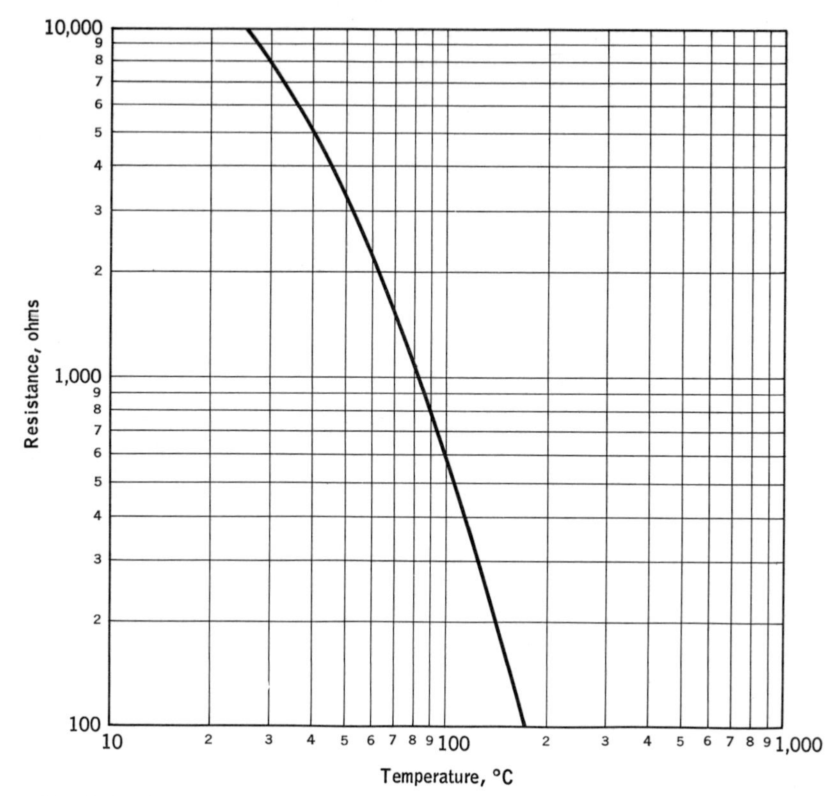

Fig. 3·6 Resistance versus temperature for a typical thermistor. (*From Bell Telephone Laboratories Monograph B 1443, Becker, Green, and Pearson.*)

It is also true that if we change the size of a wire, we also change the resistance of the wire. If the cross-sectional area of a wire becomes smaller, the current has a less favorable path and decreases in value; therefore, the resistance must have been increased. If the diameter of the wire could be increased in some manner, the wire would become a better path and the current would increase. We would then have to conclude that the resistance had decreased. Certain types of equipment used to weigh materials or measure loads or forces, as well as the rapidly changing pressures in engine cylinders, often use this principle. The use of the changing resistance of a wire will be studied in much more detail in Chaps. 10 and 12.

Thermoelectric effects. Another unusual property of materials (because much less frequently encountered in everyday life) is that, if we take two unlike metals such as copper and iron, twist an end of each around the other, and then weld the junction, we have made a very simple generator of electricity which operates when the twisted, welded junction is heated. We might be quite astonished to learn that if a current from some external source is passed through the junction, the junction becomes *cold,* thus creating a refrigerating effect. It may be that by the time you study this, refrigerators of this general type will be commercially available. The principle is not new; it was discovered by a Frenchman over 100 years ago. But it required improved materials to make it practical. Another use for the heated junction, or *thermocouple,* as we shall call it, is to provide electric power for small radio sets where commercial power is not available. A campfire can both warm and entertain you.

In Chap. 15 we shall spend a great deal of time and effort investigating the many ways in which thermocouples are employed to indicate temperature. The thermocouple is one of the best devices for measuring temperatures between 1000 and 3000°F.

The effect of pressure. Among other things, we must define or state more precisely than ever just what we mean or want to express. This will involve using certain words in ways that are quite contrary to the normal ones. For example, ask almost anyone what an "elastic material" is, and the answer will probably be "rubber." The engineer and the scientist will not accept this as the correct answer because they use the term "elastic," as we must, to mean quite a different thing. To them, an *elastic material* is one which returns almost exactly to its original shape or position whenever any force which may have caused it to bend, stretch, compress, or so on is removed. Rubberlike objects generally do not quite return to their original shapes or positions.

Because we want instruments to perform always in the same manner under the same conditions, we want to use only materials which are as

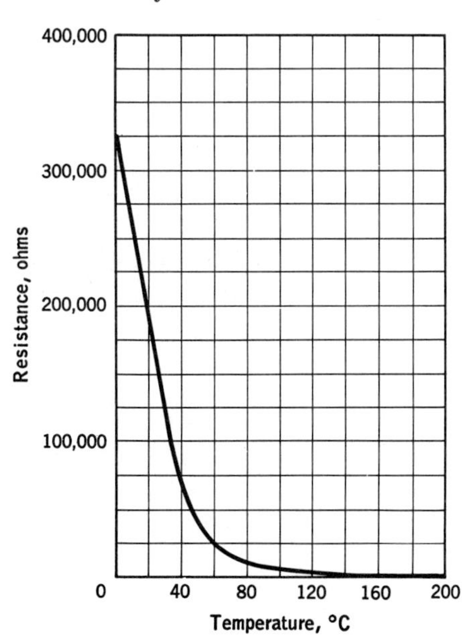

Fig. 3·7 Resistance versus temperature for a thermistor. The material is different from that represented in Fig. 3·6. Note the change in the shapes of the two curves; one is plotted logarithmically, and the other is plotted linearly. (*From Bell Telephone Laboratories Monograph B 1443, Becker, Green, and Pearson.*)

nearly perfectly elastic as possible. Steel, copper, bronze, glass, and brass are common examples of elastic materials which we use because:

1. They are able to return to their original shape after hundreds of thousands of movements under applied force.

2. The amount they bend, compress, or otherwise change physically is so nearly proportional to the applied force.

In Chap. 9 we shall learn more about the subject, but a most important principle to remember is that we can never apply any force, no matter how small, to any object without having that object change its shape or dimension in some way. The changes may be so small that we cannot detect them with our usual measuring tools, but changes are there, nevertheless. Rest a forefinger on a 6-in. shaft which is supported at two points five or six feet apart. The weight of your finger will bend the shaft. Of course, it won't be much, and you surely won't be able to see it directly. But apparatus using light waves as units of measure would readily show the amount of the deflection. This we must always remember: *forces, large or small, always cause a material to bend, stretch, compress, or otherwise alter its shape.* A spring is a common example of an elastic material which is deformed by an amount that is proportional to the load applied. This is the principle employed in simple scales.

If a material always returns to its original shape when a force on it is removed, we say that it has been stressed less than its *elastic limit.* If it does not return to its original shape, it has been stressed beyond its elastic limit. In almost all instrumentation work, extreme care is taken to ensure that the apparatus is not stressed too highly. Within the proper values we can predict and get the correct operation tens of thousands of times; but if we stress the material too highly, our predictions about performance will be quite in error and thoroughly undependable. Changes in dimension with the application of force will be found to be the heart of the operation of the many pressure-measuring devices, be they Bourdon tubes, bellows, or resistance units.

Twisting, or putting a material in *torsion* or *shear,* often occurs in instruments. The same principles apply here as in connection with tension or compression forces. Within limits, materials that are stressed by torque within their elastic limits will return to their original shapes when the torque is removed. Metal tubes are used as springs in certain instruments; the amount of twisting is directly proportional to the applied torque. In certain kinds of pressure gages, the tube functions both as a spring and as a flexible sealing member.

Changes in materials because of light. There are many materials whose resistance depends upon the intensity or color of the light falling on them. Still other materials function as tiny electric generators when light falls on them. Both types of material are used to measure and control. *Light meters* employ both principles, and many measuring circuits respond, or function in response, to changes in the electrical characteristics of their components.

Radioactive isotopes. Some materials of natural occurrence are continually undergoing a decomposition involving energy particles which are emitted from the materials. The materials may be elements such as radium and uranium or radioactive forms of elements, called *isotopes,* such as carbon 13. By radioactive decomposition, certain elements eventually become totally different elements, whereas radioactive isotopes—or *radioisotopes*—become the nonradioactive form. It takes 1,600 years for the output of particles from radium to diminish by 50 per cent. Such a time period is called a *half-life.* Other radioactive elements or isotopes *decay* at different rates and have different half-life periods. Table 19·2 lists some of the industrial uses for radioactive isotopes, most of them produced in atomic piles. Some of the particles emitted by isotopes have a great deal of energy, being able to pass through an inch of steel, while other particles are stopped by a sheet of paper. The amount of energy absorbed can be used to gage the thickness of materials or the presence or absence of materials, and it is so used by many industries. This is a very new use of an unusual property of certain materials. With time there will be great systems built around radioactive isotope transducers. In Chap. 19 will be found details for putting these characteristics to work.

Changes in color with temperature. When materials are heated above 850°F, they give off energy which is visible. The hotter the material, the greater the amount of energy radiated, and there is an accompanying change in the color of the radiation. At the lowest visible temperatures, the object is a very dull, dark red. As the temperature is increased, the amount of red becomes less and the amount of blue increases. The intense heat generated by an electric arc, the hottest man-made "fire," has a very bluish hue. The change in color is very definite as we advance from the lowest visible temperatures to the highest. Color, therefore, may be used to judge temperature. In the temperature ranges above 3000°F it becomes a very valuable tool.

There are other materials which are particularly responsive to infrared rays, which are produced by bodies whose temperature is below about 850°F. The heat which we sense as being given off by a hot stove or furnace when we stand near it, but not where the air rising from it is appreciable, is radiant infrared heat. At the other end of the spectrum we have rays which we do not see; these are the ultraviolet. Both classes of "invisible light" have their uses in industrial instrumentation.

Elastic materials. We have previously used the terms "elastic materials" and "elastic limit." We need to know more about the properties of elastic materials if we are to understand clearly just how and, perhaps of more importance, why apparatus performs.

One of the most important principles is this: whenever a force is applied to any object, there will be a change in dimension, a bending, or a twisting proportional to the force applied. We know of no exceptions to this rule. Some materials take more force than others to distort them—steel resists more than does lead; copper, far more than putty.

Instead of speaking of the total force, it is more general to find the force applied to each square inch. The force applied to each square inch, or at the rate of so many pounds per square inch, is called the *stress*. Again, this is a technical definition. Accompanying it is another term which we must use precisely: *strain*. Strain tells us how much a material has been distorted or changed in *inches per inch* because of the imposition of some force. For example, if a steel block 10 in. long and 2 in.^{2} in cross-sectional area becomes 10.1 in. when pulled with a force of 10,000 lb, we find the strain as follows:

$$10.1 - 10.0 = 0.1 \text{ in.} = \text{change in length for 10 in.}$$

$$\text{Strain} = \frac{0.1 \text{ in.}}{10.0 \text{ in.}} = 0.01 \text{ in./in.}$$

$$\text{Stress} = \frac{\text{force, lb}}{\text{area, in.}^2}$$

$$= \frac{10{,}000 \text{ lb}}{2 \text{ in.}^2}$$

$$= 5{,}000 \text{ lb/in.}^2, \text{ or psi}$$

Stress-strain diagrams. If we were to put a piece of steel, brass, copper, or some other common metal in a testing machine and pull on it, we would find that at the beginning there would be no appreciable change in the sample as the stress was increased slowly. After some period of time, we would notice that the sample was stretching but the force applied was remaining about constant. With the application of still more force we would stretch the sample to destruction. Assuming that we kept an accurate record of the force applied and the stretch of the sample, we would have acquired data which would give us a plot essentially as shown in Fig. 3·8. There are a number of important things to be learned from such a plot:

1. The section of the diagram at the beginning is essentially straight.
2. With many materials there is a fairly sharp transition from the sloping section to a section that is fairly flat.
3. The slope of the plot is essentially constant from O to EL.
4. The point above which the strain is no longer linearly proportional to the stress, point EL, we call the *proportional limit*. Up to this point, the strain is strictly proportional to the stress. From EL to YP the

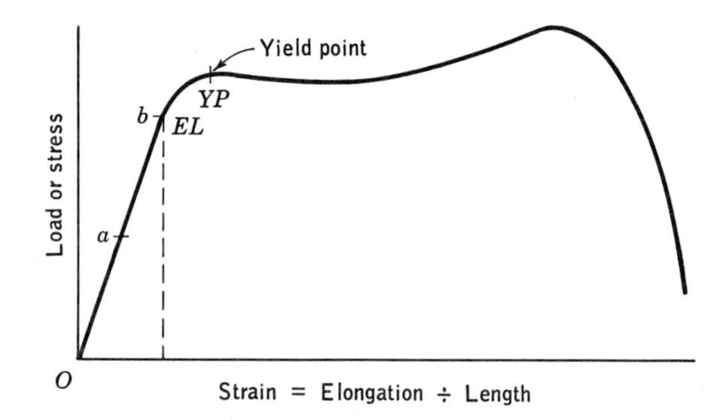

Fig. 3·8 Typical stress versus strain relationships for elastic metallic materials.

material is still within the elastic range, but the amount of strain or elongation is not easily computed. Within this range, O to YP, the material should return to its original length when the load or stress is removed.

5. When the *yield point* is reached, the sample will continue to strain without the imposition of an increased load.

6. A material stressed beyond its yield point will not return to its original length when the load is removed. For an instrument we want to have the stress below b, which is about halfway to the upper yield point. We would then expect the instrument to take a 100 per cent overload and still not be far from its original calibration. However, stresses in excess of this value might make it impossible to recalibrate the instrument for more than a limited portion of the scale. *An instrument should never be subjected to pressures greater than the designed values if the original accuracy is to be retained.*

Young's modulus. The region of the plot from O to EL, Fig. 3·8, has particular significance because it will let us determine just how much strain can be created by certain stresses. Let us assume that the sample is steel and that we have the following values:

	Stress, psi	*Strain, in./in.*
Point a	10,000	0.0033
Point b	20,000	0.0066

The slope of the line from O to E can then be found by dividing the stress by the strain:

$$\text{Slope} = \frac{\text{stress}}{\text{strain}} = \frac{10{,}000 \text{ psi}}{0.0033 \text{ in./in.}} = \frac{20{,}000 \text{ psi}}{0.0066 \text{ in./in.}}$$

This particular ratio, the slope of the line, is called Young's modulus and is generally represented by the letter E. For almost all kinds of steel we find the ratio to be about 30,000,000 psi per inch of strain. For brass and copper, the value of E is about one-half that for steel.

Table 3·4 Elastic Materials

Material	Young's modulus *	Stress at elastic limit, psi
Aluminum, rolled	10,000,000	25,000
Glass, silica	10,000,000	
Brass, annealed	14,000,000	16,000
Bronze	14,500,000	40,000
Copper	18,000,000	10,000
Steel	29,000,000	Depends upon type of steel
Concrete	2,000,000	

* Pounds per square inch per inch of elongation per inch of length.

Because Young's modulus = stress/strain = E, we now have a convenient expression which relates stress to strain. Let us try to get a better picture of what Young's modulus means and how we can use it easily to tell what strain will result from any given stress.

In Fig. 3·8 it will be noted the strain was directly proportional to the stress (Hooke's law) and that this straight-line relationship was used to compute the value of Young's modulus. For an actual sample of steel, point EL would probably correspond to a stress of 40,000 to 60,000 psi, and with a corresponding strain of 0.0013 to 0.002 in./in. Let us imagine that the straight-line portion of the curve is extended until the strain becomes 1 in. In other words, if we had started with a sample 1 in. long, it would have been stretched 1 in. for a total length of 2 in. Or if we had started with a length of 3 in., the sample would have become 6 in. long at the end. If we had kept the proper scale, we would find that *at the rate at which we started, the stress would be 30,000,000 psi to create this strain of 1 in./in.* Of course, we know that long before we reach even 100,000 psi with most steels, we will have exceeded the proportional limit and the upper yield point. Also, the sample will perhaps have been broken. Therefore, we must use Young's modulus in some such manner as this:

$$\frac{\text{Stress given}}{\text{Young's modulus}} = \frac{\text{strain}}{1 \text{ in.}}$$

or

$$\text{Strain} = \left(\frac{\text{stress}}{\text{Young's modulus}}\right)(1 \text{ in.})$$

$$\text{Stress} = \frac{(\text{strain})(\text{Young's modulus})}{1 \text{ in.}}$$

For example, what strain will be produced by a stress of 30,000 psi when the sample is steel?

$$\text{Strain} = \left(\frac{\text{stress}}{\text{Young's modulus}}\right)(1 \text{ in.})$$

$$= \left(\frac{30{,}000 \text{ psi}}{30{,}000{,}000 \text{ psi/in.}}\right)(1 \text{ in.}) = 0.001 \text{ in./in.}$$

Similarly, determine the strain if a force of 28,000 lb is applied to a piece of brass 10 in. long and of 2 in.^{2} cross section.

$$\text{Strain} = \left(\frac{28{,}000 \text{ lb/2 in.}^2}{14{,}000{,}000 \text{ psi/in.}}\right)(1 \text{ in.}) = 0.001 \text{ in./in.}$$

The total change in length is equal to the product of the strain and the length:

$$\text{Increased length} = 10 \text{ in.} \times 0.001 \text{ in./in.} = 0.010 \text{ in.}$$

EXAMPLE 3·10 It was observed under test that a piece of brass 20 in. long and with a cross-sectional area of ¼ in.^{2} was elongated 0.030 in. What was the force applied to the sample?

SOLUTION

$$\frac{(\text{Stress, psi})}{E}(1 \text{ in.}) = \text{strain, in./in.}$$

$$\text{Strain} = \frac{\text{change in length}}{\text{original length}} = \frac{0.030 \text{ in.}}{20.0 \text{ in.}} = 0.0015 \text{ in./in.}$$

Now we know the values of the strain and of E for brass:

$$\left(\frac{\text{Stress}}{E}\right)(1 \text{ in.}) = \text{strain, in./in.}$$

$$\frac{\text{Stress}}{14{,}000{,}000 \text{ psi}} = 0.0015 \text{ in./in.}$$

$$\text{Stress} = (14{,}000{,}000 \text{ psi})(0.0015 \text{ in./in.}) = 21{,}000 \text{ psi}$$

But

$$\text{Stress} = \frac{\text{force}}{\text{area}} \quad \text{or} \quad \text{Force} = \text{stress} \times \text{area}$$

$$\text{Force, lb} = (21{,}000 \text{ psi})(\tfrac{1}{4} \text{ in.}^2) = 5{,}250 \text{ lb}$$

Actually, 5,200 lb is as close as we should calculate the answer, because the value of E is not known with great precision. Recall that we get this value by predicting the force to double the length of the sample (this is a strain of 1 in. for 1 in. of the original material), provided that the material kept on stretching at exactly the same rate at which it started to stretch. Our graph really took us only about 1 per cent of the way, so our prediction may not be very good. It is much like starting off to travel 100 miles. After the first mile is behind us, we predict that we will keep on for 100 times the distance and time and in the direction that we started. Would you consider this to be a precise prediction?

EXAMPLE 3·11 How much will a piece of aluminum 20 in. long be stretched if it has a cross-sectional area of ½ in.² and is pulled with a force of 5,000 lb?

SOLUTION. We must first find the stress, which is equal to force ÷ area:

$$\frac{5{,}000\text{ lb}}{\frac{1}{2}\text{ in.}^2} = 10{,}000\text{ lb/in.}^2\text{, or psi}$$

$$\text{Strain, in./in.} = \frac{\text{stress}}{E}$$

$$= \left(\frac{10{,}000\text{ psi}}{10{,}000{,}000\text{ psi/in.}}\right)(1\text{ in.}) = 0.001\text{ in./in.}$$

$$\text{Total } \textit{change} \text{ in length} = \text{original length} \times \text{strain}$$

$$= (10\text{ in.})(0.001\text{ in./in.})$$

$$= 0.01\text{ in.}$$

The relationships in Example 3·11 can be used to tell us the change in length when we know the stress and the modulus of elasticity. Similarly, if we know that a certain material has had its dimensions changed, we can calculate the force required to change the dimensions—this might be the force developed by steam or water pressure, or it might be a mechanically generated force.

Spring characteristics. In instrumentation, springs have many shapes. Some are more easily recognized than others. First, let us define a spring as any elastic material that undergoes a distortion, stretching, or change of position which is directly proportional to the applied force. There is the conventional wire spring used for either tension or compression, and there is the Bourdon tube which is the heart of so many temperature and pressure measurements. The tube which seals certain types of flow-measuring equipment is a spring; the stiffness of the spring determines the range of the meter. Undoubtedly there are many more examples, but these three will suffice.

The basic law of the spring states that so long as the material operates over the linear portion of its range, the distortion, or stretch, or angular movement is directly proportional to the applied force or torque. For example, a coiled spring requires a force of 10 lb to stretch it 2 in. The *spring rate* is equal to the force divided by the distance.

$$\text{Spring rate} = \frac{\text{force, lb}}{\text{distance, in.}} = \frac{10\text{ lb}}{2\text{ in.}} = 5\text{ lb/in.}$$

Thus, to stretch the spring $\frac{3}{4}$ in., the force would be:

$$(\text{Spring rate})(\text{distance}) = (5 \text{ lb/in.})(\tfrac{3}{4} \text{ in.}) = 3.75 \text{ lb}$$

Or, if the force were given as 25 lb, then the deflection or stretch would be found from the relationship

$$\frac{\text{Force, lb}}{\text{Spring rate}} = \text{stretch}$$

$$\frac{25 \text{ lb}}{5 \text{ lb/in.}} = 5 \text{ in.}$$

Hysteresis. Because there are no absolutely elastic materials, no material will return precisely to its starting point when the deforming force is removed. That is because the molecules have an internal friction. The force of internal friction is far greater in some cases than others. Steel, glass, bronze, and brass develop only small internal frictional forces for small displacements; so meters and instruments made from these materials indicate almost the same values when the forces are increasing as they do when the forces are decreasing. On the other hand, meters or instruments made of rubber, despite what you think you know about rubber, would not have this property. Because the steel, brass, and glass so nearly duplicate their performance with increasing and decreasing forces, we say that they have a very small *hysteresis loss*, while that for rubber is relatively large. The hysteresis loop for a material also represents the amount of energy or work required to make the material go through a cycle of pressure or force changes.

It will be noted that the materials in Fig. 3·9 represent quite faithfully the relationship between force and distortion or strain. Gages made of such materials should be quite accurate, and they should be independent of increasing or decreasing values. But if a gage were made of a material similar to rubber, it would register a very wide difference between increasing and decreasing values and it would not return to zero when the pressure was removed.

Aside from the inaccuracies of reading caused by hysteresis, instrument designers try to avoid using materials with appreciable hysteresis

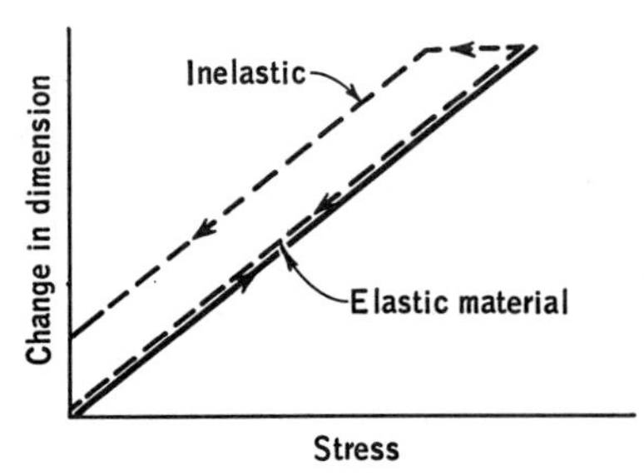

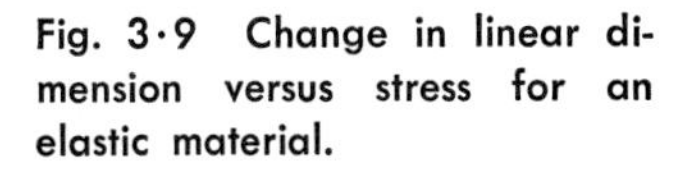
Fig. 3·9 Change in linear dimension versus stress for an elastic material.

because they are more likely to fail with repeated flexing or cycling. If you repeatedly bend a piece of metal in order to break it, you will notice that the metal at the bend becomes warm to touch. Most of the work that you do is used to overcome the internal friction. Springs can be stressed repeatedly, and at high rates, but little heating occurs because of the small amount of internal friction. In summary, materials with appreciable hysteresis are not normally employed in instruments because:

1. Inaccuracies are introduced.
2. The operating life is likely to be short.

Resistance to corrosion. The desirable characteristics of instruments are many; among them are long life, constancy of calibration, ease of adjustment, and repeatability. These features will not be realized if the parts corrode or stick.

Industrial plants and the air near them may be contaminated by corrosive fumes; moisture in various amounts may be present; and dust and dirt may also contain contaminants which aid corrosion. Many instruments therefore have to be made of materials which are quite resistant to chemical attack or can be made so by plating or other treatment. Corrosion resistance should not, however, be obtained at the price of accuracy. Unfortunately, there may be times when a compromise will have to be made: longer life at the expense of some accuracy.

We are all familiar with the rapidity with which steel and iron parts rust; they must therefore be protected in some manner. Particles of rust can jam bearings, clog nozzles, and otherwise interfere with normal operations. Brass and bronze are relatively inert to many atmospheric contaminations, but plating or painting is generally done to enhance these properties and to improve appearance.

Where acids and other active reagents are encountered, some of the stainless steels will be used. In other cases, it may be found that parts of monel or other nickel-bearing materials will be used.

SUMMARY

1. Because very few measurements are made directly, we infer one value after determining another. Pressure measurements often depend upon the change in length of an elastic material, while temperature determinations may depend upon a change in resistance or change in volume. In order that we can make the correct interpretations or make the best use of the various types of transducers, we must know what happens to materials under various conditions.

2. Almost all metals undergo changes in length and volume when their temperature is changed. The change per degree is not uniform, and average rates of change may be used over a short temperature span if high accuracy is to be obtained. If the changes in length are undesired, then it may be necessary to compensate for this behavior.

3. Liquids, as a class, also experience changes in volume with changes in temperature. If the liquid is confined within an elastic housing, pressures can be created.

The pressures are related to the amount of the expansion, and thus to the change in temperature which created the volume change.

4. Temperature is a measure of molecular and atomic energy and does not tell us how much heat there is in a substance. Heat transfers take place from high temperatures to low. Because so many materials change their physical dimensions, as well as other characteristics, determination of temperature is very important. Because temperature cannot be measured directly, it must always be inferred or determined by measuring some other property or quantity.

5. There are two general pairs of temperature scales. The absolute temperature scales, the Kelvin and the Rankine, have their starting points at absolute zero, where atomic activity is assumed to be zero. All gas-pressure measurements involving temperature must employ absolute temperatures. The centigrade and Fahrenheit scales are based on the freezing and boiling points of water, with the number of divisions purely arbitrary. The Fahrenheit degree represents five-ninths the temperature span expressed by the centigrade degree.

6. Gas-pressure measurements are among the most important measurements, not only in connection with industrial operations, but also in connection with many phases of instrumentation.

7. Almost all metals experience changes in electrical resistance as the temperature changes. There is such a close correlation between the two changes that some of our most accurate temperature measurements are made with resistance thermometers. Metals have positive coefficients of resistance versus temperature, with the coefficient being quite small.

Certain organic materials and some oxides undergo decreases in resistance with increasing temperature. Many of them are termed "thermistors," which as a class have negative temperature-resistance coefficients. Their rate of change of resistance with temperature is several times that of metals.

8. If heat is applied to a junction formed by joining two dissimilar metals, there will be created a potential which is proportional to the difference in temperature of the junction and the other ends of the wires. With wires of definite composition and manufacture, there is a very definite relationship between the temperature difference and the potential which is created. Consequently, the relationship is widely used not only for the measurement of temperature but for many other applications in which heat may be obtained or produced.

9. Some materials experience changes in resistance when subjected to light of different intensities and wavelengths. Others produce potentials which are a measure of the light intensity. Still others are capable of emitting electrons under the influence of light. All of these effects or properties are employed in various types of instruments.

10. Certain materials undergo a spontaneous disintegration which progresses at a definite logarithmic rate. The intensity of the radiation produced by the disintegration varies with the material. The emitted energy can be absorbed by other material. The time required for the emitted energy to decrease by 50 per cent is called the half-life. This time may vary from microseconds to hundreds of years.

11. An elastic material is one which returns almost precisely to its original shape when a distorting force is removed. Steel, brass, and glass are examples of highly elastic materials. Rubber is one of the most inelastic of the elastic materials. Every material undergoes some change of shape or dimension whenever subjected to a force or stress. No matter how slight the distortion, every force creates strain.

For all instrumentation work we should be most particular that no elastic part is stressed over the elastic limit, or that it is operated in the nonlinear portion of the stress-strain curve, unless suitable calibration steps are taken.

12. The spring characteristics of materials are of great importance in many pressure or pressure-sensing devices. Regardless of the shape, most springs have deflection rates which are directly proportional to the applied load. Moreover, the strain is virtually independent of the way in which the load is applied—increasing or decreasing. Highly elastic materials which make good springs have very small hysteresis effects.

QUESTIONS

3·1 What properties of materials are of particular interest in instrumentation work?

3·2 What are common examples of linear expansion with temperature? Of volume? Are there practical applications of these effects? In what equipment or instruments?

3·3 When a coefficient of expansion is given, is it an average value, or do all expansions within the span take place at the specified rate?

3·4 How can an expanding fluid create a pressure?

3·5 Should the pressure of a gas decrease or increase with an increase in temperature? With a decrease in temperature? What other condition must hold true for your answer to be strictly correct?

3·6 Why should changing the volume of a gas affect its pressure? What is the relationship between volume and pressure?

3·7 Of what use are the absolute-temperature scales?

3·8 Why should there be a difference between gage pressure and absolute pressure?

3·9 Why are absolute pressures used for gas pressure-volume relationships?

3·10 What is Charles's law? Boyle's law?

3·11 What is the expression that considers both temperature and volume changes in connection with the pressure of a confined gas?

3·12 What do temperature changes do to the resistance of most metals? Of what significance might this be?

3·13 What is an elastic material? List some good examples of highly elastic materials.

3·14 Of what significance are elastic materials in instruments?

3·15 How small a force may be applied to a piece of steel without deforming it?

3·16 What are some of the materials whose resistance changes with the intensity of illumination?

3·17 What are radioisotopes?

3·18 What is a half-life? Why do we use the term?

3·19 What is meant by strain?

3·20 What is the meaning of Young's modulus E?

3·21 On what part of the stress-strain diagram should instruments work? Are there any exceptions?

3·22 What are the shapes that a spring may take? Compare their relative extensions in normal use.

3·23 What is meant by spring rate?

3·24 What is meant by hysteresis? Is it a good thing to design an instrument so there will be a large amount of hysteresis?

PROBLEMS

3·1 To complete an electrical temperature alarm circuit, a piece of copper 5 in. long must expand 0.010 in. beyond its normal length at 70°C. What increase in temperature is required to complete the circuit?

3·2 How much more will a 10-in. piece of brass expand than a 10-in. piece of invar for an increase in temperature of 100°F? (Invar cannot be heated much above 350°F without its coefficient changing to one greater than aluminum.)

3·3 If a 2-in.-long piece of brass is to expand 0.002 in., what change in temperature will be required?

3·4 How long would a piece of glass have to be to expand as much as a piece of steel for the same temperature change? Give answer as a percentage.

3·5 A steel rail is 33 ft long. How much does its length change during the year when its temperature changes from +150°F to −20°F? Assume the rail is unrestrained.

3·6 Railroads now use continuous rails ¼ mile long. What would be the change in length, if free, of a ¼-mile rail for a temperature change of 200°F? What do you suppose accounts for the fact that the rail occupies the same space all the time, regardless of the temperature?

3·7 A gallon of water just fills a container. What part of this volume will overflow if the liquid is heated 100°F? 100°C? Assume no change in the volume of the container.

3·8 If the liquid is confined within a container of steel which has adequate strength, what happens when the temperature of the liquid increases? Has this any application in instrumentation?

3·9 A piece of aluminum 100 ft long is to be heated 1000°F. What will be its change in length?

3·10 If a tank were built of copper and had dimensions 10 by 10 by 8 ft, what would be the change in volume of the tank for a temperature change of 200°F? 200°C?

3·11 How much more would a steel tube that is 10 in. long at 100°F elongate than a pyrex glass tube that is 20 in. long at 100°F for a temperature change of 200°F? 200°C?

3·12 A steel rail is 33 ft long at 60°F. What will be the change in length if the rail is permitted to expand freely from −100°F to +200°F?

3·13 An aluminum rocket 10 ft long at 70°F is expected to have temperatures of 1000°F and −400°F. What will be its maximum and minimum lengths?

3·14 An increase of 200°F is equivalent to what change in degrees centigrade? A temperature of 200°F is equivalent to what centigrade temperature change?

3·15 An increase of 200°C is equivalent to what change in degrees Fahrenheit? A temperature of 200°C is equivalent to what Fahrenheit temperature change?

3·16 A temperature of 932°F is equal to what temperature on the centigrade scale? 168°F?

3·17 What temperature on the Fahrenheit scale corresponds to 500°C? −200°C?

3·18 500°F has what Rankine equivalent?

3·19 500°C has what Kelvin equivalent?

3·20 500°K is equal to what Rankine temperature? 500°R is equal to what Kelvin temperature?

3·21 Find the ratio of a Kelvin to a Rankine degree. Of a Rankine to a Kelvin degree.

3·22 Convert 2000°C to degrees Rankine. To degrees Kelvin. To degrees Fahrenheit.

3·23 Convert 2000°F to degrees Kelvin. Degrees Rankine. Degrees centigrade.

3·24 A metal container has a volume of 64 ft^3. To what size would it have to be reduced (*a*) to double the gage pressure? (*b*) to double the absolute pressure? Does the original pressure in the container affect your answer?

3·25 A gas at 200 psig is put into a container 4 times as large. What will be the new absolute and gage pressures?

3·26 A boy using a bicycle pump completes nine-tenths of the stroke before the compressed air equals the pressure in the tank. What is the tank pressure? Assume no temperature effects and no pump leakage.

3·27 If air which fills a 1-ft^3 container at 15 psi is allowed to fill a volume 10 times as great, find the absolute and gage pressures.

3·28 An oxygen tank has a volume of 10 ft^3. If it first contained oxygen at 200 psig and later its pressure was 100 psig, how many cubic feet of gas at atmospheric pressure had been used?

3·29 A tank holds both air and water. The tank has a volume of 40 ft^3 and contains 30 ft^3 of water. The pressure gage reads 50 psi. How many cubic feet of water could be moved before the pressure would drop to 20 psi? Assume the gage on top of the tank and no absorption of air by the water.

3·30 If a tank has 20 lb of a gas at 127°C and a pressure of 185.3 psig, what will be the pressure if the centigrade temperature is doubled? If the volume is cut to one-half of the original?

3·31 Gas produced in a chemical process emerges with a temperature of 1000°F and a pressure of 85.3 psig. If this hot gas is produced at the rate 100,000 ft^3/hr under these temperature and pressure conditions, how large a tank would be required to hold an hour's supply if the storage temperature were 300°F?

3·32 A gas thermometer holds 4 in.^{3} of helium at 270°C and 200 psig. What will be the pressure level at 500°C if the metallic housing does not change volume?

3·33 What would be the pressure at −250°C of the gas thermometer of Prob. 3·32?

3·34 A gas thermometer develops a pressure of 114.7 psia at −200°C. Plot the pressures developed at −100°, 0°, 500°, and 1000°C. Assume that there is no change in the volume of the system and that the gas acts as a perfect gas.

3·35 If an air tank has a gage pressure of 100 psi, what will be the absolute pressure when the gage pressure is doubled? Halved?

3·36 If the absolute pressure in an air tank is 200 psi, what will be the gage pressure when the absolute pressure is doubled? When the original gage pressure is doubled?

3·37 For air under normal atmospheric pressure, what will be the gage pressure when the absolute pressure is tripled?

3·38 For a "perfect" vacuum, what are the gage and absolute pressures?

3·39 What will be the gage and absolute pressures that must be applied to one side of a piston of 100 in.^{2} if the force created is to be 10,000 lb? Assume normal atmospheric pressure on the other side.

3·40 How much will the gage and absolute pressures be changed if all the gas in one cube is transferred to another cube having sides half as long as the sides of the first one? Twice as long? Assume 800 psia as the original pressure.

3·41 If a force of 100 lb is applied to a rod having a length of 100 in. and there is a change in length of 1.0 in., find the strain. If the cross-sectional area is 0.1 in.^{2}, find the stress.

3·42 If a force of 100,000 lb is applied to a steel block 2 in. square and 10.0 in. long, the steel changes in length by 0.0083 in. Find the stress and strain under these conditions.

3·43 A steel wire 20 ft long has a load of 400 lb hung from one end. Find the stress if the cross-sectional area is 0.2 in.^{2}, and find the strain if the change in length because of the load is 0.5 in.

3·44 A hollow metal post 10 ft long has a cross-sectional area of 4 in.^{2} If its length decreases by 0.10 ft when a load of 50 tons is applied, find the stress and the strain.

3·45 A load of 50 tons is carried on a steel column having a cross-sectional area of 10.0 in.^{2} Find the average stress.

3·46 How much will a steel wire 30 ft long and 0.30 in. in diameter stretch when 400 lb is hung on it?

3·47 How much will a steel rod 100 ft long and 0.05 in. square be stretched by a force of 1,000 lb?

3·48 The strain in a block of steel is used to weigh a tank full of chemicals. If the block of steel is 10 in. tall and 3 in. square, what will the tank weigh if the strain is to be 0.001 in./in.? Assume that half of the weight of the tank falls on the block of steel.

3·49 How much larger in area must an aluminum block be than a steel block if the strains in the blocks are to be the same for any given load?

3·50 The maximum allowable stress in a particular piece of steel is 30,000 psi. What maximum change in length will occur when a load of 300,000 lb is applied to a piece of steel 10 in. high and 20 in. in cross-sectional area?

3·51 What cross-sectional area of brass and aluminum would have to be used in Prob. 3·50 to obtain twice the change in length?

3·52 What are the relative sizes of steel, brass, and aluminum required to produce equal strains for any given load?

3·53 For equal cross-sectional areas of material, what are the relative loads on steel, brass, and aluminum to provide the same strains?

3·54 What stress is required in brass to cause a strain of 0.001 in.? In steel? In aluminum? What stresses to create strains of 0.002 in./in.? 0.0005 in./in.?

3·55 A force of 100 lb compresses a spring 2 in. Express the spring rate in pounds per inch and pounds per foot.

3·56 What would be the spring rate of a bellows that has an effective area of 10 in.^{2} and expands 1 in. for a pressure of 10 psi?

3·57 What spring rate is required to produce a 2-in. movement of a spring when 500 lb is applied?

3·58 If a bellows has a spring rate of 40 lb/in., what effective area should the bellows have so that 20 psi will produce a movement of ¼ in.? ½ in.?

3·59 If each of the two sizes of bellows in Prob. 3·58 is so placed in turn that it just touches a spring whose constant is 100 lb/in., how far will each bellows expand against the spring when the pressure is 50 psi?

3·60 A piece of steel 100 in. long is heated 100°F. What will be the increase in length? If a piece of brass of the same length is also heated 100°F, how do the changes in length compare?

3·61 A sensitive switch requires a movement of 0.002 in. How much of a temperature change is required if the actuating element is 5 in. long? What should be the length if the change in temperature is limited to 10°C? Material is aluminum.

3·62 If the actuating element for the switch of Prob. 3·61 need move only 0.0005 in. because of a system of levers, what are the two answers?

3·63 If a piece of railroad track is 33,000 ft long on a day when the temperature is 70°F, what will be the change in length from winter to summer when the temperature of the rail goes from −20°F to +140°F?

3·64 A gage block of steel is 1.000000 in. long at 68°F. What will be its error if its temperature rises 10°F? Drops to 60°F?

4 Basic Principles

This chapter is principally concerned with two topics; the first is a consideration of the meanings and methods of determining weight, density, and specific gravity; and the second is a study of certain hydraulic and pneumatic principles which can be used in making various industrial measurements.

WEIGHT, MASS, AND SPECIFIC GRAVITY

Weight. The *weight* of an object is a measure of the forces of attraction between the object and any other objects. Because of the great mass of the earth and its relative closeness, weight is often defined as a measure of the mutual pull of the earth and the object. However, from an observation of the ocean tides, we know that both the sun and the moon exert forces upon the earth and the objects on it. For most practical purposes, however, we shall consider that only the earth and the object are concerned in measuring the weight of the object.

The weight of an object can be determined quite easily in a number of ways:

1. By hanging it on a calibrated spring. The extension of the spring indicates the weight, and this is the principle of ordinary scales.
2. By employing the method of equal torques and a known force such as some known weight.
3. By measuring the strain in a material when the object to be weighed is supported by the strained material. This is similar to (1), but the strain is so small that auxiliary equipment must be used to detect it, whereas with the springs generally used, the amount of movement is quite easily measured.
4. By applying hydraulic or pneumatic pressures to maintain an original position after the subject weight has been applied. Certain types of load cells employ these principles.

The weight of common bulk materials in the English system is generally expressed in terms of pounds per cubic foot. It may be expressed

Table 4·1

Material	*Weight, lb/ft³*	*Material*	*Weight, lb/ft³*
Aluminum, pure	168	Gray cast	439–445
Brass	510–542	Lead	710
Carbon	125–144	Lithium	39
Copper	554	Mercury	849
Cork	15.6	Osmium	1,400
Germanium	341	Platinum	1,336
Gold	1,203	Silver	660
Ice	55–57	Tungsten	1,174
Iron:		Uranium	1,165
Pure	491	Water (max density)	62.4

in any other unit of weight and any other unit of volume, but pounds per cubic foot will customarily be employed. Table 4·1 shows how the weights of common materials compare for equal volumes of 1 ft^3 each. Another name for the pounds per cubic foot unit is *weight density.* A second term, known simply as *density,* is used to indicate the *mass per cubic foot.* These two terms must not be confused, because they refer to entirely different things. The term *mass* will be defined later in this chapter.

Specific gravity. Because of its common occurrence, the water is often used as a reference. When we compare the weight of a given volume of any material with that of an equal volume of water, under standard conditions, we are determining the *specific gravity* of the material with respect to water. Specific gravity may be expressed as a fraction:

$$\frac{\text{Weight of a given volume of the material}}{\text{Weight of an equal volume of water}} = \text{specific gravity} \qquad (4\cdot1)$$

For example, in Table 4·1 the weight of 1 ft^3 of cast iron is approximately 450 lb. The weight of an equal volume of water is 62.4 lb. The specific gravity of the iron is then equal to

$$\frac{450 \text{ lb/ft}^3}{62.4 \text{ lb/ft}^3} = 7.22 \qquad \text{dimensionless} \qquad (4\cdot2)$$

It will be noted in Eqs. (4·1) and (4·2) that the units in the numerator and denominator are the same. Therefore, the specific gravity is a dimensionless unit and indicates only the ratio of the weight of a given volume of material to the weight of an equal volume of water. Materials having specific gravities less than 1 are lighter than water, whereas those having specific gravities greater than 1 are heavier than water. It will be found in Chap. 21 that the specific gravity of a liquid will be used as a measure of its purity or concentration.

Mass. All of us are quite familiar with the term "weight" and how to use it. But another term closely related to weight is often misunderstood or used incorrectly. That term is *mass*. As we use the term in industry and engineering, the mass of an object indicates the force necessary to accelerate the object one foot per second per second; this is often written as 1 ft/sec^2. In the English or engineering system, an object which weighs 32.2 lb on the earth in the general latitude of New York City will require a force of 1 lb to accelerate it 1 ft/sec^2. An object which weighs twice as much will require 2 lb of force applied to it to accelerate it at the same rate. An object which weighs 16.1 lb will require that ½ lb of force be applied to it to accelerate it 1 ft/sec^2.

In the form of an equation, it can be stated that

$$\text{Force} = KWa$$

where K = proportionality constant
W = weight, lb
a = acceleration, ft/sec^2

Now if we know from experiment and observation that a force of 2 lb will accelerate an object weighing 64.4 lb at the rate of 2 ft/sec^2, then we can determine the value of K:

$$2 \text{ lb} = K(64.4 \text{ lb})(1 \text{ ft/sec}^2)$$

and K must therefore be equal to 2/64.4 or 1/32.2 ft/sec^2. This is the value of the gravitational acceleration for freely falling bodies in the general latitude of New York. Now if we group K and W together to form a single term, we have

$$\frac{1}{32.2}W \quad \text{or} \quad \frac{W}{32.2}$$

When we divide the weight W by the gravitational constant, we produce a unit of mass known as the *slug*. Thus, an object which here on earth weighs 32.2 lb has a mass of 1 slug; if it weighs 96.6 lb, then it has a mass of 3 slugs. The mass of any object will tell us the force required to give the object an acceleration of 1 ft/sec^2, neglecting all frictional effects. Therefore, an automobile which weighs 3,220 lb has a mass of 100 slugs and must have a force of 100 lb applied to it to accelerate it 1 ft/sec^2. If we wish to accelerate this car at the rate of 10 ft/sec^2, then we must apply 1,000 lb, and so on.

All that we need remember about mass is that the mass of any object tells us the force required to accelerate that object at the rate of 1 ft/sec^2; to get 3 times the acceleration will require 3 times the force.

It might be pointed out in passing that, while the weight of an object may change, the mass of an object does not change appreciably for speeds up to a few thousands of miles per second. But as the speed of an object approaches that of light, the mass of the object increases. This is just another way of saying that resistance to acceleration increases with extremely high speeds. In the usual types of instrumentation we will not encounter any situations in which we have to consider the change in mass because of speed, but in certain designs of mass spectrometers (Chap. 18) change of mass must be considered.

Our chief interest in the mass of objects probably will come in the determination of liquid and gas flow, when we must know the mass of the material to be accelerated through the orifice or venturi tube.

There are other units of mass besides the slug, but the other systems and units are not normally required for the solution of flow problems in this country.

Earlier in this chapter, we referred to the term "density." In engineering work, we should understand *density* to mean the *mass per cubic foot* or the *slugs per cubic foot.* We must never confuse *weight density* with *density.*

HYDRAULIC AND PNEUMATIC PRINCIPLES

Until the 1940's, when electrical and electronic devices began to be used with greater and greater frequency, the history of industrial instrumentation was centered around hydraulic and pneumatic principles. Thermometers, pressure systems, level systems, and control systems have operated mainly through the use of hydraulic and pneumatic techniques. The principles are few in number and are quite easily mastered.

When a liquid or a gas is confined within a container, a force will be exerted upon the walls and bottom in the case of liquids and upon all surfaces in the case of gases. In many instances we are concerned not with the total force exerted, but with the force per unit area. The force per unit area is called *pressure.* It may be expressed mathematically as:

$$P = \frac{F}{A}$$

EXAMPLE 4·1 A block of steel which weighs 400 lb is supported on a wooden block 10 in. square. Find the pressure, in pounds per square inch and pounds per square foot, with reference to the block.

SOLUTION

$$P_1 = \frac{F}{A} = \frac{400 \text{ lb}}{(10 \text{ in.})(10 \text{ in.})} = \frac{400 \text{ lb}}{100 \text{ in.}^2} = 4 \text{ lb/in.}^2\text{, or psi}$$

$$P_2 = \frac{F}{A} = \frac{400 \text{ lb}}{(10 \text{ in.})(10 \text{ in.})/(144 \text{ in.}^2/\text{ft}^2)} = \frac{400 \text{ lb}}{0.695 \text{ ft}^2} = 576 \text{ lb/ft}^2\text{, or psf}$$

In instrumentation work we will normally be expressing the pressure in pounds per square inch or pounds per square foot. However, when it comes to low-pressure measurements, the pressure may be expressed in terms of the height of a column of liquid required to establish a condition of pressure equilibrium. Flowmeters and differential pressure instruments may express their pressures in terms of so many inches of water or mercury, but these may be easily converted into our conventional unit of pressure of pounds per square inch.

Pressure due to gravity. Because of the gravitational pull of the earth, all objects have weight, and this pull is always in the dirction of the center of the earth for all practical purposes. The bottom of a tank, or a platform, will then be subjected to a force which is equal to the gravitational pull upon all the material which is being supported. If the load is spread uniformly over the supporting area, then the pressure can be determined as indicated previously.

If we consider a tank filled to a depth of 1 ft, then we have a pressure on the bottom because of the height of the column of liquid being supported by each unit area. Now, if we fill the tank to a depth of 2 ft, then there will be a column twice as high to be supported by each square inch or square foot, so the pressure will have been doubled. We can continue to fill the tank, and we will find that the pressure at the bottom of the tank is directly proportional to the weight of the column of liquid above it. We might also reason this way: if we start at the surface where there is no pressure because of the weight of the liquid and move a pressure gage down in the tank, the gage should register pressures which are directly proportional to the weight of the column of water above it and thus proportional to the depth to which it is moved. Stated another way, the pressure indicated by the instrument should be directly proportional to the height of the column of liquid above the gage.

The weight of liquid above any fixed point in a tank is equal to the product of the volume and the weight density, and if we call the weight and the force synonymous, then

$$W = F = (\text{volume})(\text{weight density})$$

but pressure equals force ÷ area, so

$$\frac{F}{A} = \frac{(\text{volume})(\text{weight density})}{\text{area}}$$

$$P = \frac{F}{A} = \frac{(\text{area})(\text{height})(\text{weight density})}{\text{area}}$$

$$= (\text{height})(\text{weight density})$$

Care must be exercised to make sure that the proper units are used.

If the height is in feet and the weight density is in pounds per cubic foot, then the pressure will be in pounds per square foot. If the height is in inches, then the weight density must be in pounds per cubic inch if the answer is to be in pounds per square inch.

EXAMPLE 4·2 Find the pressure, in pounds per square inch and pounds per square foot, at the bottom of a tank 10 ft deep which is filled with water.

SOLUTION

$$P = hD$$

$h = 10$ ft, $D = 62.4$ lb/ft^3

$$P_1 = (10 \text{ ft})(62.4 \text{ lb/ft}^3)$$
$$= 624 \text{ lb/ft}^2, \text{ or psf}$$

To get the answer in pounds per square inch,

$$P = hD$$

where $h = (10 \text{ ft})(12 \text{ in./ft}) = 120 \text{ in.}$

$$D = \frac{62.4 \text{ lb/ft}^3}{1{,}728 \text{ in.}^3/\text{ft}^3} = 0.0361 \text{ lb/in.}^3$$

$$P = (120 \text{ in.})(0.0361 \text{ lb/in.}^3)$$
$$= 4.32 \text{ lb/in.}^2, \text{ or psi}$$

EXAMPLE 4·3 If a cylindrical tank 20 ft tall is filled with a liquid having a specific gravity of 0.80, what pressure, in pounds per square inch, will be created at the bottom?

SOLUTION

$$P = hD$$

where $h = (20 \text{ ft})(12 \text{ in./ft}) = 240 \text{ in.}$
$D = (0.0361 \text{ lb/in.}^3)(0.80) = 0.0288 \text{ lb/in.}^3$

$$P = (240 \text{ in.})(0.0288 \text{ lb/in.}^3) = 6.7 \text{ lb/in.}^2, \text{ or psi}$$

EXAMPLE 4·4 If the tank in the example above had an area of 20 ft^2, what force would be developed against the bottom?

SOLUTION

$$F = PA$$

where $P = 6.7 \text{ lb/in.}^2$
$A = (20 \text{ ft}^2)(144 \text{ in.}^2/\text{ft}^2) = 2{,}880 \text{ in.}^2$

$$= (6.7 \text{ lb/in.}^2)(2{,}880 \text{ in.}^2) = 19{,}300 \text{ lb}$$

EXAMPLE 4·5 A column of mercury is 100 in. high. What would be the height of a column of water which would develop the same pressure at the bottom of the column?

SOLUTION. From Table 4·1 we find that the specific gravity of mercury is 13.6. Therefore, a column of mercury of any given height will establish a pressure at the bottom 13.6 times as great as that of a column of water of the same height. Consequently, it will take a column of water 13.6 times as high as the mercury column to establish an equivalent pressure:

$$h = 100(13.6) = 1{,}360 \text{ in., or } 113.3 \text{ ft}$$

It is because a very tall column of water can be replaced by a relatively short column of mercury that mercury is used for so many pressure measurements and calibrations.

EXAMPLE 4·6 A column of mercury 20 in. high will establish what pressure at the bottom of the tube in which it is contained?

SOLUTION

$$p = hW_{\text{den}} = (20 \text{ in.})(13.6)(0.0361 \text{ lb/in.}^3) = 9.82 \text{ lb/in.}^2\text{, or psi}$$

(The value of 0.0361 lb/in.³ was found in Example 4·2.)

Transmission of pressures; Pascal's law. One principle that is employed repeatedly in industrial instrumentation design is Pascal's law, which states that whenever an external pressure is applied to any confined fluid at rest, the pressure at every point in the fluid is increased by the amount of the external pressure. Another statement of this principle might be that, in a confined static fluid, pressures are transmitted equally and undiminished in all directions.

We find applications of Pascal's law in thermometer and differential pressure designs where the forces which are generated at one point are:

1. Transmitted to a remote point for actuation of some indicator
2. Used to balance some other force

When a thermometer bulb undergoes a change in temperature, assuming that the design is of the filled type, then the ensuing change in internal pressure is transmitted throughout the system. With a liquid-filled hydraulic system, the transmission of any pressure change takes place at an extremely fast rate, and a condition of static equilibrium is soon established again.

U-tube manometer. If we take a piece of glass tubing and bend it into the form of a **U,** then we have a very simple manometer. The manometer is one of the most accurate ways of measuring relative pressure, and it depends for its operation on some of the simple principles previously discussed in this section. In Fig. 4·1 we have such a simple bent tube filled with some liquid so that, with no pressures other than atmospheric, the levels of the liquids in each leg have equal elevations above the reference line. Now if we look at points P_1 and P_2, which also have equal heights above the reference line, then these points also have equal pressures, because any difference in their pressures would be corrected by the adjustment of the liquid columns to equalize the pressures. Each point is also subjected to the pressures created by the weights of equal heights of columns of liquid above them. The same situation must prevail for points P_3 and P_4.

Changing the relative sizes of the two legs of a manometer will have

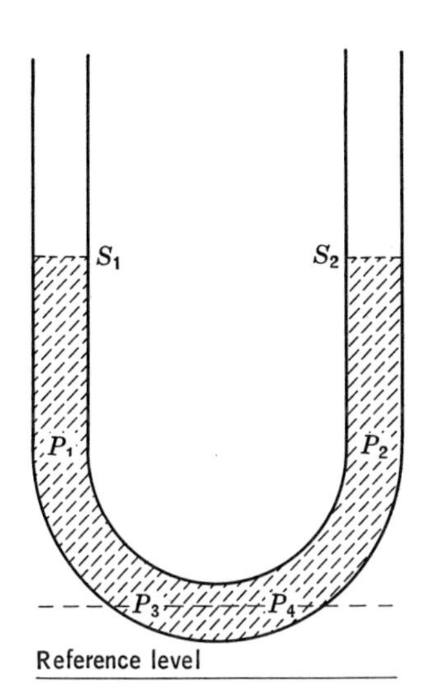

Fig. 4·1 A **U** tube illustrates conditions of pressure balance for any set of points which are at equal heights above some reference, even though the two surfaces of the liquid are not of the same height.

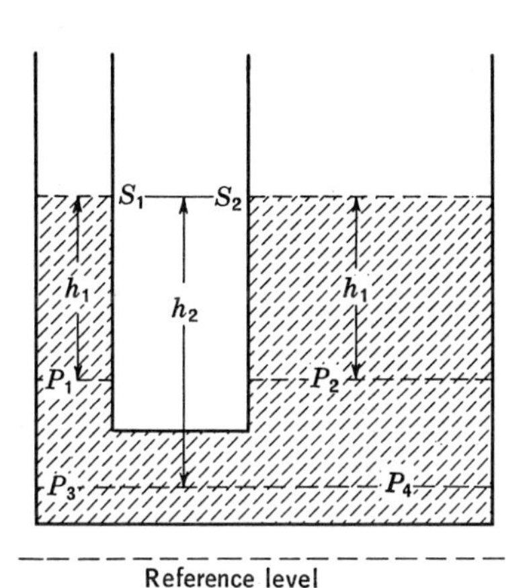

Fig. 4·2 Pressure equality of points in a **U**-tube manometer at the same height above the reference is independent of the size and shape of either leg.

no effect on the pressures established at corresponding points in the liquid, Fig. 4·2. Under stable conditions, *all points in the liquid at equal distances above the reference or datum line must be subjected to the same pressures.* ***U****-tube manometers are pressure-balance systems.*

In Fig. 4·3, we have introduced an external pressure above one leg while leaving the pressure above the other leg unchanged. This added pressure forces the liquid down in the one leg and up in the other, as is shown in Fig. 4·3. If we examine the pressures at points P_a and P_b, we find that

$$P_a = P_{\text{ext}} + P_c$$

$$P_b = P_1 + P_c$$

where P_c is the pressure created by the column of liquid between S_3 and P_a, and P_1 is the pressure created by the column which is above S_3 in the right leg + P_{atm}. If we subtract P_c from each of the above equalities, and remember that $P_a = P_b$ because they are at equal heights above the reference, then

$$P_{\text{ext}} = P_1 = \text{pressure of column } h_1 + P_{\text{atm}}$$

$$P_{\text{above atm}} = P_1 - P_{\text{atm}}$$

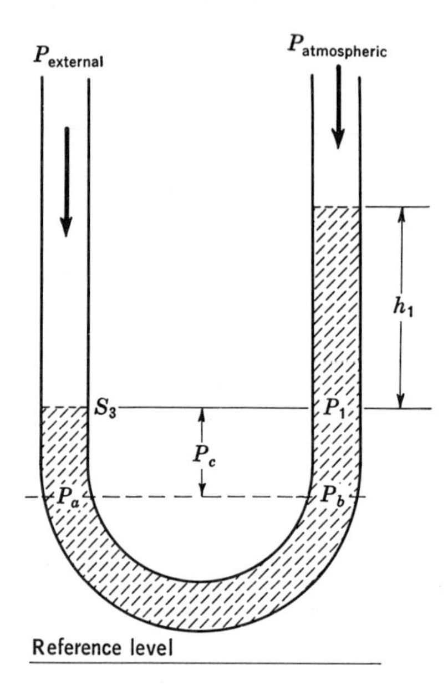

Fig. 4·3 When there is a pressure differential, the liquid rises in the leg of the manometer subjected to the smaller pressure.

The extra height of liquid which is required to establish pressure equality then becomes a measure of the pressure with respect to atmospheric pressure. The pressure which is greater has the shorter column of liquid. If the reference pressure is not atmospheric pressure, then the column of liquid represents the pressure differential. For any given pressure differential the height of h_1 will vary with the liquid: the greater the specific gravity, the shorter the column. Mercury is often used because of the relatively large differential pressure which can be measured with a given column of liquid, as compared with an equivalent column of water.

SUMMARY

1. The weight of an object is a measure of the material attraction exerted by the object and the other parts of the universe, principally the earth. The weight of an object may vary with the location of the object. All objects weigh less as they are moved farther away from the earth, for example.

2. The weight density of an object or material indicates the pounds per cubic foot of the object or material.

3. The mass of a given object is a measure of its resistance to acceleration. So far as we know, the mass of an object remains constant regardless of the position of the object in the universe; moving the object farther from the earth will reduce its weight, but it will not affect its mass. All of this assumes that the object is not moving at speeds in excess of 1,000 miles or so per second. The mass, or the resistance to acceleration, increases with speed until, as the speed of light is approached, the resistance to acceleration approaches an infinite value. From Einstein's equation

$$M = \frac{M_0}{\sqrt{1 - (V/186{,}000)^2}}$$

where M_0 = original mass

M = new mass

V = speed, miles/sec

4. (*a*) The specific gravity of a material other than a gas is a comparison of the weight of the material with the weight of any equal volume of water. The weight may be expressed in any units so long as both the material and water weights are expressed in the same units. Similarly, the volumes may be of any desired value, but they must be equal. (*b*) The specific gravity of gases is obtained by comparing the weight of the gas with that of air.

5. At any given point in fluids, liquids, and gases, pressures within closed systems are transmitted equally and undiminished in all directions.

6. (*a*) The force exerted upon each square inch or square foot is the average pressure and is generally expressed only as the pressure. (*b*) A pressure is the ratio of the force applied to the area.

7. Pneumatic and hydraulic transmission of pressures with the usual instrument does not normally involve any loss in pressure.

8. In a confined fluid, the pressures developed at points which are at equal elevations with respect to some reference point will be the same. Any differences in pressures are automatically self-correcting. This is the basic principle of the liquid manometer.

9. The shape and the relative dimensions of the manometer will have no effect upon the *pressure,* for equal pressures and depths of fluid, but the forces developed will depend upon the area.

QUESTIONS

4·1 What must be true of the density if we state that the pressure at any depth is directly proportional to the depth?

4·2 A U-tube manometer is open at one end and connected to a pressure chamber at the other. Four liquids which can be used to fill the manometer have specific gravities which range from 0.50 to 13.6. What conditions would determine which of the liquids was used?

4·3 Because of a misunderstanding, the two legs of a U-tube manometer were made of tubes of different sizes, one having four times the cross-sectional area of the other. Which leg would you connect to the pressure system and which leg would you leave exposed to atmospheric pressure, assuming that one leg would use atmospheric pressure for a reference? Why?

4·4 What are the differences between weight density and mass density? What are the conditions that generally require that we know the mass of an object?

4·5 What should be the specific gravities of liquids used in low-pressure manometers? For high-pressure manometers?

4·6 The same liquid is used to fill four containers to the same depth. The containers have unequal areas at the bottom. What do you know about the pressure at the bottom? The force on the bottom of each container?

4·7 The pressure at the bottom of a tank is to be measured, but because of the way the tank was installed, it is impractical to drill a hole in the bottom. Assuming that the liquid in the tank is a clear, nonviscous liquid, where and how might you make this measurement?

PROBLEMS

4·1 A block of metal having a volume of 10 ft^3 weighs 3,000 lb. What are its weight density, its mass per cubic foot, and its specific gravity?

4·2 A liquid has a specific gravity of 2. What is its weight density? Its mass per cubic foot?

4·3 1,000 ft^3 of a certain liquid is contained in a cubical tank. If the specific gravity is 0.8, what force is exerted on the base of the tank? What are the weight density and the mass density of the liquid?

4·4 If 100 gal of a process liquid weighs 500 lb, what is the specific gravity of the liquid? If the liquid contracts on cooling, what effect will this have on the specific gravity?

4·5 If the specific gravity of a flowing liquid is 0.6, what will be the weight of 100 gal of the liquid? 100 ft^3?

4·6 A gasoline tank is filled to a depth of 30 ft. What is the pressure halfway down? At the bottom? Assume specific gravity is 0.80.

4·7 Draw a graph showing the relationship between the pressure and the depth for the gasoline tank in Prob. 4·6. If you had a sensitive gage, would it provide a means for determining the amount of gasoline in the tank?

4·8 A square tank 20 by 20 ft is to be filled to a depth of 15 ft with a liquid having a specific gravity of 1.10. What will be the force on the bottom of the tank? If this load could be supported equally at three points, what would be the force at each point?

4·9 Assuming that air at 100 psi can be supplied to a piston to support the total load in Prob. 4·8, how large would the piston have to be?

4·10 If the amount of liquid in Prob. 4·8 is cut in half, what will be the load to be supported by the piston in Prob. 4·9?

4·11 Can you draw a graph relating the air pressure against the piston in Prob. 4·9 to the depth of liquid or the weight of liquid in the tank of Prob. 4·8? If you can draw such a graph, will it provide a way of "weighing" the tank's contents?

4·12 Some skin divers carry a small pressure gage calibrated to give them the depth. What calibration would be needed to provide for a depth of 100 ft? Assume an interval of 10 ft and sea water.

4·13 Compute the mass densities for the materials listed in Table 4·1.

4·14 Prove to your own satisfaction that whether a manometer tube is vertical or tilted will make no difference provided the vertical distance is measured above the reference point.

4·15 Why is static equilibrium required for Pascal's law to be true?

4·16 Why should fluids be different from solids when it comes to transmitting forces?

5 Energy and Force Systems

ENERGY

Energy is often defined as a capacity for doing work. *Work* must also be defined in its technical sense of a force acting through a distance. Unless both force and distance are involved, there is no work. Work is customarily defined in engineering practice in terms of foot-pounds. It is independent of time: no matter how long it takes to do a certain task, the amount of work, or foot-pounds, is the same. For example, you push a cart for 10 ft and a force of 25 lb is required. How much work have you done?

$$\text{Work} = \text{force} \times \text{distance}$$
$$= (25 \text{ lb})(10 \text{ ft}) = 250 \text{ ft-lb}$$

But a force of 100 lb exerted against a building which doesn't move appreciably gives us the following:

$$\text{Work} = \text{force} \times \text{distance}$$
$$= (100 \text{ lb})(0) = 0$$

No matter how tired we might get, we have done no work unless we have moved something or exerted a force through some distance.

We become concerned with the time required when we want to determine the horsepower. By definition, 1 hp requires that 550 ft-lb of work be done in 1 sec.

We may have energy in several forms:

1. *Chemical energy* exists in coal or dynamite because the inherent capacity for moving objects is there. Whether or not we use this energy is of no particular significance.

2. Energy which is available because of position is called *potential energy*. Water in a reservoir or lake above some hydraulic turbines is capable of flowing through the turbines and developing power. A car on

a hill, if allowed to coast down a smooth road, might be able to go part way up a hill on the other side of the valley. Theoretically, any object which is higher than some reference point possesses potential energy with respect to that point. Work can be done in going from the higher elevation to the lower as energy is given up. Potential energy represents energy which is banked or stored in an object such as a spring.

3. A third form of energy is called *kinetic energy,* or energy of motion. Any moving object, large or small, possesses energy because of its motion. Work must have been performed to get the object moving at that speed, and the object has acquired an amount of energy equal to the work done on it, less any work against friction.

One of the most important principles is that energy can be converted from one form to another or can be stored and used at some other time. Also, there may be a repeated cycle of energy transformations from potential to kinetic and back to potential, as in a watch, clock, or pendulum.

Potential energy. The implication of the words "potential energy" is that the energy *may* be available, and the implication may also be that the original state of the object may call for an elevation higher than that of any subsequent position. Whatever the original conditions, work was done at some time or other in order that the object might have a source of energy. Energy must always be obtained at the expense of something or someone; it is never free in the sense that it is never paid for. Potential energy is of the type that can be banked or stored, waiting to be employed in some way. A stretched spring also has energy stored in it because work was done to stretch it; that is, an energy deposit was made in the spring. Some of that energy can be reclaimed when the spring returns to some less stressed condition.

EXAMPLE 5·1 A spring that has a spring rate of 100 lb/in. is compressed 10 in. How much work energy is stored in the spring?

SOLUTION

$$\text{Work} = (\text{force})(\text{distance})$$

$$= \frac{0 + (10\ \text{in.})(100\ \text{lb/in.})}{2^*} \times \frac{10\ \text{in.}}{12\ \text{in./ft}}$$

$$= 418\ \text{ft-lb}$$

* The average force must be used, inasmuch as the force applied varies from zero at the beginning to a value of 1,000 lb at the end of the compression.

EXAMPLE 5·2 A 100-lb piece of steel is lifted by a crane and placed on a rack 20 ft above the ground. How much work was done in lifting it? What is its potential energy while on the rack?

SOLUTION

$$\text{Work} = (\text{force})(\text{distance})$$

$$= (100\ \text{lb})(20\ \text{ft}) = 2{,}000\ \text{ft-lb}$$

$$\text{Potential energy} = (\text{weight})(\text{distance the object could fall})$$
$$= (100 \text{ lb})(20 \text{ ft}) = 2{,}000 \text{ ft-lb}$$

These two answers are the same in value. Work was done to lift the piece of steel; the potential energy of the steel increased with the work done to lift it. The higher the steel is to be raised, the more work must be done on it. But at the same time, the potential energy will increase an equivalent amount.

Kinetic energy. The energy that an object has because of its speed represents the work done to set it in motion. Mathematically, the energy is expressed by

$$\text{Kinetic energy} = \frac{1}{2}\frac{WV^2}{g}$$

where W is the weight, in pounds; g is the gravitational acceleration with a value of 32.2 ft/sec^2; and V is measured in feet per second. With the units indicated, the kinetic energy will be found in foot-pounds.

EXAMPLE 5·3 If we ignore friction, how many foot-pounds is required to make a car weighing 3,220 lb go 50 fps?

SOLUTION

$$\text{Kinetic energy} = \frac{1}{2}\frac{(\text{weight, lb})(V, \text{ ft/sec})}{g}$$
$$= \frac{1}{2}\frac{(3{,}220 \text{ lb})(50 \text{ ft/sec})^2}{32.2 \text{ ft/sec}^2}$$
$$= \frac{1}{2}\frac{(3{,}220 \text{ lb})(2{,}500 \text{ ft}^2/\text{sec}^2)}{32.2 \text{ ft/sec}^2} = 125{,}000 \text{ ft-lb}$$

One of the very interesting characteristics of these two forms of energy, kinetic and potential, is the ease with which one form can be changed into the other form. A child on a swing, going back and forth, has only potential energy at the highest points of movement, when the swing is just ready to start down. As the swing comes down, the energy of position becomes less but the kinetic energy, which was zero at the topmost point, increases because the speed is increasing. At the lowest point in the arc the potential energy is zero for that particular path and set of conditions, but the kinetic energy is at its maximum. As the swing starts to move higher, it will exchange kinetic energy for potential energy and finally come to a momentary halt. If there were no friction or other losses, the swing would reach the same maximum height each time.

A similar situation involves the pendulum in a clock. It continually exchanges kinetic energy for potential energy, and vice versa. The hairspring and balance wheel in your watch repeatedly go through this interchange of energy; potential energy in the spring is converted into kinetic energy of the balance wheel and then back to potential energy as the hairspring becomes stressed in the other direction.

We are particularly interested in such exchanges and conversions of energy because, in certain measurements, we deliberately establish conditions which will convert *pressure energy* to *velocity energy* and then back to *pressure energy*. Unless there is friction, and there is always some, the total energy remains constant; we merely change its form. Stating it in financial terms, you may have $1,000 in cash which you may invest in United States Government Savings Bonds. Aside from any time limitations which there are in cashing the bonds, you can convert from one asset to the other at will, and so it will be with most of the energy systems which we shall use. That one form of energy can be converted to another is a very important principle.

Law of levers. Almost every instrument, indicator, recorder, or controller uses levers of one form or another. Levers are also used in almost every piece of machinery. Perhaps we think of a lever most often as being a straight piece of material which is pivoted at one point, but the forms of levers are extremely varied. They range from straight or bent forms to the wheel, which may be said to be made up of a very large number of levers which must act at a point.

Levers are used for a number of purposes:

1. To convert some force acting through a distance to a different force acting through some other distance
2. To increase movement at the expense of force, or vice versa
3. To create one moment equal to another moment
4. To reverse direction

One of the principles of the preceding section is the basis for the law of levers. That principle is that without any friction, the work or energy delivered to a mechanism will come out undiminished in value but perhaps in a different form. This is the principle of the conservation of energy. Inasmuch as work has been defined as the product of a force and the distance through which the force acts, measured in the direction of the force, we can say:

$$(\text{Force 1})(\text{distance 1}) = (\text{force 2})(\text{distance 2}) \tag{5·1}$$

But, as in Fig. 5·1, the distance moved in the direction of the force is directly proportional to the distance from the fulcrum. Thus,

$$D_1 = h(\text{moment arm 1}) \tag{5·2}$$

$$D_2 = h(\text{moment arm 2}) \tag{5·3}$$

By dividing expression (5·2) by (5·3), we get

$$\frac{D_1}{D_2} = \frac{\text{moment arm 1}}{\text{moment arm 2}} \tag{5·4}$$

Then the moment arm ratio can be substituted for the *distance*, and becomes

$$\frac{\text{Force 1}}{\text{Force 2}} = \frac{\text{moment arm 2}}{\text{moment arm 1}} = \frac{D_2}{D_1} \tag{5·5}$$

When expressed as a product,

$$(\text{Force 1})(\text{moment arm 1}) = (\text{force 2})(\text{moment arm 2})$$

By definition, moment = (force) (moment arm), where the moment arm is defined as the perpendicular distance from the point of rotation to the line of action of the force. Therefore,

$$\text{Moment 1} = \text{moment 2}$$

From Eq. (5·5) we get

$$\frac{\text{Force 1}}{\text{Force 2}} = \frac{\text{distance 2}}{\text{distance 1}}$$

or

$$(\text{Force 1})(\text{distance 1}) = (\text{force 2})(\text{distance 2})$$

but,

$$\text{Work} = (\text{force})(\text{distance})$$

Therefore,

$$W_1 = W_2$$

EXAMPLE 5·4 A force of 10 lb is applied 10 ft from the fulcrum of a plank 30 ft long, Fig. 5·2*a*. What force can be applied at the other end of the plank and through what distance will this unknown force move if the 10-lb force moves down 2 ft?

SOLUTION

$$(\text{Force 1})(\text{moment arm 1}) = (\text{force 2})(\text{moment arm 2})$$

$$(10 \text{ lb})(10 \text{ ft}) = (\text{force 2})(20 \text{ ft})$$

$$100 \text{ lb-ft of moment} = (20 \text{ ft})(\text{force 2})$$

$$5 \text{ lb} = \text{force 2}$$

Using the principle of the conservation of energy, any loss of potential energy or of work done that occurs on one side of the fulcrum will equal the gain in energy on the other side (if there is no friction), Fig. 5·2*b*. Then

$$(10 \text{ lb})(2 \text{ ft}) = (5 \text{ lb})(\text{distance})$$

where "distance" is the actual distance moved when measured in the direction of the applied force. Then

$$20 \text{ ft-lb} = 5 \text{ lb (distance)}$$

$$\text{Distance} = \frac{20 \text{ ft-lb}}{5 \text{ lb}} = 4 \text{ ft}$$

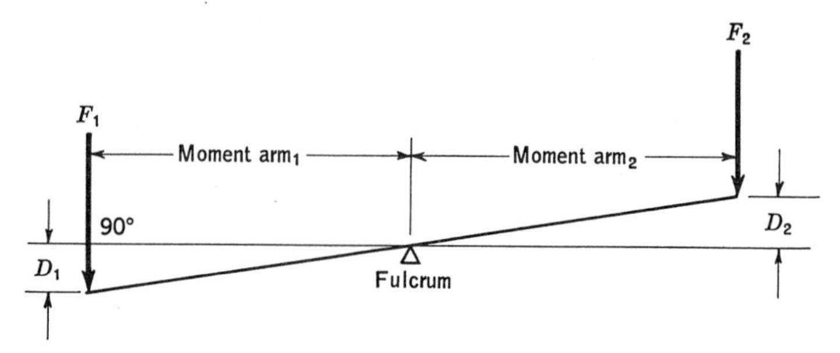

Fig. 5·1 When an object is not rotating about some point about which it is free to rotate, we must conclude that the moments tending to produce counterclockwise motion are just equal to those tending to produce clockwise motion. If such a stable condition is disturbed by slowly forcing F_1 down, then the loss of potential energy by one body is just equal to that gained by the second body.

Figure 5·1 can also be used to illustrate the *principle of equal moments.* In all instruments, any movements are so slow and of such limited value that appreciable motion is not involved. Thus, we always deal with *balanced systems.* A balanced system generally means that no acceleration is involved. A *moment* is defined as a force acting through a distance to *try* to produce rotation or turning. This turning effort is generally expressed in *pound-feet,* whereas work is expressed in *foot-pounds.* In the case of work, there must actually be movement through a distance, but in the case of moment there may not be any actual movement. An attempt to produce rotation, even if unsuccessful, is the product of the force and the perpendicular distance from the point of rotation to the line of action of the force. In Example 5·4 the original force is 10 lb acting through a *moment arm* of 10 ft. This does not mean that the 10-lb force is exerted through a distance of 10 ft; actually, the distance is 2 ft in the direction in which the force is acting. These are most important distinctions. Torque tells how hard we are trying to turn something. Actually, we may accomplish nothing. For example, suppose that a pipefitter pulls on the end of a wrench with a force of 100 lb, that the wrench has a handle 2 ft long, that the pipe does not move, and that the wrench does not slip. Find the torque developed and the work done.

$$\text{Torque} = (\text{force})(\text{moment arm}) = (100\text{ lb})(2\text{ ft}) = 200\text{ lb-ft}$$

$$\text{Work} = (\text{force})(\text{distance}) = (100\text{ lb})(0\text{ ft}) = 0\text{ ft-lb, or no work}$$

If the pipefitter moves the end of the wrench 3 in., or ¼ ft, then the work done will be

$$(100\text{ lb})(\tfrac{1}{4}\text{ ft}) = 25\text{ ft-lb}$$

It will be noted that we have used the units foot-pounds and pound-feet. There will be little trouble if we remember that foot-pounds represent work. *Work* means that the force actually moved something, not merely tried to do so. Pound-feet, on the other hand, represent an attempt to turn or rotate something; whether or not we succeed is unimportant to the definition. These two uses of the units are not invariable, but generally it will be found that energy or work is expressed in foot-pounds, while pound-feet will be reserved for bending, turning, or twisting efforts. Even more important, *pound-feet never suggest or imply convertibility to foot-pounds, or vice versa.*

Another basic principle is that every turning effort must be resisted by one of the same magnitude. Thus, the pipe in the example given above had to resist with a moment, or torque, of 200 lb-ft. As a child, you have verified this principle many times. Let two children of unequal weight try to use a seesaw. To make it balance, they adjust the position of each child with respect to the fulcrum so that the lighter child is farther out. If the heavier child weighs 100 lb and is 6 ft from the fulcrum at the time of balance, where will the lighter child, weighing 80 lb, have to sit?

To determine mathematically where the lighter child will have to sit to balance the heavier child, we need to state one of the fundamental relationships:

$$\text{Clockwise moment on left} = \text{clockwise moment on right}$$

$$(\text{Force})(\text{moment arm on left}) = (\text{force})(\text{moment arm on right})$$

$$(100 \text{ lb})(6 \text{ ft}) = 600 \text{ lb-ft} = (80 \text{ lb})(\text{distance})$$

Fig. 5·2 Figures for Example 5·4.

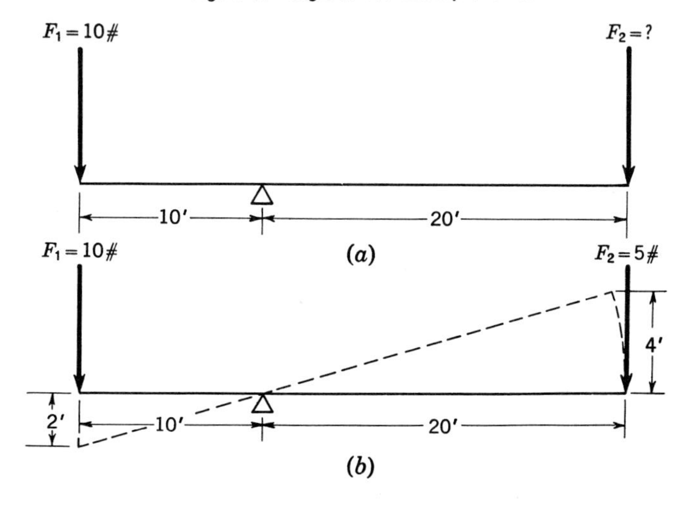

By dividing each side of the equation by 80 lb, we find the distance to be 7.5 ft.

Later, when we have force-balance and torque-balance problems to consider in transmitters and pressure seals, we shall have many occasions to employ these principles. We shall find many designs in which the fulcrum is movable and a fairly small force which is variable within limited values is used to balance the system.

The *mechanical advantage* of a mechanism or component is the ratio of the force delivered to the force applied to the device. In Fig. 5·2*a* a force of 10 lb is balanced by a force of 5 lb. Then

$$\text{Mechanical advantage} = \frac{\text{force out}}{\text{force in}} = \frac{5\text{ lb}}{10\text{ lb}} = \frac{1}{2} \qquad \text{dimensionless}$$

In this instance, the output force was one-half of the input force, but the distance through which it moved was twice the distance through which the input force moved.

When levers are used only to produce changes in direction in which the force is applied, all the basic rules still apply with respect to the equality of the torques and the work done.

Gears are special lever designs; the moment arms are the radii of the gears. All of the familiar principles of levers apply here also:

1. The tendency of a gear train to turn clockwise, for example, must be equal to the resisting counterclockwise effort. That is,

$$\text{(Force) (distance), clockwise} = \text{(force) (distance), counterclockwise}$$

2. If there is an increase in movement of the driven gear with respect to the driving gear, there will be a proportional reduction in the force available. That is,

$$\text{(Force 1)(distance 1)} = \text{(force 2)(distance 2)}$$

In the case of the Bourdon tube gages, for example, the great magnification of the movement of the tube will mean that the force available at the pointer is extremely small.

3. Ignoring friction, the work-energy delivered to the gear train equals the work-energy delivered by the gear train, or

$$\text{(Force in)(distance in)} = \text{(force out)(distance out)}$$

EXAMPLE 5·5 A driving gear having a radius of 1 in. mates with a second gear having a radius of 3 in., Fig. 5·3. If a torque of 100 lb-in. is applied to the smaller gear, find the force developed by the small gear. What torque will the large gear develop? If the small gear teeth move through an arc of 30 in., how much work will be done? What will be the work output of the large gear?

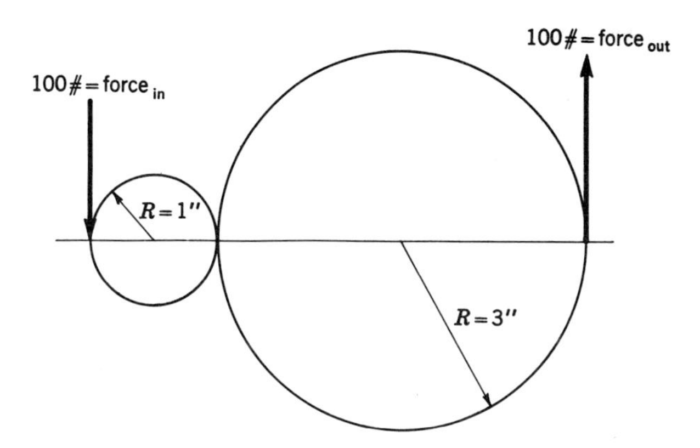

Fig. 5·3 Figure for Example 5·5.

SOLUTION. A torque of 100 lb-in. means that the product of the force and the radius of the moment arm is

$$FR = 100 \text{ lb-in.}$$

whence

$$F(1 \text{ in.}) = 100 \text{ lb-in.}$$

$$F = 100 \text{ lb}$$

The small gear pushes on the large gear with a force of 100 lb, and the large gear pushes back with the same force. But the torque developed by the large gear is

$$\begin{aligned}(\text{Force})(\text{moment arm}) &= (\text{force})(\text{radius})\\ &= (100 \text{ lb})(3 \text{ in.})\\ &= 300 \text{ lb-in}\end{aligned}$$

$$\begin{aligned}\text{Work in} &= (\text{force})(\text{distance moved in the direction of the force})\\ &= FR = (100 \text{ lb})(30 \text{ in.})\\ &= 3{,}000 \text{ in.-lb}\end{aligned}$$

$$\begin{aligned}\text{Work out} &= (\text{force})(\text{distance moved in the direction of the force})\\ &= (100 \text{ lb})(30 \text{ in.}) = 3{,}000 \text{ in.-lb}\end{aligned}$$

The large gear has to develop an output force of 100 lb to balance the input force. The output gear will also move through an arc of 30 in. length. (Because of the different radii, the larger gear will turn through a different angle, when compared with the smaller gear, but the teeth of each gear must turn through *equal distances*. It is for this reason that 30 in. is used in calculating both work in and work out.)

If friction is neglected, the input work must equal the output work. We find that these two quantities are equal, so the answers should be correct.

Force on pistons. There are many instruments in which a pressure is exerted on a piston or bellows. Sometimes the piston will be in the valve motor. Controllers of the pneumatic type are likely to have several bellows which may be said to be special types of pistons which are sealed and which have very little friction associated with their movement. The frequency and the importance of the use of pistons requires that we learn some of the fundamentals of operation. Pistons are employed in torque- and force-balance systems.

A piston is essentially nothing but a surface or area against which some pressure or force is permitted to act and which can move within certain limits. In instruments, the amount of movement generally is small, and in many designs it may approach zero because of the great sensitivity of the design. The total force applied to a piston is equal to the product of the area and the pressure, provided the two have corresponding units.

$$\text{Force, lb} = (\text{pressure, lb/in.}^2)\ (\text{area, in.}^2)$$

For example, if a pressure of 100 psi is permitted to act on the top of a piston having an area of 10 in.^{2}, the force developed could easily be computed by substituting in the above expression.

$$\begin{aligned}\text{Force} &= (100\ \text{lb/in.}^2)(10\ \text{in.}^2)\\ &= 1{,}000\ \text{lb}\end{aligned}$$

A small pressure exerted against a large area can create a large force. A high pressure directed against a small area, as in the hydraulic brake systems of cars and control systems, can also develop a large force. The force developed, less any frictional effects, is equal to the product of the pressure and the area against which it is exerted.

Pistons of the sort used in automobile or steam engines are not likely to be found in instruments, but bellows and diaphragms are used to convert pressures to forces. Some applications call for movement of the end of the piston, while others create force systems which are balanced and in which the net movement of the piston may be less than a thousandth of an inch.

HYDRODYNAMICS

We have previously considered some of the forms of energy and how transformations may be made from one form to the other, oftentimes with little or no loss of energy in the transformation. We are already acquainted with the ease with which a child on a swing converts his energy of motion to one of position at the top of the movement, and vice versa.

A very common place for these interchanges of energy to occur is in the flow of gases and liquids or where materials are placed under pressure because of the weight of the materials above them or because they are confined within a tank or some piping system. If an object moves without friction and without the application of any force, then the energy which it had before the movement should be a measure of its energy after the movement. There should have been no loss or gain in energy under these conditions.

We know that if an object is simply dropped and falls without friction, the object is continuously exchanging its potential energy for kinetic energy. And the sum of its instantaneous kinetic and potential energies must be equal to the original potential energy with respect to some fixed point. Now if we were to have a tank 100 ft high full of water, we might take 1 lb of that water at the top and enclose it within a bag which could be completely sealed. If we were then to move that bag very slowly to the bottom of the tank, the water would have lost all of its potential energy with respect to the ground, and because of its very slow rate of descent, its kinetic energy at all times would have been essentially zero. How then can we account for this loss in potential energy with no corresponding gain in kinetic energy? Inasmuch as we cannot *lose* it, the energy must be present in some other form. It is now accounted for by a *pressure energy,* because the pressure at the bottom of the tank can now apply such a force that the pound of water would escape through an opening with the speed it would have had if it had fallen frictionlessly to that level.

Figure 5·4 illustrates a very common fact that if we have holes in a tank, the liquid within the tank will escape, and the liquid at the bottom

Fig. 5·4 When an escaping jet of liquid is subjected to the same pressure as the surface of the liquid, the speed of the escaping jet is directly proportional to the square root of the vertical distance between the surface and the point of escape. If the surface of the liquid has upon it a pressure different from that on the emerging jet, this will affect the speed of discharge.

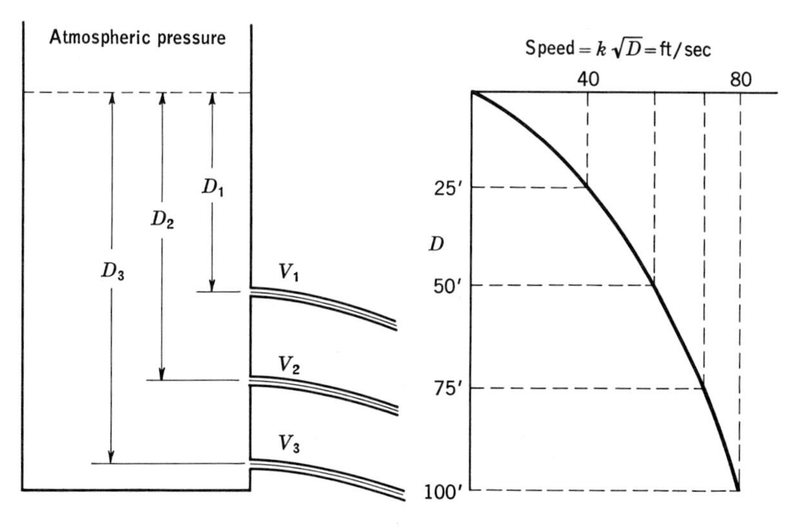

of the tank will escape with the greatest speed. If we assume that all of the potential energy lost will be converted into kinetic energy and that it makes no difference how the pound of water gets from the original high level down to any other point, then the speed will have to be the same whether the pound of water falls freely to that point or issues from an opening, because all of the energy is being converted into kinetic energy.

If this 1 lb of water were to fall 100 ft, it would lose 100 ft-lb of potential energy and would acquire 100 ft-lb of kinetic energy. Under these conditions, we can compute its speed at that, or at any other, point.

$$\text{Gain in KE} = \text{loss in PE} = 100 \text{ ft-lb} \tag{5·6}$$

$$\frac{1}{2}\frac{WV^2}{g} = 100$$

$$\left(\frac{1}{2}\right)(1)\left(\frac{V^2}{32.2}\right) = 100$$

$$V^2 = 2(32.2)(100)$$

$$V = \sqrt{2(32.2)(100)}$$

$$= 80 \text{ fps}$$

At any other point between the top and the bottom, the pressure would be proportional to the depth and the liquid would have acquired some other speed which would be proportional to the square root of the height. Where there is no friction or other loss which absorbs energy, we may state that

$$V = \sqrt{2gh} \tag{5·7}$$

If we now perform a simple mathematical step, we can change Eq. (5·7) so that it becomes

$$V = \sqrt{\frac{2dgh}{d}} \tag{5·8}$$

where d is our mass density in units which are consistent with h. But the product dgh gives the pressure at any point which is h units of height below the surface. Therefore, the equation can be converted into another useful expression:

$$V = \sqrt{\frac{2P}{d}} \tag{5·9}$$

The value of h may also be obtained from the relationship

$$h = \frac{P \text{ lb/ft}^2}{d \text{ lb/ft}^3} \tag{5·10}$$

It must be fully realized that the d in expression (5·8) is in slugs per cubic foot or some other appropriate unit of mass. But in the relationship shown in Eq. (5·10), d is expressed in terms of *weight density*. These distinctions must be carefully kept in mind if serious errors are to be avoided.

If we substitute P/d for h in Eq. (5·7), we find that

$$V = \sqrt{\frac{2gP}{d}}$$

Inasmuch as the height of the column of liquid can be found from the ratio of P to d, we can say that speed of the issuing fluid is directly proportional to the square root of the height of the column.

EXAMPLE 5·6 At what speeds will water escape from a leaky water tank if there are holes 25, 50, 75, and 100 ft below the surface? Assume ideal conditions so far as frictional and viscosity effects are concerned.

SOLUTION

$$V = \sqrt{2gh}$$

$$V_{25} = \sqrt{2(32.2)(25)} = 40 \text{ fps}$$

$$V_{50} = \sqrt{2(32.2)(50)} = 56.5 \text{ fps}$$

$$V_{75} = \sqrt{2(32.2)(75)} = 69.2 \text{ fps}$$

$$V_{100} = \sqrt{2(32.2)(100)} = 80 \text{ fps}$$

CHECK. In falling through a distance of 50 ft, for example, a weight of 1 lb has lost 50 ft-lb of potential energy. Did it gain that amount of kinetic energy?

$$\text{Kinetic energy} = \frac{1}{2}\frac{WV^2}{g} = \frac{1}{2}\frac{(1 \text{ lb})(V \text{ ft/sec})^2}{32.2 \text{ ft/sec}^2} = 50 \text{ ft-lb}$$

$$V^2 = 50(2)(32.2) = 3{,}220 \text{ ft}^2/\text{sec}^2$$

$$V = 56.6 \text{ ft/sec, or fps}$$

The answers are well within the accuracy of the slide rule used for their computation.

If we were to plot the speed of the liquid versus depth of the opening, we would find that the function is far from linear. Doubling the *head* does not double the speed.

Bernoulli's principle. If we have water flowing through a series of sections of pipe, all of which are at the same level but of different cross-sectional area, then the speed of the liquid in each section will be different. Assuming that the liquid is incompressible, then the pounds per second or gallons per minute through each of the various sections must be the same, because we cannot store any of the liquid through compression. Inasmuch as there is no loss of potential energy because all sections of the pipe are at the same elevation, the total energy possessed by each cubic foot or pound of the liquid must remain constant.

However, the speeds through each section are different. Therefore, the kinetic energy possessed by each unit of material will vary with the square of the speed. Any difference or change in the kinetic energy must then be obtained at the expense of the pressure energy which the material possesses. In a piece of frictionless pipe, or in a very short section of pipe, the pressure energy plus the kinetic energy is a constant:

$$\text{Pressure energy} + \text{kinetic energy} = \text{constant}$$

Consequently, any change in speed will be accompanied by a corresponding change in the pressure energy; an increase in the one will require a decrease in the other. Figure 5·5 shows how these variations of pressure occur with speed and change in cross section.

Before we derive the energy relationships for liquids or fluids flowing through pipe of variable cross-sectional area, we might try to get another picture of *pressure energy*. Previously, we considered what would happen to a given amount of water as it was lowered slowly from the top to the bottom of a tank. We concluded that as the water lost potential energy, it gained pressure energy which could be used to give the

Fig. 5·5 For idealized flow of a liquid under pressure, the pressure will be greatest where the flow rate is the least, and vice versa. Contractions in a flow stream will result in a decreased pressure; an increase in the cross section of the stream will increase the pressure.

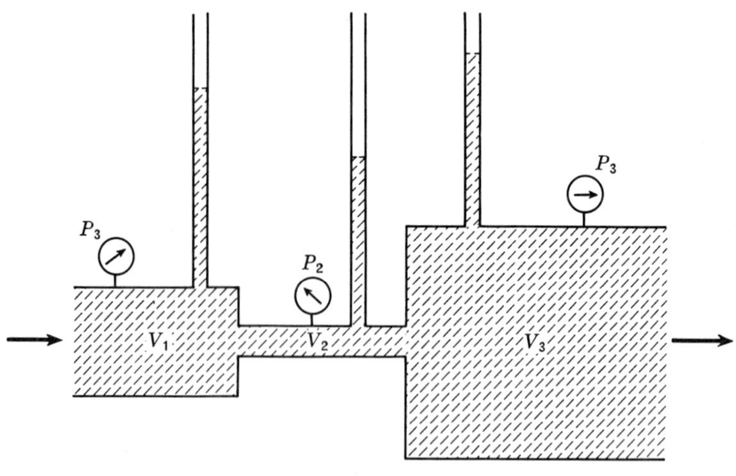

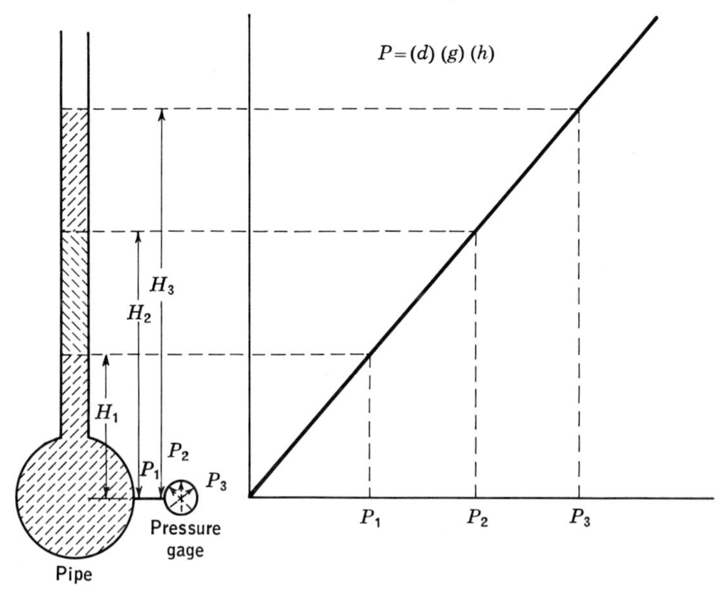

Fig. 5·6 There is a linear relationship between the pressure and the height of a column of liquid which it will support.

water a definite speed if it were allowed to escape from the tank under pressure. Now, it also takes the same kind of energy, foot-pounds, to lift an object against gravity. Therefore, we might reverse the process and notice how the body would lose pressure energy and gain potential energy if the given amount of water were raised from the bottom to the top of the tank. Thus, we might say that the pressure energy could be converted into potential energy and consequently is capable of pushing the liquid to such a height that the new potential energy would be equal to the original pressure energy.

In Fig. 5·6, we indicate that pressure P_1 is capable of lifting the liquid to some height H_1. The pressure created by a column of water of height H_1 will be exactly equal to the original pressure P_1. For the ideal conditions of no friction and no change of elevation,

Kinetic energy 1 + pressure energy 1

= kinetic energy 2 + pressure energy 2

$$\frac{1}{2}\frac{WV_1^2}{g} + mgh_1 = \frac{1}{2}\frac{WV_2^2}{g} + mgh_2 \tag{5·11}$$

but $W/g = m$, so

$$\tfrac{1}{2}mV_1^2 + mgh_1 = \tfrac{1}{2}mV_2^2 + mgh_2 \tag{5·12}$$

By dividing through by m, we get

$$\tfrac{1}{2}V_1^2 + gh_1 = \tfrac{1}{2}V_2^2 + gh_2 \tag{5·13}$$

We know that $h_1 > h_2$ and $V_2 > V_1$; therefore,

$$\tfrac{1}{2}V_2^2 - \tfrac{1}{2}V_1^2 = gh_1 - gh_2 = g(h_1 - h_2)$$

$$V_2^2 - V_1^2 = 2g(h_1 - h_2) \tag{5·14}$$

But in a pipe in which the flowing fluid is incompressible, equal volumes pass all given points in the same period of time. Therefore

$$A_1V_1 = A_2V_2 = K = \text{rate of flow} \tag{5·15}$$

and

$$V_1 = \frac{A_2V_2}{A_1} \tag{5·16}$$

Substituting this for V_1 in (5·14), we get

$$V_2^2 - \left(\frac{A_2V_2}{A_1}\right)^2 = 2g(h_1 - h_2) \tag{5·17}$$

$$V_2^2 - \frac{A_2^2V_2^2}{A_1^2} = 2g(h_1 - h_2) \tag{5·18}$$

$$V_2^2\left(\frac{A_1^2 - A_2^2}{A_1^2}\right) = 2g(h_1 - h_2) \tag{5·19}$$

$$V_2^2 = \frac{2g(h_1 - h_2)}{(A_1^2 - A_2^2)/A_1^2} = A_1^2\,\frac{2g(h_1 - h_2)}{A_1^2 - A_2^2} \tag{5·20}$$

For any given values for $A_1 - A_2$,

$$V_2 = K\sqrt{2g(h_1 - h_2)} = \text{flow rate in restriction section} \tag{5·21}$$

It must always be remembered that in a pipe with variations in size, the speed will be the greatest where the pipe is smallest, but the pressure will be the greatest where the speed is the least.

In pipe flow when there is friction, there will be a loss of head or pressure because some of the pressure energy is used to do work in overcoming the drag of the pipe friction. The speed of the fluid being the same at all points in a pipe of constant diameter, the kinetic energy of each pound must be the same; therefore, all frictional losses are at the expense of the potential or pressure energy. This is indicated in Fig. 5·7.

Example 5·7 If water at 100 psi enters a pipe having a cross-sectional area of 10 in.² and then passes through a pipe of 2.5-in.² cross section with a pressure in that section of 90 psi, what is the rate of flow in the restricted section? Neglect the effect of friction. Use Fig. 5·8. Kinetic energy is in foot-pounds; pressure energy is in

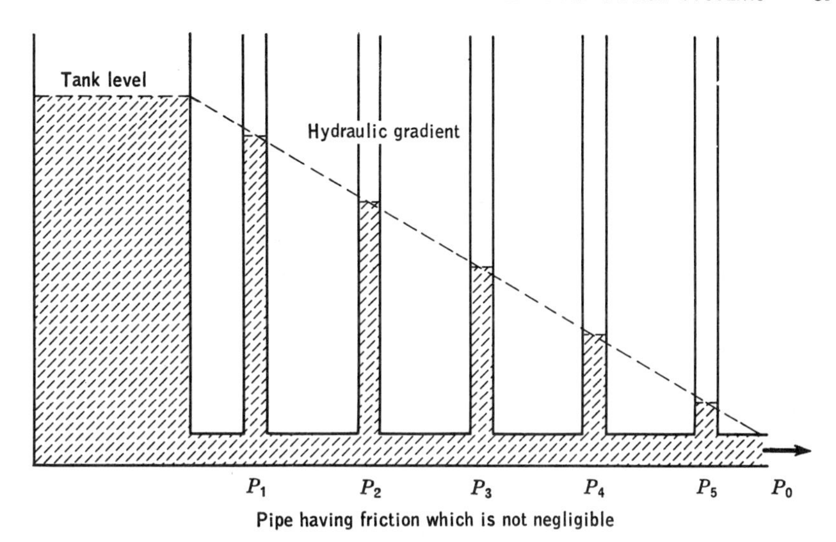

Fig. 5·7 In a pipe of constant cross-sectional area and significant friction, the pressure drop increases with the distance.

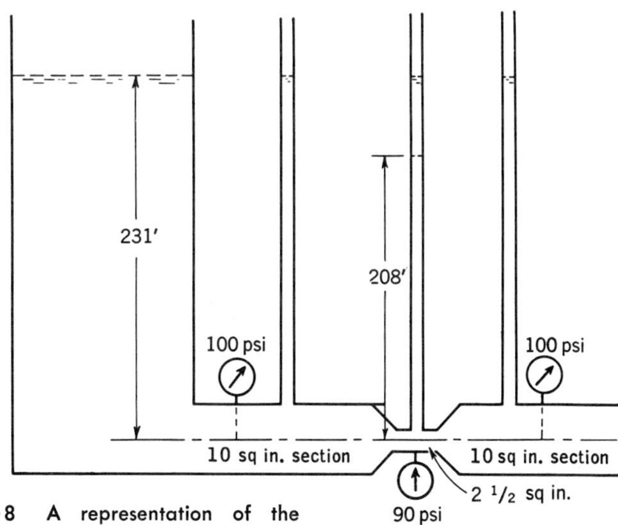

Fig. 5·8 A representation of the conditions in Example 5·7.

foot-pounds, and h is measured in feet; areas must be in terms of square feet; speed must be in feet per second; pressure must be in pounds per square foot.

SOLUTION

$$A_1 = \frac{10}{144}\text{ ft}^2 = 0.0695\text{ ft}^2 \qquad P_1 = 100(144) = 14{,}400\text{ psf}$$

$$A_2 = \frac{2.5}{144}\ \text{ft}^2 = 0.0174\ \text{ft}^2 \qquad P_2 = 90(144) = 12{,}960\ \text{psf}$$
$$= 13{,}000\ \text{psf approx}$$

Since 90 psi is equivalent to 13,000 psf and 1 ft^3 of water = 62.4 lb,

$$h_2 = \frac{13{,}000\ \text{lb/ft}^2}{62.4\ \text{lb/ft}^3} = 208\ \text{ft}$$

Since 100 psi is equivalent to 14,400 psf,

$$h_1 = \frac{14{,}400\ \text{psf}}{62.4\ \text{lb/ft}^3} = 231\ \text{ft}$$

$$V_2 = A_1\sqrt{\frac{2g(h_1 - h_2)}{A_1^2 - A_2^2}} = 0.0695\sqrt{\frac{2(g)(231 - 208)}{0.00485 - 0.003\ \text{(approx)}}} \qquad \text{from (5.20)}$$
$$= 0.0695(552) = 38.4\ \text{ft/sec}$$

$$\text{Volume} = \text{area} \times \text{speed} = (0.0174\ \text{ft}^2)(38.4\ \text{ft/sec}) = 0.67\ \text{ft}^3\text{/sec}$$

$$\text{Weight per second} = (0.67\ \text{ft}^3\text{/sec})(62.4\ \text{lb/ft}^3) = 41.7\ \text{lb/sec}$$

Example 5·8 If water enters a series of pipes having the cross-sectional areas shown in Fig. 5·9 and the initial pressure is 100 psi when the flow is 10 fps in the first section, what are the pressures at each other section? Ignore any frictional effects.

Solution. Energy at the first section equals energy at the second section, and so on.

Fig. 5·9 Schematic representation of the conditions in Example 5·8.

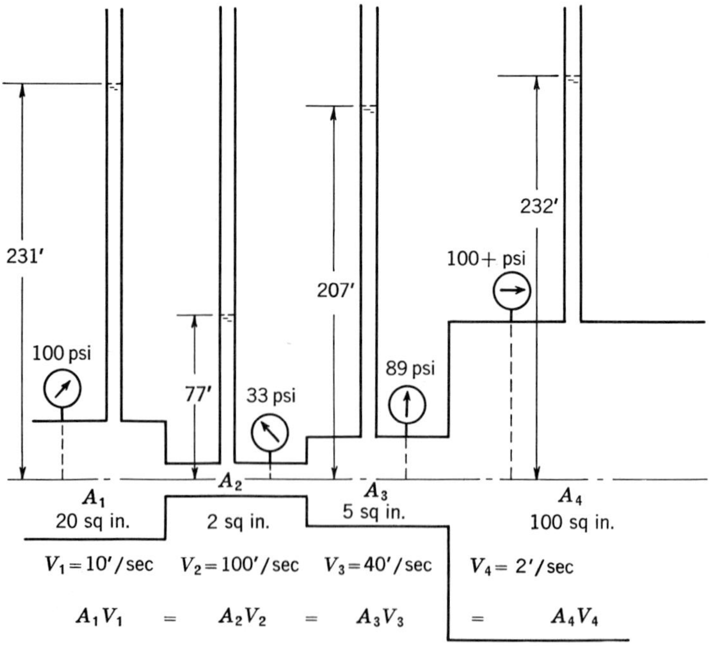

Pressure energy 1 + kinetic energy 1 = pressure energy 2 + kinetic energy 2 = · · ·

"Pressure energy" must be understood to mean the ability to lift a pound of water to a given height, which then becomes potential energy. As we are using the terms, pressure energy is capable of being converted into potential energy.

$$100 \text{ psi is created by a column } \frac{(x \text{ ft high})(62.56/\text{ft}^3)}{144 \text{ in.}^2/\text{ft}^2} = (0.433 \text{ psi/ft})(x \text{ ft})$$

$$100 \text{ psi, equivalent height} = \frac{100 \text{ psi}}{0.433 \text{ psi/ft of height}} = 231 \text{ ft}$$

$$\text{Potential energy per pound of water} = (231 \text{ ft})(1 \text{ lb}) = 231 \text{ ft-lb}$$

$$\text{Total energy per pound of water} = \text{PE} + \text{KE}$$

$$= 231 \text{ ft-lb} + 1.55 \text{ ft-lb} = 232.5 \text{ ft-lb}$$

Then at area 2, if KE = 155.5 ft-lb, PE = 232.5 − 155.5 = 77.0 ft-lb/lb or a height of 77 ft. This represents a pressure of

$$(77 \text{ ft})(0.433 \text{ psi/ft} = 33.3 \text{ psi})$$

Then for section 3, if KE + PE = 232.5 ft-lb/lb and

$$\text{KE} = \frac{1}{2}\left(\frac{1}{32.2}\right)(40^2) = \frac{1{,}600}{64.4} = 24.9 \text{ ft-lb/lb}$$

then

$$\text{PE} = 232.5 - 24.9 = 207.5 \text{ ft-lb/lb}$$

which indicates a height of 207.5 ft or a pressure of (207.5 ft)(0.433 psi/ft) = 89.6 psi.

For section 4

$$\text{KE} = \frac{1}{2}\left(\frac{1}{32.2}\right)(2^2) = 0.06 \text{ ft-lb/lb}$$

$$\text{Total energy} = \text{PE} + \text{KE} = \text{PE} + 0.06 \text{ ft-lb/lb} = 232.5 \text{ ft-lb/lb}$$

$$\text{PE} = 232.5 - 0.06 = 232.4 \text{ ft-lb/lb}$$

$$\text{Height} = 232.4 \text{ ft}$$

$$\text{Pressure} = (232.4 \text{ ft})(0.433 \text{ psi/ft}) = 100.4 \text{ psi}$$

Equation (5·11) furnishes the relationship for solving certain problems. If we have a flow of water through a piece of frictionless pipe having the four diameters shown in Fig. 5.9, and with the input pressure 100 psig when the flow V_1 is 10 fps, then we should be able to compute the pressure at each of the other sections.

CHECK FOR EXAMPLE 5·8

$$V_2^2 = \frac{(A_1^2)(2g)(h_1 - h_2)}{A_1^2 - A_2^2}$$

$$100^2 = \frac{(20/144)^2(64.4)(231 - h_2)}{(20/144)^2 - (2/144)^2}$$

$$= \frac{(400/144^2)(64.4)(231 - h_2)}{400/144^2 - 4/144^2}$$

$$10{,}000 = 400(64.4)(231 - h_2)\left(\frac{1}{396}\right)$$

$$\frac{10{,}000(396)}{400(64.4)} = 231 - h_2 = 154$$

$$h_2 = 77 \text{ ft}$$

which checks, because a column of water of height 77 ft creates a pressure of 33.3 psi.

SPECIAL TYPES OF PNEUMATIC AND HYDRAULIC AMPLIFIERS

Pneumatic amplifiers. The term *amplifier* will suggest to many a complex electronic device found in radio and television sets, but in the process control industry, and in certain types of pneumatic instrumentation, there are many common amplifiers which are quite simple and which are based on a few elementary principles, many of which have already been studied.

A wrench may be said to be part of a torque amplifier; a hammer may also be said to be a force amplifier. There is another type of amplifier, however, which is at the heart of the pneumatic measuring and control systems: the movement of a small baffle close to an orifice can produce forces which are hundreds or thousands of times as great as the force required to actuate the baffle. Figure 5·10 shows the main parts of such

Fig. 5·10 This schematic of a pneumatic amplifier employs all of the components required for a device of this sort. The shapes and materials may change with the design, but all the functions are the same, and the parts function in the same way.

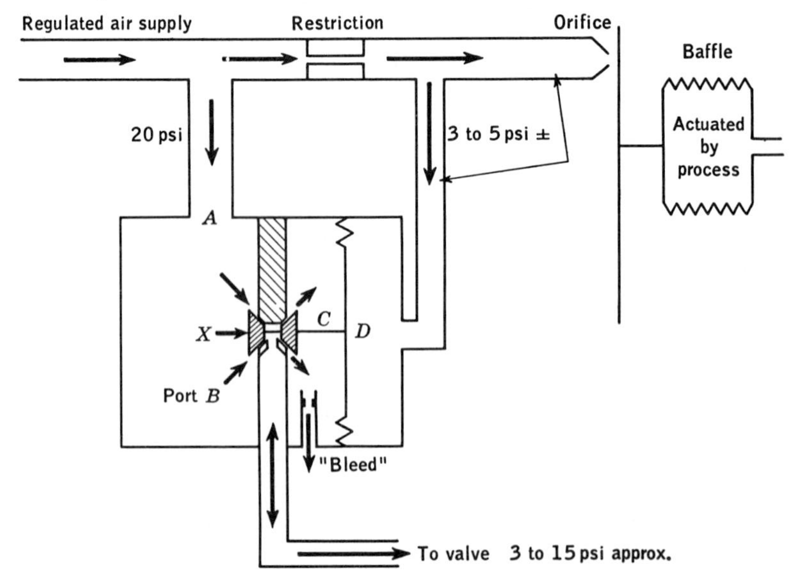

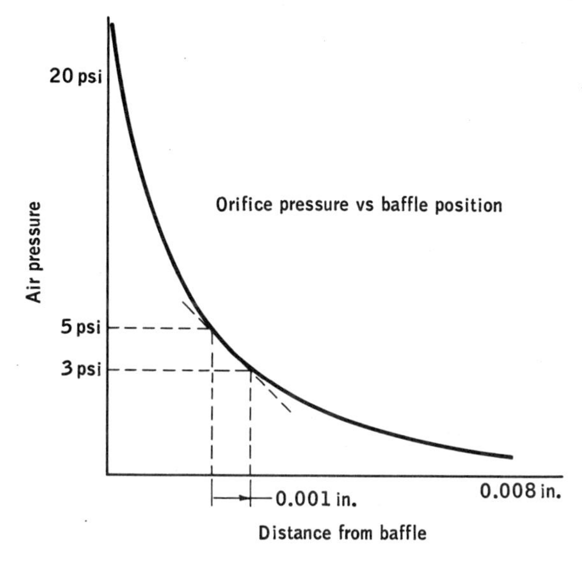

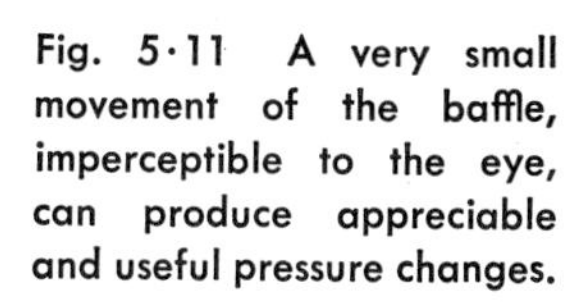
Fig. 5·11 A very small movement of the baffle, imperceptible to the eye, can produce appreciable and useful pressure changes.

an amplifier. In the tube shown, there is a small orifice or restriction which will reduce the pressure at the gage to, let us say, 1 psi when the air can escape from the orifice at its greatest rate. The orifice opening is quite small, being 0.020 to 0.040 in. in diameter. If the baffle is moved to the left toward the orifice, it will be found that as the baffle gets closer and closer to the orifice, the gage pressure will increase.

If we plot the pressure against the position of the baffle, we get a plot approximately as shown in Fig. 5·11. If the orifice were closed, the gage pressure would reach the line value of 20 psi. But with the orifice flow unimpeded, the pressure does not drop to zero.

It will also be seen on careful inspection that there is a section of the curve that is reasonably straight. This section shows that there is an easy correlation between distance and the pressure. It will also be noticed that the region where the straight-line section occurs requires a movement of the baffle of only 0.001 to 0.002 in. A very small movement of the baffle creates a relatively large pressure change. This relationship of pressure and baffle travel is the heart of most pneumatic instruments and controllers.

If we take the simple parts shown in Fig. 5·10, we have a complete pneumatic amplifier. The air relay consists of a casting or stamping with suitable ports so that air at 20 psi, for example, can be admitted at port A. Port B lets air at pressures of 3 to 15 psi go to the valve motor or instrument. At C is a mechanical connection to X which is moved by diaphragm D. When the X is moved to the right, it restricts the input air. This reduces the amount of air to the instrument or valve motor.

With little air being admitted, and air escaping at C, the pressure must fall. If, however, X is moved to the left, the input line is wide open, so the pressure will increase in the output line. It should be understood that limited movement of the baffle may be the result of a change of pressure in a Bourdon tube or the movement of some beam or other part of an instrument. Moreover, the force required to move the baffle approaches zero; thus, moving the baffle imposes no material physical load on the sensing or measuring system.

For many commercial instruments and controllers, a movement of 0.001 to 0.003 in. by the baffle changes the pressure to the relay from 5 to 3 psi, approximately. This pressure range is related to the linear section of the displacement–output pressure curve, Fig. 5·11. The controlled output pressure of the air relay will generally vary from 3 to 15 psi for a change in the input pressure of 2 psi. Other ranges of pressures and baffle travel may be obtained. Power-plant equipment often employs higher pressures because of the more massive equipment to be controlled.

This type of pneumatic amplifier should be thoroughly understood, because it is used so often in industry in instruments and controllers. Almost all pneumatic controller action in simple form will be concerned with what we use to control the baffle movement. Level, pressure, flow, temperature, and allied control processes can be easily understood if this baffle amplifier is *mastered*. Similarly, temperature, pressure and differential pressure transmitters employ this relay-amplifier in one form or another.

The measuring devices which respond to temperature, density, level, and pressure will, in many instances, do nothing more than position the baffle with respect to the orifice. The output pressure of the amplifier will then be used as a measure of the variable.

Hydraulic amplifiers. In its behavior a hydraulic amplifier is very similar to its pneumatic counterpart. The controlled pressures are higher in the hydraulic equipment, and the actuating medium is virtually incompressible. When large pieces of equipment are to be manipulated or moved to precisely defined places, the hydraulic unit will have certain advantages.

Because of the high values of pressure used with controls of this type, there is a requirement for tight packing around the rod linking the pilot piston PP to the sensing element SE, Fig. 5·12. Therefore, more force will be required to move PP than was required to move the baffle in the pneumatic system. Consequently, this hydraulic amplifier cannot work with as minute inputs as can the pneumatic units.

In the neutral position, the pilot piston PP closes the ports to the

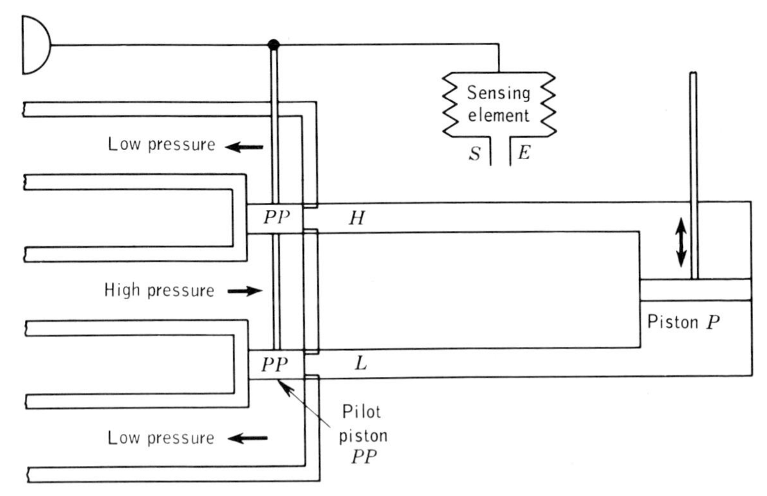

Fig. 5·12 Schematic of hydraulic amplifier system. Hydraulic systems can develop extremely large forces, but they are not so sensitive to small forces as are pneumatic systems, which require a baffle movement of only about 0.001 in.

large piston *P*. If, however, the sensing element moves the pilot piston up, it will uncover the high-pressure port so that fluid will flow in at *H* and out at *L*. This will move piston *P* down, thus moving the valve or damper connected to it. So long as there is an opening at *H* and an outlet at *L*, the piston will be forced down. When the sensing transducer no longer calls for a change in the valve opening, pilot piston *PP* will move up, closing both ports at *L* and *H* and, in effect, locking the piston in position, because the fluid on each side of the piston is virtually incompressible.

SUMMARY

1. *Work* is the product of a force exerted through a distance. There must be both a force and a distance through which the force is exerted. "Trying" does not count; there is no measure of "effort."

2. *Energy* is a measure of the capacity for doing work.

3. Energy exists in many forms. We are concerned generally with energy of (*a*) motion, which is called kinetic; (*b*) position, which is called potential; and (*c*) composition, which is called chemical.

4. Energy can be transformed from one form to another and may be transformed repeatedly, particularly from kinetic to potential, and vice versa.

5. No more energy can be removed from a system than was there originally, plus any energy added. No energy is free; it is always obtained by decreasing one or more of the various forms in which energy may exist. Aside from certain atomic reactions, energy may not be created or destroyed, but is merely converted into some other form.

6. Levers may be used to (*a*) increase movement, but at the sacrifice of force;

(*b*) increase force, but at the sacrifice of movement; or (*c*) change the direction in which a force is applied or transmitted.

7. The fundamental relations for levers are:

$$(\text{Force 1})(\text{distance 1}) = (\text{force 2})(\text{distance 2})$$

but

$$(\text{Force})(\text{distance}) = \text{work}$$

so

$$\text{Work 1} = \text{work 2} \qquad \text{if there is no friction}$$

where distance means the actual amount of motion in the direction in which the force is applied.

8. (*a*) Torque means the amount of twisting effort, and it is equal numerically to the product of force and moment arm, or perpendicular distance from the point of rotation, or fulcrum, to the point of application of the force. (*b*) Torque normally is expressed in pound-feet, a unit of measure that is in no way synonymous with, or equivalent to, foot-pounds. (*c*) Applying a force to a lever creates a torque or twisting effort, but no work is done unless the lever actually moves. This is one time that trying does count. Trying to move a lever creates torque, but not necessarily work.

9. Many instruments employ torque balance for their operation: the clockwise movement created by a pressure or by pressures is counterbalanced by torques of equal magnitude. Just as for each force there must be an equal and opposite force, so for each torque there must be an equal and opposite torque.

10. (*a*) Liquid escaping without friction from an orifice in the side of an open vessel will leave with the same speed which it would have acquired in falling freely through the vertical distance from the surface to the orifice. (*b*) If the liquid is subjected to an external pressure, then the escaping liquid, in the ideal case, will have the same speed that it would acquire in falling freely a distance equal to the height of the fluid plus the distance equal to the equivalent pressure head.

11. In a system in which the energy per unit of material is constant, any increase in one energy is at the expense of some other energy. As fluid passes through an orifice, its speed increases and so does its kinetic energy, but its pressure energy drops. Therefore, in normally confined systems of fluids, increases in flow rate are accompanied by decreases in pressure. The total values of the energies remain constant for idealized frictionless conditions.

12. Almost all pneumatic transmitters and controllers use the change in orifice pressure as a baffle is positioned close to the orifice.

QUESTIONS

5·1 What is the difference between kinetic energy and potential energy? Will 1 ft-lb of one kind of energy do as much as 1 ft-lb of the other?

5·2 Why does an earth satellite continue to travel at high speed for a long time after the propellant is used up? Why doesn't it stop within a short time?

5·3 What particular uses do we make of the conversion from kinetic energy to potential energy, and vice versa, in the field of instrumentation?

5·4 Discuss the differences between foot-pounds of work and pound-feet of torque.

5·5 Can foot-pounds be converted into pound-feet, and vice versa?

5·6 How do work and torque differ in their definition and meaning?

5·7 How do the distances used in computing work and torque differ? Why should this be?

5·8 In what direction must the force be exerted if work is to be accomplished? If torque is to be produced?

5·9 (*a*) What are the ways in which levers are used? (*b*) How many of these ways are employed in common instruments?

5·10 (*a*) How does a gear resemble a lever? (*b*) How does a wheel resemble a lever?

5·11 What is the advantage of using fluids to transmit pressures? What are the fundamental principles involved?

5·12 What is true of the potential and kinetic energies which an object has, provided that no energy is added or removed?

5·13 What use do we make of energy relationships when we measure flow?

5·14 What energy conversions are made in the flow measurements studied?

5·15 Are the flow measurements of Prob. 5·14 based on volume or mass?

5·16 Where, in a closed system of flowing fluid, will the pressure be the least? Why?

5·17 Where, in a closed system of flowing fluid, will the pressure be the greatest? Why?

5·18 Draw a typical curve to show the relationship between baffle points and orifice pressure.

5·19 List the main parts of a pneumatic relay.

5·20 What are the primary reasons for using pneumatic relays?

5·21 Do pneumatic and hydraulic systems have similar characteristics?

PROBLEMS

5·1 How much potential energy would a 100-lb block gain or lose if it were raised vertically 100 ft? If it were moved horizontally for 100 ft?

5·2 A force of 10,000 lb is required to pull a truck up a hill. If the vertical movement is 50 ft, how much work is done?

5·3 A large instrument bellows has a spring rate of 100 lb/in. What work, in foot-pounds, would be stored in the compressed bellows if the compression were 2 in.? (HINT: The average force must be used.)

5·4 An 8-oz ball must rise how far to acquire 200 ft-lb of potential energy?

5·5 How fast must a 3,220-lb car travel if it is to acquire 200,000 ft-lb of kinetic energy?

5·6 How high would the car in Prob. 5·5 have to be raised to acquire the same numerical amount of potential energy?

5·7 If the car in Prob. 5·5, were to fall freely from that height, how much of each energy would it have after it had fallen halfway?

5·8 A child who weighs 100 lb is on a swing. If the child were raised 1 ft above the lowest point and then released, how fast would he be going as he passed through the low point?

5·9 How fast must a 32.2-lb object be moving if it is to have 100 ft-lb of kinetic energy?

5·10 If none of the kinetic energy of Prob. 5·9 is wasted, how high will this kinetic energy lift the object?

5·11 A 100-lb weight is 100 ft above the ground. Find both its potential energy and kinetic energy when it has fallen (*a*) 25 ft, (*b*) 50 ft, and (*c*) 100 ft. What generalization can you make about the results?

5·12 A ball which weighs ½ lb is thrown vertically with a speed of 100 fps. How high will it go? What is its potential energy at the highest point? What are the kinetic and potential energies when it is either halfway up or halfway down?

5·13 If the child in Prob. 5·8 had been lifted 1 ft directly above the lowest point on the arc and dropped, how would his speed compare with that in Prob. 5·8?

5·14 A rope is wound around a shaft 6 in. in diameter. What torque is applied to the shaft if the rope is pulled with a force of 200 lb?

5·15 How far from the fulcrum must a force of 100 lb be applied if it is to develop a torque of 400 lb-ft? Where would a 200-lb force have to be applied to produce an equal and opposite torque?

5·16 A differential pressure gage has a net force of 10 lb applied to one end of a member 10 in. long which may be pivoted at any point. Where would the fulcrum have to be located so that a force of 3 lb applied to the other end would produce an equal and opposite torque?

5·17 If the differential pressure gage in Prob. 5·16 had a net force of 4 lb instead of 10 lb applied, where would the pivot point have to be?

5·18 A differential pressure transmitter has a net force of 10 lb acting on the diaphragm, which acts 2 in. from the fulcrum. Where will a force of 2 lb have to be applied to balance the 10-lb force?

5·19 A pen arm is 6 in. long and is pivoted ½ in. from one end. If a Bourdon tube is linked to the pen arm ¼ in. from the pivot, how much will the pen move for a ⅛-in. movement of the Bourdon?

5·20 The force developed by a bellows is 50 lb. It pushes on one end of an arm 12 in. long. Where must the pivot be if a force of 10 lb is providing an equal and opposite torque?

5·21 If the bellows of Prob. 5·20 had an effective area of 4 sq in. and an applied pressure of 5 psi, where would the pivot have to be to create a condition of torque balance?

5·22 A force of 10 lb is applied to a plank, 4 ft to the left of the fulcrum. If the plank is weightless, how far will a 5-lb weight at the torque equilibrium point be raised when the 10-lb weight is pushed down 6 in.? What PE will the 10-lb weight lose? What PE will the 5-lb weight gain?

5·23 Repeat Prob. 5·22 but with the 5-lb weight replaced by a 1-lb weight.

5·24 A pen is so connected as to increase the Bourdon movement by a factor of 10. What is the mechanical advantage? (HINT: If the movement is increased, what must be true of the force?)

5·25 If the mechanical advantage of a device is greater than 1, what must be true of the output force and distance with respect to the input force and distance?

5·26 If the mechanical advantage of a device is less than 1, what must be true of the output force and distance with respect to the input force and distance?

6 Heat and Heat Transfer

Factors affecting heat transfer. Inasmuch as almost all temperature measurements involve the transfer of heat energy, it becomes very important that we understand the factors which may influence this exchange of heat energy. It is not only in temperature measurements that we become concerned with such problems but also in certain techniques used for analysis of gases, and the measurement of air velocity, as well as high-vacuum measurements. Heat energy can be transferred to or from an object (1) by conduction, (2) by convection, and (3) by radiation. As a general rule, we want all of our measurements to reflect, almost instantaneously, any changes in the variables which they are measuring. Any factor which will delay the transfer of energy will consequently delay the response of the instrument. Errors caused by delays in response are termed *dynamic errors* to distinguish them from *static errors,* which are caused by inaccuracies in the calibration, friction, and lost motion in the linkages. Dynamic errors are not constant; they depend on the speed with which the variable is changing and on the amount of energy to be transferred in either one direction or the other with respect to the measuring device.

We might state it as axiomatic that in instrumentation we are interested in knowing the instantaneous value of the variable, and not its value 10 sec or 10 min earlier. While this book is concerned only with measuring the variable, the associated problem of control to maintain fixed or predetermined values of the variable requires that the measurement reflect the variable as faithfully as possible. The earlier that corrective action can be taken, the smaller is likely to be the deviation from the desired value. What, then, are some of the factors which affect the exchanges of energy? The most important ones are:

1. The type of fluid in which the sensing element is placed
2. Any film which tends to insulate the element from the fluid

3. The speed of movement of the fluid with respect to the sensing element
4. The material of which the sensing element is made and the mass of the sensing element
5. Any loss of energy from the sensing element to adjacent materials
6. Special bulbs
7. Location—*dead time*

Let us then consider each of the above factors as it affects the operation of thermometric devices.

The type of fluid. The word "fluid" in this case includes both liquids and gases. From personal experience, we have observed that materials can be heated much more rapidly when they are in contact with a liquid than when they are in contact with a gas, such as air, at the same temperature. A cake of ice plunged into water at 100°F will melt many times more rapidly than it will if placed in air at the same temperature. Perhaps one explanation is that each cubic inch of the water will store or give up far more energy than will an equal volume of the gas for the same temperature change. The faster the heat can be transferred to the ice, the greater will be the melting rate.

One very important factor which must always be considered is that *there must always be a difference in temperature before any thermal energy can be transferred from one object to another.* As the difference between the fluid and the sensing bulb becomes smaller and smaller, a liquid has more heat to transfer than does the gas for the same temperature difference.

Perhaps this is the place to point out that, on the basis of a mathematical expression, one object can never attain the temperature of a second. For example, on a cold winter day, bring into your kitchen, which has an assumed temperature of 70°F, a piece of iron which has a temperature of 0°F. If we could very carefully and accurately measure the temperature of the piece of iron, we would get a series of readings which, when plotted, would give us the graph shown in Fig. 6·1. You will note that as time went on, the change in temperature per minute became less and less until, after 2 hr had elapsed, there was almost no change in temperature. We would have encountered that same sort of thing if we had dropped the piece of iron into a pail of water which was kept at 70°F, but the rate of change would have been very much greater. As the temperature of the piece approached that of the air, a smaller and smaller temperature differential caused the heat energy to be transferred from the warmer to the colder object.

$$\text{Heat transferred by conduction} = K(T_h - T_1)$$

where K depends upon the surfaces, the units of heat, area, and so on; T_h is the higher temperature; and T_1 is the lower temperature.

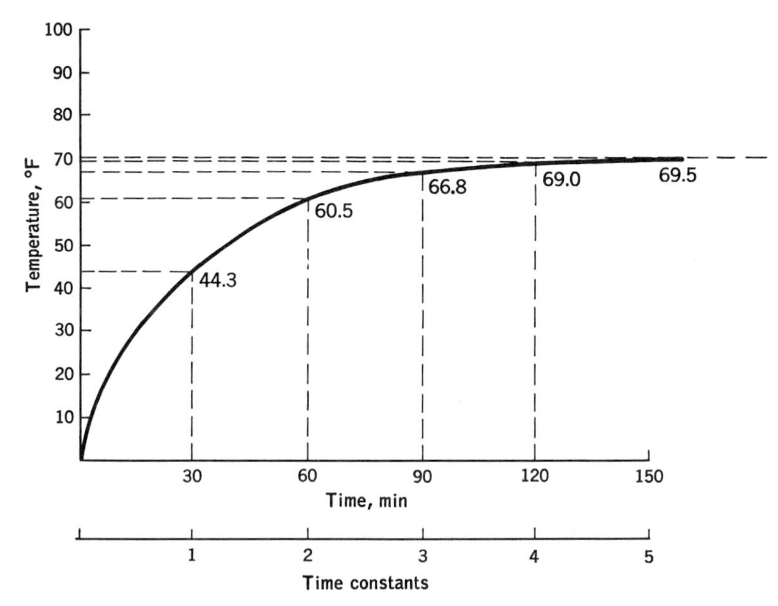

Fig. 6·1 When a step change of temperature is involved, more than five or six time constants will be required to bring the indicated temperature to within 1 per cent of its final value.

Another expression for this rate of heat transfer is

$$\text{Rate of heat transfer by conduction} = K(e^{-Rt})$$

where K = original rate, cal/sec or Btu/sec
e = base of natural logarithms
R = a constant dependent upon the conditions
t = time expressed in the same units as are used for K and R

The value of the factor e^{-Rt} approaches 0 as time increases indefinitely. Thus, as the rate of heat transfer decreases logarithmically, so must the rate of change of temperature follow, because it always takes a fixed number of heat units to create a given temperature change.

It may seem that we are making a great deal of fuss about a theoretical point—everyone of us "knows" that an object brought into a room will attain the temperature of the room. But what sort of device were we using to tell us that the two were at the same temperature? Probably some very crude device such as our hands. Now, for all practical purposes in many instances, these two do arrive at "identical temperatures." But in industry today, we are not satisfied with having a sensing unit within one or two degrees of the surrounding space. With such temperature differentials, the rates of heat transfer may be moderate. But for those situations in industry in which we now want to measure to 0.1°F, the time required to effect the same transfer of heat energy will probably be far greater:

the smaller the temperature differential, the lower the rate of heat transfer. In the example used here, the heat to be transferred may be only one-tenth as much, but there is only one-tenth the initial temperature difference, and this becomes still smaller as the sensing element approaches more and more closely the true temperature.

Insulating films. Any sensing bulb, and any other material also, will have attached to its surface a film of gas or fluid of some sort. This film adheres quite tenaciously, and it acts as a series resistance in the electrical case. It should be recalled that one of the best heat insulators is a blanket of dry *motionless* air. Storm windows are installed to preserve a film of air which has very little motion with respect to the warm inner pane. The wall and ceiling insulation which is installed in homes, industries, cold-storage warehouses, and so on is primarily nothing but material which will trap extremely small amounts of air which is kept from transferring heat to other objects by convection currents. If a film of air or gas will act as an insulator in your attic, it will function in the same way when it surrounds a thermometer bulb.

In some cases, the insulating film may also include coatings of grease or sludge. These materials also are likely to be relatively poor conductors of heat, with the result that the bulb is further insulated from the fluid with which it must exchange heat energy.

Speed of the fluid. If we have a piece of ice in water which is not stirred, it will melt fairly slowly. The reason for this is that the water close to the ice soon assumes a temperature nearly that of the ice. With only a small temperature difference, the rate of transfer of heat becomes small. The only circulation of the water is then created by the differences in the density established by the temperature differences. Now if we stir or agitate the water, water which has not been cooled by its proximity to the ice will be brought into contact with the ice. There will now be a greater temperature difference between the water and the ice with the result that the ice will melt at a greater rate. As the rate of stirring or agitation goes up, the film of relatively still water which surrounds the ice will become thinner and thinner. Thus, the transfer of heat is speeded up for two reasons:

1. The insulating water film is reduced in thickness.
2. Warmer water is continually brought to the ice.

With instruments located in air or gas, the same situation prevails: there will be a coating of the gas which surrounds the bulb and which insulates the bulb from temperature changes. The film cannot be completely dissipated even with speeds of 1,000 to 2,000 fps. Much of the energy which is required by very high speed planes is used to overcome the friction between the air through which the plane is passing and the film of air which adheres to the plane's surfaces. Despite the great rate at which cooling air is moving over the plane, the frictional forces set up between this layer of air which adheres to the surfaces and the moving air can create temperatures of well over 1000°F. Therefore, we have to accept the presence of such insulating films. But they can be reduced in thickness if the air or gas is moving at a relatively high speed over the sensing bulb's surface.

The bulb material and its mass. In order that the temperature may be sensed with the minimum transfer of energy, which would speed up its response, the ideal sensing bulb would have a wall approaching zero thickness and a mass which would also approach zero. But practical considerations preclude such an ideal solution. In thermometers which employ the filled-liquid system, particularly mercury for temperatures up to 1000 and 1200°F, the liquid must be subjected to high pressures to prevent it from boiling; pressures of 2,000 to 3,000 psi are not unusual. Consequently, the bulb has to be strong enough to stand the pressures which will be created by the confined liquid. The bulb actually acts as a spring which is stretched by the expanding liquid, and in return the bulb exerts upon the confined liquid a pressure which is then transferred to the liquid in the capillary and thus to the pressure-sensing element.

In the ideal case, we would make the sensing bulb of silver or copper because of the high rates at which those metals can conduct heat energy. But the cost of silver largely restricts its use, while the chemical activity of copper seriously limits the number of places where it can be used. Stainless steels are quite generally used for sensing bulbs because of their relative chemical inertness. But most of the stainless steels employed for these purposes are not good conductors of heat. Inasmuch as we have to have an inert material for the bulb, we must then accept the reduced thermal conductivity as the lesser of two evils. Consequently, the bulb then must be selected with the following in mind:

1. Adequate physical strength to withstand the internal pressures
2. Sufficient resilience to withstand repeated stretching without loss of calibration
3. Chemical stability in the materials with which it is used
4. Adequate volume so that the variations in the volume of the pressure-responsive element and the capillary will not adversely affect the accuracy of the unit

For the Class II thermometers which use the vapor pressure developed by the confined liquid, it is not necessary to change the temperature of all of the liquid; only the temperature at the interface between the liquid and the vapor need be changed. The small amount of energy that is necessary is responsible in some degree for the fact that this class of filled thermometers has the smallest sensing bulb and, consequently, the greatest response rate for the filled systems.

On the other hand, the transfer of heat to a confined gas requires a maximum amount of surface to effect the heat exchange. Therefore, the volume of a bulb on a gas thermometer is large to:

1. Speed up the rate of transfer of heat to the confined gas.
2. Reduce to a minimum any temperature-pressure effects which occur in portions other than the bulb itself. One rule is that the volume of the bulb shall be at least 8 times that of the rest of the system. Another rule sometimes limits the span to one-third of the range.

In certain types of resistance-temperature bulbs, the winding is made to respond at a maximum rate by using a silver core which can absorb and transfer heat with the minimum temperature differential.

The comments which have been made about the temperature bulb also apply to thermocouples:

1. The well or protecting tube is in many ways just like the bulb of the pressure units. The selection of the material is made on the same basis: adequate strength for the pressures which are present and chemical resistance to the ambient conditions.
2. The mass of the well should be kept to a minimum consistent with the other demands.
3. The amount of wire which forms the hot junction should be a minimum. A butt weld will have the minimum amount of material which must be heated or cooled. It is to be preferred to a twisted junction.

When a well is provided for use with thermometers and thermocouples, it must be realized that the response is greatly reduced for at least two reasons:

1. The mass of the system is greatly increased. There may be no way to avoid the use of a well: the thermocouple or thermometer must be removable without shutting down the system, and the pressures may be such that they would injure the apparatus. Or it may be that chemical inertness cannot be obtained except through use of a well.
2. The heat necessary for functioning of the temperature-sensing system must now pass through an additional thermal resistance. This will increase the temperature differential between the active element and the body whose temperature is sought, as well as result in a very significant increase in the time of response of the device.

The thermal element should never be protected by a well when it is to operate a controller, unless there is no other way around the problem. The increased time lag is one of the major problems in satisfactory controller action.

Although it is reasonably obvious, it must always be remembered that the thermocouple or thermometer will respond only to its own temperature, and not to the temperature of the surroundings, unless the two have the same value. Every reasonable attempt must be made to ensure that the two have an insignificant difference in temperature.

With certain thermocouples, particularly those using platinum and its alloys, there may be a ceramic gas-tight tube surrounded by a protecting tube. In this case, there may be a very serious difference between the temperature of the furnace and that of the responding element. Not only is there more material to heat in this instance, but every extra tube provides that much more resistance to the transfer of heat through it and through the two insulating films which are at its surfaces.

Loss of energy to adjacent materials. Unless there is sufficient penetration by the sensitive element into the area whose temperature is sought, there may be such a transfer of heat away from the bulb that the bulb never reaches the temperature of its immediate surroundings. For example, if a bulb were in steam but the end of the bulb and its capillary were not protected by suitable insulation, then the temperature of the capillary end of the bulb would be not that of the steam, but intermediate between the steam and the surrounding air. Inasmuch as many steam-temperature-measuring devices respond to *average* temperatures of the expanding mediums, any deviation of the *average* from the steam temperature will create an error. It will be noted in Chaps. 14 and 15 that there are specified depths of insertion for thermocouple wells which are dependent upon the way that the heat is transferred to the thermocouple. Every attempt must be made to ensure that there is a suitable minimum heat loss through the bulb and tubing. Such heat losses will also increase the delay in the response of the device to temperature increases.

At the higher temperatures, heated objects receive and send out much energy by radiation. If a thermocouple well is in a spot where it can radiate energy to a *cold wall,* for example, it will register a temperature lower than the true value because of its excessive radiation. Under such conditions, provision should be made to *shield* the thermocouple from the cold area.

The converse is also true. If a thermocouple is placed where radiant energy from a flame, or even the flame itself, is allowed to heat the device, then the indicated temperature will be greater than it should be. Much thought must therefore be given to the placement of temperature-sensing elements to make sure that the energy which the unit receives is representative of the place where it is located and that it measures only what it is supposed to measure.

Special bulbs. When air and gas temperatures are to be measured, particularly when the temperatures are not high, then long, averaging capillary-type bulbs may be used. This provides a maximum surface for the transfer of heat and a thinner wall than could be used if the bulb were in conventional form.

If a similar type of bulb is used with vapor-pressure devices, then the device becomes a *hot-spot indicator.* Such a unit can then monitor large areas and give a response to the highest temperature.

Other factors affecting location. Even though we have located our temperature bulb at a point where the temperature is representative of what we want to measure and there is no undesirable loss of energy to some other part of the system, there is one other factor which must be considered if a satisfactory installation is to be obtained. Not only must the temperature be correct, but its value must be secured at the earliest possible moment, especially for those installations where the device not only monitors the temperature but also actuates a controller. It has been mentioned before that everything possible must be done to speed up the time of response, but in those instances we were thinking of the response of the thermal element. At the moment, we are concerned with those situations in which there is a delay in reaching the sensitive element. For example, suppose that the temperature of a liquid leaving a heat exchanger is to be maintained at some predetermined value. The liquid is flowing through the pipe at the rate of 10 fps. Then for every 10 ft that the temperature bulb is removed from the discharge point of the exchanger there will be a delay of 1 sec. Move the bulb to a point 200 ft away because that simplifies the problem of reading and puts the bulb close to the operator and you will have introduced a delay of 20 sec into the response of the thermometer. (We will assume that the pipe is so well insulated that there is no change in the fluid temperature along the pipe.) The following, therefore, is one of the cardinal rules: *Always place the sensing bulb close to the point where you can first make the measurement.* This always requires the use of some judgment. When you want to measure the temperature of some fluids which are being mixed, or some liquid which is being heated by live steam, the sensing element must be far enough downstream from the point to ensure that adequate mixing has taken place and that representative temperatures are being measured.

Radiation Pyrometers. Radiation pyrometers are not subject to many of the problems which beset lower temperature measurements, but there are certain factors which are worthy of note here:

1. Although the transfer of heat by radiation is essentially instantaneous, some of the radiated heat may never reach the instrument because of the absorption by invisible gases or water vapor and other products of combustion. Consequently, any absorption may present a serious error, particularly when true temperatures rather than comparative temperatures are sought. Dust and dirt also absorb energy.

2. The field of view of the radiation unit must be at the temperature which is indicative of the measurement desired. If some cold area is included in the field, then the indicated temperature will be lower than it should be.

Electrical analogies. The flow of heat from the furnace source through the walls of the thermometer bulb to the liquid or gas which fills the

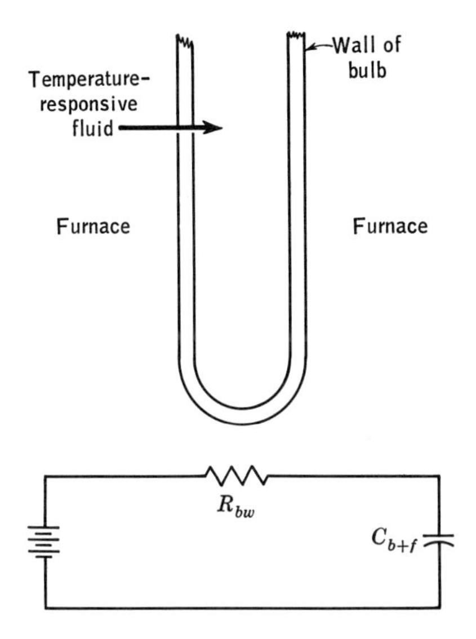

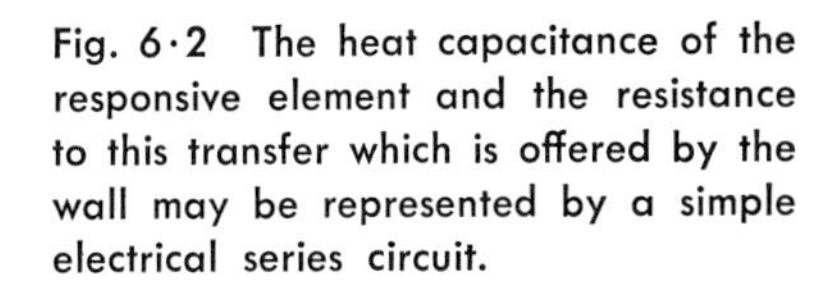
Fig. 6·2 The heat capacitance of the responsive element and the resistance to this transfer which is offered by the wall may be represented by a simple electrical series circuit.

bulb may be compared with a simple series electric circuit, Fig. 6·2. The flow of heat from the furnace or pipe to the temperature-sensing bulb is created by the temperature differential; electric current flows because of a potential differential. A temperature differential is required to cause heat to be transferred through the walls of the sensing element. We find, likewise, a difference in electrical pressure required to cause current to flow through anything which impedes the flow of electrons.

For any given size and construction of thermometer a definite amount of heat is required for a temperature change of one degree; the more heat that is added, the greater is the change in temperature of the bulb and its fill. In the electrical system, we find that adding so many electrons to a capacitor of given size will increase the back pressure or voltage by a definite amount; the more electrons added, the greater will be the back pressure, and in direct proportion. The heat-storage capacity of the temperature-sensing element may be said to be equivalent to the current-storing capabilities of the capacitor.

If the heat to be transferred to the bulb must first pass through the walls of a well and then through an intervening air space before reaching the bulb, then there is a somewhat more complex electrical equivalent which must be studied, Fig. 6·3.

Figures 6·2 and 6·3 are of the same type and may be analyzed in the same way. The three resistances in series may be treated as a single equivalent resistance, which then makes the two circuits electrically identical. However, this is an approximation.

From previous work, we know that the capacitor cannot become fully charged instantaneously when there is some resistance in series with it. And the time required for the potential across the capacitor to reach

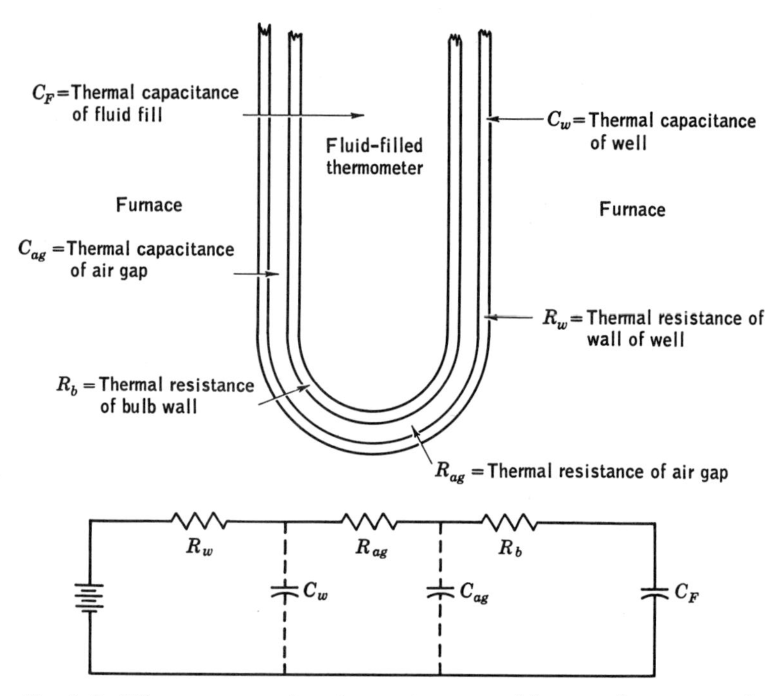

Fig. 6·3 Where a responsive element is protected by a well, an exact determination by electrical analog is quite complex. For many solutions the circuit represented by the solid lines may be adequate.

some definite fractional part of its final value depends both on the size of the capacitance, in microfarads, and on the value of the resistance, in megohms. The product of R and C will give us the time in seconds to reach 63.2 per cent of the final value. If we can determine the time for any such combination to reach 63.2 per cent of a step change, we then know the time constant of the equivalent circuit. We do not need to know the size of the resistor and the capacitor; we are interested primarily in their product.

But from a study of this analogy, it is easy to see that anything which can be done to decrease the resistance of the circuit will decrease the time constant and hence increase the speed of response. In a similar fashion, anything which can be done to diminish the thermal capacitance will also speed up the response. And high speed of response is likely to be obtained only when both values are made as small as possible.

Where we have the thermometer or thermocouple inside a well, a truer representation of the electrical equivalent circuit is as shown in Fig. 6·4. It will be noted that there must be a capacitor to represent each material which can store heat and which is not connected to the same part of the thermal circuit.

A study of the electric circuit may also reveal why no temperature-sensing element can respond in zero time. To have such performance would require that there be no resistance to the transfer of heat and that the thermal capacity of the bulb also be zero. Inasmuch as every finite amount of material requires a finite amount of heat to change its temperature, zero thermal capacity can never be attained. We may approach it quite closely with some extremely fine thermocouple and thermopile constructions, but we can never attain it.

It should be pointed out that an exactly analogous situation develops when we measure pulsating pressures. Oftentimes a needle valve is placed in the line between the gage and the source of the pulsating pressure. In addition, a small tank may be connected in parallel with the gage (see Fig. 9·46). The needle valve performs the function of a resistor, while the tank and the gage act as a capacitor. Thus the pressure at the gage will not follow the pressure peaks any more than a thermometer would follow the temperature peaks. A pressure gage working under these conditions will not follow the pressure, and it will develop a dynamic error even though it is in perfect static calibration.

Response of thermal bulbs. The data for this section were prepared by and obtained from E. W. Jensen, F. C. Kiesling, Gerald Akins, and

Fig. 6·4 When the thermal capacitance of the well may not be neglected, the equivalent electrical circuit may employ two capacitors.

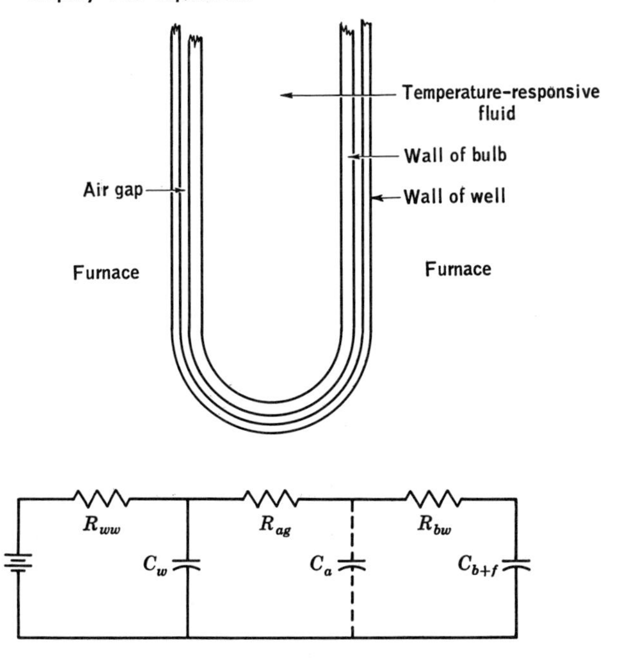

D. W. St. Clair, all of the Kodak Park Works of the Eastman Kodak Company.

Let us suppose that you have been asked to check the calibration of a recorder measuring the temperature of 170°F air in a duct. You arrive on the job with your best mercury etched-stem thermometer and place it near the recorder bulb. How long should you wait before reading the thermometer in order to measure the duct temperature within 1°F?

Another time you are asked to check the same air temperature and find that the etched-stem thermometer cycles ±5°F every 2½ min. How many degrees is the air temperature actually changing?

The answer to both of the above questions depends on the type of installation, Fig. 6·5, and on the response of the temperature-measuring system which, in turn, can be defined by a single number, the *time constant.* If the time constant of the etched-stem thermometer is 60 sec when in air traveling at 500 fpm (the velocity in the duct), it would be necessary to wait 276 sec (4.6 min, or 4.6 time constants) before making a final reading of the thermometer in order to measure the air temperature within 1 per cent or 1°F (from Table 6·1). In the second example, the thermometer was cycling +5°F to −5°F every 2½ min. Knowing the time constant of the bulb (60 sec) and the period of the cycle (5 min), the actual air temperature variation is found to be ±8°F (from Fig. 6·10).

The time constant of a thermal system is most easily illustrated by examining its response when measuring the temperature of a bath which is changing at a constant rate, Fig. 6·7*b*. Heat must flow into the thermal bulb to raise its temperature, and heat flow, in turn, must be accomplished by a temperature difference. This difference is called the *dynamic error* and depends on the heat-storage capacity of the bulb and the resistance of the bulb surface to heat flow. This temperature error causes the measured temperature to lag behind the actual temperature by a certain time T. This is the time constant of the thermal system for the particular conditions of the test.

The most common method of determining the time constant is to move the thermal bulb suddenly from one constant temperature to another. The response of a system to this change is shown in Figs. 6·6 and 6·7. It can be shown mathematically that the time required for the measured temperature to change 63.2 per cent of the difference between the two temperatures is the same time constant as that shown in Fig. 6·7*b*. Note

Table 6·1

Recovery, per cent	63.2	90	95	99	99½
Time	$1.0T$	$2.3T$	$3.0T$	$4.6T$	$5.0T$

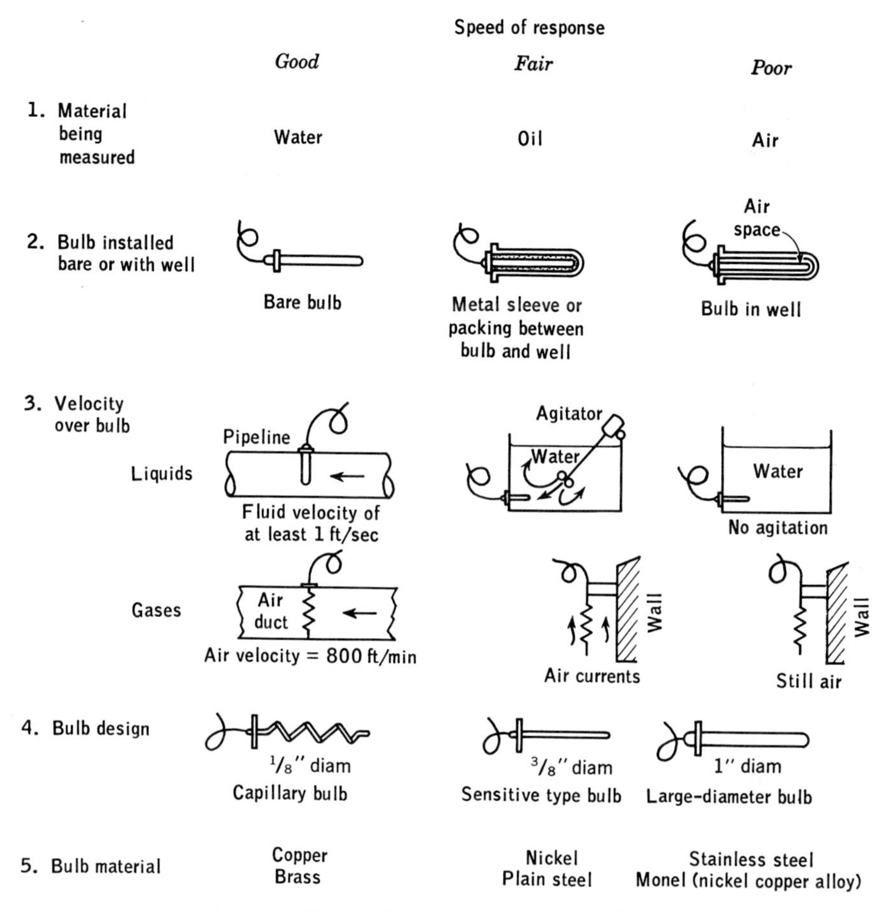

Fig. 6·5 Factors influencing the speed of response. (*Data and figure by E. W. Jensen, F. C. Kiesling, Gerald Akins, and D. W. St. Clair, Kodak Park Works, Eastman Kodak Co.*)

that the time constant for a certain bulb depends also on the composition and amount of agitation of the fluid being measured. This is explained further on page 523, and time constants for typical thermal systems are given in Figs. 6·8 and 6·9 and Table 6·2.

The recovery curve shown for a sudden temperature change is typical of that obtained when calibrating a thermal-measuring system by changing the bulb suddenly from one temperature to another. Note that a very definite length of time is required before the measured temperature reaches a certain percentage of the total change.

The time required to reach a desired percentage of the total temperature difference is related to the time constant T in Table 6·1. For the example given on page 88, the time constant was 60 sec, and the difference between duct and room temperature was 170°F − 70°F = 100°F.

Table 6·2 Time Constants for Typical Thermal Systems in Water at a Velocity Greater than 200 FPM

Item	Bulb diam, in.	Bulb material	Actuating medium	Construction	Time constant, sec
1	1	Stainless steel	Gas	Tube system	9.2
2	3/4	Stainless steel	Gas, mercury	Tube system	6.5
3	11/16	Stainless steel	Vapor	Tube system	6.1
4	1/2	Stainless steel	Mercury	Tube system	4.1
5	1/2	Stainless steel	Liquid filled	Tube system	6.0
6	3/8	Stainless steel	Vapor	Tube system	4.0
7	3/8	Stainless steel	Liquid filled	Tube system	5.0
8	1/4	Glass	Mercury, organic liquid	Etched-stem thermometers	8.0
9	1/4	Stainless steel	Bimetallic	Dial thermometer	7.0
10	3/16	Copper	Gas, vapor	Tube system	0.85
11	0.10	Chromel-constantan	Thermocouple		0.7

The time required to come within 1°F of the duct air temperature (99 per cent recovery) would be 60 sec $\times$ 4.6 = 276 sec.

When the fluid temperature is cycling, the measuring system again lags behind the actual temperature depending on the value of the time constant. The actual fluid temperature may be determined from curve *A*, Fig. 6·10 if the time constant of the bulb and the period of the cycle (the time between repeated peaks) are known. For the example on page 88, the time constant was 60 sec and the period was 5 min. Then the ratio of time constant to period is 1 min $\div$ 5 min = 0.2. Using curve *A*, Fig. 6·10, the ratio of measured temperature to actual temperature is found to be 0.63. The actual temperature is therefore cycling 5°F/0.63, or approximately $\pm$8°F. Examples of the relation of measured temperature to actual temperature for various thermal systems are shown in Fig. 6·11.

In order that the charts shown in Figs. 6·10 and 6·11 may be used effectively, it will be well to work out a number of problems. They will illustrate the significance of some of the terms and the units which must be used.

EXAMPLE 6·1 Determine the amplitude ratio value for curve *C*, Fig. 6·11.

SOLUTION

$$\frac{X}{\theta_0} = \frac{1}{\sqrt{1 + (2\pi f T)^2}}$$

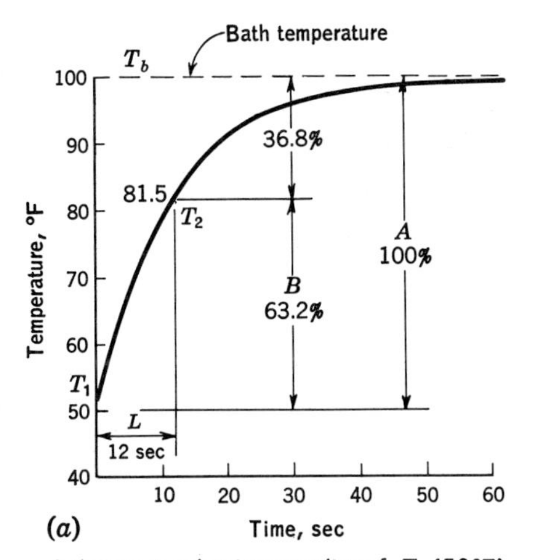

(*a*)

A themometer showing a reading of T_1 (50°F) is plunged into a bath of temperature T_b (100°F) and the temperature is recorded at intervals of 5 seconds

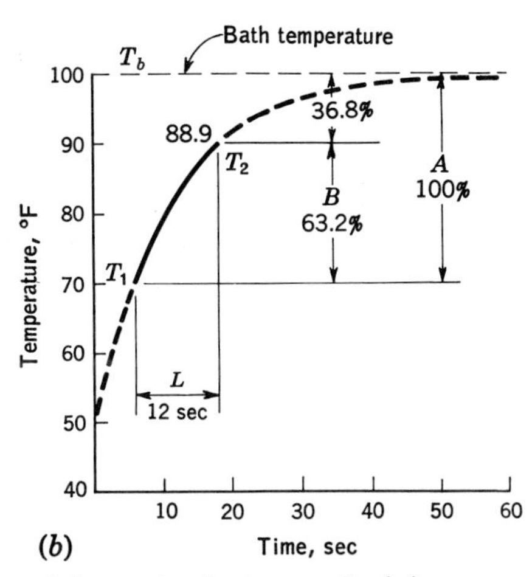

(*b*)

A thermometer showing a reading below 70°F is plunged into a bath of temperature T_b (100°F). At a predetermined temperature such as T_1 (70°F) the time required to reach T_2 (88.9°F) is recorded as the response time of the thermometer

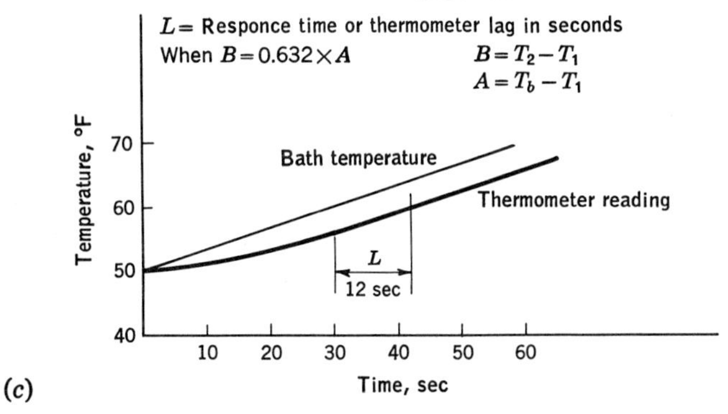

(*c*)

The same thermometer lag, " L " will apply if the temperature of the bath is changing at a constant rate. The thermometer reading will lag the actual bath temperature by the time " *L* "

Fig. 6·6 Temperature response of black-filled etched-stem thermometer in still water bath. (*Data and figure by E. W. Jensen, F. C. Kiesling, Gerald Akins, and D. W. St. Clair, Kodak Park Works, Eastman Kodak Co.*)

The frequency and the time must be both expressed in the same units of time. The air temperature is cycling at the rate of 1 cycle every 5 min [or 1 every (5 min) (60 sec/min) = 300 sec]. But 300 sec/cycle means $\frac{1}{300}$ cycle/sec. The time constant for the capillary bulb is 20 sec, so

$$2\pi fT = (2\pi)(\tfrac{1}{300})(20) = \frac{40\pi}{300} = 0.418$$

$$(2\pi fT)^2 = 0.418^2 = 0.175$$

$$\frac{X}{\theta_0} = \frac{1}{\sqrt{1+0.175}} = \frac{1}{\sqrt{1.175}} = \frac{1}{1.085} = 0.92$$

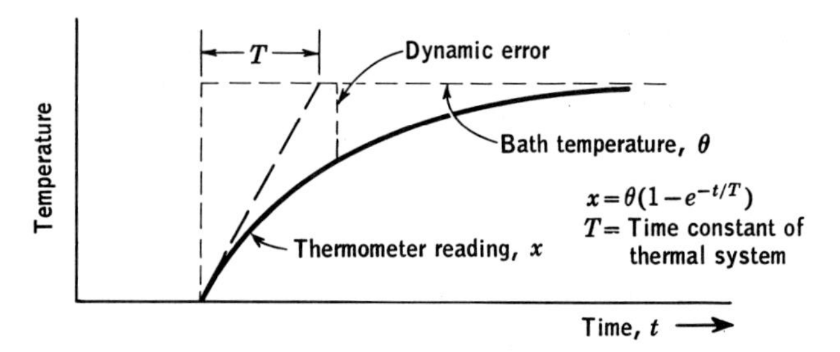

(a) Response to a sudden and sustained temperature change (step change)

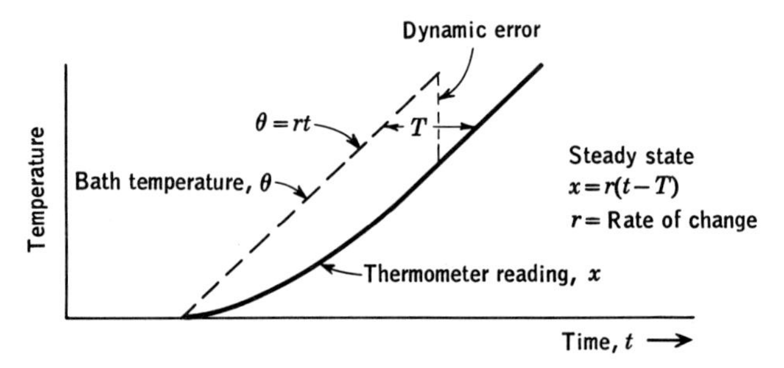

(b) Response to a suddenly applied uniformly changing temperature (ramp change)

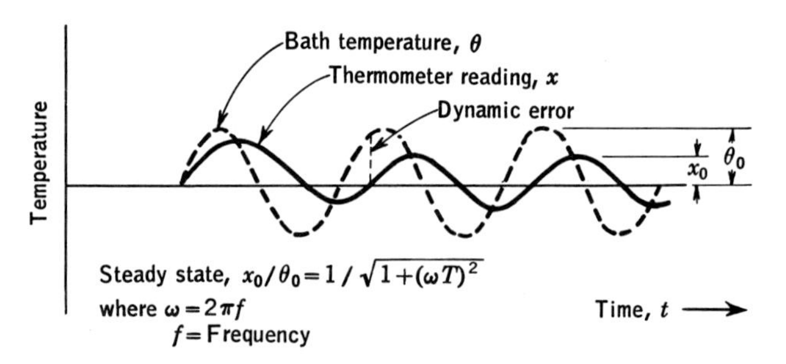

(c) Response to a cycling temperature (sinusoidal)

Fig. 6·7 Response of single-capacity thermal systems to various types of temperature changes. (*Data and figure by E. W. Jensen, F. C. Kiesling, Gerald Akins, and D. W. St. Clair, Kodak Park Works, Eastman Kodak Co.*)

So if the true temperature has a sinusoidal variation of $\pm 20°$, then the *indicated* variation would be $(\pm 20°)(0.92) = \pm 18.4°$.

EXAMPLE 6·2 Determine the amplitude ratio value for curve E, Fig. 6·11.

SOLUTION

$$\frac{X}{X_0} = \frac{1}{\sqrt{1 + (2\pi f T)^2}} \qquad f = \frac{1}{300} \text{ cps} \qquad T = 160 \text{ sec}$$

$$2\pi f T = (2\pi)(1/300)(160) = 3.35$$

$$(2\pi f T)^2 = 3.35^2 = 10.55$$

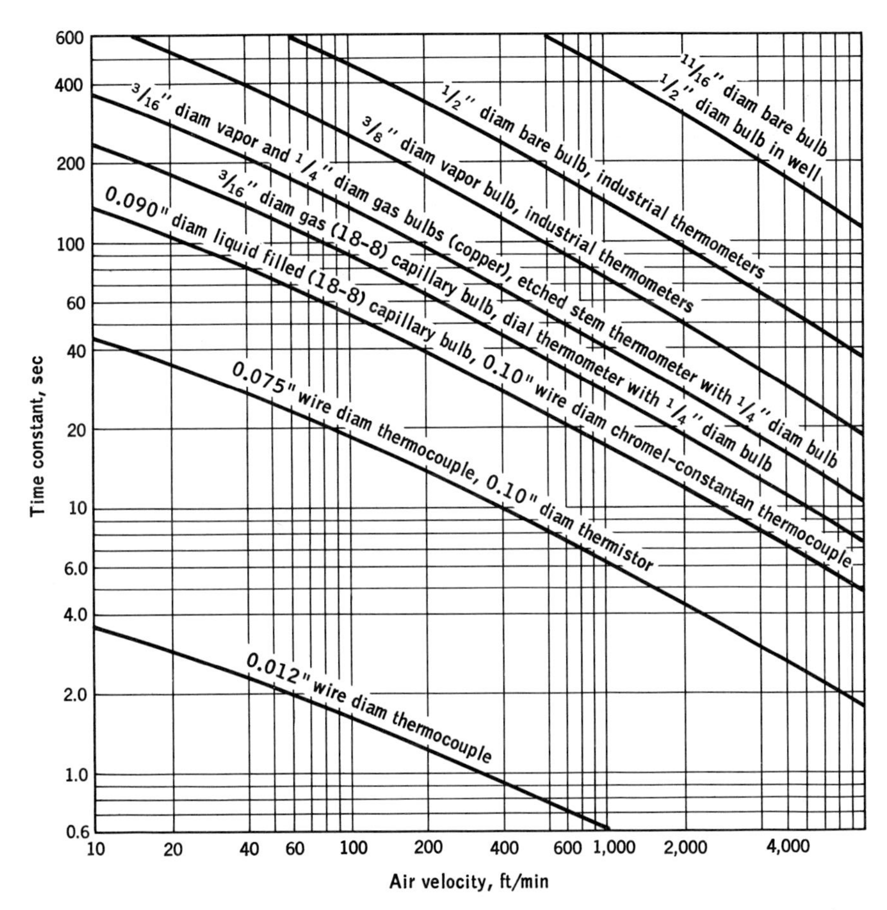

Fig. 6·8 Time constants in moving air. (*Data and figure by E. W. Jensen, F. C. Kiesling, Gerald Akins, and D. W. St. Clair, Kodak Park Works, Eastman Kodak Co.*)

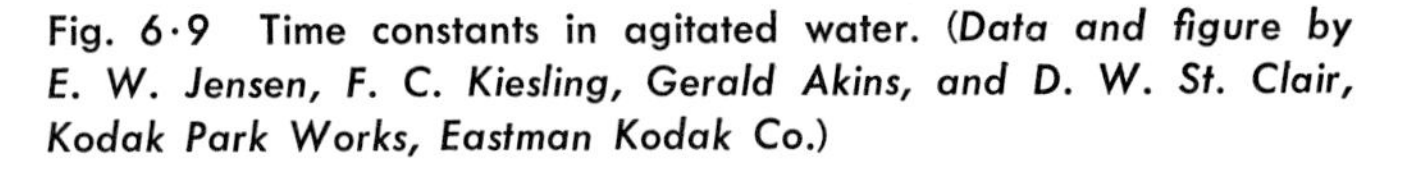
Fig. 6·9 Time constants in agitated water. (*Data and figure by E. W. Jensen, F. C. Kiesling, Gerald Akins, and D. W. St. Clair, Kodak Park Works, Eastman Kodak Co.*)

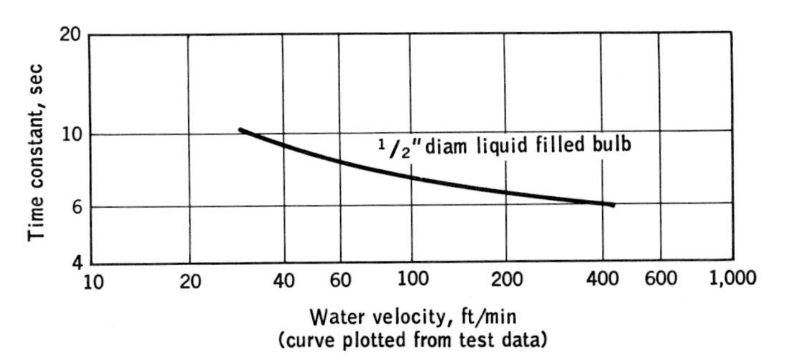

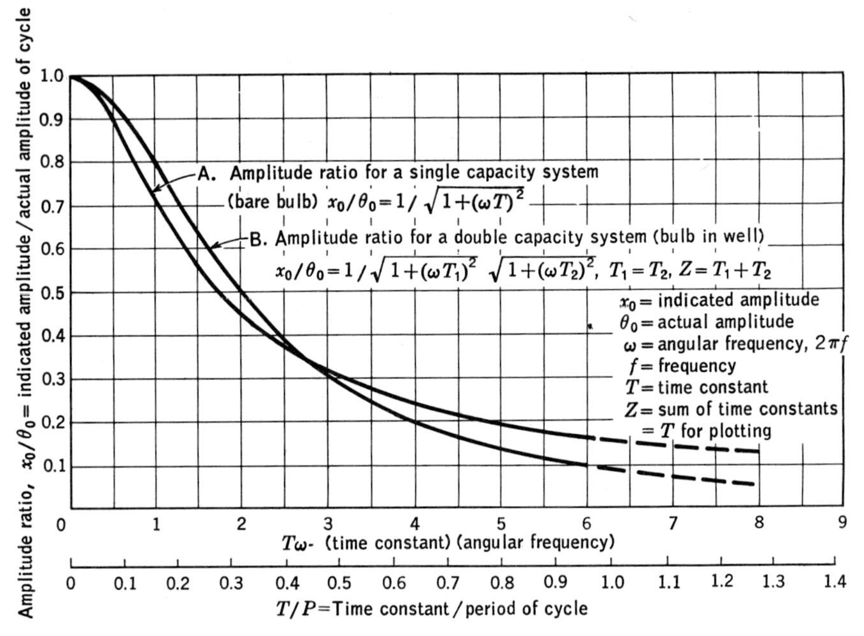

Fig. 6·10 Amplitude ratio for cycling temperature, single- and double-capacity systems. *(Data and figure by E. W. Jensen, F. C. Kiesling, Gerald Akins, and D. W. St. Clair, Kodak Park Works, Eastman Kodak Co.)*

$$\frac{X}{\theta_0} = \frac{1}{\sqrt{1 + (10.55)}} = \frac{1}{\sqrt{11.55}} = \frac{1}{3.68} = 0.272$$

Therefore, if the true temperature variation $= \pm 20°$, then the indicated variation $= (\pm 20)(0.272) = \pm 5.44°$.

To find the time by which the maximum reading of the thermometer lags the true maximum value, the terms of the denominator of the fraction $1/\sqrt{1 + (2\pi f T)^2}$ furnish the value for the tangent of the angle.

For Example 6·1:

$$\text{Tangent of the angle of lag} = \frac{2\pi f T}{1} = \frac{0.418}{1} = 0.418$$

$$\text{Angle of lag} = \tan^{-1} 0.418 = 22.7°$$

In terms of time delay, this means

$$\left(\frac{22.7°}{360°}\right)(300 \text{ sec}) = 18.8 \text{ sec}$$

For Example 6·2:

$$\text{Tangent of angle of lag} = \frac{2\pi f T}{1} = \frac{3.35}{1} = 3.35$$

$$\text{Angle of lag} = \tan^{-1} 3.35 = 73.4°$$

The maximum reading will then lag behind the maximum true value by the fraction

$$\left(\frac{73.4°}{360°}\right)(300 \text{ sec}) = 61 \text{ sec}$$

EXAMPLE 6·3 Determine the amplitude ratio value when two capacitances are involved, Fig. 6·10. Let true cyclic change $= \pm 10°$ every 2 min $= 120$ sec; then $f = 1/120$ sec. Let $T_1 = 40$ sec for well and let $T_2 = 20$ sec for bulb.

SOLUTION

$$\frac{X_0}{\theta_0} = \frac{1}{\sqrt{1 + (2\pi f T_1)^2}\sqrt{1 + (2\pi f T_2)^2}}$$

where

$$2\pi f T_1 = 2\pi(1/120)(40) = 2.09$$

$$(2\pi f T_1)^2 = 4.4$$

$$2\pi f T_2 = 2\pi(1/120)(20) = 1.045$$

$$(2\pi f T_2)^2 = 1.1$$

Fig. 6·11 Response of typical thermal systems to cycling air temperatures. (*Data and figure by E. W. Jensen, F. C. Kiesling, Gerald Akins, and D. W. St. Clair, Kodak Park Works, Eastman Kodak Co.*)

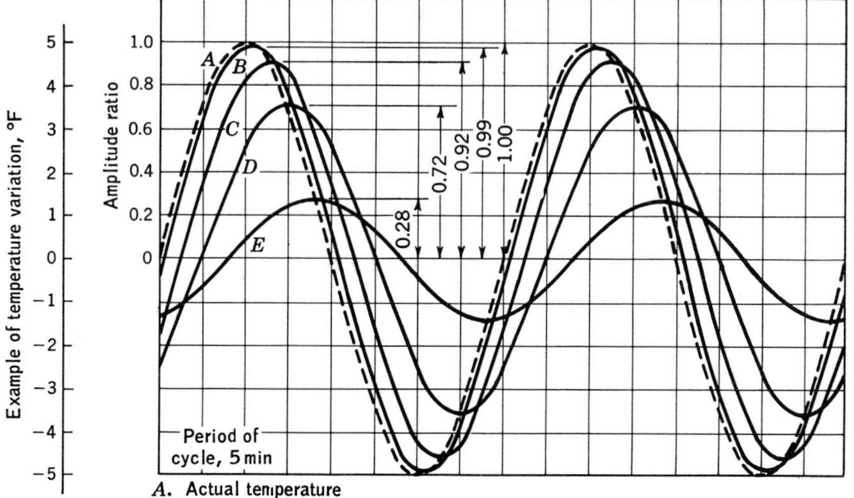

A. Actual temperature
B. Temperature reading of a thermistor, $T = 7$ sec
C. Temperature reading of a capillary bulb, $T = 20$ sec
D. Temperature reading of an etched stem thermometer, T = 46 sec
E. Temperature reading of a 1/2 in. cylindrical bare bulb, T = 160 sec

Example: If a thermal system having a time constant of 160 seconds in air is inserted in a duct whose temperature is cycling 10°F in a period of 5 minutes the temperature reading will vary only 2.8°F

$$\frac{X_0}{\theta_0} = \frac{1}{\sqrt{1 + 4.4}\sqrt{1 + 1.1}} = \frac{1}{\sqrt{5.4}\sqrt{2.1}} = \frac{1}{\sqrt{11.3}} = \frac{1}{3.36} = 0.297$$

$$\text{Angle of lag} = 77.2^\circ = \tan^{-1}\frac{4.4}{1} = 4.4$$

$$\text{Angle of lag} = 47.7^\circ = \tan^{-1}\frac{1.1}{1} = 1.1$$

$$\overline{124.90^\circ} = \text{total lag for the two elements}$$

The instrument will then register $(\pm 10^\circ)(0.297) = 2.97^\circ$ with the maximum value occurring

$$\frac{(124.9^\circ)(120 \text{ sec/cycle})}{360^\circ/\text{cycle}} = 41.6 \text{ sec}$$

after the true or actual maximum.

Let us check value given by curve, Fig. 6·10.

$$T_1 = 40 \text{ sec}$$

$$T_2 = 20 \text{ sec}$$

$$\overline{T_1 + T_2 = 60 \text{ sec}}$$

$$\omega = 2\pi f = (2\pi)\left(\frac{1}{120}\right) = \frac{2\pi}{120}$$

$$T\omega = Z\omega = 60\left(\frac{2\pi}{120}\right) = \pi = 3.14$$

From chart, for $T\omega = 3.14$, $X_0/\theta_0 = 0.29$, which checks the value previously found by computation.

Pressure pulsations. If the applied pressure were 100 ± 20 psi every 2 sec (assume a sinusoidal pressure variation with time), and the time constant for the gage is 1 sec, then

$$\frac{X}{X_0} = \frac{1}{\sqrt{1 + (2\pi fT)^2}} = \frac{1}{\sqrt{1 + [(2\pi)(1/2)(1)]^2}} = \frac{1}{\sqrt{1 + 9.9}}$$

$$= \frac{1}{3.3} = 0.303$$

or gage will read $100 \pm (0.303)(20) = 100 \pm 6.06$ psi

where $\frac{X}{X_0}$ = ratio of indicated value to true value

f = frequency, cps

$= \frac{1 \text{ cycle}}{2 \text{ sec}}$

T = time constant for the gage, sec

SUMMARY

1. Heat may be transferred from one object to another by conduction, convection, and by radiation. It may be that only one of these methods is in use at any one time, or it may be that all three methods are being used simultaneously to add heat to or remove heat from an object.

2. Normal heat transfers always cause the hot object to lose heat and the colder object to gain heat. In this case, the terms "hot" and "cold" are relative ones and have no definite connection with a specific temperature.

3. A certain period of time is always required for heat to be added to or removed from an object. Consequently, *all temperature-sensing devices will require definite times to undergo specific temperature changes,* with any given set of physical conditions. No matter how small the change in temperature is to be, a period of time will always be required to effect it. There never can be an instantaneous temperature change.

4. The time of response of temperature-sensing devices can be reduced provided that the device is located where there is the greatest movement of fluid (liquid, air, or gas) over its surface. Pockets and stagnant conditions should always be avoided.

5. The *time constant* tells the time required for the device to register 63.2 per cent of a *step change.* When there are cyclic changes in the temperature, any temperature-sensing device will lag behind the actual temperature in reporting that there is a change going on. In addition, it is quite probable that the indicated change is much smaller than that which is actually taking place.

6. Because of the requirement that the transfer of heat takes time, an instrument which is in perfect static calibration will be in error under cyclic or transient conditions. Such an error is a *dynamic error,* and all temperature-sensing devices are subject to it.

7. The use of a thermometer well will greatly increase the time of response of the sensing system. The greater the number of *resistances* placed in series, the slower will be the responses of the system.

8. In addition to the periods of time required to transfer heat to the thermometer, there is the time required to bring to the sensing element the material undergoing the temperature change. If the fluid is having its temperature changed at point X but it is 100 ft to the sensing bulb at point Y, there will be a transportation lag, or dead time, equal to the time required for the material to travel the 100 ft. If the purpose of the measurement is accurate control of temperature, then the sensing element in most cases will be located as close as possible to the point where the temperature is being changed. There are some exceptions to this rule, but in general, the distances must be kept as small as possible.

QUESTIONS

6·1 If a cyclic change is taking place every 10 sec but the time constant of the system is 10 sec, what percentage of the cyclic change will be registered? (Use Fig. 6·10.)

6·2 If the time constant were to be doubled, what would happen to the response?

6·3 If you were to extend the plot for Fig. 6·9 to a value of 1 fpm, what value for the time constant would you expect? Justify your answer.

6·4 Beyond what point is there little benefit in increasing flow in agitated water? Would you expect other liquids to behave similarly?

6·5 What accounts for the difference between the time constants for the ½-in.-diameter bare bulb and the same bulb in a well, Fig. 6·8?

6·6 What factors tend to increase the time constant of a thermometer system?

6·7 What factors tend to decrease the time constant of a thermometer system?

6·8 To what degree can you control the time constant of a system? Explain fully.

6·9 What are the ways of transferring heat energy? Which are most important at 200°F, 100°F, 3000°F?

6·10 What is meant by the time constant of a thermometer or thermal system?

6·11 How many time constants must one wait to get an answer within 1 per cent of a step change if a thermometer is inserted into a hot oven? An ice bath?

6·12 If the temperature of an oven is cycling at a true rate of ±10° every 40 sec, will any thermometer show the ±10° variation? If the time constant of the thermometer is 300 sec, what will be the apparent cyclic variation?

PROBLEMS

6·1 Using the table of constants (Table 6·2) determine the time constant for (*a*) item 1, when the speed is 40 fpm; (*b*) item 2, when the speed is 100 fpm; and (*c*) item 5, when the speed is 30 fpm. Assume all time constants change at same rate as shown in the graph, Fig. 6·9.

6·2 When the flow rate of the liquid is 50 fpm, how long will it take for item 8, Table 6·2, to be within 1 per cent of the correct value for a step change?

6·3 Repeat Prob. 6·2 for item 11, Table 6·2.

6·4 How much longer will it take for item 9, Table 6·2, to indicate 99 per cent of a step change than item 11?

6·5 Tabulate the factors that might account for the difference in values in Prob. 6·4.

6·6 How long will be required for a ½-in.-diameter bare bulb to reach 95 per cent of its final temperature when the air flow is 1,000 fpm? 4,000 fpm?

6·7 Use Fig. 6·8. Extrapolate the graph for the ½-in.-diameter bulb in a well for a flow of 200 fpm. How long will it take for 98 per cent of a step change in temperature to be indicated?

6·8 How much longer, in seconds, will it take an 11/16-in.-diameter bare bulb to reach 99.5 per cent of a step change in air at 4,000 fpm as against 1,000 fpm?

6·9 How much longer will it take a ½-in. bulb in a well to reach 98 per cent of the final value at 4,000 fpm than the same bulb without a well?

6·10 If an electric furnace has its heating elements switched on and off automatically every 2 min—60 sec on and 60 sec off—what is the true maximum temperature deviation if the sensing bulb has a time constant of 100 sec and the indicated temperature changes are ±5°F?

6·11 By how many seconds will the maximum indicated value lag behind the true value in Prob. 6·10?

6·12 If the sinusoidal cyclic variations in temperature of an air duct are ±10°, where the cycle is 100 sec long, what must be the maximum time constant for the thermometer if it is to register at least 80 per cent of the temperature changes?

6·13 In Prob. 6·10 what can be the maximum time constant if the maximum indicated value is not to lag the true value by more than 10 sec?

7 Electrical Circuits and Devices

In order to make the presentation of this material more thorough and complete, six experiments have been included in this chapter. They are relatively simple and easily performed. Because of the great importance of the principles which they present, they should be performed whenever possible. In any event, the principles and the discussion should be thoroughly mastered before proceeding to the subsequent chapters, which will employ all of these principles.

In Chap. 3 we found that when most metals are heated they are not as good conductors of electricity as they are when they are cold: the hotter they become, the poorer is their performance as conductors. This is but one of the many electrical characteristics with which we shall have to become familiar.

In Chap. 4 we studied Ohm's law. We found that the electrical pressure is measured in volts or in millivolts, which are one one-thousandth as large as volts. When you purchase electric light bulbs, you generally specify 120 volts, but when you buy flashlight batteries, you may specify a rating of 1.5 volts.

The unit of current, or electron flow, is the ampere. We may define it by saying that a current flow of 1 amp requires that 6.3×10^{18} electrons per second pass any given point. This seems like a fantastic number, but it is only about 15 per cent larger than the number of electrons moving past any given point in the familiar 100-watt electric light bulb. There are two smaller units of current which customarily are used in instrumentation: the milliampere, $\frac{1}{1000}$ amp, and the microampere, $\frac{1}{1000000}$ amp. In certain industrial measurements, measurements are made in micromicro amperes.

POTENTIOMETRIC DEVICES

Potentiometers. We will also have occasion to make very careful voltage measurements when we correlate the very small voltage generated by thermocouples with the temperature differences which created them. In order to make precise small voltage measurements, we use a *potentiometer* in most circuits in which thermocouples are the temperature transducers. A potentiometer consists primarily of:

1. A resistance element made of many turns of carefully spaced wire
2. An arm which can contact any of the turns

If, for example, the voltage applied to the potentiometer at terminals a and a', Fig. 7·1, is 1.018 volts and there are 1,018 turns on the potentiometer, then we have the same number of turns as millivolts (1.018 volts = 1,018 mv). We might say that it takes an electrical pressure of 1 mv to cause the current to flow through each turn. Then it would take 2 times this voltage to cause the current to flow through 2 turns; 10 times as much to flow through 10 turns, etc. Thus we can relate the voltage to the number of turns. To measure the voltage between a and P', all we need is some simple way to count the number of turns between the two points.

We could, of course, have half as many turns or twice as many turns as the number of millivolts and the plan would be the same. If we had 10 times as many turns as millivolts, we could get an answer of 0.1 mv per turn. By dividing each turn, we could measure still smaller parts of a millivolt.

What may appear to be a very odd value for the voltage of the battery is actually the electromotive force furnished by a *standard cell* used to furnish a known value to a high degree of accuracy. We shall learn more about this in Chap. 15.

If potentials are to be measured with good accuracy, no current can flow through the potentiometer to an external circuit. Connecting an ordinary voltmeter, as shown in Fig. 7·1b, would load the potentiometer

Fig. 7·1 (*a*) A potentiometer provides a way to obtain some desired fractional part of the applied battery voltage. (*b*) For the position of the pointer *P* to be directly related to the potential, the meter must not take an amount of current which will change the potential drop through the potentiometer.

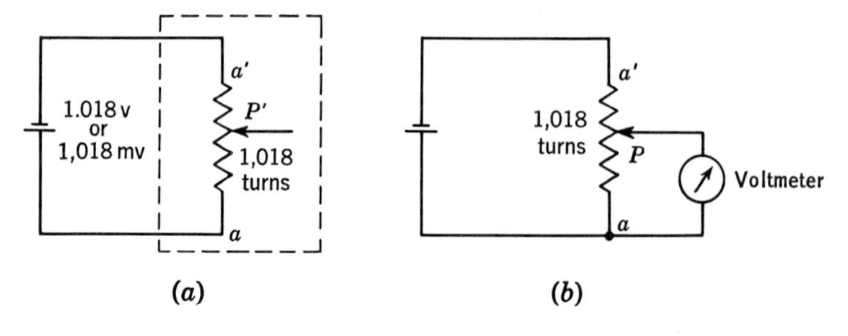

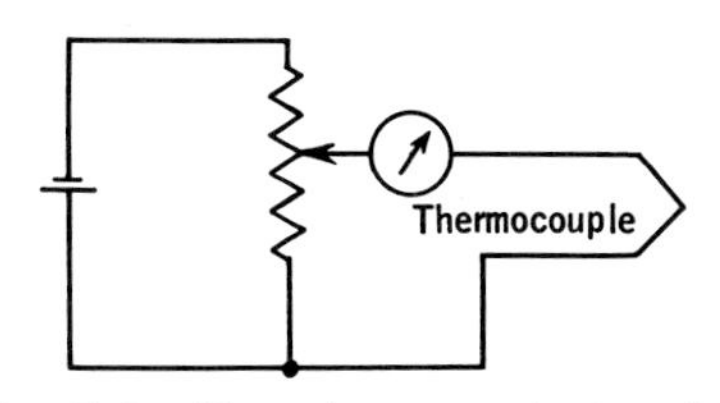

Fig. 7·2 If a thermocouple is substituted for the battery B′, Fig. 7·3, the temperature may be determined by finding the point where the two potentials are equal.

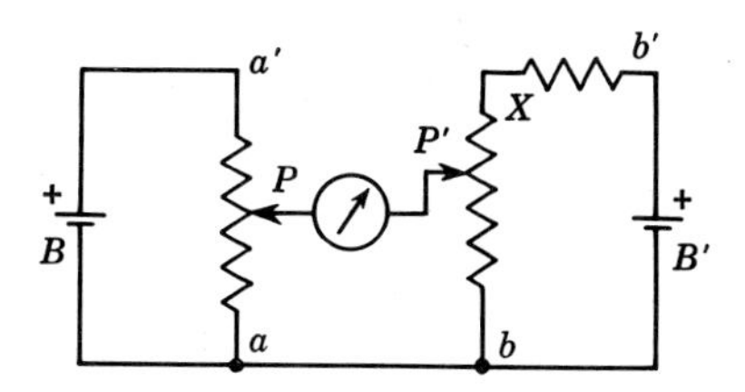

Fig. 7·3 The sensitive meter will indicate 0 when pointers *P* and *P′* are at the same potential. By knowing the value of *P*, we also know the value of *P′*.

to such a degree that the answer would not be close enough for calibrating purposes. Special vacuum tube voltmeters could be used because they do not require signal current for their operation. The chief use for the potentiometer is to create a voltage whose value can be determined without the use of voltmeters, but which will be used to cancel out another voltage. In this manner, the second voltage is measured.

For instrumentation work, our chief interest generally is in the millivolts produced by thermocouples, Fig. 7·2. But for the sake of example and clarity, a battery B' and two resistors in series are shown in Fig. 7·3. We will assume that we want to know the number of millivolts electrical pressure between P' and b.

If we connect a sensitive meter such as galvanometer, as shown in Fig. 7·3, between the potentiometer circuit and battery B' and its resistors, the following will happen:

1. If P' has a higher potential than P, with respect to a and b which are at the same electrical level (they are connected to the same conductor), then a current will flow through the meter. The direction of movement of the pointer will show that P' has the higher potential or level.
2. If P has a higher potential, the meter pointer will try to rotate in the opposite direction.
3. If P and P' are at the same level, there will be no current flow to cause the meter to move. (There must be a difference in electrical level before a current will flow.)

If, when we complete the circuit, the movement of the meter pointer indicates that P is lower in value than P', then we move the potentiometer contact P toward a'. Soon we note that the meter is returning to its zero position. When it does, P and P' have the same value. Because we know the value of P, we also know the value of P'. Inasmuch as no meter current flows as the time of balance, the meter circuit can be used over relatively long distances without difficulty.

Wheatstone bridge. The circuit which carries this name is one used very frequently for the determination of the resistance of some circuit

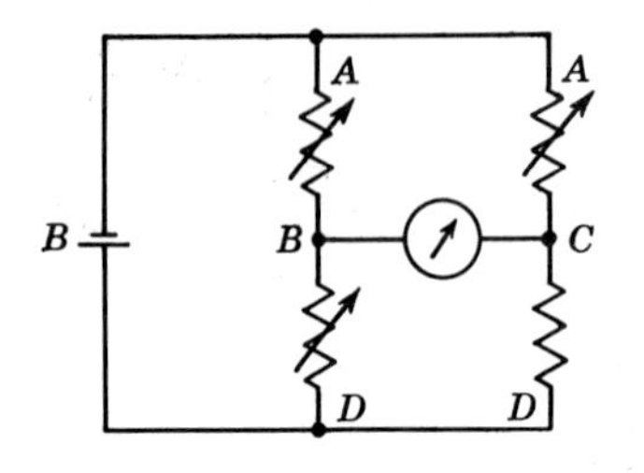

Fig. 7·4 A schematic representation of a Wheatstone bridge.

or component. In simple schematic form, it is represented in Fig. 7·4. It might be said to be another form of potentiometer, inasmuch as points *B* and *C* must have the same potential before the resistance can be measured or evaluated. We should keep this resemblance to a potentiometer in mind, because we will encounter it again when we study certain commercial equipment.

In Fig. 7·4 a battery *B* furnishes current to two paths, each having two resistors in series. A sensitive galvanometer, similar to the one used in the potentiometer, is connected between points *B* and *C*. If we make the resistance from *A* to *B* equal to that from *B* to *D* (the symbol ↗ through a resistor indicates that it is controllable in value) and we adjust the resistance from *A* to *C* until the meter indicates *balance* because its current is zero, the resistor *X* will have a value equal to the resistance from *A* to *C*. It can be shown that the following relationship is true:

$$\frac{R_{AB}}{R_{BD}} = \frac{R_{AC}}{R_{CD} \text{ or } R_X}$$

In this manner, we can ascertain all of the more usual values of resistance that we are likely to encounter in the instrumentation field. Because resistor values change with temperature, strain, and chemical concentration, we shall have many reasons to refer to the Wheatstone-bridge circuit. It is at the heart of many measuring devices. It may be used with alternating currents as well as with direct-current battery circuits.

ELECTRON TUBES

The industrial instrumentation field would be quite different from what it is today if it were to exist only with mechanical-pneumatic or mechanical-hydraulic elements. The vacuum tube, radio tube, or valve, as it is called, has created revolutions here just as it has in the communications and entertainment businesses.

Someone has remarked that if mechanical engineers were to create a communications system, they would undoubtedly use nothing but the simple mechanically operated drum. Chemists, on the other hand, would probably want to use nothing but smoke signals. The electrical engineers, obviously, would use nothing but a complex electrical system. But after a comparison of the systems and the capabilities of each, most of us have decided to use the more intricate system because of its better performance.

So it is with industrial electronic measuring and control systems. Their operating advantages outweigh their disadvantages. Most of the components which go into these systems are quite simple, substantial, and not difficult to understand so far as their function is concerned. Of all of the components employed, the electronic tube is the most important.

Radio tubes (or vacuum tubes) are able to accomplish many tasks which would be difficult or impossible to do without them. But in their operation, they are basically simple. In their simplest form, these tubes are rectifiers, that is, they permit current to pass in one direction but will not pass a current moving in the opposite direction; therefore, they convert alternating current to a pulsating current. How they do this will be more evident after studying Fig. 7·5. In schematic form we have represented the two-element tube, or diode. It consists of an anode, or plate, which must be connected to a positive potential for its operation. Electric energy heats the cathode and causes electrons to be boiled off, just as boiling water creates water droplets and particles which fill the space over the water. Some of the water vapor falls back into the liquid, while some escapes. Similarly, some electrons go back to the cathode, while others move away. At the cathode, the electrons which are made to move at higher speeds because of their temperature can be pulled away by the electric field created by the charged anode. As these electrons move toward the anode, others move through the circuit to replace them at the cathode.

If we had an exhaust fan running above a vat, the amount of vapor removed by the fan would depend upon:

1. The speed of the fan. This is analogous to the potential, or charge, of the anode, or plate.

2. The temperature of the liquid. This corresponds to the temperature of the cathode: the higher the temperature, the more vapor created. The

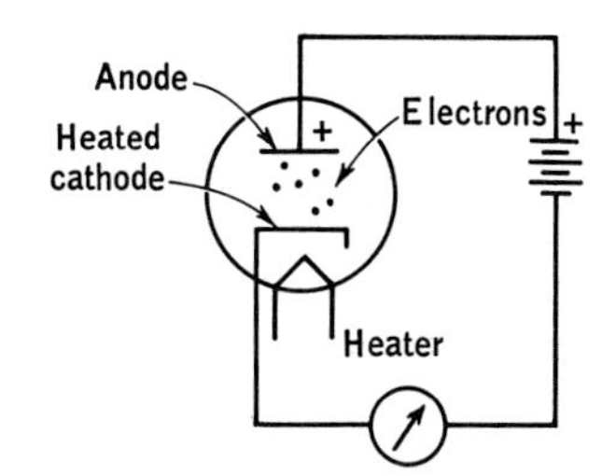

Fig. 7·5 A diode has a cathode which furnishes electrons and a positively charged anode which attracts electrons. The passage of the electrons around the circuit can be measured by a milliammeter.

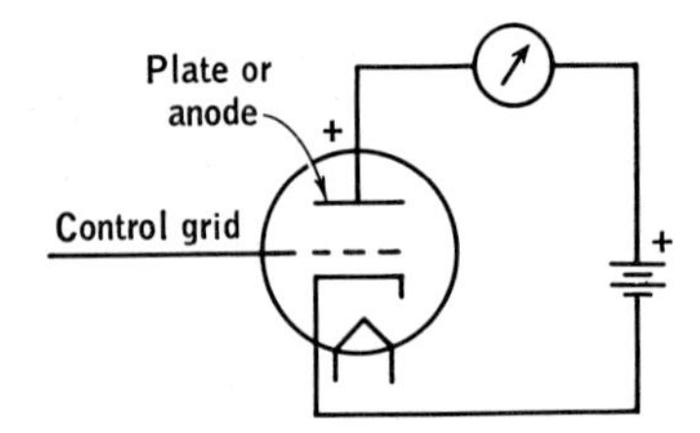

Fig. 7·6 A triode differs from a diode in that a third element, a grid, has been placed between the cathode and the anode. The addition greatly increases the ability to modulate or control the flow of electrons through the tube.

boiling and liberating of the water depends upon the amount of heat energy supplied. This is the situation with respect to the electrons freed by the heated cathode.

Historically, this device has been around a great many years. It is often called the Fleming valve to honor the man who first put it to work. Edison had noted the flow of current from heated filament to anode, but he never used the phenomenon. It was in 1907 that the critical part of this device was invented by Lee De Forest. What he did was to introduce a third member, the grid, into the tube, Fig. 7·6. This was a most unique member that he added, because it became possible to amplify a signal without requiring any significant input energy. The grid functions in the usual industrial circuit without having any current flow in it; it accomplishes everything through the potential of the signal which is impressed on it. The grid normally is a screen of fine wire placed between the cathode and the plate. The positively charged anode exerts an attractive force on the electrons, while the negatively charged grid drives the electrons back toward the cathode. If the grid has a small negative charge, some of the electrons will get through it and reach the plate. Making the grid less negative will let still more electrons reach the plate. The change in voltage on the grid modulates or controls what would otherwise be a steady flow of electrons, such as is found in a rectifier with a constant cathode temperature and fixed anode voltage.

One may think of the grid as a sort of fence, the size of whose openings may be controlled by the signal applied. A negative grid signal tends to close the opening; a positive signal tends to enlarge the openings.

The grid can respond to signals changing at the rate of hundreds of millions of times each second. Hence, it can repeat instantaneously, for all practical purposes, all information given to it. High speed of response is one of its great features. The fact that it requires no current, only voltage, means that extremely minute signals are enough to cause it to function. It also has the ability to amplify very weak signals so that they become powerful enough to operate relays, motors, valves, and associated apparatus. By putting extra grids into the tubes, certain characteristics can be changed, but they are not of primary concern at this time.

One other ability we should keep in mind is that these tubes can function as generators of alternating currents, some of which reverse their polarity hundreds of millions of times each second. At the highest frequencies this electric energy acts much as light waves; it tends to travel in straight lines, is reflected, and it can be focused.

Some of these capabilities of tubes will be studied later as we study general types of commercial designs.

Simple vacuum tube characteristics and applications. Inasmuch as the vacuum tube voltmeter is used so often in instrumentation work, it might be well to study it in some detail. The principles to be noted here will be found in many pieces of equipment. Only slight modifications in the circuit will change a voltmeter into a liquid-level gage, a relay, a temperature controller, and so on. We should keep in mind that in instrumentation, the vacuum tube will be used most often to:

1. Function as a generator of alternating current of fixed or controllable frequency.

2. Control the current in the plate circuit through the size and polarity of the signal applied to the grid. Because the plate current can be varied, we can:

a. Vary the voltage across a resistor in the plate circuit. Because this voltage can be made larger than the input signal, the tube functions as an amplifier.

b. Consider the tube to function as a variable resistor. With a constant voltage applied to the plate, a variable grid voltage produces a changing current. Such a relationship implies a variable resistance. Unlike a conventional variable resistor, this variable resistance has no movable parts.

Experiment 7·1 uses the circuit shown in Fig. 7·7. With no signal or voltage applied between points a and b, the meter will indicate a steady value of current when the cathode is heated. If points a and a' are now connected, it will be found that the meter will move as the potentiometer contact a' is moved. By connecting a voltmeter between a' and b',

Fig. 7·7 One type of vacuum tube voltmeter.

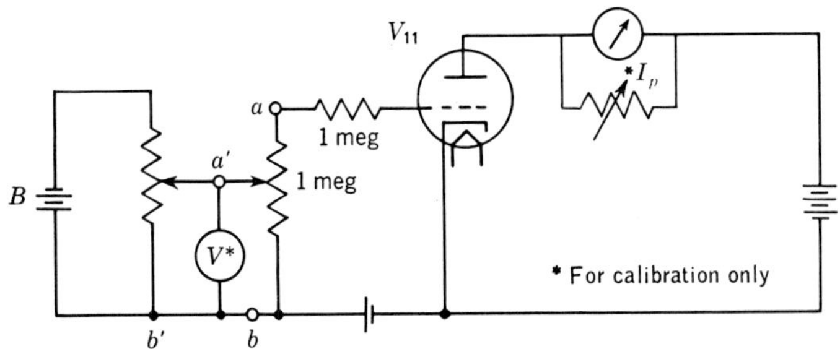

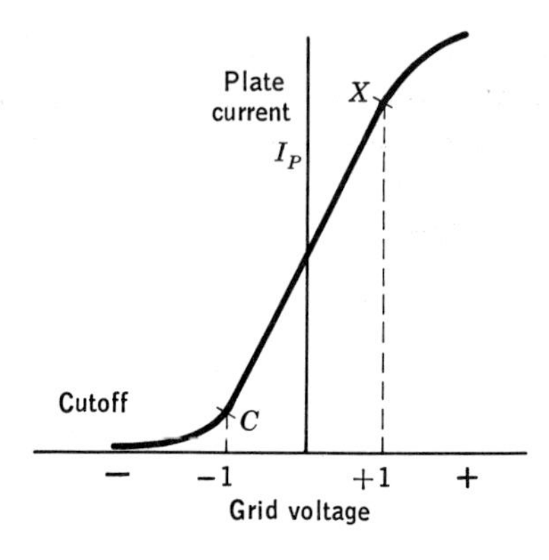

Fig. 7·8 The plate current in a triode is greatly increased by positive grid potentials and decreased by negative ones.

we can find the relationship between the applied signal and the meter current I_p. If we reverse the connections to battery B, we will find that the currents will be smaller than before. The values of voltage $a'b'$ and current I_p probably will look like the sketch in Fig. 7·8. It will be noted that a section of the curve is essentially straight but, from X on, an increase in the positive grid signal produces only a very small increase in the plate current. This is a condition of *saturation:* almost all of the electrons boiled off by the cathode are being attracted to the anode. Consequently, no more can go to the anode, no matter how easy the path is made. But as the grid is made more negative, the current again becomes more nonlinear and approaches *cutoff*. Both the cutoff and the saturation regions are normally to be avoided for measurement work but not necessarily in control work.

To extend the range of the voltmeter, a series of resistors can be so connected that only a small percentage of the signal is used. Such a group of resistors is called a *voltage divider*. If we assume that the voltage to the grid need change only ½ volt in order to cause the meter to move from its zero value to its full-scale value, then that is all that should ever be applied to the grid. Consequently, we will want to take such a portion of the input signal that the portion which is applied to the grid will be only ½ volt, approximately. Figure 7·9 shows schematically such a divider. If we have an assumed input voltage of 10 but we need only ½ volt to actuate the meter, then we need only a small fraction, or one-twentieth, of the original input voltage. The string of series resistors has the same current in each part. So while the voltage in any one part is to the voltage in any other part as the product of the resistance and current in one section is to the product of those values in the second section, the currents are common in one of the ratios and nullify each other:

$$\frac{V_1}{V_2} = \frac{IR_1}{IR_2} = \frac{R_1}{R_2}$$

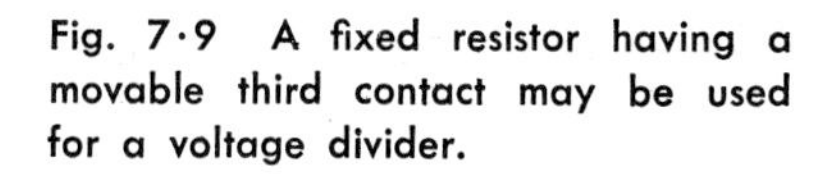
Fig. 7·9 A fixed resistor having a movable third contact may be used for a voltage divider.

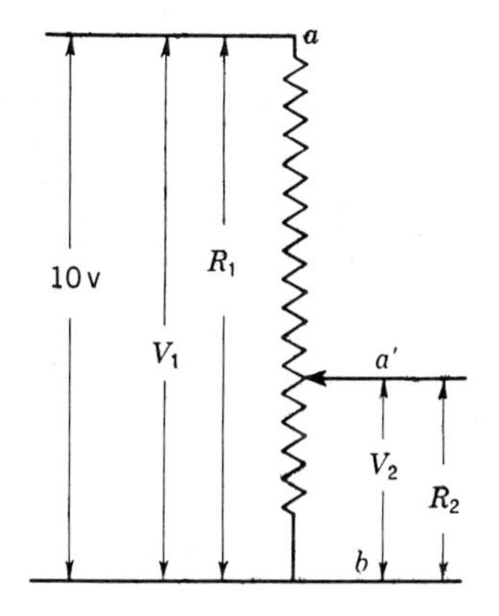

Therefore, we can generalize and state that the voltages in such a series string are to each other as the ratios of their resistances. Then, if the desired voltage is one-twentieth of the input voltage, the resistance from a' to b should be one-twentieth of the resistance between a and b. In this manner, we can measure a voltage far beyond the linear range of the vacuum tube voltmeter circuit. All that we need do is take some known fractional part of the input voltage and then multiply our answer by the reciprocal of that value. When we position the contact on the voltage divider so that it includes only one-fifth of the voltage-divider network, the multiplying factor is 5. If it includes two-fifths of the circuit resistance, then the factor is $5/2$, and so on.

In the vacuum tube voltmeter, the input circuit resistors generally are 1 million ohms each (or 1 megohm), with 10 of them in series to give easy multiplying factors. Other combinations are used, but the total input resistance to the divider is so high that for most practical purposes no current is assumed to flow in the divider, and none flows in the grid circuit of the tube.

For improved stability and performance which makes the meter reasonably independent of small line-voltage changes, two tubes (often in the same glass or metal envelope) may be used. One is used primarily for measuring and one for reference. It might be well at this point to consider one tube as a variable resistor and the reference tube as a fixed resistor. In the somewhat simplified circuit shown in Fig. 7·10 we have a Wheatstone bridge with the resistors arranged just as they are in Fig. 7·4. The chief difference is that two triode vacuum tubes have replaced two of the more conventional resistor elements. There is one other slight difference in the vacuum tube bridge: There is no provision for bringing the meter to zero; the amount of unbalance current is then taken as a measure of the bridge unbalance. Such use of a Wheatstone bridge is quite common.

If we take tube B to be the reference tube, no signal is impressed on

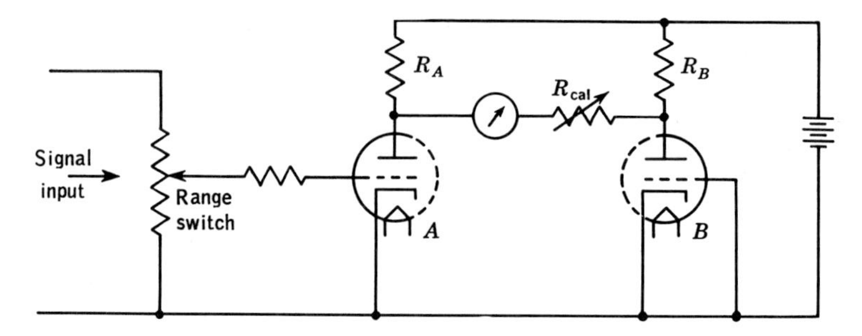

Fig. 7·10 Vacuum tubes may be made to function as fixed or variable resistors in a Wheatstone bridge circuit.

its grid because the grid is connected to the cathode through the common ground; therefore the resistance of the tube can be assumed to remain constant during normal operation. However, there will be a plate current in tube B, because a plate current will exist even when the grid and the cathode have the same voltage (see Fig. 7·8). This current in tube B will also create a voltage drop in resistor R_B (recall that it always takes a potential drop to force current through a resistance), so as the current through any resistor varies, so will the voltage across its terminals vary proportionally. If resistors R_A and R_B have the same resistance, the voltages drops in the two can be equal only when they are passing equal currents, and, of necessity, each tube would likewise be passing the same current. But if the tubes are not passing the same currents, then unequal voltages will be developed at the terminals of the two resistors. It is this inequality of voltages which will then cause current to flow through the meter and cause it to deflect. The resistor in series with the meter is adjusted to make the meter go full scale for some specified input signal.

We might summarize the vacuum tube voltmeter characteristics by saying:

1. The input resistance is high—generally several megohms. The very small current which the voltmeter takes will be of no significance except in circuits having resistances of many megohms. For extremely high resistance circuits, it may be necessary to consider the loading effect of the meter.

2. Applying the wrong voltage for a particular range setting generally will not damage the meter because of the saturation characteristics of the tubes.

Photoelectric effects. It has been noted that some materials undergo a change of resistance when light strikes them. This radiant energy may or may not be in the visible spectrum. Some detectors of radiant energy will operate on the infrared rays generated by temperatures as low as

100°F. You can easily sense that radiated energy when you stand close to a house-heating radiator which may have a temperature of 140°F.

We use the change in resistance of these materials to actuate vacuum tubes. If the materials are made part of a voltage divider, the voltage applied to the grid of the tube can be made to vary widely. Photocells of such materials may undergo a resistance change of the order of a million times when they go from an illuminated condition to a dark one. Let resistor *a-b* in Fig. 7·11 be the photosensitive resistor. Assume that its resistance when dark is 10,000,000 ohms. Assume also that resistance of *a-b* is 1,000 ohms when illuminated. Let the value of *b-c* be to 100,000 ohms. Then

$$\frac{V_{bc}}{V_{ac}} = \frac{R_{bc}}{R_{ac}}$$

When the terms are transposed, we get

$$V_{bc\ \text{dark}} = \frac{(V_{ac})(R_{bc})}{R_{ac\ \text{dark}}}$$

If the known values are then substituted

$$V_{bc\ \text{dark}} = \frac{(10 \text{ volts})(100{,}000 \text{ ohms})}{(10{,}000{,}000 + 100{,}000 \text{ ohms})}$$

$$= \frac{1}{10.1} \text{ volts}$$

This small signal applied to the grid of the tube will be inadequate to increase the plate current to such point that the relay will be energized.

Now, if the photoelectric cell is illuminated to a value of 2 ft-c (foot-candles) or more, we will find that the resistance of the photocell has dropped to about 1,000 ohms. Restating the relationships for emphasis:

$$\frac{V_{bc}}{V_{ac}} = \frac{R_{bc}}{R_{ac}}$$

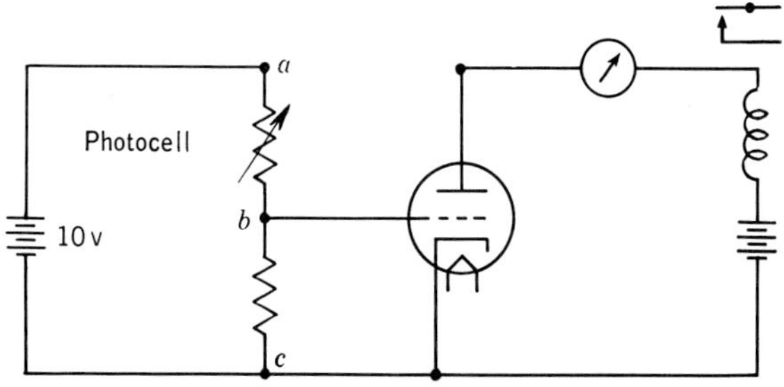

Fig. 7·11 A change in the resistance of the photocell, caused by controlling the intensity of the light falling on it, can control the tube current.

With the terms transposed,

$$V_{bc} = \frac{(V_{ac})(R_{bc})}{R_{ac}}$$

$$V_{bc\ \text{light}} = \frac{(V_{ac})(R_{bc\ \text{light}})}{R_{ac\ \text{light}}}$$

$$= \frac{(10\ \text{volts})(100{,}000\ \text{ohms})}{1{,}000\ \text{ohms} + 100{,}000\ \text{ohms}}$$

$$= \frac{(10\ \text{volts})(100{,}000\ \text{ohms})}{101{,}000\ \text{ohms}}$$

$$= 10- \quad \text{volts}$$

A signal of this magnitude will cause the current in the tube to increase greatly, thereby operating the relay and completing the circuits connected to the relay contacts. In this manner, many circuits are controlled with no moving parts in the voltage-dividing network. In a small fraction of a second, the change in light signal is communicated to the relay. Counting circuits, fire-detection circuits, and sequence circuits can be made to function in this manner.

Time-delay circuits. Just as it takes time for a temperature-sensing system, or other detecting system which involves a change in energy, to respond to the new conditions, so it takes time to "fill" certain components of an electrical system. Let us examine first, the mechanical equivalent. If two tanks are connected as shown in Fig. 7·12, but with tank B empty and valve V closed, we find that the level in tank B rises slowly after the valve is opened and rises at an ever slower rate as the level in tank B approaches that in A. (Assume that the level in A does not change appreciably.) Theoretically, tank B will never get filled to the level of tank A, but the difference will be so small that after a time interval which is 5 or 6 times as great as that of the *time constant*, there will be only a small difference in level.

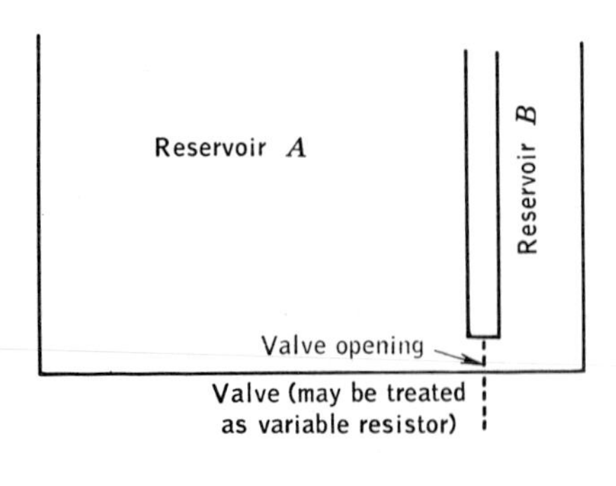

Fig. 7·12 The time required to fill tank B is governed by the size of B—this is analogous to capacitance—and the size of the valve opening, that is, resistance.

It will be fairly evident that if tank *B* is made larger, it will take more time to fill it. Similarly, the resistance of the valve will affect the time required to fill tank *B*. The size of tank *B*, or its gallons or cubic feet per foot of depth, is called its *capacitance C*. Capacitance should never be confused with *capacity*. A broad, shallow tank holding 100 gal will have a much higher capacitance than will a tall slender tank also holding 100 gal. For a given valve position, or valve resistance, and size of tank, the product of the capacitance and the resistance will give the time necessary to fill the tank to 63.2 per cent of its capacity, provided that the tank is of uniform diameter. This same product will be found mentioned in connection with the time required for a thermometer to respond to 63.2 per cent of the temperature change.

To return to our electrical system, a capacitor behaves in some ways as an electrical basket or bucket which is filled with electrons. In place of a mechanical valve to control flow, we use a resistor. If the switch, Fig. 7·13*a*, is closed when there is no charge on the capacitor *C*, the microammeter which measures the flow of current will indicate a high initial surge of current which decreases with time until it apparently reaches zero. At the same time, the voltage across the "bucket" is approaching the voltage of the battery, Fig. 7·13*b*. Again, it seems clear that if the electrical bucket is made larger, it will take longer to fill it. Similarly, if the resistor has a higher value, it will take longer to fill the bucket.

In the electrical case, the product of the resistance, in millions of ohms (megohms), and the capacitance, measured in microfarads, will tell the time in seconds for the voltage across the capacitor to rise to 63.2 per cent of its final value. At the same time, the charging current has fallen to 36.8 per cent (100 per cent − 63.2 per cent) of its original value.

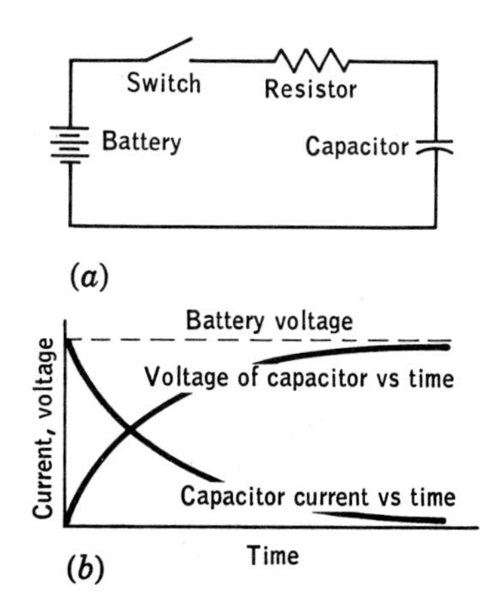

Fig. 7·13 When a resistor is in series with a capacitor, both the charging current and the potential across the capacitor vary with time.

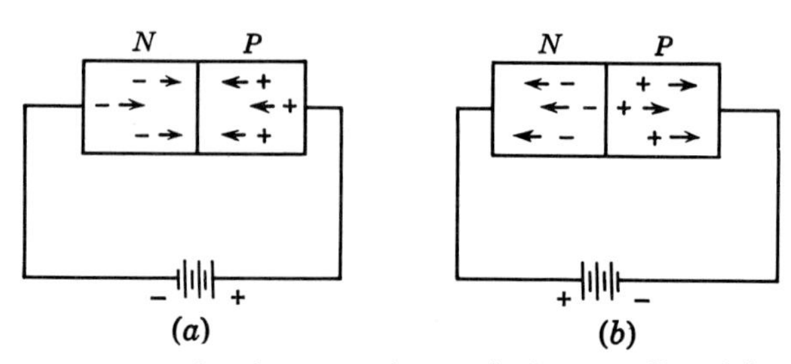

Fig. 7·14 The charges within a diode are affected by the polarity of the applied potential. (*a*) If the applied potential causes the charges to move toward their common junction, then the circuit will conduct. (*b*) If the potential causes the charges to move away from their common junction, the circuit is effectively open-circuited, and almost no current flows.

With various combinations of resistance and capacitance, a wide range of times can be obtained. Thus, if we want to delay a response, we can use such pairs of capacitors and resistors that their product will give the desired time. We will find that air reservoirs and needle valves controlling the movement of air in and out of tanks will display exactly the same time-delay characteristics. Such combinations may be used to smooth out a pulsating signal so that the sensing element will respond to the average value instead of the instantaneous peak. Time constants, whether found in electrical circuits, temperature measurements, or pulsation damping systems, have a very important bearing on performance. Any device with a long time constant may present problems in control because of the lag of the measured property.

Electrical capacitors have other qualities which should be kept in mind:

1. They can distinguish between steady signals and changing ones by passing the changing signal and holding back the steady one, except, as we have noted, that a steady voltage will build up across a capacitor, but the current will drop to zero.

2. A capacitor connected in parallel with a resistor will tend to bypass a changing current around the resistor, and this will help to keep the voltage across the resistor constant. This is one type of filtering action.

3. Combinations of capacitors, inductances, and resistances determine the frequencies at which circuits will function and the frequencies which will be generated in oscillators (see Chap. 8).

TRANSISTORS

It will be recalled that diode rectifiers function because unlike charges can be driven toward a common junction where the charges combine, Fig. 7·14*a*. Here the flow of charged particles constitutes a current. Alternatively, the unlike charges may be driven away from the junction, thereby creating a region with no combining particles, Fig. 7·14*b*. This in effect constitutes an open or high-resistance circuit with very little current flowing. The number of electrons available to cross a junction then depends in part upon the polarity and magnitude of the applied potential.

We have a somewhat similar situation with respect to the transistor which, in its simplest form, has three members, Fig. 7·15. We might carry this terminology further and show the relationship to the ordinary triode vacuum tube:

Triode	*Transistor*
Cathode	Emitter
Grid	Base
Plate	Collector

If the emitter and the collector have an excess of electrons in their crystal structure, they are known as n-type materials; if an excess of positive charges, as p-type materials. At the present time, two elements are commonly used for transistor manufacture: germanium and silicon. They must first of all be purified to extreme degrees, their necessary

Fig. 7·15 Representation cf charges in an n-p-n type of transistor. (*a*) If the base section is thick, it effectively prevents the movement of electrons from the emitter to the collector. (*b*) If the base section of an n-p-n transistor is made very thin, the electrons entering it from the emitter come under the influence of the positive potential applied to the collector. (*c*) Inserting a signal into the base-emitter circuit can change the electron output of the emitter. (*d*) By using an enlarged base section, and considering that it consists of two parts, we now have two diodes. One diode is biased for conduction and the other is biased against it.

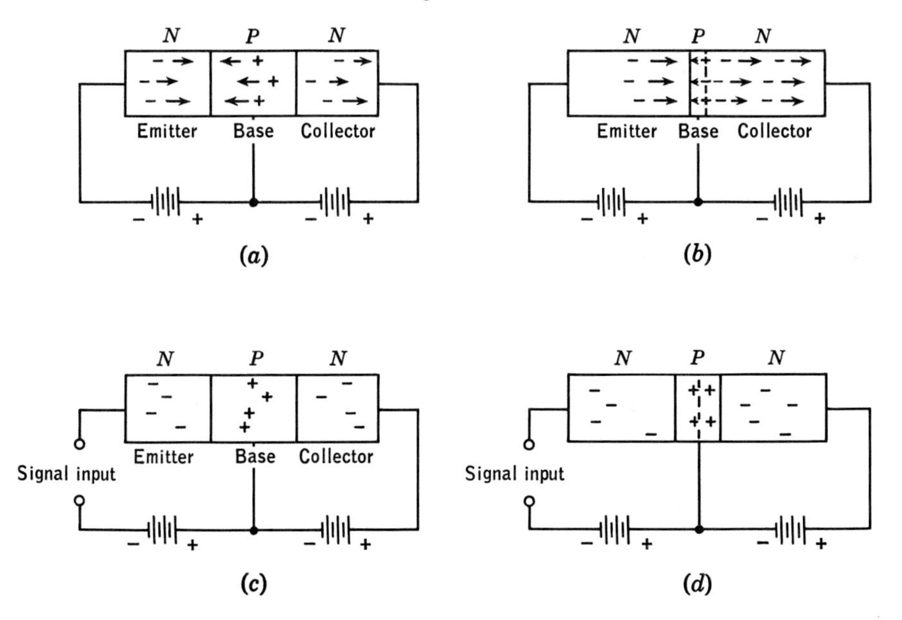

purity being of the order of 1 part in 10 billion. Then they are deliberately contaminated with a material—perhaps 1 part in 20 million—which will fit into their crystal structure and provide either an excess of electrons, which will create n-type material, or a deficiency of electrons, which will create p-type material.

In Fig. 7·15*c* we have injected a signal into the base-emitter circuit. It should be noticed that the distance between the emitter and the collector is occupied by the base and that the thickness of the base measured between the emitter and collector, may be as little as 0.0003 in. Let us go one step further and divide the base into two sections, as is illustrated in Fig. 7·15*d*. We now have what appears to be two diodes. The emitter-base diode has its biasing battery applied in such a manner that the positive components in the base will be repelled by the positive potential and driven toward the emitter. At the same time, the negative charge of the battery will drive the electrons away from the battery and in the direction of the positively charged base, which will attract the electrons. As the magnitude and polarity of the signal change, the net voltages applied to the emitter-base will increase and decrease, and so will the flow of electrons and positive charges, or *holes*. One of the cardinal rules of transistor operation is that the biasing of the base and the emitter must always be such as to create a current flow in the diode which we are considering. Should the net voltage go to zero or even reverse itself, the flow will remain at zero.

At the same time, the other diode which we have created has the battery applied in such a way that the electrons in the collector are drawn to the positive terminal of the battery and the positive charges are attracted away from the base to the negative battery terminal. This means that the current carriers are going away from their common junction, which then creates a condition of almost no current flow in the collector-base circuit. There is essentially an open circuit at the common junction of the second diode. What is needed if this section of the transistor is to carry current is a new supply of electrons into the base, and this will have to be in the immediate vicinity of the collector in order that the positive field may attract the electrons to the positive battery terminal.

As it happens, we do have this new supply of electrons conveniently close: the emitter-base diode section is transmitting electrons from the emitter into the base. Now if the positive potential of the collector can be made adequately large, by comparison with the emitter-base voltage, then it can furnish a greater attracting potential than can the base to which it was originally attracted.

If we use a very commonplace analogy, it may be easy to visualize what is going on to produce this transistor behavior. Many people are enticed into entering a store because of advertising of some sort or another; it appeals to them in terms of price, material, design, or some other quality. Consequently, they enter the store and make definite moves to see the merchandise, say, to the left of the entrance. But just as they get into the store, they find that a most attractive display on the right: there are color, commotion, and lights, and things seem to be going on. (This may all have been part of the merchandising stunt: get the people into the store with a loss leader of some sort and then sell them something else at a far higher price.) So our would-be purchaser turns to the right to see what is going on, and he never does get back to his original destination.

Our transistor is based on similar trickery: use any means at all to get the electrons out of the emitter and into the base, where they will fall under the spell of the higher collector potential. The only use for the base is to entice these electrons to such positions that the electric fields can set up what is desired in the way of electron or positive-charge flow. Because the emitter will send to the base a number of electrons which is directly proportional to the base-emitter potential and most of the electrons go on to the collector circuit, the current in the collector circuit is directly proportional to the applied signal.

It may help some if we use another analogy to explain the transistor's behavior. If a number of steel ball bearings or other spherical iron objects are resting on a table top which is perfectly level, or slightly tipped up to the right, Fig. 7·16, then the gravitational pull merely pulls the balls against the table top. But if we place a permanent magnet on the opposite side of a slot in the table top, then the magnetic force might pull some of the steel balls toward the magnet. But as soon as the balls reach the edge of the slot, the gravitational pull, which previously could accomplish nothing, is able to exert its full force on the balls and accelerate them downward. If we assume that the magnetic pull required to cause the balls to roll over the smooth surface is small when compared with the earth's pull, then most of the balls will fall and only a very few will reach the magnet. This almost exactly parallels what the two electric fields do in the case of the transistor. The base can initiate

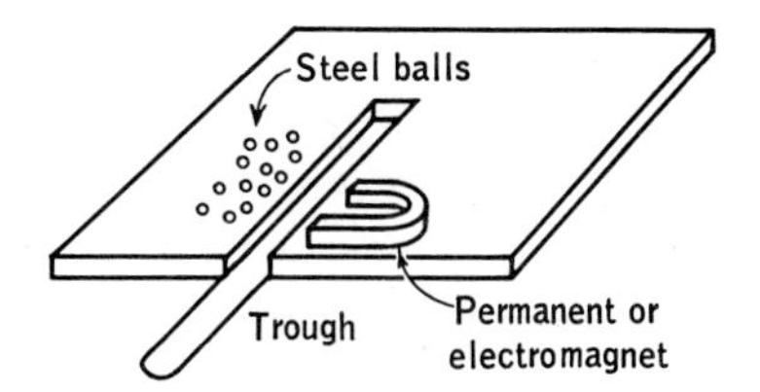

Fig. 7·16 Transistor action is somewhat analogous to the action of steel balls on a smooth level surface when the balls are subjected to two forces: gravity and magnetism. The primary function of one force is to bring the balls (or electrons) to the point where they become subject to the second force.

the movement of the electrons into the base section, but when they arrive there, they find another field which is too powerful, when compared with the field which first attracted them, to resist.

To extend the analogy a bit further, let us replace the permanent magnet with an electromagnet. Then we have essentially the same behavior as before, but with this difference: as we vary the current in the magnet winding, the effect of the magnet will vary, and presumably it will pull more or fewer steel balls to the edge of the table. In this manner, we have a controllable flow down a trough. But if the steel balls now are replaced by electrons, or other charged particles which can be attracted and moved by electrostatic fields, then the particles can be brought to such point that, as they move across the gap in response to the field, they find a second electrostatic field which diverts them from their original destination. Now if some of these electrons had approached the "edge" with some initial speed of their own, they might have had enough energy to reach the first field before the second field could pull them away. As a matter of fact, a certain number of electrons *will* complete their motion through the base and not through the collector as we desire.

In actual transistor design, every attempt is made to make the physical base narrow or to locate the emitter and the collector terminals so that they are physically close. In this manner, the collector can exert the maximum effect.

Now, if the steel balls in our analogy were given a random motion because some energy had been imparted to them, some of them might reach the slot because of their own motion. If any significant number were to do so, the electromagnet would no longer be effecting control over the number of balls going over the edge of the slot. We have an analogous condition when the electrons have such thermal energy because of their temperature that the number acted on by the base is no longer controlled by the emitter-base potential. If the random motion of the electrons causes large numbers to enter the base when there is no base-emitter signal, then the device is not satisfactory. This is precisely the sort of thing that happens when transistors get hot or above their rated temperature.

Comparison with vacuum tubes. It has been stated that the three-electrode and the three-terminal transistor are much alike in their functioning, but one distinction must always be kept in mind. Most vacuum tubes, when used for simple Class A amplifiers, respond only to potential variations and do not take any current for their operation; they are operated primarily by the potentials which are applied to the grids. But transistors require for their operation that electrons or holes (positive charges) be injected into the transistor material. Thus, one transistor is *current-operated,* while the vacuum tube is *potential-operated.*

The use of the word "common" in the illustrations means that the same electrode functions simultaneously in two different circuits. For example, in Fig. 7·15*d*, the base is common.

One of the most commonly used terms in connection with triode tubes is the *voltage-amplification factor,* usually designated by mu, μ. Some tubes are designed for very high amplifications—of the order 60 to 100—and relatively small plate currents, while others are designed for voltage amplifications of the order of 3 or 4, but with plate currents of the order of 100 to 200 ma.

The corresponding terms in transistor technology are concerned with current relationships. The ratio of the current change leaving the collector to that leaving the emitter is called *alpha,* α. It may be written as:

$$\frac{\text{Change in collector current}}{\text{Change in emitter current}} = \frac{I_c}{I_e}$$

This ratio will always be less than one for junction transistors of the type which we have been considering. Inasmuch as some electrons always end up in the base circuit after having left the emitter, the collector current can never equal the emitter current. But the smaller the base current is with respect to the emitter current, the more *efficient* might the transistor be said to be. Stated another way, with the base current representative of the signal input, the greater the ratio of the current going to the collector for each unit current in the base, the greater is the response of the transistor.

For the common emitter, or grounded emitter, there is a relationship which is a comparison of the collector current changes with respect to changes in the base. For this type of circuit, the ratio

$$\frac{\text{Change in collector current}}{\text{Change in base current}} = \frac{\Delta I_c}{\Delta I_b} = \beta$$

Another relationship which is often used to designate the same ratio is obtained from

$$\beta = \frac{\alpha}{1 - \alpha}$$

It will be noted that as α gets closer and closer to 1, the value of β increases indefinitely. It is now possible to obtain transistors which have betas as high as 200.

Typical circuits. We shall consider some of the more typical circuits which are likely to be of value in our study of instrumentation.

The measurement of small current values of 0 to 1 milliammeters is often desired because of the sturdiness of the instruments. Figure 7·17 shows a simple schematic for the purpose. Although the input current may be only a few microamperes, the input potential required to cause this

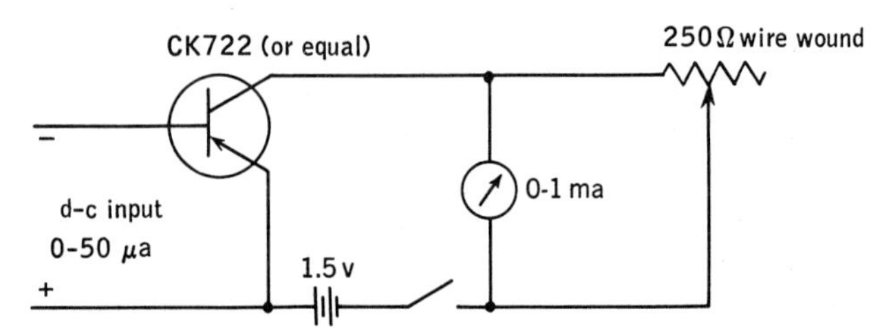

Fig. 7·17 The use of a transistor makes it feasible to measure microamperes when using the more rugged and cheaper milliammeter.

current flow is not an insignificant quantity. Consequently, we cannot use transistors to detect potential differences of the order of a small fractional part of a millivolt; the voltage of the input signal is likely to be of the order of 10 to 20 mv.

Conventional transistor action causes an input of a few microamperes to result in a collector current of 1 ma. For this reason, we shall have to use more complex circuits to tell when the potential applied to the circuit is close to zero in value.

If we want to eliminate the effect of the steady current in order to study more effectively the changes in current, then we may use the circuit shown in Fig. 7·18. If we want a greater current amplification in order that smaller input signals may be measured, then we may make use of another transistor characteristic which makes it possible to couple the output of one transistor directly into the other. Such a coupling arrangement is economical of parts and helps to minimize the possibility of failure. It will be noted that the two transistors have unlike designations, one being p-n-p and the other being n-p-n, Fig. 7·19*a*.

Transistorized voltmeter. One of the most useful instruments for checking voltages is an electronic voltmeter. The circuit is not unlike the circuits described previously. The chief variation in this circuit is in the use of a group of resistors (Fig. 7·19*b*) which are connected in series with

Fig. 7·18 By employing a bridge type of circuit, the steady current can be balanced out. The meter then responds only to the current created by the input signal.

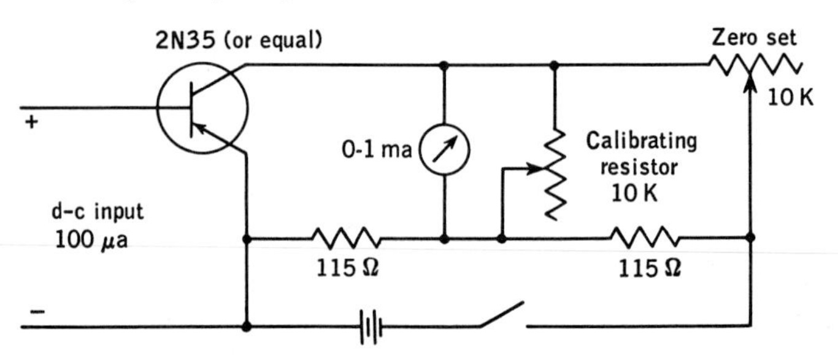

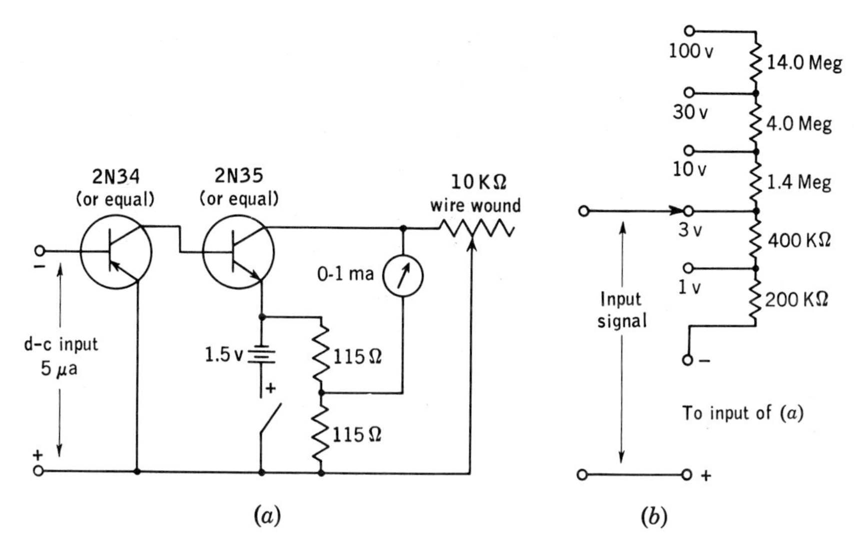

Fig. 7·19 (a) By using complementary transistors, it is possible to obtain relatively high current amplifications when employing simple circuits having few components. (b) This transistorized voltmeter requires an input signal of 5 μa in order to obtain a 1-ma signal in the meter circuit. The input resistors are chosen to furnish the 5-μa signal current with the indicated potential.

the input voltage so that the maximum current should never exceed 0.000010 amp, or 10 μa. It will be noted that where the input potential has been increased by a factor of 10, the resistance of the input circuit has been increased by the same factor. Therefore, under all normal operating conditions, regardless of the range selected, the currents in the first and second transistors will have the same ratio. The position of the selector switch will indicate by what value the output current should be multiplied in order to indicate properly the input signal current.

The range may be increased or decreased provided that care is taken to keep the correct ratio of input potential and resistance of the entire input circuit.

Geiger counter. Later we shall have occasion to study the effects of ionizing the gas confined within certain tubes. All of the essential elements of a commercial device are found in Fig. 7·20. In this circuit, the function of the transistor is quite unlike that of any transistors previously studied. This transistor is capable of converting a steady direct current into a modulated direct current which appears in the secondary as an alternating current. This circuit is called an *oscillator* because of the variations which now occur in it. Ability to function as an oscillator means that direct current may now be converted into an interrupted current without employing any movable components. This type of

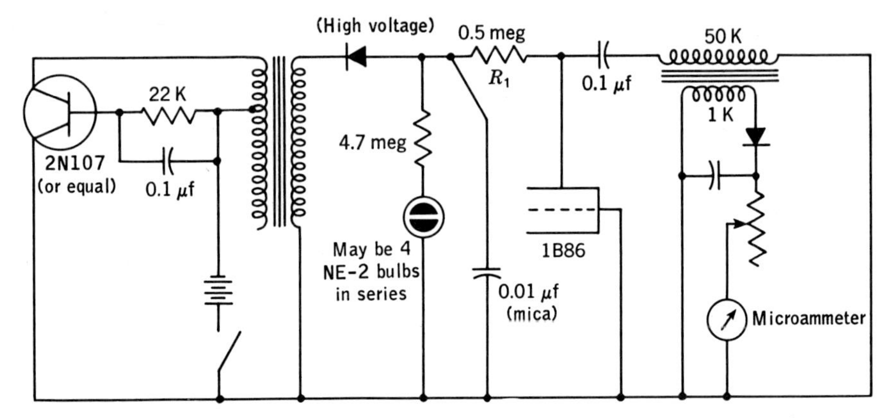

Fig. 7·20 Transistorized Geiger counter. In this example the transistor is used to convert the low-voltage direct current into alternating current which may be transformed, and then rectified for use with the ionization tube 1B86.

chopper operation will be of greater and greater significance as the performance of steady-state devices like the transistor is improved.

The high voltage output of the secondary is then rectified and fed to a voltage-regulating section, which consists of a series of neon bulbs. (Most gaseous discharge tubes maintain the same potential across their terminals during their operation, without regard for the current. Thus as the current increases but the voltage remains the same, the apparent resistance of the device decreases.) These tubes will pass just enough current to maintain the potential across their terminals constant. With a regulated value of voltage available to the Geiger tube 1B86, the ionizing of the gas is independent of the voltage and dependent only upon the number of ionizing particles which penetrate the tube. As the tube current varies with the amount of ionizing going on, so will the potential drop across resistor $R1$ vary. This change in potential will be applied to the step-down transformer, which will deliver its rectified secondary current to a microammeter.

Power output stages. The output signals of power amplifier systems used for positioning servomechanisms in instrumentation equipment generally require that the signal be large in terms of current rather than large in terms of potential. The type of output transformer selected is designed to match the current requirements of the motor with the current output abilities of the transistors. This type of circuit is simple but very effective.

BRIDGES

It is believed that the material relating to bridges will be better understood if it is presented in the form of simple experiments which can be easily performed.

Fig. 7·21

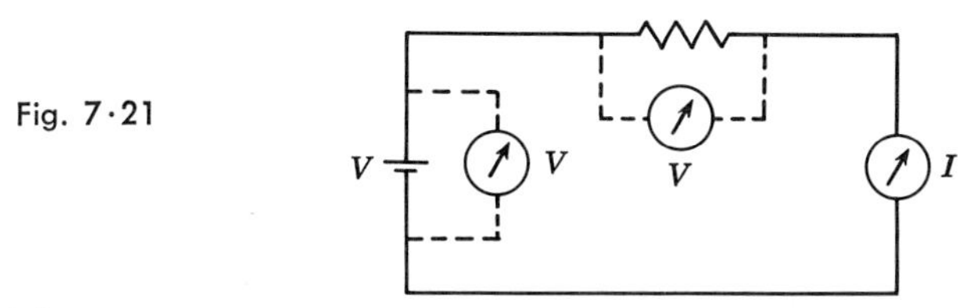

Experiment 7·1

Subject. Measurement of voltage and current; calculation of resistance for series and parallel circuits.

Theory. The resistance of a direct-current circuit is the ratio of the electrical pressure to the flow, or *pressure flow.* Using the customary units, we have pressure in volts and current in amperes. The unit of resistance is then the *ohm.* Expressed in the usual way, the relationships are volts/amperes = ohms. This relationship holds for an entire circuit, part of the circuit, or any component. If we know any two of the trio, we may compute the third one. The voltmeter V, Fig. 7·21, is used to measure the pressure drop in the circuit or part of the circuit that we are interested in. The current is measured by the ammeter, which is so placed that all the current from that part of the circuit in which we are interested will pass through it. If in Fig. 7·21, $V = 10$ volts and the current $I = 2$ amp, then the resistance of that part of the circuit will be found as follows:

$$\frac{\text{Volts}}{\text{Amperes}} = \text{ohms}$$

$$\frac{10 \text{ volts}}{2 \text{ amp}} = 5 \text{ ohms} \qquad \text{the value of } R$$

If we convert Fig. 7·21 into a *series* circuit, Fig. 7·22, the current must go through two or more parts in succession. Whatever current goes through one part must go through all the rest. Let $V_1 = 3$ volts, $V_2 = 6$ volts, and $V_3 = 1$ volt. Then

$$R_1 = \frac{3 \text{ volts}}{2 \text{ amp}} = 1.5 \text{ ohms} \qquad R_2 = \frac{6 \text{ volts}}{2 \text{ amp}} = 3 \text{ ohms}$$

$$R_3 = \frac{1 \text{ volt}}{2 \text{ amp}} = \tfrac{1}{2} \text{ ohm}$$

Fig. 7·22

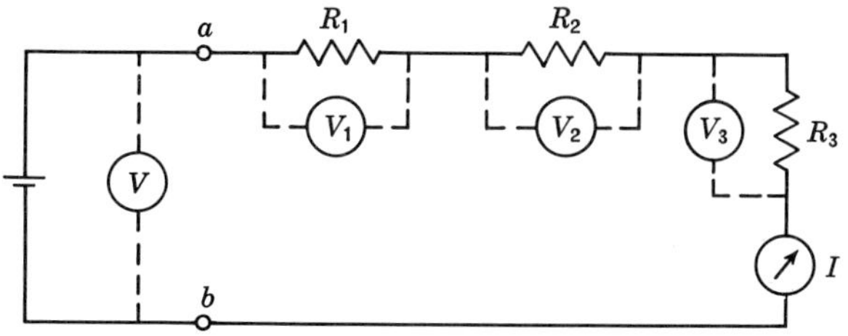

We would find that $V = 10$ volts by direct measurement. Also, $V_1 + V_2 + V_3 = 10$ volts. Therefore, we may state: One characteristic of a series circuit is that *the sum of the voltages across the various series components is equal to the applied voltage V.*

$$R \text{ for the entire circuit} = \frac{10 \text{ volts}}{2 \text{ amp}} = 5 \text{ ohms}$$

But $R_1 + R_2 + R_3 = R_{\text{total}}$

$$1.5 \text{ ohms} + 3 \text{ ohms} + 0.5 = 5 \text{ ohms}$$

Therefore, we may conclude that *the resistance of a series circuit is equal to the sum of the resistances of the individual series resistors.*

Very often the resistances are used in parallel, as are the various appliances in your home. The resistance of such a circuit is found in a similar manner. In a parallel circuit all branches or paths have the same electrical pressure or voltage, but the current in each path is governed by the resistance of that one branch only; the others do not affect it at all (theoretically).

Assume in Fig. 7·23 that the current taken by the three branch circuits is 10 amp when the battery develops a potential of 10 volts. The resistance of the entire circuit is then 10 volts ÷ 10 amp = 1 ohm. By closing the switch to just one branch circuit at a time, we find that the current to $R_1 = 2$ amp, to $R_2 = 3$ amp, and to $R_3 = 5$ amp. Then the values of each branch resistor can be found:

$$R_1 = \frac{10 \text{ volts}}{2 \text{ amp}} = 5 \text{ ohms} \qquad R_2 = \frac{10 \text{ volts}}{3 \text{ amp}} = 3.3 \text{ ohms}$$

$$R_3 = \frac{10 \text{ volts}}{5 \text{ amp}} = 2 \text{ ohms}$$

It will be seen that with parallel circuits, we cannot add the branch resistances, because the resistance of the entire circuit is only 1 ohm, although none of the branches has a value so low. R for the entire parallel circuit, based upon the resistances of the components, may be found by assuming some value of voltage applied to branches and then computing the current for each branch. All of the branch currents are then added

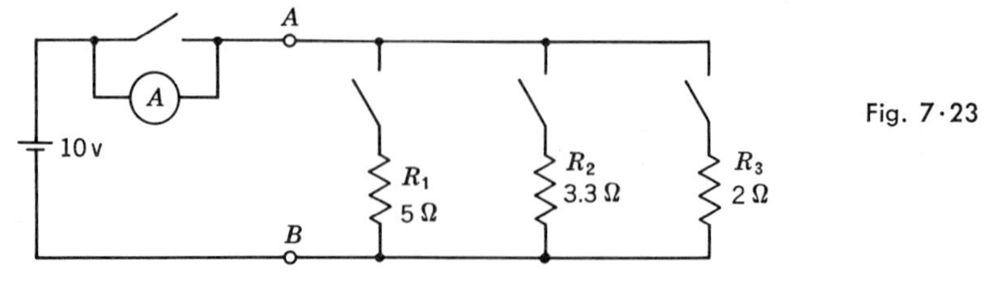

Fig. 7·23

and their sum is divided into the assumed voltage. Oftentimes 1 volt is assumed; then the value found when the resistance is divided into the voltage is called the *conductance*. However, any value of voltage may be assumed; just remember to use the same value for all parts of any one group of parallel paths. For example, for the three paths of Fig. 7·23 assume that the voltage from A to B is 100 volts. Then:

$$\text{Current in } R_1 = \frac{100 \text{ volts}}{5 \text{ ohms}} = 20 \text{ amp}$$

$$\text{Current in } R_2 = \frac{100 \text{ volts}}{3.3 \text{ ohms}} = 30 \text{ amp}$$

$$\text{Current in } R_3 = \frac{100 \text{ volts}}{2 \text{ ohms}} = 50 \text{ amp}$$

$$100 \text{ amp} = \text{total current}$$

$$\text{Resistance} = \frac{\text{pressure}}{\text{current}} = \frac{100 \text{ volts}}{100 \text{ amp}} = 1 \text{ ohm}$$

This checks the value previously found for the entire circuit.

In general, assume a value for V which is numerically greater than the largest value of R. Then the problem of decimal values will be minimized.

Summary. In series circuits: (1) The current is the same throughout the series portion of the circuit. (2) The sum of the individual resistances equals the series resistance. (3) The sum of the voltages across the various parts equals the applied voltage.

In parallel circuits: (1) The voltage applied to each part is the same. (2) The current in each branch depends upon resistance of that branch. (3) The total current equals the sum of the individual branch currents.

Apparatus

1 storage battery, 6 volt, or 4 no. 6 dry cells
3 resistors, 10, 20, and 30 ohms
1 ammeter, 0–1 amp
1 voltmeter, 0–10 volts

Procedure

1. Construct a series circuit like that shown in Fig. 7·22. Using clip leads on the voltmeter, use the voltmeter to determine V_1, V_2, and V_3. Also record the current.

2. Construct a parallel circuit like that shown in Fig. 7·23.

   a. With the knife switch in each branch open, the ammeter current should read zero.

b. Close one switch and read and record the branch current. Repeat for the other two branches.

c. Close any two branch switches and read and record the total current.

d. Close all three branch switches and read and record the total current.

Results

1. Find the series resistances and the resistance of each component in step 1.

2. Find the parallel resistance, the resistance of each branch, the resistance of two branches in parallel, and the resistance of three branches in parallel.

3. Compute and compare the calculated results with the observed values.

4. Compare the sum of the component voltages with the applied voltage for the series circuit.

5. Compare the sum of the individual branch currents with the total, or line, current.

What conclusions can you draw about the character of series and parallel circuits? Where are you likely to encounter series and parallel circuits?

Experiment 7·2

Subject. Galvanometer construction and galvanometer applications.

Purpose. To become familiar with the galvanometer as the basic electrical measuring device in industrial instrumentation.

Apparatus. Galvanometer, zero center; series and parallel resistors.

Theory. When two magnetic fields exist in close proximity, they produce a mutual reaction such as a push, pull, or a turning or twisting action. It makes no difference whether the magnetic fields are created by permanent magnets, electric currents, or by some other means.

The D'Arsonval meter is the almost universally accepted instrument for measuring small currents. The D'Arsonval design employs bearings and has the coil turn against a hairspring, whereas a still more sensitive arrangement eliminates bearing friction because the rotating element is suspended by a conducting wire which is extremely elastic. Because the strain in the wire suspension is directly proportional to the angular rotation, for small angles, the restoring torque is likewise proportional to the angular displacement.

Both designs have had wide use in industrial instrumentation. But at the moment we are concerned mainly with the more rugged meter which uses jeweled bearings, and has hairsprings to carry current to the coil, which may rotate through angles of 90 to as much as 270°.

For any given meter, it takes only so much current in the coil to produce the torque required for the full 90° (more or less depending upon the individual meter). Therefore, because a meter requires only a certain definite current, all uses of the meter should be such as to allow only this value of current to pass through the meter when maximum deflection is desired. More than this current could damage the meter because of extra heat or damaged hairsprings. Therefore, we may place resistance in series with the coil to control the meter-circuit current. When we use the galvanometer to measure the small signals developed across the points of a Wheatstone bridge, we want the greatest response of the meter, so we want no extra resistance in the meter circuit. But if we want to use the same galvanometer to measure much larger potentials, we must put such extra resistance in series with the meter that the current will be just enough to produce full-scale movement with the maximum potential to be measured. We can apply Ohm's law here; therefore, if the resistance of the meter is 20 ohms, the current required for full-scale deflection is 0.0005 amp, and the potential to be measured is 50 mv, or 0.050 volts, then

$$I = \frac{E}{R}$$

$$0.0005 \text{ amp} = \frac{0.050 \text{ volts}}{R}$$

$$R = \frac{0.050}{0.0005} = 100 \text{ ohms}$$

With 100 ohms as the required meter-circuit resistance, we must add 80 ohms in series with the 20 ohms of the meter to bring the total to 100 ohms in order to convert the galvanometer to a millivoltmeter with a range of 0 to 50 millivolts.

For any other voltage rating, the same procedure would be followed. To convert this specific meter into a voltmeter which reads 100 volts, full scale, we repeat the process. Again,

$$I = \frac{E}{R}$$

$$0.0005 \text{ amp} = \frac{100 \text{ volts}}{R}$$

$$R = \frac{100 \text{ volts}}{0.0005 \text{ amp}} = 200{,}000 \text{ ohms}$$

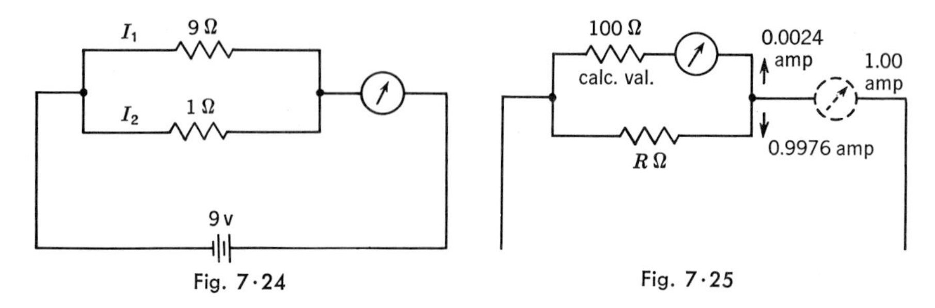

Fig. 7·24

Fig. 7·25

Theoretically, we should subtract 20 ohms from this answer, but normally we would not do it or even try to do it. The answer would be 200,000 ohms.

From these two examples, it will be seen that a galvanometer can be made into a voltmeter of almost any desired range. All that we need to do is place the correct resistance in series with the galvanometer to limit the current to the rated value.

If we want to measure current in the ampere or milliampere range, we find that the current available exceeds that required to move the meter full scale. All that we need now is to cause only a portion of the current to pass through the coil and to bypass the rest of it around the coil in some manner. Suppose that we want to measure only one-tenth or one-hundredth of the total current because that is all that the meter can pass safely or that we can read on the scale. If we can arrange such a division of currents, then all that we need do is multiply our measured value by the appropriate 10, 100, or whatever the appropriate factor might be. But in all these cases in which we measure only a factional part of the total current, we must know what the fraction is, as in Fig. 7·24.

$$I_1 = \frac{V}{R_1} = \frac{9 \text{ volts}}{9 \text{ ohms}} = 1 \text{ amp}$$

$$I_2 = \frac{V}{R_2} = \frac{9 \text{ volts}}{1 \text{ ohm}} = 9 \text{ amp}$$

$$I_{\text{total}} = 1 \text{ amp} + 9 \text{ amp} = 10 \text{ amp}$$

We find that the current in branch 1 is one-tenth of the total so long as the resistance of the branch is 9 times that of the parallel resistor; it will be one-hundredth of the total current if $R_1 = 99$ ohms when R_2 is 1 ohm. In this way, by dividing the current into two parts, we can measure any desired fraction of the total current. All that we need do is multiply the answer by the proper factor.

Procedure. To convert a galvanometer into a voltmeter.

1. To a 12-volt battery, connect a galvanometer in series with a decade

resistor. Before completing the circuit, make sure that the resistor inserts about 10,000 ohms into the circuit. Then adjust the decade resistor to make the meter read full scale. Record the value of resistance required. With a good voltmeter, check the value of the battery voltage.

2. Repeat, but with a value of 6 volts (or 4 volts; the value is of no great importance). Again, record the value of the decade resistor for full-scale meter deflection.

3. Compute the current required for full-scale meter movement and the resistance of the meter thus:

$$(I_{\text{meter}})(R_{\text{meter}} + R_{\text{decade 1}}) = 12 \text{ volts (use } \textit{your} \text{ true value)} \qquad (1)$$

$$(I_{\text{meter}})(R_{\text{meter}} + R_{\text{decade 2}}) = 6 \text{ volts (use } \textit{your} \text{ true value)} \qquad (2)$$

For this case, let us assume that $R_{D1} = 4{,}900$ ohms, and that $R_{D2} = 2{,}400$ ohms. Then, dividing Eq. (1) by Eq. (2), we get

$$\frac{(I_{\text{meter}})(R_{\text{meter}} + R_{D1})}{(I_{\text{meter}})(R_{\text{meter}} + R_{D2})} = \frac{12 \text{ volts}}{6 \text{ volts}} = \frac{2}{1} = 2$$

Therefore

$$R_{\text{meter}} + R_{D1} = 2(R_{\text{meter}} + R_{D2})$$

$$= 2R_{\text{meter}} + 2R_{D2}$$

If $R_{D1} = 4{,}900$ ohms and $R_{D2} = 2{,}400$ ohms, then

$$R_{\text{meter}} = R_{D1} - 2R_{D2} = 4{,}900 - 2(2{,}400) = 4{,}900 - 4{,}800 = 100 \text{ ohms}$$

$$\text{Current for full-scale deflection} = \frac{E}{R} = \frac{12.00 \text{ volts}}{(4{,}900 + 100) \text{ ohms}} = 0.0024 \text{ amp}$$

4. Using the calculated full-scale current, compute the resistance that must be connected in series with the meter resistance to make the pointer move full scale for 50 mv, 1 volt, and 100 volts. Using these values in decade resistor, check for full-scale deflection.

5. To cause this meter to function as an ammeter, proceed as follows: For the example meter being used, the meter requires only 0.0024 amp for full-scale deflection, but if 1 amp is to be measured, then only 0.0024 amp must pass through the meter movement and the rest must pass through the resistor in parallel with the meter. Consequently, 0.9976 amp must pass through the parallel, or shunt, resistor. This is indicated in Fig. 7·25. Because the circuits are in parallel, their potentials are equal, therefore

$$(100 \text{ ohms})(0.0024 \text{ amp}) = (R)(0.9976 \text{ amp})$$

$$R = \frac{(100 \text{ ohms})(0.0024 \text{ amp})}{0.9976 \text{ amp}} = 0.2405 \text{ ohms}$$

Compute the value of resistance that must be placed in parallel with your galvanometer to make it read full scale with 1 amp in the metered circuit.

6. Determine the value of resistance required to make the galvanometer read full scale for 0.100 amp in the metered circuit.

Results

1. By what numerical value must the galvanometer scale value be multiplied to obtain the true values for 50 mv, 1 volt, 10 volts, and 100 volts? (This value is known as the scale constant.)

2. Compare the computed values of series resistances with the per cent full-scale reading obtained. Did you always get the full-scale deflection?

Conclusions. Can you successfully convert a galvanometer into a millivoltmeter, a voltmeter, and an ammeter?

Experiment 7·3

Subject. Resistance measurements using a Wheatstone bridge.

Purpose. (1) To teach the theory and operation of a Wheatstone bridge. (2) To learn how to construct a very simple slide-wire bridge for home or laboratory use.

Theory. The general theory of the Wheatstone bridge is given in the theory section of the chapter and will only be reviewed.

1. When the galvanometer reads zero, points a and b, Fig. 7·26, are at the same electrical level with respect to either point c or point d. Current flowing in the galvanometer circuit will show that these points are not at the same electrical level. To obtain a flow of current, there must be a difference in potential. For this part of the experiment, do not connect a battery from x to y.

2. When the bridge is balanced,

$$\frac{R_1}{R_2} = \frac{R_3}{R_4} \quad \text{or} \quad \frac{R_1}{R_3} = \frac{R_2}{R_4}$$

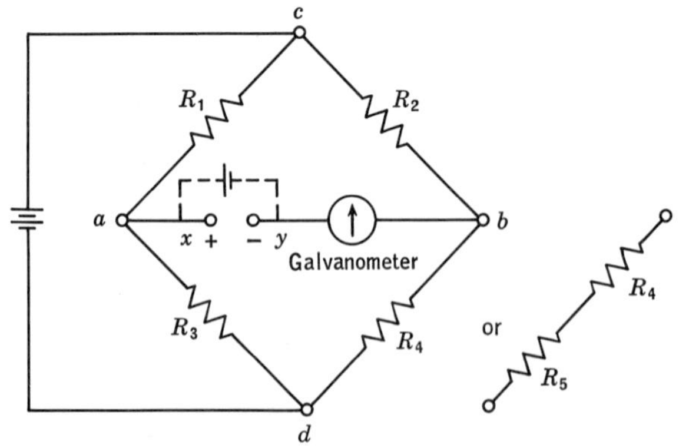

Fig. 7·26

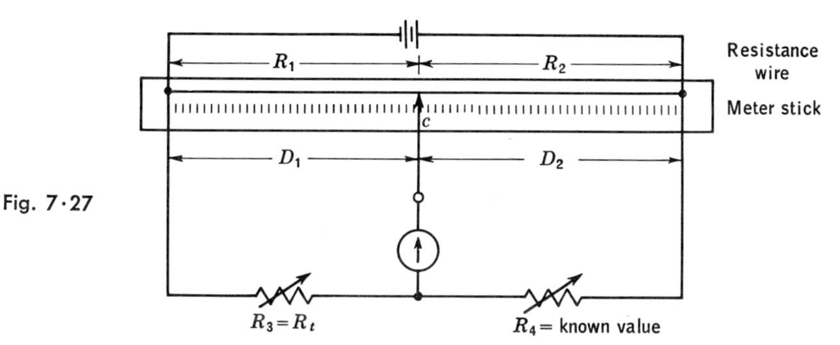

Fig. 7·27

3. By making R_2 variable, we can change the ratio of the bridge to suit our measurements.

   a. If R_1 is made large with respect to R_2, then R_3 will be proportionally larger than R_4 at the time of balance.

   b. If R_1 is made smaller than R_2, then R_3 will be proportionally smaller. By changing these ratios, the one bridge can be used to measure resistances much smaller and much larger than the resistances in the bridge itself.

4. The voltage applied to the bridge in part controls the sensitivity.

5. The currents in the bridge should be small enough so that the resistances do not change their values because of self-heating.

Procedure A

1. Connect components as shown in Fig. 7·26. Make only R_2 a variable resistor. Let R_3 be a spool of copper wire. Let $R_1 = R_4$.

2. Balance the bridge. Read and record R_1, R_2, and R_4.

3. Place R_3 in a container of oil and heat the oil slowly on a small electric hot plate. Record temperature of oil and R_1, R_2, and R_4 for each 10° temperature rise of the oil. The maximum temperature of the oil should not exceed 120°C. The oil preferably should be of the type used for quenching steel.

Procedure B (Optional). A Wheatstone bridge may be constructed very simply for home or laboratory use by using the parts shown in Fig. 7·27. A piece of resistance wire is fastened to a board about 40 in. long. To simplify the determination of the ratios involved, a meter stick should be used, in which case the length of the wire should be just 1 m. Or the distance between the ends may be divided into 10, 20, or 50 equal parts. Let R_4 be a variable resistor whose values are known. If R_4 has but one value, the bridge is balanced by adjusting the position of *c*. Otherwise, the bridge is operated by initially moving contact *c* to the middle of the bridge and R_4 is then varied to bring the bridge to balance. Should it not be possible to balance the bridge, contact *c* is moved in such a direction as to provide a usable ratio. If it is known that R_3 is greater than

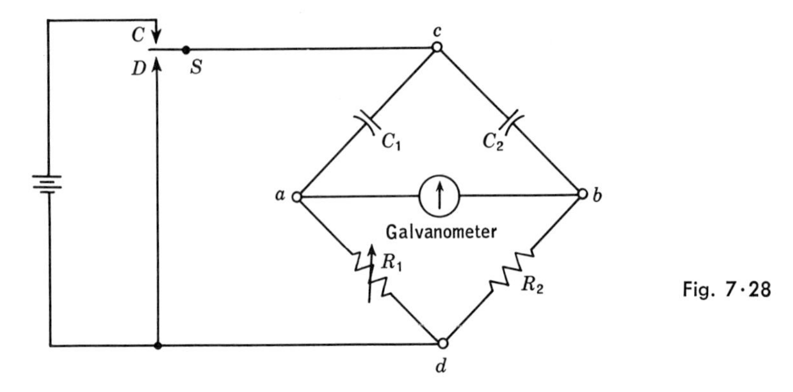

Fig. 7·28

R_4, then D_1 will have to be greater than D_2. For a trial, make D_1 twice D_2 and try again to balance the bridge. If attempt is not successful, make D_1 3 or 4 times as great as D_2. Then follow the rest of the procedure A.

Results

1. Calculate the values of R_3 when at room temperature and at all the other recorded temperatures.
2. Plot the coil resistance (vertically) against temperature (horizontally).

Conclusions. Make conclusions about the following:

1. Can the bridge be used for resistance measurements?
2. Does the copper change resistance with temperature?
3. Could the resistance for copper be used to infer the temperature?

This basic circuit is used for almost all accurate resistance thermometers and resistance measurements. Therefore, it should be mastered. Learn to know not only its operation but also its limitations and advantages.

Experiment 7·4

Subject. Capacitance-type Wheatstone bridges.

Purpose. To learn the theory, circuit, and operation of a capacitance bridge.

Theory. All Wheatstone bridges that have been studied to this point have used only resistors in their four branches. At the time of balance, the flow of current through each of the two legs of the bridge produced two points which had the same steady potential. This equality of potential was indicated by zero deflection of the indicating meter.

Another version of the Wheatstone bridge employs two resistors and two capacitors, as indicated in Fig. 7·28. When switch S is closed at C, assume that the switch had previously been connected to point D and that the capacitors were completely discharged as a result; current will flow through resistors R_1 and R_2 while capacitors C_1 and C_2 are being charged. If we assume that the galvanometer G registers no deflection when switch S is closed at C, then the potential drops across the two resistors must have been equal.

$$I_1R_1 = I_2R_2 \tag{1}$$

$$\frac{I_1}{I_2} = \frac{R_2}{R_1} \tag{2}$$

But the current required to charge each of the capacitors is directly proportional to the capacitance, or size, of the capacitors; the larger capacitor requires the larger charging current. For a properly balanced circuit, the charging current in one branch will always be a fixed percentage of the current in the other branch. It may be 2 times as much, 3 times as much, or only half as much, but the currents will always be directly proportional to the capacitance of the capacitors. Therefore, inasmuch as

$$I_1 = KC_1 \qquad \text{and} \qquad I_2 = KC_2$$

we may write them as fractions:

$$\frac{I_1}{I_2} = \frac{KC_1}{KC_2} = \frac{C_1}{C_2} \tag{3}$$

Because the ratio of the currents is the same as the ratio of the capacitances, we may substitute the capacitances for the currents in Eq. (2), or

$$\frac{C_1}{C_2} = \frac{R_2}{R_1} \tag{4}$$

By adjusting the value of R_1, we can produce the conditions of balance and then determine the relative values of the capacitances. Provided that one of the capacitances is known, the value of the second can be computed.

It is also possible to construct a bridge in which the resistors have fixed values and one of the capacitors is variable in value. Equation (4) still holds true. Certain commercial bridges employ the variable capacitors to balance the bridge rather than the variable resistance because of certain mechanical advantages:

1. The capacitor has no wearing parts (essentially).
2. The balance is not obtained by "jumps" in the voltage; instead, it is continuously variable.

By varying the resistance of R_1, this particular bridge can be used to measure capacitance. If we vary the capacitance, then we can compute the resistance.

As switch S is positioned first at D and then at C and so on, the capacitors become charged and discharged. Any unbalance in the bridge will be indicated by the galvanometer. Such an indication of unbalance will require an adjustment of the variable element.

In commercial practice, we will find that alternating current is supplied to the bridge instead of the intermittent current which is supplied through the switch. However, the bridges all use the same principles. Commercial applications will be found in some of the Foxboro designs. The charging currents of the capacitors are in the microampere range, so with this design the standard cell may be used as the constant-voltage source. With normal use, it should last at least ten years.

Procedure

1. Using the basic circuit shown for this experiment, connect one known and one unknown capacitor into the circuit as shown in Fig. 7·28. Inasmuch as no amplifiers are to be used, choose capacitors in the range of 1 to 6 microfarads, paper or better. Under no conditions use electrolytic capacitors.
2. Determine the value of the unknown capacitor. Use Eq. (4).
3. Measure several other high-grade capacitors.
4. Using small fixed capacitors, determine the smallest change in capacitance that you can detect with your bridge.

Results

1. Determine the values of the capacitors.
2. Compare the measured values with the values marked on the capacitors.
3. What was the smallest capacitance change that could be detected?

Discussion

1. In what ways would you try to make the bridge more sensitive?
2. What are the advantages and disadvantages of capacitance bridges used to measure resistance? Assume that the resistance changes are small and are created by changes in temperature.
3. Discuss commercial designs which employ capacitance bridges for resistance or temperature measurements.

Experiment 7·5

Subject. Electronic bridge circuits.

Purpose. To become familiar with electronic methods of balancing bridges.

Theory. In one sense, a Wheatstone bridge is a potentiometer: its operation depends upon bringing two points to the same electrical level. In Fig. 7·29 it is points a and b which have the same potential, with respect to either point c or point d, at the time of balance. If we have a piece of copper wire connected from x to y when we balance the bridge,

Fig. 7·29

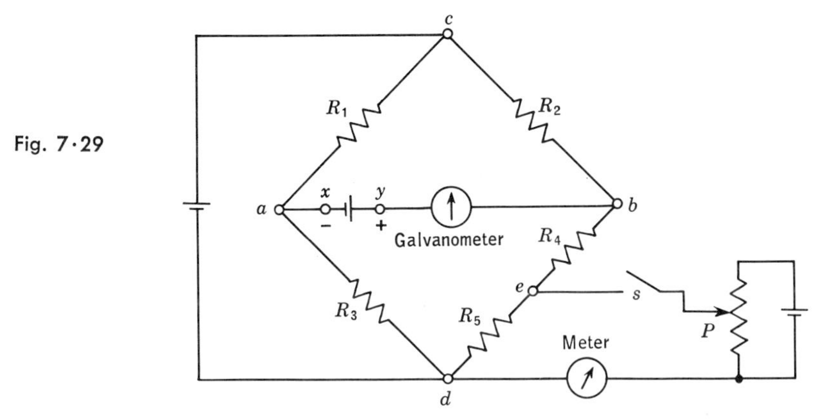

we have a conventional bridge. But if we remove the wire and add a small direct-current signal of 10 to 50 mv from some external source, the bridge will no longer be balanced and the galvanometer will indicate a flow of current. This external signal will try to raise point *y* above *b*, or lower *b* below *y*. To restore the balanced bridge condition, we can raise *b* by sending a current from some external source through a portion of the resistor, *ed* in this case. Additional current through *ed* will increase its potential and thus add to the potential of *b* with respect to *d*. In Fig. 7·29 if we control the extra current in *ed* by positioning the potentiometer rheostat *P*, point *b* can be made higher or lower than point *a*, or equal to point *a*.

The purpose of this extra circuit is to permit us to balance the bridge without having any movable parts in the bridge proper. Balancing motors, problems of inertia, overdrive, and maintenance are eliminated. In commercial designs the section indicated by the potentiometer rheostat is replaced by an amplifier whose input signal comes from the bridge-balance circuit.

Instead of injecting a signal into the bridge at *x-y*, we could leave this as it is in the conventional bridge. Then we might let R_3 change in value as it would if it were the coil of a resistance thermometer. Figure 7·30 shows such a modified circuit. This is the basic theory behind the apparently complex circuits used by certain commercial equipment. Taylor's 700 T and Swartwout's, along with others, use these principles. If you understand this theory, then you understand the basic performance of the commercial items. Your chief problem now will be to identify the parts. Experiment 7·6 will aid you, because it is even closer in construction to the commercial item.

Because the extra voltage contributed by R_5 is directly proportional to the current, the current tells us the millivolts added to balance the bridge. The current can then be used to "measure" the signal at *x-y* or

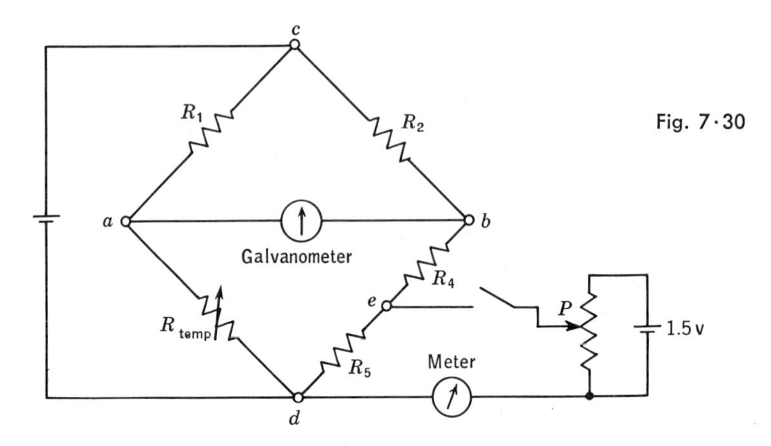

Fig. 7·30

the change in R_T. The temperature then will be "read" on a milliammeter or converted into some other type of signal by a transducer.

Apparatus. Resistors R_1 and R_2, about 100 ohms; R_3, about 100 ohms; R_4, about 200 ohms; R_5, about 10 ohms (except possibly for procedure 2); Nesco galvanometer; potentiometer rheostat; and batteries.

Procedure

1. Set up bridge in Fig. 7·29 and balance with copper wire across x-y.

2. Insert thermocouple leads at x-y instead of copper wire. Gradually heat the thermocouple over a hot plate; adjust the distance to control the temperature. Read the current required to balance the bridge. Be sure that the polarity of the thermocouple is as indicated; otherwise, balance cannot be obtained with the indicated polarities.

3. Replace resistor R_3 with a temperature-responsive resistor and convert to Fig. 7·30. Heat R_T as in Experiment 7·2. Record the current and the temperature.

Results

1. Determine whether the bridges can be balanced as indicated.

2. Determine whether current can be used to "read" temperature through a suitable calibration.

Procedure (option 1)

1. Connect bridge according to Fig. 7·31. Use a high-gain tube. (A pentode might be used instead of the indicated high-μ triode. Resistors R_5 and R_4 will have to be chosen to work with the tube which is used.) Balance the bridge by using a variable resistor for R_5. Make resistor R_{temp} a temperature-sensitive unit such as a spool of copper wire. Heat the coil of wire in oil to about 100°C. Note and record both the deflection of the galvanometer and the plate current for each 10° interval.

2. Replace the vacuum tube circuit with the potentiometer-balancing circuit of Fig. 7·30. *Do not change the values of any resistor.* Heat the coil of wire again and balance bridge by manual adjustment of potentiometer P. Record current and temperature at 10° intervals.

Fig. 7·31

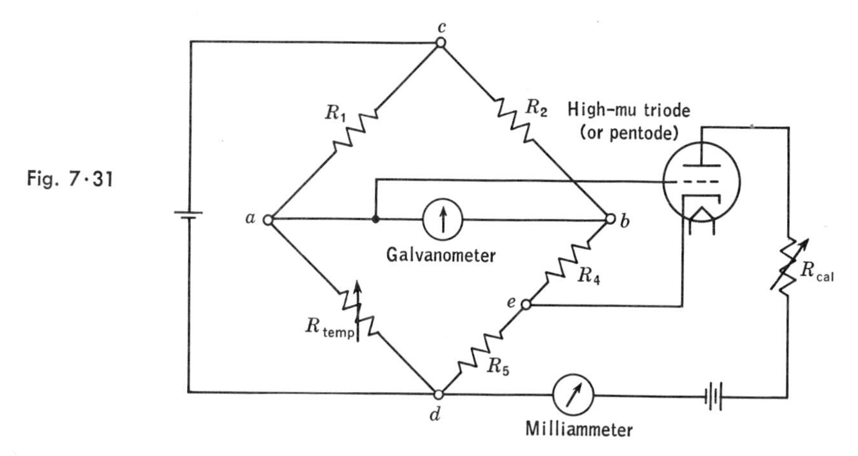

3. Replace the galvanometer with the Nesco unit (which is a galvanometer whose pointer controls the light which falls on a photocell) and connect as shown in Fig. 7·32. Again, *do not change the values of any resistors.* Heat the same copper coil to about 100°C. Read and record the temperature and the plate current.

Results

1. Tabulate temperature versus correcting current for each of the three circuits.

2. Plot correcting current versus temperature for each of the three circuits.

3. Did the vacuum tube give the same correcting current as did the manual adjustment? Why? Explain in detail.

Procedure (option 2). Build an amplifier with a gain of about 100,000 and connect according to Fig. 7·32. Repeat the appropriate parts of option 1.

Fig. 7·32

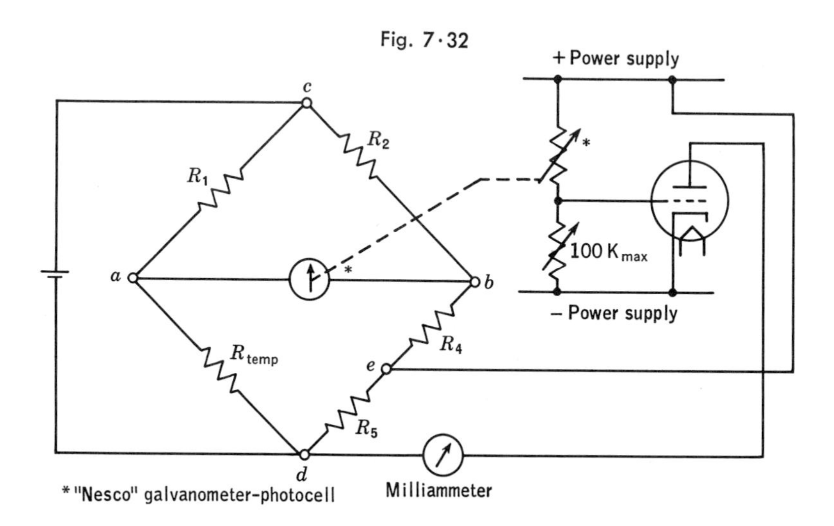

Fig. 7·33

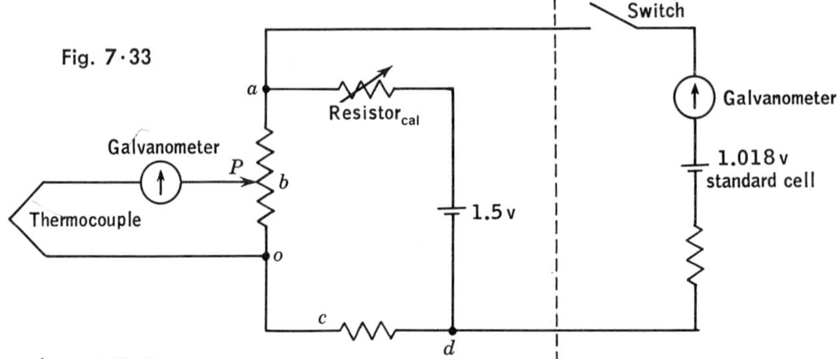

Experiment 7·6

Subject. The potentiometer.

Purpose. To become acquainted with the basic potentiometer circuit used for temperature and potential measurements and the theory of operation of the potentiometer.

Theory. In Experiments 7·2 and 7·3 the Wheatstone bridge was used as a sort of potentiometer. Values were adjusted or controlled to bring two points to the same electrical level or potential. A somewhat different type of potentiometer can be constructed as shown diagrammatically in Fig. 7·33. The section to the left of the dotted line is the potentiometer section; that to the right is used only for calibrating. Two galvanometers are shown, but by proper switching only one is actually required.

If variable contact P is adjusted until the galvanometer shows zero deflection, then the emf of the thermocouple is exactly equal to the voltage between points b and o. By calibration, the potential from a to o is established. The voltage from b to o is a definite percentage of ao, the percentage being determined by the position of P.

In this way, the emf of the thermocouple can be determined. The temperature difference which created the emf can be determined from tables. If desired, the potentiometer can be calibrated in terms of temperature instead of emf. This is the basic type of potentiometer used for emf determination by many instrument makers.

If the voltage from a to d at the time of balance is 1.018 volts, then, by proper choice of resistors ao and cd, the voltage difference between terminals a and o can be made any value that is desired. For example, the output of a thermocouple rarely will exceed 50 mv (0.050 volts). Therefore, we might decide to make the voltage from a to o equal to this value. Because at balance there is no current in the thermocouple circuit, the same current will flow in ao and cd. The voltage relationships then will be:

$$\frac{V_{ab}}{V_{ad}} = \frac{(I)(R_{ab})}{(I)(R_{cd})} = \frac{R_{ab}}{R_{ad}}$$

We know from our calibration that the voltage between a and d equals the voltage of the standard cell. Therefore for a desired value of 50 mv for V_{ao},

$$\frac{0.050 \text{ volts}}{1.018 \text{ volts}} = \frac{R_{ao} \text{ ohms}}{R_{ad} \text{ ohms}}$$

For the sake of making a simple fractional relationship, assume that $R_{ad} = 2{,}036$ ohms. Then when the substitution is made,

$$\frac{0.050 \text{ volts}}{1.018 \text{ volts}} = \frac{R_{ao} \text{ ohms}}{2{,}036 \text{ ohms}}$$

$$R_{ao} = \frac{(0.050)(2{,}036)}{1.018} = 100 \text{ ohms}$$

The standard cell is of a particular design which always gives a fixed value of potential, 1.018 volts under standard conditions. Sometimes a standard cell is used in connection with a regulated power supply. Bristol's No-Bat permits the standard cell to replace the operating battery which so often is employed. With this design, there is no requirement for calibration, inasmuch as the correct voltage is always used. Under no condition should the standard cell ever be called upon to deliver more than a few microamperes. Any other use will ruin it as a standard source of potential for comparison purposes. To meet this low current drain requirement, Bristol's potentiometers are of a high-resistance type.

In Fig. 7·34, a different type of potentiometer is used. Here the current flowing through a resistor develops an emf: $V = IR$. If R_{ab} is known and I can be measured accurately, then the value of IR is known. Thus, the product of current and resistance at the time of balance indicates the value of the emf developed by the thermocouple. Or the milliammeter can be calibrated in terms of the temperature difference when the emf is balanced.

The next step in commercial practice is to provide an amplifier which will make the circuit automatically self-balancing. Any unbalanced signal is fed to the amplifier; the output then controls the current through the resistor ab to reestablish a value of emf at ab equal to that of the thermocouple. Chapter 15 points out some of the practical considerations.

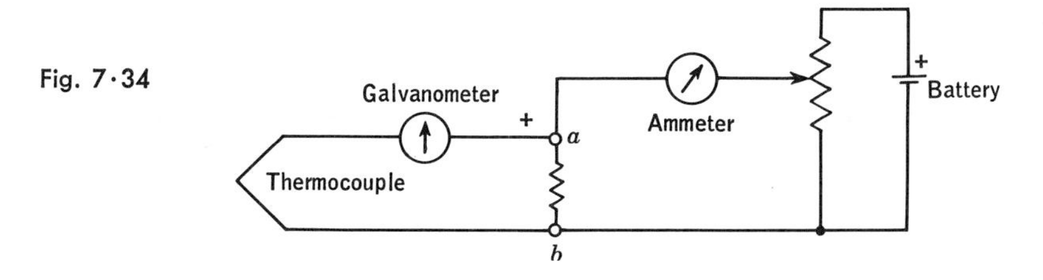

Fig. 7·34

Apparatus. Apparatus called for in Fig. 7·33 to the left of the dotted line is required. $R_{ao} = 100$ ohms; $R_{cd} = 1{,}900$ ohms; $R_{\text{cal}} = 0$–$1{,}000$ ohms; $E_{\text{bat}} = 1.5$ volts; millivoltmeter; galvanometer.

Procedure

1. Connect the apparatus as shown in Fig. 7·33 to the left of the dotted line.

2. For a potentiometer P, calibrate a good wire-wound resistor at eight or ten positions. Make a dial, or get one graduated with 100 divisions. With resistor *cd* approximately equal to 2,000 ohms, adjust the calibrating resistor until the current through the circuit will cause the potentiometer to provide a potential of 50 mv.

3. Then heat the thermocouple slowly. Adjust the position of the movable contact to keep the galvanometer at the *zero* position.

Results. Record the potentiometer position for balance, determine the emf in millivolts.

1. If the thermocouple can be heated in a furnace where the temperatures are known, did your thermocouple develop approximately the correct emf?

Discussion

1. Why do many of the commercial circuits for a potentiometer provide for standardizing the circuit?

2. What are the advantages of continuous standardization versus intermittent standardization?

3. Why are millivoltmeters not always used in commercial instruments to determine the 50-mv potential? What are the reasons for using other methods?

4. Draw a circuit in which only one galvanometer is required for calibration and operation.

SUMMARY

1. Very accurate measurements of potential, when they are required, are generally made by comparing one potential with another. When two voltages are known to be equal and one of the two is known, the other is known also.

2. The measurement of potentials by the comparison of a known and an unknown emf eliminates the flow of current during the final measurement. With no current flowing, the resistance of the circuit is of little importance.

3. When a current of known value flows through a resistor whose resistance is also known, the potential difference across the terminals can be determined. Thus, by reading a calibrated ammeter, the potential which is developed by the resistor is known. If this is the value which is just equal to the unknown potential which is being compared with the known resistor, then the unknown emf is also known. In this manner, potentiometers need not have any variable resistors or slide wires in the measuring portion of the circuit.

4. Rectifiers of the hot-cathode type have two essential parts: (*a*) the cathode, which is heated or otherwise treated so that it will emit electrons, and (*b*) an anode, or plate, which is positively charged to bring to it the negative electrons which are emitted by the cathode.

5. The addition of a grid, or grids, completely changes the performance of a diode. Vacuum tubes of this sort have noteworthy characteristics: (*a*) So long as the grid is maintained negative with respect to the cathode, no signal current is taken by the device. Because no current is required, the device functions without power being furnished to it. A change in the grid potential is all that is required for operation when the tube functions as an amplifier. (*b*) Because the grid potential can control the plate current, it can also control the potential developed by a resistor in the plate circuit. Because the potential developed is greater than the input signal, the device becomes an amplifier. By placing several stages in series, amplifications of millions of times are possible. (*c*) Because the input signal can control the plate current, the tube also functions as a variable resistor which has no moving contacts or parts. In many bridge circuits, and elsewhere, the tubes are used primarily as variable resistors. (*d*) When properly connected with other components, vacuum tubes having three or more elements can be made into oscillators, or generators of alternating current. The frequencies may reach values as high as several billions per second. (*e*) For linear operation there are well-established characteristics and limits.

6. The use of voltage-divider networks makes it possible to employ only a small fraction of some input signal. If the signal applied to a vacuum tube voltmeter is but one-hundredth that of the input, then the measured signal must be multiplied by a factor of 100. Vacuum tube voltmeters may measure a wide range of values, but the actual measuring circuit may be limited to a maximum value of ½ volt or less.

7. The use of photoelectric (variable-resistance type) cells in a voltage divider makes it possible to control voltages easily.

8. Electrostatic capacitors are capable of storing energy. The time required for them to acquire a full charge depends upon their size and the resistance of the circuit to which they are connected. When the size is expressed in microfarads and the resistance is measured in megohms, the product of the two terms will give the *time constant,* in seconds, required for the capacitor to reach 63.2 per cent of its final voltage, if being charged, or to have lost 63.2 per cent of its original potential.

9. Circuitwise, capacitors can differentiate between direct current and alternating currents: they pass the alternating current and block the direct current. When employed with inductors, they form frequency-responsive units which can pass or block any portion of the frequency spectrum normally used.

10. Transistors employ some of the basic components of a three-element tube: an anode, which is called the *collector;* an *emitter,* which corresponds to the conventional hot cathode; and a *base,* which functions in a manner similar to the grid.

Unlike the vacuum tube, which is operated by a potential on the grid, the transistor functions because of current furnished by the input circuit. Tubes operate from sources of potential, while transistors must have a source of current.

11. When current is supplied to the base-emitter portion of a transistor, electrons or holes are made available and can then be attracted by the collector as they pass through the base section.

The successful operation of a transistor requires, among other things, that the electrons or holes which are injected into the base are injected only in response to the electrical signal, and not by thermal agitation of the materials.

12. Because of their different input requirements, transistors and vacuum tubes are not directly interchangeable. Each one must be used where it is best suited to the task to be performed.

13. Electrical meters respond to current flow. Consequently, a meter will function either as an ammeter or a voltmeter provided the circuit constants, or those of the

meter, limit the current through the active portion of the meter only to that required to produce full-scale deflection.

14. A voltmeter is a relatively high-resistance device, while an ammeter is a low-resistance device.

15. A Wheatstone bridge is one of the standard means of measuring electrical resistance. At the same time we should recognize that it is also a potentiometric device, because we are trying to place two portions of the circuit at the same potential. We recognize this condition when the current through the meter is zero and the meter has no pointer deflection.

16. Various means may be used to balance a bridge. Capacitors function just as well in alternating-current circuits as do resistors whose value may be varied.

17. By controlling the current through a portion of one resistor, it is possible to balance a bridge without having any variable parts in the sensitive portions of the bridge circuit proper. Resistance as well as small potentials may be measured very effectively by such means.

QUESTIONS

7·1 What types of measurement require the use of potentiometers?

7·2 Why cannot one use a very sensitive voltmeter when making thermocouple readings? Or if it is possible, what precautions must be taken to ensure accurate readings?

7·3 How does a Wheatstone bridge operate? Is it correct to say that its correct operation is independent of the applied voltage? When and where might the bridge voltage be of significance?

7·4 What are the relative advantages and disadvantages of the electronic type of potentiometer? What instruments that you know of employ potentiometers of this type?

7·5 In what important ways do vacuum tubes and transistors resemble each other?

7·6 In what important ways are vacuum tubes and transistors quite different?

7·7 Is the difference between vacuum tubes and transistors of any importance in their applications in instrumentation?

7·8 What is the significance of the value of 1.018 volts, Fig. 7·1?

7·9 If the battery, Fig. 7·6, were to be reversed, what would be the effect upon the operation of the tube?

7·10 The grid-biasing battery, Fig. 7·7, applies a positive potential to the grid of the tube. What is the effect of reversing this battery?

7·11 What is the result of using a grid voltage which is greater than $+1$ and less than -1, Fig. 7·8? Might such use be desirable in instrumentation work? Under what conditions might it be necessary to stay within these limits?

7·12 Tell in your own words just how a voltage divider works.

7·13 What is the purpose of the tubes in Fig. 7·10? Why are two tubes used?

7·14 What advantages and disadvantages are involved when a photocell is used as part of a voltage divider, Fig. 7·11? Is there any practical significance?

7·15 If the valve, Fig. 7·12, is almost closed, why will the level in reservoir A always be higher than that in B?

7·16 In what way is the liquid level in Fig. 7·12 related to the level in a capacitor which is being charged?

7·17 How must the base-emitter section of a transistor be biased?

7·18 How must the emitter-collector section of a transistor be biased?

7·19 How do the required biasing arrangements of Questions 7·17 and 7·18 dictate the operation of a transistor?

7·20 Will a transistor operate if the bias of the base-emitter section is reversed from the normal?

7·21 What is the effect of high ambient temperatures on transistor operation?

7·22 What are the characteristics of series circuits?

7·23 How do parallel circuits differ from series circuits with respect to both current and voltage?

7·24 Do you know of any instrumentation problems or conditions in which series circuits are employed? Parallel circuits?

7·25 What are the basic components of a moving-coil galvanometer?

7·26 Just how do the galvanometer components function in producing meter movement?

7·27 How can a galvanometer be converted into a voltmeter? An ammeter?

7·28 How much more current is there in the meter coil when the meter is reading a full-scale value of 150 volts instead of a calibrated full-scale value of 10 amp? What generalization can you make about the meter?

7·29 What relative value of resistance must be inserted in the coil circuit of a galvanometer to use the meter as a voltmeter?

7·30 What relative value of resistance is placed in parallel with the coil circuit of a galvanometer to use the meter as an ammeter?

7·31 What determines the accuracy of measurements which can be made by using the bridge shown in Fig. 7·26?

7·32 What advantages or disadvantages are there to using capacitance bridges instead of resistance bridges?

7·33 What advantages or disadvantages exist if a bridge is balanced by inserting a potential instead of varying a resistance?

7·34 What advantages are there when a bridge is balanced, Fig. 7·29, by positioning a potentiometer rheostat instead of varying one of the bridge resistors?

7·35 What is the effect of a change of temperature upon the bridge, Fig. 7·29?

7·36 If the bridge battery, Fig. 7·31, were to be reversed, would the system function satisfactorily? (Assume that the tube has an adequate amplification.)

7·37 In what significant ways do the circuits shown in Figs. 7·30 and 7·31 differ? Which would give the better performance?

7·38 What is the purpose of the standard cell? What is the purpose of the calibrating resistor, Fig. 7·33?

7·39 How can current be used to balance a potentiometer? Can such a potentiometer be more accurate than one using a standard cell? What must be true of the circuit constants and elements if current is to be used as a measure of potential? Why do we not use *standards* when we measure current?

7·40 What advantages are there in balancing a bridge with current if a remote indication of resistance or potential is desired?

PROBLEMS

7·1 If a change in the grid potential of a triode is 1 volt and it changes the potential across a resistor in the plate circuit by 50 volts, find the true amplification factor of the tube. What would be the amplification factor if the voltage change had been 20 volts? 100 volts?

7·2 With a fixed plate voltage of 100 volts, 0.010 amp exists in the plate circuit, which has no other resistors, the applied grid signal being -2 volts. When the grid potential is changed to -1 volt, the plate current rises to 0.030 amp. What is the resistance of the tube for each value of grid potential? How much does the resistance change with a grid change of $+1$ volt?

7·3 A voltage-divider network consists of four resistors in series: 10,000, 20,000,

30,000, and 40,000 ohms. If 10 volts is applied to the circuit, what percentage of the applied voltage will appear across each resistor?

7·4 If the 10,000-ohm resistor were replaced by two other series resistors, one of 1,000 ohms and the other of 9,000 ohms, what voltages would be established between the initial terminal and the terminals of the other resistors? Arrange the resistors in order of their resistance, with the smallest at the beginning. Do these values give you anything which might be used conveniently in an instrument?

7·5 Four resistors of 100, 200, 300, and 400 ohms are connected in series across a source of 100 volts. What current will flow and what will be the potential across each resistor?

7·6 If the four resistors of Prob. 7·5 were to be connected in parallel with 100 volts applied, what current would exist in each resistor? What would be the total current and the resistance of the entire circuit?

7·7 If four 1,000-ohm resistors are connected in parallel across a potential of 100 volts, what current will the group take? What is the resistance of the combination?

7·8 What is the equivalent resistance of ten 10,000-ohm resistors in parallel? Can you prove it?

7·9 A photocell which has a dark resistance of 1,000,000 ohms is in series with a resistor of 10,000 ohms. If the light resistance is 90,000 ohms, what will be the voltage across the fixed resistor for both conditions if 10 volts is applied to the circuit? If the grid of a triode amplifier is connected as in Fig. 7·10, will this provide an adequate signal to make an appreciable change in the plate current?

7·10 A meter requires 0.001 amp for full-scale deflection. The resistance of the movement is 100 ohms. What resistors should be placed in series with the coil if it is desired to measure 1 volt, 10 volts, 50 volts, 100 volts, 1,000 volts? (Assume full-scale deflection.)

7·11 What is the minimum voltage which will provide full-scale deflection with no added resistance in the coil circuit of the meter in Prob. 7·25.

7·12 If the coil had a basic sensitivity of 0.0001 amp for full-scale deflection and a coil resistance of 1,000 ohms, find the series resistance required to meet the requirements of Probs. 7·10 and 7·11.

7·13 If the meter movements of Probs. 7·10 and 7·12 were to be used to measure current, what parallel resistance would be required for currents of 0.001 amp, 0.010 amp, 1.000 amp, 5 amp, and 10 amp?

7·14 In Fig. 7·26, $R_1 = 1{,}000$ ohms, $R_2 = 4{,}000$ ohms, and $R_3 = 250$ ohms at the time of bridge balance. What is the value of R_4?

7·15 If in Fig. 7·27 the values of D_1 and D_2 are 60 and 40 units, respectively, find the value of R_3 when $R_4 = 100$ ohms.

7·16 If the ratio of R_1 to R_2 is 100, find the ratio of C_2 to C_1, Fig. 7·28.

7·17 If in Fig. 7·29 the input signal at x-y is 0.100 volt, what added current will have to flow through R_5 if it is to balance the bridge? $R_5 = 100$ ohms. Assume that the bridge was balanced before the 0.100-volt signal was inserted. If the polarity of the input signal were reversed, what current would be required?

8 Alternating-current Circuits

Electric energy may be generated in many ways:

1. The heating of the junction of two dissimilar metals will produce a potential which is proportional to the difference in temperature between one end of the wires and the other. If a closed circuit is provided, a current which will be proportional to this temperature difference will flow. This is an inefficient way of generating electric energy at the present time.

2. Two dissimilar metals placed in an acid solution will create an electromotive force which is a function of the metals. This is a fairly common phenomenon. Dry cells and storage batteries are very common examples.

3. Radiant energy falling on certain materials can dislodge electrons and in this manner can create potential differences. Photographic exposure meters generally employ cells of this type.

4. Moving a conductor in a magnetic field will generate a voltage which is proportional to the strength of the field and the speed with which the conductor is moved. This is the manner in which most electric energy is created today. Whether the conductor moves in the magnetic field or whether the field moves past the conductor makes little or no difference. In machines of small capacity the field is likely to be stationary, but in the large machines employed by the public utilities for power generation it is customary for the field to move while the active conductors remain motionless.

This is the basic theory back of the recent designs of magnetic flowmeters. The flowing liquid functions much as conductors do and generates a potential which is directly proportional to the speed of the liquid.

When a conductor is moved past a magnetic pole, or the pole is moved relative to the conductor, the magnitude of the potential generated depends upon:

1. The magnetic intensity of the field
2. The length of the conductor
3. The speed of the conductor with respect to the field

Stated mathematically

$$e = KBLV$$

where e = potential developed in the conductor
B = magnetic flux density, lines/in.^{2} or lines/cm^2
L = active length of the conductor, in. or cm
V = relative speed, ips or cm/sec
K = proportionality constant

EXAMPLE 8·1 A conductor 10 in. long is passing at 10 fps through a magnetic field in which B = 1,000 lines/in.^{2}. What potential will be induced? (This is a problem similar to that found in the design of certain magnetic flowmeters.)

SOLUTION

$$e = KBLV$$

$K = 10^{-8}$ when the answer is desired in volts. Without K the answer is in abvolts.

$$e = (10^{-8})(1{,}000)(10)(10) = 10^{-3} \text{ volts} = 1 \text{ millivolt}$$

One other factor not considered here is that the polarity of the potential developed in any given conductor will depend on whether the conductor is near a north pole or a south pole. If a given end of the conductor is called plus when the conductor is being moved past a north pole, then that same end will have the opposite, or minus, polarity when the movement is past a south pole. In all common types of generators, the potential developed by each conductor changes its polarity each time the conductor passes a pole of opposite polarity. Figure 8·1 shows what happens to the polarity of a conductor as it is moved under the four poles of a small generator. At a point between the poles the conductor does not cut flux but moves parallel to the field. At this point the voltage goes through zero and then reverses its polarity.

It will be noted that on one side of the conductor the two magnetic fields are acting in the same direction. This convention indicates that at that point they are of the same polarity. Experiments and experience have taught us that such similar, or *like*, poles repel each other and try to increase the distance between them. On the other side of the conductor we will find that the fluxes go in opposite directions and thus are of *unlike* polarity. Unlike poles always attract each other. Thus, we find forces created whenever two or more magnetic fields react with each other. These forces may cause a conductor to move, as in the case of a motor, or they may oppose movement as in the case of a generator, which requires that we force the conductor to move through the magnetic field.

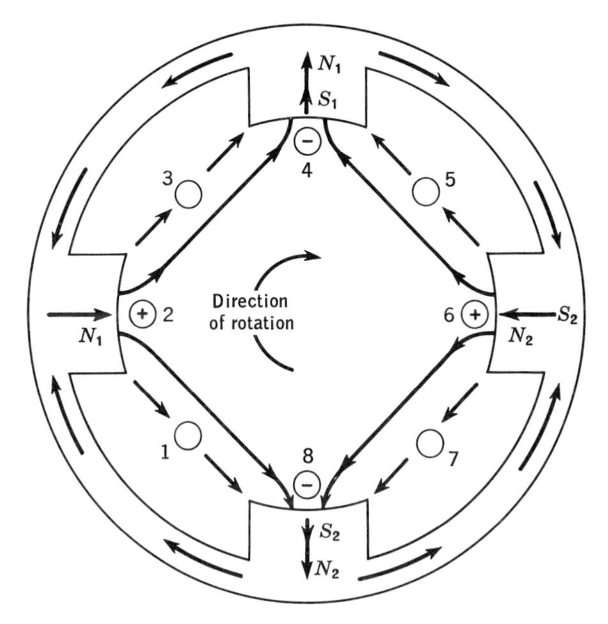

Fig. 8·1 For any given direction of rotation of a conductor under the poles of a motor or generator, the polarity of the induced voltage changes each time the conductor moves from one pole to the next.

At a later time, we may want to use Lenz's law to tell us what is happening in certain circuits in which we want to determine the polarity of certain conductors or windings. We may safely say the following: The magnetic fields created by currents resulting from induced voltages will always oppose any increase in these fields, but once established, they will oppose any reduction.

ELECTRIC POTENTIAL AND MAGNETIC FIELDS

Lenz's law. One of the most important laws of basic electricity is called Lenz's law, which states that whenever a conductor has a potential induced in it, the current which flows because of this potential will establish such a magnetic polarity as to oppose movement of the conductor through the field. This is a law which also deals with the conservation of energy in that it states that we must do work on the conductor if the conductor is to produce electric energy. Here is another situation in which we may convert from one type of energy into another; we deliver mechanical energy to the generator, which converts it into electric energy which may be later converted into mechanical or heat energy.

Figure 8·2 shows the assumed direction of the magnetic field around a conductor which is carrying current. It will be noted that reversing the direction of current flow will change the direction of the magnetic field. Figure 8·3 shows the assumed flow of magnetic flux. The north pole is defined as the pole from which the flux leaves, whereas the south pole is the pole at which the flux in the circuit enters the magnetic ma-

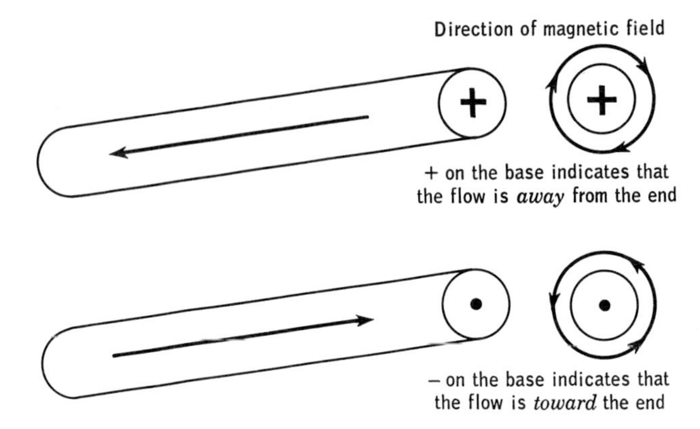

Fig. 8·2 The direction of the current flow determines the direction of the magnetic field around the conductor.

terial. It is assumed that there is an internal magnetic circuit between the north and south poles. Where there are more than two poles, the same assumptions are made.

Many electrical types of tranducers use a "motor" which in many ways resembles a loudspeaker. (See Chaps. 10 and 15.) The same forces acting on a conductor must be considered in a design of this type.

The force developed depends upon:

1. The length of conductor in the magnetic field
2. The strength of the magnetic field in the air gap
3. The current

In the form of an equation,

$$\text{Force} = KLBI \qquad \text{lb}$$

where K = proportionality constant = $1/(1.13)(10^8)$
L = length, in. = (6 turns)(5 in./turn, assumed) = 30 in.
B = flux density, lines/in.^{2} = 5,000 lines (assumed)
I = current, amp = 0.010 (assumed)

$$\text{Force} = \frac{30(5{,}000)(0.010)}{1.13(10^8)} = \frac{3(5)(10^2)}{1.13(10^8)} = \frac{13.3}{10^6}\ \text{lb}$$

Direct current. If we create electric potentials by chemical means, we find that when a circuit is completed from one plate to the other through some circuit element such as a resistor, motor, or meter, the direction of current is always in one direction. This unidirectional flow of current we call *direct current.* In any given circuit, the current need not remain constant in magnitude, but if it always sends electrons through the circuit in the same direction, then it is termed direct current.

In such common creators of direct current as automobile generators, the individual conductors have their polarities reversed each time they pass another magnetic pole. However, the reversal of potential and current in each conductor is not allowed to reach the external circuit. By means of a mechanical *rectifier* or switch called a *commutator,* the circuits to the various conductors are switched or transferred in such a

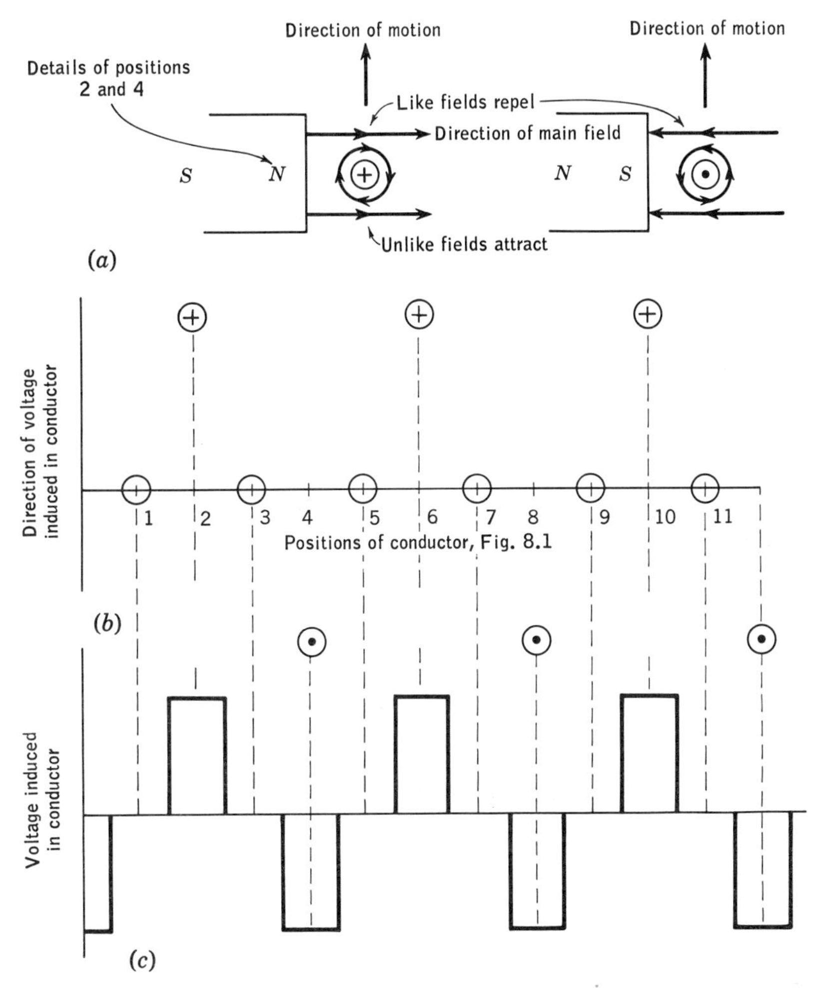

Fig. 8·3 (*a*) The induced emf in a conductor will be such that any ensuing current will create a magnetic field which will oppose further motion of the conductor through the field. (*b*) A potential is induced in a conductor only when it is cutting flux. The polarity of the induced potential reverses as the conductor moves from pole to pole. (*c*) If the poles are all of the same strength, the emf will have the same numerical value at each pole, but its polarity will change. Rotational speed is assumed constant.

manner that the current in the external circuit always flows in one direction, regardless of what is happening in the various conductors.

Historically, direct-current generators and motors are our oldest type of mechanical-electrical converters. But it must be remembered that the flow of current in the individual conductors is not unidirectional but instead reverses each time the magnetic flux changes. In ease of control, direct-current motors far excel any other type. We will find in industry an increasing use of motors of this type where the speed or torque of the motor or the voltage of the generator is controlled by deviations from a desired value. But in terms of easy change of potential, amplification of signal, or distance of transmission, the direct-current system leaves something to be desired.

Alternating current. We will note that the potential of the conductor shown in Fig. 8·3*c* keeps reversing the direction of current flow in its conductors, or we might say that the current alternates in its direction of flow, whence the name *alternating current.* Over the years, alternating-current systems have become the predominant systems in many parts of the world because:

1. Alternating-current power can be transmitted farther at lower cost than can direct-current power.

2. It is a very easy matter to change the voltage of alternating-current systems through the use of static devices which are known as *transformers.*

3. Stable high-gain amplification is easier to achieve with alternating-current systems than with direct-current systems. This is of significance in many electronic designs.

4. Alternating current makes easy circuit isolation possible.

In most industrial applications, every attempt is made to ensure that the voltage and current waves which are produced are sine waves; Fig. 8·4 shows how such an electric current varies with time. Our chief interest

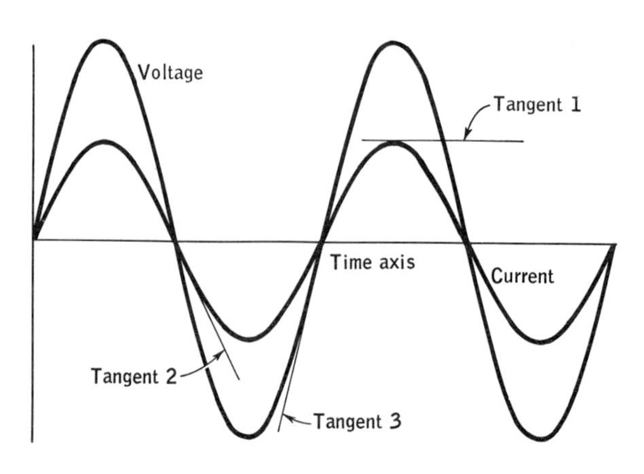

Fig. 8·4 Of particular significance are the tangents to the current wave, which give us a measure of the rate of change of current and also of the flux which may be changing simultaneously with it.

in the sine wave is that it is the only waveform whose amplitude (or value) can be changed by a transformer without changing the shape of the wave. Square waves, rectangular, or triangular waves impressed on the input of a transformer will have an output form which does not even faintly resemble the original shape. With a sine wave, on the other hand, there will be no substantial change in the shape of the wave, regardless of the number of transformations, provided that the transformers are of the usual commercial quality and the device is operated within its voltage and current ratings.

We should pay particular attention to Fig. 8·4. The two slopes which are marked tangent 1 and tangent 2 are of particular interest. Many phenomena connected with alternating currents are based on the *rate* at which the current changes, not on the value of the current at that time. In the graph of the current wave, it should be noted that where the current is a maximum, the tangent (tangent 1) is horizontal, which indicates that the current at that particular moment is neither increasing nor decreasing. Consequently, anything which depended on how fast the current is changing at that point would have a zero value.

Quite in contrast to this condition is the tangent to the current curve when the current is going through zero. Mathematically, we can prove that the tangent to the curve has its greatest value at the zero point, indicating that, at that point, any value which depended upon the rate of change of current would be a maximum. These are very important principles:

1. When the value of the current wave is a maximum, the rate of change of current is zero.
2. When the current wave is going through zero, the rate of change of current is a maximum.

When we study transformer action, we will have to employ these principles if we are to understand what is going on. It will be noted that nothing has been said about the rate of change of the voltage curve. Normally, it is the current which creates magnetic fields and is otherwise responsible for the performance of alternating-current equipment; it is the rate of change of current which creates most of the phenomena that we will study.

Self-induced voltages. If we had a coil of copper wire wound on a hollow core of cardboard or fiber and connected a galvanometer as shown in Fig. 8·5*a*, a deflection of the galvanometer would take place when we tried to insert the indicated permanent magnet. For the polarities indicated, the current flowing in the meter circuit would flow through the coil in such a direction as to create a magnetic pole opposing the entrance of the north pole; this is in accordance with Lenz's law. When the magnet stopped its movement into the core, the current in the meter circuit would

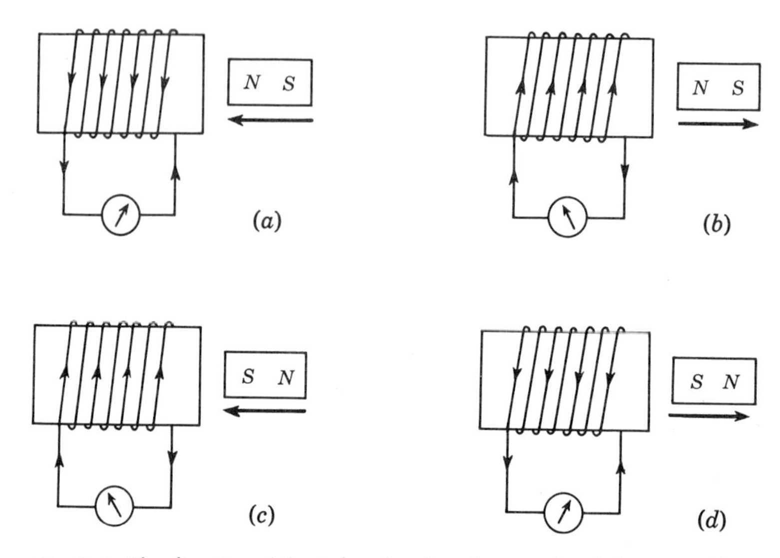

Fig. 8·5 The direction of the induced emf conforms to Lenz's law regardless of the type of conductor or the design of the magnetic field.

drop to zero, indicating that a stationary condition would not be opposed. But if we now wanted to withdraw the magnet, as shown in Fig. 8·5*b*, the meter would deflect in the other direction. With a reversal of the coil current, the polarity of the magnetic poles would reverse and we would find attraction instead of repulsion of these poles. This too is in accordance with Lenz's law, because the action is to maintain the flux which had previously been established in the coil. In all cases of induced voltages, we will find this opposition to an increase in the flux which exists in a circuit; but once the flux has become established, there will be an equally great reluctance to let it decrease.

The magnitude of the self-induced voltages may be determined quite readily.

$$\text{Average induced emf} = \frac{(\text{no. of turns})(\text{change in flux})(10^{-8})}{\text{time required for flux change}}$$

$$= \frac{N(\Delta\phi)(10^{-8})}{(\Delta t)}$$

where $\Delta\phi$ = change in flux
Δt = time required for the change

EXAMPLE 8·2 If 10,000 flux lines link or thread 2,000 turns, find the emf created when the flux goes from 0 to 10,000 lines in 0.01 sec.

SOLUTION

$$\text{Average emf} = \frac{(2{,}000)(10{,}000 - 0)(10^{-8})}{0.01} = 20 \text{ volts}$$

If the flux changes from 5,000 to 10,000 lines in the same time,

$$\text{Average emf} = \frac{(2{,}000)(10{,}000 - 5{,}000)(10^{-8})}{0.01} = 10 \text{ volts}$$

Because the flux does not undergo as great a change, the average emf will be less. We shall meet this situation again when we study magnetic amplifiers, the action of which depends upon controlling the amount of flux change that can occur.

If the same flux change were to occur in 0.001 sec, then

$$\text{Average emf} = \frac{(2{,}000)(10{,}000 - 5{,}000)(10^{-8})}{0.001} = 100 \text{ volts}$$

In Fig. 8·5*c* it will be noted that inserting the south pole produced an opposite current flow from that produced by inserting a north pole. It should be observed that inserting a north pole produces the same current flow as retracting a south pole. But every attempt to change the flux linking a circuit by moving the permanent magnet will produce an induced voltage whose resulting current flow will try to oppose any further change in the flux.

If we now connect the same coil to a battery, Fig. 8·6*a*, then, when the switch is closed, the changing magnetic field created by the increasing current will create a voltage of self-induction which will oppose any further increase in the current. Motion-picture cameras can easily record that the current increases relatively slowly to its final value. But when the switch is opened, Fig. 8·6*b*, the decreasing flux which accompanies the diminishing value of current will create a potential which acts to maintain the current at its existing value. The direction of the mag-

Fig. 8·6 When the switch is closed, the induced emf opposes the applied battery potential. But when the switch is opened, the polarity of the induced emf is reversed and tries to maintain the current.

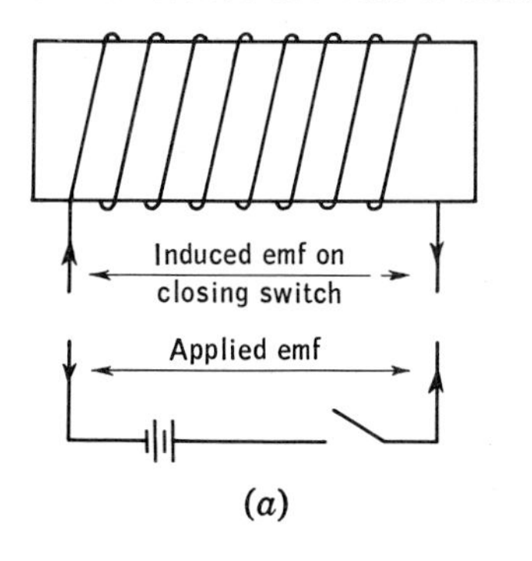

(*a*)

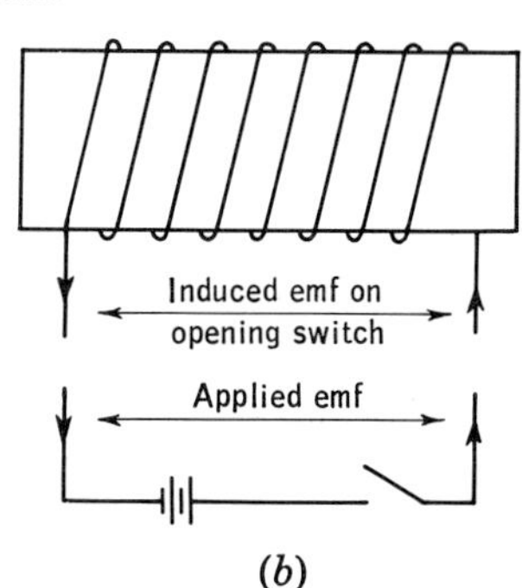

(*b*)

netic fields created by the current can be checked by the actions of a compass needle.

It should be noted that the end of the compass which points north is a *north* pole. It is attracted by a *south* pole which is located at our north magnetic pole in northern Labrador. We have every reason to believe that the magnetic representation which we call *flux* enters the earth at our north magnetic pole and leaves it at the south magnetic pole. By our definition of a north pole—the pole at which the flux leaves the magnet—our north magnetic pole is located near the south geographical pole and not near the north geographical pole.

In many ways, the direction of magnetic flux intensity on the earth's surface was largely an academic question until the use of magnetic mines in World War II made it a matter of immediate concern to every steel ship which is magnetized by ambient earth's field.

A changing magnetic field induces a voltage in a circuit or coil. It makes no difference whether the changing magnetic field is created by some moving external permanent magnet or whether the current flowing in the coil is changing and consequently is changing the magnetic flux. *It must always be remembered that wherever we have a current flow we have a magnetic field created by that current flow.* The coil cannot tell the source of the changing magnetic field, so it will experience an induced voltage within itself which will *always try to avoid any change in the current and the flux linking the circuit.*

This self-induced voltage, called a voltage of self-induction, is proportional to the amount of flux linking the circuit and the rate of change of the current.

Before we try to see how the voltage of self-induction can control the current, in part, let us study the circuit shown in Fig. 8·7. If we start with the circuit shown, then the ammeter will register 1 amp so long as potentiometer contact P remains at a. But if we moved the contact toward b, then the voltage available to send the current through the resistor is no longer 10 volts, but is 10 volts minus the voltage between P and a. When point b is reached, the voltage inserted by battery B will be exactly equal to that of battery A and the net voltage on the circuit will be zero. Consequently, the current will also be zero. Control of the

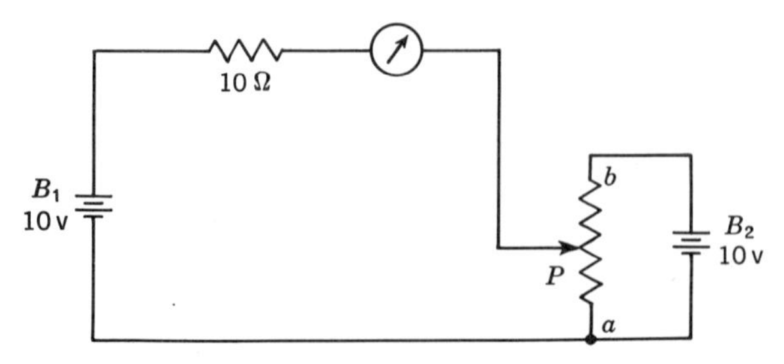

Fig. 8·7 In addition to the use of variable resistances to control the amount of current in a circuit, we may employ a potential whose value may be varied.

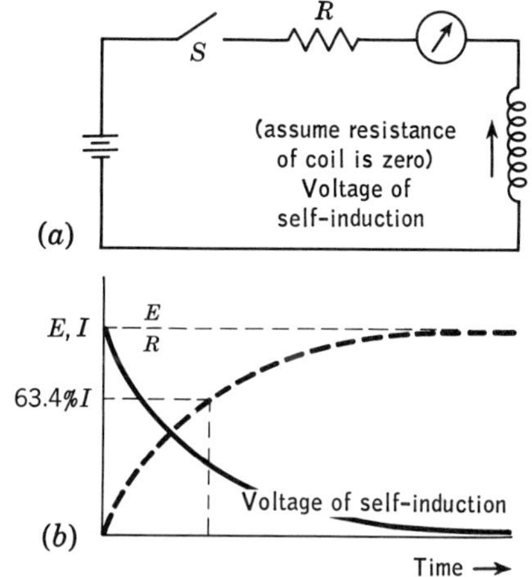

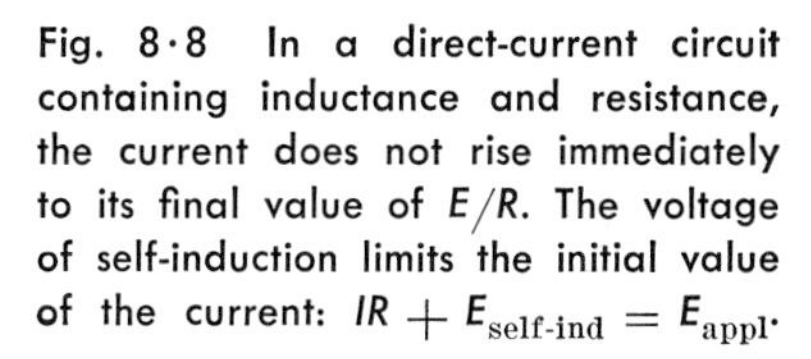
Fig. 8·8 In a direct-current circuit containing inductance and resistance, the current does not rise immediately to its final value of E/R. The voltage of self-induction limits the initial value of the current: $IR + E_{\text{self-ind}} = E_{\text{appl}}$.

current can be obtained by inserting a voltage which acts in opposition to the applied voltage.

If we now modify the circuit so that we have a coil replacing the potentiometer and battery B, Fig. 8·8*a*, and we close switch S, as the current in the circuit starts to build up, the changing flux in the coil creates a voltage of self-induction which will do the same thing that the battery and potentiometer did in the circuit of the preceding paragraph; the voltage of self-induction acts to prevent an increase in the current. With direct-current circuits, this opposing voltage is a transient condition as is shown in Fig. 8·8*b*.

In direct-current circuits having magnetic components, the current builds up to a steady value given by the ratio E/R, which is true for all direct-current circuits. It will be observed that the voltage of self-induction falls from a value equal to E toward zero. The time in seconds required for the current to reach 63.2 per cent of its final steady value is called the time constant of the circuit.

Inductance. Any circuit component in which there is a change of the magnetic field when the current in the circuit changes is said to have inductance. The unit of inductance is the *henry*. A coil is said to have an inductance of 1 henry when there is a voltage of self-induction of 1 volt when the current in the circuit is changing at the rate of 1 amp/sec.

Another way to define the unit of inductance is to consider that the product of the lines of flux change and the number of turns on the coil through which the flux expands must give us an average product of 100,000,000 when the current is changing at the rate of 1 amp/sec. One hundred million flux linkages per second will create a voltage of 1 volt. This is one of the definitions of a volt.

We may express the effect on current flow of the induced voltage by stating

$$I = \frac{E_{\text{appl}} - E_{\text{cemf}}}{R} \quad \text{or} \quad \frac{E_{\text{appl}} - E_{\text{self-ind}}}{R}$$

$$= \frac{E}{R}(1 - e^{-RT/L})$$

$$IR = E - Ee^{-RT/L}$$

where E = applied voltage

Ee = self-induced potential

When the switch is first closed, the current is zero. So $E = Ee^{-RT/L} = Ee^{-TC}$. Thus the induced cemf is just equal to the applied potential when the time is zero, and the value of the time constant is also zero. But any quantity raised to the zero power has a value of 1.

However, as time increases and the number of time constants also increases, the value of Ee^{-TC} decreases. The *net* voltage acting on the circuit increases with time until, after six or seven time constants, the net potential is essentially that of the battery, and

$$I = \frac{E - 0}{R} = \frac{E}{R} \qquad \text{Fig. 8·8}b$$

Variable-voltage direct-current circuits. If we establish a circuit such as is indicated in Fig. 8·9 and start with the potentiometer contact at o and if P is moved with fairly uniform motion toward a, the changing current in the circuit will create a more or less constant voltage of self-induction in coil C because the current is changing at a relatively uniform rate. Consequently, at no time will the current in the circuit be equal to the battery voltage divided by the resistance. Instead, it will be equal to the difference between the applied battery voltage and the induced voltage divided by the circuit resistance.

$$I = \frac{E_{\text{bat}} - E_{\text{self-ind}}}{R}$$

We might then move the contact, again at a uniform rate, from a to b. This time the voltage of self-induction will try to maintain the current, and the voltage of self-induction must be added to that provided by the battery. The higher the rate at which we try to vary the potentiometer setting, the greater the voltage of self-induction. Therefore, with an input of variable voltage, the circuit current is governed not entirely by the applied voltage and the resistance, but by the difference between the applied voltage and the voltage of self-induction plus the resistance.

If, instead of trying to move the potentiometer pointer manually at a high rate of speed to secure a larger voltage of self-induction to limit the

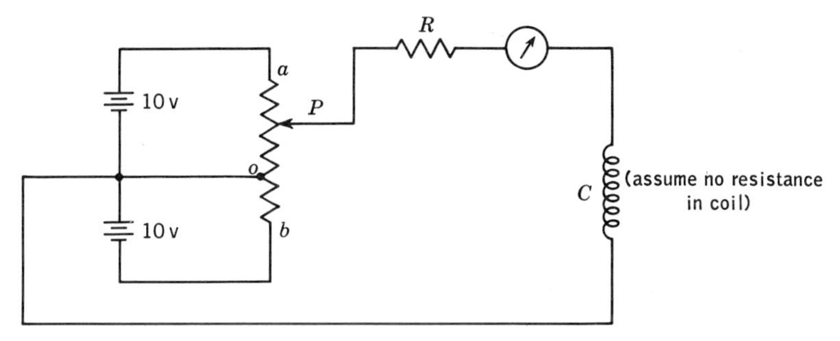

Fig. 8·9 Movement of contact *P* between points *a* and *b* applies potentials of differing magnitudes and polarities. This simulates an alternating-current system whose frequency depends upon the speed with which pointer *P* is moved.

current to a small value, we were to apply to this circuit a voltage which automatically gave us the desired voltage variation, we would find that we were dealing with an alternating-current system. Here the variations in the applied voltage and the ensuing current changes produce induced voltages which seriously limit the value of the current. We cannot apply Ohm's law to an inductive circuit which has a variable current. Ohm's law applies only to *steady-state* conditions.

Transformers. If we consider the coil and its magnetic core represented in Fig. 8·10 and we apply an alternating voltage to the coil, then the current which flows in the circuit will create a flux in the indicated direction. This same flux will also create our voltage of self-induction which, if the magnetic circuit is good and the resistance of the coil is small, will be almost as large as the applied voltage, but it never can quite equal or exceed the applied voltage under normal operating con-

Fig. 8·10 The changing flux that creates a voltage of self-induction within the primary winding will also create an emf in any other winding through which the flux passes. This induced potential will vary directly with the number of turns. With the same flux threading both primary and secondary windings, the emf per turn is the same.

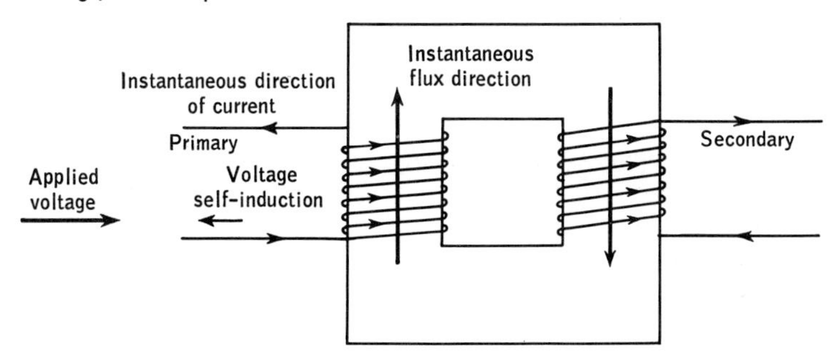

ditions. This flux circulates throughout the magnetic circuit. If we now put on the magnetic core a second coil having the same number of turns as the first coil, then the changing flux which creates our voltage of self-induction in the one coil will induce an equal voltage simultaneously in the second coil. The coil to which energy is supplied is called the *primary;* the second coil, in which magnetic energy creates the voltage, is called the *secondary.*

We now state some fundamental transformer relationships without proof:

1. The induced voltage in the secondary will vary at the same rate as the current flowing in the primary. This means that the frequency of the output voltage wave is equal to the frequency of the input voltage.

2. Because the same flux links both primary and secondary, in good designs for power purposes the induced voltages will be directly proportional to the number of turns in each. That is $V_p/V_s = N_p/N_s$.

3. There may be more than one secondary winding. The voltage induced in each winding will be in direct proportion to the number of turns in the winding.

4. Because the transformer inherently is a device of high efficiency, the power delivered to the primary may be considered to be the power delivered by the secondary. Another statement which will need defining later is that the product of the voltage and the current of the primary will equal the product of the voltage and current of the secondary.

$$V_pI_p = V_sI_s$$

relates the "apparent power" in both sets of windings. Further,

$$V_pI_p \cos \theta = V_sI_s \cos \theta$$

where $\cos \theta$ is called the *power factor.* Unlike direct-current circuits, in which the product of current and voltage always gives the power, alternating-current circuits require a multiplying factor because the current and voltage do not always act in unison.

5. Voltages may be stepped up or down with almost equal facility for all common voltage and current ratings.

6. Transformers often carry *polarity* markings to indicate the terminals which have the same *instantaneous* voltages. In many circuits it is important that certain points have their maximum or minimum values at the same instant. The polarity markings of a transformer indicate these terminals.

7. A transformer is used to change the *impedance* of a circuit or some part of a circuit.

EXAMPLE 8·3 If the voltage of the primary is 120, what will be the secondary voltage if the primary has 1,000 turns and the secondary has 10,000 turns? The induced voltages are assumed to be equal to the applied voltages, and the voltage of each winding is proportional to the number of turns.

SOLUTION

$$\frac{V_p}{V_{s1}} = \frac{N_p}{N_s}$$

$$\frac{120 \text{ volts}}{V_{s1}} = \frac{1{,}000 \text{ turns}}{10{,}000 \text{ turns}} = \frac{1}{10}$$

$$V_{s1} = (120 \text{ volts})(10) = 1{,}200 \text{ volts}$$

EXAMPLE 8·4 If a third winding contained 100 turns, what would the voltage in it be?

SOLUTION

$$\frac{V}{V_{s2}} = \frac{1{,}000 \text{ turns}}{100 \text{ turns}}$$

$$\frac{120 \text{ volts}}{V_{s2}} = \frac{1{,}000 \text{ turns}}{100 \text{ turns}} = 10$$

$$V_{s2} = \frac{120 \text{ volts}}{10} = 12 \text{ volts}$$

EXAMPLE 8·5 If there are but two windings, one on the primary and one on the secondary, find the current in the secondary when the primary current, Example 8·4, is 10 amp.

SOLUTION

$$(\text{Volts})(\text{amperes})_{\text{prim}} = (\text{volts})(\text{amperes})_{\text{sec}}$$

$$(120 \text{ volts})(10 \text{ amp}) = (1{,}200 \text{ volts})(I)_{\text{sec}}$$

$$I = 1 \text{ amp} = \text{secondary current}$$

To determine the polarity of a transformer, the following test is made. Even subscripts or odd subscripts indicate the same instantaneous polarity, Fig. 8·11*a*.

1. Measure primary voltage V_1. (This does not need to be rated alternating current voltage for this winding.)
2. Measure secondary voltage V_2.

Fig. 8·11 Schematic of the circuit to be used when checking transformer polarities.

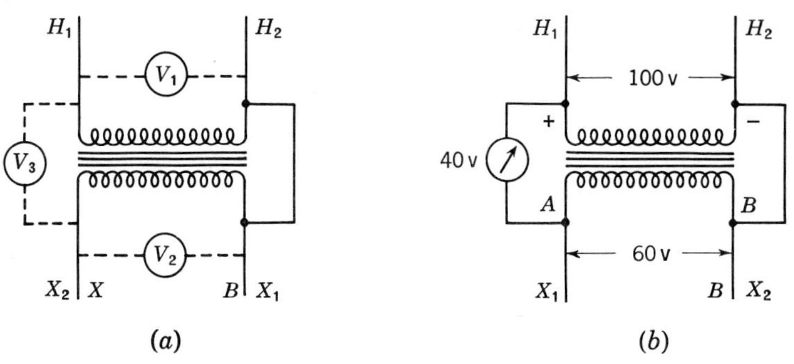

3. Connect a wire between one side of the primary and one side of secondary as shown.

4. Measure V_3. If V_3 is greater than V_1 (it may be equal to $V_1 + V_2$), then that means that the two voltages are being added +, −, +, −. Then, assuming H_1 is plus, H_2 is minus, point B must be plus, and terminal X must be minus. We have assumed H_1 to be plus, so the secondary, designated by X, will be X_1 at the plus terminal or line connected to the primary. When V_3 is greater than V_1, the transformer is known as an *additive polarity transformer.*

Suppose that when the voltages are measured, we get the following, Fig. 8·11*b*: $V_1 = 100$ volts, $V_2 = 60$ volts, and $V_3 = 40$ volts. Because the arithmetical sum of the voltages is less than the primary, then the polarities must be +, −, −, +. The primary is marked as shown. The terminal B of the secondary must have the same polarity as H_2; therefore, it is X_2. This transformer is known as a *subtractive polarity* transformer.

RESISTANCE, REACTANCE, AND IMPEDANCE

Resistance. In the study of direct-current circuits, we learned that whenever current passes through a resistor, heat is generated; the electric energy is transformed into heat energy. But only those resistors which are physically connected by wire are heated. When we have alternating-current equipment, some of the electric energy may be converted into magnetic energy which in turn creates voltages and currents in parts of the circuit to which there is no physical wire connection. But the energy so dissipated as heat has to come from the original circuit and might be said to be "charged up to it." Thus, the alternating-current resistance of a circuit may be many times its direct-current value. For other conditions, the two resistances may be of equal value, but the *alternating-current resistance can never be less than the direct-current resistance.* For very high frequencies, the current tends to crowd to the outside of the wire, creating the effect that the wire has a smaller cross-sectional area than it really has. For extremely high frequencies, hollow conductors are used because the current would not use the center of the material in any case. For such installations, a solid conductor represents a waste of material. Whatever power is required for a circuit can be said to be equal to the product of the current squared and the alternating-current resistance. (This statement is made about circuits which have but one source of voltage or power.)

Inductive reactance. Inductive reactance is an entirely new term to most of us; it is related to the voltage of self-induction. We may think of resistance as limiting current flow through the use of a "brute force method." But we found that the current in a circuit could be limited quite easily if we applied a voltage in opposition to the original voltage.

It will be recalled that the voltage of self-induction does that very thing. The voltage of self-induction depends upon the usual variables:

1. How many turns are linked by the changing flux?
2. How much flux links or cuts these turns?
3. How fast is the current changing, or what is the frequency?

The first two variables often are combined in the single term *inductance.*

A coil is said to have an inductance of 1 henry when a current of 1 amp changing at the rate of 1 amp/sec will create an induced potential of 1 volt. Now if the current changes at the rate of 10 amp/sec and none of the circuit conditions are changed, then the induced voltage will be 10 times as great, or the induced voltage will be 10 volts. Stated as the electrical engineer would state it,

$$\text{Voltage of self-ind} = (\text{rate of change of current})\,(\text{inductance})$$
$$= 2\pi fLI$$

Dividing each side by the current, we obtain

$$\frac{\text{Voltage of self-ind}}{I} = \frac{2\pi fLI}{I}$$

$$\frac{\text{Voltage of self-ind}}{\text{amp}} = 2\pi fL = \text{inductive reactance}$$

The inductance must be expressed in *henrys.* The rate of change of current is $2\pi f$, where f is the frequency, in cycles per second, and $\pi = 3.1416$.

The voltage of self-induction, then, is directly proportional to the frequency of the power supply. In general, the higher the frequency, the smaller will be the transformer for any given voltage rating.

EXAMPLE 8·6 Find the voltage of self-induction, $2\pi fLI$, for a 60-cycle power supply where the inductance is 2 henrys and $I = 1$ amp.

SOLUTION

$$V_{\text{self-ind}} = (2\pi)(60)(2)(1) = 754 \text{ volts}$$

If $I = 10$ amp, then

$$V_{\text{self-ind}} = (2\pi)(60)(2)(10) = 7{,}540 \text{ volts}$$

Impedance. We have previously noted that resistance helps to limit the current in a circuit, and that the passage of current through a resistance always results in the generation of heat. Voltages of self-induction can also limit current flow, but without the creation of heat. If both resistance and a voltage of self-induction occur at the same time in a circuit, we will have to combine their effects. But because they are of different origins, their combined effects may not be obtained simply by adding them together arithmetically. To obtain the correct value for their sum, we must add them *vectorially* at right angles.

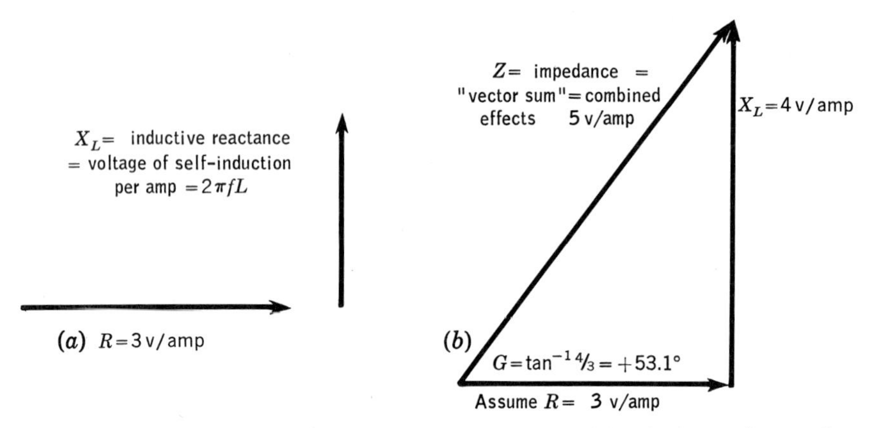

Fig. 8·12 To obtain the impedance in a series circuit containing both a resistor and a coil, the resistance and the inductive reactance must be combined vectorially in the indicated manner.

In earlier chapters, we have defined resistance by the voltage necessary to force 1 amp through a noninductive circuit. Let us retain this definition and also define inductive reactance as the voltage of self-induction when the current is 1 amp. The impedance of a circuit is then the combined vector values of two quantities required to obtain a current of 1 amp. The impedance also represents the voltage necessary to overcome the resistance and the voltage of self-induction for each ampere of current flow.

Figure 8·12*a* shows how we may represent the individual vectors. Vectors always tell us two things about any quantity: the size of the object or property and the direction in which it is acting. Unless both the size and the direction are known, we have no vector. The resistance is always represented along the horizontal axis, while the inductive reactance, $2\pi fL$, is always plotted vertically. Their combined values then are represented by the hypotenuse of the right triangle which they form, Fig. 8·12*b*. The vector sum of the resistance and the inductive reactance can never be equal to the arithmetical sum of the two; it must always be obtained by taking the square root of the sum of the squares of the values as shown. By the pythagorean theorem

$$Z^2 = R^2 + X^2$$
$$= 3^2 + 4^2 = 25$$
$$Z = \sqrt{25} = 5 \text{ volts/amp}$$

Because it represents a vector, the hypotenuse exists at some particular angle. In this case the angle is 53.1° because its tangent is $4/3$.

To indicate that these two values are to be added vectorially, the following symbolism is often used:

$$Z = R + jX \qquad \text{volts/amp}$$

The letter j indicates that X is to be measured at right angles to R. The positive value of j is used to express inductive reactance, and the vector must always be drawn up to represent this value. If we use the negative value for j, then the vector must be drawn down and represents *capacitive reactance.*

To indicate both the magnitude and the angle of a vector we may employ the *polar form.* In polar form we express both the magnitude and the angle of the vector. Thus, in the example under consideration, the hypotenuse has a magnitude of 5 units and a direction given by the angle whose tangent is X/R. In this case the value of the tangent is $4/3$, or 1.33, and the angle is 53.1°.

$$Z = R \pm jX = \text{magnitude} \underline{/\theta}$$
$$= 3 + j4 = 5\underline{/53.1^\circ}$$

Capacitive reactance. We have previously encountered capacitors which acted as electrical buckets which could be filled or emptied at different rates, depending on the resistance of the circuit. If we were to place a sensitive pressure gage at the bottom of a water bucket, we would find that the pressure at the bottom increased directly with the amount of water in the bucket. The more the bucket was filled, the greater the pressure would be. Similarly, as the electrical bucket is filled, it builds up a back pressure which limits the flow of current into it. But we might define this pressure as the back pressure created by forcing 1 amp, for some definite time, into the bucket. This term is usually called the *capacitive reactance* and is designated by X_C to differentiate it from its opposite number, which is *inductive reactance,* X_L.

The value of X_C, measured in terms of the back pressure or counter voltage per ampere, is usually determined from the expression

$$\text{Capacitive reactance} = X_C = \frac{1}{2\pi fC} \qquad \text{volts/amp}$$

where C is measured in farads and f is measured in cycles per second.

When a series circuit has both resistance and capacitive reactance, both must be represented as in Fig. 8·13*a*, with the capacitive reactance vector lagging 90° behind the resistance vector. Their sum is shown in Fig. 8·13*b*.

Capacitive reactance is always drawn *vertically down* because a negative phase angle is required to produce the *leading* angle of current

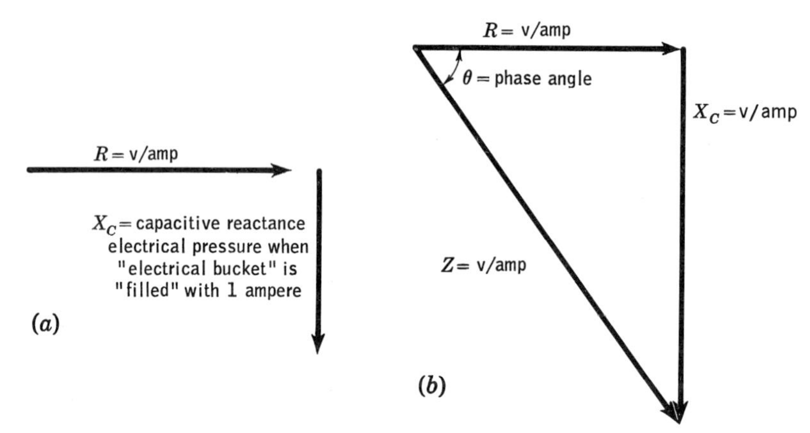

Fig. 8·13 When a series circuit is composed of a resistor and a capacitor, the resistance and the capacitive reactance must be combined as shown in order that their combined effects, the impedance, may be determined.

which characterizes capacitive circuits. To represent a condition which produces a leading current with an impedance having a negative angle may seem to be unreasonable. However, it will be found later in the chapter that the condition is governed by the mathematical relationships. The negative impedance angle must be used if a leading current is to be obtained from the particular algebra which we use.

Accordingly, inductive reactance is represented with a positive angle, drawn vertically up, because a positive angle for the impedance is required to produce the lagging current which exists in inductive circuits.

It will be noted that the capacitive reactance is inversely proportional to both the size of the "container" and the rate at which the container is being filled. It is quite easy to acquire some image of just what is going on if we will recall that dumping a quart of water into a bathtub would create only a very small pressure on the bottom, but dumping a quart of water into a tube ½ in. in diameter would create a far higher pressure. The same amount of water will create pressures which are inversely proportional to the cross-sectional areas of the containers. So it is with the electrical counterpart: a given amount of electricity will establish a far higher electrical pressure (voltage) in a small capacitor than it will in a large one.

The effect of frequency will require quite a different analogy. Let us consider Fig. 8·14. If we assume that the volume A is equal to the indicated maximum volume B, then as we move the piston from its extreme right-hand position to its extreme left-hand position, it will just raise the level of the liquid from XX to $X'X'$.

Now if we assume that it takes 1 min to move 10 qt of liquid from the lower level up to $X'X'$, we might call that the basic unit of resistance to filling. In this analogy the indicated *resistance* is not to be thought equivalent to the ordinary resistance of a circuit. If we turn the crank at a higher rate so that we raise the level to $X'X'$ twice in a minute, then the apparent resistance will be only half as much, because it takes only half the time to do the same work that was done when the crank was turned at the rate of 1 rpm.

Now if we increase the rate of turning of the crank to 10 rpm, then it will take only one-tenth of the time originally required—*all for the same amount of fluid moved from XX to $X'X'$*. If we can do the same amount of work in one-tenth the time, it appears that the faster we fill the upper chamber, the "easier" it becomes. At least our resistance seems to be less because we do the job in a shorter period of time. *But the same effort is required each time to raise a fixed volume of fluid through a fixed distance.* So if we can do a given amount of work in less and less time, it seems to be easier, or its resistance is less. If we define our resistance to filling as equal to

$$\text{Resistance} = \text{time/volume change}$$

then by this definition, as we move the crank faster, the time required for any given volume change becomes less, and the apparent resistance decreases.

All of this is one attempt to show the effect of frequency upon the capacitive reactance. *The more often we can move one ampere into the*

Fig. 8·14 The effects of capacitive reactance may be simulated to show that the counterpressure varies inversely with the frequency.

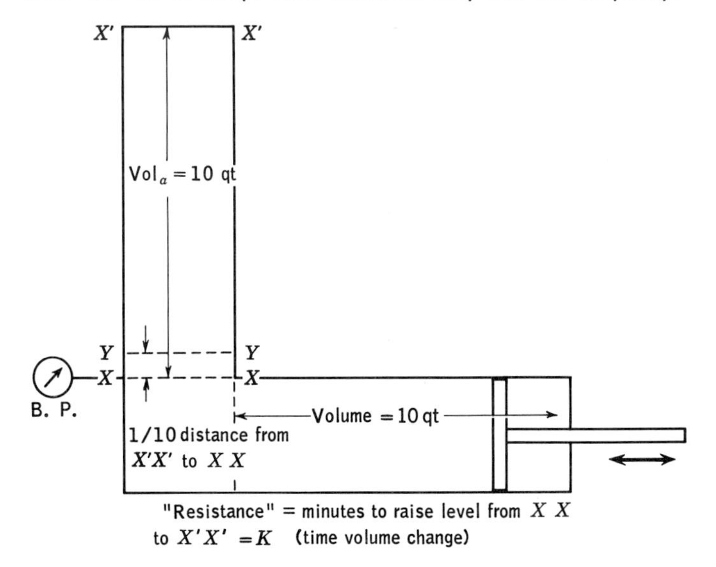

capacitor in any given period of time, the smaller is the capacitive reactance. Another way to look at it is this: If we decrease the stroke of the pump so as to raise the liquid level from XX to YY, where this distance is only one-tenth of the distance from XX to $X'X'$, but run the pump at 10 rpm, then we will raise the same volume of water above XX in 1 min as in the case when there was one full stroke per minute. But the pressure caused by the column of water raised above XX to YY would be only one-tenth of that occurring with the full stroke. So if we wish to define our resistance as the back pressure caused when a given volume of water per minute is raised above XX, then this back pressure will go down as we decrease the stroke of the pump but increase its speed proportionately. Perhaps this second analogy is better: the back pressure caused by raising the level above XX becomes smaller the faster the piston is operated, *provided that the same quantity or volume of fluid is moved in any given period of time.*

Thus, the faster we try to "fill" our capacitor, the less current will we have to put into the capacitor each time the polarity changes in order to get a certain number of electrons to move into the capacitor each second. The fact that these electrons may leave the capacitor with each alternation need not bother us any more than it did when we were considering how much fluid we moved above XX; it had to be the same fluid each time, made up of the same molecules. But we had no objection to counting those particles more than once, and so we should have no hesitation about counting the electrons more than once.

One definition of an ampere states that it is at the rate of so many electrons per second passing any given point. The definition says nothing about counting the same electron more than once, or even counting it for each direction of travel. Notice that the ampere is defined as a rate: 1,000 for 1 sec is at the same rate as 500 for $\frac{1}{2}$ sec, 100 for $\frac{1}{10}$ sec, 10 for $\frac{1}{100}$ sec, or even 1 for $\frac{1}{1000}$ sec.

CIRCUITS

Series circuits. In practice we shall find that, more often than not, we do not have some potential applied to a device which contains only resistance or inductive reactance or capacitive reactance. There is far more likely to be a combination of two or three of these elements. If the current goes through each element in turn, then it is a series circuit, and it will be quite easy to handle.

In Fig. 8·15a the current is equal to the input circuit voltage divided by the impedance.

$$Z = R + j0 = R \text{ volts/amp} = R\underline{/0^\circ}$$

In this case we could form no impedance triangle because there is no reactance; so the sum of the resistance and the reactance is equal to the

Fig. 8·15 (*a*), (*b*) The current in an a-c circuit is obtained by dividing the voltage by the impedance, both quantities being treated as vectors. (*c*), (*d*) A series circuit which includes resistance, inductive reactance, and a capacitive reactance requires the use of a vector solution of the type indicated. (e) The vector representation of the voltages which exist at the terminals of the components shown in (c).

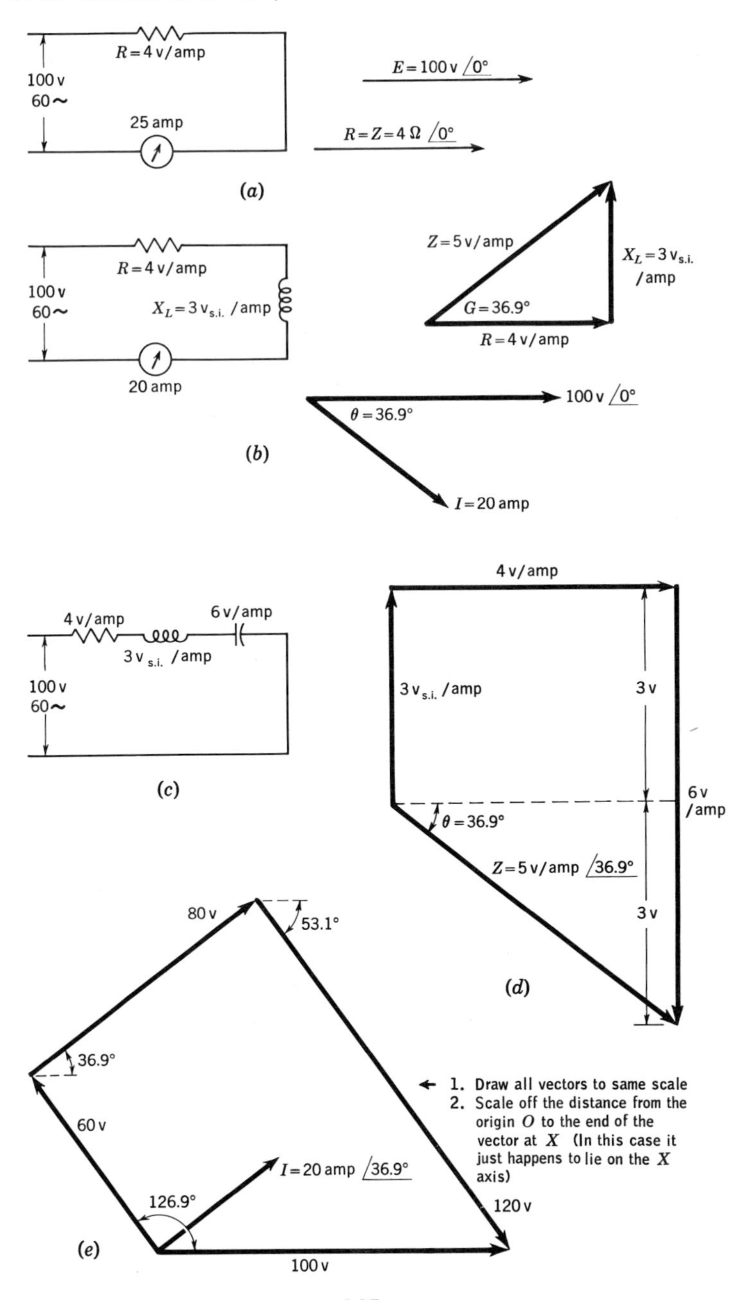

resistance only. It also lies in the same direction that we have assumed for the resistance. Let us also assume that the voltage acts in a direction which is horizontal and to the right. Now

$$I = \frac{V\underline{/\theta_1}}{Z\underline{/\theta_2}} = \frac{100\underline{/0}}{4\underline{/0}} = 25\underline{/0 - 0} \text{ amp} = 25\underline{/0}$$

The angle of zero indicates that this current is directed horizontally to the right.

In Fig. 8·15*b* we have placed an inductive reactance in series with the resistor of the preceding example. We are to determine the current and the phase angle. The phase angle is the angle between the current and the voltage, measured in degrees. As before

$$Z = R + jX_L = 4 + j3 = \sqrt{4^2 + 3^2} = \sqrt{25} = 5\underline{/\tan^{-1} 3/4}$$

$$= 5\underline{/36.9} \text{ volts/amp}$$

If we assume, as we did before, that the voltage vector lies along the horizontal to the right and therefore has an angle of zero degrees, then

$$I = \frac{E\underline{/\theta}}{Z\underline{/\theta}} = \frac{100\underline{/0°} \text{ volts}}{5\underline{/36.9°} \text{ volts/amp}} = 20\underline{/0 - 36.9°} \text{ amp}$$

$$= 20\underline{/-36.9°} \text{ amp}$$

Twenty amperes at some angle is known as a *polar* quantity and is represented by a magnitude at some angle; 100 volts at an angle of 0° is a polar quantity. When we divide and are using polar quantities, the magnitudes are divided in the customary manner, but the angles are treated as though they were exponents in the usual algebraic division; where the exponent of the divisor must be subtracted from the exponent of the dividend.

When we want to multiply polar quantities, we multiply the numerical quantities in the usual way, but we add the angles algebraically as though they were exponents. The numerical value of the angle can be obtained from a table of tangents. The tangent of an angle, for right triangles, is equal to the *side* opposite the angle divided by the *side* adjacent to the angle. In this case, the tangent is equal to ¾, or 0.75. The angle then is 36.9°.

In Fig. 8·15*c* we have added one more circuit component: a capacitor in series with the resistor and the inductor. The value of the impedance is found in a manner similar to that used previously. This time there are two reactances which must be combined vectorially with the resistance to determine the impedance.

To simplify the drawing and to show the parts in better relationship to each other, the inductive reactance was plotted first, then the resistance, and finally the capacitive reactance. Note that when the sign of the reactance is *plus*, the vector is drawn *up*, and when the sign is negative, the vector is drawn *down*. The vector sum is then equal to

$$Z = R + jX_L - jX_C = R + j(X_L - X_C) \qquad \text{Fig. } 8\cdot15d$$

$$= 4 + j(3 - 6) = 4 - j3$$

$$= 5\underline{/\tan^{-1} - \tfrac{3}{4}} = 5\underline{/-36.9°}$$

$$I = \frac{E}{Z} = \frac{100\underline{/0°}}{5\underline{/-36.9°}}$$

$$= 20\underline{/0 - (-36.9)°} = 20\underline{/36.9°}$$

Without trying to prove or justify the statement, let us accept that inductors cause the current to lag behind the voltage, but capacitors cause the current to lead the voltage. When we have each type of reactance in a series circuit, the one which is numerically larger will prevail and the over-all circuit will take on the characteristics of this "larger" quantity. In this example, the capacitive reactance exceeds the inductive reactance, and the circuit as a whole behaves as though 3 ohms of capacitive reactance were in series with 4 ohms of pure resistance.

Series resonant circuits. If in the preceding example the inductive reactance and the capacitive reactance had been numerically equal, then the impedance would have consisted of only a resistance term, and the impedance then would be 4 ohms, or 4 volts/amp. This particular condition when the two reactances are numerically equal we call *resonance*. The characteristics of a circuit at that time are important and should be learned. Resonant circuits are used in many ways. The characteristics of resonant series circuits are as follows:

1. The inductive and capacitive reactances are numerically equal.
2. The impedance is a minimum and is equal to the resistive term only. Therefore, the current is a maximum.
3. The phase angle of the circuit is zero.
4. The voltage across each circuit component is a maximum for any given input circuit voltage.
5. The voltage of self-induction of the inductor is just canceled by the counter voltage in the capacitor.

If we consider the voltage across each component where we have both inductive and capacitive reactances which are not equal, we get the following:

$$V_L = (3 \text{ volts/amp } \underline{/90°})(20 \text{ amp } \underline{/+36.9°})$$

$$= 60 \text{ volts } \underline{/90° + 36.9° = 126.9°}$$

$$V_R = (4 \text{ volts/amp } \underline{/0°})(20 \text{ amp } \underline{/36.9°})$$

$$= 80 \text{ volts } \underline{/0° + 36.9° = 36.9°}$$

$$V_C = (6 \text{ volts/amp } \underline{/-90°})(20 \text{ amp } \underline{/36.9°})$$

$$= 120 \text{ volts } \underline{/-90° + 36.9°}$$

If these voltages are then drawn, as indicated in Fig. 8·15*e*, their vector sum is 100 volts at an angle of zero. This is the value of voltage which was assumed. Oftentimes in dealing with series circuits, we place the current along the reference axis, where the angle will be zero because the current is the only thing which is common to each circuit element.

It is very important that we learn to draw these vector diagrams. With them, the solution of the current and voltage relationships is quite easy; without such a "map," we can experience only difficulty and trouble. The vector diagram should be your guide in establishing the values and the angles, and vice versa.

Series resonance is desired when we want to pass one signal but exclude other frequencies. For example, tuning a radio causes one frequency to furnish a strong signal while all others are attenuated.

Parallel circuits. In dealing with parallel circuits, we have to remember only that each branch has the same voltage but that the current in any one branch will be governed by its own impedance, not by that of its neighbors. The current for each branch will be determined just as we did for the series circuit shown in Fig. 8·15*c*. Having determined each of the branch currents, our only problem then is to add them *vectorially*. As with the voltage in the series case, the values cannot be added arithmetically, *they must always be combined vectorially*.

If we take the very simple situation shown in Fig. 8·16*a*, then we have only "pure" reactances in two of the paths: one is a capacitance with no resistance in the circuit, and the other has inductive reactance only. It should be realized that these are taken for study purposes only. Capacitors may have such small losses that their resistance may be said to be zero, but inductors will almost always have some resistance.

Let us compute the current in each branch and then see what the ammeter should read when all switches are closed at the same time, Fig. 8·16*a*.

$$I_L = \frac{E\,\underline{/\theta_1}}{Z\,\underline{/\theta_2}} = \frac{100 \text{ volts } \underline{/0°}}{(0 + j10) \text{ volts/amp}}$$

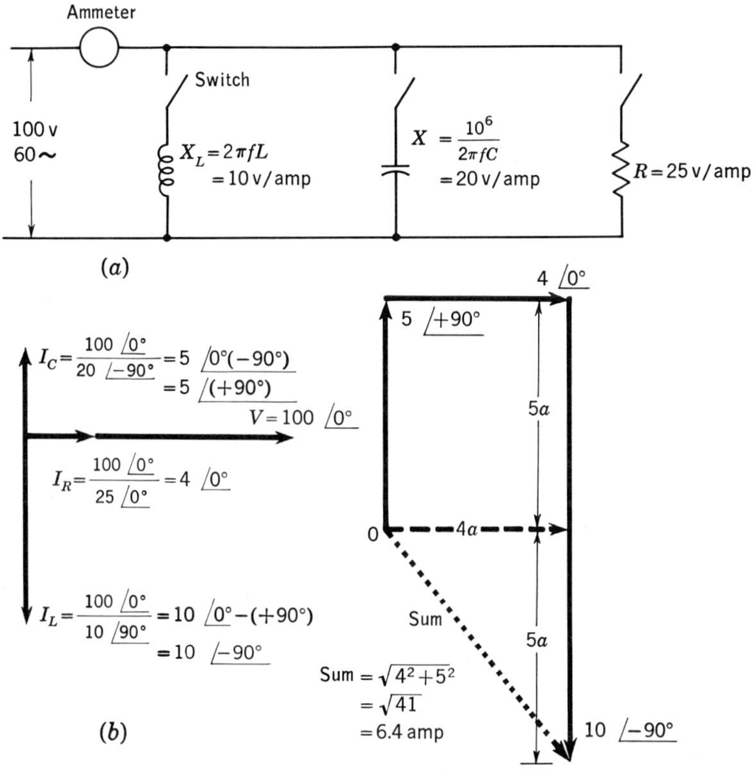

Fig. 8·16 (a) When parallel circuits are employed, the total current is the vector sum of the individual branch currents. (b) To simplify the figure, the capacitive current (leading by 90°) is taken as the first current vector; then the resistive current; finally the lagging inductive current.

$$= \frac{100 \text{ volts } \underline{/0°}}{10\underline{/+90°}} = 10 \text{ amp } \underline{/0° - (+90°)} = 10\underline{/-90°} = 0 - j\,10$$

$$I_C = \frac{100\underline{/0°}}{20 \text{ volts/amp}\underline{/-90°}} = 5 \text{ amp } \underline{/0° - (-90°)}$$

$$= 5 \text{ amp } \underline{/+90°} = 0 + j5 \text{ amp}$$

$$I_R = \frac{100\underline{/0°}}{25 \text{ volts/amp } \underline{/0°}} = 4 \text{ amp } \underline{/0° - 0°} = 4 \text{ amp } \underline{/0°}$$

$$= 4 \text{ amp} + j0$$

$$I_{\text{total}} = \text{vector sum}$$

$$= 4 + j5 - j10 = 4 - j5 = 6.4\underline{/-57.35°}$$

The ammeter should read 6.4 amp.

IMPEDANCE-MATCHING TRANSFORMERS

Transformers are often used to match impedances. This means that an attempt is made to match the current and voltage requirements of the source with the current and voltage requirements of the device. For example, a motor is designed for operation at 25 volts and normal current requirements are 50 amp. A 100-volt supply is available, but the current is limited to 15 amp. How can we operate the motor? If the motor were applied directly to the line, it might run much too fast, if it were of the series type, and the current would far exceed the 15-amp line rating. But if we use a transformer to reduce the potential to 25 volts,

$$V_1 I_1 = V_2 I_2$$

$$100(I_1) = 25(50)$$

$$I_1 = 12.5 \text{ amp}$$

which is well within the allowable line current. The supply line could furnish high voltage and low current values, but the device required a low voltage and a large current. By the use of a transformer we have been able to run the motor.

To use a somewhat different analogy, when you wish to start an automobile, you are faced with a similar problem. The motor can furnish a high speed but can develop a fairly small torque. To get the car underway, we need a large torque at low speed. To meet these conflicting conditions, we insert a transmission between the motor and the differential. In the low position we reduce the speed and increase the torque delivered to the rear wheels. When in second gear, the speed is not decreased so much, but the torque is also greater than engine torque. In high gear the output of the engine is fed directly to the wheels, which under these conditions need high speed but little torque.

The gear shift or automatic transmission in a car is an impedance-matching device or mechanical transformer which tries to match the output of the engine with the requirements of the load. The same engine needs to furnish high torque at low speed when starting and low torque at high speed when running. The transmission makes it possible to have both.

EXAMPLE 8·7 A servomotor to drive a potentiometer requires 40 volts and 0.2 amp. The output of the amplifier is limited to 0.50 amp at 200 volts. Can we select a transformer that will meet the maximum limitations of each device?

SOLUTION. $V_1 I_1$ should be equal to $V_2 I_2$:

$$200(I_1) = 40(0200)$$

$$I_1 = 0.040 \text{ amp}$$

which is within the current rating of the amplifier.

The impedances of the primary and secondary circuits are in the ratio of the square of the turns ratio, or transformation ratio. If the voltage ratio is greater than 1, then the turns ratio squared will be more than 1; if the turns ratio is less than 1, then the turns ratio squared will be less than 1.

The impedance of the high-side winding, Example 8·9, is

$$\frac{E}{I} = \frac{200 \text{ volts}}{0.040 \text{ amp}} = 5{,}000 \text{ volts/amp, or ohms}$$

The impedance of the low side is

$$\frac{E}{I} = \frac{40 \text{ volts}}{0.20 \text{ amp}} = 200 \text{ volts/amp, or ohms}$$

$$\frac{5{,}000 \text{ volts/amp}}{200 \text{ volts/amp}} = 25 = n^2$$

$$5 = n = \frac{200 \text{ volts}}{40 \text{ volts}} = \text{transformation ratio}$$

SATURABLE-CORE REACTORS AND MAGNETIC AMPLIFIERS

Two of the most important components found in instrumentation today are the saturable-core reactor and its very close relative, the magnetic amplifier. Both of these devices function without any moving parts; they depend entirely upon the behavior of magnetic materials and suitable rectifying elements which control the amount of direct current which flows in the various circuits.

If we take one of the very simple reactors, which is merely a coil of wire wound on a magnetic core, and connect it to a suitable source of alternating-current power, we expect that very little current will flow. (By "suitable" we mean that the voltage of self-induction for the particular frequency involved can at least equal the applied voltage.) The changing current in the winding has produced a magnetic field which by its rate of change with time creates within the winding a voltage of self-induction which is almost equal in magnitude to the applied voltage and is opposite in polarity, with the result that the net voltage applied to the coil is almost zero. Under such conditions one would expect very little current to flow.

If we assume, for the moment, that the voltage applied is 100 volts and that the voltage of self-induction is 98 volts for the frequency of 60 cps, then we know the maximum voltage of self-induction which can be produced with this core is in the vicinity of 98 volts. We will assume that for applied voltages in excess of 100 volts, the voltage of self-induction will not increase significantly. If, for example, the applied voltage is increased to 200 volts, then the voltage of self-induction, which remains substantially unchanged, and the applied voltage will have a difference of 102 volts instead of the 2 volts which was previously the case. We are ignoring the fact that these additions and subtractions should be of a vector nature. If we realize that there are some differences in the numerical values, but remember the numerical relationships, no great harm will have been done.

If we use the arithmetical difference as quite adequate, then we have a ratio of voltage differences of $^{102}\!/_{2}$, or 51. Thus instead of having a negligible current as with 100 volts applied, there will now be a very large current which will be determined largely by this voltage difference and the impedance of any other part of the circuit which might be in series with the coil and whose voltage of self-inductance had originally been quite insignificant. For rated voltages and below, we find

$$V_{\text{appl}} - V_{\text{self-ind}} = \text{something greater than zero}$$

Assuming values and using arithmetical relationships instead of vectors:

100 volts, applied − 98 volts, assumed voltage of self-induction = 2 volts

But if the applied voltage exceeds the rated voltage, then the voltage of self-induction will not approximate the applied voltage. If

200 volts, applied − 98 volts, assumed voltage of self-induction = 102 volts

for the conditions assumed, there will be $^{102}\!/_{2}$, or 51, times as much current if the coil has 200 volts applied instead of 100 volts. *The current is dependent upon the difference of two potentials; one is the applied emf, and the other is the voltage of self-induction,* Fig. 8·16*b*.

We find that this is the sort of thing which happens when a transformer which is designed for 120 volts is applied to a 240-volt circuit. On its specified rated voltage of 120 volts, the transformer is able to generate a countervoltage which limits the current to a very nominal value for which the transformer was designed. Now if the transformer is applied to this higher potential, there is no longer an effective control of the current by the magnitude of the counter emf. So much current will flow through the primary winding because of the large difference in the applied and counter potentials that the transformer will be destroyed through self-heating brought about by the excessive primary currents. It must be realized that none of the primary current is caused by any load on the secondary, which we will assume is open-circuited.

If we had any easy way of varying the applied voltage to such a re-

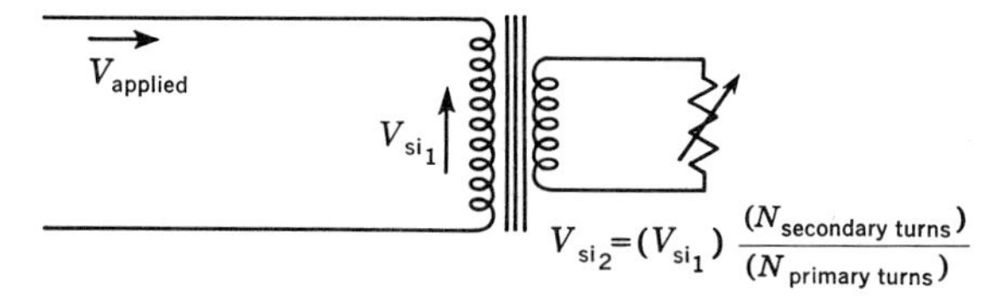

Fig. 8·17 Changing the resistance of the load in the secondary circuit will change the impedance of the transformer. As it does so, the primary current changes.

actor, it would give us one way of controlling the current more or less steplessly. But we are generally limited by fixed input potentials. Under such conditions, we must find some other way to control our output current which will bear some definite ratio to the input current.

Let us therefore consider the schematic circuit indicated by Fig. 8·17. This is a conventional transformer with a variable resistance in the secondary circuit. With no current in the secondary circuit, and with the same coil and core for a primary as we had previously considered, the applied voltage and voltage of self-induction are essentially equal and very little current flows in the primary winding—just enough to create the flux which produces the voltage of self-induction.

But if we now reduce the resistance of the secondary load to some amount so that a current of 1 amp, for example, exists in the load resistor, then there will have to be an additional current of 1 amp in the primary winding. (Let us make the illustration easier by assuming that there are as many turns on the secondary winding as on the primary, and we will use arithmetical addition when we know that these quantities are vector magnitudes.)

Just how did we get the extra current in the primary to match the secondary current? The flow of current in the secondary came about because of the relationship of the secondary voltage and the resistance of the load. This current flowing in the winding produces a demagnetizing effect which reduced the flux produced by the primary. With the reduction in the flux linking the circuit there had to be a reduction in the voltage of self-induction in the primary; this caused a greater difference between the applied voltage and the voltage of self-induction. With a greater potential difference, more current would result. Thus the current flowing in the secondary coil demagnetizes the core and in this manner controls the amount of current flowing in the primary circuit.

Changing the value of the resistance of load in the secondary circuit, Fig. 8·17, will change the impedance of the transformer. It also constitutes one way of using the current in one circuit to control the current in another circuit which is electrically isolated. The secondary current

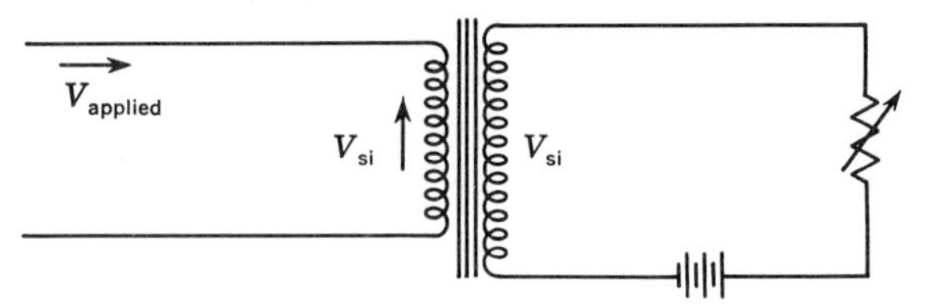

Fig. 8·18 A small direct current flowing through many turns can create a demagnetizing magnetomotive force which can also bring about large changes in the current in the primary winding of the unit.

demagnetizes the core slightly, thereby reducing the counter emf. As it does so, more current will flow in the primary circuit.

If the control of the primary current is the only purpose of the variable resistance in the secondary, then this is a very wasteful way to operate. All of the energy going into the resistor produces nothing but heat; this is costly and is to be avoided in practice.

Now let us replace the secondary coil with one having many more turns, but located on the same core. Let us also connect a battery and a variable resistor in series with the secondary. Now the core, Fig. 8·18, would be aware only that a demagnetizing effect of so many ampere-turns had been applied to it. Whether it came from some of its own current or whether the effect was produced by some external current would be of no concern to it. If we were to produce the same number of demagnetizing ampere-turns with the battery as we had produced with the previous load current, then the current in the primary would show a similar increase in magnitude, and this is essentially what we want to do.

There are some very real objections to the circuit which we have just used. In order to keep the control current small in value, we used many turns on the control winding. But the control winding is linked by the same flux that links the primary. Consequently, there will be induced in the control winding a voltage that is equal, approximately, to the applied voltage multiplied by the ratio of the secondary turns to the primary turns. It would be easy for this potential to exceed a thousand or more volts, and this could be dangerous to equipment and personnel. To find a way of overcoming this very practical problem, let us consider Fig. 8·19.

Let us divide the primary winding and the core into two equal cores which have a potential rating of 50 volts each. Then with the assumed direction of primary current flow, the flux in the two cores will be as indicated. It will be noted that the control winding now links both cores, and the direction of the flux is opposite in the two cores. The net result is that the average value of the flux is zero; whatever voltage is induced in the control winding by the flux in one winding is canceled by an

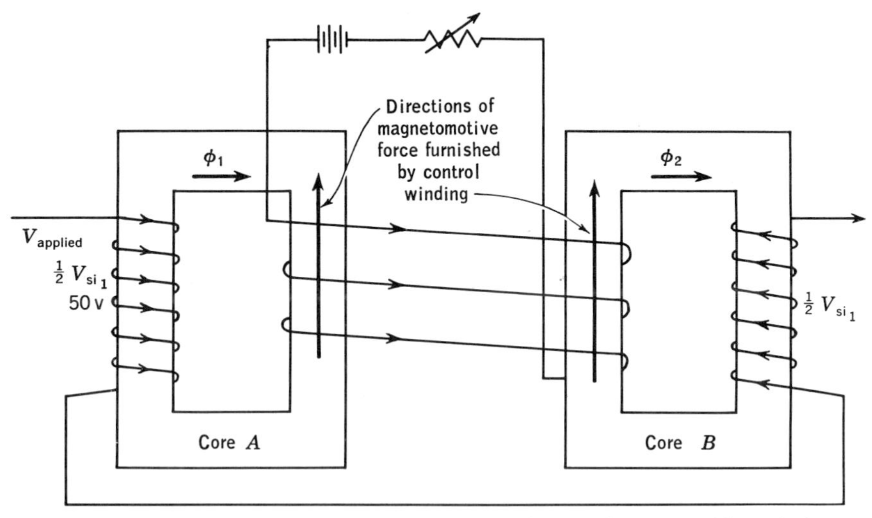

Fig. 8·19 The use of two cores and associated windings makes it possible to balance out the voltage induced in the control winding of core A by having an equal and opposite potential induced in the control winding of core B. For the directions assumed for the currents and the turns, core B will saturate first.

equal and opposite voltage induced by the flux from the other winding. For the assumed direction of the control current, the magnetic effect upon core A will be to demagnetize the core, while in core B the control current will increase the magnetic effect.

If we find that the magnetic material has the idealized characteristic shown in Fig. 8·20, the effect of the control current is to:

1. Add demagnetizing ampere-turns to core A, bringing it to a and the flux to b. This makes possible a large change in flux, and consequently, a large voltage of self-induction.

2. Add ampere-turns to c in core B and reduce the possible flux change so that there will be a very much smaller voltage of self-induction. That is, the effect in core B is opposite that in core A.

Let us assume that without the control winding, the flux change would have been quite adequate to produce a voltage of self-induction of 50 volts. If we now recall that each of these coils can produce a voltage of self-induction of 50 volts, then if the condition of core B is such that it can no longer produce any change in its flux and hence in its flux linkages, there will be a sudden increase in the load current because there is no longer a counter emf which is almost equal to the applied voltage. The increase in the load current caused by the saturation of core B also results in saturating core A almost at the same time because the ampere-turns required to saturate the material are only a very small fraction of the ampere-turns which are produced by the load current.

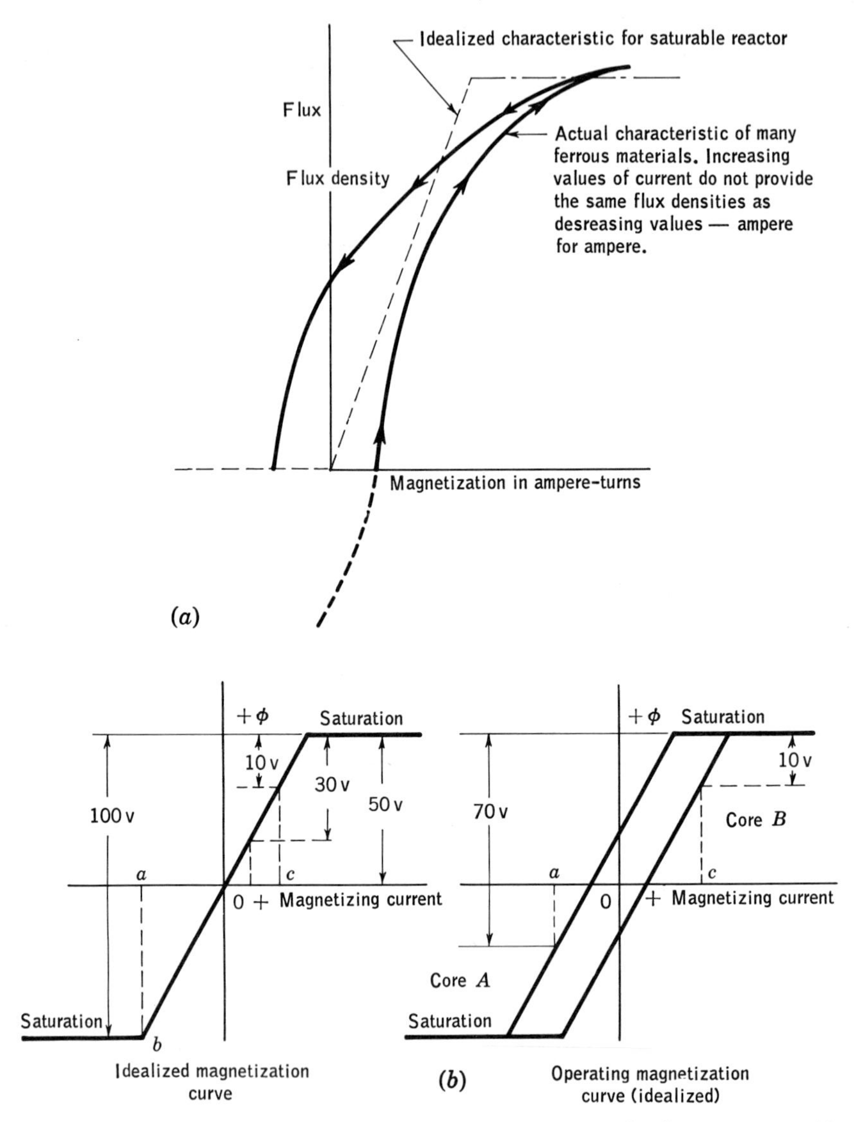

Fig. 8·20 (*a*) The ideal saturation characteristic can be quite closely approximated by some of the materials now available. (*b*) By "biasing" the core, we can increase or decrease the number of ampere-turns which must be added by the load before the core saturates.

For the particular portion of the cycle which we are studying, core *B* is the one which is going to initiate the action. As the applied voltage goes through the first half of its cycle, with the direction of voltage and current as assumed, the current in the load circuit will lag 90 electrical degrees behind the applied voltage, because that is one of the characteristics of circuits which are predominantly inductive and that is what the core is until the moment of saturation. That the current lags so far behind the potential which produces it means that we arrive at the positive maximum value of the current when the potential is going through zero and that the current is zero when the applied voltage is a maximum. (See Fig. 8·21*a*, which is based on no biasing current.) So long as the current in core B is negative—during the first half of the alternation—then the magnetizing current and the control current are working against each other. But during the next quarter of a cycle, the magnetizing current and the control current will be aiding in their magnetic effects.

If in Fig. 8·21*a* the core is not saturated when the magnetizing current reaches its maximum at point x, then the core never will saturate. If it happens that the core just reaches saturation with the idealized magnetic material of Fig. 8·20*b*, then there will be no change in the load current inasmuch as the applied voltage is at that particular moment going through zero. But if we now add to the magnetizing current a control current which has the magnitude indicated in Fig. 8·21*b*, then the core will saturate earlier in the alternation, at point y. The voltage applied to the load will be almost equal to the value of the sine wave between points a and b; the drop in the reactor will be negligible under these conditions and the line voltage being applied to the load will now cause a very sharp increase in the load current, Fig. 8·21*c*. This sharp increase in the conductivity of the circuit is termed *firing,* because the sharp transition from little conduction to maximum conduction takes place very quickly and is suggestive of the behavior of a thyratron.

As the control current is made more and more positive, the firing takes place earlier and earlier in the alternation. When the magnetizing current is negative, it is working in opposition to the control current, but as the magnetizing current becomes less negative, it permits the control current to control the firing anywhere during the first 90 electrical degrees. When the magnetizing current is positive, it adds to the effect of the control current to establish the firing point.

Because the load current may be many times the control current, devices of this sort may also be termed magnetic amplifiers. To be useful in instrumentation work, they must be as nearly linear as possible in order that the input signal may be faithfully reproduced in the output circuit.

So far as the terminology is concerned, saturable reactors are made to handle relatively huge quantities of power. But magnetic amplifiers are generally rated in terms of watts or some other fairly small unit.

The linearity obtained and the general performance of the device is quite largely dependent upon the type of magnetic material used and its magnetic history. Some of the materials are so sensitive to any pres-

Fig. 8·21 (*a*) Because the magnetizing current lags 90° behind the applied voltage, if saturation is obtained just as the magnetizing current passes through its maximum value, no added current will flow. At that moment the applied emf is zero. (*b*) When *control ampere-turns* are added to the ampere-turns of the normal magnetizing current, saturation occurs before the magnetizing current has reached its maximum positive value and when the voltage wave is greater than zero. By advancing the time at which saturation accurs, there will be a potential available to cause current to flow through the load. (*c*) By increasing the value of the control ampere-turns, saturation of the core occurs earlier in the cycle. Because the net circuit voltage suddenly increases from almost zero to some significant amount, the load current suddenly increases.

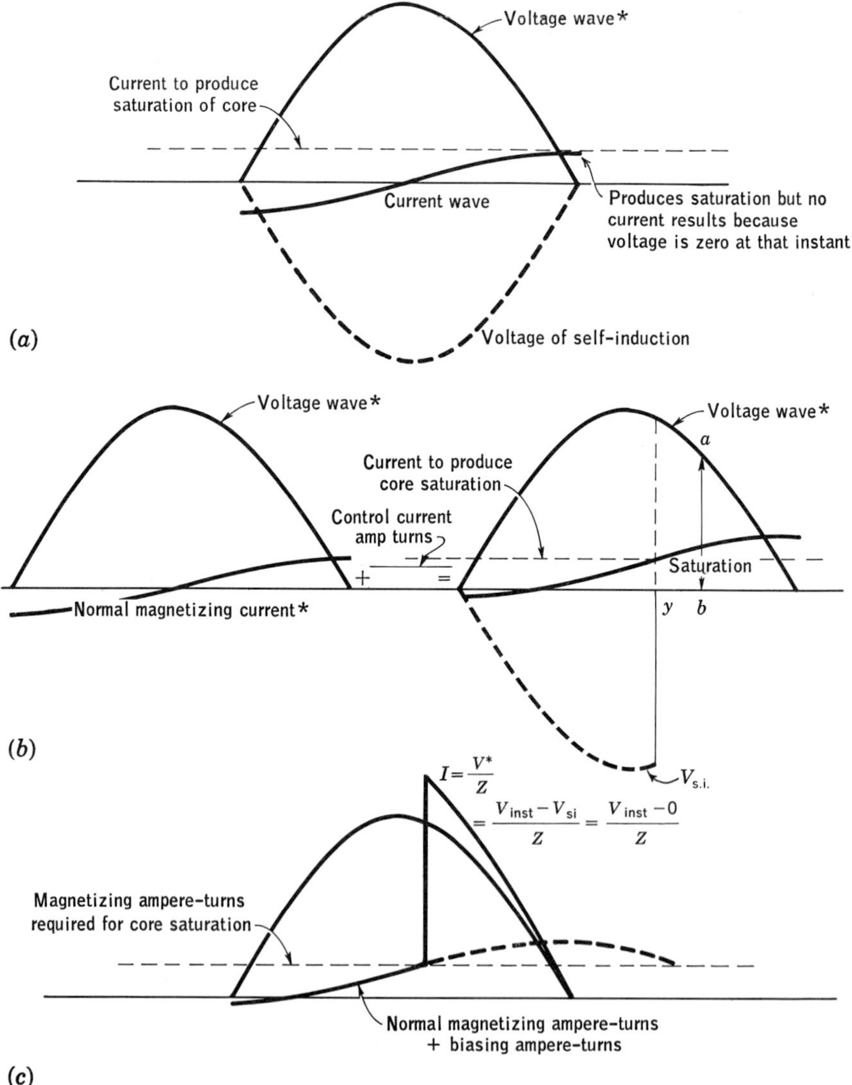

sure which will produce significant strains that the entire magnetic characteristic can be changed when the core material is even slightly deformed. Some materials are so responsive to small magnetic fields that they are completely saturated by the very weak earth's magnetic field. Such devices must therefore be carefully shielded from the effects of stray magnetic fields.

SUMMARY

1. Electric energy may be generated in a variety of ways: (*a*) by heating one junction of two dissimilar metals and cooling the other junction, (*b*) by the chemical action of an acid on two dissimilar metals, (*c*) by causing radiant energy to fall on photocells, or (*d*) by moving a conductor in relation to a magnetic field or vice versa.

2. When a conductor is moved in a magnetic field, the magnitude of the potential may be expressed by

$$E = KNBAPS$$

where K = proportionality constant
N = number of series conductors per path
B = flux density
A = area of a pole
P = number of poles
S = number of revolutions per unit of time

3. The emf induced in a conductor has a polarity which depends upon the pole whose flux is being cut; for any given direction of rotation the emf induced when the conductor is under a north pole is opposite to that induced when the same conductor is under a south pole. Except for a very special type of generator which finds but limited use, all generators induce alternating voltages in their conductors and then use a mechanical switch, called a commutator, to convert the alternating emf into a direct or unidirectional potential.

4. Surrounding every current-carrying conductor there is a magnetic field which generally will be proportional to the current. If the magnetic field is created by an alternating current, then the changing magnetic flux may induce emf's in any conductor which senses the changing magnetic field.

5. The current which results from any induced voltage will take such a direction of flow that it will oppose any change in the magnetic field. This is Lenz's law.

6. In the field of industrial instrumentation, alternating currents find many applications. Some of the most important reasons are the following: (*a*) With alternating currents, transformers may be used. They provide easy ways of changing voltages, providing electrical isolation, and matching the current and voltage requirements of one device to those of another. (*b*) Alternating-current amplifiers have certain advantages over direct-current amplifiers. For long-time stability of output, the alternating-current amplifiers are generally used. (*c*) By comparing the polarity of the error signal with the polarity of the supply voltage, it is possible to sense the direction in which any correction should be made to reestablish equality of signals or the balance of bridges.

7. Transformers and capacitors respond to the rate of change of the current. Provided that the potential and the current are both sine waves, then the current and potential waves will (*a*) change at their maximum rate when they pass through zero, and (*b*) have their minimum, or zero, rate when the potential and current

waves have their maximum values. For most normal uses of transformers and capacitors, we must always remember that their performance or response depends upon how fast the current or flux is changing, and not on its absolute value.

8. Whenever a changing magnetic flux threads or cuts a conductor, a potential will be induced. The magnitude of this potential depends upon (*a*) the amount of flux linking the turns or coil, (*b*) the number of turns of wire linked by the flux, and (*c*) the rate at which the flux is changing. The higher the rate at which the current, and hence the flux, changes, the greater the value of the emf. Some equipment is made to operate on frequencies as high as 400 cps because the induced voltages will be much higher than they would be for frequencies of 60 cps, assuming the same amount of flux.

Inductance is one property of a circuit linked by flux. If the current changes at the rate of 1 amp/sec and the induced voltage is 1 volt, then the circuit has a self-inductance or an inductance of 1 henry. Any change in the flux linking an inductive circuit will create a voltage of self-induction. This voltage of self-induction will always be of such a polarity as to act in opposition to the applied voltage.

9. The voltage of self-induction can limit the current in a circuit just as effectively as a resistance can and without the creation of any wasted heat.

10. Although resistive effects and voltages both tend to limit the flow of current in any given circuit, their effects are not directly additive. They must be combined vectorially, with the reactive, or self-induced, voltages acting at right angles to the resistive voltages. If the inductive effects predominate, the reactance effects are plotted as positive vertical values; if the capacitive effects predominate, the reactance must be plotted as negative vertical values. If the capacitive and reactive values are equal, the circuit functions as though it were composed of resistances only.

The vector sum of the resistive and reactive effects is termed the *impedance* of the circuit. It indicates the magnitude of the potential required to force 1 amp through the circuit. Dividing the potential of the circuit by this value will then indicate the circuit current. Because the impedance is a vector quantity, the current may be shifted with respect to the voltage which creates it.

11. If in a series circuit, the inductive and capacitive effects are numerically equal, the effects cancel and the current will be a maximum. This is called *series resonance.*

12. If in a two-branch parallel circuit, the inductive current of one branch is just equal to the capacitive current of the other, the condition is known as *parallel resonance* and the current to the two branches is a minimum. This condition is used to block or exclude currents of a particular frequency and thus permit us to select one of several frequencies that occur in the same circuit.

13. The polarity markings of transformers are used to indicate terminals which have the same instantaneous polarity. This is of particular significance in instrumentation when it is necessary to compare instantaneous potentials and when it is necessary to know how one potential compares with another in polarity.

14. In instrumentation, transformers have a particular function of matching the current and potential ratings of amplifiers, tubes, and transistors to those of the servomotor or other load to which they may be connected.

QUESTIONS

8·1 What are the ways of generating electricity? Which way or ways can create only small potentials? Which are of significance in instrument work?

8·2 Arrange the answers to Question 8·1 in terms of their probable usefulness in instrumentation.

8·3 State Lenz's law. What is the connection between the law and "perpetual motion"?

8·4 What are some of the reasons for using alternating rather than direct current? Direct current rather than alternating current?

8·5 What are the factors that govern the magnitude of the induced emf?

8·6 What would be the effect on an automobile ignition system of slow and fast circuit interruptions?

8·7 How do we define and distinguish between north and south magnetic poles?

8·8 What is invariably associated with every flow of current?

8·9 Is there any relationship between the time constant of an electrical circuit and the time constant of a temperature-measuring device?

8·10 What might be the advantages of controlling the flow of current by using the voltage of self-induction rather than a variable resistance?

8·11 Of what particular significance is the slope of the tangent to the current wave in a representation of an alternating-current wave?

8·12 Of what practical significance to instrumentation is it to know the instantaneous polarities of electrical quantities?

8·13 Of what use might the knowledge of polarities be in other fields than instrumentation?

8·14 How could you use polarity markings on a transformer to get a voltage that was the sum or the difference of two available secondary voltages?

8·15 Just what are the relationships among resistance, reactance, and impedance?

8·16 What are the differences between inductive and capacitive reactances? Why do we distinguish between them?

8·17 What are the characteristics of series circuits? Of series resonant circuits? Of parallel resonant circuits?

8·18 Why might we like certain telemetering circuits to operate near the resonant point? Draw a sketch to illustrate, then study briefly some of the systems used in telemetering. Explain any discrepancies.

8·19 Compare and contrast the characteristics of series and parallel resonant circuits.

8·20 What would be the advantages of using higher frequencies (higher than 60 cps) for instrumentation equipment? Are there any drawbacks?

8·21 What are some of the differences between a transformer used to connect vacuum tubes to a suitable load, such as a motor, and a transformer used with transistors for the same purpose?

8·22 Discuss how an automobile transmission, or any other speed-torque-changing equipment for that matter, functions to match "impedances."

PROBLEMS

8·1 A conductor 20 in. long is moving at 100 fps where the flux density is 10,000 lines/in.^{2}. What will be the emf under these conditions?

8·2 If the speed in Prob. 8·1 is doubled, what will be the value of the emf? If the flux density is then doubled, what will be the emf?

8·3 If we assume that the conductor of Prob. 8·1 represents a portion of the fluid flowing in a magnetic flowmeter, write an equation which would represent the electrical signal when the quantity of fluid is Q cfm and the pipe has a cross-sectional area of A in.^{2}. What will be the difference in the expression if Z is expressed in cubic feet per second?

8·4 An electrical tachometer which is to indicate the speed of certain equipment has 1,000 conductors in series in a path. There are four poles, each with an area of 2 sq in. What will be the value of e if the flux density is 1,000 lines/in.^{2} and each conductor is 2 in. long and if the speed is to vary from 100 to 2,000 rpm? Plot the answers on graph paper.

8·5 Show the direction of the current in the conductor if it is (*a*) functioning as part of a motor, (*b*) functioning as part of a generator. What will happen if the motion is reversed?

8·6 What can be done to increase the output of the transducer in Example 8·2?

8·7 For the transducer shown in Example 8·7, what is the effect of increasing the turns to 1,000 and doubling the diameter of the armature?

8·8 Draw a sine wave on cross-section paper, using an amplitude of 3 in. and a length of 4 in. for the complete cycle. Determine the slope of the tangents to the curve at 0°, 30°, 60°, and 90°.

8·9 If a transformer is to decrease the voltage by a factor of 50, what is the ratio of the primary to secondary turns? If the voltage of the secondary were to be increased by the same factor, what would be the ratio?

8·10 A transformer is rated at 120 volts on the primary. It has 1,200 turns. How many turns are required for a secondary voltage of 6.3 volts? 700 volts?

8·11 A transformer which is rated at 100 volt-amperes can carry what primary and secondary currents if the voltages are 100 and 10 volts, respectively?

8·12 A transformer is rated at 100 volts on the primary and 25 volts on the secondary. What would V_3 of the polarity test sketch read if the leads are brought out for a subtractive polarity connection? For an additive polarity?

8·13 A transformer has two secondary windings, one rated at 6 volts and one rated at 5 volts. Sketch the two windings. If the transformer is of the subtractive-polarity type, indicate the polarity markings and show the connections to furnish 11 volts and 1 volt.

8·14 If the supply voltage is 100 volts, what other connections would have to be made in the transformer of Prob. 8·13 if one wanted 111 volts? 106 volts? 101 volts?

8·15 A resistance of 4 ohms is in series with an inductive reactance of 3 ohms. Determine the impedance and the phase angle. Determine the current if the voltage is 100 volts. Sketch the phase angle.

8·16 Repeat Prob. 8·15 but with a capacitive reactance.

8·17 Determine the impedance vector diagram and the current if the impedances of Probs. 8·15 and 8·16 are combined in series. In parallel.

8·18 A coil having an inductance of 0.1 henry and a resistance of 40 ohms is connected to a 100-volt line. Determine inductive reactance, impedance, vector diagram, current, and power. $f = 60$ cps.

8·19 Assuming that the characteristics of the material do not change, how much higher voltage could you use on a given transformer if one is permitted to use a frequency 10 times the rated frequency?

8·20 If the rated frequency of a transformer could be increased 10 times, what might you do with the number of turns and size of core if the voltage rating remains the same?

8·21 A servomotor to balance a potentiometer requires 100 volts and 0.20 amp. What must be the rating of the amplifier and the turns ratio of a transformer if the amplifier can supply 0.080 amp at 300 volts?

8·22 Determine the ratio of the impedances for the high and low sides of the transformer when supplying the motor in Prob. 8·21.

8·23 A transistor can supply 2 amp at 30 volts. What would the impedance ratio of a transformer have to be to furnish this power to a motor rated at 120 volts? Assume that the motor is rated at 60 va.

8·24 (*a*) If a mechanical gear train steps down the speed by a factor of 5, what is the ratio of the output torque to the input torque? How does this factor compare with the mechanical advantage of the gear train? (*b*) Then divide input and output speeds by the torques expressed in terms of the input torque to get an equivalent impedance. What is the ratio of the "impedance"? (*c*) Is there a relationship to the impedance of the electrical case?

8·25 What is the inductive reactance of a coil having an inductance of 10 henrys when $f = 10$ cps? 20 cps? 60 cps?

8·26 If $R = 1{,}000$ ohms, find the impedance for each frequency in Prob. 8·25.

8·27 A capacitor has a rating of 10 μf. What will be its reactance at 10 cps? 20 cps? 60 cps?

8·28 If the reactances of Probs. 8·25 and 8·26 are placed in series, what will the combined reactances be for 10, 20, and 60 cps?

8·29 If a resistance of 300 ohms is placed in series with the reactances of Prob. 8·28, compute the impedance.

8·30 A transformer has a rating of 10 amp and 100 volts for its primary rating. What will be the secondary currents if the secondary voltage is 10 volts? 20 volts? 50 volts?

8·31 If the secondary voltage of a transformer is 100 volts and the current is 10 amp, what is the secondary impedance? What is the apparent impedance of the transformer and load if the primary voltage is 10 volts? 100 volts? 1,000 volts? What generalizations can you make about a transformer relative to the line and load impedance ratios?

8·32 When 100 volts is applied to the primary of a transformer, 20 volts is induced in the secondary. When a polarity test is made, the voltmeter reads 80 volts. Of what type is the polarity? How should the primary and secondary terminals be marked? Draw a sketch and show the four terminals.

8·33 Could the transformer of Prob. 8·32 be connected to provide either 20 or 120 volts for a secondary load? 20 or 80 volts? What precautions would have to be observed because we have eliminated the electrical isolation that insulates the secondary, in many cases, from the primary or line voltage?

8·34 What forces, in pounds, will be produced in a pressure-current transducer when the flux density = 10,000 lines/in.^{2}, there are 1,000 turns having a diameter of 3 in., and the current is 0.100 amp?

Part II — Instrumentation in the Process Industry

9 Pressure Measurements, Mechanical Transducers

Perhaps the most common industrial measurement made is that of pressure. It covers an extremely wide variation in values: from one-millionth that of a gentle breeze to pressures so great that materials are permanently distorted, water is compressed, and extremely strong vessels or pipes are required to confine or limit them. We cannot see a force or a pressure, which is the force per unit area. Despite the fact that a force is weightless, can be transmitted at high speed, and accomplishes all sorts of things, we find it difficult to define it. We know that forces move objects, accelerate them, bend objects, distort them, and otherwise change them.

Because of this dilemma, we shall "measure" forces by what they do. Chapter 3 relates the magnitude of the force and the distortion which results. It cannot be emphasized too much that *the application of any force, or pressure, will always produce a deflection, a distortion, or some change in volume or dimension, no matter how small or large the force.*

It should be pointed out here that certain types of level measurements, flow measurements, and temperature measurements are really pressure measurements. Therefore, it is of great importance that this chapter be completely understood before going on with the other measurements. The force types of instrument are all alike; they differ mainly in the way the pressure is generated.

Division of pressure instruments. The range of the pressure instruments is so vast, Fig. 9·1, that their construction is of many types, being dependent upon the pressure range used. In order to make our study as direct and coherent as possible, we have divided all pressure instruments into two general classes: those employing mechanical means primarily and those which rely upon some electrical phenomenon or relationship for their proper functioning. Consequently, Chap. 9 is concerned with the first group, while Chap. 10 discusses the second group. In order that the continuity of the subject may be retained, the summary, the ques-

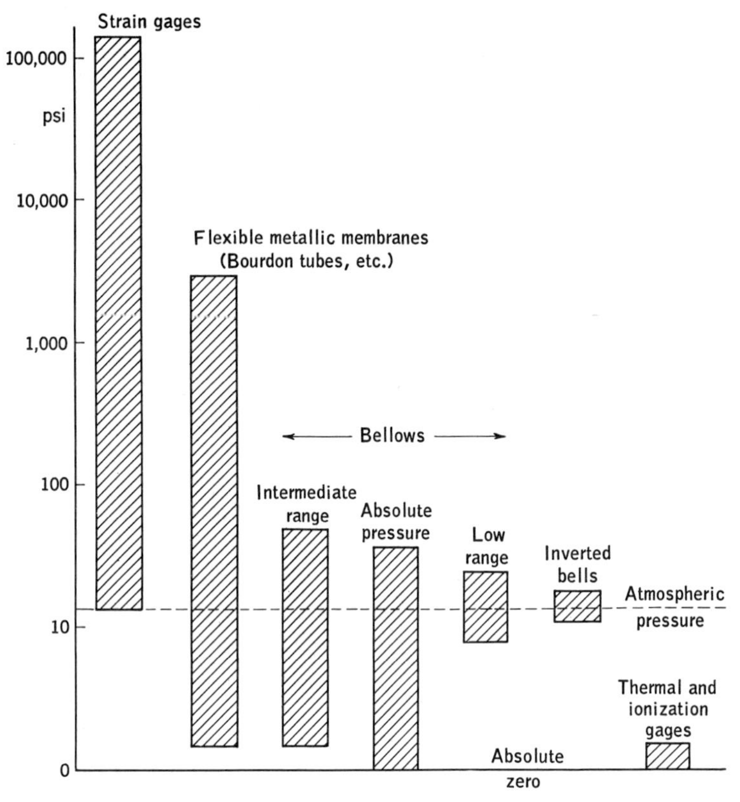

Fig. 9·1 Ranges of elements and gages. Pressure gages measure vacuums as great as 10^{-11} mm of mercury and pressures as great as 150,000 psi. This figure indicates approximate ranges and spans of the various devices for measuring pressure. (*Minneapolis-Honeywell Regulator* Co.)

tions, and the problems are not separated into two distinct groups but are treated as an entity at the end of Chap. 10.

For our analysis the following divisions of mechanical-type instruments could be made in the general order of their pressure ranges:

1. Large-tube McCloud vacuum gage
2. **U**-tube manometer, liquid column
3. Well-type manometer
4. Limp diaphragm
5. Metal diaphragm and bellows
6. Bourdon tube
   a. Spiral
   b. Helix

Still another division might be made by considering not the pressure ranges, but how the measurement is made:

1. Comparing an unknown pressure with a known pressure or force
   a. **U**-tube manometer
   b. Well-type manometer
   c. Limp diaphragm
2. Inferring the pressure from the measured deformation of an elastic material (see Chap. 3)
   a. Metal diaphragm or bellows
   b. Bourdon tube
      1) Spiral
      2) Helix

Group 1 is primarily a low-pressure group, whereas group 2 can measure pressures of only a fractional part of a pound per square inch or go as high as several hundred thousands of pounds per square inch. In most instances, the two groups find quite different uses.

PRESSURE-BALANCE GAGES

Chapter 5 states that there are many ways to balance forces. There are position balances, torque balances, and force balances with which we find unknown forces by determining the forces necessary to counteract their effect. In the field of liquids, we can also use pressure balances. In general, these gages are for the lower pressure ranges, where the pressure range may be from a few inches of water to several pounds per square inch.

It may not be generally recognized that all common types of mechanical pressure gages (to differentiate them from electrical designs) are fundamentally differential pressure gages—that is, they are designed to measure the difference of two pressures. Quite often it is not well understood that the conventional Bourdon spring responds to the difference between the internal force and the force created by the atmosphere. In a similar way, all diaphragm and bellows gages respond to the difference between the internal and external forces. With liquid manometers, it is quite easy to see that two different pressures are being applied. Our analysis then will be based on the assumption that pressure-measuring device is of the differential type, but for the special case when the external pressure is zero the gage will respond to *absolute pressure.* If the pressure applied to the outside of the bellows or tube or to one surface of a liquid manometer is other than atmospheric or zero pressure, then the instrument is what is commonly called a differential pressure gage.

Manometer theory. One of the simplest of the pressure-balance types of manometers is the *liquid-column pressure gage,* in which the pressure created by a column of liquid is used to balance the unknown pressure. If no pressure other than air pressure is applied to either leg of the manometer, Fig. 9·2, then the surfaces will be at the same height above some reference. But if air, gas, or some liquid exerts a force down on the

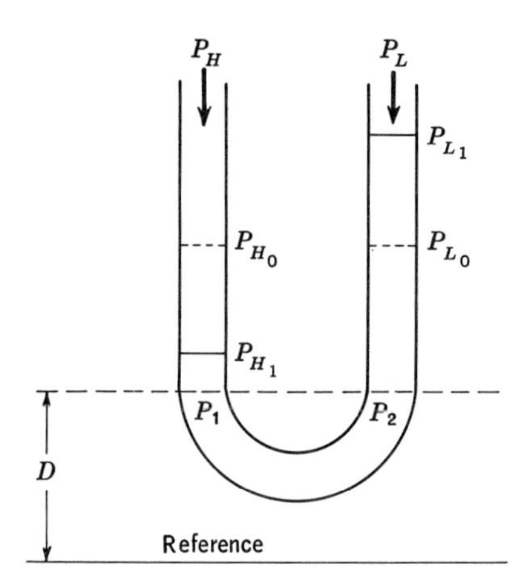

Fig. 9·2 The U-tube manometer. The U-tube manometer is a *pressure-balance device*. Any difference between P_H and P_L will cause enough liquid to move from the high-pressure leg to the other leg to bring about pressure equalities at all points at equal distances above some reference, such as P_1 and P_2.

left column, for example, liquid will be forced up in the right leg until the static balance of the pressures is again achieved. Because of the manometer construction, the columns must always alter their heights so that, at any given distance D above the reference, the pressures in the two legs must be equal. If they are not equal, there are forces available to shift the liquids to bring about the equality. It will be recalled that one of Pascal's principles states that forces at a point in a liquid will be transmitted equally and undiminished in all directions.

By using a heavy liquid, such as mercury, in the right leg, a pressure of many pounds per square inch can be created by a column only a few feet high. If a very light liquid is used, the pressure created will be proportionally less. So long as the reference liquids used are homogeneous, the pressure created by the column can be stated thus: The pressure is proportional to the density and the difference in heights of the liquid columns. That is,

$$\text{Pressure} = KdH' \qquad \text{lb per unit area}$$

where K = proportionality constant which corrects for units and other factors which do not enter directly into the expression

d = density, lb/in.^{3} or lb/ft^3 (the choice of unit of measure will affect the value of K)

H' = height of the surface of the higher column above the surface of the lower column

If water is used in the manometer and the columns have a height difference of 100 in., then the pressure applied is

$$\begin{aligned} P &= KdH \\ &= \left(\frac{1 \text{ ft}^3}{1{,}728 \text{ in.}^3}\right)\left(\frac{62.5 \text{ lb}}{\text{ft}^3}\right)(100 \text{ in.}) \\ &= 3.62 \text{ lb/in.}^2\text{, or psi} \end{aligned}$$

If d is given in pounds per cubic inch, then the expression becomes:

$$P = 1\left(\frac{\text{lb}}{\text{in.}^3}\right)(\text{in.}) = \frac{\text{lb}}{\text{in.}^2}$$

Through the use of accurate weighing systems, the weight densities of liquids can be measured accurately. Consequently, we can determine pressure most accurately by the manometer method—it is the basic or standard method of measuring pressure. All other pressure standards are referred to the liquid-column manometer. Under the best laboratory conditions, pressure can be measured with accuracies as high as one part in 10,000.

It should be noted that the area of the column has nothing to do with the pressure, because the pressure on a unit area has nothing to do with pressure on the total area. In connection with this type of manometer we should therefore talk of pressure balance rather than force balance, because quite a different concept is involved. Force is the product of a pressure and an area:

$$\text{Force} = \left(\frac{\text{lb}}{\text{in.}^2}\right)(\text{in.}^2) = \text{lb}$$

(It should be especially noted that the units used in a formula or expression must also be correct if the final units are to be so. These units are treated as algebraic units. The determination of the correct units in complex systems is called *dimensional analysis*. All answers to problems and questions should meet this test; it is not enough that the numerical answers alone be right. It makes a tremendous difference whether the answer is in pounds per square foot or pounds per square inch.)

Where the pressures are very low, the **U**-tube manometer may be

Fig. 9·3 Slant-tube manometer. Any given pressure difference between P_H and P_L will always create the same difference in *vertical* level for any given instrument and liquid. By using a slant scale, the liquid will have a greater movement over the scale for the same vertical travel. This facilitates reading.

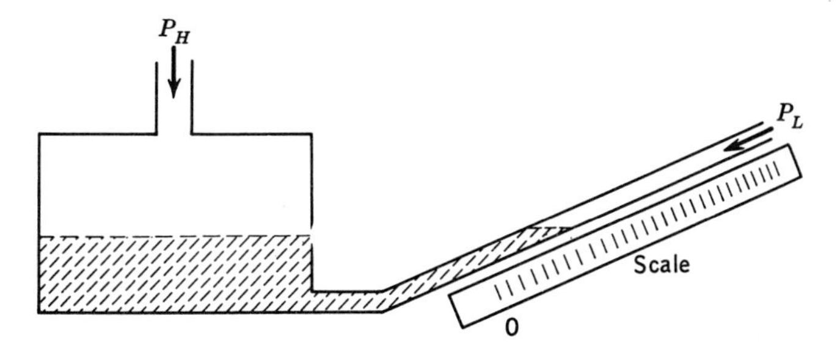

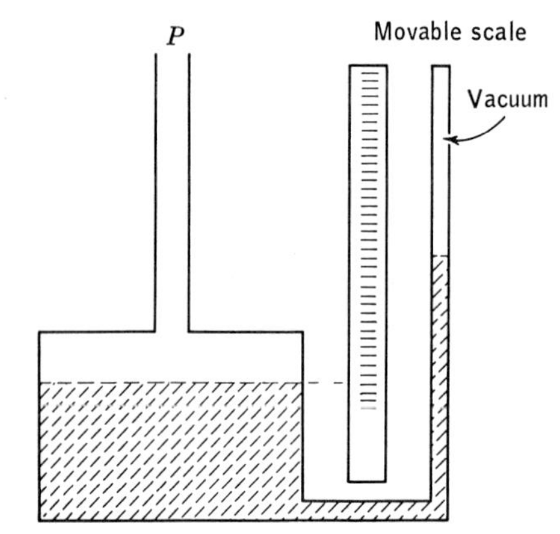

Fig. 9·4 A U-tube manometer can be converted from differential measurements to absolute pressure by evacuating and sealing one leg.

modified in construction. As shown in Fig. 9·3, the use of the slanting leg will stretch the scale by a factor of 4 for the dimensions used.

If a sealing liquid is used to isolate the gas or liquid from contact with the liquid in the manometer, then corrections will have to be made for that part of the pressure contributed by the sealing liquid. In all types of manometers, the use of sealing liquids will require corrections unless the use has been anticipated in the design.

The absolute-pressure vacuum gage. The measurement of absolute pressure is almost identical with the measurements previously considered. Not only is the top of the tube above the standard reference liquid sealed off, Fig. 9·4, but a vacuum as nearly perfect as possible is created above the sealed liquid. This removes the pressure of the atmosphere, which is 14.7 psi under standard conditions of atmosphere, elevation, and temperature (sea level and 68°F). The terms "absolute pressure" and "gage pressure" are defined and derived in Chap. 3. Absolute pressure is understood to mean the total pressure, which includes any pressure which might be contributed by the atmosphere. Gage pressure is understood to mean a pressure in excess of that contributed by the weight of the atmosphere. The abbreviations for these terms are psia and psi for pounds per square inch absolute and pounds per square inch, respectively.

The well-type manometer. In the U-tube manometer (Fig. 9·5) the application of pressure causes the liquid in one leg to go down while that in the other leg goes up. We are interested in the differences in height, so there is no fixed reference. This tends to make the measurement of the height more difficult than it would be if one surface could be maintained at some fixed level. If it could be, our scale could be read from the reference or be adjusted to compensate for any deviation. Such a fixed reference becomes of greater importance as we desire to introduce transducers of one sort or another and resort to mechanical reading instead of reading by some individual.

The well-type manometer meets the mechanical reading requirement

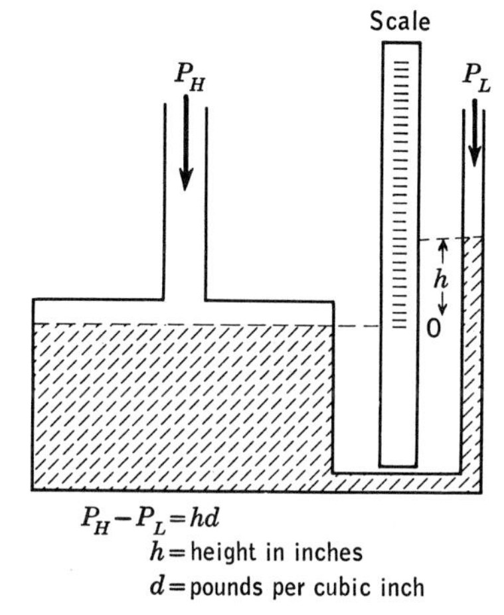

Fig. 9·5 The well type of manometer. This is a modification of the **U**-tube manometer. With a large area for the well, the level changes slightly over the normal pressure range. If P_L = atmospheric pressure, then P_H = gage pressure.

by having a relatively large reservoir for the reference leg and a small-bore tube for the indicating column. The liquid displaced from the reservoir must be moved into the tall column. If the area of the reservoir is 10 in.^{2} but that of the tube is but 0.01 in.^{2}, then a column 100 in. high will require enough liquid to lower the reservoir only 0.1 in. If the bore of the tube were made smaller, the reservoir would be lowered a smaller amount. However, there are very practical reasons for not making the tube too small.

The barometer. A barometer (Fig. 9·6), is a well-type absolute-pressure gage whose range is from zero pressure, absolute, to atmospheric pressure. The readings are generally in millimeters of mercury; high vacuums are not measured with this sort of gage. To obtain correct values for the height of the column, the surface of the liquid in the well is brought to the index by changing the volume of the reservoir. In very accurate work, correction would be made for temperature effects.

FORCE-BALANCE GAGES; NO ELASTIC DEFORMATION

Bell-type pressure gages. This type of gage, Fig. 9·7, is used extensively in industry to measure low pressures, with a range generally from 1 in. of water to 10 in. of water. By proper design to minimize friction, this gage can be made responsive to the smallest pressure variations normally encountered in industry, except for those pressure measurements which might be termed high vacuum. Because many flow measurements are inferred from low-pressure measurements, this same instrument will be met again in Chap. 11.

The theory of the bell-type gage is extremely simple. When no pressure above atmosphere or other reference pressure is applied, the bell just rests on the bottom. The counterbalance is so adjusted that with no extra pressure under the bell, the bell will just sink. As pressure is applied to the inside of the bell, the difference between the external atmos-

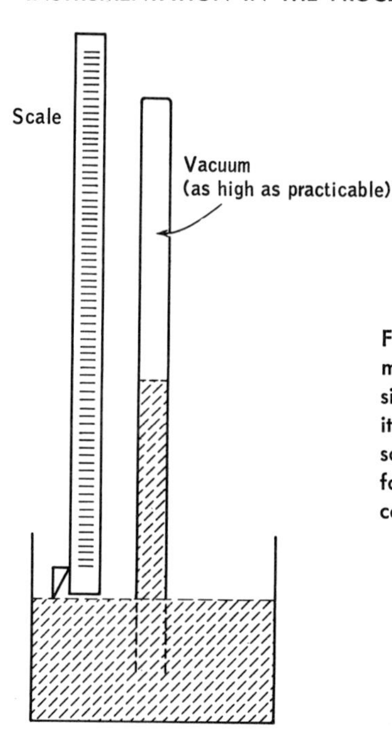

Fig. 9·6 A barometer is a special well manometer in which the low-pressure side has zero pressure (or as close to it as practicable). By adjusting the scale so that its index is always at the surface of the liquid, the height of the column is ascertained.

Fig. 9·7 A bell-type manometer. A manometer of this design can measure differential pressures if suitable connections are made at *A*. If atmospheric pressure only exists above the bell, then the measurement is that of gage pressure. But if the space above the fluid is evacuated to produce a high vacuum, then the gage will measure absolute pressure.

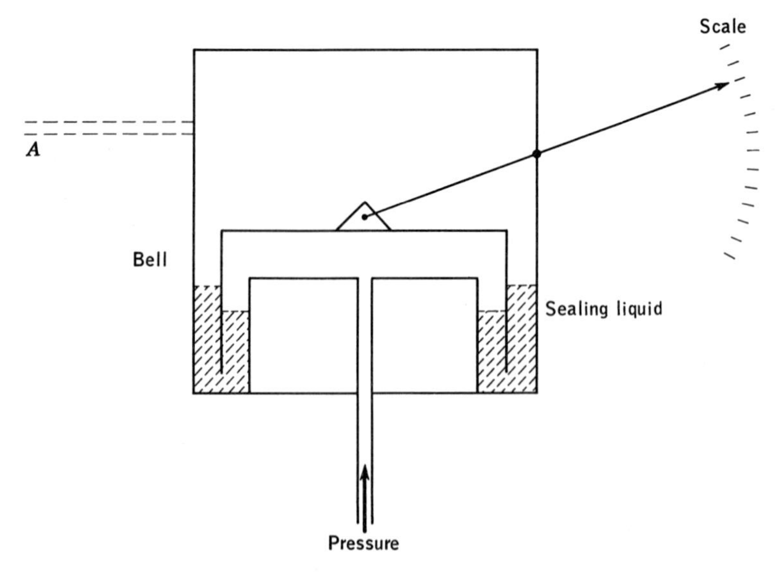

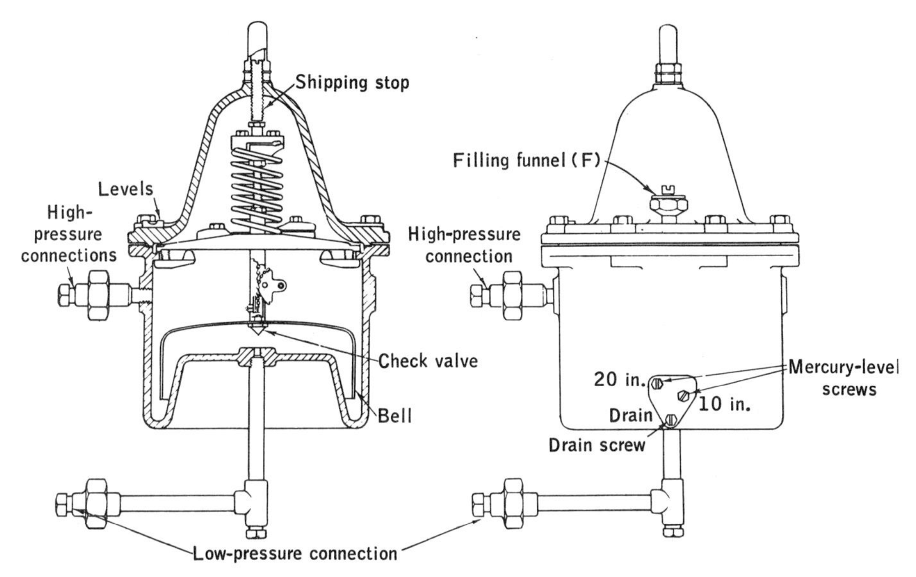

Fig. 9·8 With this design of the bell-type instrument, most of the bell will be out of the liquid when the differential pressure is zero. Increasing the differential will then force the bell down against the action of the spring and the lower pressure. Whether the pressure forces the bell up or down is of no concern in studying the characteristics of this type of instrument. (*The Foxboro Co.*)

pheric pressure and the inside pressure becomes a lifting force. As the bell leaves the liquid, it loses some of its buoyant force. It will rise until the forces balance again: the gravitational pull (weight) plus the exterior force must equal the internal lifting force plus the buoyant forces.

By proper mechanical design, the range can be governed. Actual design will employ springs, lever, and weights to establish the force systems. Figure 9·8 is used to show the operating principles; Fig. 9·9 is a section of an actual instrument.

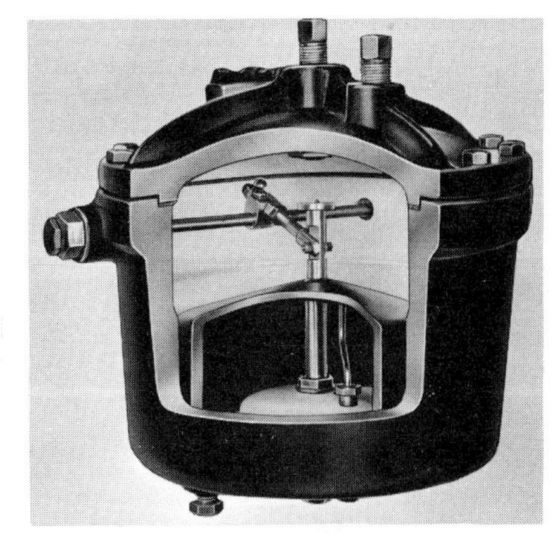

Fig. 9·9 The bell type of manometer is particularly adapted to measuring gases which are flowing under low pressures. For purposes of efficiency it is desired to reduce to a minimum the pressure losses brought about by metering devices. The normal ranges are 0–1 to 0–10 in. of water. (*American Meter Co.*)

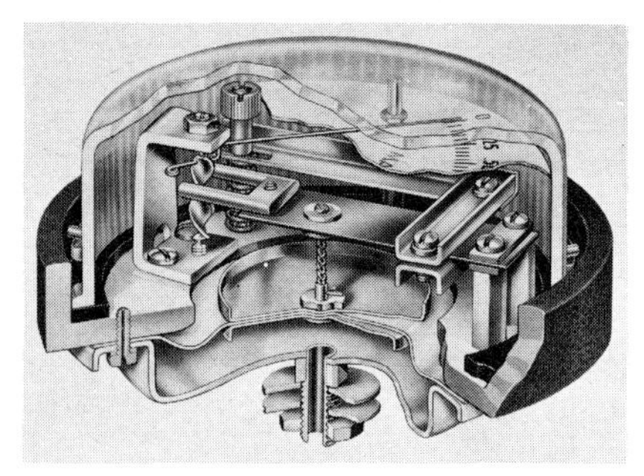

Fig. 9·10 This limp-diaphragm gage employs a unique design to convert linear motion to rotary motion, with no physical connection to the rotating element. As the **U**-shaped magnet is moved by the pressure, the soft-iron helix which carries the pointer rotates so as to keep the air gap reluctance to a minimum. (*F. W. Dwyer Mfg. Co.*)

Limp-diaphragm gage. There are many occasions in industry when it is necessary to measure pressures equivalent to those established by a column of water only a few inches high but, for reasons of room, ease of servicing, durability, and other operational conditions, the use of the liquid-filled manometer is not desired. For such installations, another type of force-balance gage has been designed. The gage is truly a force-balance unit. The pressure to be measured is applied to a piston which is sealed against leaks by a very flexible, limp material which takes almost no force to deform. The force against the piston is used to deform the range spring. As the spring is deformed, the pointer to which the spring is mechanically connected is moved. Figures 9·10 and 9·11 indicate how the commercial versions are constructed.

Reaction-type low-pressure gage. A totally different way to measure low pressures mechanically involves principles that have not previously

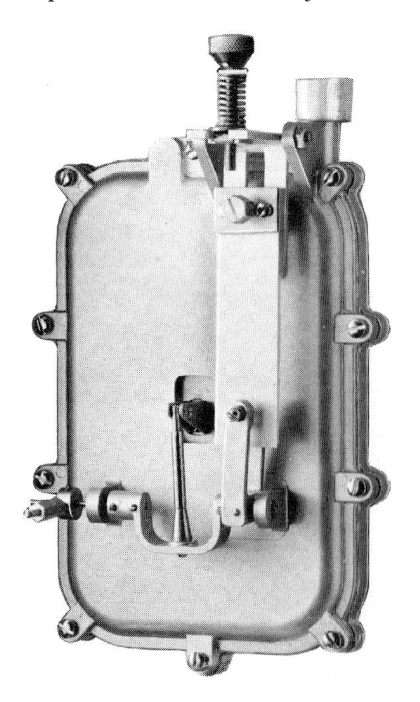

Fig. 9·11 The low-pressure gages used in powerhouses to measure draft often are of the limp-diaphragm type. Despite the low differential pressure of a few inches of water, the forces available to position pen or pointer are large because of the large diaphragm area. (*The Hays Corporation*)

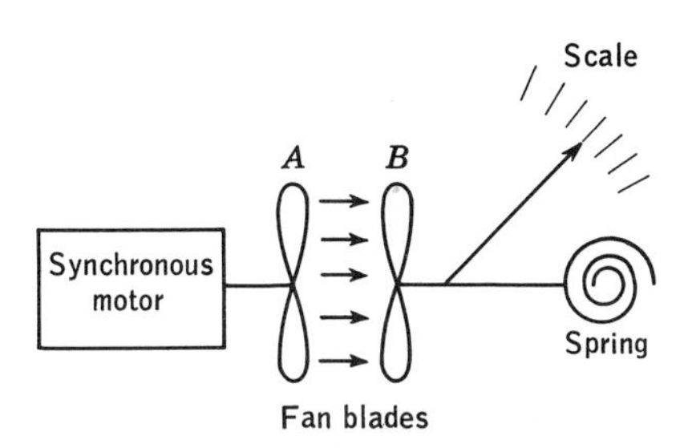

Fig. 9·12 A reaction type of pressure gage. The number of molecules of air or gas which are driven against the receiver fan blades *B* will create a force directly proportional to the number of particles present and hence proportional to the gas pressure. By using a constant-speed synchronous motor, the particles should always have the same additional speed imparted to them.

been considered but are quite basic, simple, and easily understood. For example, we have all seen fans rotate only because moving air was striking the fan blades. This is the basic principle involved in the reaction-type low-pressure gage. If we create a positive movement of air by using fan A, Fig. 9·12, to direct the movement of air against the blades of the second fan B, the second fan will try to rotate because of the force of air striking the blades. If we use a spring to restrain the motion of B, then we will get a rotation proportional to the force with which B is struck by the air.

Now let us put this arrangement in a room where the air pressure, and hence the number of molecules per cubic foot, has been reduced to half of the normal atmospheric pressure. With only half the number of molecules striking the fan blades, we might expect to develop only half the striking force. If the pressure is reduced to one-fourth of the original pressure, only one-fourth of the original number of molecules should strike the blades of B and thus develop only one-fourth of the original force and torque.

If propeller B is restrained by a very delicate hairspring and friction has been reduced almost to zero, then only a very slight torque will be required to produce rotation. This slight torque can be produced by having a relatively small number of air or gas molecules strike B. Thus, the amount of rotation of B can be used to obtain a measure of the number of molecules present and hence a measure of the pressure, which is directly proportional to the number of molecules per unit volume provided the temperature is constant (Chap. 3). It will be seen that liquids and diaphragms have been replaced by forces generated as one object strikes another and both objects undergo accelerations.

FORCE-BALANCE GAGES; ELASTIC DEFORMATION

In Chap. 3 it is repeatedly emphasized that any material will be deformed or distorted when *any* force, no matter how small or great, is applied to it. If all materials react in this manner, the amount of deformation or movement or distortion can be used as a measure of the force, and consequently of the pressure which created the force. Experi-

Fig. 9·13 Pressure elements which employ the deformation of elastic materials. The pressure-sensitive elements which employ flexible membranes take many shapes. A **C** type of Bourdon tube is limited in its movement; therefore, it uses a gear train to provide rotation of the pointer. The spiral and helical elements are capable of providing satisfactory movement for many instruments without having to resort to gear trains. The diaphragm and the bellows elements are employed most often with low-pressure systems. (*The Foxboro Co.*)

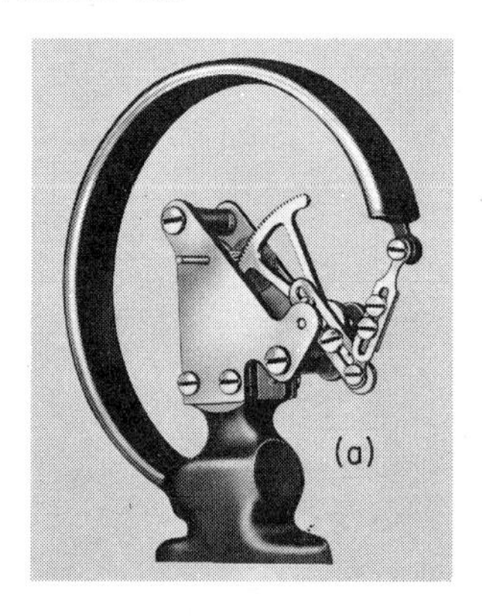

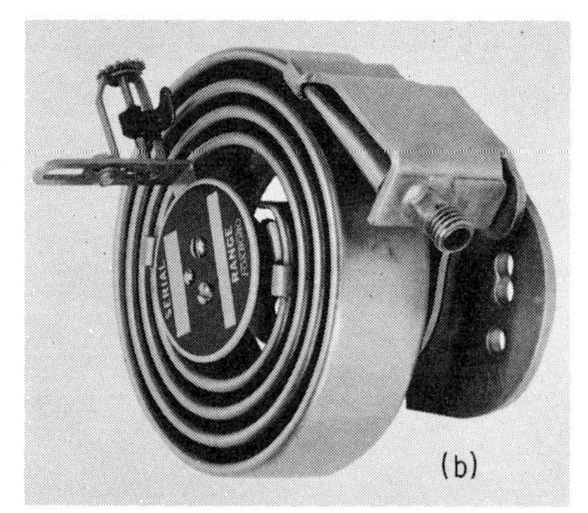

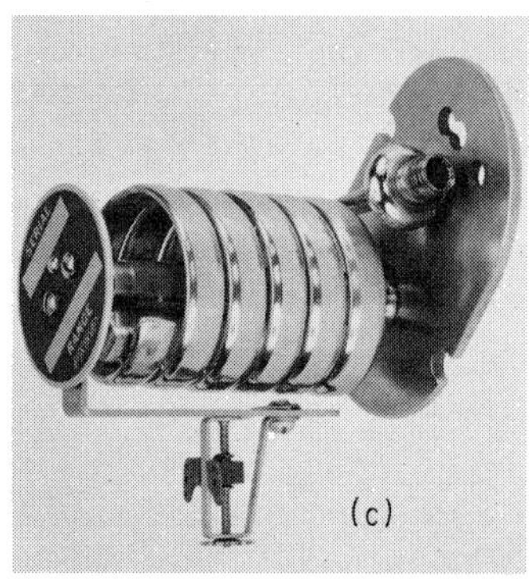

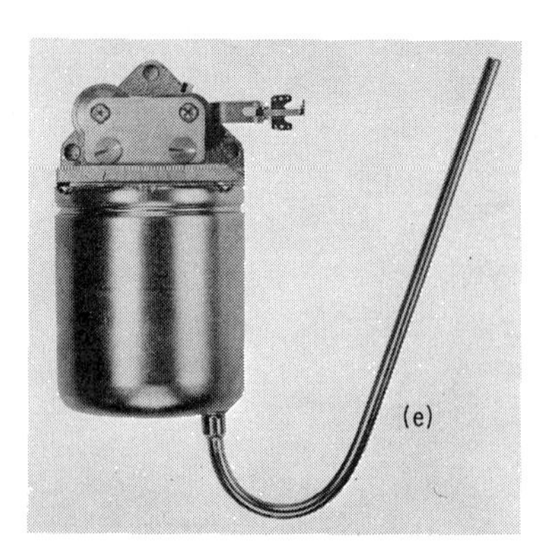

ence indicates that the most useful applications of elastic materials in pressure gages are (Fig. 9·13):

1. As a bellows
2. As a metallic diaphragm
3. As a Bourdon spring: (*a*) spiral and (*b*) helix

The above sequence is also in order of pressure ranges from low to high. Diaphragm pressure gages are used by the millions for trucks, tractors, and other heavy-duty applications. For industrial measurements, their main use is in differential pressure measurements where the movement of the diaphragm is generally small.

At the start, it should be realized that the stresses required to produce a strain of 0.001 in./in. in steel would be about 30,000 psi. The stresses generally created in most instruments are of lower value, so the strains are proportionally less. With the exception of the bellows, the change in dimension of the elastic member is quite small, and can be indicated, in almost all instances, only through the use of gear trains having a very high amplification factor.

The metallic diaphragm and the Bourdon tube also "use up" practically all of the force generated within them in changing their own dimensions. A Bourdon tube should have almost identical movements with and without its gear train. Because almost all of the pressure variations supplied to the gage are used in deforming the elastic member, all gear trains or other indicating elements must have an absolute minimum of friction. Friction in any part will prevent an accurate indication; it will tend to cause low readings when the pressure is increasing and high readings when the pressure is decreasing.

All three of the forms of elastic materials listed above function as springs. From personal experience, we know that a spring changes its dimensions with the applied force. Friction from any source will produce erroneous results. For example, we know that if the pointer attached to a spring rubbed against the scale, the indications would be incorrect because the friction would work against the positioning of the spring. The springs have queer shapes perhaps, but their behavior in all respects follows that of springs in that highly elastic materials have:

1. Deformations proportional to the applied force
2. Very little hysteresis
3. Small strains, only a small fraction of a thousandth of an inch per inch

Consequently, metallic diaphragms and Bourdon tubes should never be expected to do any work, that is, exert any appreciable force through a distance, if any degree of accuracy is expected of them. The elimination of friction to the greatest degree practicable is characteristic of every design that uses these components.

Fig. 9·14 Bellows and their associated elements may take many forms. The pressure, the length of stroke, the operating conditions, and the desired life will dictate the design. (*Robertshaw-Fulton Controls* Co.)

The bellows. A metallic bellows, Fig. 9·14, is a series of circular parts so formed or joined that they can be expanded axially by pressure but not appreciably in any other direction. Because the materials can be of fairly light gage and may have large effective areas, they can develop far larger forces than are needed to strain them through a desirable distance. Figure 9·15 is a nomograph which shows how large deflections cut down on the cyclic life of a bellows.

Inasmuch as it is desirable to limit the travel of the bellows, a range spring is generally employed so that the bellows works against it. Both members function as high-grade springs, so all linear relationships are retained.

The type of service will determine the material to be used, and the material probably will dictate the method of fabrication of the bellows. Fabrication generally employs one of the following:

1. Turning from solid stock
2. Soldering, or welding, stamped annular rings
3. Rolling tubing
4. Hydraulically forming drawn tubing

By increasing the number of convolutions in the bellows, the axial movement or stroke can be increased. Increasing the diameter of the bellows will increase the force for a given pressure or, for a given diameter, make it possible to measure small pressures. Bellows, therefore, are

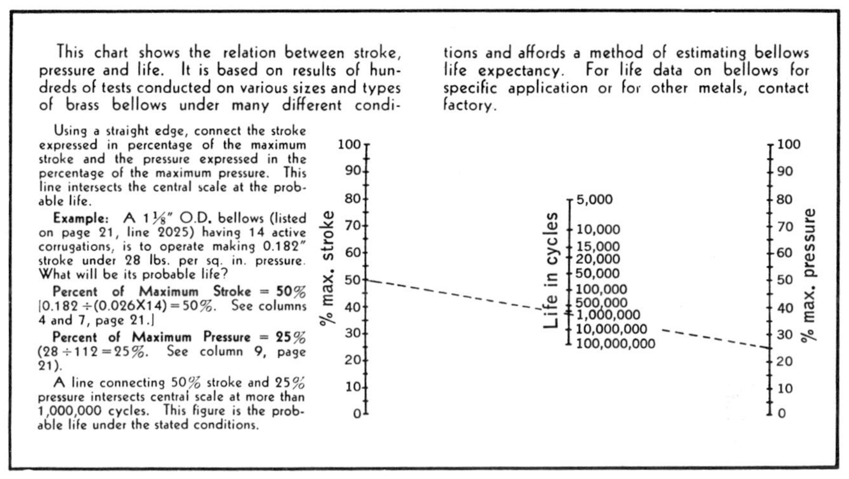

Fig. 9·15 Nomograph relating life in cycles, per cent maximum pressure, and per cent maximum stroke. The pressure to which the bellows is subjected and the length of the stroke will govern the life expectancy of a design. For any given set of conditions, the anticipated life may be determined from the nomograph. (*Robertshaw-Fulton Controls* Co.)

used in the span from 6 to 8 oz/in.^{2} to 50 to 75 psi. In practice, the useful stroke normally is limited to 5 to 10 per cent of the length of the bellows.

Choice of materials for metallic bellows. The type of service will govern the material used. There are many factors to be considered:

1. Strength, i.e., the pressures to be used
2. Hysteresis
3. Fatigue, i.e., the cycles that will constitute the normal life of a unit
4. Corrosion resistance of the material or means of keeping corrosive contaminants from reaching the bellows
5. Ease of fabrication, i.e., the possibility of making a bellows out of the materials that satisfy conditions (1) to (4)

Application of bellows in pressure gages. By using bellows with large areas and many convolutions, very low pressures will create appreciable forces which may be balanced by the spring action of the bellows alone or by using a spring to reinforce the spring rate (resistance) of the bellows. By proper choice of material, convolutions, and calibrating spring, pressures of fractional parts of 1 oz/in.^{2} to 50 to 75 psi may be measured. This construction is quite adequate for simple pressure gages. It should be noted that atmospheric pressure is acting on the outside of the bellows so the bellows responds only to pressures in excess of atmospheric. In its simple form, Fig. 9·16, the bellows is really a differential type of instrument.

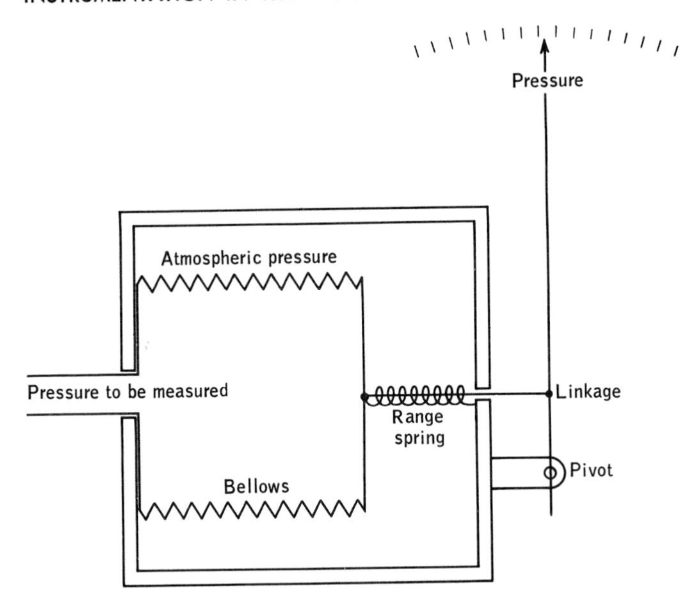

Fig. 9·16 Simplified section of a bellows-type pressure gage for low-pressure use.

Absolute-pressure gage. When measuring low pressures the variations in the atmospheric pressure may be objectionable. To overcome this difficulty, the bellows may be enclosed in an airtight housing from which practically all air has been removed, Fig. 9·17. Under these conditions, the bellows will have pressure only on the inside and will respond to the one pressure only.

Differential pressure gage. Often when making pressure measurements, particularly those related to gas and liquid flow, we are interested only in a differential pressure, not in the pressures themselves. This is true because the pressure differential is used to infer the flow rate.

If in place of a vacuum, a pressure of 50 psi were applied to the outside of the bellows, Fig. 9·18, then it would take pressures greater than 50 psi to balance the external force. We will assume that P_2 will always exceed P_1 if the meter is to read upscale. Therefore, all values of P_2 which are less than 50 psi could not be read, or would be *suppressed*. Such construction could produce a span of 50 to 60 psi, or 50 to 70 psi, and so on. A large scale for a small span facilitates easy reading. It should be recognized, however, that it does little good to find a pointer position with a very high degree of accuracy when the inherent accuracy of the instrument does not warrant any such degree of precision. The use of short spans may be useful in getting repeated pressure conditions rather than precise values.

The basic principles of the absolute-pressure gage indicate that it is

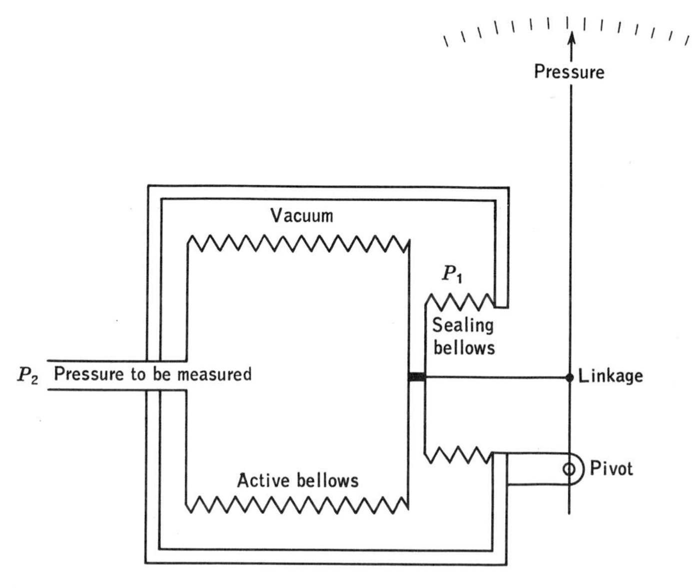

Fig. 9·17 A section through a schematic of an absolute-pressure gage.

really a differential pressure gage in which the pressure on the low side has been reduced to zero.

For the higher pressure ranges, the bellows is unsatisfactory for reasons of mechanical strength and is replaced by the Bourdon tube, which has

Fig. 9·18 Schematic of a differential pressure gage employing bellows.

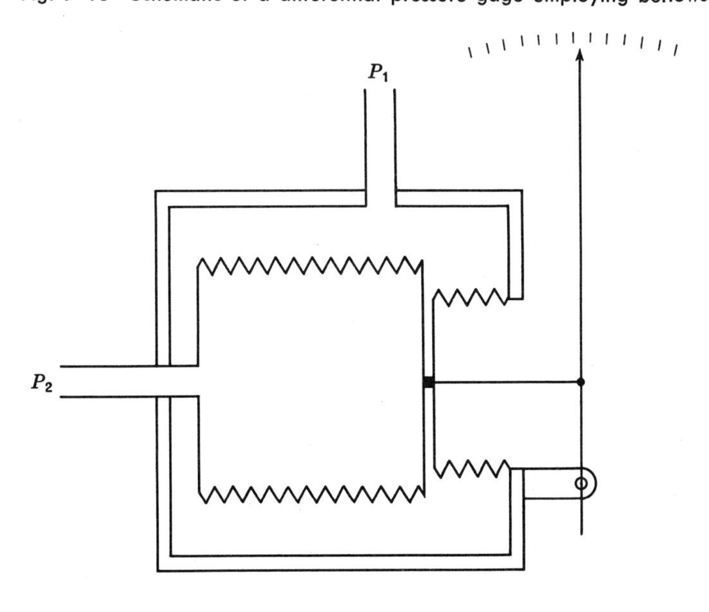

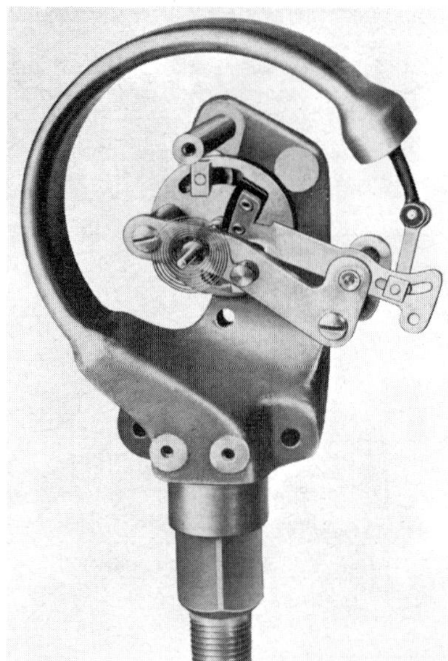

Fig. 9·19 Bourdon-tube pressure gage. The principal components are:

1. One-piece drop forging
2. Socket and tube welded together
3. The Bourdon tube, made from material suitable for the liquid and pressure
4. A forged tip which closes the tube
5. The connecting link, which transmits the motion of the tip of the tube to the linear-rotary motion unit
6. Linear-rotary Helicoid unit (Fig. 9·24)

(*American Chain and Cable, Helicoid Gage Division.*)

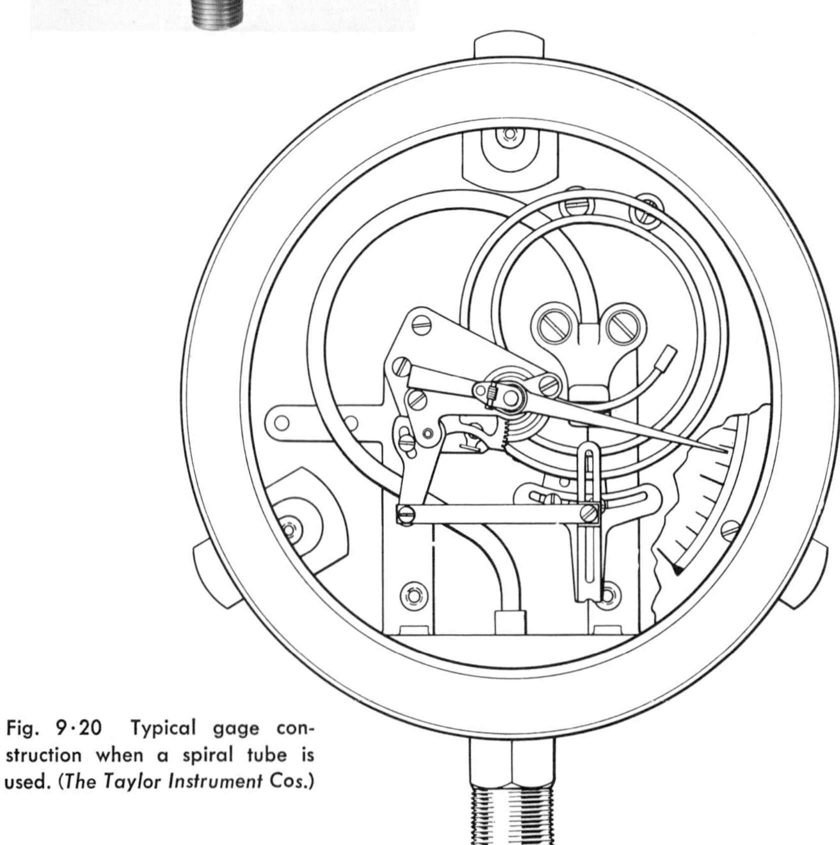

Fig. 9·20 Typical gage construction when a spiral tube is used. (*The Taylor Instrument Cos.*)

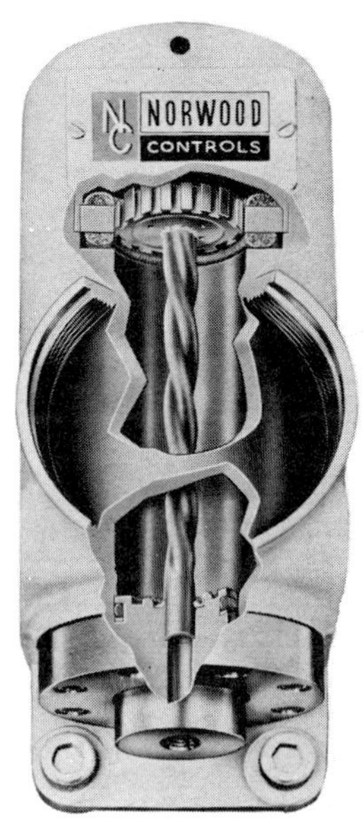

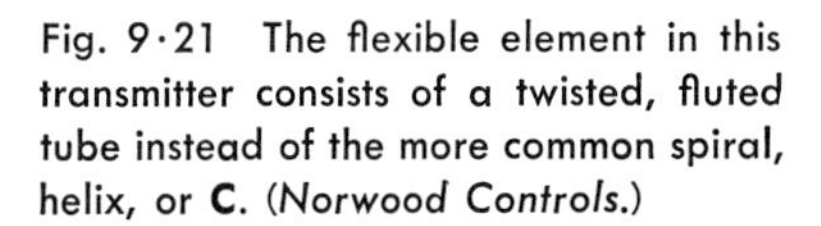
Fig. 9·21 The flexible element in this transmitter consists of a twisted, fluted tube instead of the more common spiral, helix, or **C**. (*Norwood Controls.*)

a greatly reduced area for the pressure to work against and a far higher spring rate than the bellows has.

The Bourdon tube pressure gage. The basic design of the Bourdon tube was produced in France over a hundred years ago, and it has been changed very little since then. The original patent in 1852 described it as a curved or twisted tube whose transverse section is not circular. The application of internal pressure causes the tube to unwind, or straighten out. The movement of the free end is transmitted to a pointer or other indicating element. This spring is so stiff that almost all of the internal force is used to deform or stretch the material. Unlike the bellows, no external spring is used to alter or control the movement of the free end. Because virtually no external force is available to actuate the indicating mechanism, the pointer and its gear train are made to have a minimum of friction.

Figures 9·19, 9·21, and 9·22 show a conventional Bourdon tube, a very rigid twisted tube used with an electrical transducer, and a helical unit to increase the rotation and the mechanical rigidity of a high-pressure unit. Figure 9·23 shows how the conventional gear train differs from a helical movement which contains no gears, Fig. 9·24.

When it is desired to get more movement than can be obtained from a single turn of a Bourdon tube, the tube may be constructed as a spiral, Fig. 9·20, or helix. The heavy-duty helix, Fig. 9·22, has the movement required to position a pointer or pen without the use of delicate gears.

Superficially, the theory of the Bourdon tube is very simple, but in practice its analysis can be very complex and somewhat beyond the scope of this book.

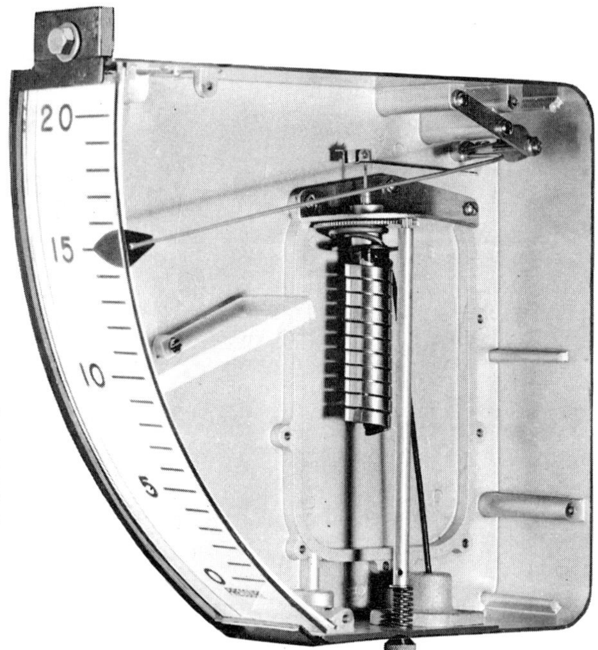

Fig. 9·22 The helical-spring member is not restricted to use with circular charts or round-faced dials. (*The Hagan Corporation.*)

Fig. 9·23 The pinion and sector are employed in many gages to amplify the movement of the Bourdon tube.

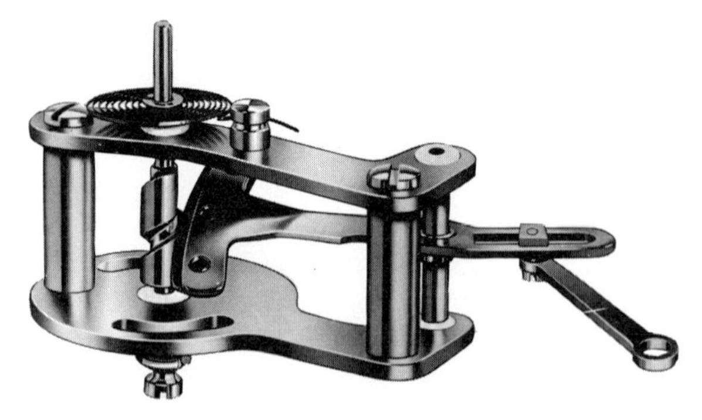

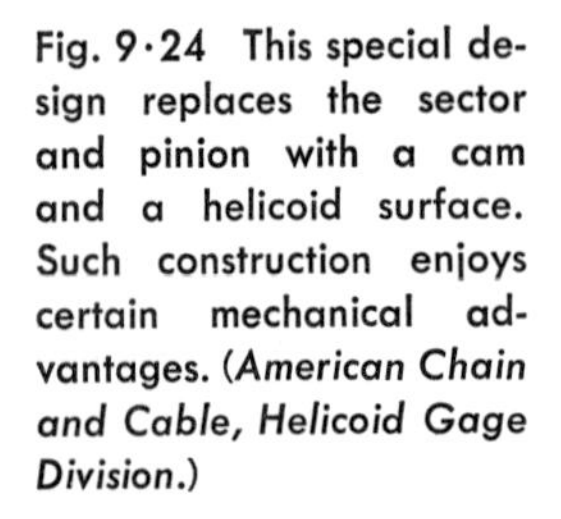

Fig. 9·24 This special design replaces the sector and pinion with a cam and a helicoid surface. Such construction enjoys certain mechanical advantages. (*American Chain and Cable, Helicoid Gage Division.*)

Choice of Bourdon tube materials. The pressure range, the number of operating cycles, formability, and the material which exerts the pressure on the inside of the tube will dictate the material to be used in tube construction. Phosphor bronze, beryllium copper, steel, chrome-alloy steels, and stainless steels are generally used. For the lower pressure ranges, phosphor bronze is generally preferred; where corrosion is a problem, stainless steel may be employed. Where stainless steel will not meet the gage requirements but protection against corrosion or viscous liquids is necessary, an appropriate fluid or membrane is used to protect the gage. The uses of such seals and the precautions to be followed in their employment will be found in the Installation and Application Notes at the end of this chapter.

COMMERCIAL DIFFERENTIAL PRESSURE GAGES

While the direct measurement of pressure is one of the most common industrial measurements, oftentimes there are conditions that require that the differences in two pressures be determined. There are two approaches to that problem:

1. We could measure each pressure and then subtract one from the other.
2. The difference might be secured through arrangement and mechanical linkages. (Certain types of temperature compensation designs do this.)

Because differential pressure gages are normally designed for pressures far higher than the differential values, they do not have much sensitivity to small pressure changes of the sort encountered in level and flow measurements.

Types of differential gages. We have already used a **U**-tube manometer for pressure measurements. This basic tube with certain modifications to improve its usefulness is used for many differential pressure measurements. Other designs use the same basic principles although their shape and dimensions may be quite different.

1. **U**-tube manometer
2. **U**-tube modifications
   a. Well manometer
   b. Inclined tube (well type)
   c. Enlarged-leg float manometer
   d. Ring manometer
   e. Tilting **U**-tube manometer
3. Bell differential pressure
4. LeDoux bell
5. Differential pressure bellows
6. Two Bourdon tubes
7. Differential pressure transmitters

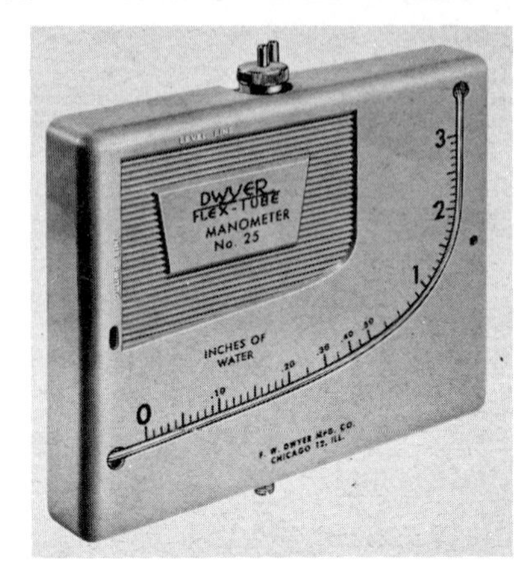

Fig. 9·25 By employing a variable slope, easy readability can be obtained at low pressures; yet significant spans are available. (*F. W. Dwyer Mfg. Co.*)

Well-type manometer. If a **U**-tube differential pressure manometer were constructed as shown in Fig. 9·5 with a large reservoir for the liquid, we would find that for small sizes of tubing for the vertical legs (not less than ¼ in.) the liquid level of the reservoir would change hardly at all. The zero of such a scale becomes relatively fixed for all except the most accurate measurements.

If we always connect the source of higher pressure above the reservoir,

Fig. 9·26 Schematic diagram of a commercial differential pressure gage using a liquid. As the liquid is moved through a large distance in the high-pressure leg, the level of the low-pressure leg is raised only a moderate amount. The movement of the float is then a measure of the differential pressure. There must be a pressure-tight seal where the shaft passes through the housing.

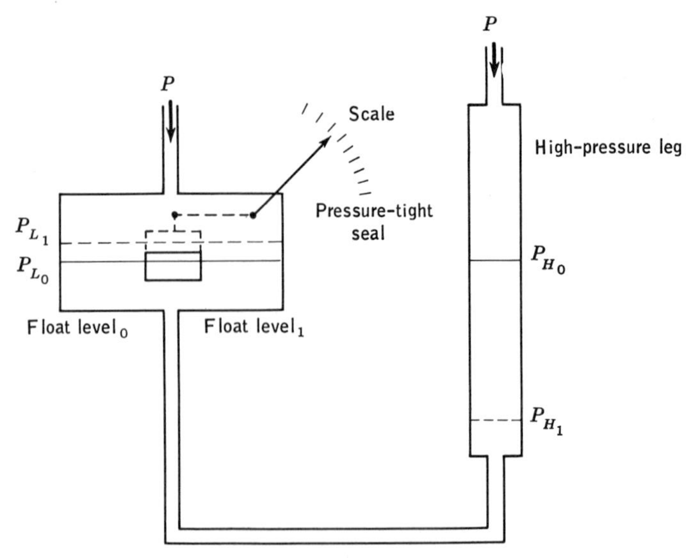

the liquid never will rise in that tube, so we can dispense with it. To simplify the reading of small pressure differentials, the tube may be inclined. This may spread out a vertical range of 2 to 3 in. so that it may be read over a slant distance of 12 to 15 in., Fig. 9·25. The vertical height that the liquid rises is the same in the inclined tube as in the conventional vertical leg, but an extended scale may help. In all such cases, the manometer must be carefully leveled or there will be very appreciable errors.

Figures 9·26 and 9·27 show how one commercial version of the **U**-tube differential pressure gage is actually constructed. A metal float is supported by the mercury in the reservoir. A higher differential pressure forces liquid down and out of the high-pressure leg; the mercury then rises in the reservoir, carrying the float with it. A shaft through a pressure-tight bushing carries the float movement to a pointer. By changing the volume of the high-pressure leg, the change in reservoir level for any given differential pressure can be controlled. By using the proper tube, pressure differences of 20, 25, 50, 100, and even 200 in. of water may be measured. By this general design, the mercury level in one chamber may move through a relatively large distance but the float has only a limited movement.

It should be remembered that although the measurements may be referred to as so many inches of water, they are almost universally made by using mercury in this design of instrument. A column of water 200 in. high would be 16.6 ft high. This would be most awkward to install. By using mercury, which is 13.6 times as heavy as water, the column can be reduced to about 14.5 in.

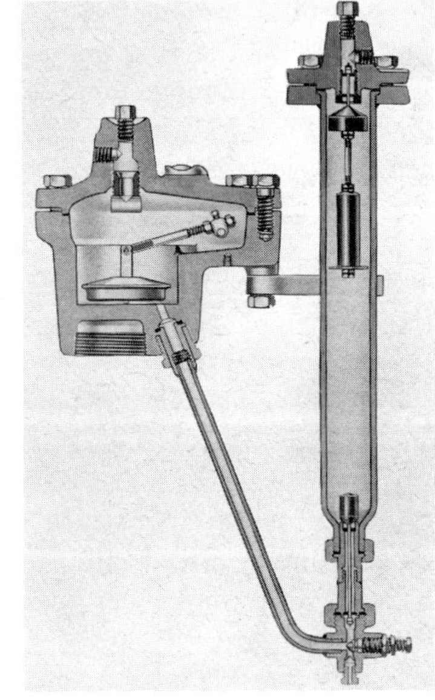

Fig. 9·27 Cutaway section of a commercial differential pressure gage. (*American Meter Co.*)

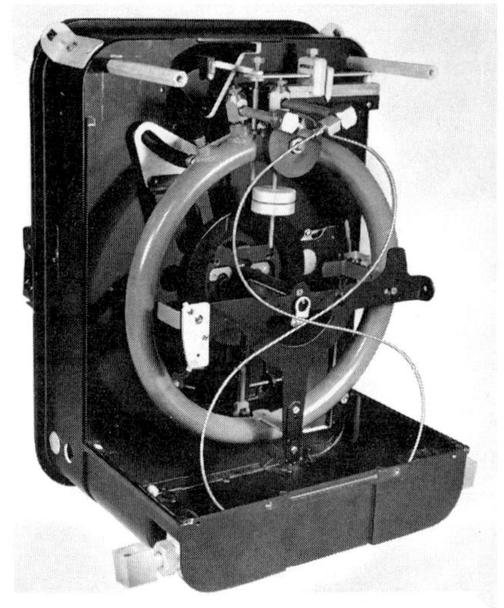

Fig. 9·28 A commercial design of a ring differential pressure gage. (*The Hagan Corporation.*)

Ring manometer. If the **U** tube is changed to a ring, we have another manometer which has a wide use in industry, Fig. 9·28. With equal pressures, the two mercury legs are level. But if the differential pressure becomes some finite amount, then the higher pressure will drive some mercury out of the high-pressure leg and into the low-pressure leg. Under these conditions, the low-pressure side will now weigh more than the high-pressure one. If the ring is balanced on a fulcrum, some other force must now be applied to maintain a condition of static balance. The magnitude of the restoring force and how it is applied is indicative of the pressure differential and characterizes the various designs.

Tilting U-tube manometer. The tilting **U** tube is just a special form of the **U**-tube or the ring manometer. In theory, it is like any other of the liquid-filled manometers. It and the ring manometer have certain advantages when we wish to convert the differential pressure into a pneumatic or an electrical signal.

PRESSURE TRANSMITTERS

Oftentimes it is desired to read pressure values at some point remote from the instrument. When the methods are to be nonelectrical, a pressure transmitter employing the basic features shown in Fig. 9·29 will be used. The pressure being measured moves a beam which is pivoted about F. An increase in the measured pressure causes the bellows P to expand, thereby moving beam B and its baffle closer to orifice O. As this happens, the pressure within the orifice housing increases. Consequently, air at higher pressure is forced into the balancing bellows BB, thereby causing it to expand and move the beam away from the orifice until a torque balance is obtained.

The range spring assists the balancing bellows BB in counteracting

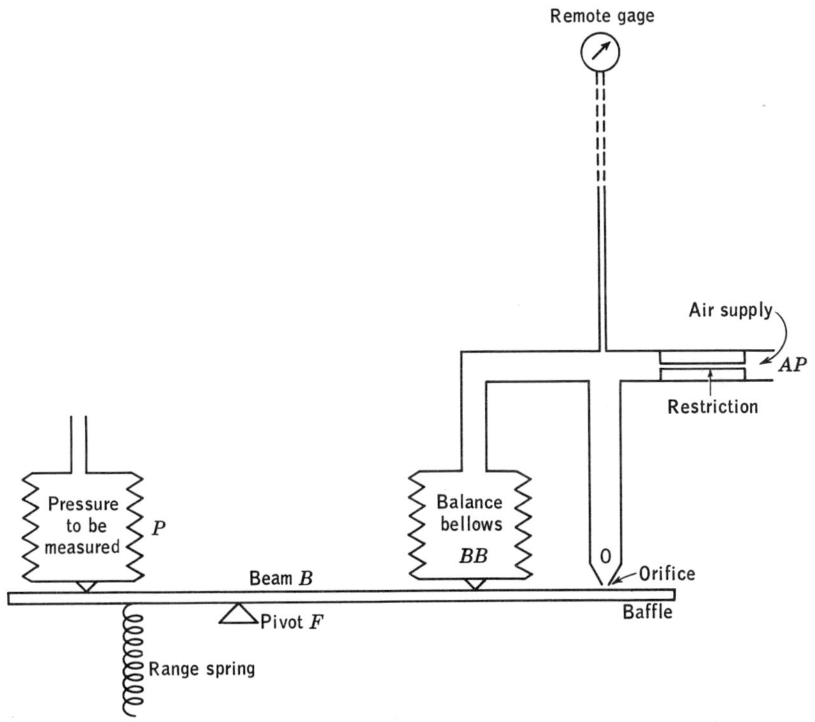

Fig. 9·29 Schematic of a pneumatic pressure transmitter. Most transmitters employing pneumatic means are based on the principles indicated here.

the movement of bellows *P*. A *torque balance* is obtained with a very small movement of the beam with respect to the orifice. A few thousandths of an inch is enough to change the pressure at the orifice from 1 to 2 psi to the maximum pressure at *AP*. By making the beam *B* long and mounting the orifice several inches from the fulcrum, it may be said that the balance is achieved with virtually no movement of the beam.

When the transmitter must also indicate the value of the variable being transmitted, then such a mechanical arrangement as is shown in Fig. 9·30 might be used. As pressure changes move the Bourdon tube, the indicating pointer will change position. As it does so, the baffle clearance is modified. The ensuing pressure changes within the system then cause the baffle to be repositioned with respect to the baffle. The pressure required to obtain equilibrium then becomes a measure of the input variable and is transmitted as being an accurate measure of the variable.

By various mechanical arrangements the movement of the baffle may be amplified 30 to 50 times, Fig. 9·31. With such construction, extremely small movements of the beam, and thus very small pressure changes in pressure, can be detected. We don't care what causes the changing pres-

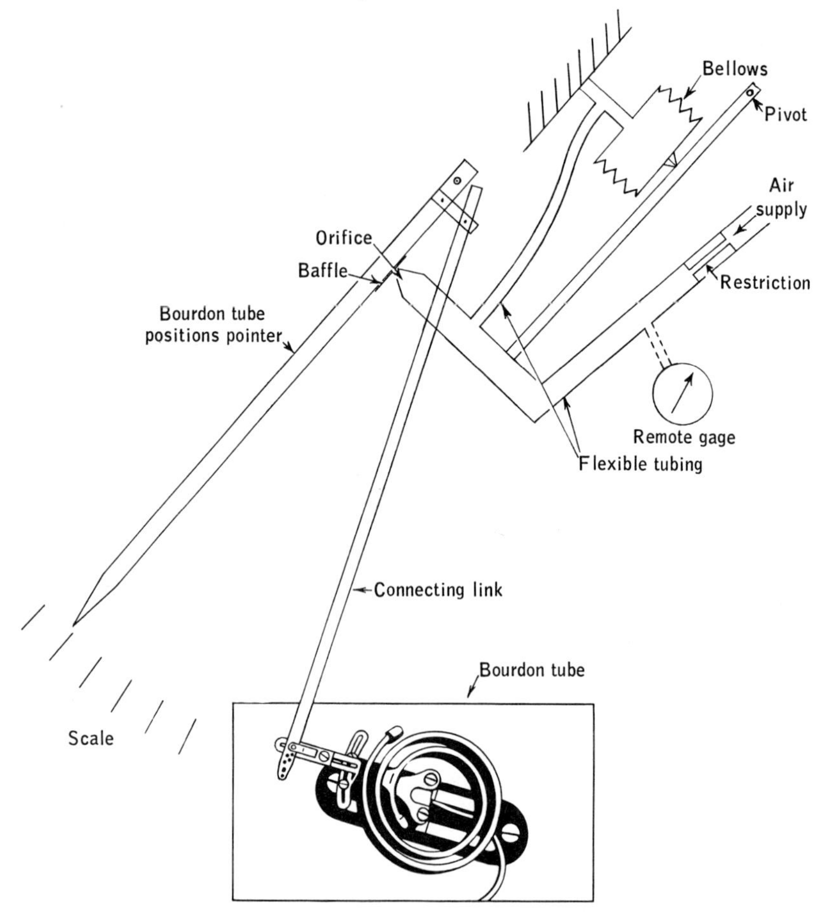

Fig. 9·30 Schematic of indicating pressure transmitter. As the pressure-sensitive element moves the pointer, the pressure at the orifice changes—decreasing when the pointer moves to the left and increasing when the pointer moves to the right. The bellows then has a change of pressure which causes the orifice to be brought to its position near the pointer to reestablish equilibrium. The pressure within the measuring system may be created by temperature effects or by liquid level, or it may result from weighing a tank or some other load. The source of the pressure is immaterial. (*The Taylor Instrument Cos.*)

sure in P; it may be an expanding liquid in a thermometer bulb or the pressure of a head of liquid at the bottom of a tank.

In Fig. 9·32 we find a pressure transmitter which employes a Bourdon spring and a filled system. The Bourdon tube spring moves the pointer to reposition the baffle with respect to the nozzle. The nozzle back pressure is then amplified in a reverse-acting relay and becomes the output pressure. This pressure also goes to the feedback bellows and acts as a restoring torque to bring the nozzle and baffle back to their original positions.

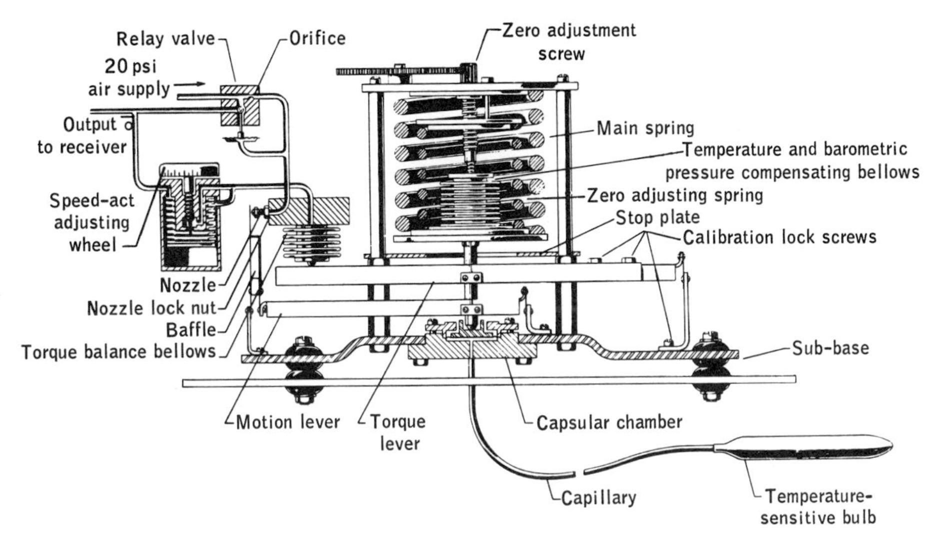

Fig. 9·31 High-sensitivity pneumatic transmitter. This transmitter will work, in basic design, with any variable which will create a pressure. Particular attention should be given to the construction and the connection to the motion lever of the baffle. This particular design provides an amplification of about 50 times without using gears. This makes it possible to sense movements of the motion lever which are of the order of 20 to 30 millionths of an inch. (*The Taylor Instrument* Cos.)

Fig. 9·32 As the movement of the Bourdon tube positions the pointer and its connected baffle, the pressure transmitted to the relay is controlled. An output pressure which is directly proportional to the relay pressure is then transmitted to the remote indicating element. At the receiver the pen or pointer will move an amount directly proportional to the transmitted pressure. (*The Taylor Instrument* Cos.)

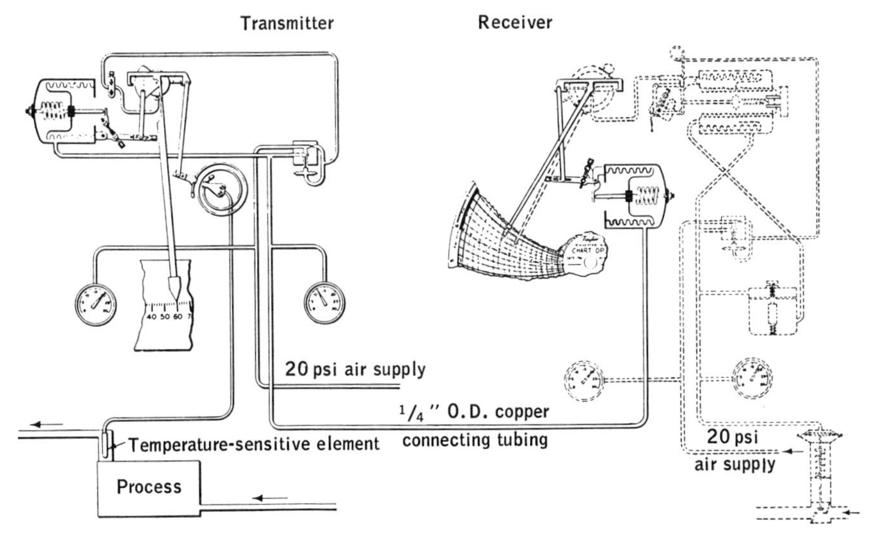

The feedback elements and the pressure-conversion components are somewhat different in Fig. 9·32. But in all cases of the pneumatic transmitters, any change of baffle position caused by the variable will initiate such action that the relative position of baffle and nozzle is maintained.

Differential pressure transmitters. If it is desired to measure and transmit differential pressures (liquid-level, interface-measurement, and flow problems are often approached by using differential pressure measurements), the transmitter is quite similar in construction to that described in the preceding section, Fig. 9·33.

In this instance, it would bc quite easy to change the range of the pressure differential by moving the fulcrum. Commercial versions of such a transmitter are shown in Figs. 9·34 to 9·38. In the commercial design, a metal diaphragm has been substituted for the bellows because it must be able to withstand pressures as high as 3,000 to 5,000 psi. Should either the high- or low-pressure lines to the diaphragm become clogged, the full line pressure would be applied to only one side of the diaphragm. In the

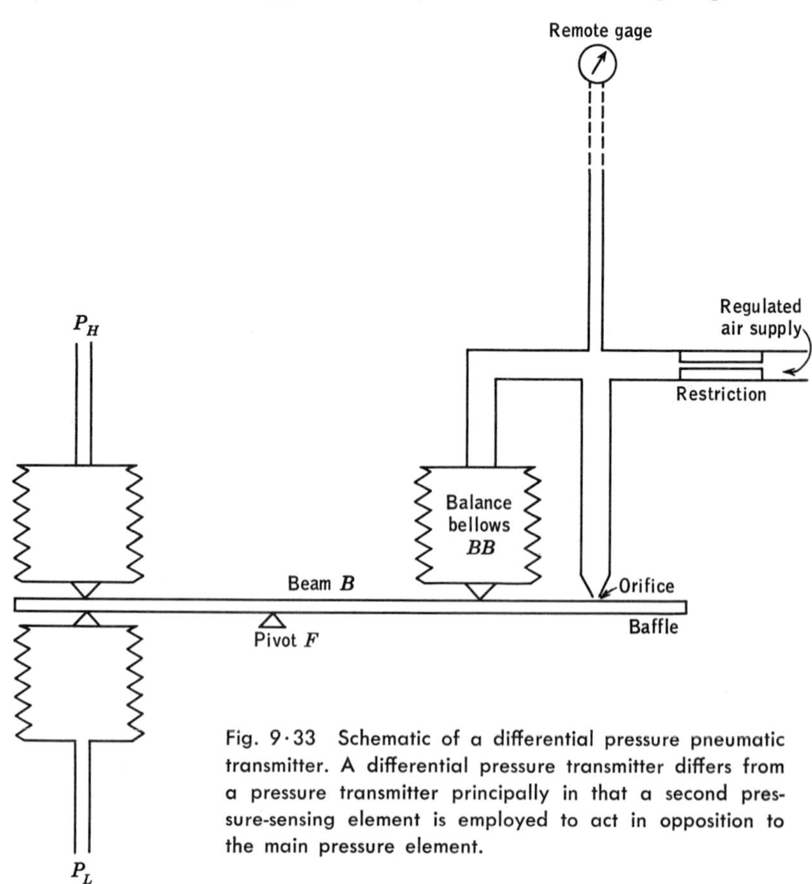

Fig. 9·33 Schematic of a differential pressure pneumatic transmitter. A differential pressure transmitter differs from a pressure transmitter principally in that a second pressure-sensing element is employed to act in opposition to the main pressure element.

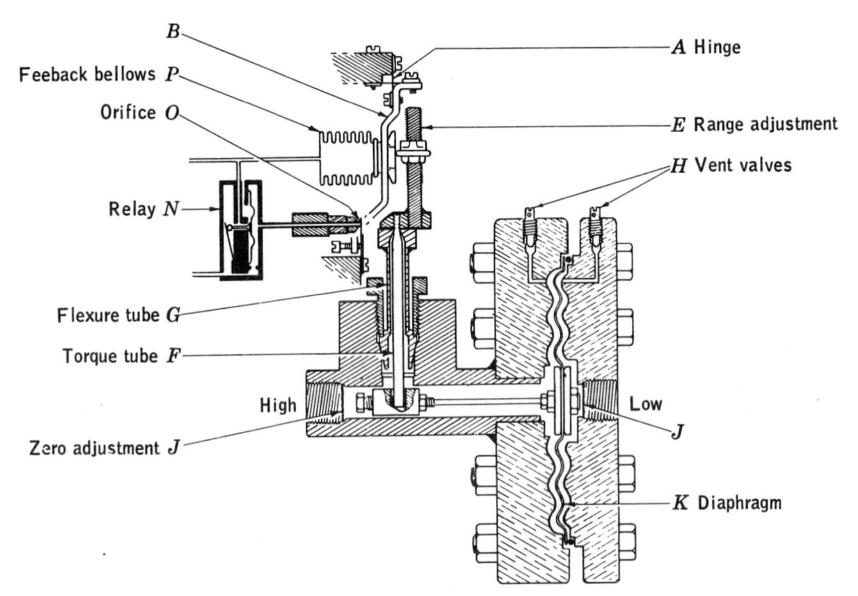

Fig. 9·34 Cutaway view of a pneumatic type of differential pressure transmitter. (*The Foxboro Co.*)

instrument shown in Fig. 9·34 the diaphragm will be forced into mating portions of the body, the body being capable of taking the forces created. The differential pressure transmitter, like its counterpart the pressure transmitter, is of the torque-balance type, with extremely restricted movement.

The torque-balance differential pressure pneumatic transmitter shown in Fig. 9·34 employs the movement of the diaphragm to the right (the high pressure is on the left) to control the air pressure to the nozzle. This controlled pressure, acting with the pneumatic relay, supplies air to the bellows. An increase in the baffle-nozzle pressure will produce an increase in the air pressure in the bellows. As the bellows pushes the hinged member away from the orifice, a condition will be reached which will produce torque equality and, consequently, equilibrium.

A somewhat similar type of torque-balance differential transmitter is shown in Fig. 9·35. In operation, the flapper-nozzle combination measures the degree to which torque equality has been achieved. If an increase in the differential pressure forces the twin-diaphragm capsule A to the right, then the force bar B rotates counterclockwise about pivot E, which is formed from a special diaphragm. As the clearance between the nozzle and the baffle is decreased, the air pressure at the nozzle is increased, and this causes the air pressure in bellows G also to increase. With an increase in the output air pressure, a greater clockwise torque about fulcrum F is generated. This continues until the two opposing

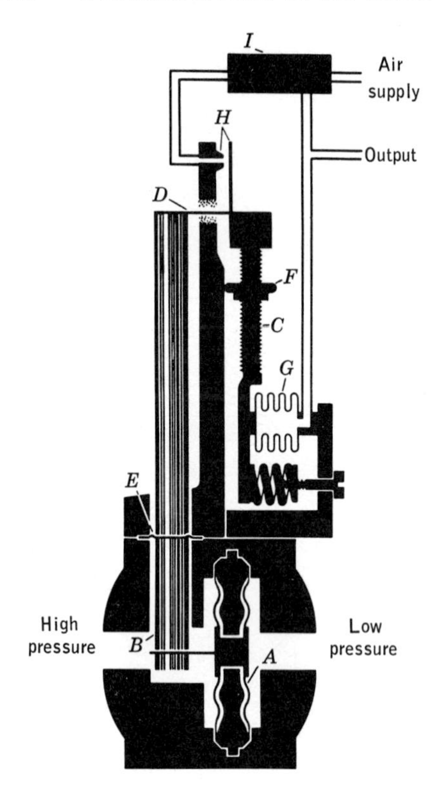

Fig. 9·35 Schematic representation of the important elements of a pneumatic differential pressure transmitter. Note the torque-balance type of feedback to ensure that the output signal is an accurate measure of the differential pressure. (*The Foxboro Co.*)

torques become equal. The construction shown in Fig. 9·36 employs the same principles, but the mechanical components differ in details.

Quite a different design of a differential pressure transmitter is sketched in Fig. 9·37. This design is of the *force-balance* type in contrast to the torque-balance principles previously considered. The measured fluid received in the process side of the transmitter exerts a force on the sensing diaphragm proportional to the differential pressure. This force moves the stack, thereby establishing the distance between baffle and nozzle. As the clearance between nozzle and baffle is decreased, the greater pressure that is developed tries to force the feedback diaphragm to the right in opposition to the force created by the differential pressure which acts on the sensing diaphragm. A condition of force balance is soon reached, with the output air pressure being a direct measure of the differential pressure. It will be noted that this design has eliminated all levers and fulcrums.

Figure 9·38 is of interest because of the very high sensitivity that can be obtained with the indicated baffle-nozzle arrangement, which can provide mechanical amplifications of many times without the use of gears or levers. Otherwise, the design is basically similar to some of the others which have been considered.

Fig. 9·36 This differential pressure transmitter is similar in operation to the others which have been discussed. It should be noted that the overrange protection is provided by two **O** rings which will trap some of the fluid and cause it to support the diaphragm. (*The Taylor Instrument Cos.*)

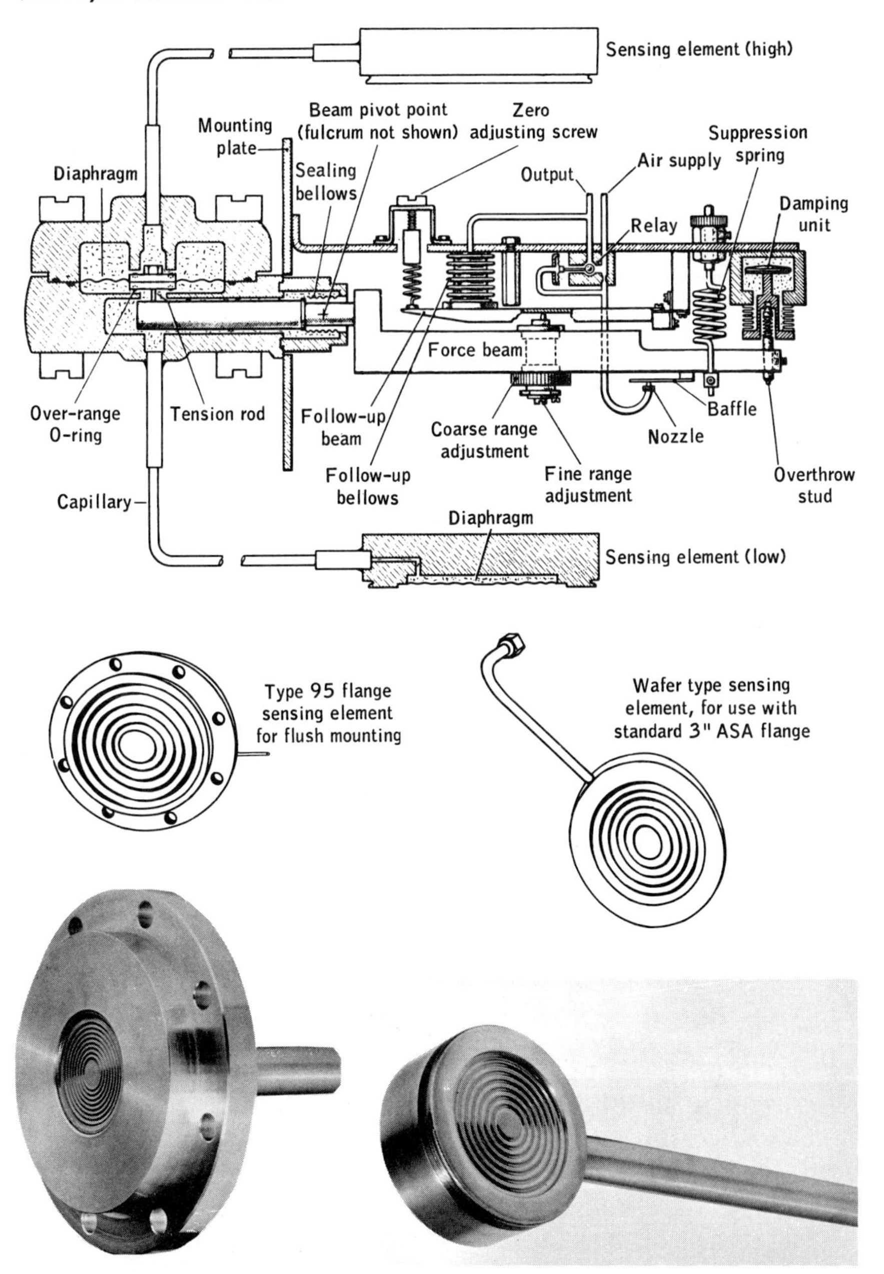

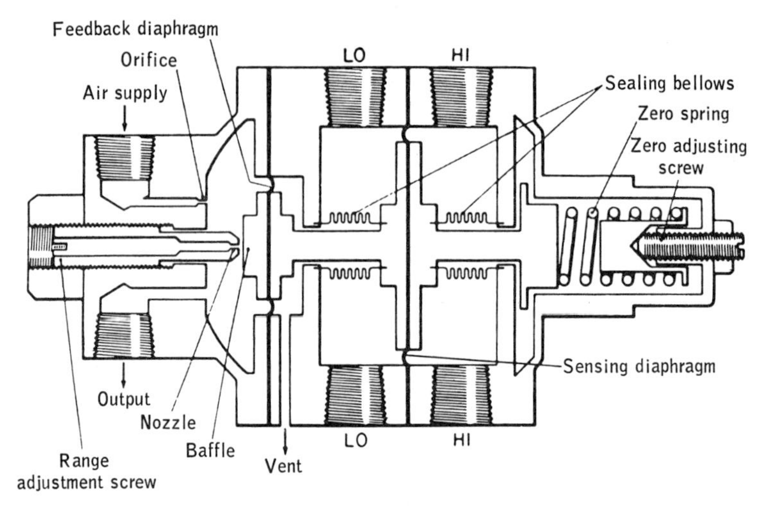

Fig. 9·37 A differential pressure transmitter in which the feedback element operates without employing levers or mechanical amplification. (*The Taylor Instrument Cos.*)

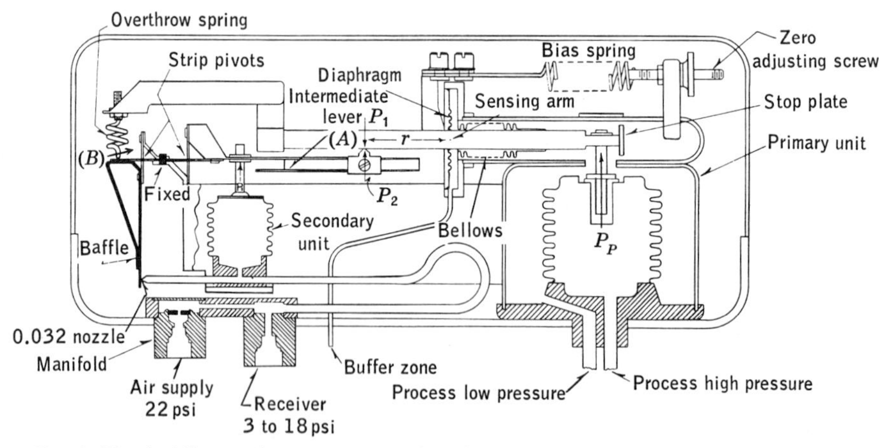

Fig. 9·38 A different design to ensure that the output pressure is a true measure of the differential pressure. (*The Taylor Instrument Cos.*)

The mechanical elements of other pressure transmitters are present in the transmitter shown in Fig. 9·39 but with changed position and appearance. An increased differential pressure will try to rotate the beam assembly clockwise about the fulcrum. For the assumed increase in differential pressure, there must be an increase in the output pressure of the escapement pilot valve. As a greater pressure is exerted against the balancing diaphragm, the greater force that is developed rotates the beam counterclockwise through use of a floating force junction. This

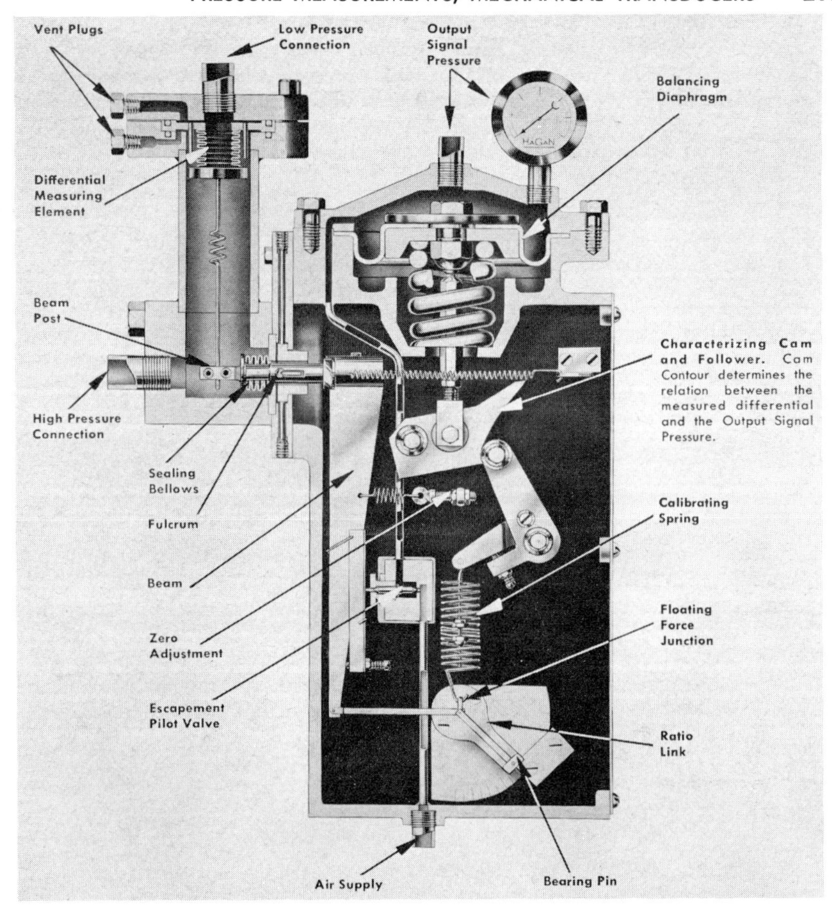

Fig. 9·39 The mechanical elements of this differential pressure transmitter are somewhat different from those previously studied. The *floating force junction* permits conditions of torque equilibrium to be established. (*The Hagan Corp.*)

counterclockwise rotation will continue until the torque created by the input signal is offset by the balancing torque which is created by the output air acting against the balancing diaphragm.

The construction of both ring and **U**-tube manometer, Fig. 9·40, simplifies the construction of differential pressure transmitters.

Pneumatic principles. In order to keep the output pressure of the transmitter linear with changes in temperature or pressure, an air relay often is used. A plot of pressure inside the orifice, that is, between the restriction and the orifice, Fig. 9·41, shows that only a few thousandths movement of the baffle will control the pressure from its maximum value (line pressure) to its minimum. There is a short distance where the curve has a section which is essentially straight. In the 3- to 5-psi range, the

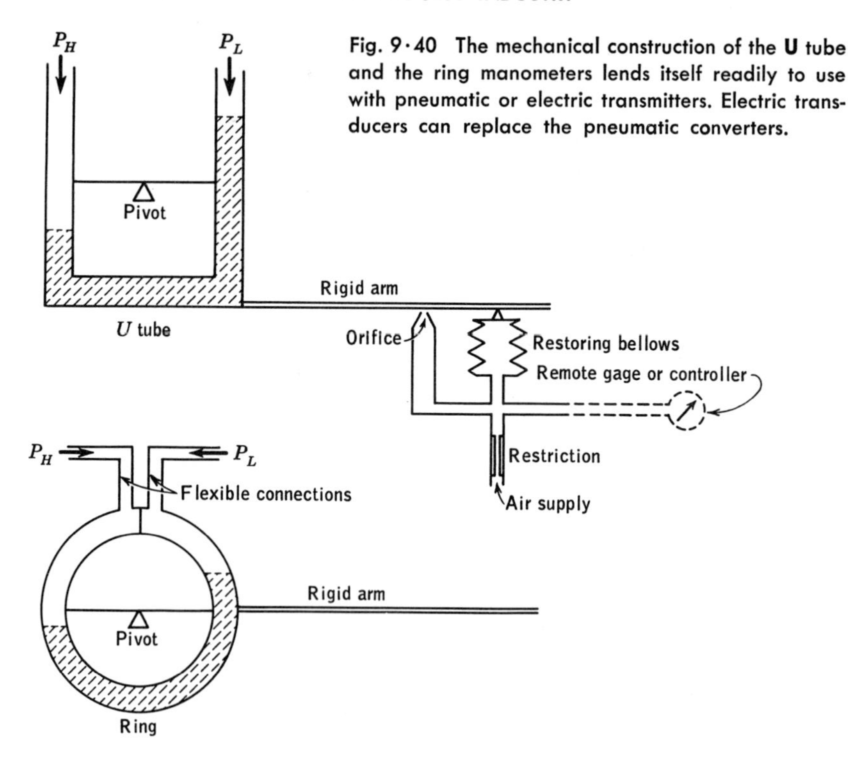

Fig. 9·40 The mechanical construction of the **U** tube and the ring manometers lends itself readily to use with pneumatic or electric transmitters. Electric transducers can replace the pneumatic converters.

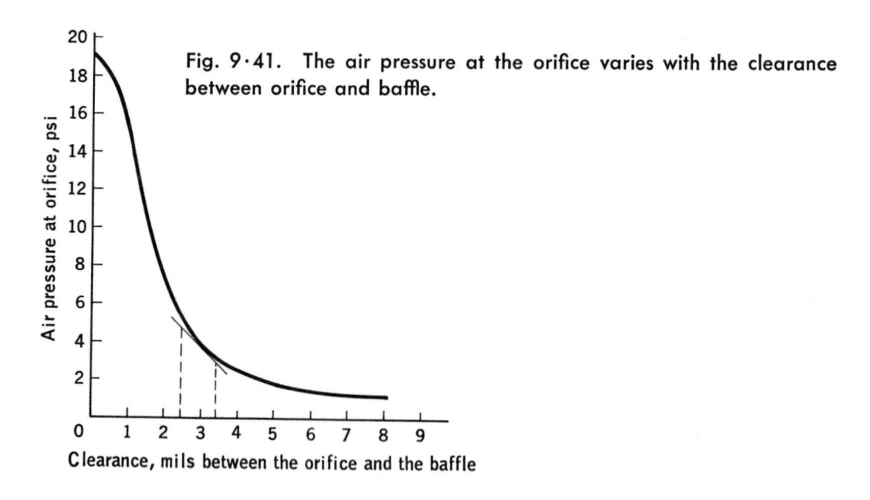

Fig. 9·41. The air pressure at the orifice varies with the clearance between orifice and baffle.

baffle movement creates a pressure change directly proportional to the change in baffle position. In order to use the small pressure change having the linear relationship, an air relay is often employed. The relay is designed to convert the pressure changes ahead of the orifice into much greater pressure changes with far larger volumes of air than can be passed through the pressure-reducing restriction, Fig. 9·42.

If we move the baffle toward the orifice, pressure P will increase. This will be noted on the gage, and it will also move the diaphragm D in the relay. As the diaphragm moves, it will push ball B up so that it partially closes the path for the air entering at H. The upward movement of the ball will also increase the clearance at E, which will permit more air to escape. The pressure gage G will indicate a reduction in pressure. This particular design is known as a reverse-acting relay. The output pressure decreases for an increase in the pressure applied to the diaphragm. A change of 2 psi at the nozzle produces a change of 12 psi at the output, so the relay may be called a pressure amplifier of a special kind. We are not trying to employ the increased pressure at the moment. Our chief concern right now is to see how a small baffle movement can be made to produce large pressure changes. How we move the baffle is of little interest; many transducers will use the same principles.

For those situations in which an increase in the value of the variable must be translated into an increase in air pressure for transmission, we

Fig. 9·42 Schematic of a reverse-acting relay. With this type of relay, an increase in nozzle pressure, caused by bringing the baffle closer to the orifice, will produce a decreased output air pressure.

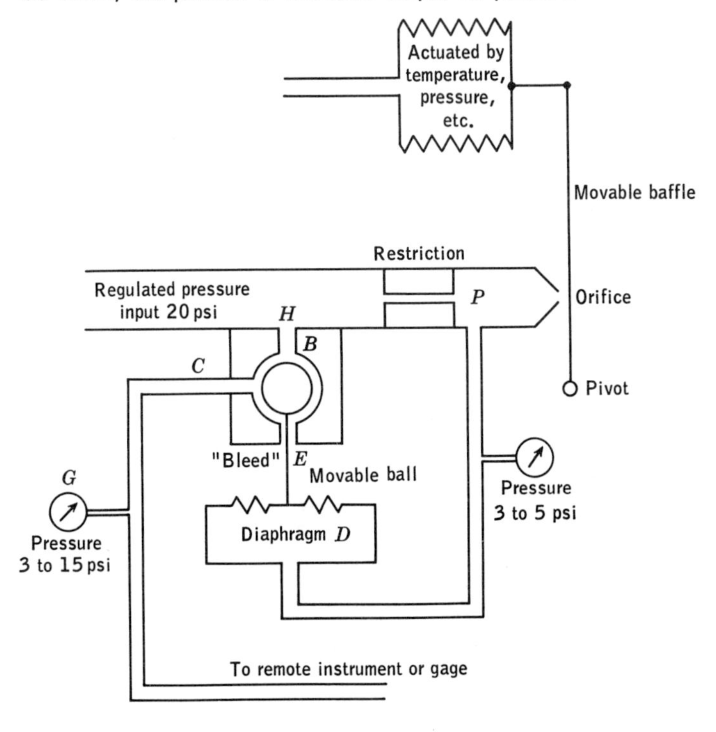

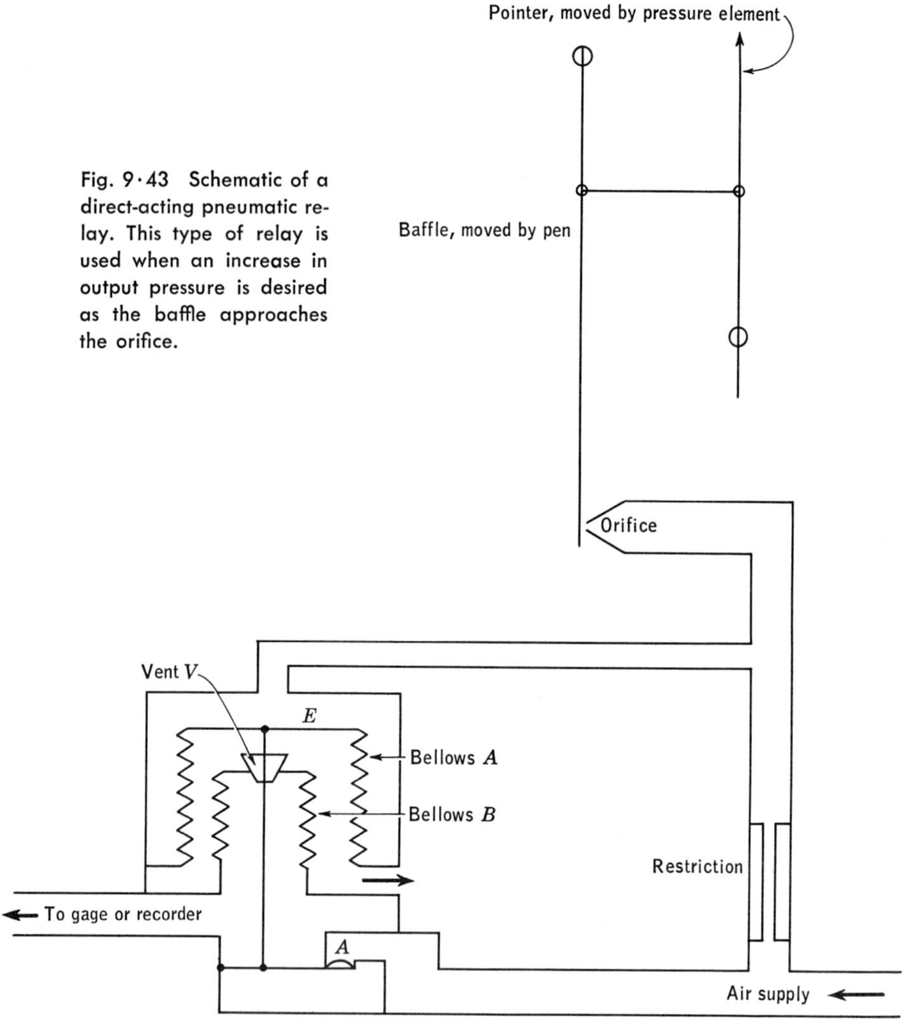

Fig. 9·43 Schematic of a direct-acting pneumatic relay. This type of relay is used when an increase in output pressure is desired as the baffle approaches the orifice.

employ a *direct-acting pneumatic relay*. All of the essential features of other pneumatic relays are found in this one, Fig. 9·43. Certain constructional features modify some of the response. Let us assume that the baffle is moved toward the orifice. This will increase the orifice pressure and push down end E of bellows A. As E moves down, it forces the linkage to open the input-air valve at A. The added air expands bellows B and causes the valve to close when the force of the orifice air is balanced by the air going to the gage or recorder.

When the baffle moves away from the orifice and decreases the pressure in bellows A, it withdraws the valve at vent V until the pressure in bellows B is correct for the pressure in bellows A.

Diaphragm-type differential pressure devices. Differential pressure-measuring devices are generally designed for conditions such that the pressures are of the order of a few inches of mercury up to about 50 psi. But there are many commercial measurements, particularly in connection with flow, such that the differential pressure may be small but the pressures which create the differential are large. For example, the line pressure may be 1,000 psi, but the differential pressure only 10 or 20 psi. For such conditions the conventional bellows construction will not work.

Figure 9·44 shows one design in which the two pressures whose differential is sought are applied to the outsides of the two sets of diaphragms. The interior is filled with a liquid which keeps the bellows from collapsing. By design and construction, one side is always designated as the high-pressure side. This ensures that the net action on the bellows will always be to move the bellows in one direction from the zero position. However, as they try to move, they actuate an arm which is connected with a pneumatic torque-balancing system which will then apply such a torque as to restore balance. The pressure which is required is then a measure of the differential pressure.

A modification of this type of differential pressure-measuring unit employs a silicone-filled twin diaphragm which is so placed in its housing that under normal operating conditions the torque-balance conditions can be established in the manner discussed. But should there be a major increase in the differential pressure, or should the pressure to one side become zero while that of the other remained normal, then the restoring

Fig. 9·44 Cutaway section of a diaphragm type of differential pressure transmitter. (*The Foxboro Co.*)

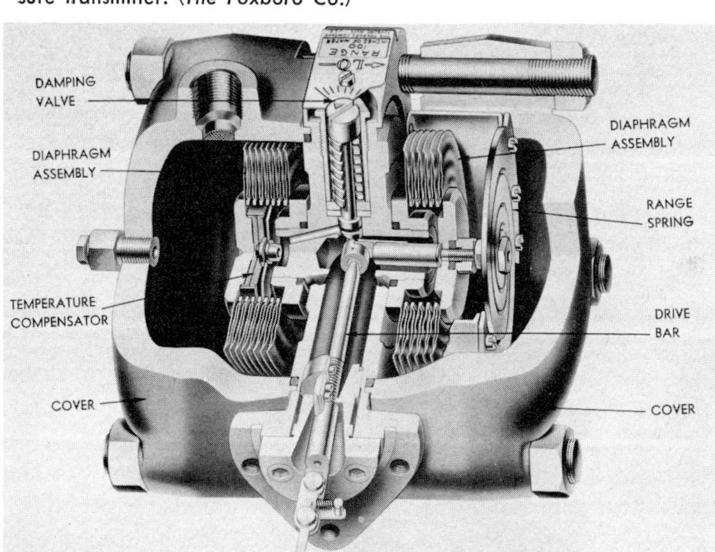

torque furnished by the air pressure in the bellows would not be adequate. Under these conditions, the diaphragm would be forced against the body with which it mates. The excessive differential pressure would then be transmitted to the body of the cell where there is adequate strength to absorb it.

It will be noted that the ring and **U** manometers lend themselves readily to use with auxiliary equipment to obtain pressures for transmission. These pressures are suitable for remote reading or for controller or valve actuation.

It will be noted that regardless of the specific design, all of the pneumatic transmitters have functioned in exactly the same way:

1. The pressure-measuring element causes movement of some sort, either toward or away from an orifice.

2. The resulting change in pressure in the orifice section is communicated to a bellows which can reposition the orifice. But because of the feedback between bellows and orifice, the behavior of the one is largely offset by behavior of the other until a condition of equilibrium has been reached.

3. The pressure required to bring the orifice close to the pressure-responsive element is a measure of the total pressure and is obtained without imposing any physical load (of any significance) on the indicating section.

It should be clearly understood that any variable which can position the baffle can be converted into a pneumatic equivalent.

The source of the pressure is also immaterial. Whether the pressure is created by temperature, level, or flow conditions will make no difference in the way the measurement is made or converted into other values.

INSTALLATION AND APPLICATION NOTES

1. Flexible copper tubing and compression fittings are recommended for most installations, particularly for very low pressures and high vacuums, when no leaks can be tolerated in gage lines or connections. Gage lines must be adequate to withstand the pressure and corrosion of the media being measured.

2. The speed of response of pressure gages is affected by the internal size and the total length of the gage line. In applications using controllers, low range or vacuum gages, the size and length of gage line must be carefully considered in order to avoid objectionably slow response. The sizes shown in Table 9·1 are recommended.

3. The installation of a gage cock and tee in the line close to the gage is recommended because it permits the gage to be removed for testing or replacement without having to shut down the system. Figure 9·45 shows such an installation. The tee is also used to facilitate filling the line.

4. Pulsating pressures in the gage line are objectionable because:

a. The constant movement of the pen or pointer will produce an unreadable signal.
b. Violent pulsations will fling the ink out of the pen.
c. Constant motion will cause excessive wear, may cause loss of calibration, and may result in early rupture of the pressure element.

Table 9·1 Recommended Gage Line Size

Pressure	Length, ft	Std pipe, in.	Copper water tube, type *K*, in.
0 to 12 in. water	Up to 10	¼	⅜
0 to 12 in. water	10 to 25	½	⅝
0 to 12 in. water	25 to 50	¾	1
12 in. water to 15 psi	Up to 25	¼	⅜
12 in. water to 15 psi	25 to 100	½	⅝
15 to 50 psi	Up to 50	¼	⅜
15 to 50 psi	Over 200	¼ and ½ *	⅜ and ⅝ *
Above 50 psi	Up to 200	¼	⅜
Above 50 psi up	Over 200	¼ and ½ *	⅜ and ⅝ *

* Use smaller line for first 50 ft nearest gage and large line for balance of length. If pressures require it, use extra-heavy pipe of internal diameter similar to standard pipe in table.

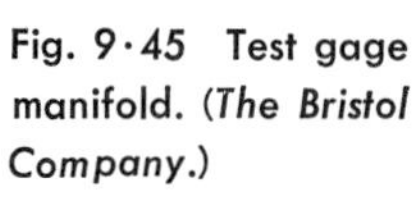

Fig. 9·45 Test gage manifold. (*The Bristol Company.*)

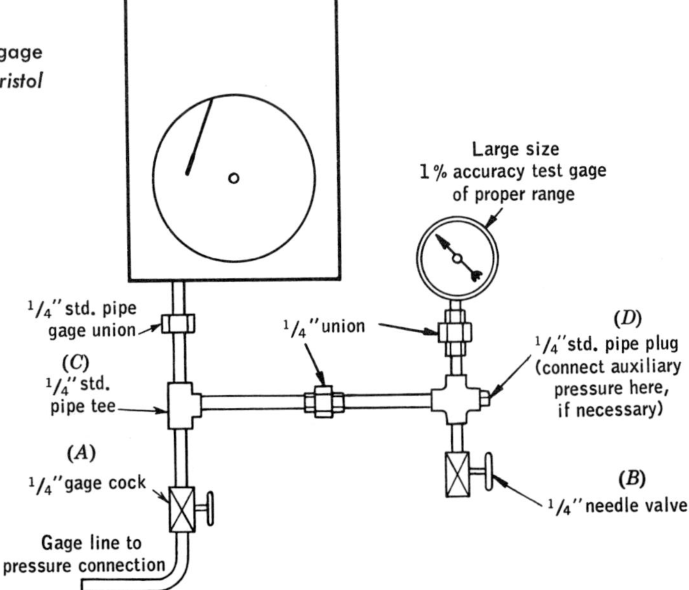

These pulsations may be damped out or minimized by adding restrictions (resistance) and capacity to the gage line. The simplest pulsation damper is a ¼-in. needle valve installed in the gage line several feet from the gage. The needle valve should be closed slowly until the pen or pointer indicates a constant pressure yet responds to significant average pressure changes. If there is not sufficient capacity in the line, a capacity tank made of 2-in. pipe and pipe caps can be installed in the line as shown in Fig. 9·46. For more violent pulsations, commercial dampeners may be required.

It will be found in the application notes dealing with differential pressure gages that the insertion of needle valves to reduce pulsation cannot be made haphazardly. Special precautions must be taken to ensure that the values being read or recorded are the correct ones.

5. When corrosive material must be gaged, the material must be excluded from the gages unless the gages are made with special noncorrosive elements and tubing. Three common methods of excluding the measured media are:

a. Liquid seals. The gage and its connecting line is filled with an

Fig. 9·46 Capacity-resistance unit for pulsation damping. The tank also serves as a condensate or sludge trap.

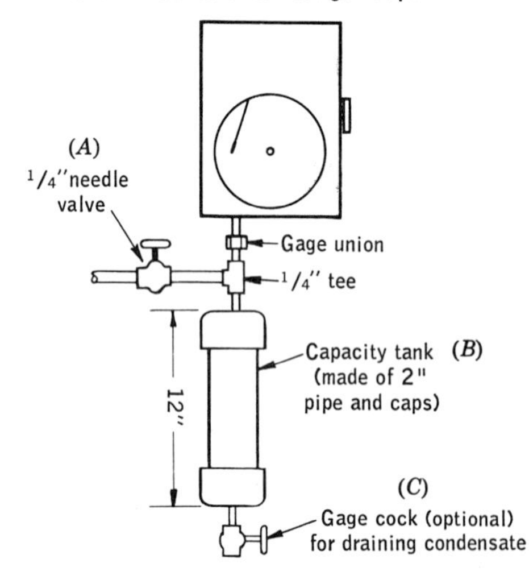

Fig. 9·47 When the working fluid must be excluded from the gage, the pressure may be transmitted to the flexible element through a fluid which is retained by a sealing member. (*American Chain and Cable, Helicoid Gage Division.*)

inert liquid which is retained in position and kept from mixing with the corrosive element in some designs through use of some sort of physical seal. The pressure (sometimes without a seal) is then transmitted (through the seal when used) to the liquid, which in turn transmits it to the gage (Fig. 9·47). Choice of the sealing fluid is based on the following properties (Table 9·2):

1) Resistance to chemical combination with the pressure medium.
2) Resistance to physical combination with pressure medium such as evaporation, amalgamation, or dissolving.
3) Specific gravity, which can keep the sealing liquid separate from the pressure medium under all existing temperatures. The sealing liquid may be either heavier or lighter than the pressure medium.

Table 9·2 Suggested Sealing Liquids for Common Pressure Media

Pressure Media	*Suggested Sealing Liquids*
Acetic acid, to 50% solution	Kerosene; mercury (very slow reaction)
Acetone	40% $CaCl_2$ solution, pH 8.5
Alum solution	Light, refined mineral oil
Ammonia	Mercury; light, refined mineral oil
Benzol	40% $CaCl_2$ solution
Brine, sodium or calcium	Light, refined mineral oil; perchlorethylene
Butane or propane gas	40% $CaCl_2$ solution
Butadiene	40% $CaCl_2$ solution
Caustic soda	Refined mineral oil; kerosene
Chlorine	Diaphragm seal—silver or tantalum, purging system
Creosote, wood-preserving	40% $CaCl_2$ solution
Ethyl acetate	40% $CaCl_2$ solution; mercury
Fuel oil	40% $CaCl_2$ solution; 50/50 Prestone or glycerine and water
Hydrochloric acid	Silver diaphragm; perchlorethylene as one component of a double seal
Lactic acid, 10%	Kerosene
Menthol	40% $CaCl_2$ solution
Milk of lime	Light, refined mineral oil
Molasses	Light, refined mineral oil
Phenol	40% $CaCl_2$ solution
Pitch or tar	40% $CaCl_2$ solution
Phosphoric acid	Monsanto's HB-40 for concentrated solutions; refined mineral oil for dilute solutions
Soda ash	Refined mineral oil
Sulfate liquor	Refined mineral oil
Sulfuric acid	Refined mineral oil
Zinc oxide fumes	Refined mineral oil; kerosene
Oleum, mixed acids, fuming nitric acid, hydrogen fluoride	Chlorpropane Liquid 170, Hooker Electrochemical Company

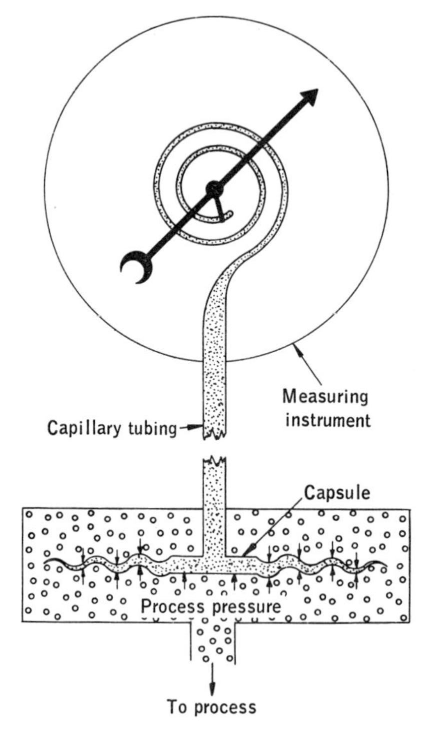

Fig. 9·48 Schematic diagram of a capsule-type pressure seal. (*The Bristol Company.*)

Fig. 9·49 A flexible member used to confine fluid within the measuring element. (*American Chain and Cable, Helicoid Gage Division.*)

4) Physical characteristics such as correct viscosity, vapor pressure, freezing point, and diffusion of the pressure medium through the sealing fluid.

b. Capsule-type seals. The gage may be furnished with a completely sealed system whereby an inert diaphragm retains the liquid in the gage and keeps the pressure medium from coming in contact with it. Figures 9·48 to 9·51 show three entirely different types of force-balance seals; one is pneumatic, while the others are primarily hydraulic. The sealing member should be cleaned at intervals because materials may tend to coat it and reduce the pressure applied to the seal.

c. Purge systems. Sticky, flocculent, or excessive amounts of solids can be excluded from gages and gage lines by a liquid- or gas-purge system such as is shown in Fig. 9·52. However, the following conditions must be met:

1) The purge fluid must not combine in an objectionable way, chemically or physically, with the pressure medium.

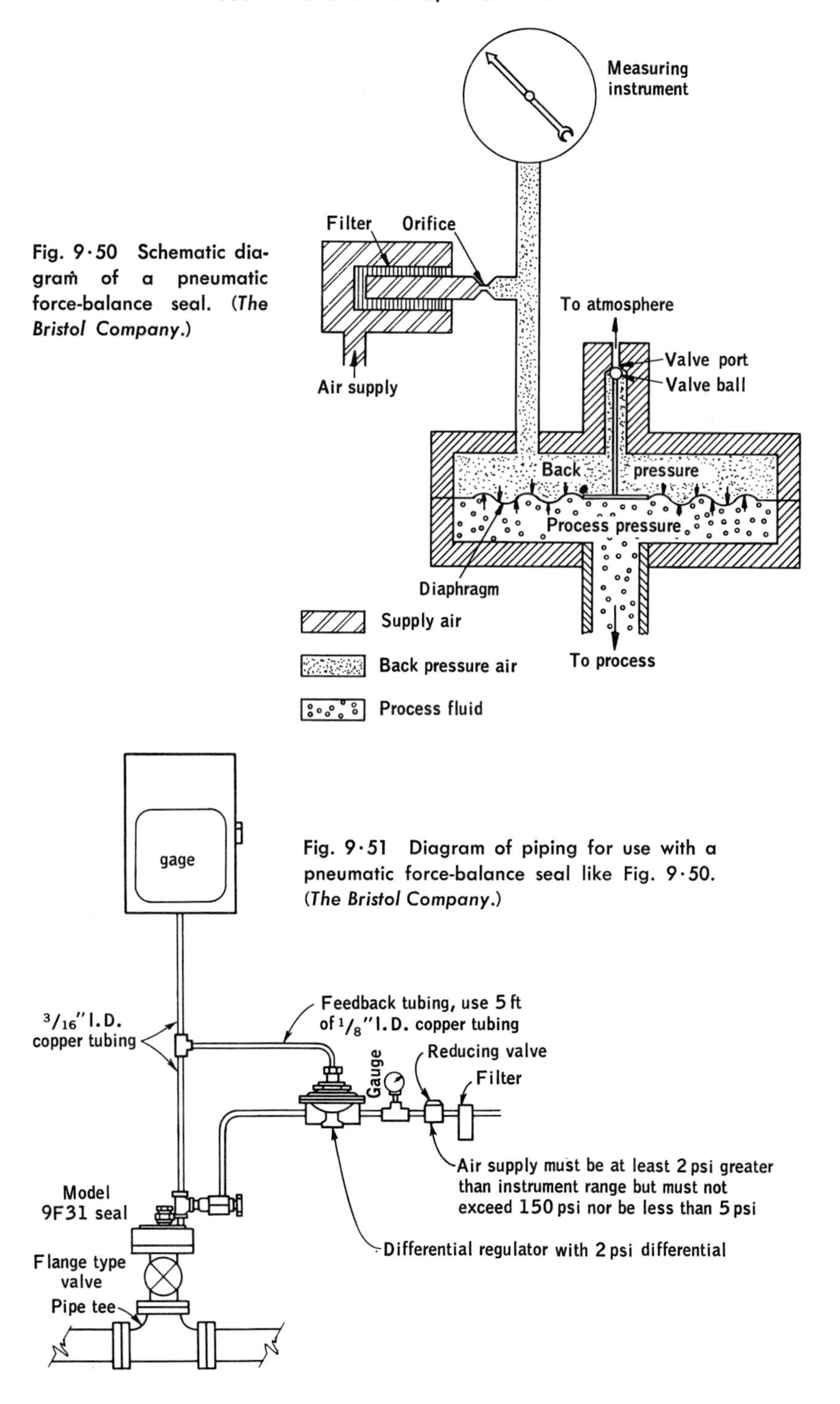

Fig. 9·50 Schematic diagram of a pneumatic force-balance seal. (*The Bristol Company*.)

Fig. 9·51 Diagram of piping for use with a pneumatic force-balance seal like Fig. 9·50. (*The Bristol Company*.)

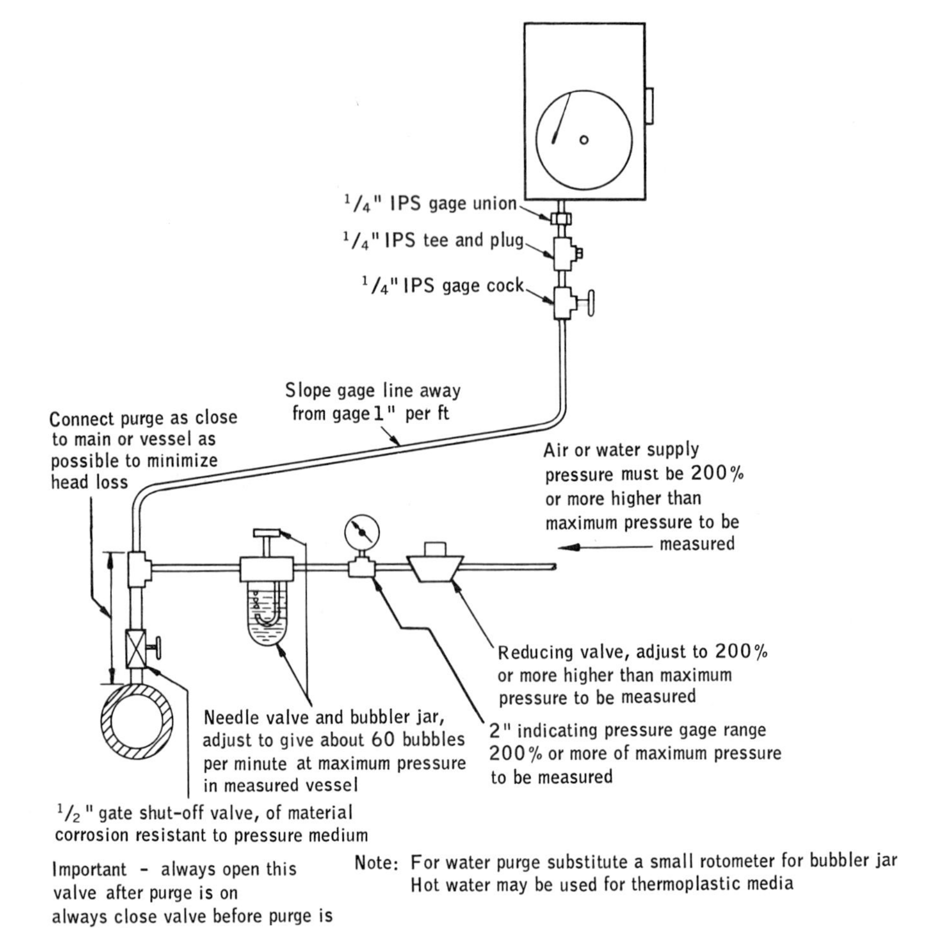

Fig. 9·52 An air or water purge system may be used to exclude pressure media which are corrosive or solidifying or which contain solid matter. (*The Bristol Company*.)

2) The purge fluid must not dilute the pressure medium to any objectionable degree.

3) In a purge system using a gas such as air, venting of gas to the atmosphere must be possible. Otherwise, the process will be filled with the purging gas.

The common purging fluids are water and air. In general, water purging can be used only where the pressure medium is a large quantity of flowing solution in which the dilution effect of pure water is negligible. Typical applications are numerous in such plants as pulp and paper and sewage disposal.

d. Purge system operation. If the flow of the purging medium is not held within fairly low limits, the pressure drop introduced by the flow will be added to that of the medium being measured. Therefore, the following should be observed if this error is to be reduced to a negligible minimum:

1) Connect the purge line into the gage line as close as possible to the vessel or main being measured, Fig. 9·53.
2) Use not less than ½-in. standard pipe between purge connection and the vessel being measured. Use a gate valve to minimize the pressure drop.

Fig. 9·53 Typical installation diagrams for capsule-type pressure seals. (*The Bristol Company.*)

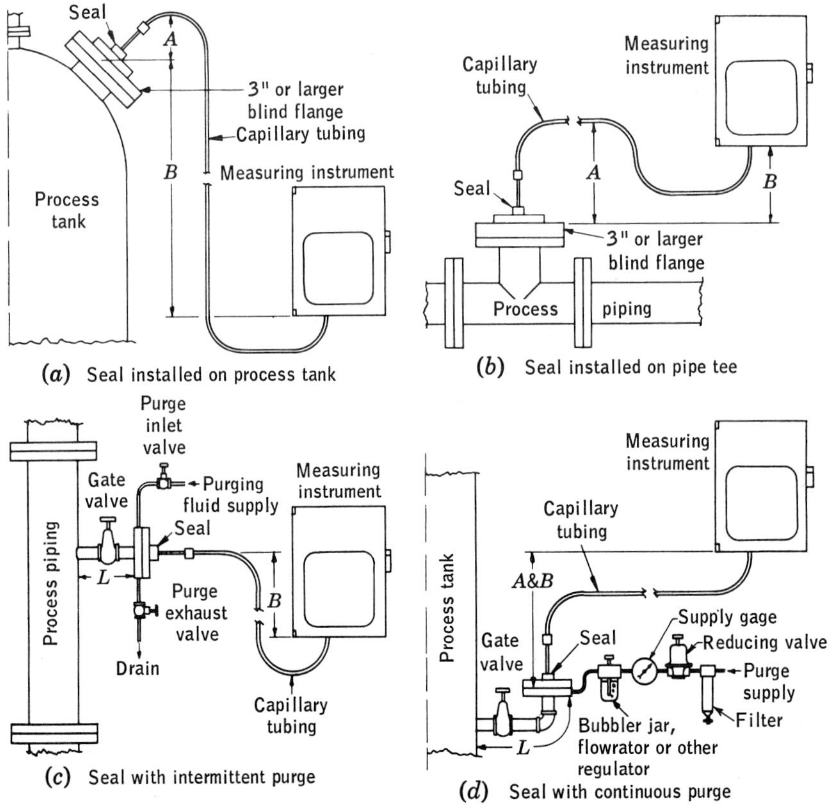

A No point in this capillary tubing may be more than 30 feet above the seal when measuring pressure or 5 feet above the seal when measuring vacuum

B Appropriate corrections must be made (espedially on low ranges) for the head of liquid in the capillary tubing

L Make the piping from the process to the seal as large and as short as possible to reduce friction and prevent clogging

3) Pressure of the purge supply should be at least 200 per cent and preferably 300 per cent of the maximum range to be measured.
4) The reducing valve must be set to deliver at least 200 per cent of maximum range. The higher the pressure delivered to the needle valve, the less the differential pressure change across it will be as the measured pressure changes and the more uniform will be the flow of the purging material. Such an arrangement will prevent excessive flow of the purging fluid when the measured pressure is low.
5) The adjustment of the purge flow rate must always be made when the pressure is a maximum. Adjust the needle valve of an air purge to give 30 to 50 bubbles per minute; for a liquid purge, the flow rate should be adjusted for about 2 oz of liquid per minute.
Caution: *Always turn on the purge system before admitting the measured medium to the system; always turn off the measured medium before shutting off the purge system.*

6. Steel is not appreciably corroded by ammonia gas or liquid. Gages with carbon-steel elements, tubing, and case connections can be directly connected to ammonia without seals.

7. *Mercury amalgamation.* Mercury amalgamates readily with copper, brass, silver solder, etc., and the amalgamation will rapidly cause failure of pressure elements containing such metals. Therefore, great care must

Fig. 9·54 Gage lines may be heated electrically to prevent solidifying or freezing of the pressure media. (*The Bristol Company.*)

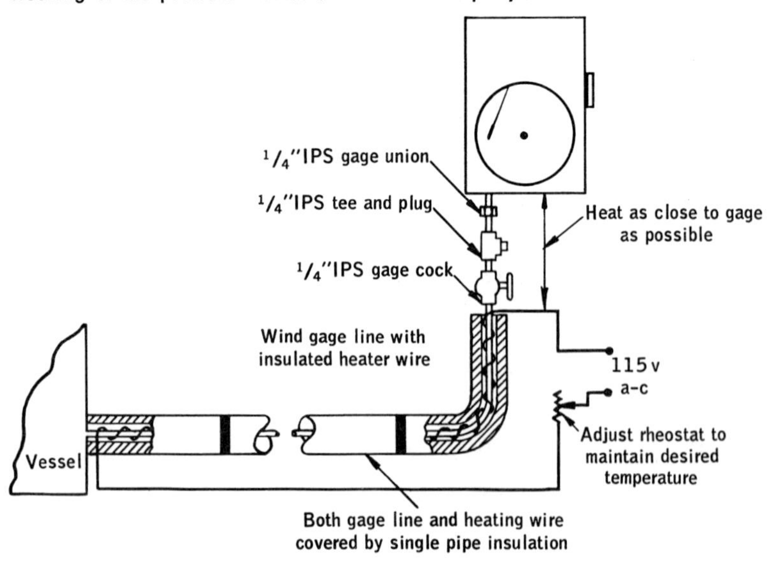

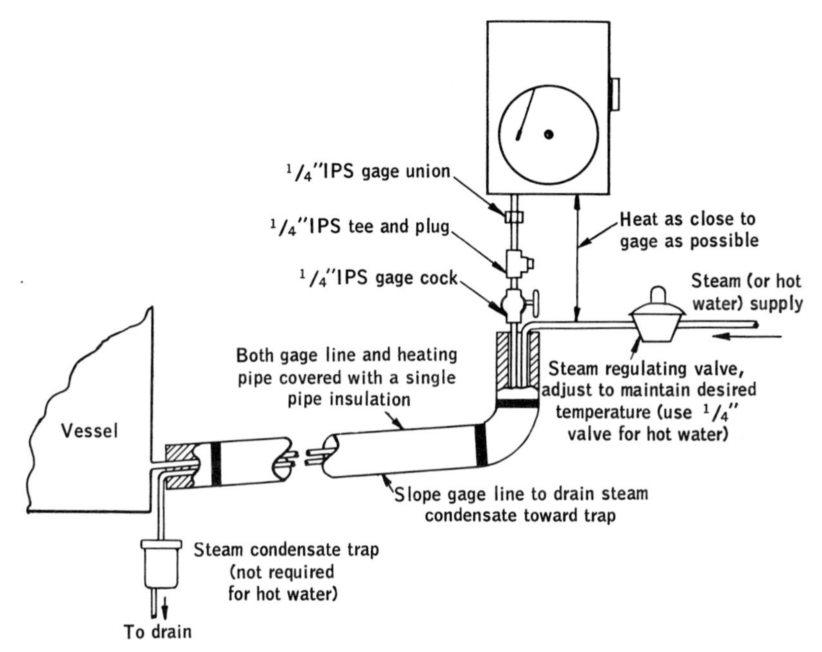

Fig. 9·55 Steam heat may also be used to prevent solidifying or freezing of pressure media. (*The Bristol Company.*)

be observed to prevent any accidental introduction of mercury or mercury vapors into any part of pressure gages except those of all-welded ferrous construction. Under many sets of conditions, the air entrapped in the pressure element and internal tubing will prevent any liquid or fumes from entering the gage.

In installations of pressure pens having mercury-manometer connections, the gage lines must be so installed that the mercury which may be accidentally blown from the manometer cannot run down into the pressure gage.

8. *Viscous or solidifying fluids.* Fluids which thicken or solidify on standing or cooling must be excluded from pressure gages and connecting lines or kept in a continually fluid state, Figs. 9·54 to 9·56.

9. *Freezing ambient temperatures.* When gages measuring either freezable liquids or gases bearing condensable moisture must be mounted outdoors or in unheated places, both the measuring system and the exposed gage line must be protected from freezing.

10. *Overrange protection.* Pressure gages must never be subjected to pressures above their scale range unless they have been protected against such damage. Even momentary overpressures may permanently deform the element or linkage and thus destroy the calibration.

A simple overrange protection for air and gas gages, which can be

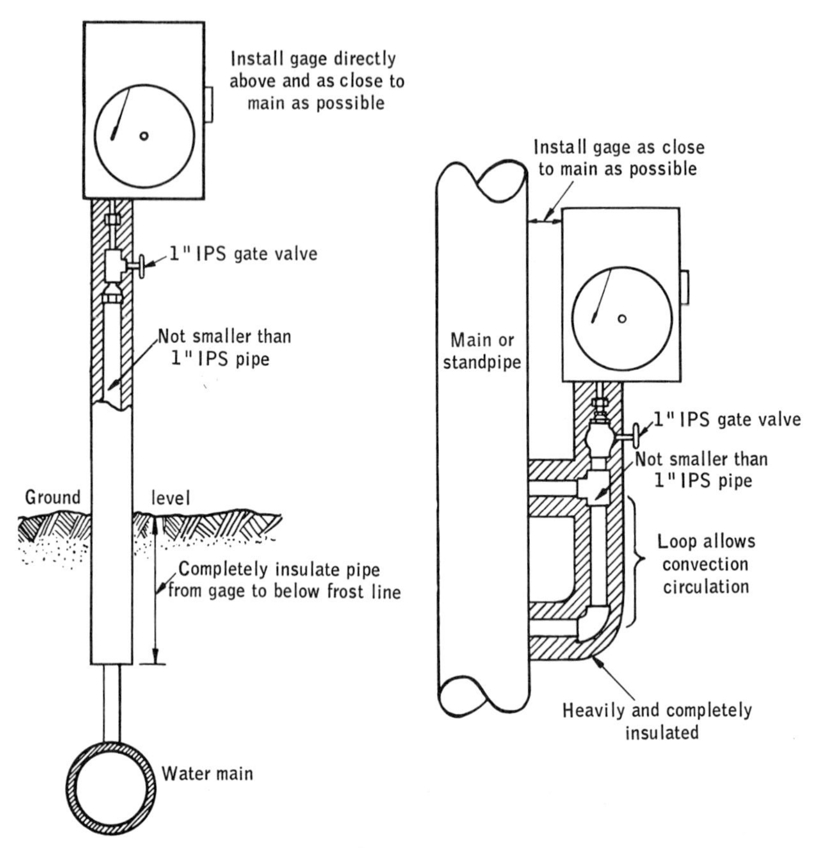

Fig. 9·56 Protection from freezing may be obtained by using insulation, along with a piping loop, to ensure continuous circulation of the fluid. (*The Bristol Company.*)

easily built and installed in the field, is shown in Fig. 9·57. The height of the sealing fluid L above the bottom end of the bubble pipe should be about 10 per cent greater than the head in sealing liquid equivalent to the maximum range of the gage, according to the following formula:

$$L\text{, in.} = \frac{(1.1)(\text{gage range, in. water})}{\text{sp gr of seal liquid}}$$

A temperature correction for specific gravity may be necessary. A sealing fluid which will not evaporate, freeze, or be contaminated by gas should be used. Do not use a viscous liquid, because the pressure element may be damaged before bubbles can escape through the liquid.

After the gage and overpressure seal are installed, close the needle valve

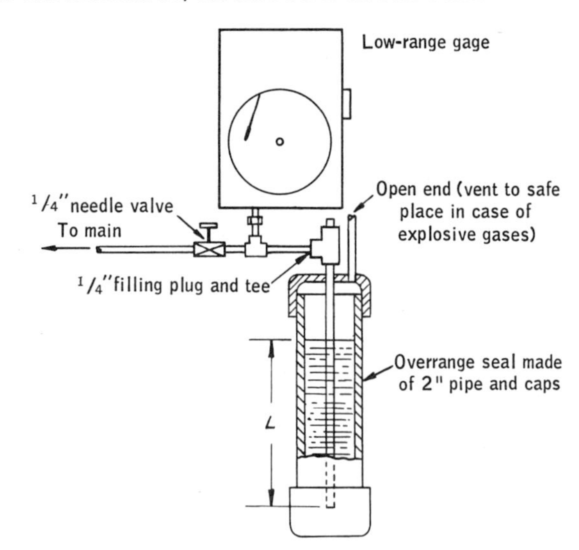

Fig. 9·57 This arrangement will provide overpressure protection for low-range instruments. (*The Bristol Company.*)

and pour sufficient sealing liquid through the filling plug to produce the correct value of L.

Open the needle just enough to give satisfactory speed of response to normal pressure changes.

11. *Gages measuring gas, air, and liquid.* When gage lines for the measurement of air and gas are installed, provision must be made for draining and venting of any condensate. Figure 9·58 is self-explanatory. As previously mentioned, when liquids are being measured, equal attention must be given to draining or filling the lines, Figs. 9·59 to 9·61.

12. *Pressure gages for hot condensable gases or steam.* The temperature of the fluids entering pressure gages should, in general, never exceed 150°F.

a. Gage mounted below connection. When the gage is mounted below the connection, Fig. 9·62, it will automatically be protected against high temperatures; the condensate in the line will form a seal. The chief precaution is to ensure that the lines are sloped properly to ensure venting and that correction is made for the head of liquid in the connecting line. It is often wise to fill the lines with water to ensure that no overheating will occur before the condensate has formed in sufficient quantities to make an effective seal.

b. Gage above the connection. When a pressure gage is installed above a steam line, a coil pipe seal *must* be installed. Before steam is turned on, enough water should be poured through the filling plug to ensure that the coil pipe is full. The gage lines should not be smaller than ½-in. standard pipe size to provide free counterflow of steam and condensate. Avoid low points in the line which can trap the condensate and create head errors.

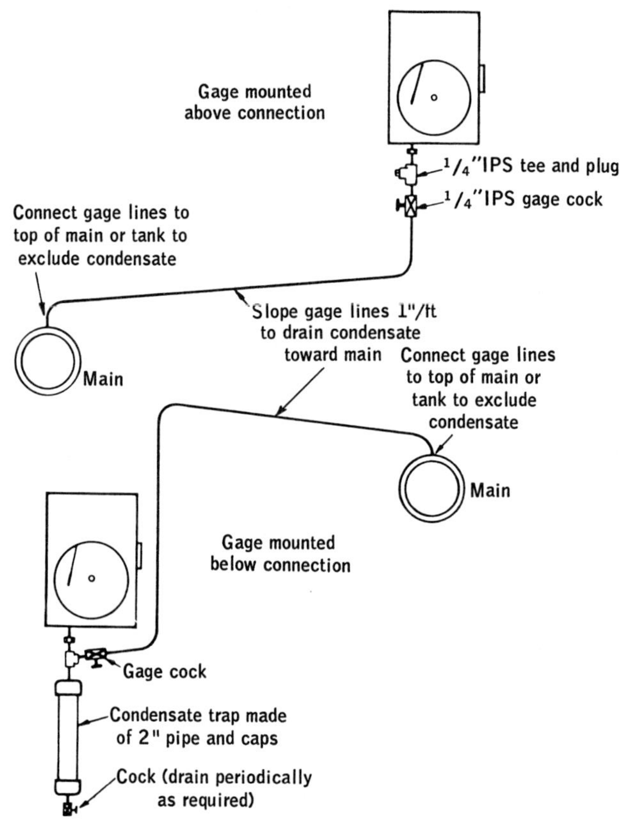

Fig. 9·58 Recommended piping arrangements for noncorrosive gases and air. (*The Bristol Company.*)

13. *Compound pressure and vacuum gages.* Great care must be taken to prevent corrosive or solidifying pressure media from being drawn into gage lines and elements when vacuum is released and pressure is applied.

14. *Case purging in explosive or corrosive atmospheres.* A pressure gage containing any electrical elements should be air purged if its ambient atmosphere may at any time contain explosive gases. Similarly, any gage mounted in corrosive atmospheres should be internally protected by purging the case.

Air purging consists in providing a constant supply of clean, low-pressure air to the instrument case, which keeps it at an internal pressure slightly above the pressure of the surrounding atmosphere. This prevents the entrance of corrosive and explosive elements.

15. A two-way gage cock should be installed just ahead of the gage. This makes it possible to remove, service, test, or replace the gage without shutting down the system.

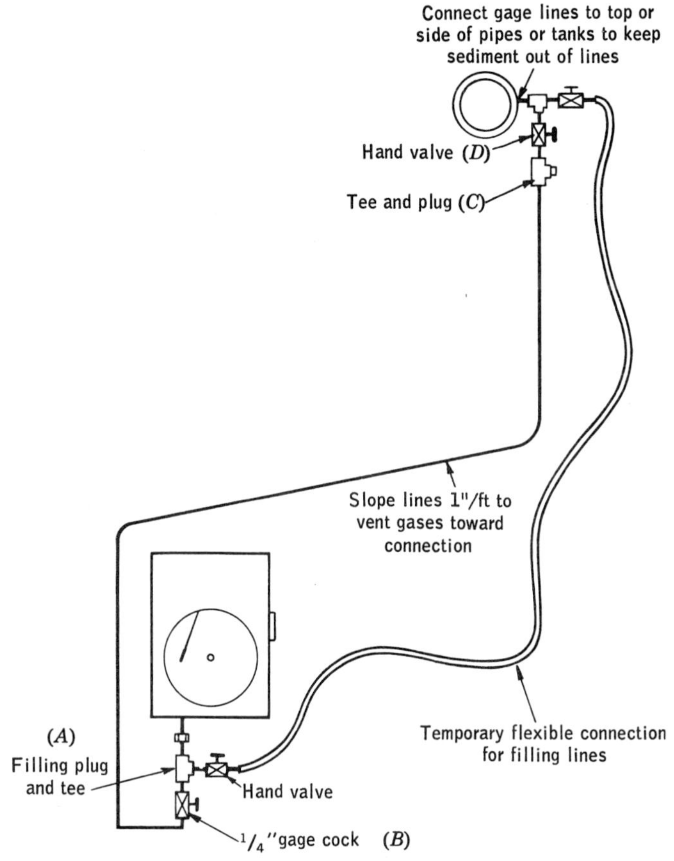

Fig. 9·59 Filling gage lines; gage below connection. (*The Bristol Company.*)

16. Pressure pulsations are generally created by reciprocating pumps, compressors, or engines. Hydraulic pressure surges should not be permitted to enter the gage. Provide protection against surges, pulsations, and so on by:

a. Using restrictions and capacity chambers to reduce pressure variations. The true average may not always be obtained; it may deviate several per cent from the mathematical average, Fig. 9·63.

b. One of the simplest dampers is a ¼-in. needle valve installed several feet ahead of the gage. A long narrow needle will produce equal flow resistance for in and out flow. The needle valve should be adjusted slowly until the gage pulsations are damped out, but the meter will still respond to significant pressure changes. If the capacitance of the gage and tubing is insufficient, additional capacitance can be added.

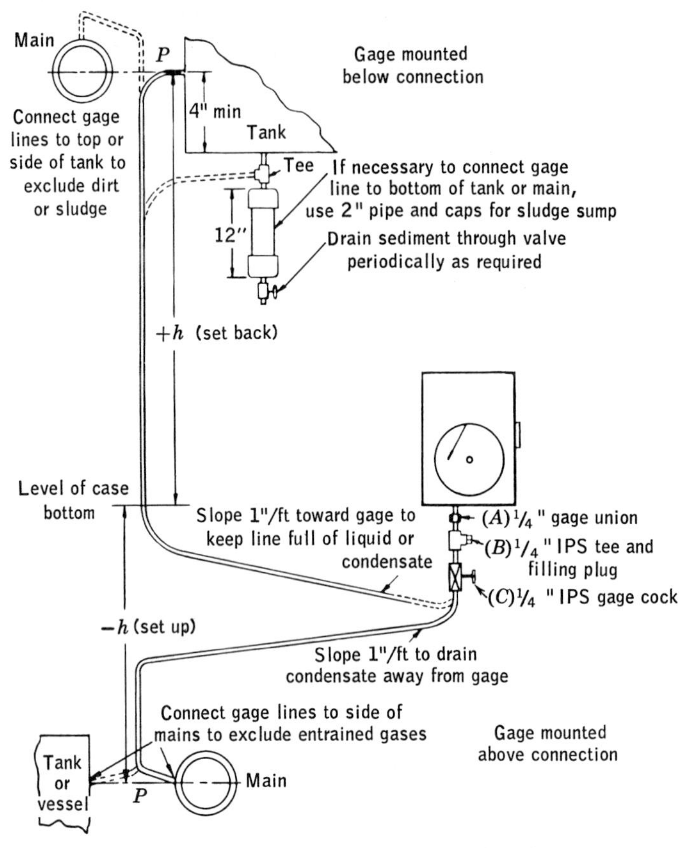

Fig. 9·60 Gage installation for noncorrosive liquids. (*The Bristol Company.*)

c. For still more severe pressure pulsations, specially designed dampers should be used.

17. All elastic-member pressure gages (Bourdon tube, bellows, diaphragms) measure the difference between the internal and external pressure. On low-range gages these errors may be serious. Gages installed in rooms ventilated by exhaust and circulating fans may have serious errors. Opening and closing the case doors of low-range instruments may cause a momentary reading variation which will be especially apparent on recording gages.

Gages should be so located as to be as free as possible from dust, dirt, corrosive fumes, vibrations, and extremes of temperature.

18. *a*. Absolute-pressure gages should be wall or panel mounted as close to the measuring point as possible if delayed response is to be minimized.

b. For low-pressure gages, there can be *no* leaks in gage lines. Copper tubing with soldered fittings is considered superior to iron or brass pipe. Whether tubing or pipe is used, ream all ends, clean inside with solvent, and blow out with clean compressed air. If brass or iron pipe is used, use pipe-sealing compounds to ensure vacuum-tight joints.

c. Good practice often includes a condensate trap, Fig. 9·46.

19. Do not subject an absolute-pressure gage to more than 10 psi above normal atmospheric pressure.

20. The pneumatic force-balance seal is a device used with pressure and liquid-level gages to seal corrosive, viscous, or solid-bearing fluids out of the gage and gage piping. Such a seal is shown in Fig. 9·50. Air from some supply is allowed to flow through a pressure-reducing orifice to both the gage and the back of the seal. When the pressure applied by the process is exactly equal to the pressure applied by the air on the back of the

Fig. 9·61 Installation and filling where gage lines go over an obstruction. (*The Bristol Company.*)

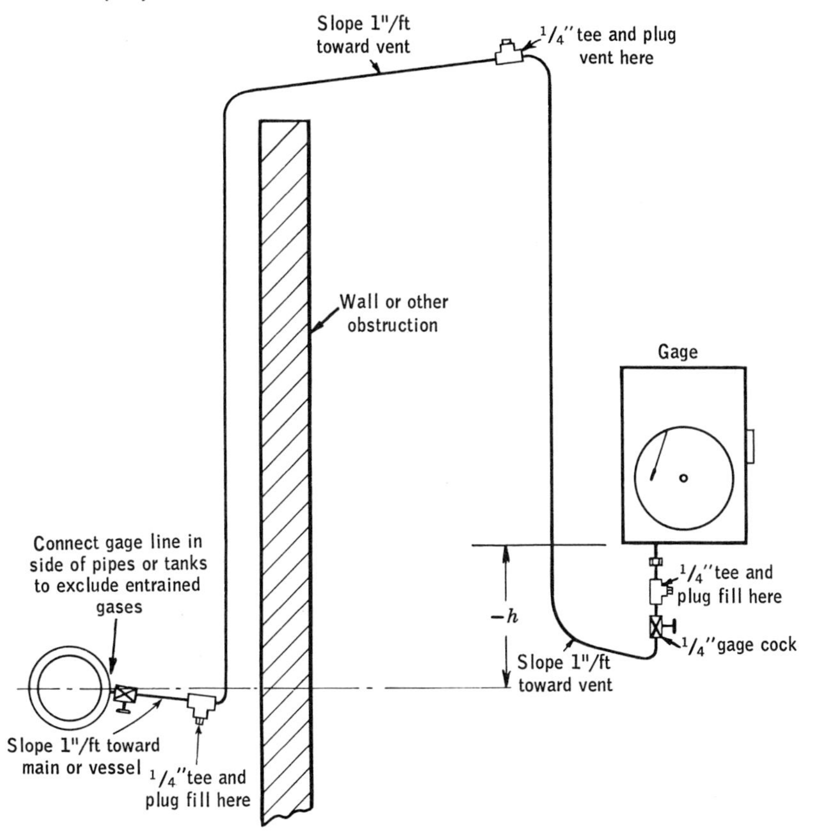

diaphragm, the valve ball will be just enough off the valve seat to maintain a condition of equilibrium. The pressure which does this is indicated by the gage. If the process pressure drops, then the ball will be moved farther away from the seat, allowing air to escape until the two pressures are again balanced. The supply air pressure must always be at least 2 psi greater than the process pressure. Some designs limit the top air pressure to a maximum of 150 psi and a minimum of 5 psi.

21. When installing the capsule type of pressure seal illustrated in Fig. 9·48, the seal should be installed as close as possible to the process.

Fig. 9·62 Gage installation for steam or hot condensable vapor. (*The Bristol Company.*)

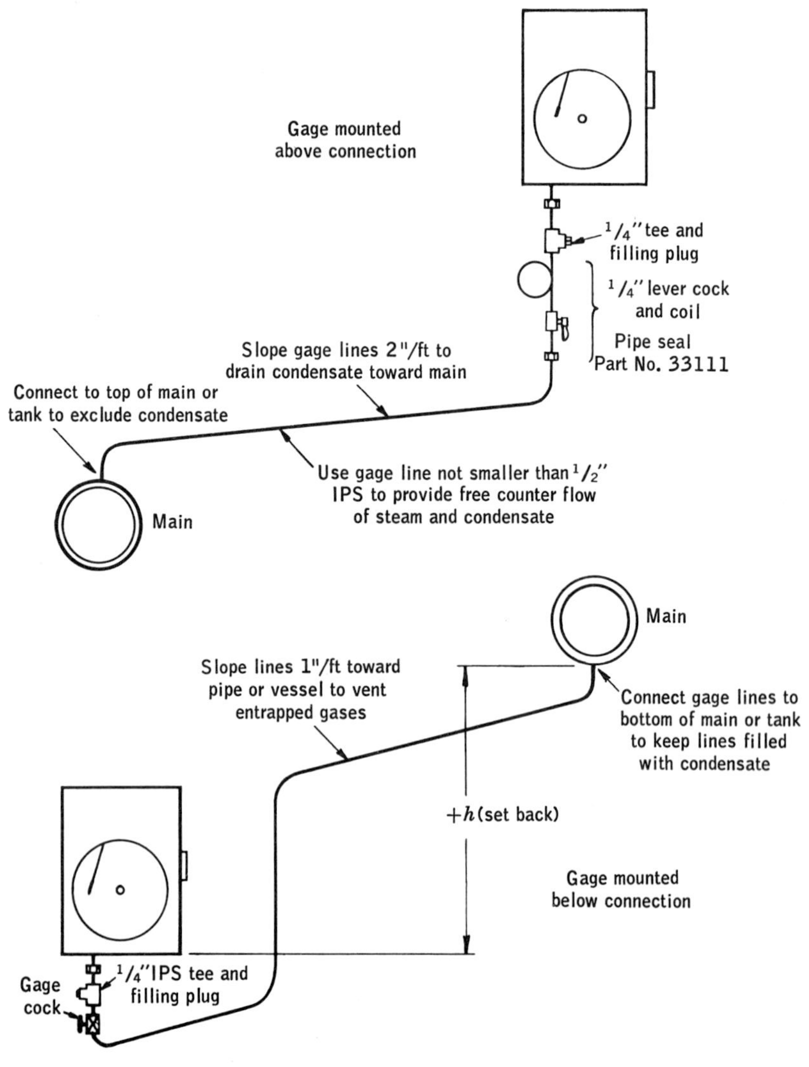

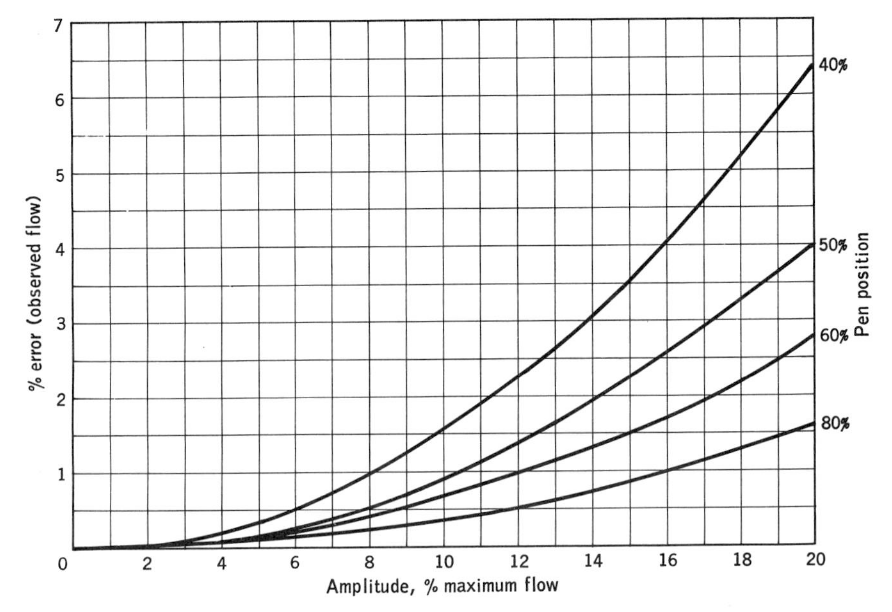

Fig. 9·63 The error in measurement which results from damping. (*The Taylor Instrument Cos.*)

However, if the process temperature exceeds that allowable with the seal, then an extension pipe of adequate size may be used.

22. *Correction for gage location.* If high accuracy is required for low-pressure measurements—especially low-range gages—and differential pressure measurements, corrections may have to be applied to the readings unless the gage and the point of measurement have the same elevation. In the case of differential pressure meters, if both legs have equal divergence from the datum level and are filled with the same liquid, or liquids of equal specific gravities, then the elevation effects will nullify each other.

a. Pressure gage located below the point of measurement. When the gage is located below the point of measurement, the indicated gage pressure will be higher than the true desired pressure by the pressure created by the column of liquid between the gage and the measured point. The correction is negative and is equal to hd, where h is the difference in heights measured in inches and d is the weight density, in pounds per cubic inch. The assumption is made that the pressure is to be in pounds per square inch.

b. For pressure gages located above the point of measurement, the same sort of correction is required but the correction must be added, inasmuch as the column of liquid creates a pressure which opposes the desired pressure. Without any correction the gage will give a *low* reading.

c. When it is desired to correct the meter indication for elevation errors, the following may be done:

1) Make sure that the line to the gage is completely filled with liquid, which may be of the same sort that fills the process pipe, sealing fluid, or condensate from the gas being measured.
2) Provide atmospheric pressure at the point of measurement.
3) Set the pointer to zero.

23. *Filling the gage lines.* Lines should be installed according to Figs. 9·60 and 9·61. It is essential that gage lines of all gages measuring liquid pressures and all gages mounted below points of connection when measuring pressures of gases with condensable vapors be completely filled with some liquid pressure medium or condensate. Trapped air at any point of gage lines may cause serious errors in pressure reading.

In general, it is best to fill the gage lines from the *lowest* point and vent them from the *highest* point. Therefore, a tee and plug should be installed at the lowest and highest points, together with the necessary shutoff valves. All horizontal runs should have a slope of about one inch per foot toward the nearest vent. When filling the lines, rap the horizontal portions gently to dislodge any air bubbles. Gravity-feed filling methods may be employed, however, if certain precautions are taken. In filling a system with water, the service mains may be used as a source of water pressure.

Gravity feed methods. An overhead tank or a cup and riser may be used in filling. If so, the downcoming feed should be at least ½-in. tubing and must be kept full at all times to prevent entrapping air. Unless a second man is available to manipulate the valve at the entrance to the piping being filled, a gate valve should be connected just below the filling cup or funnel. This will enable the operator to keep the riser full of liquid.

Pressure tank or barrel pump. The recommended method of filling is by means of a pressure tank or barrel pump. A painter's spray tank may well be used for this purpose. In constructing a pressure tank, an automobile inner tube valve is used as the air pump connection that allows the passage of air inward only. The tank is partially filled with seal liquid and then pumped up with air. The tank now serves as a source of supply under pressure, its output being controlled by a valve. A gage glass should be installed to indicate supply level, because it is important to prevent the level dropping so low as to unseal the end of the internal pipe.

If a pressure tank is unavailable, a barrel pump may be used. To prevent drawing up air, the depth should be frequently checked and kept above the suction line opening. A telescopic type of suction line is not recommended. However, if it is used, the level must not be allowed to drop below the highest telescopic joint. Otherwise, air might be sucked up through the joint. All packing glands on the pump should be tightened before using the pump for this service.

The reason for filling the gage lines with liquid is not that the air or gas will not transmit the pressure, but rather that variable amounts of liquid in the line require a variable correction of the sort called for in item 10 above. Air and gases are absorbed in time to varying degrees; consequently, one never knows just what the correction should be.

24. *Gages measuring steam or hot condensable vapors.* When the gage is mounted below the steam line, it will automatically be protected from the high-temperature steam by the condensate which will form in the line. The line should, however, be sloped up toward the steam line so that gases and vapor will eliminate themselves. Corrections must be made, of course, for any head of water or condensate in the gage line.

Gage mounted above the connection. In addition to using a coil pipe seal, the steam gage line should not be smaller than ½ in. to provide a free counterflow of steam and water. Take care to avoid low points in the line which would trap condensate and thus create errors.

25. *Testing pressure gages.* A liquid manometer is the most accurate means of calibrating pressure gages, particularly in the lower ranges. Gages which have ratings in excess of those that can be tested with manometers can be tested in either of two different ways. By the first method the gage under test and a standard test gage are connected to a common manifold. By means of a plunger which is moved mechanically by a screw, any desired pressure can be set up in the hydraulic system. This will produce identical pressures at the two gages. The accuracy of such a system is dependent upon the performance of the standard gage.

The use of a dead-weight tester, Fig. 9·64, eliminates the possibility that the reference gage may be in error. A series of weights—the number is adjustable with the desired test pressure—is supported by a piston which is positioned by the hydraulic pressure created in the system. Turning

Fig. 9·64 A dead-weight tester. (*The Foxboro Co.*)

the hand crank can force enough fluid into the cylinder which holds the piston so that the piston is lifted off its seat. When supported by the fluid, it can be rotated readily, and it will continue to rotate for some little time inasmuch as the frictional effects are very low and are at right angles to the desired test pressure. Such rotation ensures that the piston is not being supported in part by the frictional drag of fluid of the system.

Because the size of the piston is small, moderate loads can create pressures of hundreds of pounds per square inch. So long as the weights are correct and the proper test conditions are maintained, this should enable one to get accurate calibrations.

26. Table 9·3 lists gage materials and the chemicals with which it is suggested they be used.

Table 9·3 Gage Materials and the Chemicals with Which They Should Be Used

Bronze

Acetone
Acetylene
Alcohol
Beer
Benzine
Benzol
Bordeaux mixture
Butane
Butanol
Butyric acid
Calcium chloride
Calcium hydroxide
Carbon dioxide, dry
Casein
Chloroform
Coal gas
Cuprous oxide
Dextrine
Ethers
Ethyl acetate
Ethyl cellulose
Ethylene
Ethylene dibromide
Ethylene dichloride
Ethylene glycol
Eucalyptol
Formaldehyde
Freon
Gallic acid
Gas, illuminating
Gasoline
Glucose
Glycerine
Kerosene
Lacquers
Lysol
Magnesium sulfate
Methyl salicylate
Naphtha
Nickel acetate
Oil, lubricating
Oil, refined
Oxygen
Paraffin
Potassium chloride
Prestone
Propane
Pyroxylin
Salicylic acid
Sea water
Tanning liquors
Toluene
Water
Whiskey

Steel

Acetone
Acetylene
Alcohol
Ammonium carbonate
Ammonium hydroxide
Ammonium phosphate
Benzine
Benzol
Benzyl alcohol
Butane
Butanol
Calcium hydroxide
Carbon bisulfide
Carbon dioxide, dry
Chloroform
Coal gas
Cottonseed oil
Creosote, crude
Ethers
Gasoline, refined
Glucose
Glycerine
Kerosene
Magnesium hydroxide
Magnesium sulfate
Mercury
Methyl chloride
Picric acid, dry
Potassium chloride
Potassium cyanide
Potassium sulfate
Sodium carbonate
Sodium hydroxide
Sodium nitrate
Sodium sulfate
Sodium sulfide
Sodium sulfite
Sulfur dioxide, dry
Sulfuric acid, 75%
Toluene
Vegetable oils

Type 316 Stainless Steel

Acetic acid
Acetic anhydride

Table 9·3 Gage Materials and the Chemicals with Which They Should Be Used (Continued)

Alums
Aluminum sulfate
Ammonia
Bleach liquors
Butane
Calcium bisulfite
Carbon disulfide
Chromic acid
Citric acid
Copper sulfate
Ferric nitrate
Ferric sulfate
Hydrocyanic acid
Hydrogen
Hydrogen peroxide
Lactic acid
Nitric acid, pure
Nitrous acid
Picric acid
Phosphoric acid
Photographic solution
Pickling solutions
Potassium permanganate
Silver nitrate
Sodium cyanide
Sodium peroxide
Sodium phosphate
Sulfur dioxide
Sulfurous acid
Vinegar
Vegetable oils
Wines

Monel Housing
K-Monel Tube

Acetate solvents
Acetic acid, 5%
Acetone
Acetylene
Alcohol
Aluminum chloride
Aluminum fluoride
Aluminum sulfate
Ammonium chloride
Ammonium phosphate
Ammonium sulfate
Amyl chloride
Beer
Benzine
Boric acid
Brines
Calcium chloride
Carbolic acid
Carbon dioxide
Carbonic acid
Carbon tetrachloride
Chlorine, dry
Chromium fluoride
Citric acid
Cottonseed oil
Creosote, crude
Ethers
Ethyl chloride
Ethylene dibromide
Ethylene dichloride
Formaldehyde
Freon
Gasoline
Gelatine
Glucose
Glycerine
Hydrocyanic acid
Hydrofluoric acid, under 30%
Hydrofluosilic acid
Hydrogen peroxide
Hydrogen sulfide
Lacquers and solvents
Lime
Lithium chloride
Magnesium chloride
Magnesium hydroxide
Magnesium sulfate
Naphtha
Natural gas
Nitrosyl chloride
Oleic acid
Oxalic acid
Oxygen
Palmitic acid
Petroleum oils
Potassium chloride
Potassium hydroxide
Potassium sulfate
Propane gas
Resin
Salt water
Sodium carbonate
Sodium bicarbonate
Sodium bisulfate
Sodium chloride
Sodium fluoride
Sodium hydroxide
Sodium metaphosphate
Sodium nitrate
Sodium perborate
Sodium peroxide
Sodium phosphate
Sodium silicate
Sodium sulfide
Sour gas
Stannous chloride
Stearic acid
Sulfur chloride
Sulfur dioxide, dry
Sulfur trioxide, dry
Sulfuric acid, 5–75%
Tartaric acid
Tetraethyl lead
Titanium sulfate
Toluene
Trichlorethylene
Turpentine
Vegetable oils
Vinegar
Whiskey
Zinc chloride
Zinc sulfate

In order that the subject of pressure instruments and measurements may be treated as an entity, the summary, questions, and problems will be found at the end of Chap. 10.

10 Pressure Measurements, Electrical Transducers

Pressure gages which employ electrical elements in order to produce the desired indication may be classified as follows:

1. Gages which use the rate at which heat is conducted away from heated components as a measure of the pressure.
2. Gages which employ the amount of ionization of the gas as a measure of its pressure.
3. Gages which depend primarily upon the distortion or flexing of an elastic member to move or strain another elastic member.
4. Gages which depend primarily upon the distortion or flexing of an elastic member to change the capacitance, mutual inductance, or resistance of a circuit.
5. Gages which employ a change in resistance of a conductor, or conductors, to indicate the pressure.
6. Gages which employ magnetic or piezoelectric effects.
7. Gages which employ none of the above methods.

It is also true that we might classify electrical pressure gages in terms of the range of pressures measured. The first two groups listed above are essentially low-pressure instruments, whereas the remaining groups cover a range from medium vacuum to pressures of several hundred thousands of pounds per square inch. Figure 10·1 shows graphically how the different types overlap in their ranges.

Aside from the mechanical-type large-bulb McCloud, Knudsen, and Phillips gages, the electrical gages can measure lower absolute pressures than can their mechanical counterparts. They also have certain advantages if high-frequency pressure variations are desired. Otherwise, there is little to choose from between the two in terms of accuracy. The mechanical gages are all far simpler than the electrical ones. But the electrical signal often is more useful, and for that reason increased complexity of measuring components is tolerated.

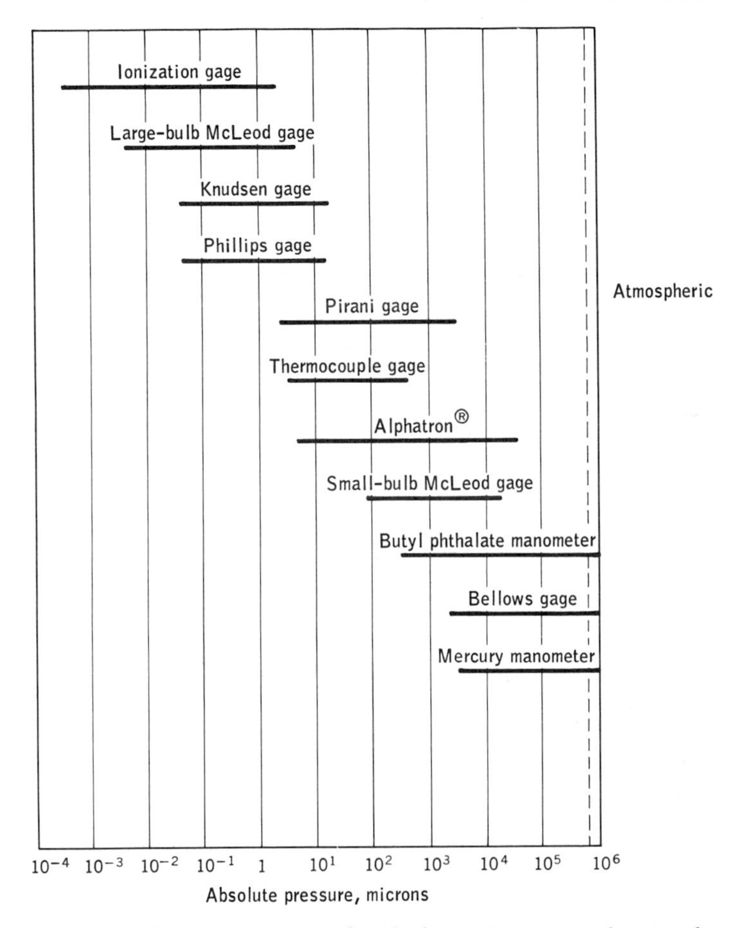

Fig. 10·1 Spans normally used with the various types of gages for measuring absolute pressure. (*Distillation Products Industries, Division of Eastman Kodak Co.*)

HEAT-CONDUCTION GAGES

The Pirani, thermocouple vacuum, and Knudsen gages all employ heated elements in order to achieve a pressure indication. The Pirani and the thermocouple type depend primarily upon the fact that the presence of a gas will help to cool a heated object. Inasmuch as the resistance of a conductor and the output of a thermocouple are both functions of temperature, differences in temperature with respect to some standard of reference are attributed to the amount of gas present.

The Pirani gage. The schematic diagram of the Pirani vacuum gage is shown in Fig. 10·2. It will be noted that the gage is essentially a Wheatstone bridge having a regulated power source to ensure constant

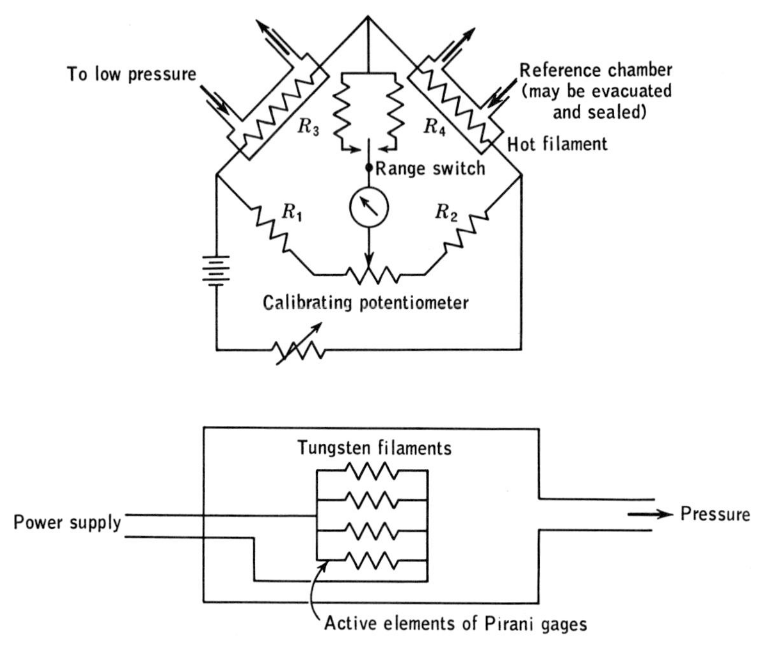

Fig. 10·2 Pirani electrical pressure transducer. This is one of the most important of the instruments to be studied. It will be found in many instruments which are not concerned with the measurement of pressure. Analysis instruments often employ the basic principles of this gage.

heat input to the measuring cell and the temperature-compensation cell. In the temperature-compensation cell, the pressure has been reduced to about one micron of mercury. Under these conditions, the energy lost by the heated wire will be transmitted by radiation or lost along the input leads. The measuring cell has the same energy delivered to it, but it can lose heat by convection and conduction to any gas which is present, in addition to the ways that the reference cell loses heat. With equal energy inputs, but with a greater cooling capacity, one expects the measuring cell to have a lower temperature which will be reflected in a lower resistance than that of the reference unit. This will create a condition of voltage unbalance which can then be measured in any one of a number of ways.

The Pirani gage is simple in theory and in construction, but its calibration depends upon the pressure of the gas which is to be measured. Such gases as argon, carbon dioxide, air, water vapor, helium, acetylene, and hydrogen may normally be used with gages of this type. Figure 10·3 shows how the performance of the instrument varies with the gas. Figure 10·4 shows that for very low pressures, the output voltage is almost a linear function of gas pressure. With increasing pressure, the response

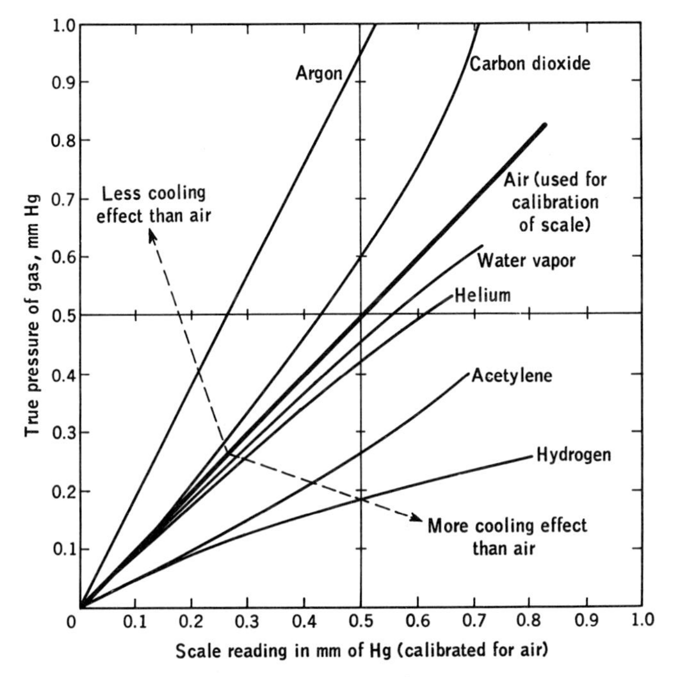

Fig. 10·3 The relative responses of different gases in a Pirani gage, compared with air as a standard. (*Douglas M. Considine, ed., Process Instruments and Controls Handbook, McGraw-Hill Book Company, Inc., New York, 1957.*)

rapidly levels off and thus effectively determines the upper limit at which the Pirani gage may be used.

The thermocouple gage. In many ways, the thermocouple vacuum gage closely resembles the Pirani gage, not only in construction but also in its pressure range and span. The essential elements of such a gage are shown in Fig. 10·5. There are two identical cells carrying resistance elements which are heated by the passage of a current through them. As before, one cell is evacuated and is used as a reference unit while the other one can be connected to the unit whose pressure is sought. The presence of any gas in the measuring cell cools the resistor below the temperature of its corresponding unit in the reference cell. If a thermocouple is attached at the middle of each resistor, the thermocouples will develop an emf which is proportional to the temperature difference of their junctions. Inasmuch as both sets of thermocouples can have equal reference temperatures, their output emf, when connected in series opposing, will reflect the difference in temperature of the hot junctions. Figure 10·6 indicates what corrections must be made when gases other than air are measured.

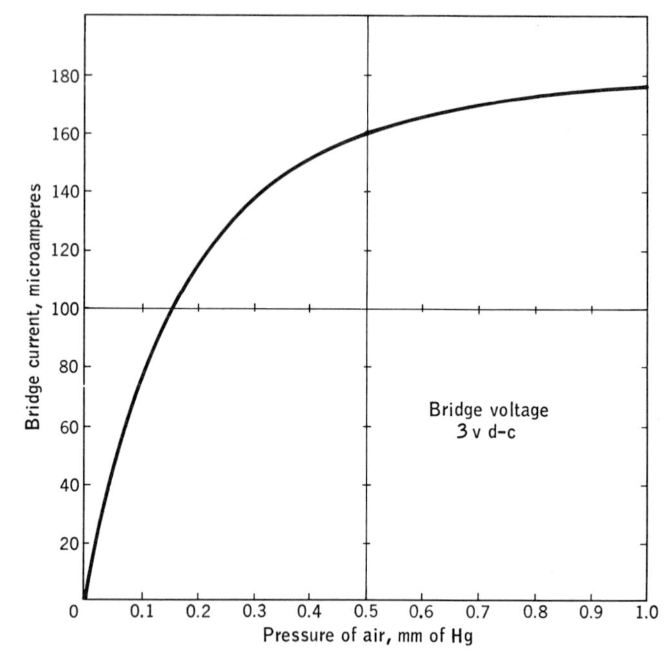

Fig. 10·4 Typical response of a Pirani gage when plotted against pressure. (*Douglas M. Considine, ed., Process Instruments and Controls Handbook, McGraw-Hill Book Company, Inc., New York, 1957.*)

Fig. 10·5 The commercial version of the thermocouple type of pressure gage employs a regulated power source to maintain sensitivity and performance. As the pressure of the measuring chamber approaches that of the sealed reference chamber, the filament temperatures approach equality. The thermocouple outputs will then be equal, and their difference voltage will approach zero. (*Minneapolis-Honeywell Regulator* Co.)

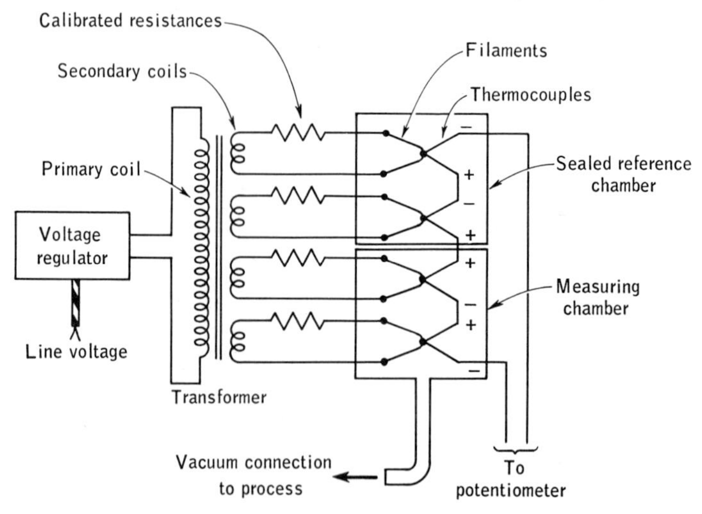

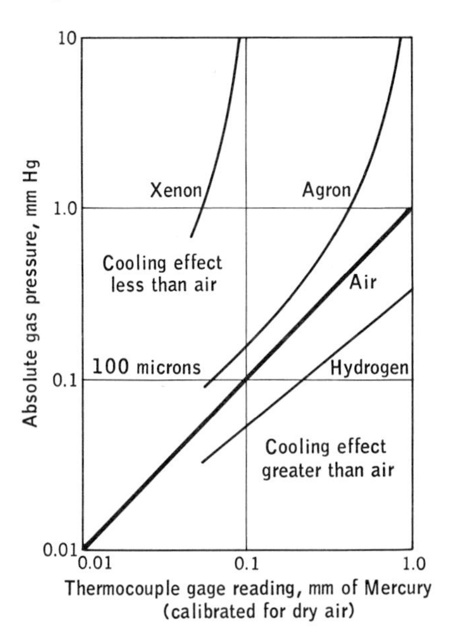

Fig. 10·6 The composition of the gas will seriously affect the accuracy of the gage if the cooling effect of the gas is materially different from that of air. (*National Research Council.*)

IONIZATION-TYPE VACUUM GAGES

There are two types of ionization gages; one is very much more responsive to low pressures than the other. Actually, the lower limit of the one is the upper limit of the other. The chief difference lies in how the gages create their ionizing effect.

The hot-filament ionization gage. In many ways, this type of gage very closely resembles an ordinary three-electrode radio tube. It has a heated cathode to furnish electrons, a grid, and a plate. But the grid and the plate do not function in the conventional manner.

In Fig. 10·7 the essential elements and their operating voltages are indicated. It will be noted that the grid is maintained at a positive potential of approximately 150 volts, while the plate is maintained at a negative potential of about 25 volts.

The high positive voltage on the grid accelerates the electrons away from the cathode. But because of their speed and the relatively wide spacing between the turns of the grid, most of the electrons continue past the grid. As they move, they encounter some of the gas molecules which may be present. If the electrons have sufficient energy when they collide with the gas, the gas molecules may be divided into positively and negatively charged particles. The positive ones will now be attracted by the negatively charged plate. The flow of ions to the plate will establish a current flow which can be detected or used in a number of ways.

The effect of the high temperature of the cathode is to dissociate the gas whose pressure is being measured. Consequently, the indicated pressures are a function of the gas composition.

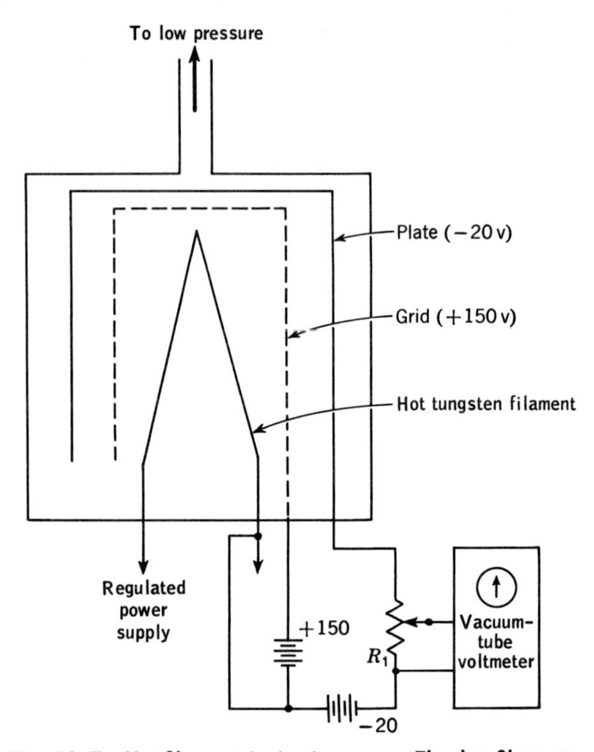

Fig. 10·7 Hot-filament ionization gage. The hot-filament ionization gage employs the electrons freed by the filament. As the electrons are accelerated toward the positive grid, any positive ions which are formed by the collision of electrons and gas molecules should be attracted to the negative plate. This will create a current through resistor R_1, which will then create a potential which will be measured by the vacuum tube voltmeter.

Radioactive-source ionization gage. The high temperature at which the filament, or cathode, must operate in order to furnish an adequate number of electrons creates a number of operating problems: (*a*) the cathode can oxidize if appreciable amounts of oxygen are present, (*b*) the gases being analyzed may be dissociated because of the cathode temperature, and (*c*) the cathode may be attacked by the gases, or it may enter into undesirable chemical combinations with them.

If a source of radioactive material is installed within the pressure chamber, Fig. 10·8, the particles given off by the radioisotope can ionize the gas molecules in much the same way that the fast-moving electrons could in the hot-filament gage. In the Alpha Ray ionization gage, the energy available from the alpha particles is much less than is available in the hot-filament gage. Consequently, the sensitivity of the Alpha Ray

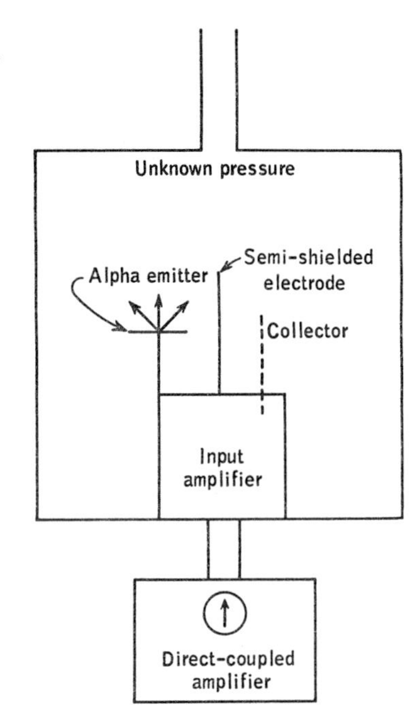

Fig. 10·8 Alphatron ionization gage. The absence of a hot filament makes possible the measurement of pressures higher than those obtained with the heated-filament design. However, the gain is obtained at the expense of sensitivity.

gage is much less than that of the other design. On the other hand, the Alpha Ray gage is not damaged by oxidation of a cathode or filament, and it can follow rapid decreases of pressure with a greater degree of fidelity. The number of ions available in this device is also dependent upon the net atomic number of the molecule.

For the measurement of pressures higher than those than can be measured by the gages previously considered, we must resort to the instruments which are primarily mechanical in nature but which can be connected to transducers of one sort or another. Some of these transducers will be very simple in principle and construction, while others will be complex on both counts.

MECHANICAL-ELECTRICAL PRESSURE TRANSDUCERS

There are many situations in which the use of the usual mechanical pressure, level, flow, or temperature gage is not desirable for any one or more of a number of reasons. The distance over which the signal must be transmitted, the range or span of desired readings, the speed of transmission, the ease with which the signal may be combined with some other value, or the type of presentation desired may preclude the use of the instruments which have previously been studied. If we do not use mechanical elements, then the chances are very great that we should employ some sort of electrical system.

In all probability, we shall find that the mechanical-electrical combination will be considerably more complex, that it will employ more basic principles, and that it will cost more. But we are willing to pay the price because of the greater utility of these systems for particular applications. In this chapter, we shall be concerned with the various ways by which pressure can be measured. This is a basic measurement, because flow and level are essentially pressure measurements, while temperature measurements also are often based on pressure changes.

Transducers. A transducer is a device for converting a signal of one sort into one of another type. Strictly speaking, a pressure gage is a transducer for converting the distortion of a Bourdon spring into an indication through use of a calibrated dial; a speedometer is a transducer for converting the rotary motion of its drive shaft into an indication of speed.

Pressure transducers. Because the measurement of pressure is of immediate concern, we shall start with the various transformations which are needed to convert pressure to some sort of usable electrical signal. Some of the more common means employ the following:

1. The change of resistance of an elastic material such as a wire
2. The change in capacitance when the distance between capacitor plates is varied or the dielectric is changed or altered
3. The piezoelectric effect: the conversion of a changing pressure into electrical voltage signals
4. The creation of a voltage by the movement of a coil within a magnetic field
5. The change in output voltage of a differential transformer as the core is moved
6. The generation of a voltage or current through the use of photoelectric effects
7. Torque-balance systems (electromagnetic)

Each method has its particular advantages and limitations. Therefore, we must learn the limitations as well as the outstanding virtues of each of the various types of transducer if we are to use them effectively. Each of the major types listed here will be studied in turn.

Resistance-change strain gages. In Chap. 3 it is pointed out that the resistance of a metallic conductor, for a constant temperature, varies directly with the length and inversely with the cross-sectional area. That is,

$$R = \frac{KL}{A} \qquad \text{ohms}$$

where K = a constant dependent upon the kind of wire
L = a length measured in the same units as those used in K
A = cross-sectional area measured in the same units as those used in K

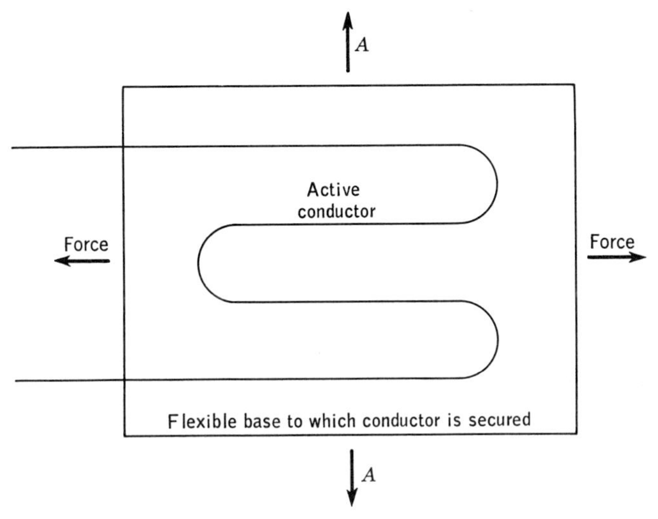

Fig. 10·9 Bonded type of resistance strain gage. The bonded type of resistance strain gage may have many sizes and shapes. For satisfactory performance, the elongation of the material must be such as to provide the maximum change in the length of the wire. Attaching the gage so that the stretch will be in the direction of arrow A will produce a negligible effect.

Inasmuch as the volume of a wire, for any given temperature, is constant, any mechanical load which would increase the stress would produce two effects:

1. The length of the wire would be increased.
2. The diameter, and hence the cross-sectional area, would be reduced.

Both of these effects will increase the resistance. The only problem, then, is to determine the best way of measuring the resistance in order to get meaningful results. The various bridges and amplifiers are considered in earlier chapters and will not be reviewed here. Whether we use a Wheatstone bridge, an alternating-current bridge, or a direct-current bridge is not of primary interest.

In theory, any wire of any size could be used to measure the strain of the sample to which it is secured. In practice, we find certain wire arrangements and dimensions to be definitely advantageous. As a result, these strain gages may be divided into (1) bonded and (2) unbonded types.

Bonded gages. The usual bonded gage, Fig. 10·9, consists of several loops of very fine wire which are cemented to a base or carrier sheet of paper or thin plastic. The wire has a diameter of about 0.001 in. and a nominal resistance, generally of about 120 ohms. The bridge current which flows through the element is about 25 ma. The base material is relatively weak; the mechanical strength of the gage must be essentially zero with respect to the sample under test. With a good bond between

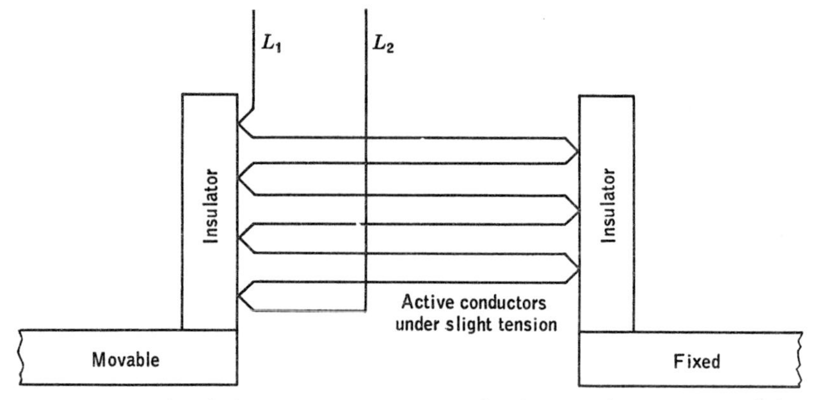

Fig. 10·10 Unbonded resistance strain gage. The shape and construction of the unbonded strain gage is governed only by the ingenuity of the user, the space available, and the amount of motion of the part.

the base and the subject, either tensile or compressive strains may be measured.

Unbonded gages. The unbonded gage, Fig. 10·10, differs from the bonded type in that the strain-sensitive resistive components are mounted on parts which have relative motion with respect to each other; this motion is independent of strain as such. The wire is applied with an initial tension so the device can measure both tension and compression. It should be understood that the wire can be strained repeatedly to about 2,000 μin./in. Such small total movement makes the installation critical; the gage must be subjected to total movements within its range.

Temperature effects. Because current flows through the strain gage element, there will be a heating effect which is proportional to the square of the current. Inasmuch as all the wires which are suitable from a resistance versus strain standpoint have an appreciable resistance-temperature coefficient, any change in resistance because of an ambient temperature change or because of current heating will have to be either considered or eliminated through any one of a number of circuits.

One of the simpler methods is to use a dummy gage through which the same current will flow and which will have the same ambient temperature as the active element. This is connected to the bridge, Fig. 10·11, so that everything but the resistive effect because of strain will have been canceled. The dummy gage, while not subjected to strain, should be close enough to the active gage so that its temperature is that of the active gage, and the material to which each gage is secured should be essentially the same in order that the heat transfer from each gage will be as nearly the same as is possible and practicable. The very small amount of power flowing through such fine wires can create very appreciable heating.

The bonded type of strain gage may be cemented to any elastic ma-

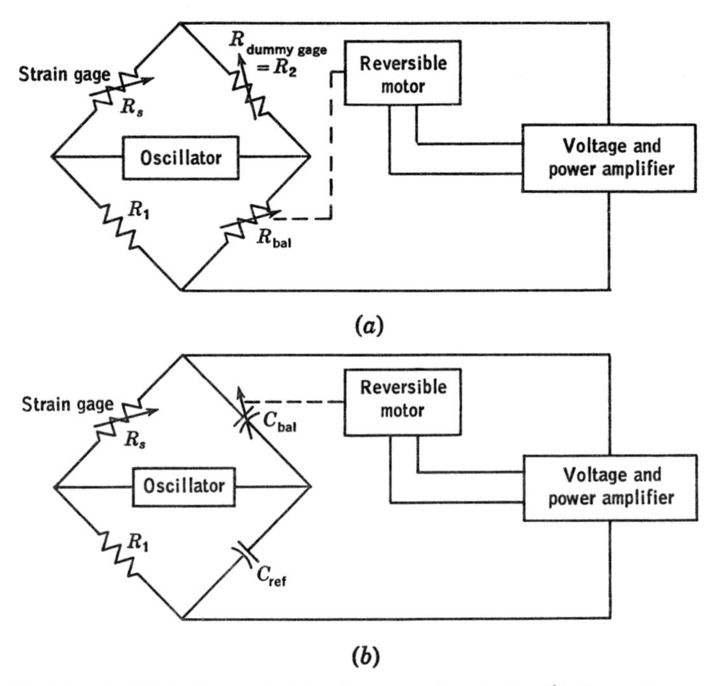

Fig. 10·11 (*a*) Resistance bridge for measuring strain. (*b*) Capacitance bridge for measuring strain. Within limits, any suitable system may be used for measuring the change in resistance of the gage, provided that it will not create serious problems of self-heating in resistors R_s and R_2. R_2 is the dummy gage.

terial whose temperatures and strains are within the limits of the gage and the cements with which it is attached. Strain gages are available in lengths of $\frac{1}{16}$ to 1 in. and with resistance from 60 to 5,000 ohms. Temperatures of 300°F can be sustained indefinitely, while for short periods of time 500°F represents a practical maximum.

The elastic medium to which the gage element is cemented may be the familiar Bourdon tube, the bellows, or a cantilever which is moved by a bellows or by two bellows working in opposition to produce a differential measurement.

Figures 10·12 and 10·13 show typical ways of employing the bonded gage. Table 10·1 shows some of the ranges and accuracies possible.

The type of servomechanism, amplifier, or meter used to balance the bridge or read its unbalance is dictated largely by the economics of the situation and the speed of response desired. For very high speed transient conditions one might use an oscilloscope and camera; for some other application a potentiometer recorder might be quite appropriate. The various types of read-out devices are discussed in detail in subsequent chapters.

Table 10·1 Ranges and Accuracy of Selected Bonded and Unbonded Gages

Sensing element	Type of gage	Ranges,* psi	Accuracy,* % full scale	Temp limit, °F
Bellows and cantilever beam	Bonded	0 to 10 0 to 2,500	±¼	−50 to +200
Bourdon tube and bellows	Bonded	0 to 10 0 to 50,000	±½	0 to +250
Bellows and diaphragm	Unbonded	0 to 0.1 0 to 10,000	±½	−50 to +225

* These accuracies and ranges should be available with either direct- or alternating-current bridges.

Variable-capacitance gages. It has been known for years that the value of a capacitor can be changed in many ways; some of the more common ones are to alter the area of the plates which are in proximity, to change the distance between plates, or to alter the value of the dielectric medium between the plates. It will have been noted in earlier chapters that there are many uses for capacitors and that a capacitor is basically nothing but two metal plates which are separated from each other by some type of insulator, be it air, oil, mica, or silk. Such a device can store small amounts of energy; it can block the passage of a direct current while permitting an alternating current to pass; it can change the

Fig. 10·12 Pressure-resistance transducer. Placing the strain gages on both sides of the cantilevered member will convert the dummy gage to an active one. This will increase the output, one unit being stretched while the other is being compressed. Any variable which can move the beam an adequate distance can employ such a design.

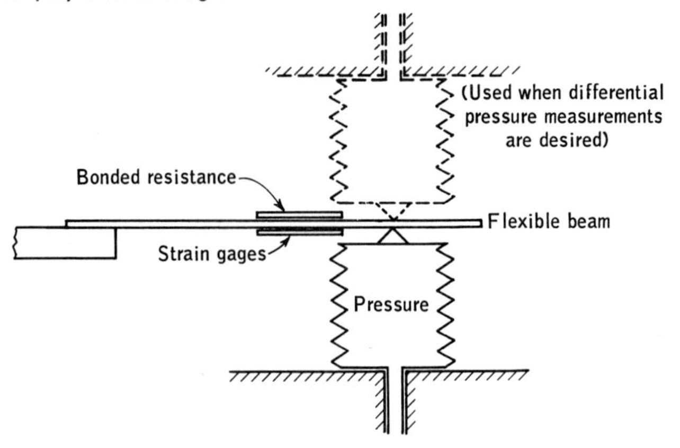

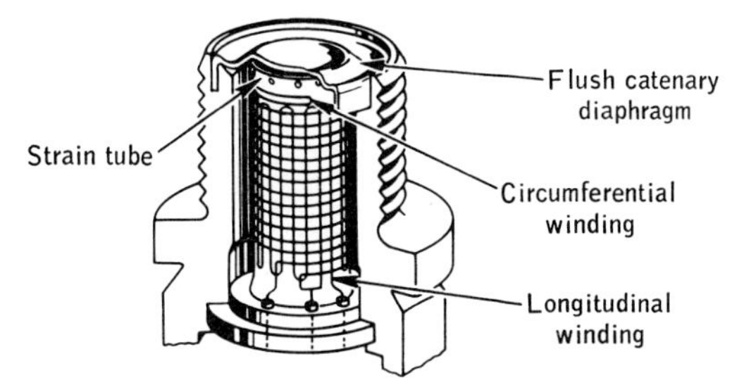

Fig. 10·13 Commercial air-cooled engine pressure transducer. By employing two separate windings, changes in the axial dimensions as well as circumferential strains can be used simultaneously to increase the output. An increase in the resistance of one unit would normally be accompanied by a decrease in the resistance of the other element. Temperature effects would also be nullified by such construction. (*Norwood Controls Division, Detroit Controls.*)

frequency of an alternating-current circuit; and it has a reactance which is inversely proportional to the frequency of the applied voltage. It is only reasonable to expect that capacitive pressure gages will employ capacitors for most of these purposes.

If only two parallel plates are used for the capacitor, the capacitance can be obtained from the expression

$$C = 0.225K \frac{A}{d}$$

where C = capacitance, $\mu\mu$f
A = plate area, in.^{2}
d = distance between plates, in.
K = dielectric constant

Should the capacitor be in the shape of concentric cylinders, then the following relationship holds:

$$C = \frac{0.614KL}{\log_{10}(b/a)}$$

where C = capacitance, $\mu\mu$f
K = dielectric constant of material between plates
L = length of cylinder, in.
b = inner diameter of outer electrode, in.
a = outer diameter of the inner electrode, in.

Figure 10·14 shows such a pressure-capacitance transducer in one of its very simple forms. The shapes of the capacitor plates and the manner in which the plates are moved relative to each other are primarily matters for the designer, who will employ the means which give him the maximum signal for the span of the variable to be measured, consistent with the other factors of dependability, long life, and so on.

By keeping plate D fixed with respect to plate P_2, the two plates form a *fixed*, or reference, capacitor with which the other one employing plates P_1 and D may be compared. If plate D were moved closer to P_1 and away from P_2, a greater change in the relative values of capacitance

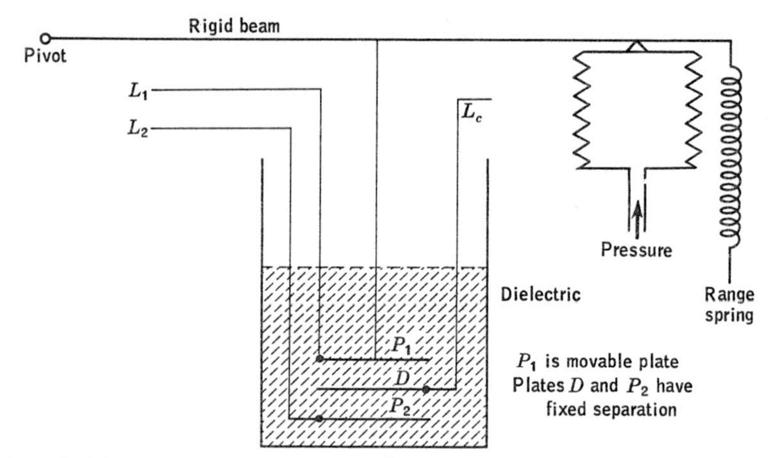

Fig. 10·14 Pressure-capacitance transducer. Changes in capacitance can control current, frequency, and power factor. By keeping the spacing between D and P_2 fixed, there is a standard value of capacitance which can be compared with the variable capacitor formed by plates P_1 and D.

would be found. Because of the mechanical arrangement of the plates, for example, it is desired to move only P_1. A suitable bridge circuit will make it easy to determine either the absolute values or the relative values of the capacitances. For example, the two capacitors could replace two of the resistors in a conventional Wheatstone bridge, and by suitable calibration, the resistance values required for balance could be correlated with the pressure which acted upon the bellows to change the position of plate P_1. It should be recognized that a Bourdon spring or some source of motion could also be used to move the beam.

Potentiometric devices. One of the simplest of the pressure-voltage transducers is shown in Fig. 10·15. This has been used for many years and under all sorts of conditions where there is sufficient force available to position the potentiometer. Normally, a bellows can develop adequate force, but a Bourdon tube is incapable of developing any significant amount of torque. If the Bourdon tube or some other elastic element such as a diaphragm is to be used to position it, the potentiometer must be of a design that has reduced the frictional drag to an insignificant amount.

Provided that adequate torque is available to perform the necessary functions, the potentiometer provides one of the best methods of converting a pressure signal into an electrical voltage:

1. The signal applied to the potentiometer may be relatively large, of the order of several volts if necessary.
2. Potentiometers which have extremely linear outputs can be obtained.
3. The use of a large output signal may obviate any requirement for high-gain amplifiers.
4. By the use of potentiometric methods, the signal can be sent relatively long distance.

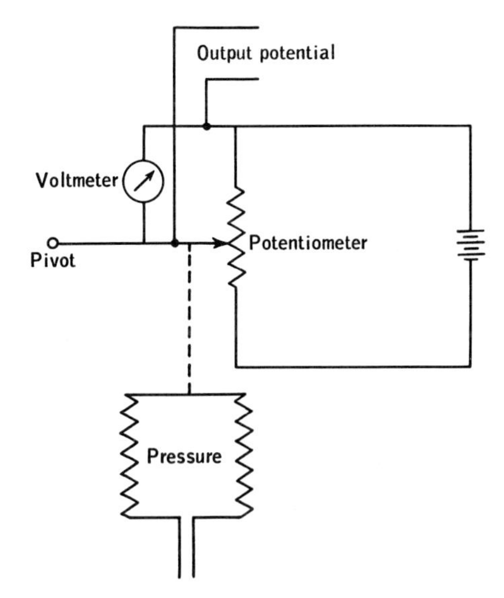

Fig. 10·15 Pressure-voltage transducer. A bellows should be able to position accurately a potentiometer having little friction. Levers may be used to match the required movement of the potentiometer with that of the bellows.

One method of obtaining a remote-reading transducer is shown in Fig. 10·16. In this instance, a potentiometer is positioned at the receiving station to maintain the galvanometer in the *zero* position. The absence of current in the galvanometer circuit indicates that the two movable contacts are at the same potential with respect to some arbitrary point, and the receiver then faithfully reflects the position of the movable contact on the transmitter.

Fig. 10·16 A potentiometric type of transmitter-receiver system. Remote-reading systems may employ such a simple, manually adjusted system or extremely complex servomechanisms. In this system, the correct value is indicated only when the potentiometer receiver is adjusted to bring the galvanometer to the center, or zero, position.

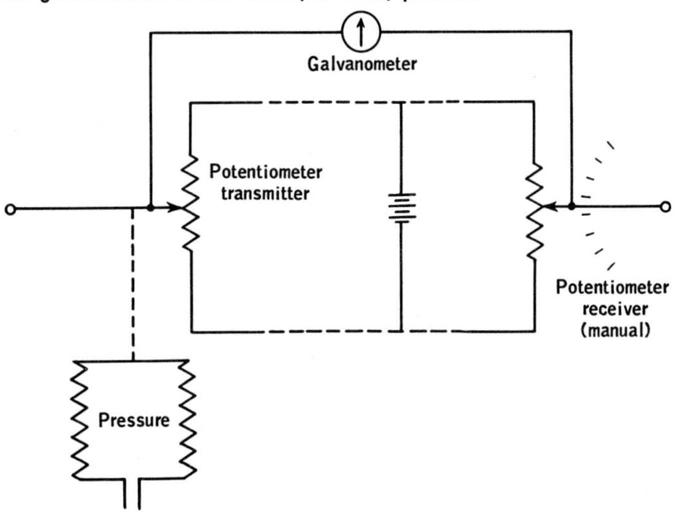

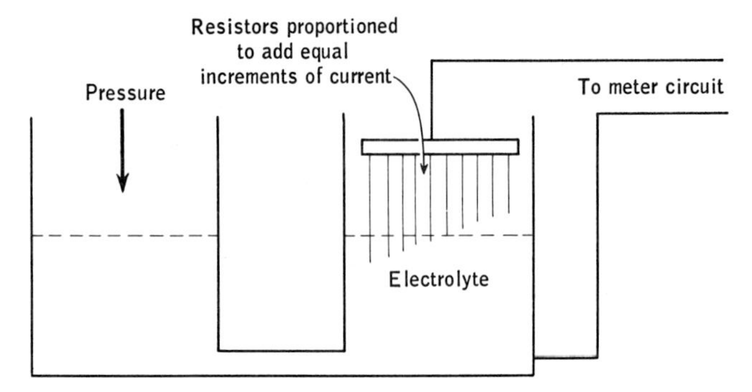

Fig. 10·17 Pressure-current transducer using electrolyte. Pressure-current transducers of this type have the great advantage of having negligible friction. But because of their mass, they cannot respond to rapid pressure fluctuations.

Manometer transducers. Another very simple way of converting pressure measurements into electrical signals is shown in Fig. 10·17. While this is represented as a pressure measurement, it could easily be a differential pressure measurement with pressure applied to each of the two surfaces. With the arrangement shown in the figure, increases of pressure will force the electrolyte down in the left-hand leg of the manometer and up in the right-hand leg. As the level of the electrolyte rises in the right-hand leg, it will touch more and more of the resistors which have been cut to such a length and so positioned that when the liquid contacts each one in turn, the current in the meter circuit will increase by essentially equal amounts. For accurate and essentially stepless meter positions, there would have to be almost a hundred of these resistors.

Mercury is often used for the electrolyte because it is an excellent conductor of electricity and, at the same time, quite an extensive pressure range can be covered with a device of modest size. If desired, almost any electrolyte may be used with either direct-current or alternating-current power supplies and meters. Polarization and electrolysis would have to be guarded against. This type of device eliminates all problems caused by mechanical movement and friction.

A simple modification of the same components produces the transducer shown in Fig. 10·18. Resistance wires are inserted in each leg of the manometer, which may measure gage pressure, absolute pressure, or differential pressure. As the level of the electrolyte changes in each of the legs because of pressure changes, the Wheatstone bridge becomes unbalanced. Balance can be restored by varying R_2. With the inherently high sensitivity of such a bridge, changes in level of less than 0.001 in. (down to 0.000030 in.) are easily determined. As in the preceding example, the electrolyte may be mercury or some other suitable liquid which

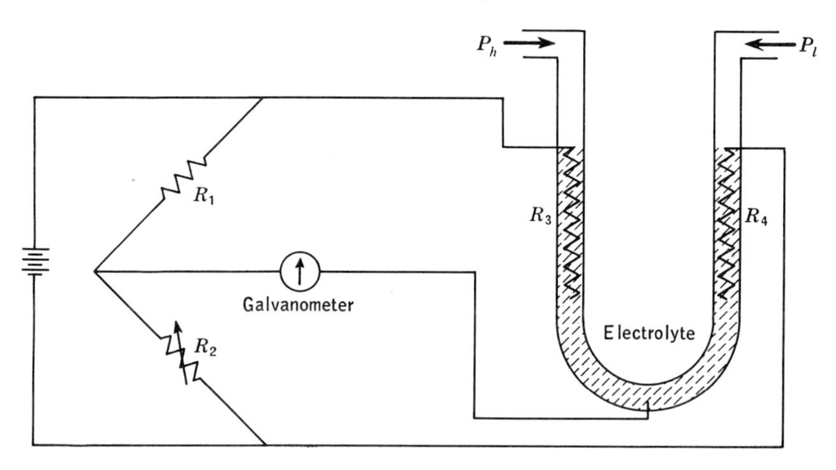

Fig. 10·18 Pressure-resistance transducer. A continuous resistor can be used in place of a series of separate resistors. Whether used for gage pressure, absolute pressure, or differential pressure measurements, this design can obtain indications of extremely small pressure changes. Differences in level of 20 to 30 μin. are possible.

will not react adversely with the resistance wires or polarize with the current.

Torque-balance pressure gages. In one design of a torque-balance pressure gage, the motion is produced by a bellows. It should be pointed out that these transducers will work with any suitable device that can apply a force of the proper magnitude. A motion or a force which can be applied through a spring, attached to the beam at such a point that it is possible to obtain a torque balance, should make the device function. An increase in pressure, for example, will rotate the rigid beam, Fig. 10·19, and move the shutter so that more light impinges on the photocell. This changes the photocell resistance and increases the grid bias in the triode tube. As a consequence, the tube resistance decreases and more current passes through the electromagnet. The opposition of the two magnetic fields drives the beam in the opposite direction until a condition of torque equilibrium has been reached.

Because of the very high sensitivity of the photocell-amplifier system, there is very little net movement of the beam. A block diagram, Fig. 10·20, shows the relationships of the various components.

Oscillator-type pressure-current transducers. A technique that has been used for many years and for many different purposes is one that requires a coil in the plate or grid circuit of an oscillating tube (one that is generating alternating currents) to deliver current to some other circuit or component. The energy may be transferred into some other winding, in

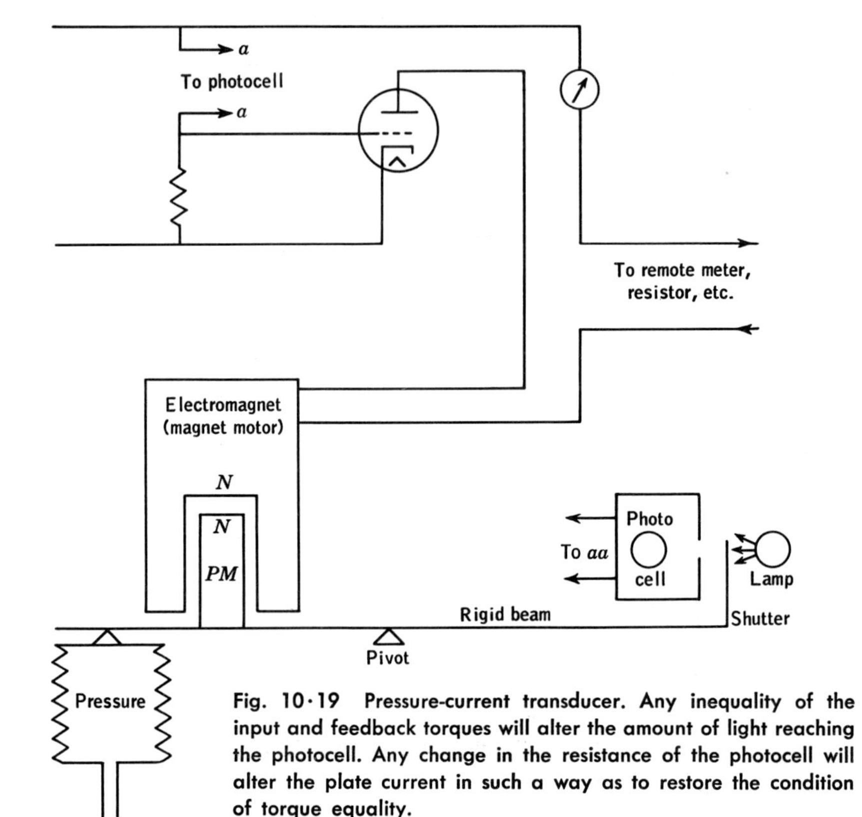

Fig. 10·19 Pressure-current transducer. Any inequality of the input and feedback torques will alter the amount of light reaching the photocell. Any change in the resistance of the photocell will alter the plate current in such a way as to restore the condition of torque equality.

which case we have transformer action, or if a conductor in the form of a short-circuited turn or a piece of metal plate is placed close to the winding, power will also be transferred to it by more or less conventional transformer behavior. In any event, if the coil has to deliver more power to the secondary winding, the additional power must come from the wind-

Fig. 10·20 A schematic representation of the essential elements of a transducer which converts an input pressure or mechanical torque to an electrical analog.

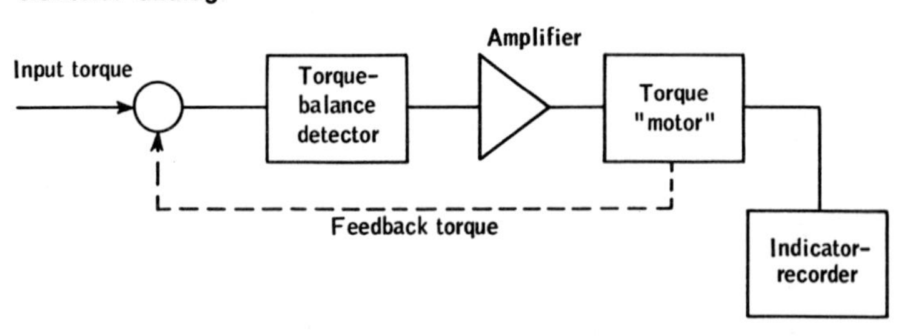

ing in the plate or grid circuit. Such an increase in power will become evident by an increase in the plate current. This increase in current may be used to do any one of a number of things: it may cause relays to operate at some predetermined values of current; it can be used to develop voltages across known resistors; it can cause electromagnets to establish forces which can be used to measure an applied force.

Figure 10·21 shows in schematic form a transducer employing some of these principles. If the pressure in the bellows increases, the beam pivots about its fulcrum and moves the metal vane (an extremely light piece of material, perhaps of the order of aluminum foil) closer to the oscillator coil. In this position, the vane has more energy transferred to it by the alternating magnetic flux which is created by the coil. Any such increase in coil energy is accompanied by an increase in the plate current which passes through the windings of the magnet-motor whose magnetic field opposes that of the permanent magnet. The mutual opposition of these two magnetic fields causes the beam to be pushed down

Fig. 10·21 Basic Microsen pressure-current transducer. Any lack of torque equality will move the metal vane with respect to the oscillator coil and in this manner control the current to the electromagnet so that a position of equilibrium is produced.

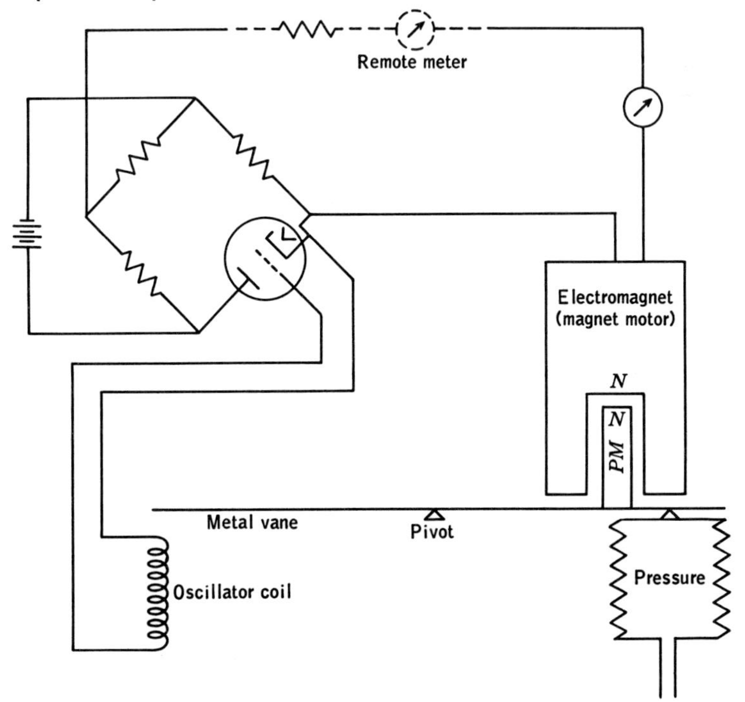

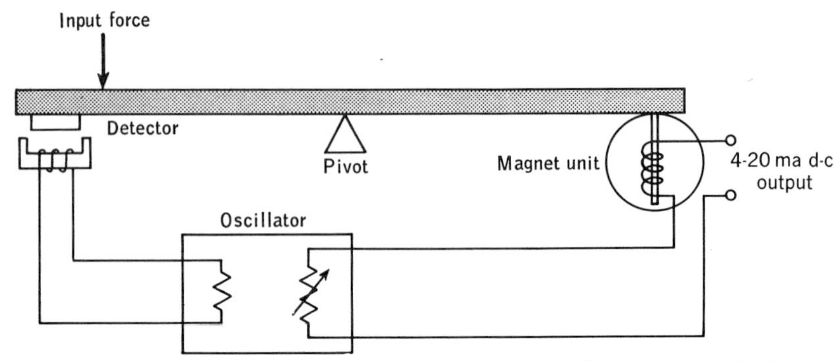

Fig. 10·22 Pressure-current transducer. Any change in the position of the beam with respect to the detector coil will vary the inductance of the oscillator circuit. This will produce a change in the output current, which must again establish a condition of torque equilibrium with the input pressure signal, for example. *(Minneapolis-Honeywell Regulator Co.)*

on the right-hand end. This moves the metal vane away from the oscillator coil until a condition of balance has again been obtained. Because of the extremely small movement of the metal vane that produces very large current changes in the tube current, there will be an almost imperceptible physical movement over the normal range of the instrument.

Any phenomenon which creates pressure changes or movements within the range of the device can be used with the device. Pressure is shown because it is so basic.

Perhaps almost any means of changing the inductance or capacitance of an oscillating circuit might be used for purposes of controlling the output current. Figure 10·22 indicates how the position of the beam controls the inductance of a circuit whose value determines the magnitude of the plate current. As the degree of oscillation decreases, the plate current rises; it is highest when all oscillations have ceased. This increased plate current then provides a feedback to the other end of the beam, thereby creating a torque which acts in opposition to the input force, which might have been created by some pressure-sensing device. When the two torques become equal, the beam will have been moved slightly from its former position of equilibrium. The current required to bring about this condition of equilibrium then becomes a measure of the input torque.

The major advantage of transducers which employ oscillators is that they can develop relatively large current changes for beam or vane movements of less than 0.001 in. In this manner, the physical conditions remain essentially unchanged throughout the operating range. For example, the magnetic pull developed by a solenoid is a linear function of current only so long as the magnetic circuit remains unchanged. If an air gap changes its dimensions, the forces developed are no longer de-

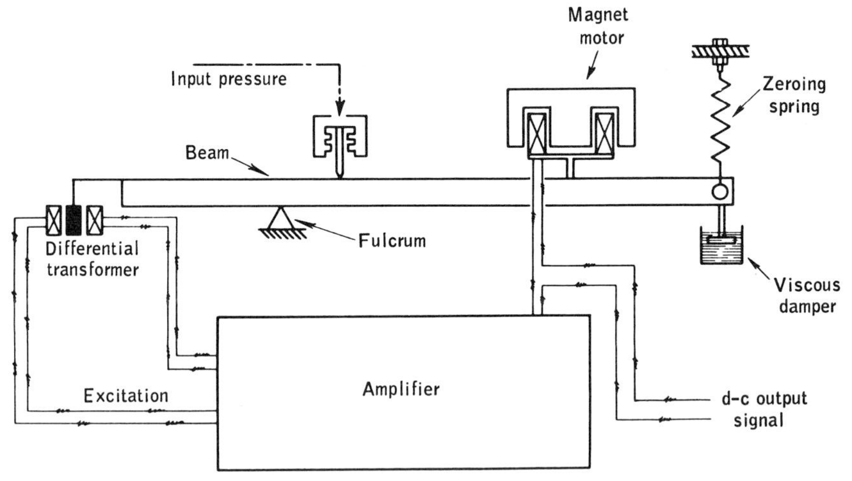

Fig. 10·23 Pressure-current transducer. The use of an alternating-current differential transformer to detect any difference in the input and output torques is somewhat of a departure from the other designs. It will be noted that the output of the amplifier is a direct-current signal which is used in the comparison magnet-motor to provide torque equality. (*Hagan Chemicals and Controls, Inc.*)

pendent only upon the current; they now must consider the increased (or decreased) flux brought about by the change in the air gaps.

The device shown in Fig. 10·23 employs still another way to obtain torque balance. In this device, a pressure bellows moves one end of a feedback beam and the other end of the beam drives a differential transformer to generate an output voltage proportional to the beam movement. The amplifier then converts the potential signal to one of current which can be used in the feedback circuit as well as in the external circuit.

ELECTROPNEUMATIC TRANSDUCERS

In industrial practice, the vast majority of power-operated valves and other final control elements are pneumatically operated. Therefore, it is of great importance that we be able to convert to pneumatic signals which control our process those electrical signals which we have been striving to obtain because transmission distances are long or because extremely high speed of transmission is necessary. One such converter is shown in Fig. 10·24. The appropriate electrical signal is fed to the coil of the electromagnet and there creates a magnetic field which acts in opposition to that of the permanent magnet secured to the rigid beam. As the opposition forces increase, the beam is forced down on the left end and up on the right. This movement of the beam brings the baffle closer to the orifice, which immediately increases the pressure within the tube. A movement of only a few thousandths of an inch will be enough to vary the pressure within the tube from almost zero, if the restriction has been

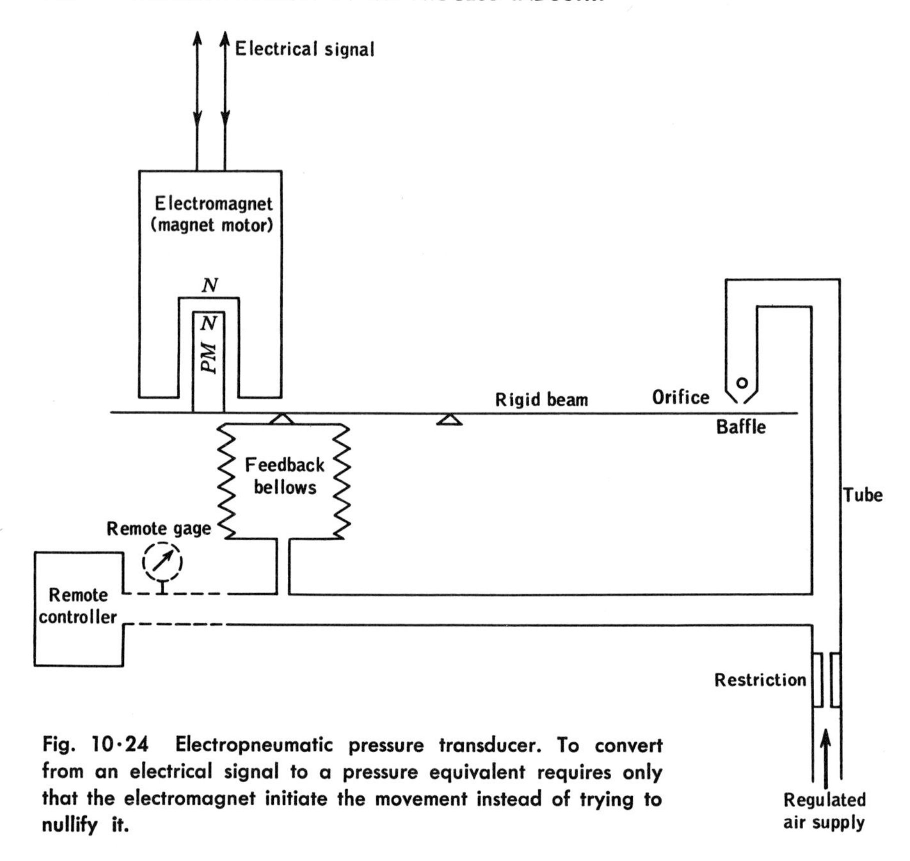

Fig. 10·24 Electropneumatic pressure transducer. To convert from an electrical signal to a pressure equivalent requires only that the electromagnet initiate the movement instead of trying to nullify it.

properly selected, to almost that of the air supply. Thus, as the baffle approaches the orifice, the pressure increase within the tube is communicated to all parts of the system, including the feedback bellows, which now acts to oppose the motion created by the electrical signal. Very quickly a position of balance has been achieved, and an air pressure which is representative of the electrical signal has been created within the measuring system.

It was noted previously that the metal vane had to move an extremely small amount in order to control the oscillator tube current; so too does the baffle have to move but a short distance in order to create the necessary torque balance.

For an even more precise translation of current into air pressure, the nozzle and baffle are modified so that any movement of the beam is effectively amplified by a factor of 40 to 50, Fig. 9·31. Consequently, if the baffle need move 0.001 in. to create the necessary change in the pressure within the tube, the rigid beam will have to move but 0.000020 in. Under such conditions, the electromagnet and the permanent magnet *PM* always

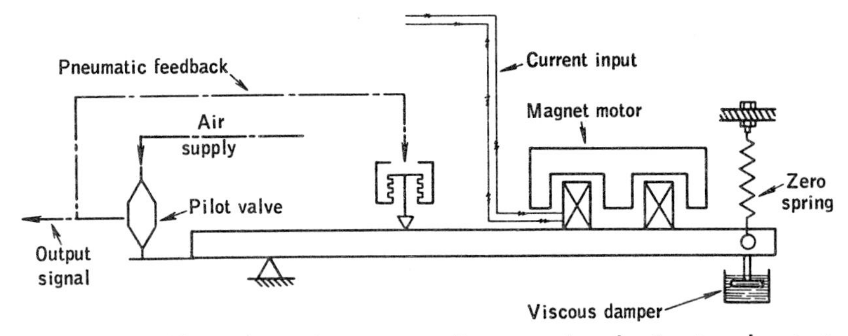

Fig. 10·25 Electropneumatic converter. By comparing the input and output torques, the equivalence of input and output signals becomes almost a certainty. (*Hagan Chemicals and Controls, Inc.*)

operate at the same point, and the force of repulsion set up between the two is a linear function of the current in the coil.

Figure 10·25 suggests an alternative design. This electropneumatic converter is a torque-balance device used for the conversion of single direct-current signals into pneumatic signals. In operation, the torque created by the magnet motor is balanced by the pneumatic feedback from the output supply system. This ensures the linear relationship between input and output signals.

Transducer-positioner. The conversion of electrical signals to pneumatic signals for purposes of operating valves or other apparatus is extremely important. There are numerous ways of making such conversions. In many cases, there will be some type of feedback so that the incoming electrical signal can be compared with the converted function in order to ensure correct transformation.

In one such design, shown in Fig. 10·26, the electrical signal of 1 to 5 ma is used to control the air delivered to an air motor which positions a valve—it might be any other device or object which is being moved or positioned. A magnetic motor is used to control the baffle position. A permanent magnet establishes equal fluxes and equal flux densities in the two paths which are in parallel with the permanent magnet. As these fluxes pass through the soft-iron armature, they create equal and opposite torques on the armature. But if a current is allowed to pass through the coil, it too will create a flux and establish a magnetic polarity making, for example, the left-hand end of the armature a north pole and the opposite end a south pole. This then will result in repulsion between the like poles and attraction between the unlike poles. This produces a clockwise rotation which is resisted by the flexure pivot and, at a slightly later time, by the springs which act as the feedback element. Let us assume that the current in the armature coil has increased, thereby producing a torque which moves the armature and the baffle, which is mounted on

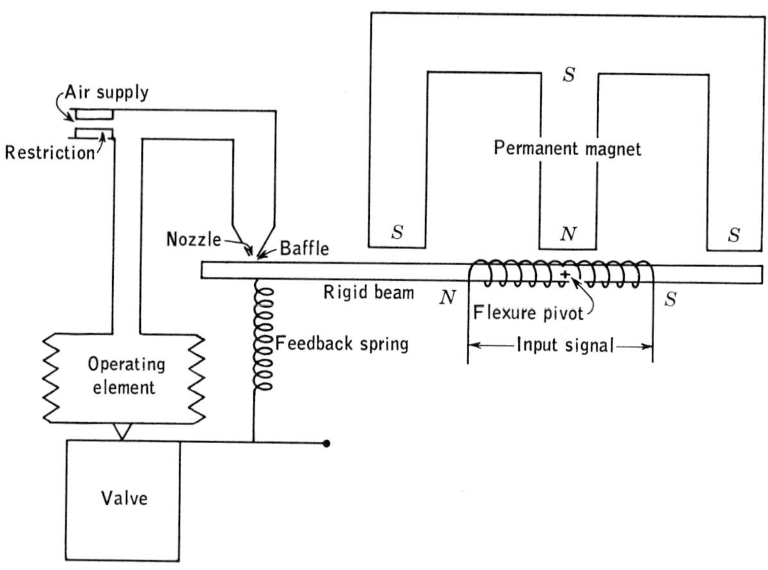

Fig. 10·26 Current-position transducer. By using mechanical feedback, the performance or position can be made to conform to the value of the input signal. Current varies between 1 and 5 ma. It never reverses direction, so increasing values of current always try to produce clockwise rotation. (*The Taylor Instrument Cos.*)

the armature, nearer to the orifice, which is above the baffle. As the baffle approaches the orifice, the pressure within the orifice and its associated pneumatic amplifier will increase. This will apply additional air pressure to the valve and drive it down. As the valve is driven down it will pull down with it the spring which is also connected to the armature and baffle. Stretching the spring will apply extra force to the armature and move it away from the orifice. In this way, a final condition of balance will have to be achieved between the torque created by the current in the armature coil and the two opposing torques established by the flexure pivot and the spring which measures the amount of movement of the valve.

The accurate conversion of one quantity into some analog of it requires that the output value be compared in some way with the input. Without such a comparison there is no certainty that the input and the analog are at all times equivalent.

Because of the high pressures which are required in many industrial processes, many of the devices previously studied for converting pressure to electrical equivalents are not satisfactory. **U**-tube manometers and bellows are restricted to the lower pressure ranges. For the higher ranges it will be necessary to employ such a differential pressure converter as is shown in Fig. 10·27. It will be noted that this is a torque-balance type of device in which a current span of 10 to 50 ma must furnish the balancing torque created by the input differential pressure. In order that the one current span will be adequate, the range wheel can be so positioned that the moment arms can be altered to permit torque balance with the one span of current, 10 to 50 ma. The balancing is strictly analogous to the situation in which one child uses a seesaw with several playmates who may be heavier or lighter than he is. If he retains one position on the board, then the fulcrum must change with each change in weight of the second child.

Any unbalance in the torques will move the laminated core from its mid-position in the differential transformer. As the core is displaced, the differential transformer output is no longer zero and is fed into a high-gain amplifier. Here the signal is converted into an output current

Fig. 10·27 High-pressure pneumatic-electric converter. When differential pressures must be converted into electrical equivalents, it may be necessary to change the form of some of the components. However, all of the principles are retained. The torque created by the output current must be equal to the torque created by the input differential pressure. The means of measuring equilibrium and obtaining it will vary with the design. (*The Foxboro Co.*)

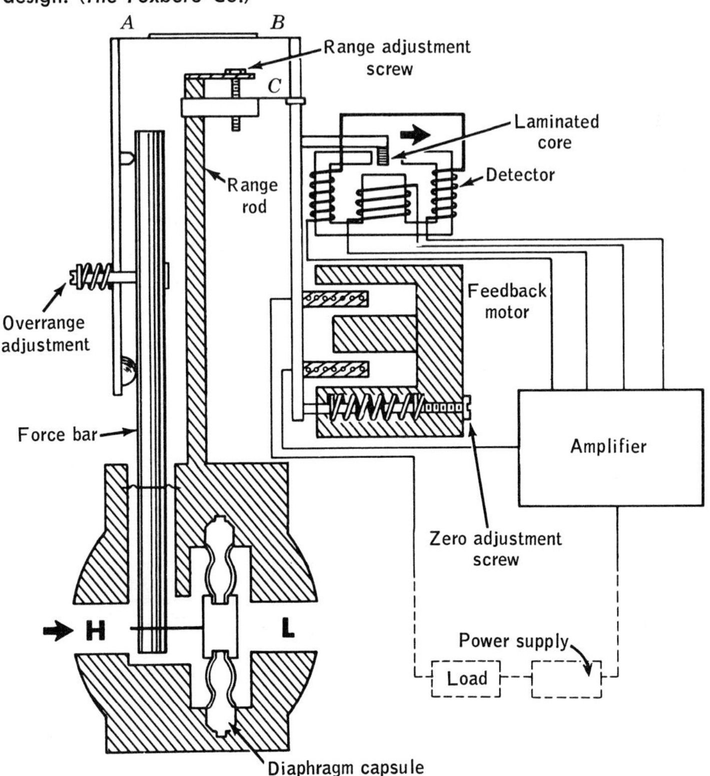

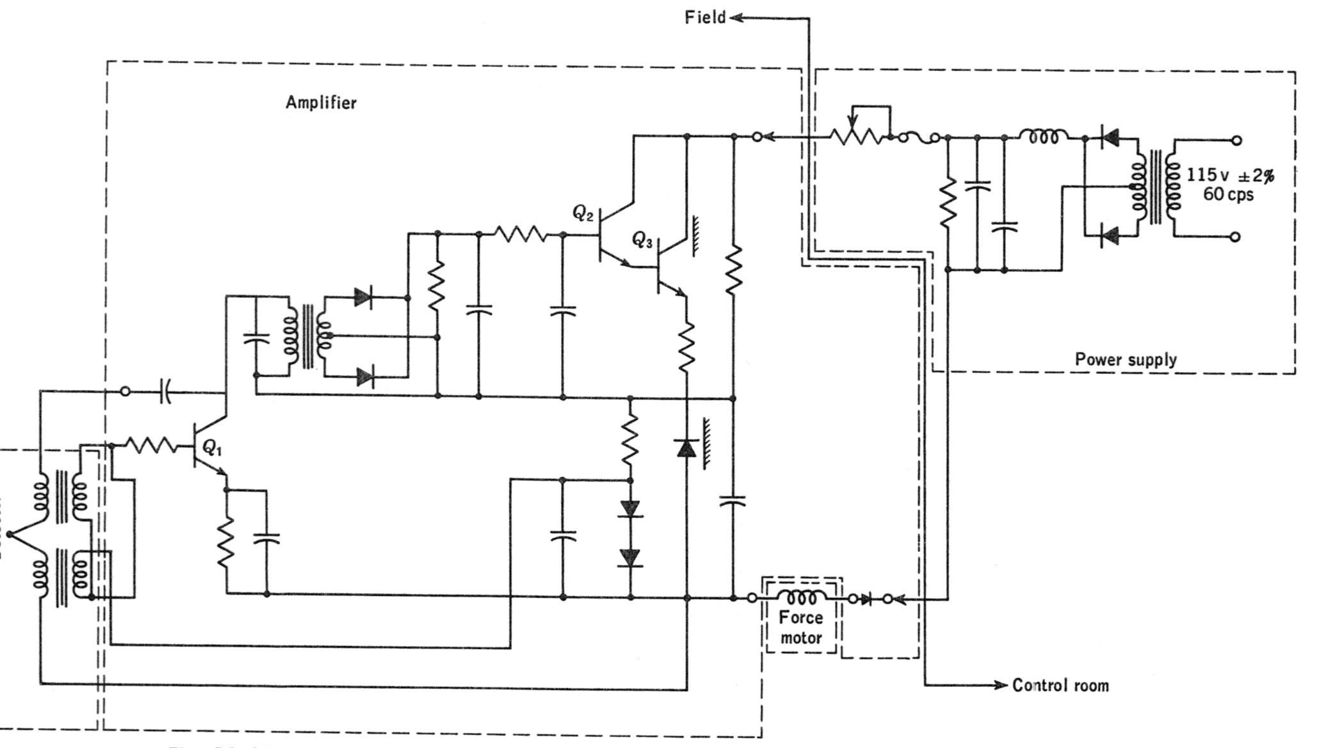

Fig. 10·28 A transistorized version of a torque-current converter. The value of the differential pressure may be determined by means of a milliammeter which can be connected in series with the line. Or the potential developed across the receiver resistor may be used as a measure of the differential pressure. (*The Foxboro Co.*)

which produces a torque in the feedback torque motor. As the current increases, the repulsion between the "like poles" increases, trying to restore the motor armature to the mid-position. It can never regain the exact center position, because to do so would mean that the output signal going to the amplifier would be zero. With no input to the amplifier, the output current to the torque motor would also fall to zero. Consequently, the torque motor can only approach the mid-position; it can never attain it. All devices with amplifiers of this sort require that there always be an error, no matter how inconsequential, for their operation. The circuit is shown in Fig. 10·28. An amplifier that requires no error for its continuous operation is considered in Chap. 15.

OTHER PRESSURE TRANSDUCERS

Magnetic-circuit transducer. For a fixed coil operating on a fixed commercial or oscillator frequency, the only variable left is that of the magnetic circuit. In contrast to air, which is a relatively poor carrier of magnetic flux, iron and many of its alloys and relatives are exceedingly fine conductors of magnetic flux; some are of the order of 200,000 times better than air. Consequently, when we substitute some of the iron family for air, more flux will be created by the coil for any given value of current. This in turn will create a greater voltage of self-induction. The voltage of self-induction can never exceed the applied voltage; if it did, the coil would have become a generator and would be delivering power without power being supplied to it. But as the voltage of self-induction increases, the difference between the applied voltage and the induced voltage becomes less, and the current decreases proportionally. The current never reaches zero, but it may come close to zero in devices having high-permeability materials and low-reluctance flux paths.

We may conclude, therefore, that as a magnetic material is inserted into a winding, more flux will link the winding for any given current in the winding. This will produce a greater voltage of self-induction which in turn will reduce the current. When the coil is connected to a fixed voltage source, the position of the magnetic core will in large part determine the current. Consequently, the core position and the current have a definite correlation and can be so used.

For a given position of the core within the coil, the current is an inverse function of the frequency. As the flux changes at a higher and higher rate with increases in frequency, the voltages of self-induction increase linearly with the frequency. This action reduces the difference between the applied voltage and the induced voltage, with the result that the current becomes smaller and smaller, as shown in Fig. 10·29*a*.

When a capacitor is placed in series with a variable inductance, Fig. 10·29*b* (the power being supplied from a source of fixed frequency and voltage), there will be one position of the core which will furnish a much

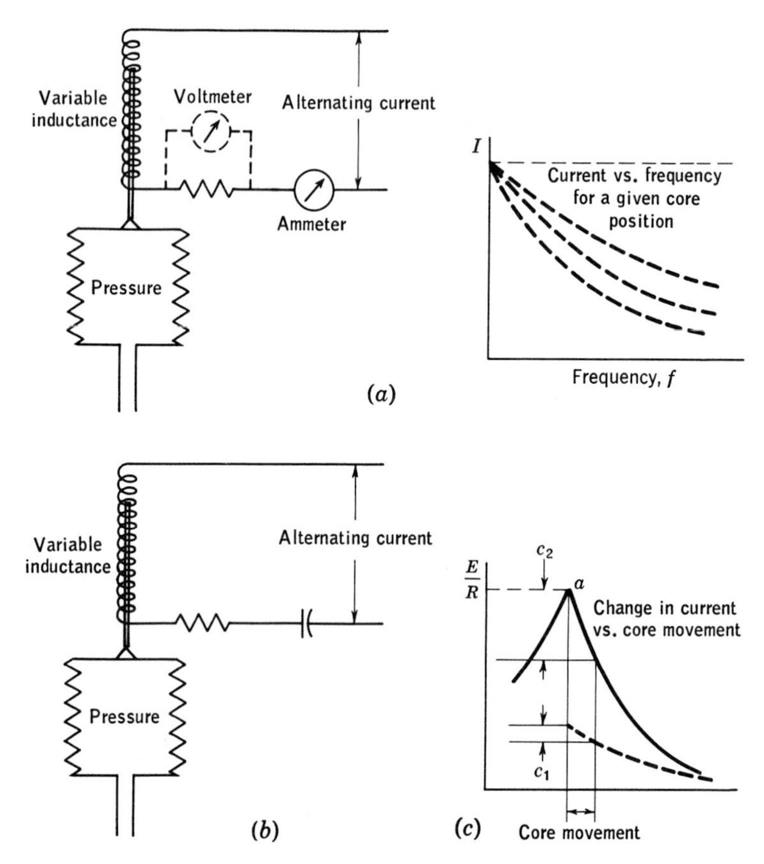

Fig. 10·29 Variable-inductance pressure-current transducers. (*a*) Because the bellows can position the core of the inductor, there will be a definite relationship between the current and the pressure which extended the bellows. (*b*) Using capacitors to *tune* the circuit will greatly increase the current change for any given core movement.

higher value of current than any other position, provided that appropriate values of inductance and capacitance have been used. Under ideal conditions, when the resistance of the coil is very small, the voltage of self-induction which is indicated by the coil voltage should be almost equal to the counter voltage created by the charge on the capacitor plates. These two counter voltages are in series with each other, but the polarities are not in series aiding. Therefore, the two voltages cancel each other and the circuit behaves as though it contained only resistance. Under such conditions, the current is determined only by the ratio of the voltage to the resistance and will be a maximum equal to E/R.

For any other position of the core, these two voltages of self-induction will not be equal, and therefore one will predominate. For the core more removed from the coil than it is for the resonant position, the capacitor predominates, but for the core inserted farther into the coil, the inductive characteristic predominates. The use of resonant circuits is well known because it provides a far faster change of current for any given movement of the core than can be obtained through the same movement of the core in Fig. 10·29*a*. Figure 10·29*b* suggests such a circuit and means of positioning the core. Figure 10·29*c* shows how much more is the change of current in the resonant circuit than it is in the other, for equal core movements. This greater change of current may be used in many ways.

There are some designs which employ not one coil and a movable core, but two or more as is indicated in Fig. 10·30*a*. If we assume that the two coils are identical or center-tapped, then for any given applied voltage, they will have identical terminal voltages only when the core is centered with respect to both of them. If the core is so moved that one coil has more of it than the other, the self-induced voltages will not be equal; therefore, the terminal voltages cannot be equal. If we now connect a meter between the center tap of the two coils and the mid-point of a resistor to which is applied the same voltage as is applied to the coils, Fig. 10·30*a*, then a current will flow and be registered on the meter, inasmuch as two points which are not at the same potential are now connected. The greater the displacement of the core from the mid-position, the greater will be the inequality of the voltages and the greater the meter indication, which can then be calibrated in terms of the core position. Under such conditions, the source of potential must be regulated for constant value.

If a second coil, similar to the one used in Fig. 10·30*a*, is connected to the same source—let us substitute the inductance shown in Fig. 10·30*b* for the center-tapped resistor—then any inequality of the magnetic-core positions will cause a current to flow through the center-tap connection. This current will act in such a manner as to position the free core again to bring about a similarity of positions. Any change of position of the transmitter core should create currents which will reposition the receiver core, and with it the indicating pointer.

It will be noted that in all these devices the magnetic core is not long enough to occupy all of the winding space at one time. Therefore, any core movement into one coil is at the expense of the other, in terms of its magnetic linkages. While this produces a greater current change than can be obtained by comparing one coil with a standard coil, an even greater effect can be obtained by applying the principles illustrated in Fig. 10·29*b*.

The Fischer and Porter Company design for a remote-reading unit is shown in Fig. 10·31. Both schematics tell the same story; one representation may be more useful than the other in showing just how an industrial design would function. By using capacitors to tune the circuits, far greater changes of current can be obtained for any given transmitter core

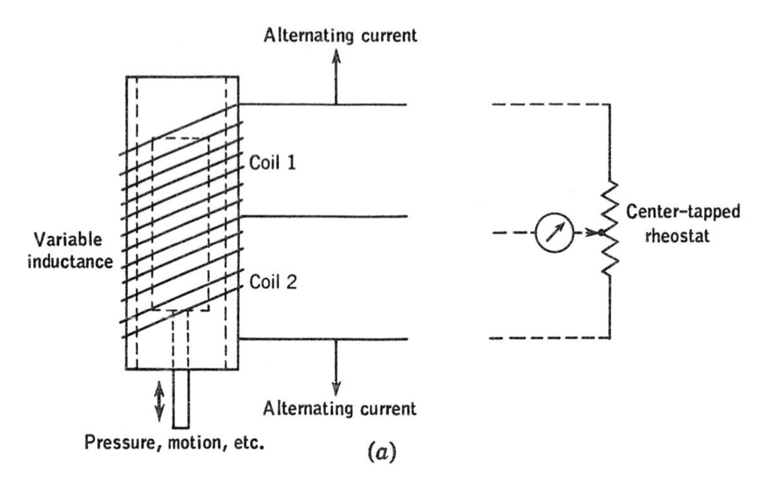

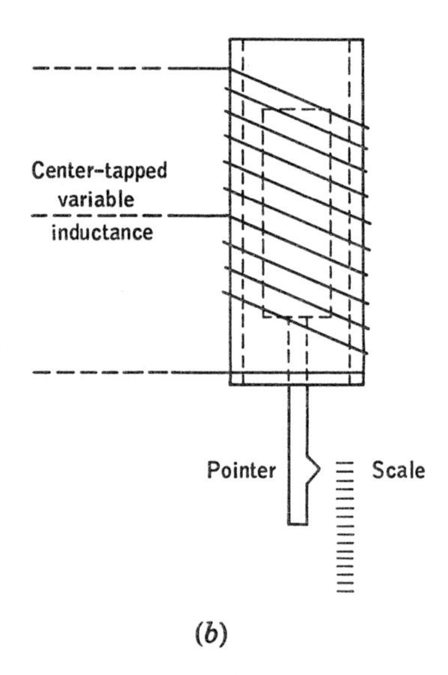

Fig. 10·30 Remote-reading pressure-motion transducers. (*a*) The position of the potentiometer at the time of current balance is a measure of the variable which has produced the position of the core. (*b*) If a second center-tapped variable inductance having a counterbalanced core is used, the circulating currents will position the core. Core position then indicates the input to the transmitter.

movement. Inasmuch as it is current that creates the magnetic field to reposition the receiver core, the larger the current change, the greater the forces which cause the receiver to reproduce the movements of the transmitter faithfully.

The differential transformer transducer. Another version of a device which employs the position of a magnetic core material to measure a pressure is shown in Fig. 10·32. This particular design and others related to it employ three or more coils. A coil that is connected to the power supply is called the primary winding. The flux linking it tends to be fairly constant in value. The two other windings are wound on each of the two outside legs of the core, one coil on each leg. Each coil has the same number of turns and is made as nearly like its neighbor as it is possible

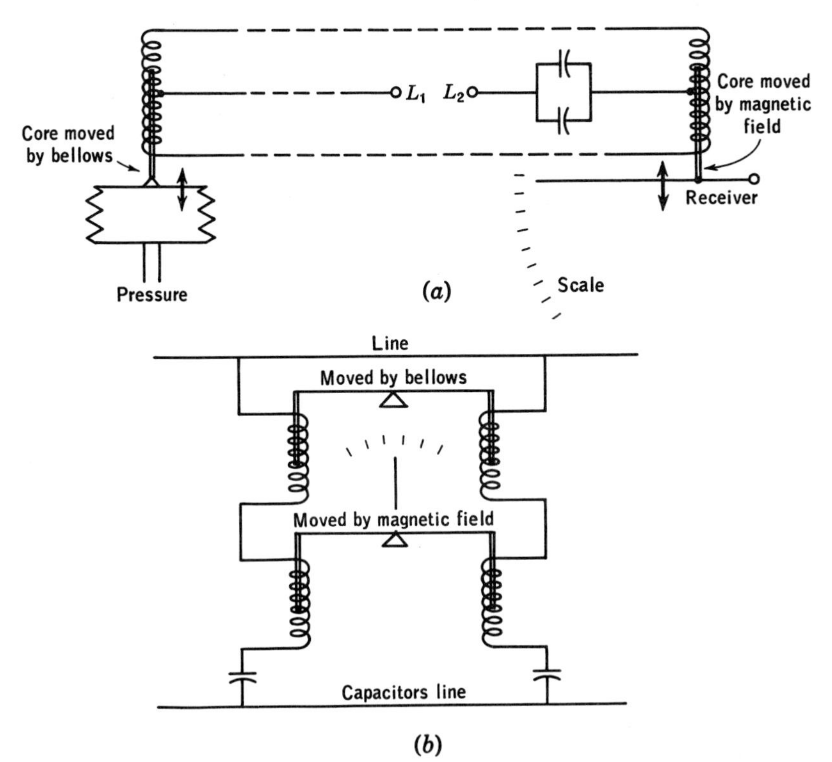

Fig. 10·31 Remote pressure indicators of the variable-inductor type. Certain commercial designs employ capacitors to increase the sensitivity of the receiver. (*The Fischer and Porter Company.*)

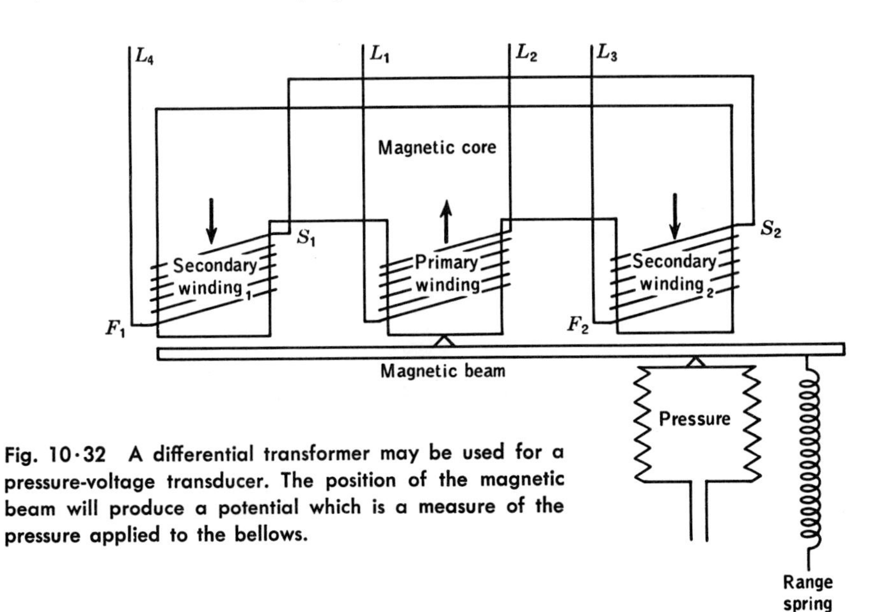

Fig. 10·32 A differential transformer may be used for a pressure-voltage transducer. The position of the magnetic beam will produce a potential which is a measure of the pressure applied to the bellows.

for it to be under commercial conditions. The movable core in this case is part of a rigid beam which is positioned by pressure, motion, and so on. If the beam is brought closer to one coil than the other, then the flux will be greater in one coil than in the other and the voltages of self-induction of the two coils will also be different. The coil having the beam closer to it will have the greater induced voltage. The two coils are so connected that their induced voltages are in series opposition and thus tend to cancel each other. The only time the two voltages will be equal and thus have a sum of zero is when each coil has the same flux linking it. With respect to the applied line voltage with which polarity must be compared, the differential voltage output of these coils will have the same instantaneous polarity when the core is nearer one coil and an opposite polarity when the core is nearer the other coil.

Figure 10·33 shows schematically the operation of an electrical type of differential pressure transmitter. Pressures are applied through the high- and low-pressure connections to opposite sides of the twin-diaphragm sensing capsule. Any difference between the two applied pres-

Fig. 10·33 A pressure or torque-to-current converter. By appropriate feedback means, the output current is compared with the input signal.

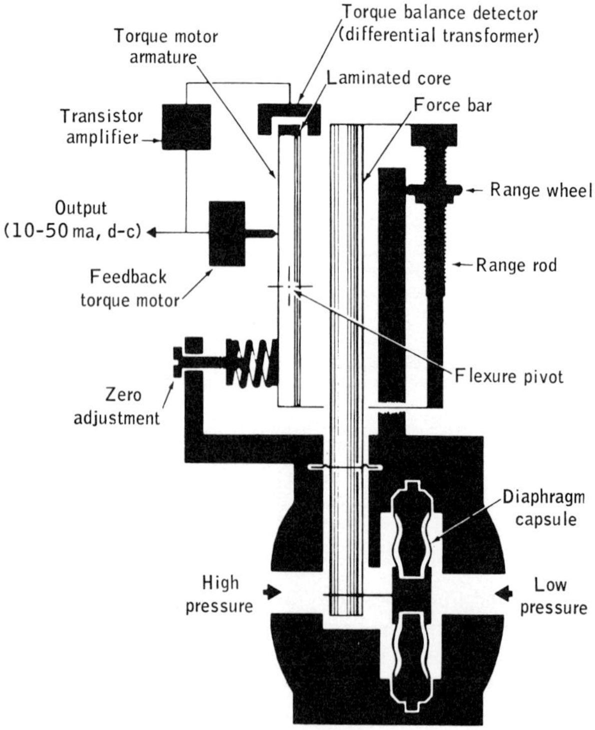

Torque balance detector (differential transformer)
Torque motor armature
Laminated core
Force bar
Transistor amplifier
Range wheel
Output (10-50 ma, d-c)
Range rod
Feedback torque motor
Flexure pivot
Zero adjustment
Diaphragm capsule
High pressure
Low pressure

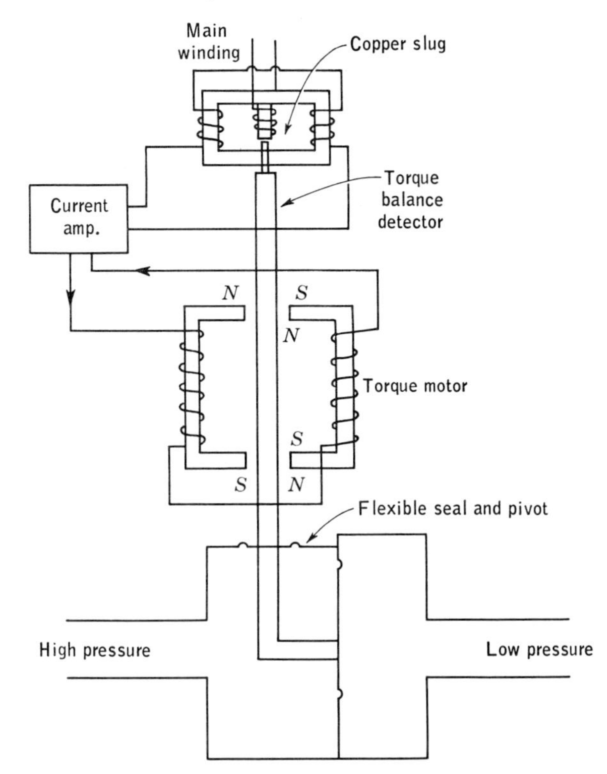

Fig. 10·34 A representation of the important elements required in one design of differential pressure transmitter which employs a differential transformer to control the current which is required for conditions of torque balance. (*The Foxboro Co.*)

sures causes the diaphragm capsule to exert a force on the lower end of the force bar. The force bar transmits this force, which is linearly proportional to the differential pressure, to an electronic torque-balance system. This system comprises a torque-balance detector (a differential transformer), a transistor amplifier, and a feedback torque motor.

In operation, the torque-balance detector compares the effects of the measured forces developed at the diaphragm capsule with the forces developed by the feedback torque motor. The transistor amplifier supplies current to the feedback torque motor to maintain a condition of torque balance between these forces. Any change in the differential pressure produces a minute movement of the detector's core. This small movement results in a change in output current, simultaneously producing a new condition of torque balance between input and output variables.

A better idea of the relationships may be obtained from a consideration of Fig. 10·34. With an increasing differential pressure, the force tends to rotate the armature counterclockwise about its pivot. This tendency is sensed by the torque-balance detector, which sends a larger signal to the transistor amplifier. This produces, in turn, an increased current output (10 to 50 ma). With an increase in the current in the electromagnet, increased force is developed by the magnetic poles. The resulting clockwise torque acts in opposition to the counterclockwise torque which initiated the increase in amplifier current. Under steady conditions, the torque produced by the current is just equal, but opposite in sense, to that created by the pressure differential.

For situations in which the differential transformer is used as a measure of the pressure, displacement, and so on, the transformer must be supplied from a regulated constant-voltage source. When the positions of two units are to be compared through use of a difference current, a regulated source is not required. Sensitivity will, however, be affected. The following characteristics are possessed by good differential transformers.

1. Their output voltage, for fixed frequency and power supply, is a linear function of their displacement from the zero position.
2. They have a wide variety of linear ranges.
3. In most cases, they require negligible force for their actuation.
4. Within the unit, there is a complete absence of friction and mechanical hysteresis.
5. There is no mechanical or electrical connection between the moving element and the transformer.
6. They are relatively small and light in weight.
7. They have a high sensitivity and can furnish relatively large signals.
8. Their output will vary with the amount of loading, but their linear relationships will be maintained.
9. There are no perceptible steps in the voltage output; stepless variation is the normal behavior.
10. They are relatively immune to shock and temperature effects.

Figure 10·35 illustrates both the construction and the variation of voltage with core position of the transformer. Aside from the mechanical details of construction, this differential transformer is the same as the one shown in Fig. 10·32. In all designs, an increase in the flux linkages of one secondary coil will be accompanied by a decrease in the flux linkages of the other.

While Fig. 10·35*b* indicates that the output will pass through zero when absolute balance is attained, the enlarged portion of Fig. 10·35*c* indicates that there will still be a small residual voltage which may be balanced out in other ways. The linearity of response when plotted against displacement (for all except the exact zero position) is what makes this transducer of so much value.

It was previously noted in the study of potentiometers that whenever a potentiometer is delivering current, its indicated calibrated values are subject to error. Figure 10·35*d* indicates that care must be exercised in choosing the instrument or device which is to read the differential voltage output. It should be noted, that unlike potentiometric devices, the load current does not appreciably alter the linear voltage relationships. It will be noted also that if increased output voltage or sensitivity is desired but a greater primary voltage cannot be employed, an increase in frequency will provide a much greater response, but not necessarily in a direct ratio of the frequencies.

One way of applying the differential transformer to pressure measurements is with a Bourdon tube, although it well might be a bellows or any other device which will provide the necessary mechanical movement. Inasmuch as the Bourdon spring is incapable of developing any appreciable force, a differential transformer makes an ideal transducer because the weight of the core can be balanced out if desired, thus leaving the tube with little or no force to exert.

The Foxboro differential transformer used for transducing pressure is more like Fig. 10·32 than Fig. 10·35. It will be noted in Fig. 10·36 that there is no gap left in the magnetic circuit so far as the secondary coils are concerned, but there is an appreciable gap in the primary coil circuit. Whereas the designs considered previously have controlled the flux linking the coils by controlling the dimensions of the air gap across which the flux has to pass, the Foxboro design controls the flux linking each coil by entirely different means. A ring of some conducting metal, such as copper, is placed around the core linking the two secondary coils. When the ring is directly centered with respect to the primary coil, equal fluxes will be diverted to each secondary coil. But as the pressure causes the Bourdon spring to deflect to some other position, the copper ring will be moved with it. If we assume that the pressure is increasing and the ring is being moved toward secondary coil E_1, then the changing flux in the core will create voltages within the ring which will cause currents to circulate in the ring. The direction of these currents will be such that their magnetic field opposes that of the field created by the primary coil current. Because of this opposition between the two magnetic fields, less flux will pass through coil E_1 than will pass through coil E_2. The total movement of the copper ring will be only a few thousandths of an inch, but it will be sufficient to control the ratios of the fluxes going to each of the secondary coils. If the pressure were to decrease, the action would be reversed, with the ring trying to prevent so much flux from going to coil E_2 and trying to force it through coil E_1. Such a mechanical design eliminates problems of counterbalancing and imposes almost zero mechanical load on the Bourdon spring.

Whereas certain of the pressure-current (or potential) transducers have

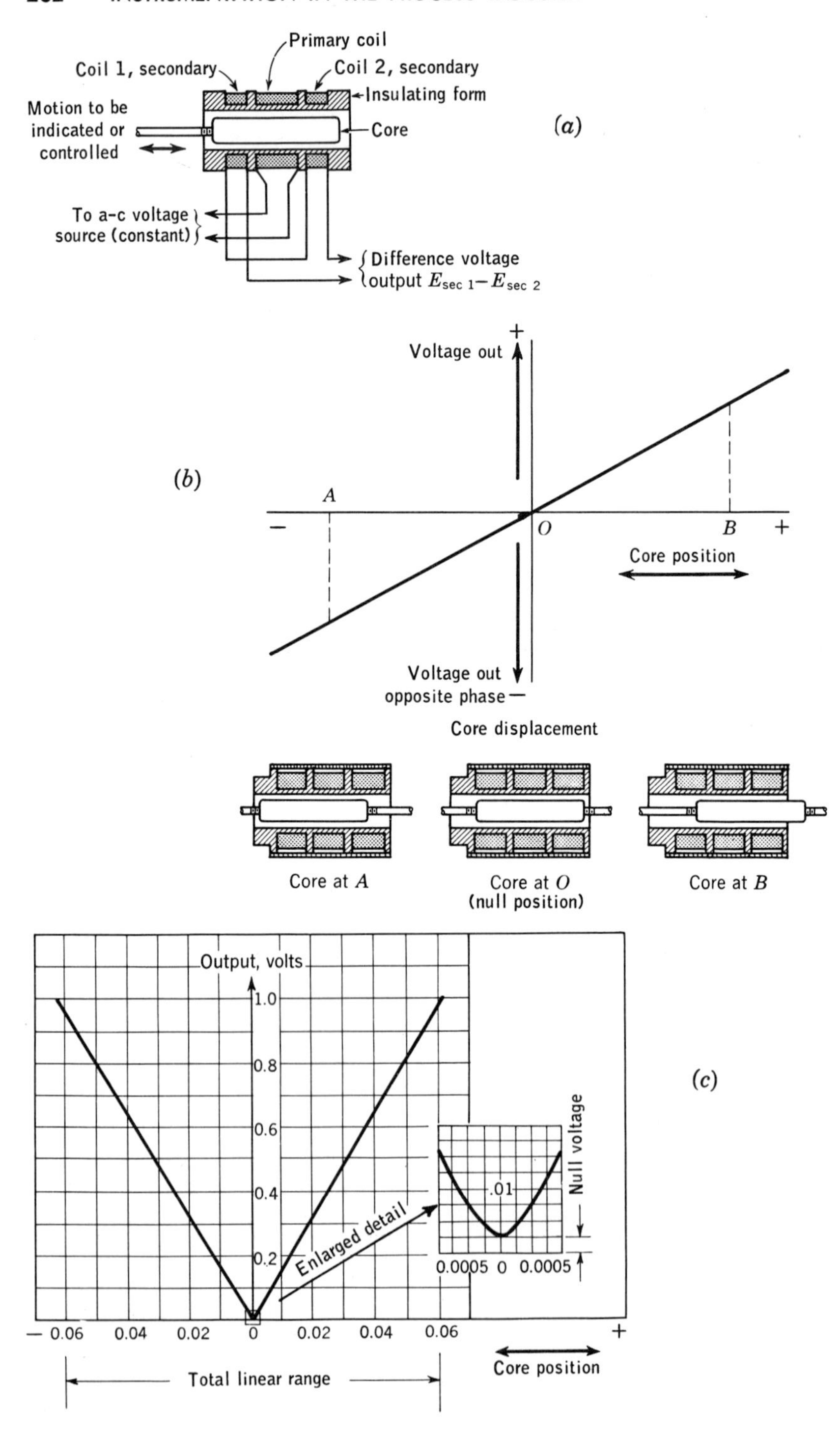
Primary coil
Coil 1, secondary
Coil 2, secondary
Insulating form
Motion to be indicated or controlled
Core
(a)
To a-c voltage source (constant)
Difference voltage output $E_{sec\ 1} - E_{sec\ 2}$
Voltage out
(b)
A
O
B
Core position
Voltage out opposite phase
Core displacement
Core at A
Core at O (null position)
Core at B
Output, volts
1.0
0.8
0.6
0.4
0.2
Enlarged detail
.01
Null voltage
0.0005 0 0.0005
(c)
− 0.06 0.04 0.02 0 0.02 0.04 0.06
Total linear range
Core position

(d)

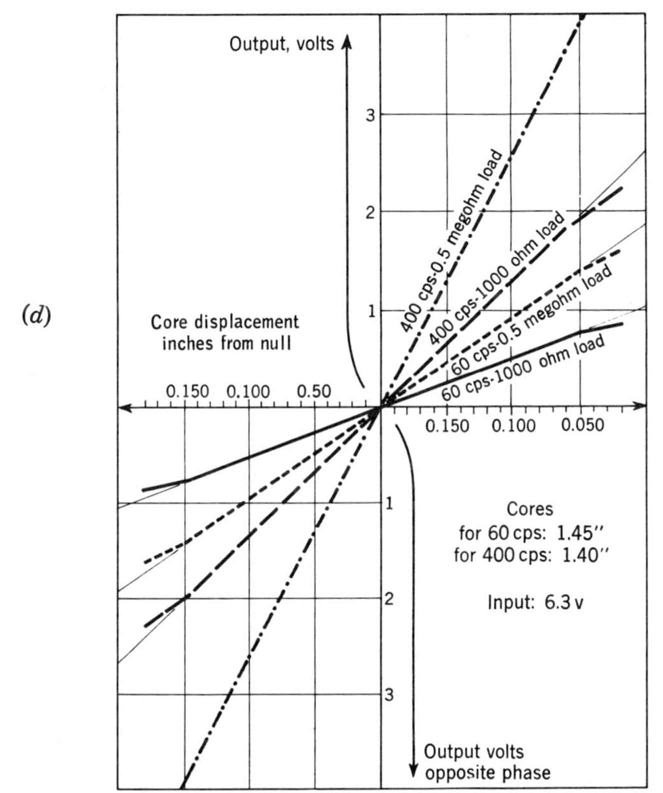

Fig. 10·35 The linear variable differential transformer. (a) The linear variable differential transformer has the essential components of other differential transformers. (b) The linear characteristics impose limited core movements. The relative positions of the core create the potentials indicated. In practice, the potential does not quite become zero. (c) The absolute magnitude of the output voltage as a function of the core position. The insert shows a magnified view of the null region. (d) The output voltage of the differential transformer is a function of both frequency and load resistance. (*Schaevitz Engineering Company.*)

had certain feedback circuits or features, the design illustrated in Figs. 10·37 and 10·38 relies upon the excellent life and performance characteristics of its bellows and differential transformer to furnish correct conversion signals.

The pressures whose difference is desired are fed into the two chambers indicated by "process." The process liquid or sealing fluid then forces the process diaphragms against their internal filling fluids. The high-pressure side forces fluid against the outside of the bellows, which would compress it, while the low-pressure side tries to expand the bellows. Attached to the free end of the bellows and moving with it is a rod which performs two functions:

1. Should excessive differential pressures be applied to the device, the

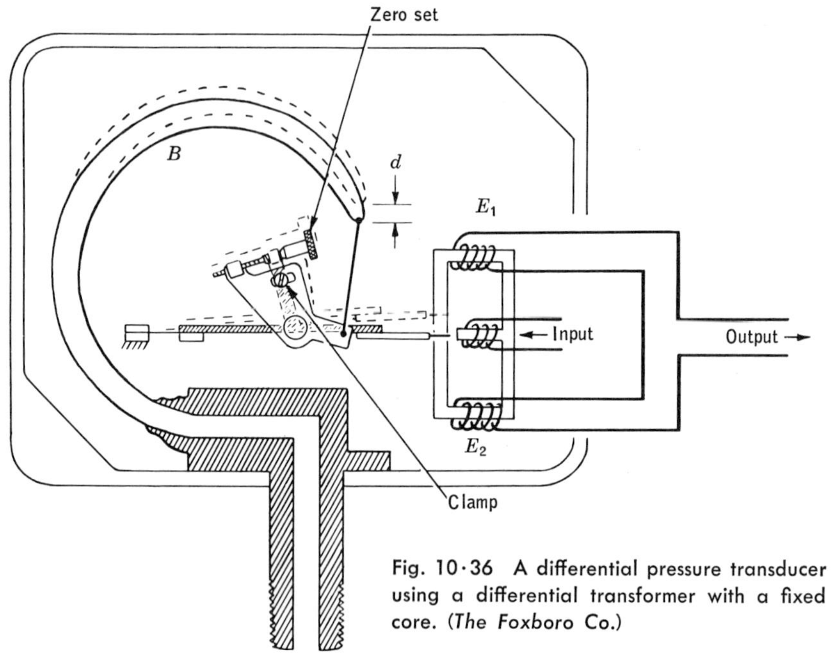

Fig. 10·36 A differential pressure transducer using a differential transformer with a fixed core. (*The Foxboro Co.*)

bellows would have excessive movement. To prevent damage to the bellows, more than normal movement will cause **O** rings to seal the filling fluid against further movement. The fluid trapped in either of the extreme positions will furnish sufficient back pressure to limit movement of the process diaphragms while, at the same time, protecting the movement of the bellows and the differential pressures applied to it.

2. As the rod is moved, it carries with it the core of the differential

Fig. 10·37 A single instrument can be used for either flow or liquid-level measurements. Such an instrument has virtually nothing to wear or change with time. Transducers which provide *potential signals* generally rely upon their sustained accuracy, but those which produce *current signals* provide for feedback so as to compare input and output signals. (*The Taylor Instrument Cos.*)

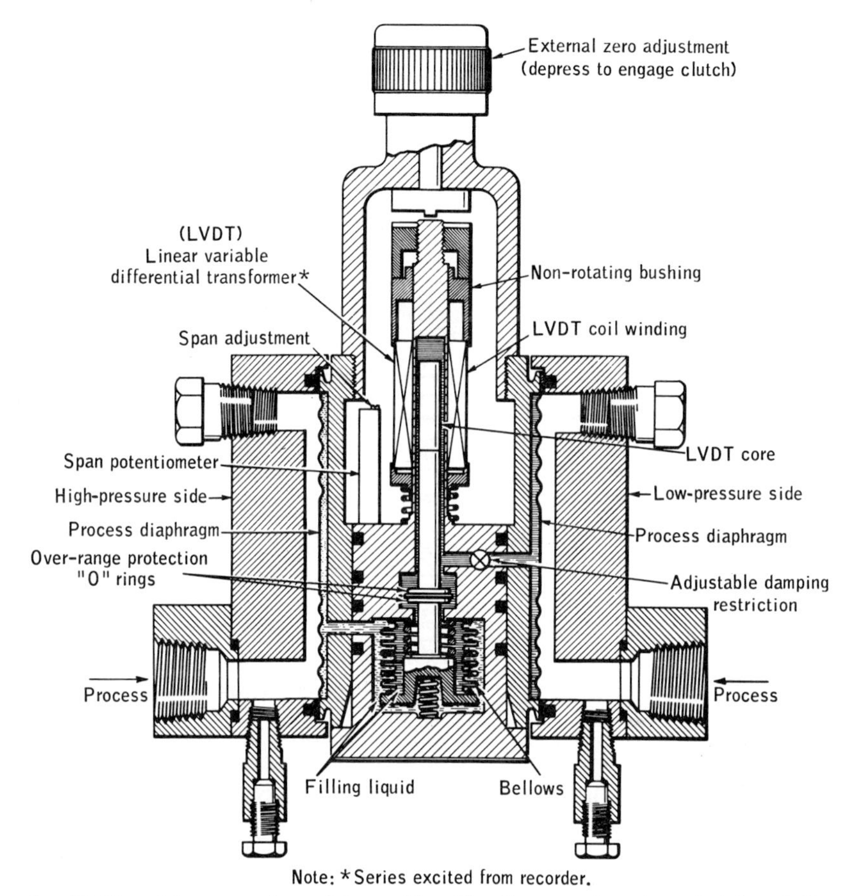

Note: *Series excited from recorder.

Fig. 10·38 Cutaway section of transmitter. The linear spring characteristics of the bellows and range springs ensure an output potential proportional to the differential pressure. It should be noted that the fluid which transmits the pressure to the bellows is kept from mixing with the process fluid—this excludes dirt, particles, or fluids which might coagulate and clog the metering element. The provision made for overrange protection should be studied. (*The Taylor Instrument Cos.*)

transformer. By using an external adjusting screw, the differential transformer may be so positioned that the unit will respond only to "forward" differential pressures, or it may be so centered that either side may be the high-pressure side, as might well be the case when there is reversed flow.

The normal output of the differential transformer is greater than that normally required. By feeding this signal into a potentiometer rheostat, any desired fractional part (less than one) of the signal may be selected for transmission to the other components. The time of response is reduced by making the volumetric displacement very small.

A different design of a pressure-current transducer is given in Fig. 10·39. The principles are relatively straightforward and will be easily

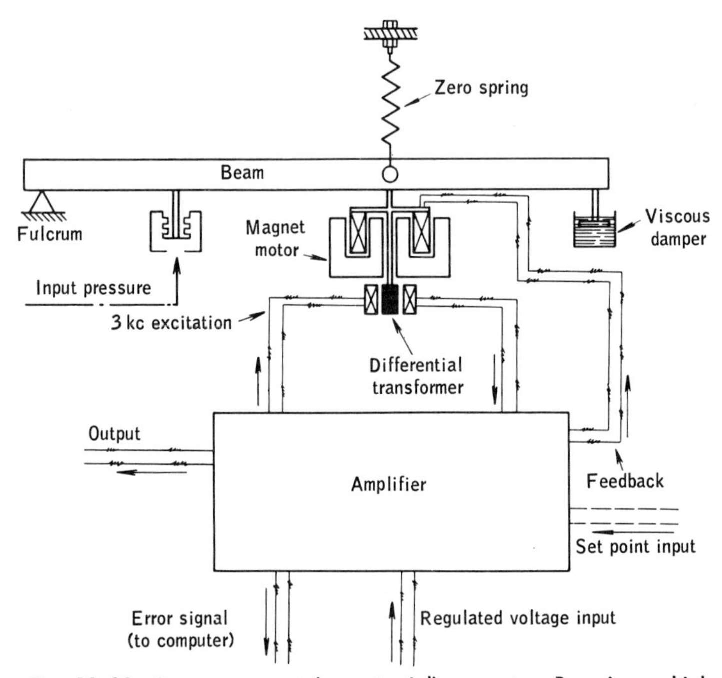

Fig. 10·39 Pressure-current (or potential) converter. By using a high frequency (3,000 cps as contrasted with the more common values of 60 or 400 cps), the output potential of the differential transformer is greatly increased for any given small displacement. The remainder of the circuit is like those previously studied. (*Hagan Chemicals and Controls, Inc.*)

understood if we analyze the behavior of the transducer. Let us assume that the input pressure is increasing. This will tend to raise the right end of the beam and produce a different output of the differential transformer. The differential transformer output is fed into the amplifier, which will increase the current output. This increased current will then flow through the coil that is attached to the beam but is located within the magnetic field of the magnetic motor. The direction of the current through the coil will be such as to produce attraction between the two magnetic fields. The attractive force will try to pull down the beam in opposition to the added upward thrust of the pressure bellows and, at the same time, decrease the output of the differential transformer until a new position of equilibrium is reestablished.

When this pressure-current (or potential) transducer is employed with a logging system (see Chap. 17), the bellows system is filled with liquid in order that the pressure system may function in the minimum time.

Calibration of differential transformer transducers. When it is desired to measure strains or to calibrate a differential transformer, the position of the core may be determined with a micrometer. Output potential versus position can be determined quite readily. When the motion produced by the primary unit is rotary, or should be converted into rotary motion, the transformer must employ a different type of magnetic circuit. Because of the shape of the movable magnetic element, the linear portion of the voltage curve is limited to an angular movement of about 50° each side of the zero position.

SUMMARY

1. Pressure measurements are the most basic industrial measurements. They are not only concerned with pressure but involved in many liquid-level, interface, density, flow, and temperature measurements.

2. Pressure measurements generally employ some of the following:

a. Pressure-balance systems in liquid-column manometers
b. Elastic materials whose distortion is a measure of the pressure
c. Force-balance and torque-balance pneumatic, electric, and hydraulic systems
d. The change in resistance of a wire with a change in dimension
e. The variation in capacitance with a change in dimensions
f. Some other quantity which is related to the strain and which is such that transducers can be used to show the relationship
g. Changes in resistance created by variations in the rates at which self-generated heat can be dissipated

3. For very low pressure measurements of the order of a few microns, electrical techniques generally are employed. For very high pressures, up to several hundred thousands of pounds per square inch, Bourdon tube elements and electric strain gages are more common.

4. The choice between electric, mechanical, and pneumatic pressure systems depends upon:

a. The pressure range and the span
b. The distance the signal is to be transmitted
c. The time required to transmit the signal
d. The availability of auxiliary instrument air, and so on
e. Compatability with other equipment
f. Availability of trained personnel to service the equipment

5. Electric strain gages can respond to signals up to 20,000 cps with only a minor drop in the response. When used with cooling systems, they can monitor pressures which change at a high rate, as in internal-combustion engines.

6. When the differential transformers are employed in connection with a mechanical type of transmitter element, a linear transducer is generally produced. The electrical signal may be employed in many ways: voltage balance, current balance, and so on.

7. Liquid-column manometers, being virtually free of any friction or hysteresis effects, have become our standards for pressure and differential pressure measurements.

8. The vast range of the mechanical type of pressure-responsive devices is summarized in Table 10·2.

9. When pressure gages depend upon the deformation of elastic materials, the gages should never be subjected to pressures which will stress the elastic material beyond the yield point or beyond the point at which the material acquires a permanent set.

Table 10·2 Industrial Pressure-measuring Instruments

Type of measurement and basic construction	Commercial accuracy, % of full scale	Maximum uncompensated ambient temp error for ±25°F change, % of span	Maximum operating temp, °F	Maximum static pressure, psi	Pressure range
Absolute Pressure:					
U-tube manometer	±½ to ±1	±¼ of reading	Ambient		5 to 50 in. Hg
Well manometer	±½ to ±1	±¼ of reading		100	6 to 125 in. liquid
Bellows	±1	±1			100 mm to 60 in. Hg abs
Diaphragm	±1	±¼	200	15 diff	6 to 760 mm Hg abs
					3 to 400 in. water
Gage Pressure:					
Bell	±½ to ±1				0.2 to 40 in. water
C-Bourdon	±½ to ±1	±¼	200		12 to 100,000 psi
Helical Bourdon	±½ to ±1	±¼	200		10 to 80,000 psi
Spiral Bourdon	±½ to ±1	±¼	200		10 to 40,000 psi
Diaphragm	±½ to ±1	±¼	200		2 in. water to 500 psi
Bellows	±½ to ±1	±¼	200		5 in. water to 800 psi
Differential Pressure:					
Inclined-tube manometer	0.01 in. liquid	±¼ of reading	Ambient	150	0.5 to 50 in. water
U-tube manometer (mercury)	±½ to ±1	±¼		250	10 to 500 in. water
Bell manometer	±½ to ±1				0.2 to 40 in. water
Bellows	±½ to ±1	±¼		500	20 to 400 in. water
Diaphragm	±½ to ±1			15	0.2 to 125 in. water
Ring manometer	±½ to ±1			1,500 to 10,000	20 to 420 in. water

10. Bourdon tubes employ all of their energy to deform the spring-like member and therefore have no capacity for applying any significant force to actuate a control or other device if the gage is to indicate correctly. Because bellows and diaphragms have relatively large areas, and can thus develop relatively large forces, they may retain their accuracy of indication while performing other tasks.

11. In selecting a material to be employed in a gage, consideration must be given to the pressures to be used, any possible hysteresis, probable life in terms of the number of cycles of flexing, the materials with which it is to be employed, and the ease of fabrication.

12. Pressure transmitters are employed when it is not feasible to extend the capillary or pressure line for the desired distance. They generally ensure accuracy of performance by comparing the input and output signals.

13. The pneumatic principles employed in the transmitters are extremely important and will be found embodied in all pneumatic controllers.

14. Pressure measurements should not be made casually if accurate and significant results are to be obtained. Attention should be given to the application notes.

15. Many different systems are available to keep liquids and slurries out of the gages. The method to use will depend on the conditions.

16. Filling the lines properly will do much to ensure accurate readings.

17. When steam pressure is to be measured and the gage is above the line, there must be an adequate connection to ensure the free flow of steam and condensate.

18. Electrical pressure gages cover about the same pressure span that can be covered by mechanical gages of one type or another. One of their chief advantages lies in the facility with which the emf or current can be converted into some other value or can be used for remote reading.

19. The principle of the Pirani gage is used in the measurement of low pressures but is also the heart of certain analytical instruments.

20. The composition of the gas has a significant effect upon the cooling of the Pirani gage resistors. Corrections must be made according to the constituents of the gases being measured.

21. A great many measurements employ the distortion of conventional elastic spring materials to change the resistance, capacitance, or position of some element. The displacement or movement of the sensing element then provides a usable signal of the desired sort.

22. Bonded and unbonded resistance-type strain gages have great inherent sensitivity, but they must be used with care. They must be applied in such a direction as to measure the desired strain. Any change in resistance because of the current in the bridge must be neutralized in some manner. The gages have a high frequency of response, but they do require the use of high-gain amplifiers.

23. If pressure can cause the elastic member to alter the spacing of two or more metal plates, changes in capacitance are created. Changes in frequency of oscillators, altered power factor of circuits, and disturbance of bridge balance may result from any pressure changes and may be used to measure pressure.

24. Remote-reading electrical pressure gages may employ potentiometric means, or they may employ circulating currents to reposition a magnetic core. The current will be a minimum when transmitter and receiver cores occupy corresponding positions with respect to their windings. The receiver core should be as nearly perfectly balanced as possible if maximum correspondence is desired.

25. When the pressure is converted into an equivalent current, most transducers will use the output current to create a torque which must be equal and opposite to that created by the input pressure signal. Except for some major mechanical or

electrical failure, this almost guarantees that the current will be a true representation of the input pressure.

26. When the pressure is converted into a potential, there generally is no comparison of input and output signals because the potential cannot create a torque. The creation of a torque requires current, and almost all of the transducers are incapable of delivering enough current to provide torque balance and at the same time maintain the desired output potential.

27. Control of the degree of oscillation of a circuit provides an extremely high rate of change of the output current for a small movement of the position-sensing device. As a consequence, the maximum span of such a device may cause the beam to move little more than a thousandth of an inch.

28. Differential transformers or other error-sensing devices may be used to sense any inequality of input and output torques. Feeding this error signal to an amplifier can provide a change in the output current of the amplifier, and thus a modification in the torque which may be compared with the input variable. *In devices of this type there must always be some small error. If the input signal were to go to zero because of perfect positioning of the core in the differential transformer, then the output current would immediately drop to zero. This would immediately create a new condition which would reposition the core to increase the output current.*

29. By using certain geometrical relationships, current-to-pressure and position-to-pressure transducers can be made to function with total movements of only a few microinches.

30. In almost every transducer, every effort is made to keep the amount of movement to a minimum so that the spring characteristics of the materials will be of no significance and so that any changes in the magnetic circuits will not materially change the reluctance.

QUESTIONS

10·1 Why are pressure measurements so important in industrial instrumentation?

10·2 What metals are generally used for the elastic elements in pressure gages? What characteristics should they possess?

10·3 What is the probable effect of pulsating pressures upon gage performance?

10·4 How might you minimize gage pulsations? Would you affect the accuracy of the gage by the methods you have suggested? Why?

10·5 Do mechanical pressure gages always employ gear trains to amplify the movement of the elastic member?

10·6 If an elastic diaphragm is used to actuate the pointer of a pressure gage, what is the probable effect of friction in the gear train?

10·7 Would friction be more or less of a problem in a gage operated by a bellows, a diaphragm, or a Bourdon tube? Why?

10·8 How much of an overrange pressure might a good gage take without serious damage? Is this true of the highest ratings?

10·9 How should pressure gages be treated? Be specific in your answer.

10·10 Why are pressure systems filled from the lowest point when it would be easier to fill them from the top?

10·11 Why and where are pressure taps connected to pipes carrying liquids and gases?

10·12 What precautions must be taken when measuring very low absolute pressures?

10·13 What does a long run of pipe to a gage do to the response rate of the gage measuring gas pressure?

10·14 Why should the needle of a needle valve used to insert resistance into a line be long and tapered?

10·15 What industrial conditions might prohibit the use of manometers with mercury?

10·16 If you want a differential pressure unit to follow rapid pressure variations, what type of instrument do you think would be the most suitable? Why?

10·17 Compare and contrast the mercury and the bellows types of manometers.

10·18 What are the desirable characteristics of liquids used to seal pressure gage lines?

10·19 What are the pressure spans for each of the major classifications of pressure instruments? Include mechanical and electrical types in your answer.

10·20 What determines what will make a good standard?

10·21 What are the standards for pressure measurements in the various spans?

10·22 What are the probable accuracies for each of the standards of Question 10·21?

10·23 In calibrating with the dead-weight tester, what precautions should be observed? What is the theory of operation of this tester?

10·24 Why are all bellows-type pressure gages differential pressure responsive?

10·25 Are Bourdon tubes also differential pressure gages? Are they so used?

10·26 Why are the usual high-pressure gages not suited for making differential measurements at high static pressures?

10·27 From the viewpoint of resistance to mechanical shock and vibration, would you consider the various types of Bourdon tubes, C type, spiral, and helical, to have any advantages?

10·28 From the viewpoint of simplicity of construction and long life, particularly where there may be vibration and pulsating pressures, is one design of Bourdon tube superior to the other?

10·29 If a pressure system is filled with a gas, what is the effect of a long pipe on the response of a pressure gage? Does the magnitude of the pressure being measured have any bearing on your answer?

10·30 What should be the characteristics of the materials used in purging pressure systems?

10·31 What are the factors which determine whether or not purging systems are required?

10·32 Why is a definite size of pipe specified for use in measuring steam pressures when the gage is above the measuring point? To get good measurements under these conditions, what must be going on in the pipe? If the pipe should become clogged with water, what would be the effect?

10·33 When using electric strain gages of the bonded type, what conditions must be maintained?

10·34 Under what conditions must one use a dummy gage? What is the purpose of the dummy gage?

10·35 When using the capsule type of seal, are there any precautions that should be observed?

10·36 Would a bellows type of pressure gage calibrated at Boston be in calibration at Denver? Explain your answer. Are there any exceptions?

10·37 Would the answer to Question 10·36 be true of all bellows gages?

10·38 Are there any particular precautions which must be observed when installing an air-seal pressure system?

10·39 What are the relative merits of mercury differential pressure gages and bellows-type differential pressure gages? Discuss in detail.

10·40 Which type of differential pressure gage can follow most closely rapid changes in the pressure differential?

10·41 Why must the line to the gage be filled with liquids when the pressure of liquids is being measured? Are there any exceptions to this rule?

10·42 What are the reasons for not installing pressure gage lines in a horizontal position?

10·43 Why is "resistance" deliberately inserted in some pressure gage lines? What precautions must be observed?

10·44 What differentiates pressure from force?

10·45 Over how large an area must a force be distributed before we can express the answer in pounds per square inch? How small an area?

10·46 (*a*) What are the various types of pressure-electrical transducers? (*b*) What are the chief uses of each type? (*c*) Name the principles upon which each type is based.

10·47 To increase the differential pressure range of a mercury manometer, would you increase or decrease the size of the high-pressure leg? Why?

10·48 Why does the kind of gas being measured have to be considered with certain of the high-vacuum gages?

10·49 What is the probable result of applying full voltage to the cathode (heater) of an ionization-type gage before the gas has been greatly reduced in pressure?

10·50 If a Bourdon tube is limited in its ability to exert any significant force, how can it position the core of a differential transformer and still retain its correct position?

10·51 Provided the load remains constant, how does the load resistance affect the output of a differential transformer? Does the value of the load resistance affect the ratio of the output voltages for any given core position?

10·52 Which is likely to require more movement, the core of the differential transformer in Fig. 10·24 or the core of the transformer in Fig. 10·32?

10·53 What type of electrical pressure transducer is best suited to high-frequency response of the order of several hundreds of cycles per second? What makes the other types less well suited?

10·54 Compare the sketch in Fig. 10·15 with Fig. 10·38, the Taylor Pneumatic Electrical Transmitter. Do you find any similarity?

10·55 What are the distance limitations, Fig. 10·16?

10·56 What practical operating conditions are required for the system shown in Fig. 10·16 to be satisfactory?

10·57 Would the equipment shown in Figs. 10·17 and 10·18 be suited to accurate laboratory work or industrial use?

10·58 What is the use of the vacuum tube in Fig. 10·19? Does the meter read average or maximum value of current?

10·59 What is the purpose of the photocell, Fig. 10·19? Would this circuit have any advantages over the use of mechanical contacts to complete the grid-cathode circuit? Assume the rest of the grid supply circuit is correct.

10·60 What would be the action of the system, Fig. 10·21, if the leads to the electromagnet were reversed?

10·61 Is the system in Fig. 10·26 one of force balance or torque balance?

10·62 In theory, would it seem feasible to replace the tube in Fig. 10·21 with a suitable transistor? What might be the advantages?

10·63 Could the transducer in Fig. 10·24 be made into a pneumatic-to-electric transducer by supplying a pneumatic signal to it instead of an electrical one? Explain your answer.

10·64 How important are the spring characteristics of the bellows, Fig. 10·26? Would your answer apply also to Fig. 10·24? What must be the conditions before the spring constant, hysteresis, etc., are of little consequence?

10·65 What are the factors which determine the value of a capacitor?

10·66 Compare the action of the transducer in Fig. 10·30 with that of the transducer in Fig. 10·31.

10·67 Does the representation of the voltage outputs of a differential output transformer, Fig. 10·35*c*, justify calling this a linear-voltage differential transformer? Are you justified in believing that this indicates the voltage-core position of all differential transformers, or only of this particular one?

10·68 Why is the indicated voltage output, Fig. 10·35*b*, different for 400 cps than for 60 cps?

10·69 Why is the rotation of a rotary differential transformer limited to a relatively small angle?

PROBLEMS

10·1 What force will be developed against each square foot at the bottom of a tank 10 ft deep if the tank is filled with water?

10·2 What will be the pressure, in pounds per square inch, at a depth of 9 ft in the tank of Prob. 10·1?

10·3 Convert 288 psf to pounds per square inch.

10·4 If a spring member has a rate of 10 lb/in., how much pressure may be applied to a limp diaphragm having an effective area of 5 in.^{2} in order to get a movement of 1 in.? 2 in.?

10·5 A column of water in one leg of a **U**-tube manometer is 50 in. high and is balanced by a 25-in. column of liquid X in the other leg. What are the specific gravity and the weight density of the unknown liquid?

10·6 1,000 in.^{3} of standard air would have to be compressed to what volume at 40°F before it could balance the pressure of a head of liquid 20 ft high? The specific gravity of the liquid is 2.0.

10·7 Using the nomograph, Fig. 9·15, what is the probable number of operating cycles possible if 30 per cent of P_{max} and 40 per cent of the maximum stroke are used?

10·8 What will happen to the life of the bellows in Prob. 10·7 if the elongation is increased from 40 per cent to 60 per cent?

10·9 An open manometer is used for measuring tank pressures. If the difference in level in the legs of the manometer is 50 in. of mercury when the atmospheric pressure is 29.5 in. of mercury, what is the tank pressure in pounds per square inch absolute? Gage? If water were used instead of mercury, for the same difference in level of the legs, what would the absolute and gage pressures be for the tank? See Fig. 10·40.

10·10 At certain times the level of a reservoir is 1,000 ft above the turbines which use its water. What is the theoretical value of the hydraulic pressure at the turbines?

10·11 A differential pressure gage is built essentially as shown in Fig. 10·41. How much will the range spring be compressed for a pressure differential of 100 in. of water, provided the spring rate is 100 lb/in.?

10·12 If the travel of the diaphragm is to be limited to $\frac{1}{16}$ in., what would the spring rate have to be, Prob. 10·11?

10·13 If a differential pressure gage were to be built essentially as shown in Fig. 10·42, what would be the spring rate of the range spring to limit the travel of the

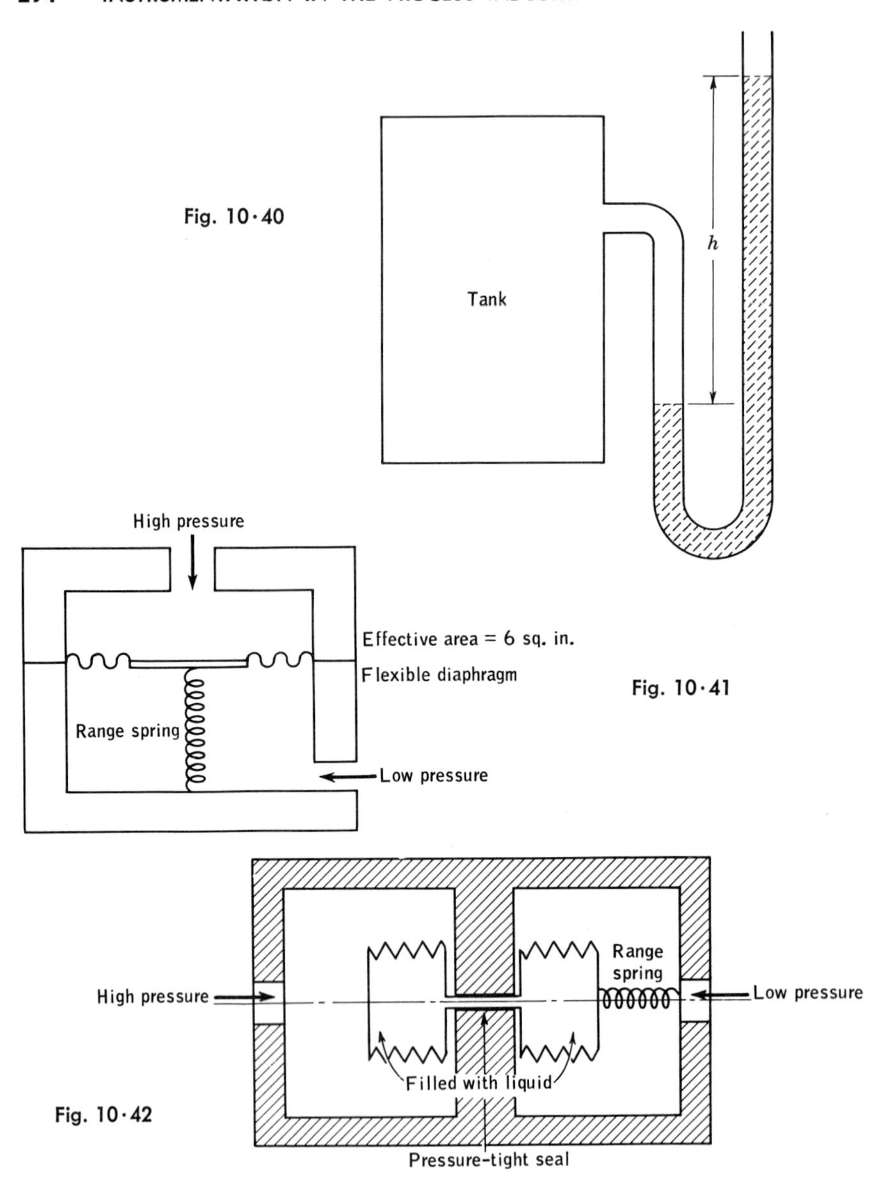

Fig. 10·40

Fig. 10·41

Fig. 10·42

diaphragm to ⅛ in. for a pressure differential of 20 in. of water? 50 in. of water? Ignore spring rates of the bellows. Effective area of each bellows is 4 in.^{2}

10·14 In a certain hydraulic press, the radius of the smaller piston is 3 in. and that of the larger is 30 in. When the force applied to the smaller piston is 200 lb, what force can be created at the larger piston? What pressure?

10·15 The density of a liquid can be determined with the apparatus shown in Fig. 10·43. It was found that $h = 100$ in. and $h' = 50$ in. What are the weight density of the liquid and the specific gravity? If the values of h and h' were interchanged, what would be the weight density and the specific gravity?

10·16 Oil of a specific gravity of 0.8 is poured into an arm of a **U**-tube manometer.

The **U** tube previously had been partially filled with water. If $h = 100$ in., find the value of h'. If mercury were used in place of the water, for the same value of h, find the new value of h', Fig. 10·44.

10·17 The three vessels shown in Fig. 10·45 all have circular bases. Compute the pressure and the force on each when all are filled to a depth of 20 in. with water. When the liquid is mercury.

10·18 If a closed manometer is used to measure the tank pressure in Fig. 10·46,

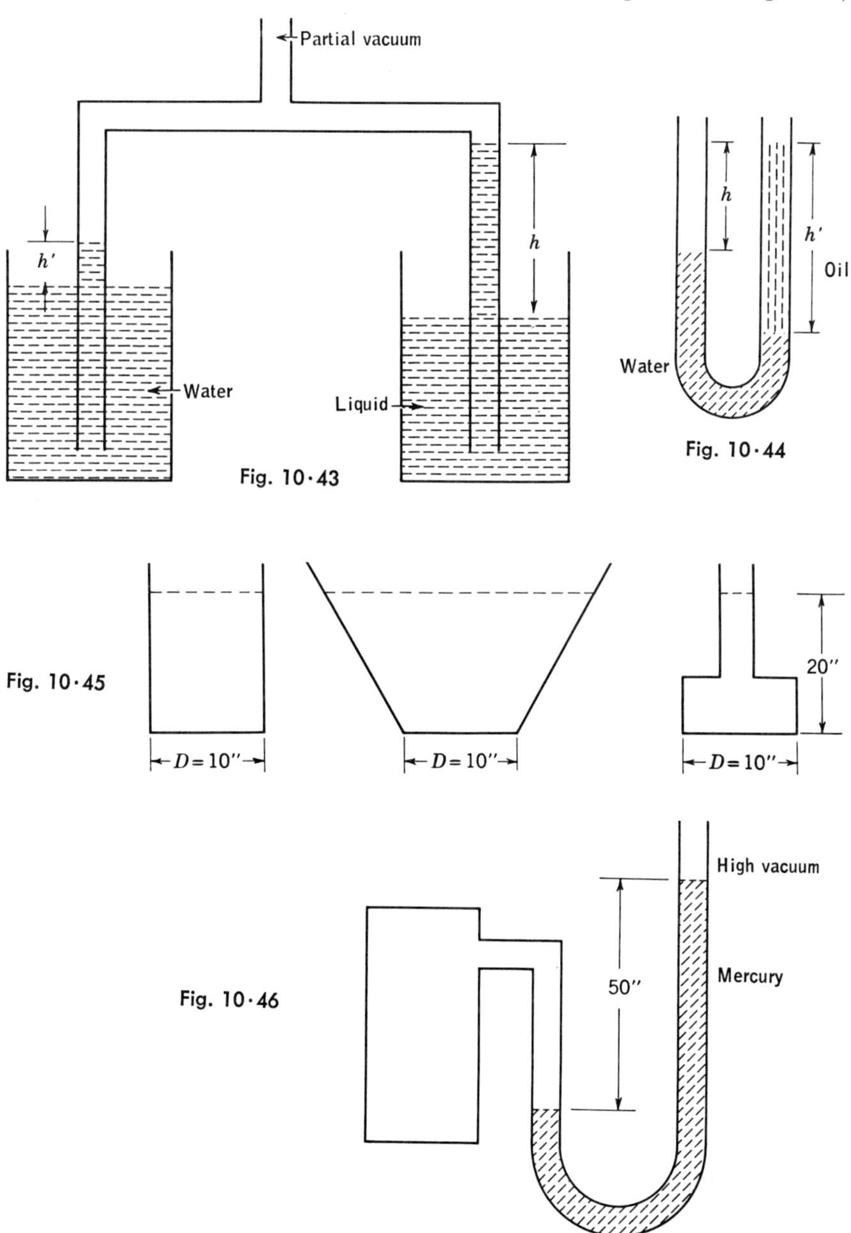

Fig. 10·43

Fig. 10·44

Fig. 10·45

Fig. 10·46

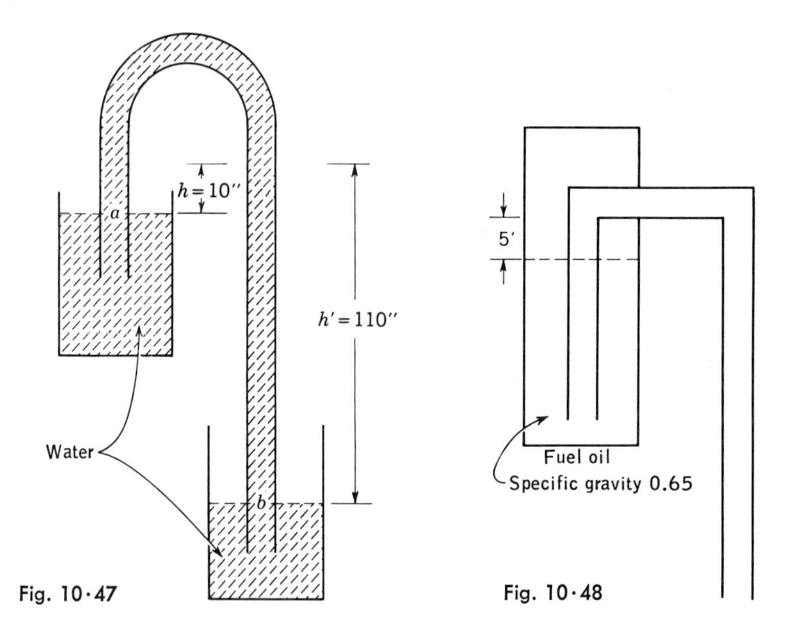

Fig. 10·47

Fig. 10·48

what will the tank pressure be when h is equal to 50 in. of mercury? 50 in. of water? (Would you be able to make such a manometer using water instead of mercury?)

10·19 For the siphon shown in Fig. 10·47, what would be the difference in pressure between points a and b for the indicated values? The assumption is made that at the moment there is no movement of the liquid in the tube. Under ordinary conditions what does the pressure difference do?

10·20 If the liquid used in the tube in Prob. 10·19 had a specific gravity of 0.8 and fairly low vapor pressure, what is the greatest length that the tube could have and still maintain itself full of liquid?

10·21 If the reservoir for a certain city were 100 ft above the lowest part of the city, what is the maximum water pressure that would be obtainable? Under what conditions might you obtain it?

10·22 A large tall tank of fuel oil, specific gravity 0.65, is to be discharged through a siphon pipe, Fig. 10·48. The siphon is 5 ft above the present level of the fuel oil. How much pressure must be created within the tank above the fuel oil to start the siphon action? How much can the tank level be lowered, theoretically, if the atmospheric pressure at the time is 30 in. of mercury?

10·23 If a barometer is completely filled with mercury so that no air space is left above the mercury, what will be the atmospheric pressure that will sustain a column of mercury 25 in. high when the tube is inverted into a bowl of mercury?

10·24 A bellows has an effective area of 2 in.^{2} and a spring rate of 100 lb/in. What will be the movement when the pressure goes from 0 to 50 psi? From 25 to 75 psi?

10·25 If a spring that has a rate of 50 lb/in. is connected to the bellows of Prob. 10·9 so as to combine the spring rates, what change in pressure will be required to produce a movement of 0.25 in.?

10·26 A **U** tube 50 in. tall is half filled with a liquid having a specific gravity of 2.00. What pressure must be applied to one side in order to push the liquid in the low-pressure leg to the top of the tube? Assume that each leg has the same cross-sectional area of $\frac{1}{16}$ in.^{2}

10·27 In Fig. 10·11*a*, $R_s = 100$ ohms, $R_1 = 200$ ohms, and $R_2 = 80$ ohms. Find the value of R_{bal}. If the frequency is changed from 500 cps to 1,000 cps, will your answer be changed?

10·28 In Fig. 10·11*b*, $R_s = 20$ ohms, $R_1 = 40$ ohms, and $C_{ref} = 1$ μf. What must be the value of C_{bal} when the oscillator frequency is 500 cps? 1,000 cps?

10·29 In Fig. 10·16 the transmitter potentiometer has a value of 300 ohms and the receiver potentiometer has a value of 200 ohms. If the transmitter contact has moved two-thirds of the way from the lowest position to the highest, or to a value of 200, where must the receiver potentiometer be positioned for balance of the galvanometer? How many ohms measured from the lower end?

10·30 If R_1 is twice R_2, Fig. 10·18, for conditions of balance, will the electrolyte be lower or higher in the left leg of the tube?

10·31 If the electromagnet, Fig. 10·24, is 4 in. from the pivot while the feedback bellows is but 3 in., what are the relative values of the forces? Of the torques?

10·32 Using Fig. 10·35*d*, how much more sensitive is a differential transformer using a 400-cycle source than one that uses a 60-cycle source?

10·33 If plates P_1 and D, Fig. 10·14, are each 10 cm square, and the spacing is 0.100 cm, calculate the capacitance if the dielectric constant has a value of 100. If the value is 1,000.

11 Liquid and Gas Flow Measurements, Mechanical Transducers

The accurate measurement of gas and liquid flow is important for many reasons:

1. Accurate measurement to obtain specific proportions is necessary.
2. The maintenance of definite rates of flow is important for maximum efficiency and production. Without accurate measurements, precise quality control is impossible.
3. Costs which are based on flow measurements will be incorrect if the measurements are erroneous. Because huge volumes of gas, steam, and liquids may have to be measured daily, a very small percentage error can amount to large sums.

Just as pressure and level measurements are mostly inferred from what happens to an elastic membrane or to a liquid, so flow measurements are largely inferred from other responses. There are only a few instances of direct measurement of flow; all the others are of the indirect or inferential type.

WAYS OF MEASURING LIQUID AND GAS FLOW

There are many ways of determining the flow rates of both gases and liquids. The type of material, its viscosity, its vapor pressure, its temperature, its conductivity, the amount of material in suspension, and the flow rate are the important factors. They will determine which of the methods is the most suitable. In one way or another, flow measurements will use some of the following:

1. Direct volume measurements
2. Positive-displacement pump measurements
3. The voltage created by a liquid moving in a magnetic field (see Chap. 12)
4. The amount of heat required to raise a liquid's temperature by a fixed amount

5. The force required to accelerate a fluid around a bend
6. The force developed by a liquid whose speed is governed by the variable cross-sectional area of the measuring tube
7. The force developed by a liquid or gas impinging on an obstruction
8. The speed of propellers in the stream of fluid (propellers turn at a rate proportional to the flow rate)
9. Vanes which can develop a torque proportional to the flow rate
10. Fluids passing through variable-area sections which will change their speed and pressure
    a. Venturi tubes in some designs
    b. A flow nozzle in some designs
    c. The orifice plate, the simplest of the devices for creating a pressure differential

The arrangements listed here, aside from the first four, are in terms of an obstruction placed in the pipe to produce the phenomena required for the measurement. Magnetic flowmeters require no obstruction in the pipe, whereas the orifice plate is an obstruction deliberately placed in a pipe to produce a differential pressure which is interpreted in terms of flow rate.

In order to simplify the treatment, mechanical-type flowmeters are treated in this chapter, and electrical systems are discussed in Chap. 12. The subject summary, questions, and problems will be found at the end of Chap. 12.

Direct volume measurements. The filling of a container is perhaps the simplest direct way to measure liquid flow. The number of times per minute, or hour, that a container of given size can be filled measures the flow rate. By keeping count, the quantity can also be determined. Such a device is of only limited use, but its principle is employed in certain types of pumps and meters. Obviously, it cannot be used to measure quantities of gas or vapor.

The quantity of liquid handled by a positive-displacement pump depends upon the volume of the cylinder filled each time and how often the filling and emptying cycle takes place. If the effective volume of the pump can be changed, or if the speed of the pump can be controlled, then the total flow and the rate can be measured; within their range, they can measure and proportion very accurately. These are some of the few measurements which can be made directly.

Quantity meters. Most of the flowmeters previously mentioned do not measure flow directly, but instead measure something associated with flow, but there is a class of meters in which fixed or variable volumes can be filled and the rate and total amount can be registered or indicated.

The tilting trap meter illustrates the principle employed. When the meter is filled, its center of gravity with respect to its pivot is such as to rotate the meter. Once the tipping action starts, the shifting of the center of gravity of the liquid aids the emptying process.

The piston meter uses a totally different measuring principle. If a fluid is admitted on one side of the piston, the fluid will push the piston to the end of its stroke. This requires that a fixed definite volume be filled. But at the moment that the piston reaches the extreme position, the inlet and exhaust valves are actuated and are reversed from their previous positions.

1. Fluid will now be permitted to enter the space on the other side of the piston and start to reverse the direction of movement.

2. The exhaust valve for the filled portion of the cylinder, will now be opened, thus permitting the fluid to flow into the outlet, while the exhaust valve on the other side is closed. This procedure is then repeated.

In certain variations of the piston meter the amount of movement of the piston before the valves are actuated can be controlled. Oftentimes the speed at which the meter functions can be controlled. The volume is then directly proportional to both the length of stroke and the speed.

$$\text{Volume} = \text{quantity} = KLN$$

where K = area of piston
L = length of the stroke
N = number of times the operation takes place

The K used for a particular calculation depends on the units of measure chosen for L and N.

Because of its construction, the piston meter is not suited to the measurement of viscous or corrosive liquids. Flow rates for this design range from about 10 to 1,000 gpm.

The nutating disk meter is one of the most common meters, being used in many domestic water systems. It resembles the piston meter in that there is a movable piston. But the movable portion rotates at an angle, and it provides not only a rotation but a sweeping motion which causes definite volumes of water to be passed from the inlet side of the disk to the outlet. Its normal flow rates are from 10 or 15 to 500 gpm. It is simple and rugged; it develops a low pressure drop; it is low in cost; and it maintains excellent accuracies of better than ± 1.5 per cent. Because it is the measuring portion of so many water meters, most people have heard the click of the disk as it wobbles its way around.

The rotary-vane meter employs a rotating member which is not centered in the pump housing but is offset by some amount. To accommodate this eccentricity, the blades of the pump must be free to move within the rotating member so as to close the variable dimension between rotating and fixed parts. In this way, each succeeding quadrant will meter the same amount of fluid and pass it under pressure to the outlet. Their design gives rotary-vane meters outstanding accuracies, of the order of 0.2 to 0.3 per cent, with limited flow rates.

Gas meters. The measurement of gases imposes several problems not only because of the difficulty in confining the gas, but also because so many of the characteristics and values of gases change with both temperature and pressure.

One method, obviously, would be to trap fixed volumes of gas and count the number of units of volume. This is exactly what is done in the gas meter shown in Fig. 11·1. Except for using an oiled-sheepskin diaphragm instead of a metal piston, the operation is exactly the same as that of the piston pump.

Inasmuch as gas meters must be tight to avoid losses, the rotary-vane meter could be used for certain operations. In practice, one is much more likely to see a modification of it in which the blades rotate but are so designed that they can change their points of contact with each other and continually change the effective volume of the compartment as they compress the gas and force it out of the meter.

Magnetic flowmeters. This is a meter whose principles have been known for over a hundred years but, because of a lack of suitable amplifiers, could not be employed practically. The development of special electronic amplifiers has made application practical. The theory is given in more detail in Chap. 12. It is enough to say here that as a conducting liquid flows through a magnetic field, it generates a very small electromotive force which is proportional to the rate of flow of the liquid. This small signal is then amplified and fed into the indicating or controlling equipment. High accuracies and the absence of pressure drop within the meter

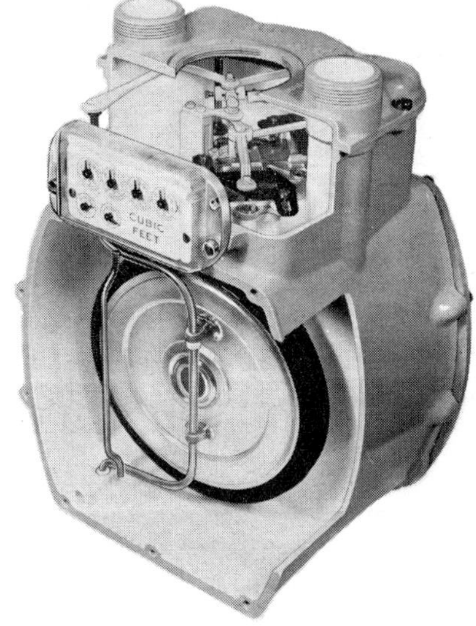

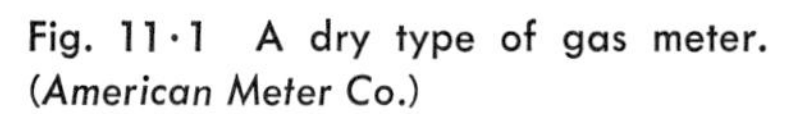
Fig. 11·1 A dry type of gas meter. (*American Meter Co.*)

are two very important factors in applying meters of this type. The absence of projections or restrictions means that liquids with all sorts of solids and slurries will not clog the meter. The pressure drop within the meter is only that of a straight section of pipe of equivalent size, a negligible amount.

Because there are no wearing parts within the fluid-flow section of the meter, the meter characteristics do not change materially with time. Thick, abrasive slurries can be measured as readily as water. One requirement must be met, however: the resistivity of the liquids must be of a suitable value. The magnetic flowmeter has certain definite advantages:

1. Density variations or changes have no effect on the accuracy.
2. The meter is immune to the viscosity of the fluid.
3. The amount of suspended solids is of no consequence.
4. The accuracy of the meter is not affected by piping arrangements either ahead or behind the meter.
5. Flow in either direction can be measured with equal ease.
6. The meter has a fast response.

Heat-input flowmeters. For any given fluid, a definite amount of heat is required to increase the temperature of each pound or gallon by some fixed amount, 1°F, for example. If the increase in temperature can be controlled accurately and the heat can be measured simply, then measuring the heat should provide a way to measure flow. The heat required can be measured quite simply by using a wattmeter to indicate the instantaneous rate at which electric power is being applied to maintain the temperature increment; the heat required is also directly proportional to the rate at which the moving fluid is absorbing heat and thus directly proportional to the flow rate itself. A watthour meter indicates the total heat required and thus registers the total flow.

If the flow of the liquid can be kept laminar—that is, free from turbulence so that there is relatively little mixing of the fluid—then the only part of the liquid which needs to be heated will be the thin layer next to the outside of the pipe. Under such conditions the power requirements will be quite small.

As with the other flow-measuring devices, the specific heat of the fluids depends upon the composition. The multiplying factor will be different for each material to be measured.

This type of instrumentation has some of the advantages of the magnetic flowmeter in that no obstruction is placed in the pipe and there are no wearing parts. The flow rate indicator and the flow totalizer are both standard electrical instruments of extreme durability.

Elbow flowmeters. So far as we know, all physical materials require forces to produce accelerations. These accelerations may be of two types:

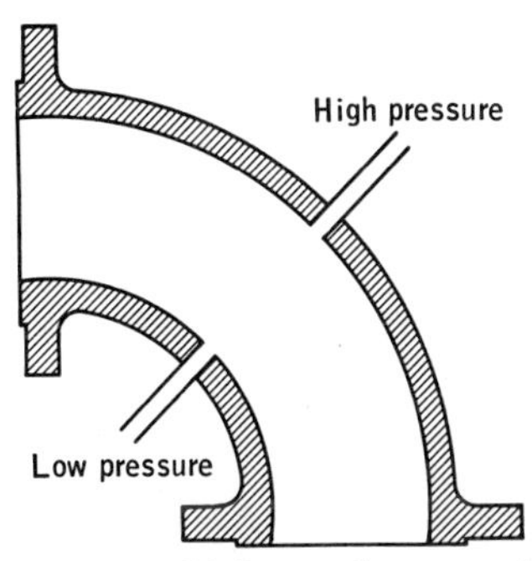

Fig. 11·2 The elbow can be used for a primary flow element. Meters which employ the elbow to establish the differential pressure are very stable over long periods of time. (*The Taylor Instrument Cos.*)

one which produces a change in speed, and one which produces a change in direction with or without a change in speed. One basic law states that the force is directly proportional to the magnitude of the acceleration and to the mass so accelerated.

Any material, such as a liquid flowing in a pipe, which is made to change direction will require that a force be applied to it. Without this force, no acceleration can take place. If we deliberately insert an elbow into a pipe, then the liquid will be accelerated as it changes direction. There will, however, be no change in the speed if the elbow has the same cross-sectional area as the pipe.

The accelerating force will be a maximum where the change in velocity is the greatest, at the "outside" face of the elbow. There will be a reduction in pressure at the opposite face on the inside of the bend. The two pressures then are the clues to the flow rate, Fig. 11·2. A differential pressure meter of any of the many designs previously studied may be employed to measure the difference of the two pressures and, by proper calibration, to indicate the flow rate.

This type of metering device provides excellent repeatability: it is not likely to clog or show much wear. Its use at the present time is not sufficiently wide that there are charts and tables available for predicting the calibration. But once calibrated, the meter is not likely to drift out of adjustment because of any changes in the metering element. Some measurements indicate that taps at the $22\frac{1}{2}°$ point are more effective than those at the 45° point.

Impact meters. Another flow-measuring device which employs virtually no restriction in the pipe is shown in Figs. 11·3 and 11·4. What it does employ is the fact that fluids, be they gases or liquids, will exert a force on an object if that object tries to impede the flow of the material. Moreover, that force will vary with the relative speed of the two objects. It will be noted that there are two sets of nozzles located at the inner periphery; half of them face upstream and half face downstream. As the fluid flows through the tube, it will create an additional pressure on the nozzles which face upstream and will decrease the pressure on those facing the opposite way.

Fig. 11·3 A recent design of flow tube. This design enjoys a large pressure differential even though the stream is but little restricted as it passes through the tube. It is used for metering water, air, steam, oil, jet fuel, hydrocarbon liquids, clay, slurries, raw sewage, brine, coke oven gas, and other materials which have proved difficult to meter. (*General Controls Co., Foster Engineering Div.*)

Fig. 11·4 Unlike most metering elements, which are seriously affected by the conditions upstream from the metering element, the Gentile flow tube shows virtual independence of the configuration of the piping. (*General Controls Co., Foster Engineering Div.*)

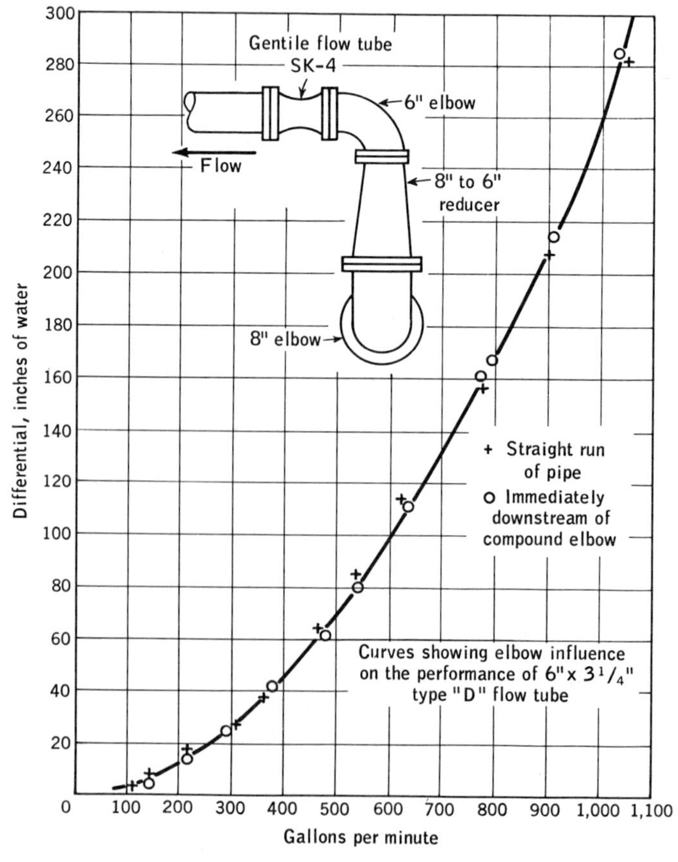

$$P_{up} = \text{static pressure} + \text{impact pressure up}$$

$$P_{down} = \text{static pressure} - \text{impact pressure down}$$

$$\begin{aligned}\text{Differential pressure} &= P_{up} - P_{down} \\ &= (\text{static pressure} + \text{impact pressure up}) - (\text{static pressure} - \text{impact pressure down}) \\ &= \text{impact pressure up} + \text{impact pressure down}\end{aligned}$$

Inasmuch as these pressures vary with the square of the speed through the tube, the expression may be condensed to become

$$\text{Differential pressure} = KV^2$$

In this design, there is no conversion from one type of energy to the other, as will be noted when we study other meters which are classed as *head meters.* Meters which employ differential pressure measurements generally restrict the flow area to increase the fluid flow rate in order to produce the measurable pressure differential.

The Venturi, nozzle, and orifice flowmeters can also be used to measure gas flow. The particular problems encountered with the measurement of gas flow are discussed in the sections dealing with these meters.

VARIABLE-AREA METERS

Variable-area constant-force instruments; rotameters. The magnetic, the heat-input, and the elbow types of flow-sensing devices do not change the size of pipe or do anything to control the flow rate at any point. The first two types are basically linear in their response. The rotameter likewise has a linear scale, in marked contrast to some of the other designs which will be studied later. But the rotameter deliberately controls the flow rate within the element.

Basically, the rotameter is a tapered vertical tube with the fluid flowing up through it, Figs. 11·5 and 11·6. The tube is smallest at the bottom and largest at the top; the velocity is then greatest at the input and a minimum at the output. The marker, called a *float,* that is used to indicate the flow rate is slightly heavier than the fluid which is being measured. Consequently, it will sink to the bottom of the tapered tube when there is no flow, but as the flow increases there will be a force directed against the marker which will lift it. By design, the fluid velocity upward decreases with height above the zero reference because of the larger size of the tube at the upper end. At some point in this stream having variable speed, the upward push of the moving fluid will just equal the downward force which makes the float sink in the still liquid. The higher the speed, the higher the marker position and the higher the indicated flow rate.

The quantity of liquid that can be measured by a rotameter is small when compared with the capacities of some of the other methods. But the rotameter has a wide range of accurate performance, the ratio of the highest flow to the lowest being about 10. If the lowest accurate value is 3 gpm, then the highest value will be 10 times 3, or 30, gpm. Another important feature is that changes in viscosity and density can be made to have little effect on the accuracy.

If the float should come in contact with the wall, there might be a tendency for it to stick. Still other operating conditions might cause the float to be pushed against the wall. To eliminate such situations, the float is sometimes guided by a wire which passes up through its center or by glass ribs on the inside of the tube. Vertical movement is free, but sidewise movement is restrained.

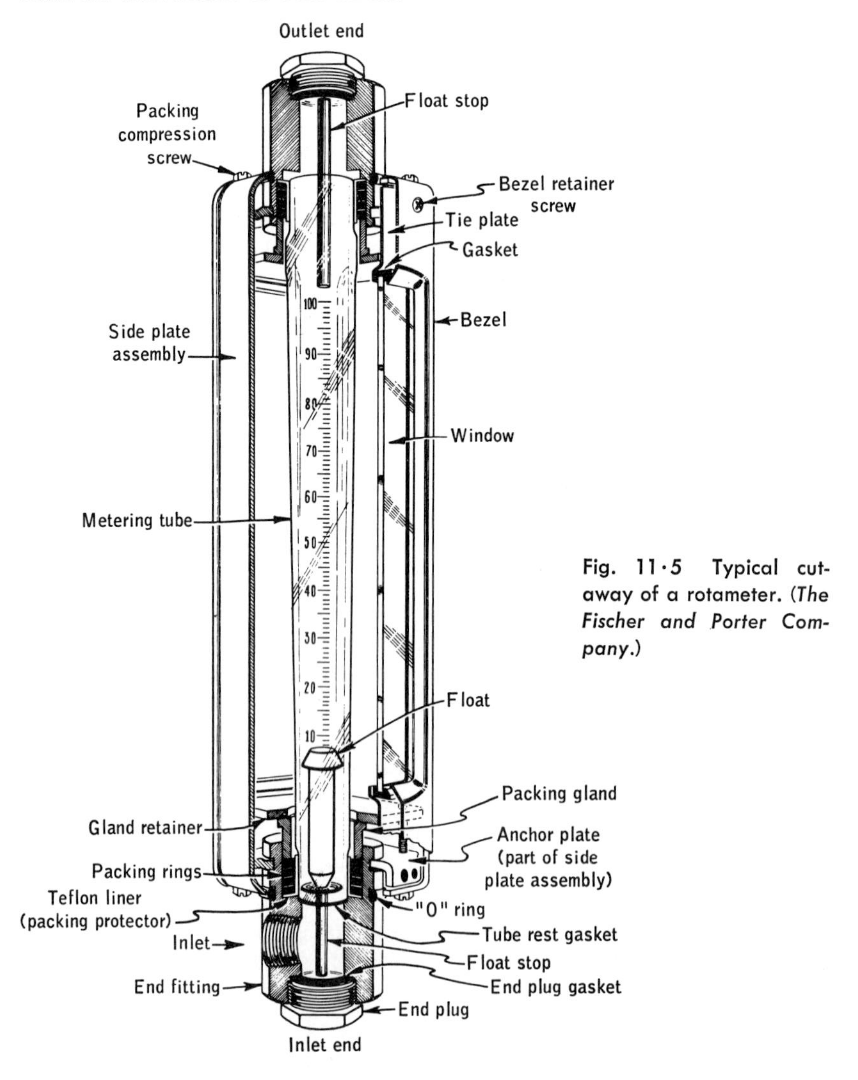

Fig. 11·5 Typical cutaway of a rotameter. (*The Fischer and Porter Company.*)

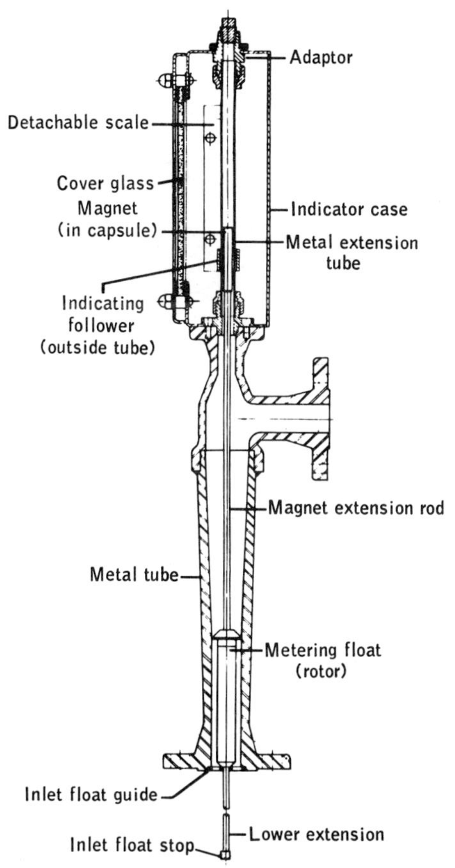

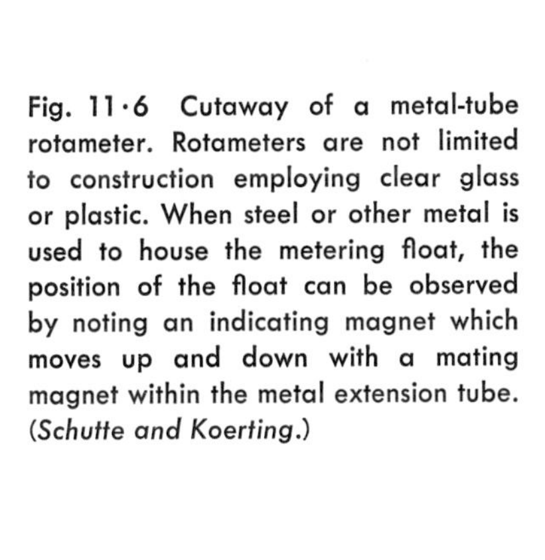

Fig. 11·6 Cutaway of a metal-tube rotameter. Rotameters are not limited to construction employing clear glass or plastic. When steel or other metal is used to house the metering float, the position of the float can be observed by noting an indicating magnet which moves up and down with a mating magnet within the metal extension tube. (*Schutte and Koerting.*)

Many means are used to facilitate reading a rotameter when the liquid is dark or opaque. In one design, ribs are installed to guide the float and leave so little clearance that the float will be visible through the thin film between it and the tube wall. (There is one major liability in such a design: foreign material may jam the float and cause it to stick. On the other hand, clogging will increase the force trying to lift the float.) Another means is to use a transparent disk which is illuminated from the rear. This creates a line of light which then indicates the reading.

Should the flow rate exceed the ratio of 10:1, it may be necessary to use two instruments such that the range of one supplements the range of the other. When large flow rates must be measured and it is desired to use the rotameter type of instrument, some of the main flow may be bypassed around the meter so that only a definite percentage of the flow is metered. The Fischer and Porter Ori-Flowrator and the Schutte & Koerting Bypass Rotameters are typical of such designs, Figs. 11·7 and 11·8. An orifice plate, which is an obstruction placed in a pipe to create a pressure change as the liquid passes through it, causes a fixed per-

Fig. 11·7 A rotameter used for large-volume flow measurements. (*The Fischer and Porter Company.*)

Fig. 11·8 A rotameter used for large-volume flow measurements. When the flow rate exceeds that of the rotameter, the rotameter may be used to measure only a small portion of the flow. A simple multiplying factor will then convert the observed flow to the true flow. (*Schutte and Koerting.*)

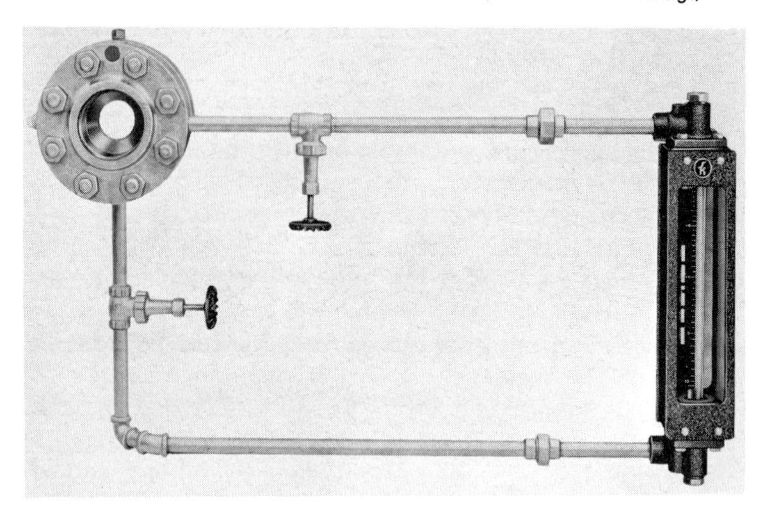

centage of the flow to go through the meter. In this manner, the linear scale of the rotameter is retained.

Variable-area constant-force instruments; cylinder and piston type. This type of meter may not appear to be in the class of meters that involve an obstruction put in the pipe, but it does properly belong with the rotameter because the principles of operation are similar. Each employs a float which is positioned by the flow; in each case the force is constant in value. In the cylinder and piston design, the float exerts a constant downward force; the difference in pressures between one side of the float, which functions as a piston, and the other side positions the piston. If the downstream flow is greater, then the pressure is reduced on the load side of the piston; the increased differential pressure then forces the piston up, thereby increasing the number or area of the openings through which the fluid can flow until the pressure differential is again balanced, Fig. 11·9.

Fig. 11·9 Variable-area meter for viscous materials. The piston acts as a valve in the variable-area meter. Any change in the difference of P_H and P_L will create a new force which will reposition the piston. The difference of the pressures must create a force equal to the weight of the piston because this piston is of the constant differential type.

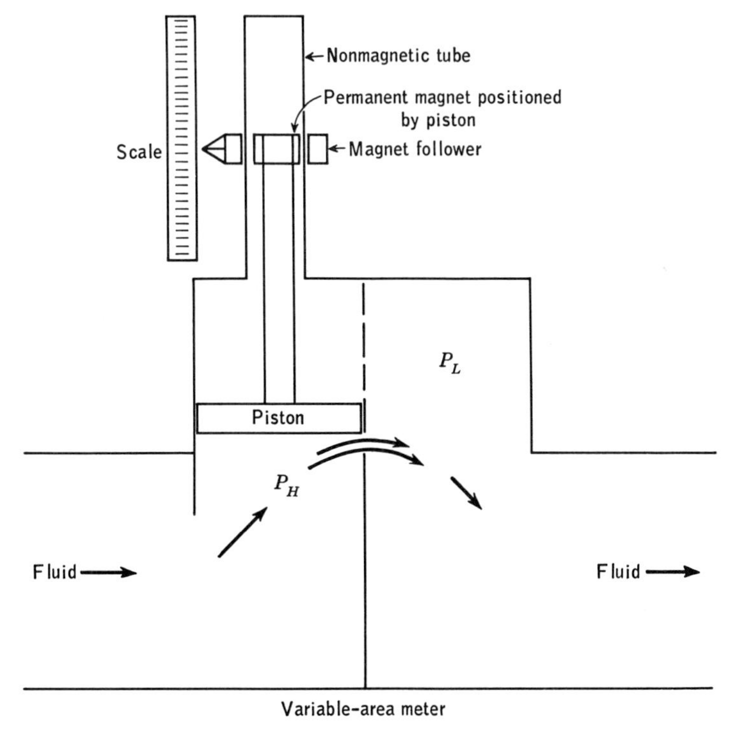

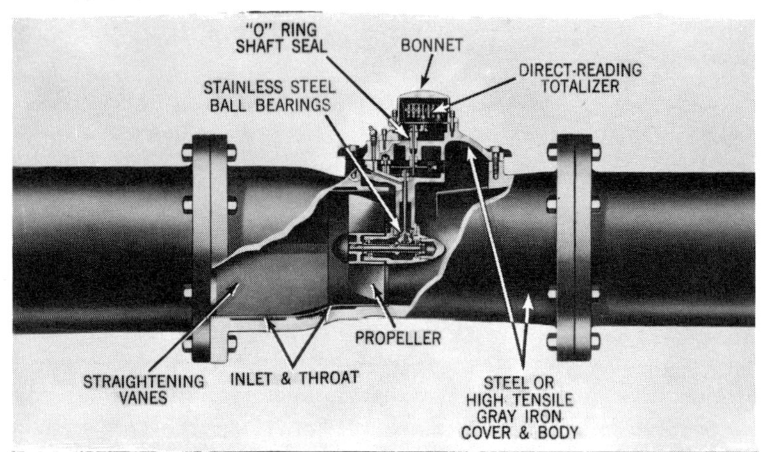

Fig. 11·10 The movement of the liquid through the pipe turns the propeller an amount which is directly proportional to the flow. The number of revolutions can be recorded on a revolution counter which can be calibrated in terms of total flow. (*B-I-F Industries, Inc.*)

The chief uses of meters of this type are for measuring (1) high-viscosity fluids such as fuel oils, tar, and chemical liquors, (2) materials which are corrosive or might clog lines, and (3) materials whose flow coefficients are not well known.

Liquid-flow velocity-turbine meters. Just as air speeds can be determined by the rate at which a propeller or other aerodynamic device is rotated, so liquid flow can be determined by the rate at which a propeller or turbine can be made to rotate. All of this presupposes that:

1. So little force will be required to turn the device that there will be no slippage.
2. The liquid will not change the flow and torque characteristics of the propeller.

In practice, devices of this type are limited to clean liquids of low viscosity. Gasoline, alcohol, and other materials of similar character give the best results. Figure 11·10 illustrates the chief features of their construction. In some designs, the functioning is entirely mechanical: the propeller turns a gear train which registers the total flow. The accuracy of such a design depends upon keeping friction to a minimum. Generally, there will be stuffing boxes unless there is a magnetic coupling which eliminates such a requirement.

The many means for converting rotary speed into electrical signals are discussed in detail in Chap. 12.

OPEN-CHANNEL FLOW MEASUREMENTS

When the flow is not under pressure and the fluid is of such nature that an open channel can be used, the measurement of the flow requires that certain modifications be made to the flow pattern to obtain the flow rate.

For a known cross-sectional area and flow rate, the flow can be quite easily determined. But with a varying depth of flow, the area is not constant and the flow rate is not easy to measure. These factors have made it necessary to restrict or modify the flow so that the areas could be determined or the flow rate would be known for certain conditions. The obstructions are generally in the shape of a dam, with either a flat surface over which the flow takes place (this is called the crest) or a **V** notch, Figs. 11·11 and 11·12. The generic name is *weir*. Figure 11·13 illustrates typical construction.

Design of the weir, rectangular notch. If accurate results are to be obtained from a weir, certain relationships of size and position must be maintained. The significant dimensions are indicated.

1. The width of the box is 3 times the width of the notch.
2. The end contractions should be 3 times the maximum head H.
3. The bottom contraction should be greater than $2H$ and preferably be $9H$.
4. A crest width of at least 6 in. should be used.
5. The head H should always exceed 0.1 ft and be less than 1 ft.
6. The head should be measured upstream at least 4 times the value of H.

Fig. 11·11 The essential parts of an installation employing a weir. The effect of the liquid discharging into the weir box must be minimized. (*"Principles and Practice of Flowmeter Engineering" by Spink, The Foxboro Co.*)

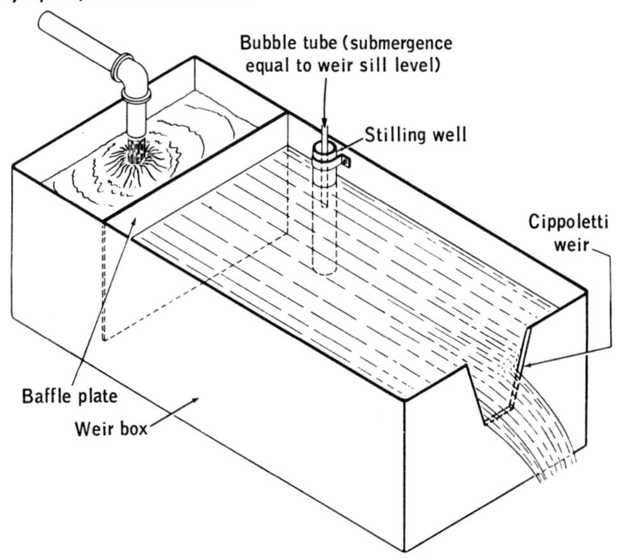

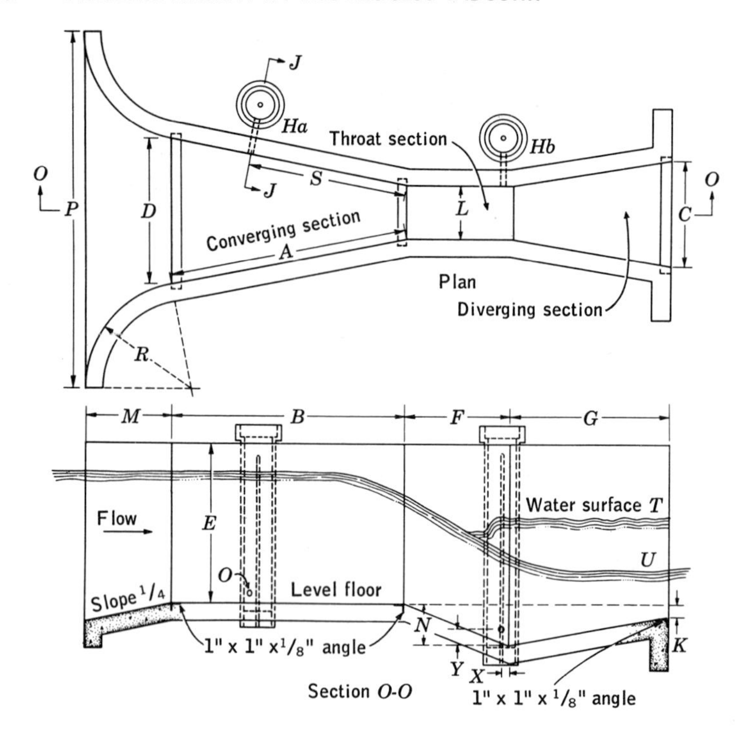

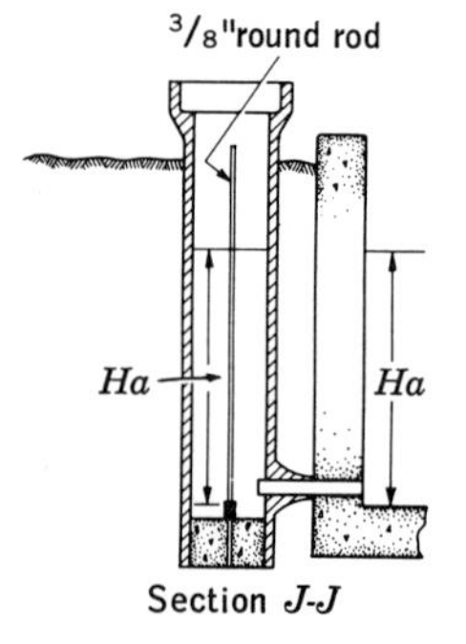

Fig. 11·12 The Parshall flume is of particular use when large amounts of material are held in suspension, as in some irrigation water. (*The Foxboro Co.*)

Note: If a recorder is used, mount gage rod outside wall and make well large enough to provide clearance for at least a 12" circle.

Design of the weir, V notch

1. The notch should be straight and sharp at the apex.
2. The area of the approach channel should be 9 to 12 times that of the notch area, for maximum head conditions.
3. The upstream edges must be sharp and straight.

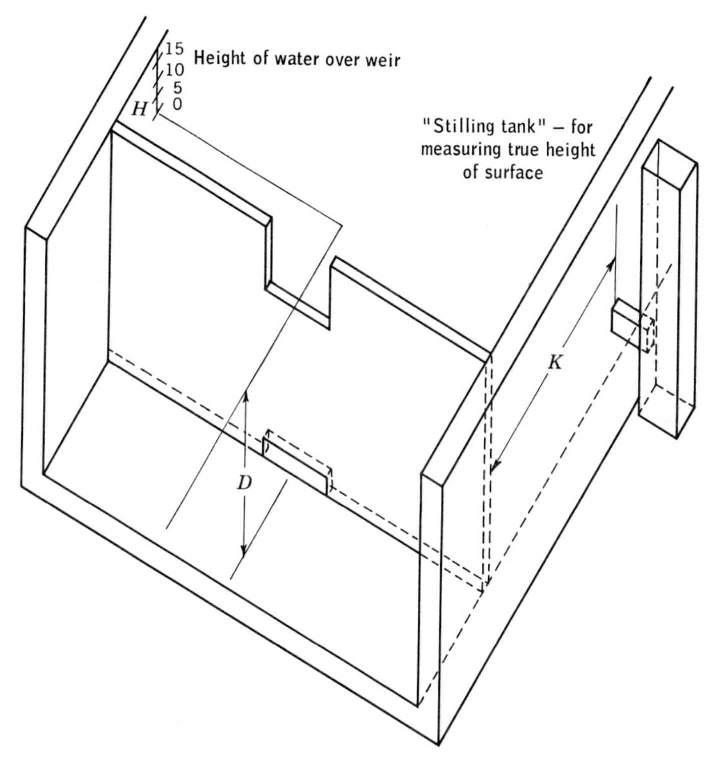

Fig. 11·13 Essential construction of a weir. The main elements are represented in this sketch. The true height of the surface of the liquid is measured in the stilling well. To ensure that the level is not influenced by the draw-down, the value of K should be at least four times the value of H. (*"Principles and Practice of Flowmeter Engineering" by Spink, The Foxboro Co.*)

Under certain types of operation the weir may be unsatisfactory, particularly when the fluid contains large amounts of suspended material, such as sewage and industrial process products. By employing certain principles of the venturi tube and using its lack of obstruction to the flow of suspended solids, a more satisfactory device than a weir can be made for industrial and commercial measurements. (Accurate laboratory flow measurements should employ a weir.) One particular design, known as the Parshall flume, is shown in Fig. 11·12. The flow is constricted both horizontally and vertically. Flow rates are obtained from Fig. 11·14.

Flow calculations. For the rectangular weir with side contraction, the theoretical flow rate is

$$Q = \frac{2b}{3} \sqrt{2gh^3}$$

Fig. 11·14 Flow curves for Parshall flumes. (*The Foxboro Co.*)

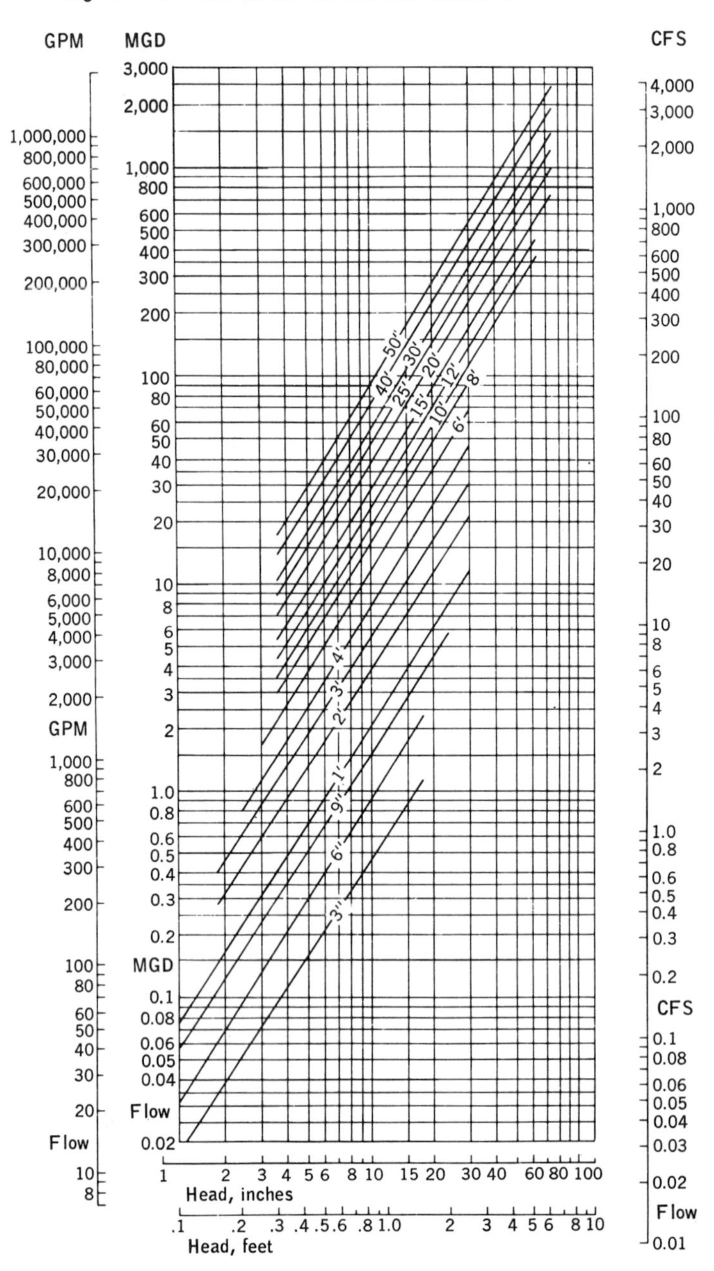

where b = width of the rectangular notch, ft
h = head over weir crest, ft

Actual flow requires use of a flow coefficient which is 0.644 ± 5 per cent:

$$Q = (0.644 \pm 5 \text{ per cent}) \frac{2b}{3} \sqrt{2gh^3}$$

Similarly, the flow through the **V** notch is

$$Q = (0.605 \pm 6 \text{ per cent}) \left(\frac{8}{15} \tan \frac{\theta}{2}\right) \sqrt{2gh^5}$$

The results may be checked by using Fig. 11·15.

The flow through the Parshall flume is expressed in almost the same manner as that for the rectangular weir; the chief difference is in the coefficient. The equation then becomes:

$$Q = (0.63 \pm 3 \text{ per cent}) \left(\frac{2b}{3}\right) \sqrt{2gh^3}$$

FLOW-CONSTRICTION HEAD METERS

In the types of flowmeters previously considered, the flow has been quite unimpeded. The measurements are obtained by measuring voltages, lifting effort, or the forces of acceleration. The majority of industrial liquid and gas flow measurements, however, use a different means to actuate the indicating elements. As in the other cases, these are inferred measurements because direct measurement of flow is difficult, if not totally impractical, especially for gases and liquids which may vaporize under certain metering conditions.

The three general subdivisions of flow-constriction meters or head meters are the venturi, the flow nozzle, and the orifice plate. All employ the same principles:

1. If the fluid is incompressible and if the flow through a pipe is constant, then at any point the product of cross-sectional area and speed must be a constant. The smaller the opening, the higher the speed must be, and vice versa.

2. If we can consider only a small section of pipe, or consider that the frictional effects are negligible, then the energy—both kinetic and potential—must be a constant at all times. Therefore, we may convert one energy to another form, but the total value is unchanged. In a pipe in which the fluid is subjected to pressure, the energies may be present in three forms:

a. Kinetic energy—the energy of motion which is represented by $\frac{1}{2}WV^2/g = \frac{1}{2}MV^2$, where M is equal to W/g.

b. Potential energy with reference to some arbitrary level.

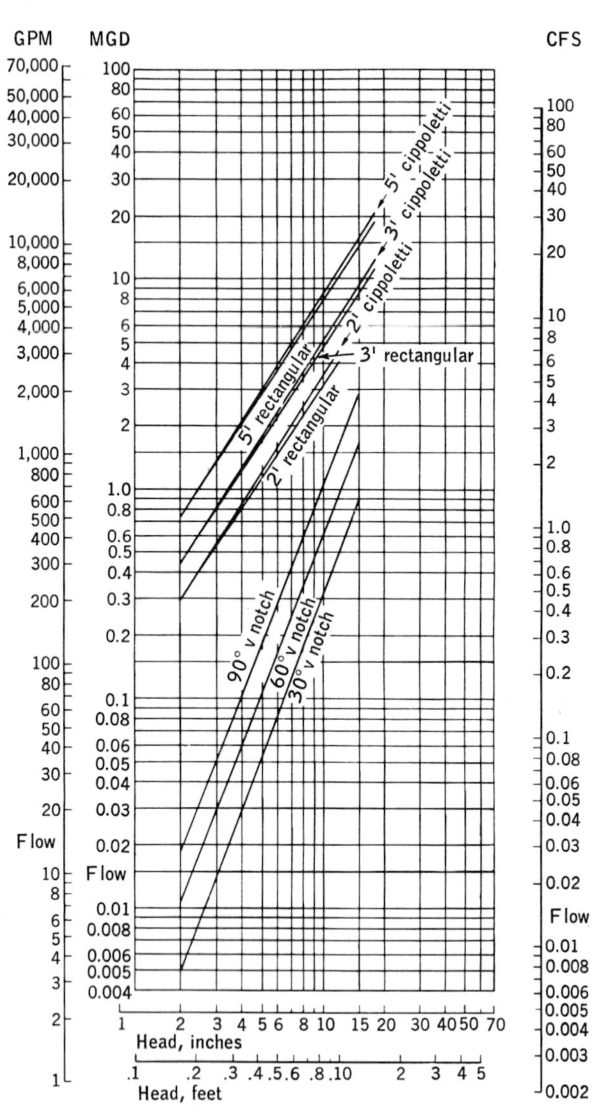

Fig. 11·15 Flow curves for weirs. (*The Foxboro Co.*)

c. Pressure energy, the energy which a liquid exchanges for potential energy as it descends very slowly through some tank. This pressure energy is equivalent to the loss of potential energy with respect to the starting level. Or it may be considered as being able to lift the fluid to its original starting point.

Thus, as the fluid proceeds through level pipe, any increase in speed will increase the kinetic energy at the expense of the pressure energy.

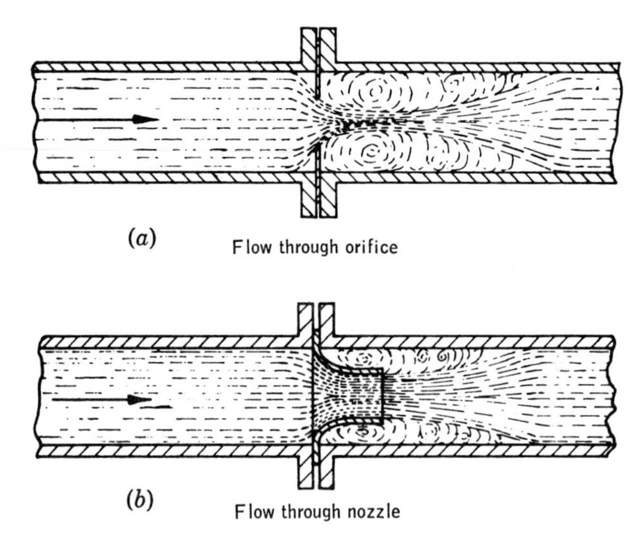

Fig. 11·16 (a) Flow patterns through orifice. (b) Flow patterns through nozzle.

Fig. 11·17 Use of flange-mounted orifice plate. The flange-mounted orifice plate can be inserted and removed without disturbing the piping or the differential pressure connections. Any increase in the flow rate of a liquid under pressure will be accompanied by a decrease in the pressure. The difference in the pressures occasioned by the changes in the flow rates becomes a measure of the flow. One of the major advantages of the orifice plate is the accurate predictability of the contraction of the flow stream.

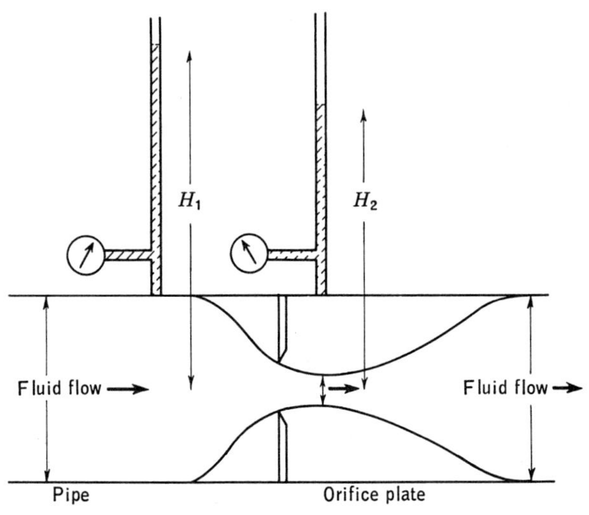

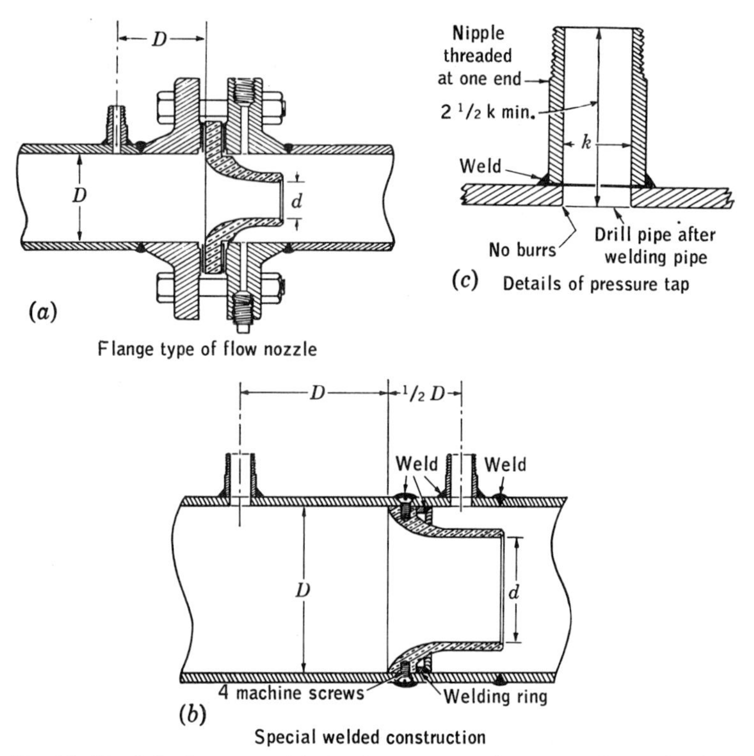

Fig. 11·18 Whether using (*a*) the flange type of flow nozzle or (*b*) the welded nozzle, particular attention must be given to the position of the (c) pressure taps, their size, and the absence of burrs. (*The Foxboro Co.*)

The potential energy with respect to some reference cannot be affected, inasmuch as we assumed that the pipe is level at this point, so there can be no change in the potential energy due to position. So if the kinetic energy gains at the expense of the pressure energy, then the pressure energy will have to decrease with any increase in speed.

It will be noted in Figs. 11·16 to 11·20 that the primary elements are quite different in their construction and performance. The efficiencies of the meters, in terms of the energy used in the metering element, is highest for the Dall tube, Fig. 11·24, and lowest for the orifice plate, Fig. 11·21. It will be noted, too, that the venturi offers a smooth path for the fluids and any suspended materials, whereas certain other designs may suffer from abrasion or the settling of solids which will markedly change the shape of the approach. The area of the stream at its narrowest part will thus depart from calculated values.

In terms of simplicity and low cost, the order is completely reversed: the orifice plate, the flow nozzle, and the venturi. So many design and calculated data are available that the orifice plate may be constructed

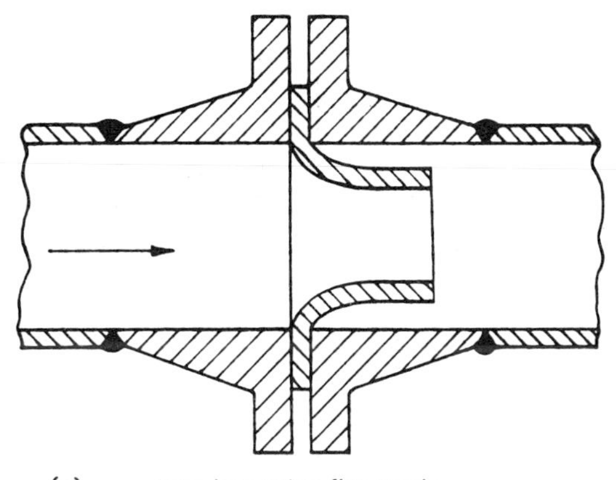

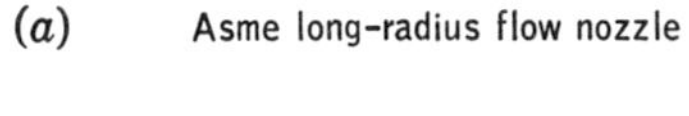

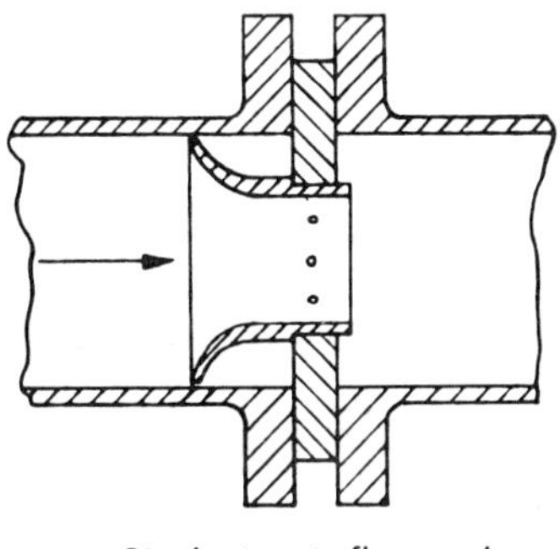

Fig. 11·19 There are several accepted designs of flow nozzles, (*a*) the ASME long-radius and (*b*) the Simplex type TG being among the more common.

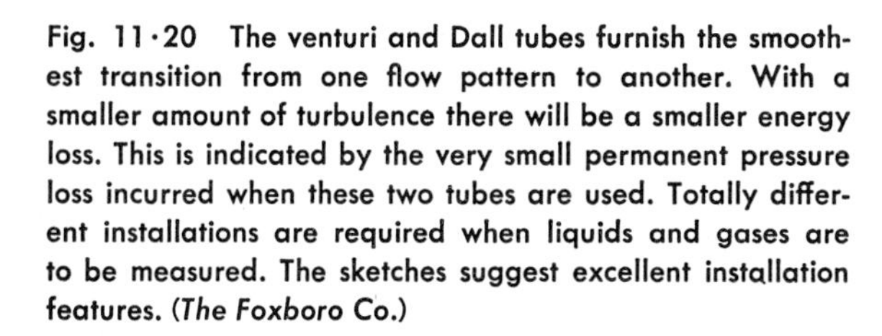

Fig. 11·20 The venturi and Dall tubes furnish the smoothest transition from one flow pattern to another. With a smaller amount of turbulence there will be a smaller energy loss. This is indicated by the very small permanent pressure loss incurred when these two tubes are used. Totally different installations are required when liquids and gases are to be measured. The sketches suggest excellent installation features. (*The Foxboro Co.*)

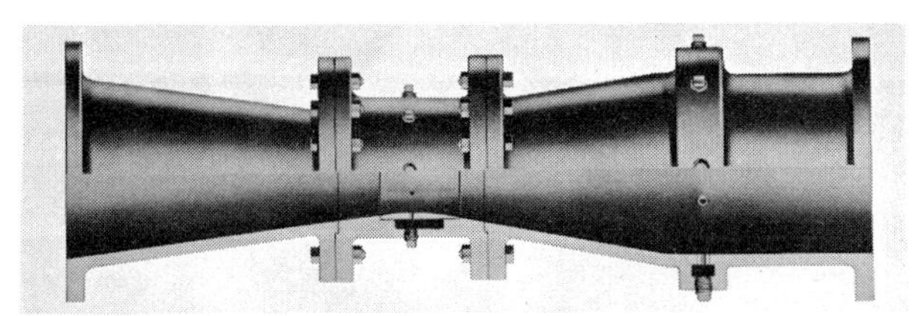

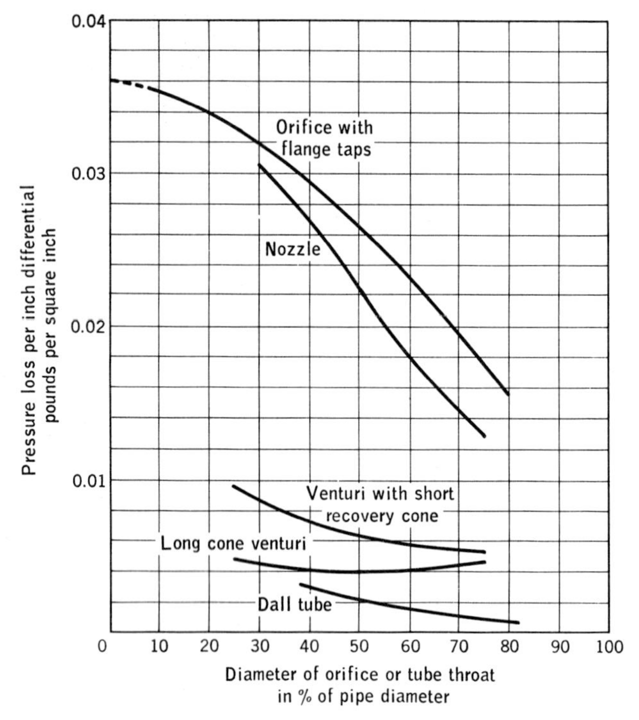

Fig. 11·21 Nozzles and orifices create relatively large pressure losses.

in any machine shop. The flow nozzles and the venturi tubes require factory construction to get the correct proportions at all points. When the saving of energy at the metering element is a factor, then the more expensive elements will be used. This becomes primarily a matter of economics.

The design of the nozzle is such as to prevent contraction of the stream as it passes through the nozzle. In practice, the stream fills about 98 per cent of the area. This accounts for its fine performance.

The flow nozzle should be inserted in a straight section of pipe. Results are improved by keeping it as far downstream as possible from any source of disturbance. Although the nozzle may be installed in any position, performance will be improved if certain precautions are observed when vapors are present or there are materials in suspension. The nozzle should be so installed that the flow will be *vertically up* when there are entrained vapors or gases; otherwise, there may be some possibility of gases accumulating. When there is a possibility that liquids with suspended materials are to be metered, the flow should be *vertically down* so that materials which drop out of suspension will be unable to alter approach conditions because there will be no good place for them to lodge and remain.

Whether using the flange type of flow nozzle or the welded nozzle, Fig. 11·18*a* and *b*, particular attention should be given not only to the position of the pressure taps but also to the size of and the complete absence of burrs in the taps.

The suggested construction of the pressure taps requires that the nipple be welded to the pipe at the desired tap location. Using a nipple drilling guide, drill a hole of dimension k through the nipple and the pipe. Before drilling, the internal diameter of the pipe nipple should be slightly less than the k dimension:

$$k = 1/2 \text{ in. for 4-in. and larger lines}$$

$$k = 3/8 \text{ in. for 3-in. lines}$$

$$k = 1/4 \text{ in. for 2-in. lines}$$

There should be no roughness inside the nipple, nor should there be any burrs or rough edges on the inside edge of the hole if accurate results are sought.

The venturi tube is used whenever the permanent loss of pressure is to be reduced to a minimum or when there are such amounts of material in suspension that orifice plates and flow nozzles are not satisfactory. Because of the very gradual transition of tube size, there is very little chance that any material can alter the approach to the tube. Moreover, the tube has unusually uniform coefficients on viscous flows.

The installation of the venturi, Figs. 11·20 and 11·21, is relatively simple; it consists in setting the venturi into the line as another piece of pipe with the shorter cone forming the inlet or upstream end. The tube may be installed in any position, horizontal, vertical, or inclined. But, as with the other types of head meters, the venturi must be installed as far downstream as possible from any source of flow disturbance, such as reducers, valves, and combinations of fittings. Typical installations are shown in Figs. 11·22 and 11·23. The most dependable readings will be obtained when pressure connections to the venturi tube are made as follows:

On horizontal and inclined piping:

To the top of the tube, for gas installations
To the side of the tube, for liquid installations
To the top of the tube, for steam installations
  when the meter is above the line
To the side of the tube, for steam installations
  when the instrument is below the line

On vertical pipes:

To any part of the tube

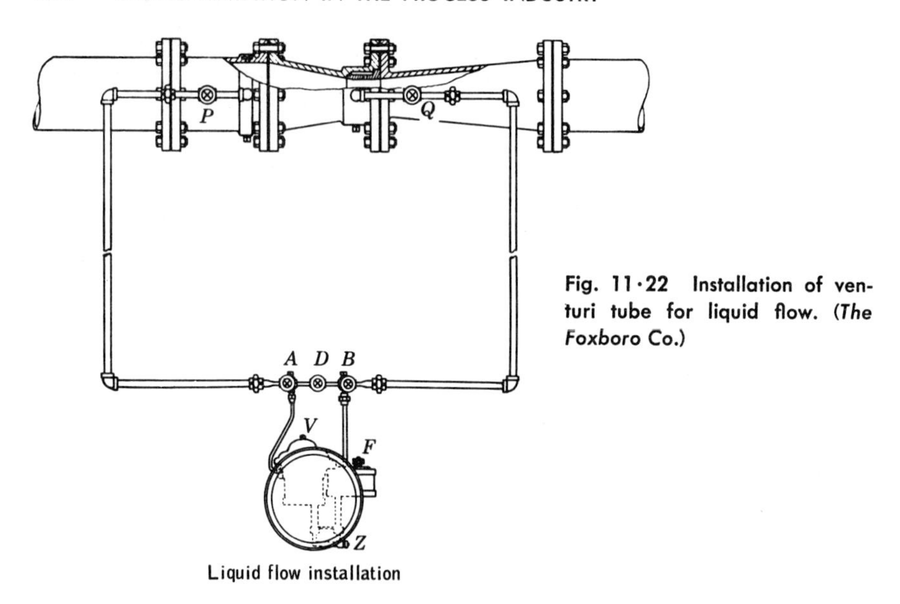

Fig. 11·22 Installation of venturi tube for liquid flow. *(The Foxboro Co.)*

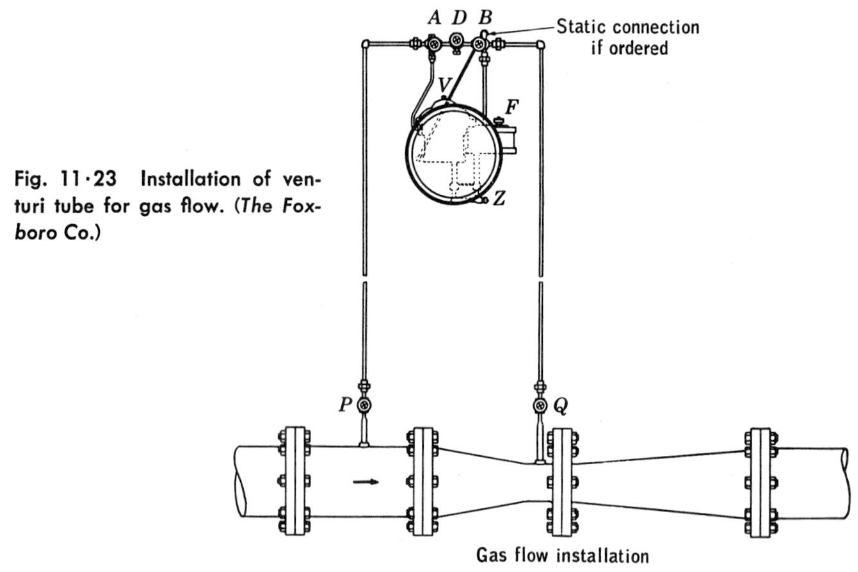

Fig. 11·23 Installation of venturi tube for gas flow. *(The Foxboro Co.)*

A modified type of venturi tube is the Dall flow tube, Fig. 11·24. It is a primary device used to measure the flow of gases, water, steam, and other fluids which do not contain solids which can settle out. This device has several important characteristics when compared with other primary elements. For a given flow rate and differential, it has a lower permanent pressure loss than any other common type. Despite this, it has a much higher differential than the usual venturi tube or flow nozzle having the same throat-to-pipe diameter ratio d/D. It is shorter, lighter, and some-

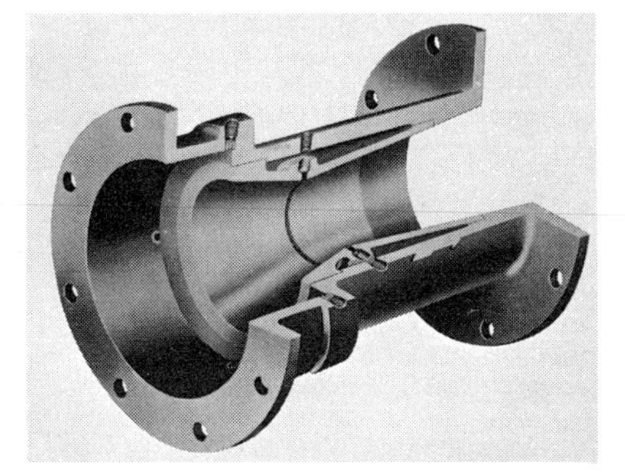

Fig. 11·24 The Dall flow tube. ***(B-I-F Industries, Inc.)***

what cheaper than a short-cone venturi of the same capacity. It was developed in England in the middle 1940's and has since undergone extensive testing. It is now being manufactured in the United States.

Despite the resemblance of the Dall tube to the venturi, there are some noteworthy differences. The Dall tube has a short, straight inlet section, the end of which abruptly decreases in diameter to the inlet shoulder; a converging cone section; a narrow annular gap or slotted throat annulus; and a diverging outlet cone. The inlet-pressure tap is directly upstream of the inlet shoulder, while the throat tap is over the annulus.

As the fluid flows through the inlet cone of the Dall tube, its velocity increases and its pressure decreases. Figure 11·25 shows that the Dall

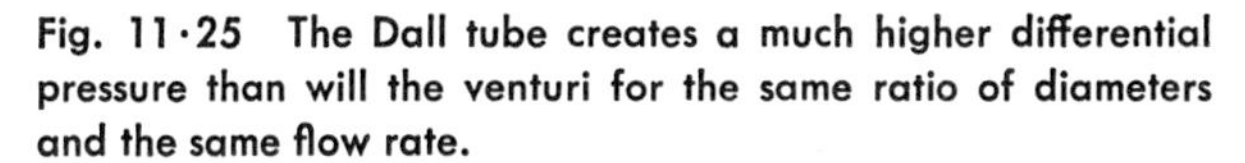

Fig. 11·25 The Dall tube creates a much higher differential pressure than will the venturi for the same ratio of diameters and the same flow rate.

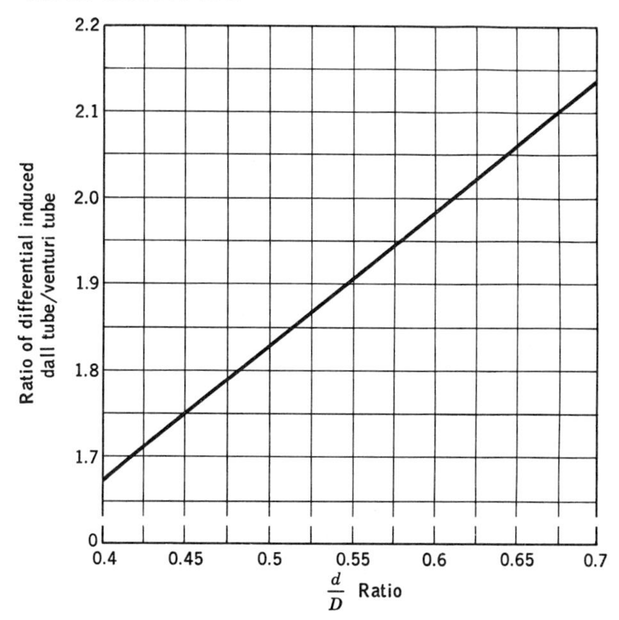

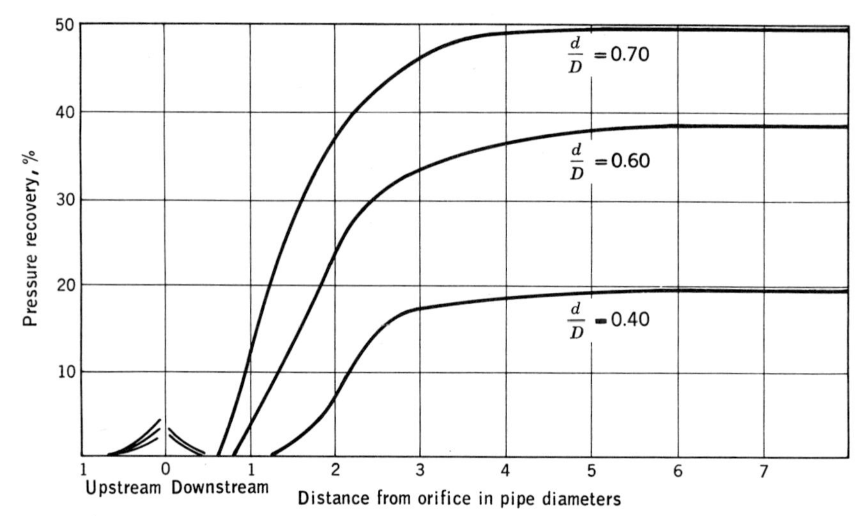

Fig. 11·26 Approximate recovery of pressure as a function of distance from orifice and of d/D ratio.

tube has a differential pressure drop almost twice that of the venturi with the same d/D ratio and flow rate.

Flowmeters which employ differential pressure measurements really are of the *mass* type, because the acceleration given the liquid, and thus the increase in the rate of flow of the liquid, varies inversely with the density of the liquid or gas.

The per cent of pressure recovery varies with distance and the ratio of d/D. Figure 11·26 indicates that the maximum pressure differential will be obtained by placing the low-pressure tap about a half pipe diameter downstream from the orifice.

The viscosity of the liquid also affects the flow rate. At low rates, the flow is viscous or laminar. With viscous flow there is virtually no turbulence or churning of the fluid. From Fig. 11·27 it will be seen that with viscous flow the speed at the center of the pipe is about twice the speed of the fluid adjacent to the walls. Most flow, however, is turbulent, with the speed of all parts more nearly alike. The speed at the center will still be somewhat greater than the average flow rate.

Reynolds number. If very high degrees of accuracy are desired when head meters are employed, then corrections will have to be made for the Reynolds number, R_D. The Reynolds number is a very important reference number in accurate flow determinations. It is used to determine the point at which the flow goes from *viscous* to *turbulent*. As the flow changes from viscous to turbulent, and vice versa, there is a very marked change in the value of the flow coefficient, but there is relatively little change with further increases in speed. The change in the flow characteristics is dependent upon three factors:

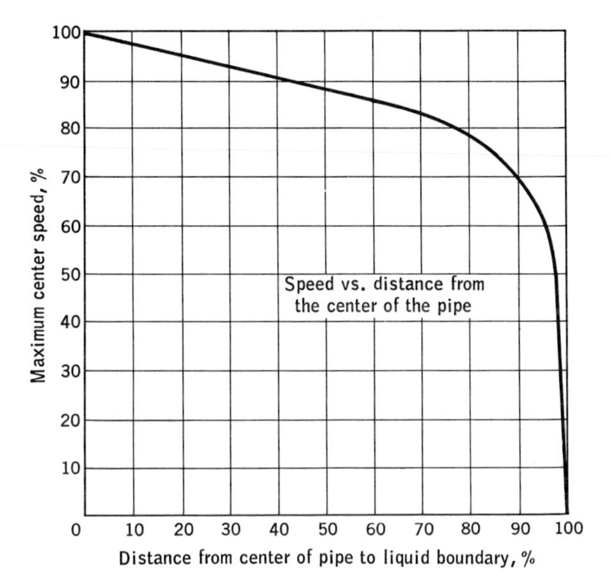

Fig. 11·27 There is a very marked reduction in the speed of the liquid as it approaches the junction with the pipe surface. Accurate flow measurements must employ *true average* values.

1. The rate of flow in pounds per unit time. The higher the rate of flow the sooner will the turbulent condition be established.
2. The pipe diameter. This is an inverse relationship because the larger the pipe, the higher must be the flow rate when turbulence occurs.
3. The absolute viscosity. The higher the viscosity the higher the flow rate for turbulence.

To ensure that you have a clear picture of the significance of the Reynolds number R_D, study Figs. 11·27 to 11·30. In the first of these figures, it will be noted that there is a wide change of speed as we look at the speeds of the fluid at different distances from the pipe wall. For turbulent flow, the average speed is approximately 82 per cent of the maximum speed.

If we now consider the second figure, we find that there is a wide range of values for average speed divided by maximum speed; they go from 0.5 to almost 0.82. It will be noted that for values of R_D greater than 40,000 there is almost no change in the value of average speed divided by maximum speed. Therefore, to ensure results that are reliable and as predicted, the values of R_D for air and water should exceed 40,000.

If the measurements were made with R_D equal to 10,000, the ratio of the two speeds would be 0.78. Consequently, the flow value would have to be multiplied by 0.78/0.82, or 95 per cent. For $R_D = 4{,}000$, the correction factor would be $0.75/0.82 = 91.5$ per cent. If possible, therefore, all flow should take place with the (*a*) speed (or pounds per unit of time) high enough, (*b*) pipe diameter small enough, and (*c*) viscosity low enough to ensure that *turbulent flow* exists and that the expected ratio of average speed to maximum speed will be realized.

Fig. 11·28 Ratio of average to maximum speed in a pipe as a function of the Reynolds number. For Reynolds numbers above 20,000 the ratio of the average to maximum speed in the center remains essentially constant. But for lower values there is a marked change in the ratio. This points up one of the chief uses of Reynolds number, the point at which the flow pattern stabilizes. (*"Principles and Practice of Flowmeter Engineering,"* by Spink, The Foxboro Co.)

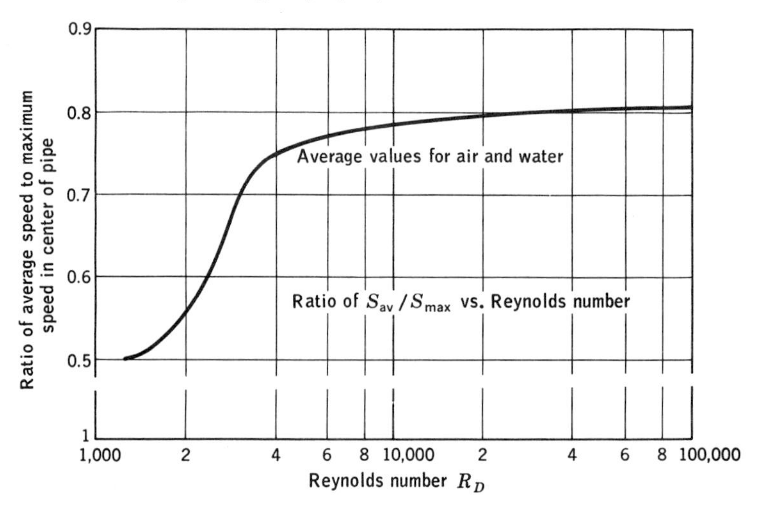

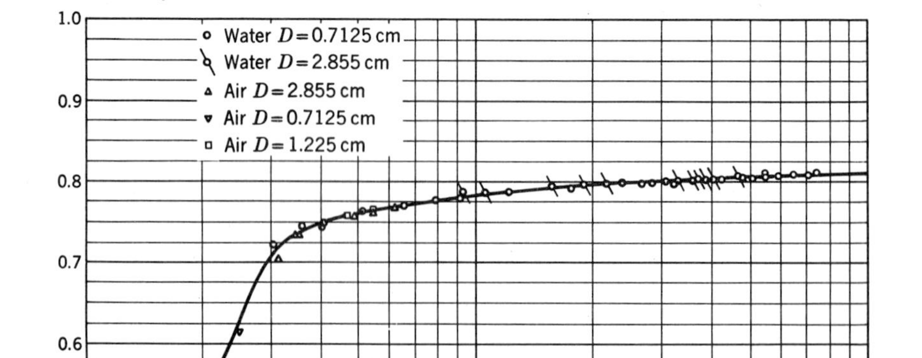

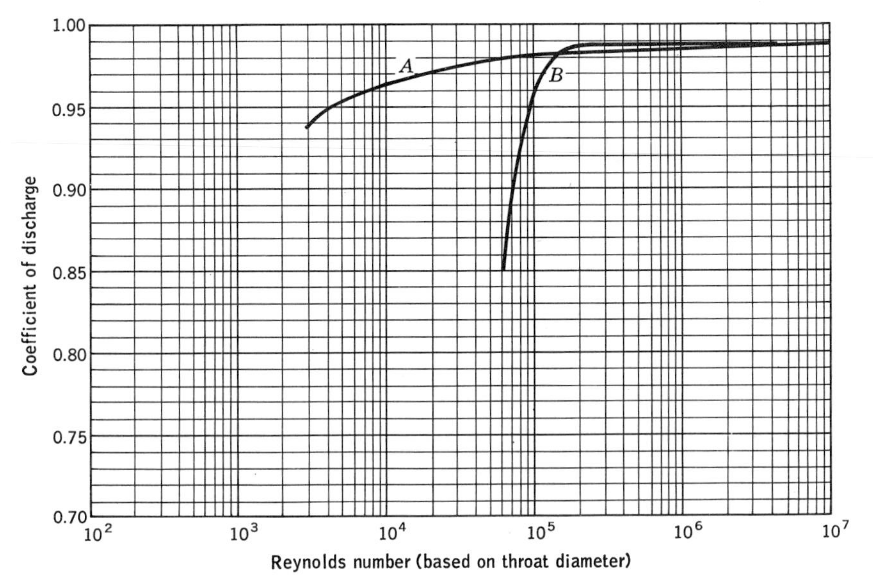

Fig. 11·29 Coefficient of discharge versus Reynolds number for two different nozzles. (*Minneapolis-Honeywell Regulator* Co.)

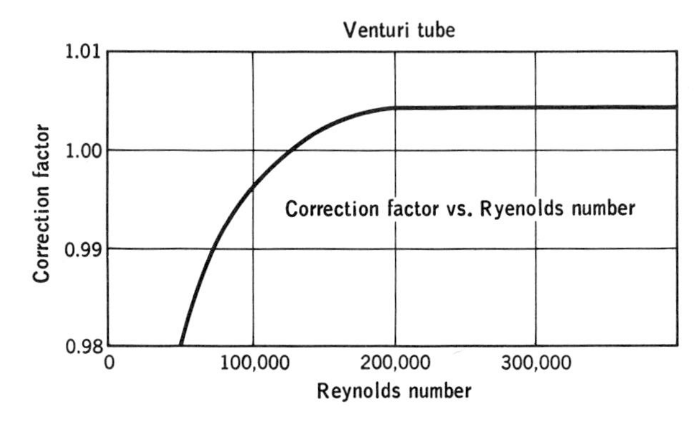

Fig. 11·30 Correction factor versus Reynolds number, venturi tube. It will be noted that only very precise calculations will require corrections for values over 100,000.

The other graphs included under Figs. 11·31 and 11·32 show how the flow correction factors vary with orifice ratio d/D, viscosity, and the Reynolds number. It will be observed also that in almost all instances that when a certain range of Reynolds numbers has been reached, all flow at higher flow rates will not require any modification of the flow factors, for any given value of d/D.

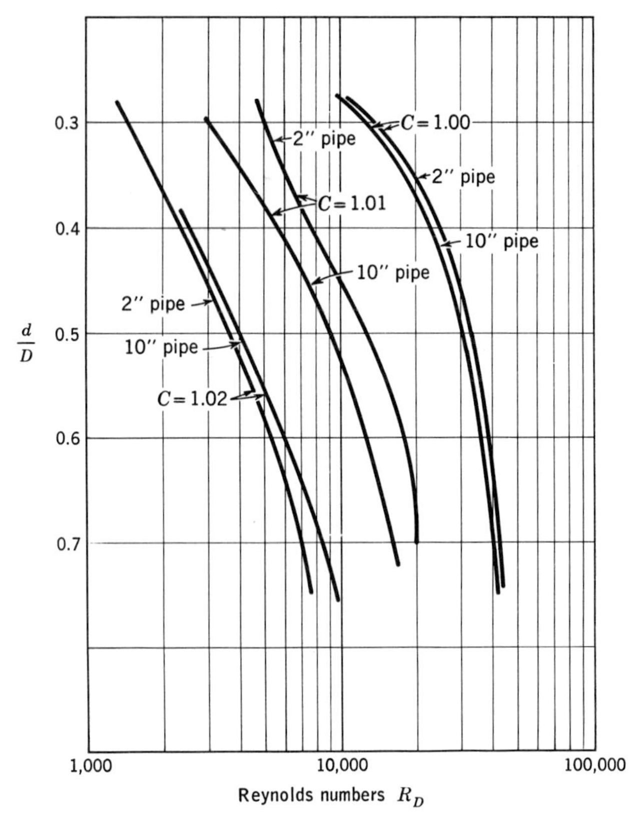

Fig. 11·31 Flow-correction factors as a function of pipe size and Reynolds number, for *D* and ½ *D* taps.

The Reynolds number is also defined as the ratio of the inertial forces to the viscous forces when the number is expressed by

$$R_D = \frac{VDS}{\mu}$$

where S = weight density
V = speed, fps
D = diameter of the pipe
μ = absolute viscosity of the material

It will be noted that the numerator of this fraction does not express a weight of material, let alone any force from some implied acceleration. In order to make this term much more meaningful and thus more significant, let us resort to some legitimate mathematical manipulations. Let us replace D with its equal $2r$ and then multiply numerator and denominator by πr. The expression then becomes

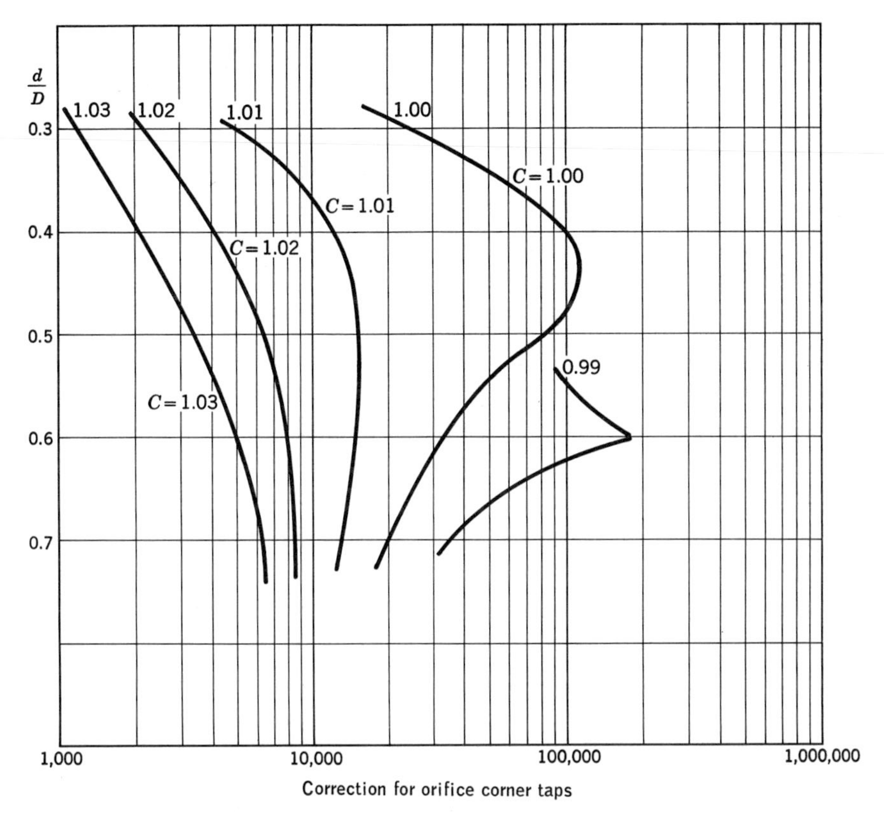

Fig. 11·32 Flow-correction factors as a function of pipe size and Reynolds number, concentric orifice with corner taps.

$$R_D = \frac{(V)(\pi r)(2r)(S)}{\pi r(\mu)}$$

But πr^2 = area of the pipe; therefore,

$$R_D = \frac{2(\text{area of pipe})(\text{speed of liquid})(\text{weight density})}{\pi r(\mu)}$$

But the product of pipe area and fluid speed produces volume per unit time, and multiplying by weight density yields pounds per second. Therefore

$$R_D = \frac{2}{\pi}\frac{\text{lb/sec}}{r\mu}$$

In order to keep our units consistent, r should be expressed in feet, and $2/\pi$ becomes K or 0.636. Then

$$R_D = \frac{0.636 \text{ flow, lb/sec}}{r\mu}$$

Therefore we may state that the Reynolds number which has been correlated with flow and the correct value of flow coefficients may be defined as directly proportional to the flow, in pounds per second, and inversely proportional to the pipe radius and the absolute viscosity.

Should the pipe size be expressed in terms of the diameter, then the coefficient becomes

$$R_D = \frac{1.272(\text{flow, lb/sec})}{d\mu}$$

where d is in feet.

If we group all of the constants together and correct for the differences in the units:

$$R_D = \frac{0.237(\text{rate of flow, lb/hr})}{(\text{pipe radius, in.})(\text{abs viscosity,(lb)(sec)/ft}^2)}$$

$$= \frac{6.32(\text{rate of flow, lb/hr})}{(\text{pipe diameter, in.})(\text{abs viscosity, centipoises})}$$

$$= \frac{6.32(3{,}600)(\text{rate of flow, lb/sec})}{(\text{pipe diameter, in.})(\text{abs viscosity, centipoises})}$$

$$= \frac{22{,}800(\text{rate of flow, lb/sec})}{(\text{pipe diameter, in.})(\text{abs viscosity, centipoises})}$$

The pitot tube. In many ways, the pitot tube, Fig. 11·33, is similar to the target meter in that part of the force or pressure developed is a measure of the impact pressure caused by stopping the movement of the fluid or accelerating it so that it follows some different path. Because of its very small area, the pitot tube measures the pressure at a point rather than the average pressure across a pipe or duct. For this reason it is often used to make a traverse of a pipe in order to find the relative speeds with respect to some reference point such as the center.

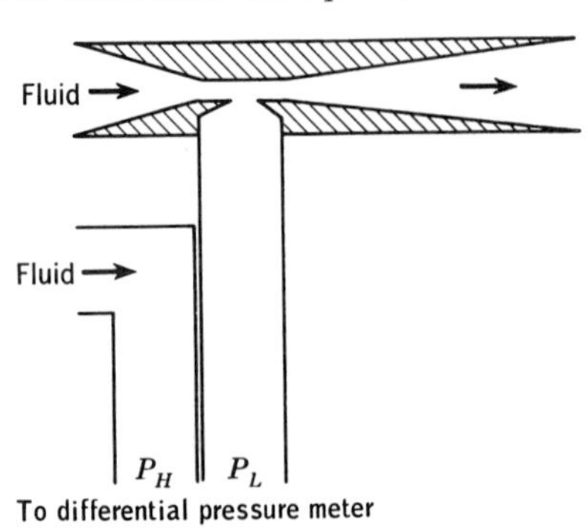

Fig. 11·33 The Zahm type of pitot-venturi.

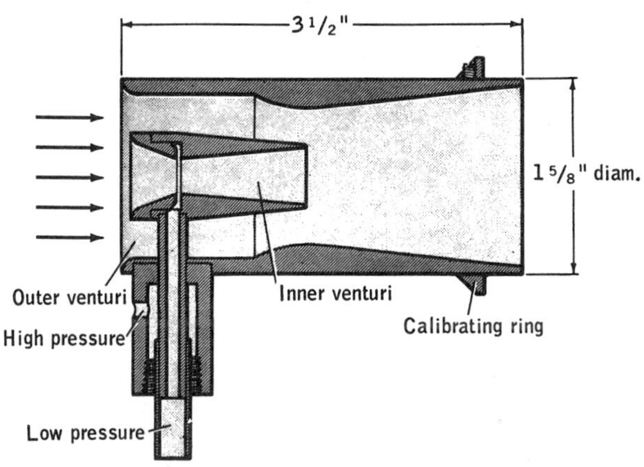

Fig. 11·34 Venturi tube construction to increase the output of a pitot tube. Single or double venturi sections may be added to a pitot tube in order to increase the pressure differential. Because such a small port furnishes the high pressure, the device is subject to plugging unless the fluids are clean. (*The Taylor Instrument* Cos.)

Aside from its deficiency in measuring only point pressures, the pitot tube is very easily fouled by foreign material in the flowing fluid. It has another shortcoming in that the pressure drops which it creates are often too small to be measured by standard differential pressure meters. In order to increase the differential pressure effects, a venturi tube is occasionally used to increase the flow at the expense of the static pressure Fig. 11·34.

Certain definite precautions must be observed if accuracies as high as ± 2 per cent are to be obtained with a pitot tube. It has been suggested that when relatively high degrees of precision are required, four conditions must be met:

1. Provide a duct diameter of 4 in. or greater.
2. Make an accurate traverse across the duct and average the readings.
3. Make sure that there is straight-line flow by having at least ten diameters of straight duct both before and after the pitot tube.
4. Install some type of device to straighten the flow upstream of the tube.

If conditions, or time, will not permit the making of a traverse, install the pitot tube in the center of the duct. Then multiply the indicated flow rate by 0.9. Such a method should give answers correct to ± 5 per cent.

An empirical relationship between flow rate and differential pressure obtained with the pitot tube is

$$\text{Air speed} = 1{,}096 \sqrt{\frac{P_v}{D}}$$

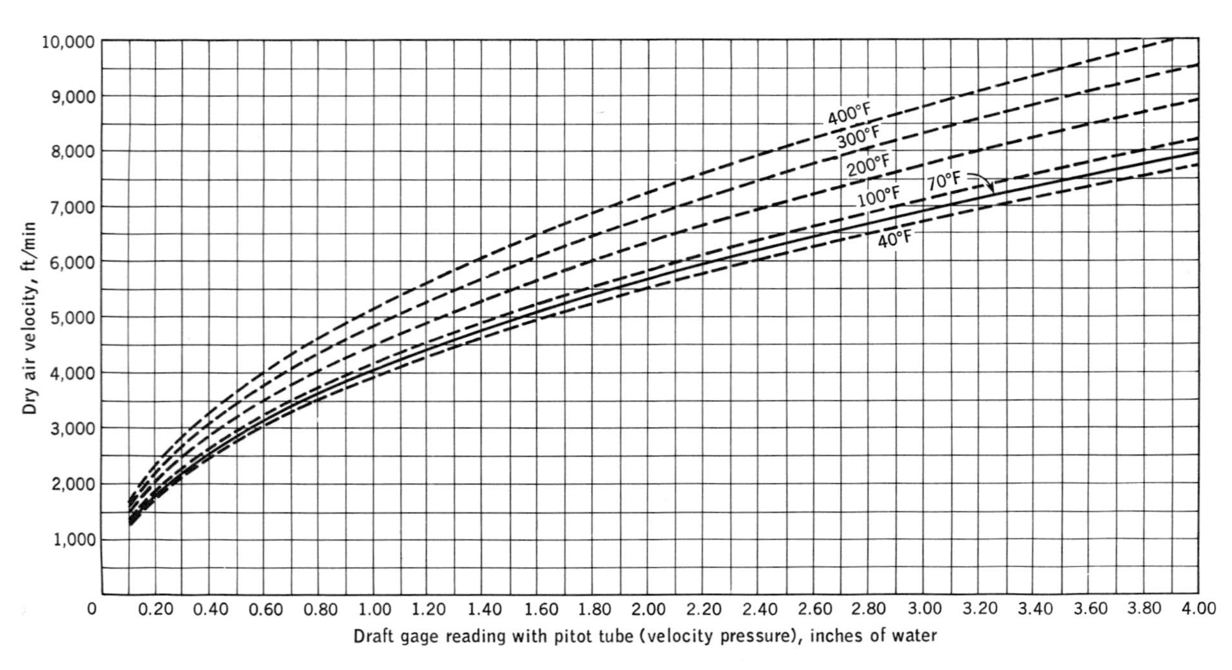

Fig. 11·35 Dry air velocity, in feet per minute, versus draft gage reading with pitot tube, in inches of water. (*F. W. Dwyer Mfg. Co.*)

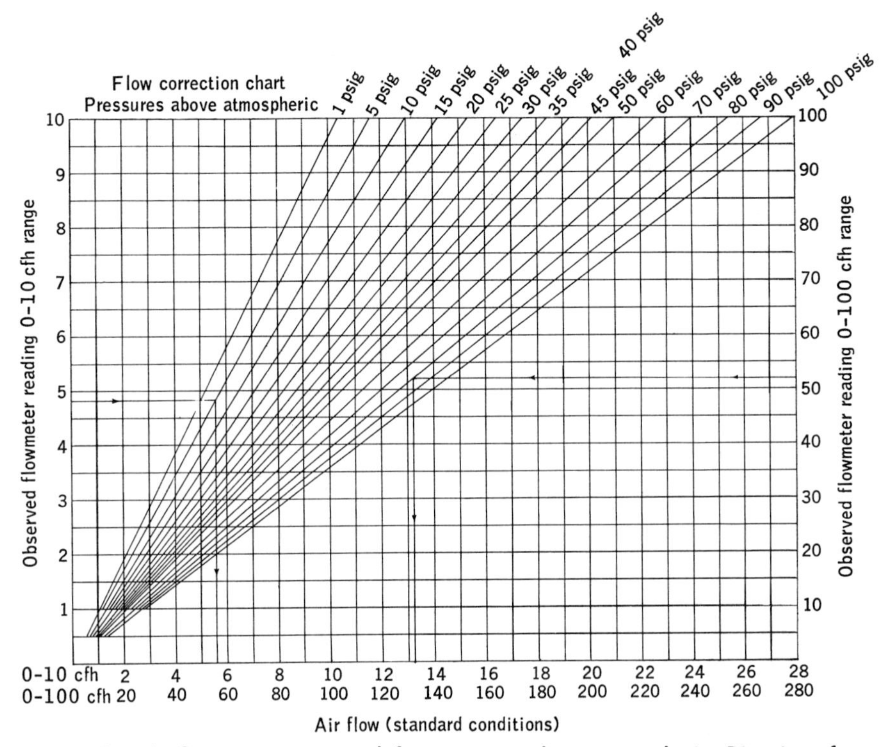

Fig. 11·36 Air-flow curves corrected for pressures above atmospheric. Directions for use of chart: (1) Find the observed flowmeter reading on the vertical axes. (2) Move horizontally to the appropriate pressure slant line. (3) Proceed vertically down to the horizontal axis to find the air flow under standard conditions. (*F. W. Dwyer Mfg.* Co.)

where P_v = air pressure, in. of water
D = air density, lb/ft.^{3}
$D = 1.325\ (P_B/T)$
P_B = barometric pressure, in. of mercury
T = absolute temperature, °R

The important relationships are presented in graphical form in Figs. 11·35 to 11·37.

THEORY OF FLOWMETERS, HEAD TYPE

Because of the nature of the material, the quantity to be measured, or some other factor, it may be concluded that the head type of meter offers the best solution to a metering problem. The main question is the selection of the primary element. Let us summarize the points in which they are alike:

1. Each one will require the same sort of differential pressure meter.
2. Each one will require the same piping arrangement and precautions.

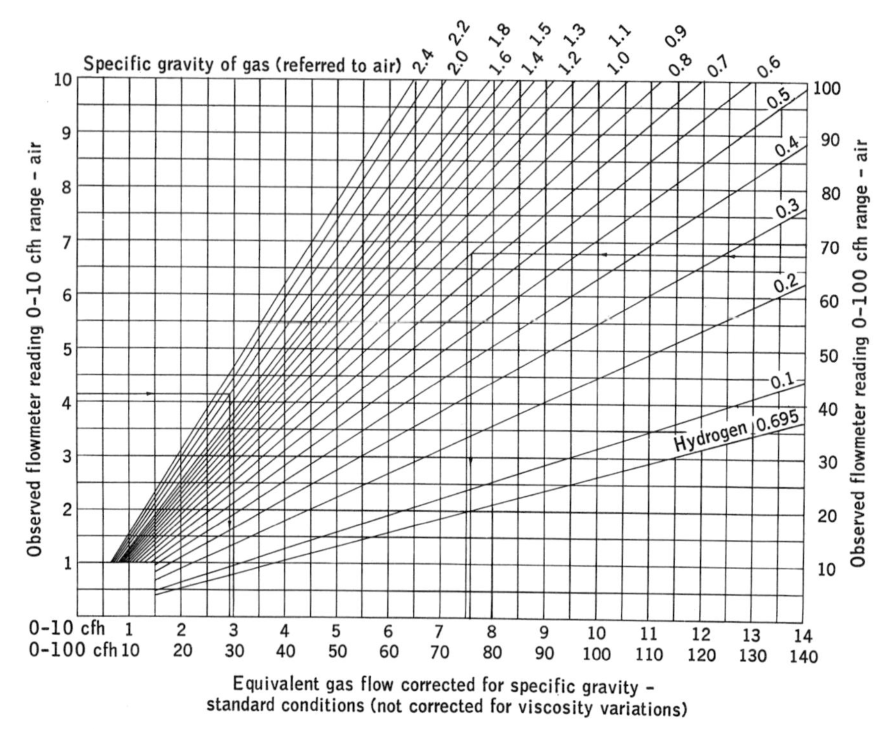

Fig. 11·37 Gas-flow curves corrected for specific gravity of gas. Directions for use of chart: (1) Find the observed flow rate on the vertical axis. (2) Move horizontally to the slant line which gives the specific gravity referred to air. (3) From the intersection, read directly below to get the corrected value. (*F. W. Dwyer Mfg. Co.*)

3. Each one will involve a pressure drop in the line.

They also have differences which may be tabulated as:

1. The pressure drop in the orifice plate is much greater than it is with either the flow nozzle or the venturi.
2. The recovery of the pressure head is least with the orifice plate.
3. The cost of the venturi is the greatest and the cost of the orifice plate is the least.
4. The orifice plate may be constructed by almost any machinist, but the other devices require factory fabrication.
5. The flowing stream fills the openings of the venturi and the flow nozzle but is contracted in passing through the plate, the reduction in area being about 40 per cent.
6. The flow coefficients for the venturi and the flow nozzle remain substantially constant over all normal operating ranges, but the coefficient for the orifice plate varies by several per cent because of viscosity and kindred effects.
7. The venturi is little affected by the presence of solids or slurries;

the flow nozzle can handle small concentrations of foreign material; but the characteristics of the orifice plate are greatly changed if solids build up in front of it.

Theory of operation. All types of head meters are based on one of the fundamental relationships concerning the energy associated with any given amount of material under certain specific conditions. For any given conditions, the amount of energy possessed by an object is fixed. That energy cannot be destroyed; it can only be transformed from one form to another. What we are primarily concerned with here is the conversion of pressure energy into speed and, conversely, the conversion of speed back into pressure energy. The term "pressure energy" is being used rather loosely in this case. It simply means that the pressure which exists is capable of creating both kinetic energies and potential energies.

By restricting the area of the flowing liquid, we will have to increase the speed of the liquid, provided that the rate of flow is constant throughout the pipe. With an increase in speed, which will cause an increase in the kinetic energy, there must be a corresponding reduction in some other energy, which in this case is the pressure energy. This analysis will be limited to the case of incompressible fluids.

For the conditions of flow of an incompressible fluid through a venturi (or other head-type primary element), Fig. 11·38, the quantity Q, per unit of time, is constant and is equal to

$$Q = A_1V_1 = A_2V_2 = A_3V_3 = A_4V_4 = A_5V_5 = A_6V_6$$

Fig. 11·38 The pressure in a straight level pipe changes with the flow rate. An increase in the speed of the liquid is accompanied by a decrease in the pressure. Much of this loss of pressure is recovered when the larger sections of pipe are reached.

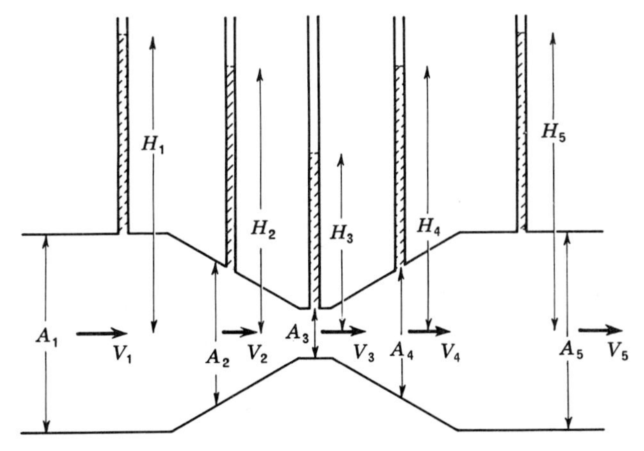

where A = cross-sectional area of the pipe
V = speed at that particular cross section

Both A and V are measured in the same system of units. Taking any two sets of relationships,

$$A_1V_1 = A_3V_3 \qquad V_1 = \frac{A_3V_3}{A_1}$$

To obtain an increase in speed, an acceleration is required. A force is always necessary if an acceleration is to result. Therefore, there must be a greater pressure at A_1 than there is at A_2, A_3, A_4, or A_5, acting from the left.

For over 150 years it has been known that in a closed pipe, such as the one shown in Fig. 11·38, an increase in flow is accompanied by a decrease in pressure at the point where the flow stream is constricted and the velocity is increased. As the pipe regains its size, the speed decreases and the pressure increases. Venturi, an Italian, measured these relationships in 1797, but over 100 years were to elapse before they were used for flow measurements. Figure 11·38 represents the fundamental relationships which occur when a constriction is introduced into the line: an increase in speed is obtained at the expense of pressure.

Bernoulli's equation shows the more complicated relationship between pressure and speed at any point in the pipe. It is useful in flow work, and in simplified form it becomes

$$\frac{P_2}{\rho} + \frac{V_2^2}{2g} = \frac{P_1}{\rho} + \frac{V_1^2}{2g} \tag{11·1}$$

which may be rewritten as

$$\frac{V_2^2 - V_1^2}{2g} = \frac{P_1 - P_2}{\rho} \tag{11·2}$$

$$V_2^2 - V_1^2 = \frac{2g}{\rho}(P_1 - P_2) \tag{11·3}$$

$$= K(P_1 - P_2) \tag{11·4}$$

$$= K'F$$

where g = acceleration due to gravity
P = static pressure, absolute
ρ = fluid weight density
V = fluid-stream speed
K and K' = proportionality constants
F = force (a change in speed requires the application of force)

The term $V^2/2g$ is common in connection with the speed of falling bodies, which may be either liquids or solids. A certain distance of fall h produces a speed such that

$$V^2 = 2gh \tag{11·5}$$

We may then rewrite Eq. (11·1)

$$\frac{P_2}{\rho} + h_2 = \frac{P_1}{\rho} + h_1 \tag{11·6}$$

Pressure can be changed into a speed, and speed can be converted into a pressure head. When a faucet is open, the water rushes out; the pressure just ahead of the faucet is *converted* into speed because the back pressure at the delivery end of the spout is only atmospheric. The pressure difference creates a force which must produce an acceleration, inasmuch as the liquid is not restrained.

It has been noted previously that

$$Q = A_2V_2 = A_1V_1 \tag{11·7}$$

where Q = flow rate

Therefore

$$V_2 = \frac{Q}{A_2} \qquad V_1 = \frac{Q}{A_1} \tag{11·8}$$

$$V_2^2 = \frac{Q^2}{A_2^2} \qquad V_1^2 = \frac{Q^2}{A_1^2} \tag{11·9}$$

But

$$V_2^2 - V_1^2 = \frac{Q^2}{A_2^2} - \frac{Q^2}{A_1^2} = Q^2\left(\frac{1}{A_2^2} - \frac{1}{A_1^2}\right) = Q^2\left(\frac{A_1^2 - A_2^2}{A_1^2A_2^2}\right) \tag{11·10}$$

and

$$\frac{2g}{\rho}(P_1 - P_2) = V_2^2 - V_1^2 \tag{11·11}$$

Therefore

$$Q^2K'' = \frac{2g}{\rho}(P_1 - P_2) \tag{11·12}$$

$$Q = \sqrt{\frac{2g}{\rho K''}(P_1 - P_2)} = K'''A_2\sqrt{\frac{2g}{\rho}(P_1 - P_2)}$$

$$= K_0A_2\sqrt{\frac{2g}{\rho}(P_1 - P_2)} \tag{11·13}$$

$$K_0 = \text{flow coefficient} = \frac{K'''}{\sqrt{1 - \beta^4}} \tag{11·14}$$

$$\beta = \text{diameter ratio} = \frac{d_2}{d_1} \tag{11·15}$$

where d_2 = diameter at stream restriction
d_1 = pipe diameter

For the liquid, ρ = a constant and the difference between P_1 and P_2 is called the head h; so the flow equation may be written

$$Q = KA\sqrt{2gh} \tag{11·16}$$

where A = area of throat, ft^2
K = velocity of approach factor

Equation (11·16) is the basis for the computations which will be studied in a separate section.

FLOW CALCULATIONS

The instantaneous volume rate of flow through any pipe or section is equal to the product of the area and the average speed:

$$\text{Volume or quantity} = \text{area} \times \text{speed}$$

If we want the answer in gallons per minute or pounds per day, then we must bring in additional factors. For example, multiplying the volume by the weight density will provide an answer in pounds, assuming that the weight density is expressed in that unit.

All of this is based on the assumption that we know the rate of flow and the area of the flowing material. Sometimes these values are difficult to obtain, and in almost no cases do we obtain them directly. We have to measure several other quantities before we can infer what the answer may be. From experience we know that the effective cross-sectional area of the liquid flowing through either a flow nozzle or a venturi tube is about equal to the area of the constriction. However, in the case of the orifice-plate meters where the orifice is simply a hole in a plate (there being a sharp edge on the upstream face of the opening), the stream undergoes a marked reduction in cross section as it goes through the opening. In general terms, the effective cross-sectional area of the stream is about 60 per cent of the orifice opening. The speed of the liquid at the point of maximum contraction is theoretically equal to the speed that the liquid would have if it were to fall freely in a vacuum for a vertical distance equal to the differential pressure head, expressed in feet for the flowing liquid, if the initial speed is negligible.

$$V = \sqrt{2gh} = \sqrt{2g\,\frac{(P_1 - P_2)}{\text{sp gr}}}$$

Fortunately, there is a vast amount of experimental data available which will aid us in ascertaining what this effective area is.

We must also make allowances for the temperatures because densities and specific gravities are both affected. The flow computations which we shall make are based upon mass, but our answers probably will be desired in gallons per minute or barrels per hour.

In order to clarify the fundamental relationships, it is proposed to determine the flow relationships for (1) water, (2) petroleum products, (3) steam, and (4) gases. From the work with Bernoulli's equation, we know that (1) there can be reversible conversions of energy from one form to another, i.e., from kinetic energy to pressure energy, and (2) for liquids confined in a pipe, and at points of equal elevation, the pressure is greatest where the speed is the least, and vice versa. From this earlier discussion, we know too that the speed of the fluid is proportional to the pressure difference necessary to force the fluid through the restriction. If the system were perfect, then the speed of the fluid through the orifice would be obtained from the expression

$$\text{Speed} = \sqrt{2gh} = \sqrt{2g\,\frac{(P_1 - P_2)}{\text{sp gr}}}$$

where g = gravitational constant, assumed equal to 32.2 ft/sec^2

h = the height, ft, of a column of fluid which would establish the differential pressure

P_1 = orifice pressure, high side

P_2 = orifice pressure, low side

In actual practice, the speed term is about 98 per cent of its theoretical value.

Our main problems, therefore, are concerned with the determination of the area of the contracted stream as it passes through the orifice plate, for example. In the case of the flow nozzle and the venturi, the design is such that the flow will completely fill the tube. (For the flow nozzle the effective area is 98 per cent of the actual area.) The fact that the flow nozzle and venturi employ the full area of the orifice accounts for the fact that they can deliver about 60 per cent more fluid, per unit of time, than can an orifice plate having the same orifice opening and the same pressure drop.

For incompressible liquids, the volume can then be expressed by the following expression:

$$\text{Volume} = C_v C_a A \sqrt{2gh}$$

where C_v = correction for actual rather than the theoretical speed

C_a = correction for the reduced area of flowing liquid

A = area of the orifice opening

In addition to these factors, we must recognize that flow is also a function of:

1. The viscosity and the specific gravity of the flowing fluid.
2. The inside diameter of the pipe.
3. The ratio of the orifice opening to the inside pipe diameter. Their effect is termed the *velocity of approach* factor.

For approximate solutions, the constant coefficients may be combined so that the flow equation becomes

$$\text{Volume} = C'\sqrt{h}$$

where h, in feet, represents the equivalent height and must consider the specific gravity of the flowing liquid.

Approximate values of the flow coefficients. If one needs only approximate values of the flow coefficients in order to determine the apparent flow, then the following will suffice:

Venturi tube

$$C = 1 \text{ (area of smallest venturi opening)}$$

Flow nozzles

$$C = 0.98 \text{ (area of smallest opening of flow nozzle)}$$

Orifice plates

$$C = 0.62 \text{ (area of orifice opening)}$$

But for the more precise flow calculations, it is necessary to consider the type of liquid, its viscosity, its speed, and the related factors as they are involved in the Reynolds number. Figure 11·41*a* shows the serious error which occurs when the velocity of approach is neglected.

Actual flow determination. Many calculations have to do with the flow of water. The graphical solution for orifice plates shown in Fig. 11·39 makes it very easy when we know the desired flow, the pipe size, and the maximum differential pressure or the pipe diameter, and the total hourly or daily flow, and an approximate value of d/D. In order to use the graph, let us assume the following conditions:

Maximum flow meter differential is 100 in. of water.

$$\frac{d}{D} = 0.7$$

Pipe size is 8 in.

1. Using the nomograph, follow to the right on the 100-in. pressure differential line until the curve for $d/D = 0.7$ is met.
2. Then move directly vertically toward the top of the nomograph until an intersection is made with the 8-in. pipe-diameter line.
3. The desired flow rate will be read on either the right- or left-hand edges: 1,730,000 gpd, or 1,200 gpm.

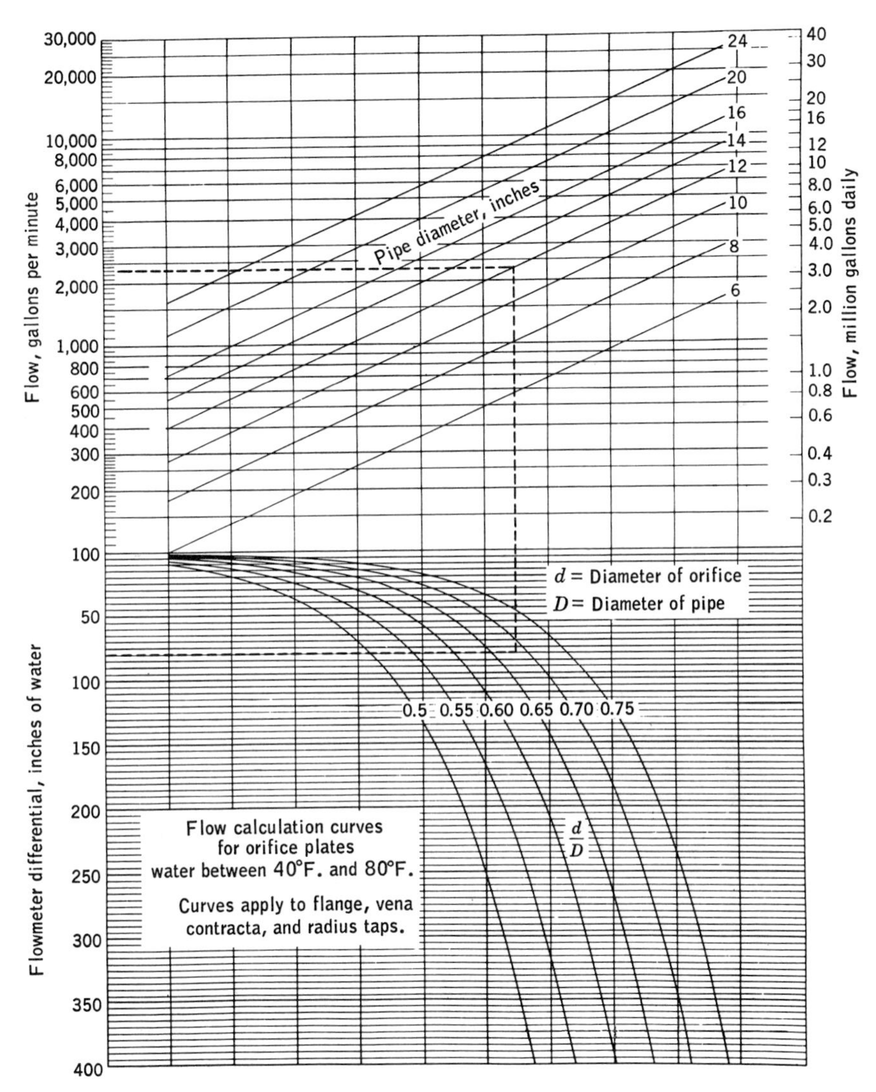

Fig. 11·39 Flow calculation curves for orifice plates. Curves apply to flange, vena contracta, and radius taps. (*The Foxboro Co.*)

It will be noted that for a 16-in. pipe, the flow for the same pressure differential is 4,900 gpm. This is what should be expected, inasmuch as a 16-in. pipe has 4 times the cross-sectional area of an 8-in. pipe.

A similar solution for flow nozzles is shown in Fig. 11·40. We may know the desired flow rate, the preferred differential pressure, and the pipe size. The only thing not determined is the ratio of d/D. If we know any three of the four quantities, then we may easily find the fourth. For example, if we desire a differential pressure of 75 in. of water, a ratio

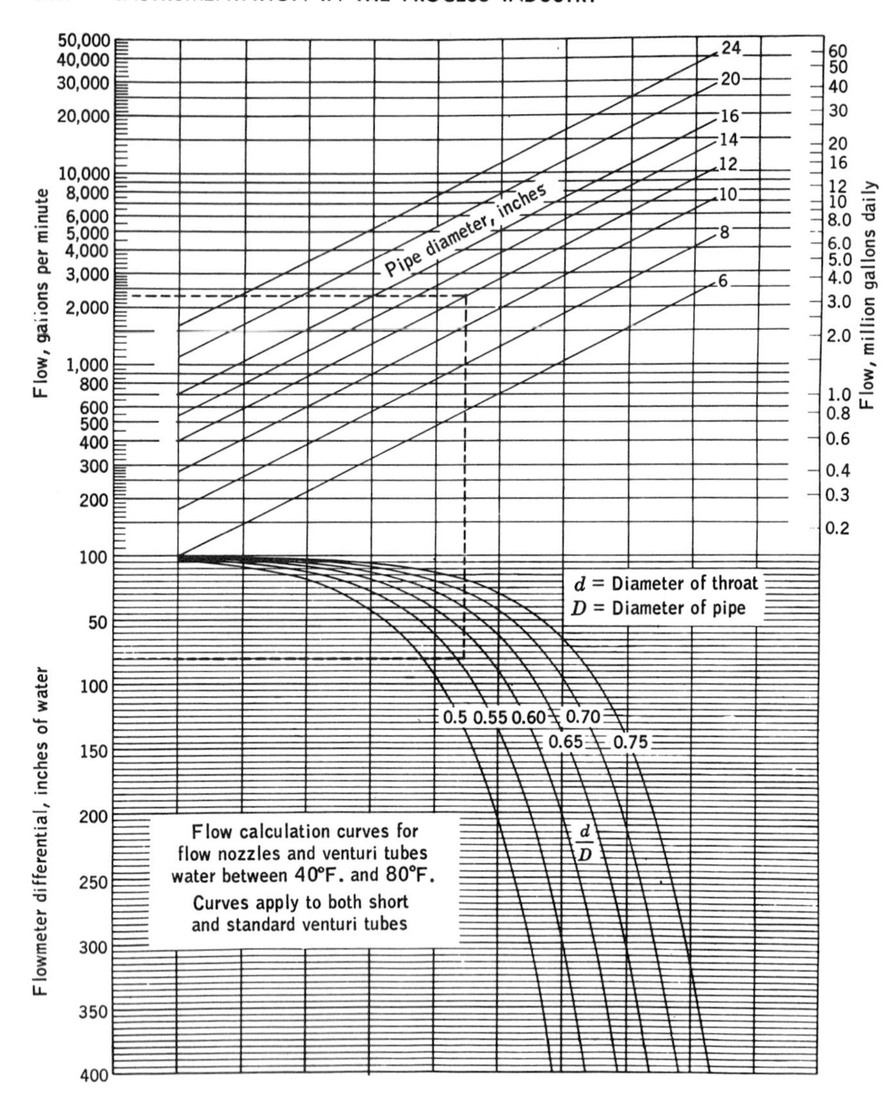

Fig. 11·40 Flow calculation curves for flow nozzles and venturi tubes. Curves apply to both short and standard venturi tubes. (*The Foxboro Co.*)

of $d/D = 0.57$, and a 12-in. pipe, what would be the rate of flow? We enter the nomograph at the desired differential pressure and proceed horizontally to the right to the correct value of d/D. We then move vertically to the specified size of pipe. We find the flow rate by moving horizontally to the left if the answer is desired in gallons per minute and to the right if in millions of gallons per day.

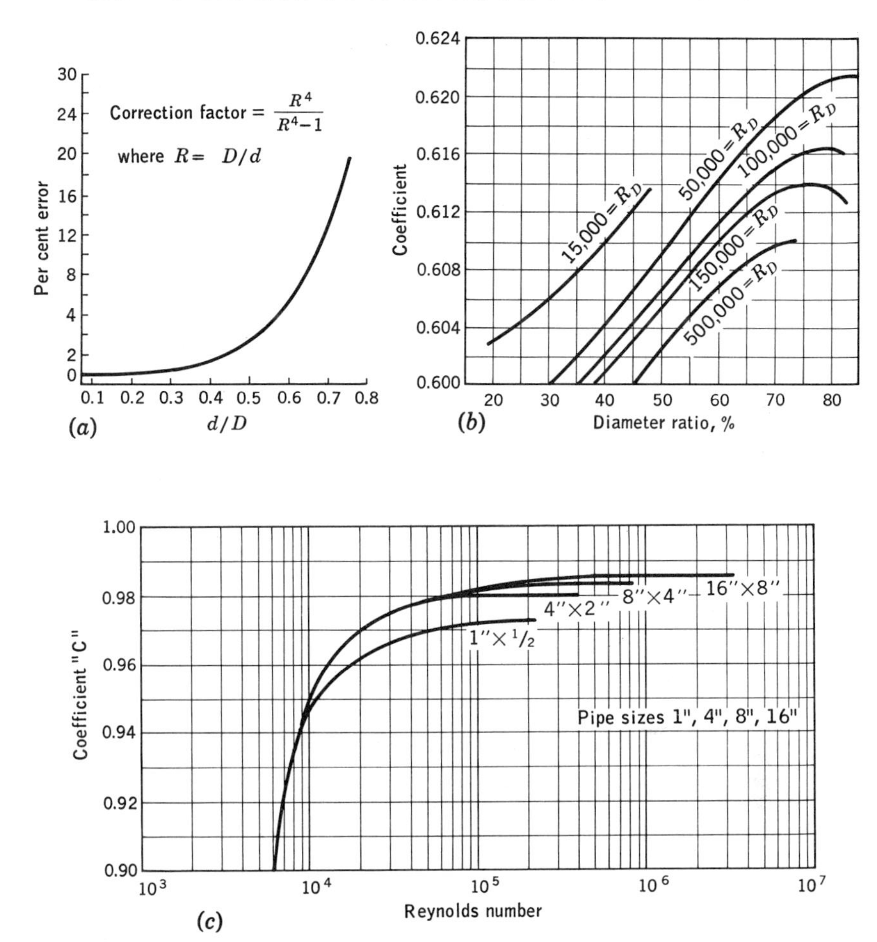

Fig. 11·41 (*a*) Percentage error versus velocity-of-approach factor. For large values of d/D, large errors are introduced by neglecting the velocity-of-approach factor. (*The Foxboro Co.*) (*b*) Variations of flow coefficients with Reynolds number and ratio of d/D. These values are for 3-in. pipe, concentric orifice, and vena contracta taps. The velocity-of-approach factor has been omitted. (*From "Fluid Meters: Their Theory and Applications," ASME Research Publication, 4th ed., 1937.*) (*c*) The venturi discharge coefficient varies with throat size and Reynolds number.

Instructions for use of nomograph, Figs. 11·42 and 11·43

$$\text{Reynolds number } R_D = \frac{(\text{constant})(\text{lb/hr})}{(\text{pipe diam})(\text{abs viscosity})}$$

$$= \frac{\text{constant}}{\text{pipe diam}} \times \frac{\text{lb/hr}}{\text{abs viscosity}}$$

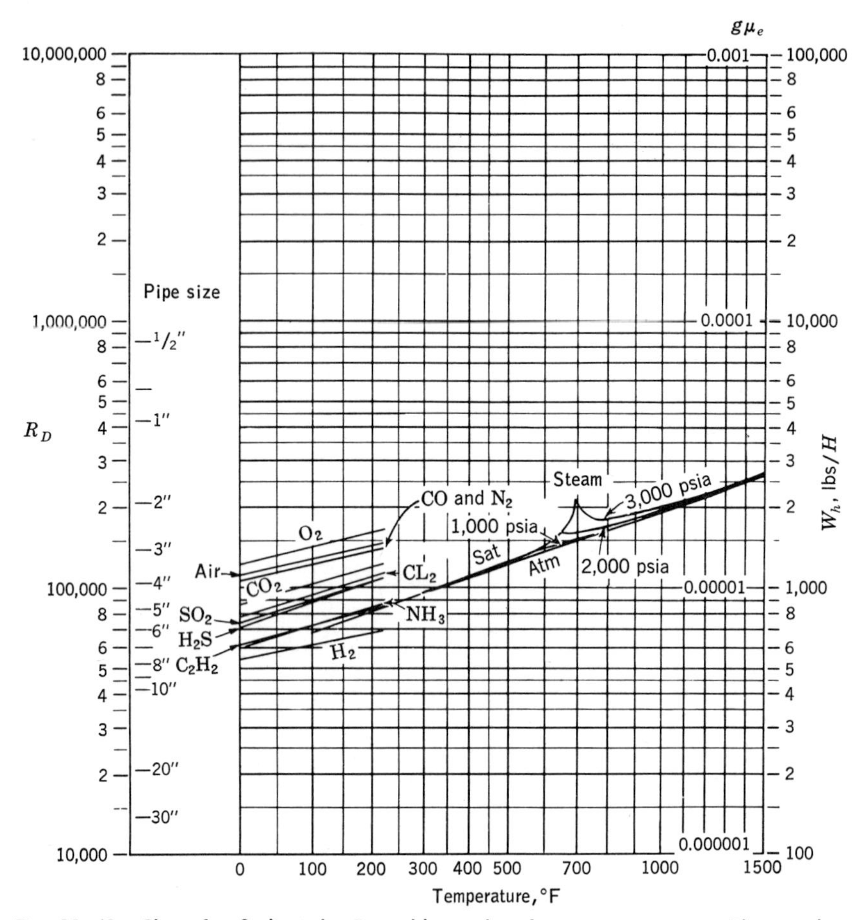

Fig. 11·42 Chart for finding the Reynolds number for steam or vapor. These values apply for any primary device or any tap location. (*"Principles and Practice of Flowmeter Engineering,"* by Spink, The Foxboro Co.)

When using logarithms, division becomes a subtraction and multiplication is an addition process.

Step 1. Taking the difference of the pounds per hour and the absolute viscosity graphs performs the necessary division. Find this graphical difference with a pair of dividers.

Step 2. Transfer this graphical difference to the left-hand column. Set one point of the dividers at the appropriate pipe size. Where the other point touches the portion designating Reynolds numbers, read the value of R_D.

SPECIAL FLOWMETER DESIGNS

The use of the elbow for measuring flow illustrates the use of a fundamental property of matter: it always takes force to accelerate any phys-

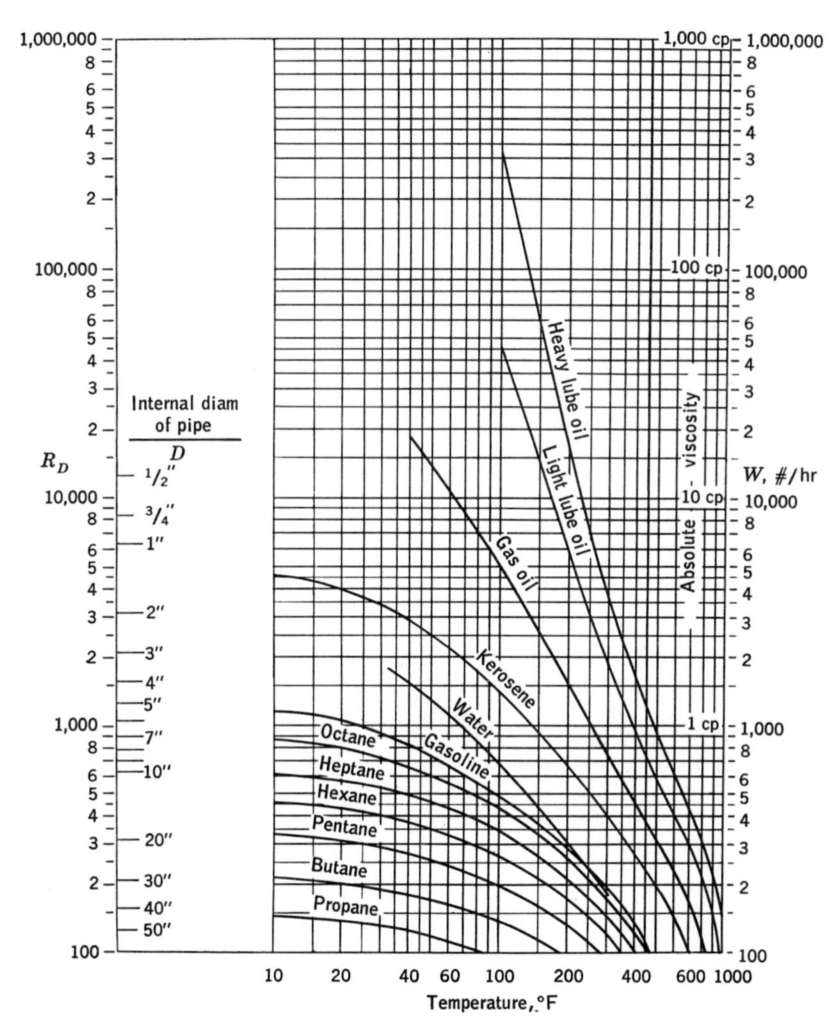

Fig. 11·43 Nomograph for finding the Reynolds number. When the pipe size, the pounds per hour, and the viscosity are known, the Reynolds number may be found. (*The Foxboro Co.*)

ical material. The associated principle is that the force is directly proportional to the acceleration which is imparted. Figure 11·44 illustrates a design which utilizes these principles directly. The disk which is situated in the middle of the flow stream will have developed on it an impact force which is proportional to the acceleration which must be imparted to the fluid flowing around it. Because the fluid would have passed straight through the pipe if the disk were not there, the fluid will require an acceleration if it is to take some other path. It is required by the ob-

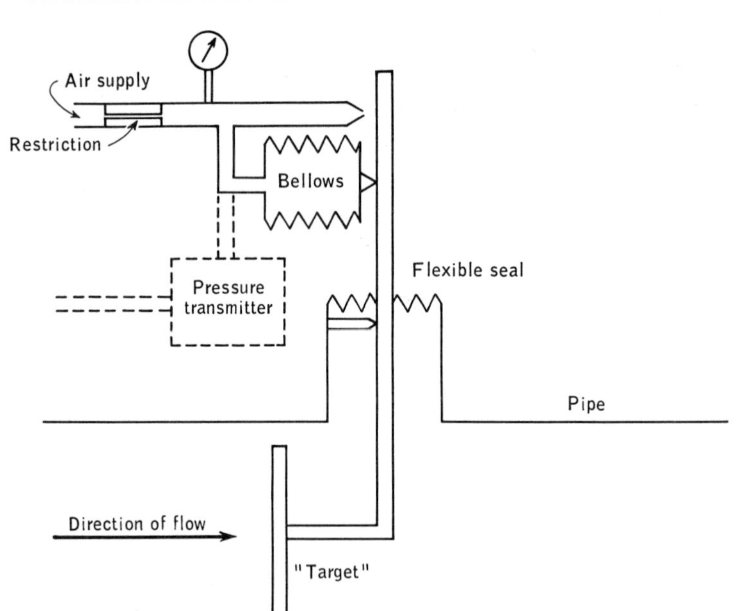

Fig. 11·44 Target type of flowmeter. This is one of the newer types of flowmeters. As the moving fluid impinges on the target, a force will be required to accelerate the material so that it will move in some other direction. The force is directly proportional to the square of the speed of the fluid. As the fluid tries to rotate the target, the upper end of the lever arm is brought closer to the nozzle. This increases the pressure in the bellows, which establishes a torque to oppose the movement of the lever arm. This type of meter has been found to give consistent, dependable service under severe conditions in which no differential pressure type of meter has been satisfactory.

struction to take another path. In this manner, force and flow may be correlated.

$$\text{Force} = AC_d d \frac{V^2}{2g}$$

where A = area of sensing element
C_d = drag coefficient
d = fluid density
$\frac{V^2}{2g}$ = velocity head

The force which is developed on the disk is then balanced by a torque exerted at the other end of the lever arm. A variable air pressure forces a bellows against the lever arm in order to establish the condition of torque equilibrium. The air pressure required to establish this condition can then be transmitted to give indications or controller actions.

FLOW INDICATORS

When there is a requirement for an indication that flow exists but it is not necessary to know the exact flow rate, devices employing movable balls, vanes, or small turbines may be quite adequate. The position of the ball or the vanes, as seen through the transparent body, will give some indication of the flow rate for those installations in which the flow rate is not critical. The rate of turning of the turbine rotor is also a rough measure of speed.

For situations in which a limited flow of purging air or gas is required, a bubbler sort of system may provide an adequate indication that there is a flow. The number of bubbles per unit of time furnishes a measure of the flow rate of the air or gas.

EXAMPLES OF FLOW CALCULATIONS

If in the ideal condition, $Q = A_2 \sqrt{2gh}$, where h is the change in pressure or differential pressure expressed in feet of equivalent head, there will have to be corrections to provide for

1. The compressibility factors Y for gases and vapors
2. Velocity of approach factors M
3. The contraction of the stream as it passes through the constriction C

If these are all grouped together, the expression then becomes $YMCA\sqrt{2gh}$

For ordinary liquids there will be no correction for compressibility, but for gases the correction must be applied. The values will be selected from Tables 11·1 and 11·2.

If the orifice opening is small, then most of the forward motion of the fluid will have been lost because of the restriction. But as the opening becomes a larger percentage of the total pipe opening, more and more of the forward motion is retained.

Table 11·1 Specific Gravities of Gases

(At 68°F and 14.7 psi abs referred to air)

Gas	*Specific Gravity*	*Gas*	*Specific Gravity*	*Gas*	*Specific Gravity*
Acetylene	0.907	Carbon dioxide	1.52	Methane	0.55
Air	1.000	Carbon monoxide	0.97	Natural gas	0.67
Ammonia	0.58	Chlorine	2.49	Nitric oxide	1.03
Argon	1.38	Ethane	1.05	Nitrogen	0.97
n-Butane	2.05	Helium	0.14	Oxygen	1.10
Isobutane	2.10	Hydrogen	0.07	Propane	1.56

The value of the correction

$$M = \sqrt{\frac{r^2}{r^2 - 1}} = \sqrt{\frac{1}{1 - (1/R)^4}} = \sqrt{\frac{R^4}{R^4 - 1}}$$

where r = ratio of pipe area to throat area
R = ratio of upstream pipe diameter to orifice diameter

See Table 11·3 for values of M.

Table 11·2 Approximate Values of the Expansion Factor Y

(Use correction is to be used with the expressions *used here* to correct for the flow of air or gases through a venturi, flow nozzle, or orifice)

Ratio of pressure		For ratios of throat-to-pipe diameter d/D:					
$\frac{P_2}{P_1}$	$\frac{P_1 - P_2}{P_1}$	0.25	0.50	0.60	0.70	0.75	0.80
		Expansion factors for venturi tubes and flow nozzles					
0.98	0.02	0.989	0.988	0.987	0.984	0.981	0.978
0.96	0.04	0.978	0.976	0.974	0.969	0.964	0.958
0.94	0.06	0.967	0.965	0.961	0.953	0.948	0.938
0.92	0.08	0.956	0.953	0.948	0.940	0.932	0.919
0.90	0.10	0.945	0.941	0.935	0.925	0.915	0.900
0.88	0.12	0.933	0.928	0.921	0.909	0.898	0.881
0.86	0.14	0.922	0.916	0.907	0.895	0.881	0.863
0.84	0.16	0.909	0.903	0.894	0.880	0.865	0.845
0.82	0.18	0.897	0.890	0.882	0.865	0.849	0.828
0.80	0.20	0.885	0.878	0.867	0.851	0.834	0.810
		Expansion factors for square-edged orifices					
0.98	0.02	0.994	0.994	0.994	0.993	0.993	0.992
0.96	0.04	0.986	0.986	0.986	0.985	0.985	0.984
0.94	0.06	0.980	0.980	0.979	0.979	0.977	0.976
0.92	0.08	0.974	0.974	0.973	0.972	0.970	0.969
0.90	0.10	0.970	0.969	0.968	0.964	0.962	0.961
0.88	0.12	0.964	0.963	0.962	0.959	0.956	0.953
0.86	0.14	0.958	0.956	0.954	0.952	0.948	0.946
0.84	0.16	0.952	0.950	0.948	0.944	0.940	0.937
0.82	0.18	0.947	0.944	0.942	0.937	0.933	0.929
0.80	0.20	0.941	0.938	0.935	0.930	0.925	0.920

Table 11·3 Velocity-of-approach Factor, Values of M

$1/R$	M	$1/R$	M	$1/R$	M
0.200	1.0008	0.450	1.0212	0.700	1.1472
0.250	1.0020	0.500	1.0328	0.750	1.2095
0.300	1.0041	0.550	1.0492	0.800	1.3015
0.350	1.0076	0.600	1.0719	0.850	1.4464
0.400	1.0131	0.650	1.1033	0.900	1.7052

Venturi tube calculations. It will be noted in Fig. 11·30 that, for the larger pipe sizes and higher Reynolds number, the discharge coefficient varies between 0.97 and 0.99. For Reynolds numbers in excess of 100,000, there is essentially no change in the coefficient.

EXAMPLE 11·1 If an 8-in. venturi is reduced to 4 in. at its constriction and the pressure differential is 100 in. of water, what is the flow rate when water is flowing? Water temperature is 70°F.

SOLUTION

$$\begin{aligned}
\text{Flow} &= YMCA\sqrt{2gh} \qquad \text{cfs} \\
&= 1(1.0328)(1.00)\left(\frac{12.5 \text{ in.}^2}{144 \text{ in.}^2/\text{ft}^2}\right)\sqrt{2(32.2)\left(\frac{100 \text{ in.}}{12 \text{ in./ft}}\right)} \\
&= 1(1.0328)(0.983)(0.087)\sqrt{64.4(8.33)} \\
&= 2.05 \text{ cfs}
\end{aligned}$$

Now let us determine whether there should be a correction for the Reynolds number. One expression for the number is

$$\begin{aligned}
R_D &= \frac{(22{,}800)(\text{rate of flow, lb/sec})}{(D, \text{ in.})(\text{abs viscosity, centipoises})} \\
&= \frac{22{,}800(2.05)(62.5)}{8(0.68)} = 123{,}000
\end{aligned}$$

For such a Reynolds number, Fig. 11·30, the coefficient is 1.00. This would increase our answer a negligible amount. But if this number were 5,000, a correction would be desirable. Let us ignore the value of the Reynolds number at this time by using the flat portion of the graph. If necessary, we can correct later.

EXAMPLE 11·2 Light lubricating oil that has an absolute viscosity of 10 centipoises at 170°F is flowing in a 16-in. venturi with a throat of 8 in. The pressure differential, always in terms of height of the flowing fluid, is 200 in. of water. What is the flow rate? Assume the specific gravity of the oil to be 0.70.

SOLUTION

$$Q = YMCA\sqrt{2gh}$$

$Y = 1$, $M = 1.0328$, $C = 0.985$, and

$$A = \frac{50.25 \text{ in.}^2}{144 \text{ in.}^2/\text{ft}^2} = 0.349 \text{ ft}^2$$

$$\sqrt{2gh} = \sqrt{2(32.2)\frac{(200)}{0.70(12)}} = 39.2 \text{ fps}$$

Then

$$Q = 1(1.0328)(0.985)(0.349 \text{ ft}^2)(39.2 \text{ fps})$$
$$= 13.9 \text{ cfs}$$

Again, let us check the Reynolds number.

$$R_D = \frac{22{,}800(\text{rate of flow, lb/sec})}{(D \text{ in.})(\text{viscosity, centipoises})}$$
$$= \frac{22{,}800(13.9)[62.5(0.70)]}{16(10)}$$
$$= 86{,}800$$

This calls for a coefficient of 0.98, or a correction of 0.5 per cent.

$$Q \text{ corrected} = 1(1.0328)(0.98)(0.349)\sqrt{2(g)\frac{200 \text{ in.}}{(12 \text{ in./ft})(0.7)}}$$
$$= 13.8 \text{ cfs}$$

Again, from the chart, no correction is required.

EXAMPLE 11·3 Steam flow. An 8-in. steam line uses a venturi with a 4-in. throat section to measure steam flow. If the pressure drop is 100 in. of water, how many pounds of steam per hour is being passed? Steam pressure is 200 psig, 100 per cent quality. Steam data from Tables 11·4 and 11·5.

SOLUTION

$$Q = YMCA\sqrt{\frac{2gh}{\text{sp gr}}}$$

Y 100 in. of water = 3.6 psi

$$P_2 = P_1 + 14.7 - 3.6 = 211.1 \text{ psia}$$

$$\frac{P_2}{P_1} = \frac{211.1}{214.7} = 0.98 \qquad \frac{D_2}{D_1} = 0.5$$

$$Y = 0.988$$

M $R = 2 = \dfrac{8 \text{ in.}}{4 \text{ in.}}$

$$\frac{1}{R} = \frac{4 \text{ in.}}{8 \text{ in.}} = \frac{1}{2} = 0.500$$

$$M = 1.0328$$

C Assume Reynolds number > 100,000, so $C = 0.982$.

A 4-in. opening = 12.56 in.^{2} = 0.087 ft^2

Table 11·4 Saturated Steam: Temperature Table *

Temp., F	Abs. press., psia	Specific volume			Enthalpy			Entropy			Temp., F
		Sat. liquid	Evap.	Sat. vapor	Sat. liquid	Evap.	Sat. vapor	Sat. liquid	Evap.	Sat. vapor	
t	p	v_f	v_{fg}	v_g	h_f	h_{fg}	h_g	s_f	s_{fg}	s_g	t
32	0.08854	0.01602	3306	3306	0.00	1075.8	1075.8	0.0000	2.1877	2.1877	**32**
35	0.09995	0.01602	2947	2947	3.02	1074.1	1077.1	0.0061	2.1709	2.1770	**35**
40	0.12170	0.01602	2444	2444	8.05	1071.3	1079.3	0.0162	2.1435	2.1597	**40**
45	0.14752	0.01602	2036.4	2036.4	13.06	1068.4	1081.5	0.0262	2.1167	2.1429	**45**
50	0.17811	0.01603	1703.2	1703.2	18.07	1065.6	1083.7	0.0361	2.0903	2.1264	**50**
60	0.2563	0.01604	1206.6	1206.7	28.06	1059.9	1088.0	0.0555	2.0393	2.0948	**60**
70	0.3631	0.01606	867.8	867.9	38.04	1054.3	1092.3	0.0745	1.9902	2.0647	**70**
80	0.5069	0.01608	633.1	633.1	48.02	1048.6	1096.6	0.0932	1.9428	2.0360	**80**
90	0.6982	0.01610	468.0	468.0	57.99	1042.9	1100.9	0.1115	1.8972	2.0087	**90**
100	0.9492	0.01613	350.3	350.4	67.97	1037.2	1105.2	0.1295	1.8531	1.9826	**100**
110	1.2748	0.01617	265.3	265.4	77.94	1031.6	1109.5	0.1471	1.8106	1.9577	**110**
120	1.6924	0.01620	203.25	203.27	87.92	1025.8	1113.7	0.1645	1.7694	1.9339	**120**
130	2.2225	0.01625	157.32	157.34	97.90	1020.0	1117.9	0.1816	1.7296	1.9112	**130**
140	2.8886	0.01629	122.99	123.01	107.89	1014.1	1122.0	0.1984	1.6910	1.8894	**140**
150	3.718	0.01634	98.06	97.07	117.89	1008.2	1126.1	0.2149	1.6537	1.8685	**150**
160	4.741	0.01639	77.27	77.29	127.89	1002.3	1130.2	0.2311	1.6174	1.8485	**160**
170	5.992	0.01645	62.04	62.06	137.90	996.3	1134.2	0.2472	1.5822	1.8293	**170**
180	7.510	0.01651	50.21	50.23	147.92	990.2	1138.1	0.2630	1.5480	1.8109	**180**
190	9.339	0.01657	40.94	40.96	157.95	984.1	1142.0	0.2785	1.5147	1.7932	**190**
200	11.526	0.01663	33.62	33.64	167.99	977.9	1145.9	0.2938	1.4824	1.7762	**200**
210	14.123	0.01670	27.80	27.82	178.05	971.6	1149.7	0.3090	1.4508	1.7598	**210**
212	14.696	0.01672	26.78	26.80	180.07	970.3	1150.4	0.3120	1.4446	1.7566	**212**
220	17.186	0.01677	23.13	23.15	188.13	965.2	1153.4	0.3239	1.4201	1.7440	**220**
230	20.780	0.01684	19.365	19.382	198.23	958.8	1157.0	0.3387	1.3901	1.7288	**230**
240	24.969	0.01692	16.306	16.323	208.34	952.2	1160.5	0.3531	1.3609	1.7140	**240**
250	29.825	0.01700	13.804	13.821	216.48	945.5	1164.0	0.3675	1.3323	1.6998	**250**
260	35.429	0.01709	11.746	11.763	228.64	938.7	1167.3	0.3817	1.3043	1.6860	**260**
270	41.858	0.01717	10.044	10.061	238.84	931.8	1170.6	0.3958	1.2769	1.6727	**270**
280	49.203	0.01726	8.628	8.645	249.06	924.7	1173.8	0.4096	1.2501	1.6597	**280**
290	57.556	0.01735	7.444	7.461	259.31	917.5	1176.8	0.4234	1.2238	1.6472	**290**
300	67.013	0.01745	6.449	6.466	269.59	910.1	1179.7	0.4369	1.1980	1.6350	**300**
310	77.68	0.01755	5.609	5.626	279.92	902.6	1182.5	0.4504	1.1727	1.6231	**310**
320	89.66	0.01765	4.896	4.914	290.28	894.9	1185.2	0.4637	1.1478	1.6115	**320**
330	103.06	0.01776	4.289	4.307	300.68	887.0	1187.7	0.4769	1.1233	1.6002	**330**
340	118.01	0.01787	3.770	3.788	311.13	879.0	1190.1	0.4900	1.0992	1.5891	**340**
350	134.63	0.01799	3.324	3.342	321.63	870.7	1192.3	0.5029	1.0754	1.5783	**350**
360	153.04	0.01811	2.939	2.957	332.18	862.2	1194.4	0.5158	1.0519	1.5677	**360**
370	173.37	0.01823	2.606	2.625	342.79	853.5	1196.3	0.5286	1.0287	1.5573	**370**
380	195.77	0.01836	2.317	2.335	353.45	844.6	1198.1	0.5413	1.0059	1.5471	**380**
390	220.37	0.01850	2.0651	2.0836	364.17	835.4	1199.6	0.5539	0.9832	1.5371	**390**
400	247.31	0.01864	1.8447	1.8633	374.97	826.0	1201.0	0.5664	0.9608	1.5272	**400**
410	276.75	0.01878	1.6512	1.6700	385.83	816.3	1202.1	0.5788	0.9386	1.5174	**410**
420	308.83	0.01894	1.4811	1.5000	396.77	806.3	1203.1	0.5912	0.9166	1.5078	**420**
430	343.72	0.01910	1.3308	1.3499	407.79	796.0	1203.8	0.6035	0.8947	1.4982	**430**
440	381.59	0.01926	1.1979	1.2171	418.90	785.4	1204.3	0.6158	0.8730	1.4887	**440**
450	422.6	0.0194	1.0799	1.0993	430.1	774.5	1204.6	0.6280	0.8513	1.4793	**450**
460	466.9	0.0196	0.9748	0.9944	441.4	763.2	1204.6	0.6402	0.8298	1.4700	**460**
470	514.7	0.0198	0.8811	0.9009	452.8	751.5	1204.3	0.6523	0.8083	1.4606	**470**
480	566.1	0.0200	0.7972	0.8172	464.4	739.4	1203.7	0.6645	0.7868	1.4513	**480**
490	621.4	0.0202	0.7221	0.7423	476.0	726.8	1202.8	0.6766	0.7653	1.4419	**490**
500	680.8	0.0204	0.6545	0.6749	487.8	713.9	1201.7	0.6887	0.7438	1.4325	**500**
520	812.4	0.0209	0.5385	0.5594	511.9	686.4	1198.2	0.7130	0.7006	1.4136	**520**
540	962.5	0.0215	0.4434	0.4649	536.6	656.6	1193.2	0.7374	0.6568	1.3942	**540**
560	1133.1	0.0221	0.3647	0.3868	562.2	624.2	1186.4	0.7621	0.6121	1.3742	**560**
580	1325.8	0.0228	0.2989	0.3217	588.9	588.4	1177.3	0.7872	0.5659	1.3532	**580**
600	1542.9	0.0236	0.2432	0.2668	617.0	548.5	1165.5	0.8131	0.5176	1.3307	**600**
620	1786.6	0.0247	0.1955	0.2201	646.7	503.6	1150.3	0.8398	0.4664	1.3062	**620**
640	2059.7	0.0260	0.1538	0.1798	678.6	452.0	1130.5	0.8679	0.4110	1.2789	**640**
660	2365.4	0.0278	0.1165	0.1442	714.2	390.2	1104.4	0.8987	0.3485	1.2472	**660**
680	2708.1	0.0305	0.0810	0.1115	757.3	309.9	1067.2	0.9351	0.2719	1.2071	**680**
700	3093.7	0.0369	0.0392	0.0761	823.3	172.1	995.4	0.9905	0.1484	1.1389	**700**
705.4	3206.2	0.0503	0	0.0503	902.7	0	902.7	1.0580	0	1.0580	**705.4**

* From B. G. A. Skrotzki and W. A. Vopat, "Power Station Engineering and Economy," McGraw-Hill Book Company, Inc., New York, 1960, abridgment of "Thermodynamic Properties of Steam," by Joseph H. Keenan and Frederick G. Keyes, John Wiley & Sons, Inc., New York, 1937.

Table 11·5 Saturated Steam: Pressure Table *

Abs. press., psia	Temp., F	Specific volume		Enthalpy			Entropy			Internal energy		Abs. press., psia
		Sat. liquid	Sat. vapor	Sat. liquid	Evap.	Sat. vapor	Sat. liquid	Evap.	Sat. vapor	Sat. liquid	Sat. vapor	
p	t	v_f	v_g	h_f	h_{fg}	h_g	s_f	s_{fg}	s_g	u_f	u_g	p
1.0	101.74	0.01614	333.6	69.70	1036.3	1106.0	0.1326	1.8456	1.9782	69.70	1044.3	**1.0**
2.0	126.08	0.01623	173.73	93.99	1022.2	1116.2	0.1749	1.7451	1.9200	93.98	1051.9	**2.0**
3.0	141.48	0.01630	118.71	109.37	1013.2	1122.6	0.2008	1.6855	1.8863	109.36	1056.7	**3.0**
4.0	152.97	0.01636	90.63	120.86	1006.4	1127.3	0.2198	1.6427	1.8625	120.85	1060.2	**4.0**
5.0	162.24	0.01640	73.52	130.13	1001.0	1131.1	0.2347	1.6094	1.8441	130.12	1063.1	**5.0**
6.0	170.06	0.01645	61.98	137.96	996.2	1134.2	0.2472	1.5820	1.8292	137.94	1065.4	**6.0**
7.0	176.85	0.01649	53.64	144.76	992.1	1136.9	0.2581	1.5586	1.8167	144.74	1067.4	**7.0**
8.0	182.86	0.01653	47.34	150.79	988.5	1139.3	0.2674	1.5383	1.8057	150.77	1069.2	**8.0**
9.0	188.28	0.01656	42.40	156.22	985.2	1141.4	0.2759	1.5203	1.7962	156.19	1070.8	**9.0**
10	193.21	0.01659	38.42	161.17	982.1	1143.3	0.2835	1.5041	1.7876	161.14	1072.2	**10**
14.696	212.00	0.01672	26.80	180.07	970.3	1150.4	0.3120	1.4446	1.7566	180.02	1077.5	**14.696**
15	213.03	0.01672	26.29	181.11	969.7	1150.8	0.3135	1.4415	1.7549	181.06	1077.8	**15**
20	227.96	0.01683	20.089	196.16	960.1	1156.3	0.3356	1.3962	1.7319	196.10	1081.9	**20**
25	240.07	0.01692	16.303	208.42	952.1	1160.6	0.3533	1.3606	1.7139	208.34	1085.1	**25**
30	250.33	0.01701	13.746	218.82	945.3	1164.1	0.3680	1.3313	1.6993	218.73	1087.8	**30**
35	259.28	0.01708	11.898	227.91	939.2	1167.1	0.3807	1.3063	1.6870	227.80	1090.1	**35**
40	267.25	0.01715	10.498	236.03	933.7	1169.7	0.3919	1.2844	1.6763	235.90	1092.0	**40**
45	274.44	0.01721	9.401	243.36	928.6	1172.0	0.4019	1.2650	1.6669	243.22	1093.7	**45**
50	281.01	0.01727	8.515	250.09	924.0	1174.1	0.4110	1.2474	1.6585	249.93	1095.3	**50**
55	287.07	0.01732	7.787	256.30	919.6	1175.9	0.4193	1.2316	1.6509	256.12	1096.7	**55**
60	292.71	0.01738	7.175	262.09	915.5	1177.6	0.4270	1.2168	1.6438	261.90	1097.9	**60**
65	297.97	0.01743	6.655	267.50	911.6	1179.1	0.4342	1.2032	1.6374	267.29	1099.1	**65**
70	302.92	0.01748	6.206	272.61	907.9	1180.6	0.4409	1.1906	1.6315	272.38	1100.2	**70**
75	307.60	0.01753	5.816	277.43	904.5	1181.9	0.4472	1.1787	1.6259	277.19	1101.2	**75**
80	312.03	0.01757	5.472	282.02	901.1	1183.1	0.4531	1.1676	1.6207	281.76	1102.1	**80**
85	316.25	0.01761	5.168	286.39	897.8	1184.2	0.4587	1.1571	1.6158	286.11	1102.9	**85**
90	320.27	0.01766	4.896	290.56	894.7	1185.3	0.4641	1.1471	1.6112	290.27	1103.7	**90**
95	324.12	0.01770	4.652	294.56	891.7	1186.2	0.4692	1.1376	1.6068	294.25	1104.5	**95**
100	327.81	0.01774	4.432	298.40	888.8	1187.2	0.4740	1.1286	1.6026	298.08	1105.2	**100**
110	334.77	0.01782	4.049	305.66	883.2	1188.9	0.4832	1.1117	1.5948	305.30	1106.5	**110**
120	341.25	0.01789	3.728	312.44	877.9	1190.4	0.4916	1.0962	1.5878	312.05	1107.6	**120**
130	347.32	0.01796	3.455	318.81	879.9	1191.7	0.4995	1.0817	1.5812	318.38	1108.6	**130**
140	353.02	0.01802	3.220	324.82	868.2	1193.0	0.5069	1.0682	1.5751	324.35	1109.6	**140**
150	358.42	0.01809	3.015	330.51	863.6	1194.1	0.5138	1.0556	1.5694	330.01	1110.5	**150**
160	363.53	0.01815	2.834	335.93	859.2	1195.1	0.5204	1.0436	1.5640	335.39	1111.2	**160**
170	368.41	0.01822	2.675	341.09	854.9	1196.0	0.5266	1.0324	1.5590	340.52	1111.9	**170**
180	373.06	0.01827	2.532	346.03	850.8	1196.9	0.5325	1.0217	1.5542	345.42	1112.5	**180**
190	377.51	0.01833	2.404	350.79	846.8	1197.6	0.5381	1.0116	1.5497	350.15	1113.1	**190**
200	381.79	0.01839	2.288	355.36	843.0	1198.4	0.5435	1.0018	1.5453	354.68	1113.7	**200**
250	400.95	0.01865	1.8438	376.00	825.1	1201.1	0.5675	0.9588	1.5263	375.14	1115.8	**250**
300	417.33	0.01890	1.5433	393.84	809.0	1202.8	0.5879	0.9225	1.5104	392.79	1117.1	**300**
350	431.72	0.01913	1.3260	409.69	794.2	1203.9	0.6056	0.8910	1.4966	408.45	1118.0	**350**
400	444.59	0.0193	1.1613	424.0	780.5	1204.5	0.6214	0.8630	1.4844	422.6	1118.5	**400**
450	456.28	0.0195	1.0320	437.2	767.4	1204.6	0.6356	0.8378	1.4734	435.5	1118.7	**450**
500	467.01	0.0197	0.9278	449.4	755.0	1204.4	0.6487	0.8147	1.4634	447.6	1118.6	**500**
550	476.94	0.0199	0.8424	460.8	743.1	1203.9	0.6608	0.7934	1.4542	458.8	1118.2	**550**
600	486.21	0.0201	0.7698	471.6	731.6	1203.2	0.6720	0.7734	1.4454	469.4	1117.7	**600**
650	494.90	0.0203	0.7083	481.8	720.5	1202.3	0.6826	0.7548	1.4374	479.4	1117.1	**650**
700	503.10	0.0205	0.6554	491.5	709.7	1201.2	0.6925	0.7371	1.4296	488.8	1116.3	**700**
750	510.86	0.0207	0.6092	500.8	699.2	1200.0	0.7019	0.7204	1.4223	598.0	1115.4	**750**
800	518.23	0.0209	0.5687	509.7	688.9	1198.6	0.7108	0.7045	1.4153	506.6	1114.4	**800**
850	525.26	0.0210	0.5327	518.3	678.8	1197.1	0.7194	0.6891	1.4085	515.0	1113.3	**850**
900	531.98	0.0212	0.5006	526.6	668.8	1195.4	0.7275	0.6744	1.4020	523.1	1112.1	**900**
950	538.43	0.0214	0.4717	534.6	659.1	1193.7	0.7355	0.6602	1.3957	530.9	1110.8	**950**
1000	544.61	0.0216	0.4456	542.4	649.4	1191.8	0.7430	0.6467	1.3897	538.4	1109.4	**1000**
1100	556.31	0.0220	0.4001	557.4	630.4	1187.8	0.7575	0.6205	1.3780	552.9	1106.4	**1100**
1200	567.22	0.0223	0.3619	571.7	611.7	1183.4	0.7711	0.5956	1.3667	566.7	1103.0	**1200**
1300	577.46	0.0227	0.3293	585.4	593.2	1178.6	0.7840	0.5719	1.3559	580.0	1099.4	**1300**
1400	587.10	0.0231	0.3012	598.7	574.7	1173.4	0.7963	0.5491	1.3454	592.7	1095.4	**1400**
1500	596.23	0.0235	0.2765	611.6	556.3	1167.9	0.8082	0.5269	1.3351	605.1	1091.2	**1500**
2000	635.82	0.0257	0.1878	671.7	463.4	1135.1	0.8619	0.4230	1.2849	662.2	1065.6	**2000**
2500	668.13	0.0287	0.1307	730.6	360.5	1091.1	0.9126	0.3197	1.2322	717.3	1030.6	**2500**
3000	695.36	0.0346	0.0858	802.5	217.8	1020.3	0.9731	0.1885	1.1615	783.4	972.7	**3000**
3206.2	705.40	0.0503	0.0503	902.7	0	902.7	1.0580	0	1.0580	872.9	872.9	**3206.2**

* From B. G. A. Skrotzki and W. A. Vopat, "Power Station Engineering and Economy," McGraw-Hill Book Company, Inc., New York, 1960, abridgment of "Thermodynamic Properties of Steam," by Joseph H. Keenan and Frederick G. Keyes, John Wiley & Sons, Inc., New York, 1937.

$$\sqrt{\frac{2gh}{\text{sp gr}}} \qquad \sqrt{2(32.2)\frac{(100 \text{ in.})}{(0.0079)(12 \text{ in./ft})}}$$

$$\sqrt{64.4(1{,}050 \text{ ft})} = 2{,}600 \text{ ft/sec}$$

$$Q = 0.988(1.0328)(0.982)(0.087 \text{ ft}^2)(2{,}600 \text{ ft/sec}) = 226 \text{ ft}^3\text{/sec}$$

But steam with 100 per cent quality at that pressure (214.7 psia) has 2.14 ft^3/lb. Therefore

$$\frac{226 \text{ ft}^3\text{/sec}}{2.14 \text{ ft}^3\text{/lb}} = 106 \text{ lb/sec}$$

$$\text{Flow} = (106 \text{ lb/sec})(3{,}600 \text{ sec/hr})$$

$$= 380{,}000 \text{ lb/hr}$$

Check for the Reynolds number. For many flow problems when water is not the fluid, it will be necessary to evaluate the flow conditions in terms of the pipe size, rate of flow, viscosity, and related factors.

Figure 11·43 provides a graphical solution for the value of the Reynolds number for a given set of conditions. This chart is most useful when the normal flow rate in pounds per hour is known. Use a pair of dividers or the edge of a straight piece of paper to measure the required distances. Determine the distance between the value of the absolute viscosity, on the right side of the chart, and the normal flow rate, in pounds per hour. Attention will have to be given to the material and temperature to ensure that the correct value of viscosity is used. Then set off the same distance, just determined, from the internal-diameter pipe size toward the viscosity scale. This will then give the correct Reynolds number.

For precise flow measurements, the Reynolds number must be considered. Its value may also be determined from

$$R_D = \frac{22{,}800(\text{lb/sec})}{(D \text{ in.})(\text{abs viscosity, centipoises})}$$

For Example 11·3

$$R_D = \frac{22{,}800(106)}{(4 \text{ in.})(0.0163)} = 382{,}000 \text{ lb/hr}$$

This value indicates that no adjustment is required. In all of our flow measurements we need to know both the actual area of the flowing stream and its true speed. The Reynolds number furnishes a clue to the corrections which should be made in order that we may effectively use the results of thousands of flow-measuring tests. The flow of a thick, viscous material will be quite different from that of very thin, watery material for the same pressure differential.

EXAMPLE 11·4 Let us assume the same conditions as for Example 11·1 except that the flow nozzle is an ASME long-radius nozzle with a diameter ratio of 0.428. The ratio for an ASME long nozzle is 2.33:1.

SOLUTION

Y 1(for noncompressible liquid)

M $R = \frac{D}{d} = \frac{8}{3.42} = 2.33$

$$\frac{1}{R} = 0.428$$

$M = 1.017$ by interpolation

C Assume $R_D = 200{,}000$, so $C = 0.985$.

A $9.2 \text{ in.}^2 = \frac{9.2 \text{ in.}^2}{144 \text{ in.}^2/\text{ft}^2} = 0.0638 \text{ ft}^2$

$\sqrt{2gh}$ $\sqrt{2(32.2)\frac{100 \text{ in.}}{(12 \text{ in./ft})(1)}} = \sqrt{64.4(8.33)} = 23.13 \text{ fps}$

$$\text{Flow} = 1(1.017)(0.985)(0.0638 \text{ ft}^2)(23.13 \text{ ft/sec})$$
$$= 1.47 \text{ ft}^3/\text{sec}$$

Checking for the value of the Reynolds number,

$$R_D = \frac{22{,}800(\text{rate of flow, lb/sec})}{(D \text{ in.})(\text{abs viscosity, centipoises})}$$
$$= \frac{(22{,}800)(1.47)(62.3)}{(8)(0.96)} = 272{,}000$$

No correction is needed.

Orifice plate calculations. The flow calculations for an orifice plate are shown in the following example.

EXAMPLE 11·5 If a flow of 300,000 lb/hr of kerosene is to be sent through a 3-in. pipe, what pressure differential in inches of water will be created with corner taps if the orifice plate opening is 75 per cent of the pipe diameter? Assume temperature of 140°F.

SOLUTION

$$\text{Flow, cfs} = \frac{300{,}000 \text{ lb/hr}}{(3{,}600 \text{ sec/hr})\ [(62.3\ 6/\text{ft}^3)(0.8 \text{ sp gr})]} = 1.65$$
$$= YMCA\sqrt{2gh}$$

Y 1(for incompressible fluid)

M For $D = 0.75$, $M = 1.2095$, from Table 11·3

C From Fig. 11·32. C requires the Reynolds number, which is obtained from Fig. 11·43. For a viscosity of 1 centipoise for the kerosene at 140°F, $R_D = 620{,}000$. Then since $d/D = 0.75$, $C = 0.609$.

A $A = 7.068 \text{ in.}^2$; assume pipe ID = 3.000 in.

$$\frac{7.068 \text{ in.}^2}{144 \text{ in.}^2/\text{ft}^2} = 0.0491 \text{ ft}^2$$

$\sqrt{2g} = \sqrt{2(32.2)} = 8.03$

$$1.65 \text{ ft}^3/\text{sec} = 1(1.2095)(0.609)(0.0491)(8.03)(\sqrt{h})$$
$$= 0.289\sqrt{h}$$
$$\frac{1.65}{2.89} = \sqrt{h} = 5.7$$
$$h = 32.4 \text{ ft of kerosene}$$
$$= 32.4(0.80) = 25.9 \text{ ft of water}$$
$$= 311 \text{ in. of water}$$

EXAMPLE 11·6 If the differential pressure were to be limited to 100 in. of water, what would be the hourly flow in terms of gallons per hour and pounds per hour for 3-in. pipe, $d/D = 0.75$, and a temperature of 140°F?

SOLUTION

$$\text{Flow} = YMCA\sqrt{2gh}$$

Y 1

M 1.2095 (as before)

C 0.6 Assume that this is true and, after determining approximate flow, compute the Reynolds number and determine if a correction is required.

$$\text{Flow} = 1(1.2095)(0.6)(0.0491 \text{ ft}^2)(8.03)\sqrt{\frac{100 \text{ in.}}{12(0.80)}} \text{ cfs}$$
$$= 0.286\sqrt{10.4 \text{ ft}}$$
$$= 0.925 \text{ cfs}$$
$$\text{Hourly flow rate} = (0.925 \text{ ft}^3/\text{sec})(3{,}600 \text{ sec/hr})(62.3 \text{ lb/ft}^3)(0.80)$$
$$= 165{,}000 \text{ lb/hr}$$

Check for Reynolds number: From Fig. 11·43 $R_D = 450{,}000$; from Fig. 11·41 with $d/D = 0.75$ and $R_D = 450{,}000$ the flow coefficient $C = 0.609$. The flow rate then becomes 165,000 (0.609/0.60) = 167,500 lb/hr.

$$\text{Gallons per hour} = (167{,}500 \text{ lb/hr})\left(\frac{1}{8.35 \times 0.80}\right)$$
$$= 25{,}100$$

Note: 1 gal of water weighs 8.35 lb.

EXAMPLE 11·7 How many pounds of saturated steam per hour will flow through a 6-in. line if $d/D = 0.60$? The differential pressure is not to exceed 400 in. of water. Steam temperature is 400°F, pressure is to be 247.3 psia. Use flange taps and orifice plate.

SOLUTION

$$\text{Flow} = YMCA\sqrt{2gh} \quad \text{cfs}$$

Y compressibility factor. From Table 11·2

$$\frac{P_1 - P_2}{P_1} = \frac{247.3 - 14.9}{247.3} = 0.945$$

For $d/D = 1/R = 0.600$

For pressure ratio of 0.97 and $d/D = 0.60$, $Y = 0.965$

M 1.0719

C 0.60, ignoring viscosity effects

A $\dfrac{28.88 \text{ in.}^2}{144 \text{ in.}^2/\text{ft}^2} = 0.201 \text{ ft}^2$

$\sqrt{2g}$ $\sqrt{2(32.2)} = 8.03 \text{ ft}^{1/2}/\text{sec}$

$$400 \text{ in. of water} = 33.3 \text{ ft}$$

$$\frac{33.3}{\text{sp gr at 400 psig}} = \text{equivalent height}$$

From steam tables, Table 11·4, 1 lb saturated steam at 400°F and 247.3 psia occupies 1.863 ft³, so each cubic foot weighs 0.537 lb. Water weighs 53.5 lb/ft³. Specific gravity of steam = 0.537/53.5 = 0.01.

Therefore $$\text{Equivalent height} = \frac{33.3 \text{ ft of water}}{0.01} = 3{,}330 \text{ ft of gas}$$

to establish the same pressure, and $\sqrt{3{,}330} = 57.7$.

$$\begin{aligned} \text{Flow} &= (0.965)(1.0719)(0.600)(0.201)(8.03)(57.7) \\ &= 56.5 \text{ cfs} \\ &= (56.5 \text{ cfs})(0.537 \text{ lb/ft}^3)(3{,}600 \text{ sec/hr}) \\ &= 109{,}000 \end{aligned}$$

Check for correction for Reynolds number: From Fig. 11·42 $R_D = 7{,}000{,}000$. From table of orifice coefficients, allowing for velocity of approach, $C = 0.605$. No correction is needed.

12 Flow Measurements, Electrical Transducers

With the principles that have already been studied and by using any one of the various types of transducers with which we are now acquainted, there are many measurements which can now be studied profitably.

MAGNETIC FLOWMETER

One of the simplest flowmeter designs in terms of the principles only is shown in Figs. 12·1 to 12·7. It is based on principles which have been known for many years, but no use could be made of the design until potential devices were available which had great sensitivity and which would take practically no current from the measuring circuit. Unless the device could be made to operate on potential only, any flow of current through the very high resistance of the fluid would use up so much of the emf that little or none would be left for the measuring circuit.

Theory. To review the principles again, whenever a conductor moves through a magnetic field, Fig. 12·3, there will be induced in the conductor a potential which is directly proportional to the strength of the field, the length of the conductor, and the speed of the conductor. This statement is independent of the resistance of the conductor. Therefore, all conductors that have the same length and are moving in the same field at the same speed should develop equal potentials, without regard for the material.

$$\text{emf} = KBLV$$

where K = proportionality constant
B = flux density
L = length of the conductor
V = speed of the conductor through the magnetic field

The speed of the liquid close to the pipe wall will be somewhat lower than the speed of the liquid at the center of the pipe, so the emf pro-

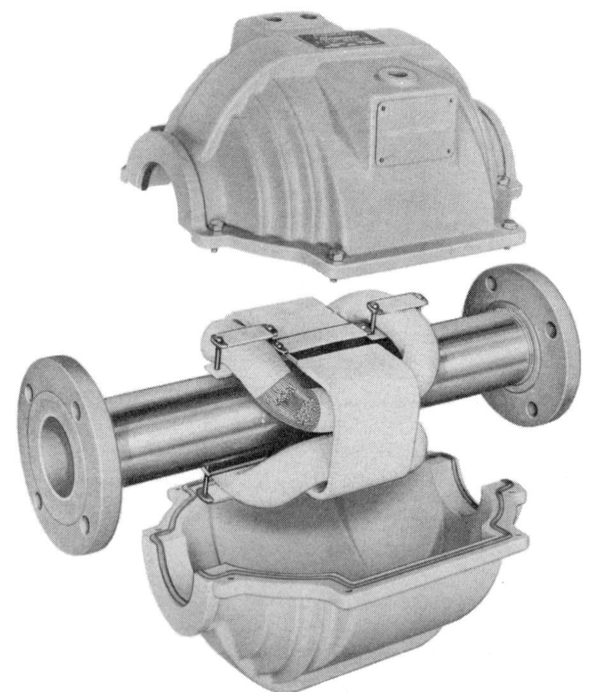

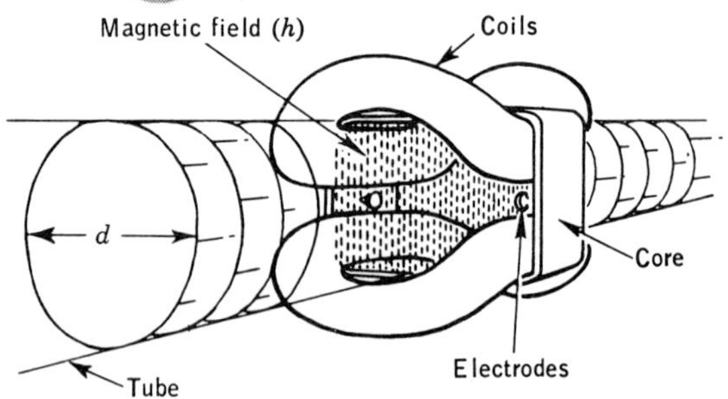

Fig. 12·1 As fluid passes through a magnetic field it generates a potential which is proportional to the rate of flow. The interior of the pipe is smooth and does not introduce a pressure drop. (*The Foxboro Co.*)

duced by devices of this sort is dependent on the average and not the maximum or minimum speed of the fluid.

Operation. In practice, the inside of the pipe carrying the fluid must be insulated so that the potentials created in the flowing liquids do not become short-circuited by the pipe and consequently reduce the external potentials almost to zero. In effect, the only portion of the flowing liquid which contributes to the emf is that which is directly in the path of the two insulated electrodes.

Because of the high resistance of the liquid, there is a requirement that there be no current flow when the balance point is achieved and the readings are to be taken as correct. For there to be no current flow, the emf of the flowing liquid may be connected in series opposition with the out-

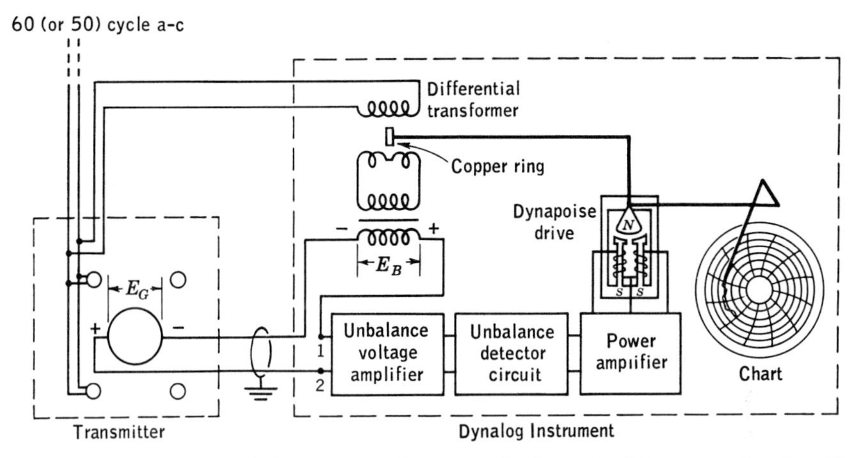

Fig. 12·2 The output of the magnetic flowmeter is determined by comparing it with the potential created by a differential transformer. (*The Foxboro Co.*)

put of a differential transformer, as in Fig. 12·2. So long as the two emf's are equal and their sum is zero, there will be no input to the unbalanced voltage amplifier. But if the flow should increase so that the sum of the two potentials is no longer zero, then there will be an input to the voltage amplifier which will create an input to the power amplifier and thus to the solenoid motor which, in turn, will move the core of the differential transformer to restore the equality of the two potentials. Moving the core of the transformer will also move the indicating pen. The device is

Fig. 12·3 Schematic of operation of the Fischer and Porter magnetic flowmeter.

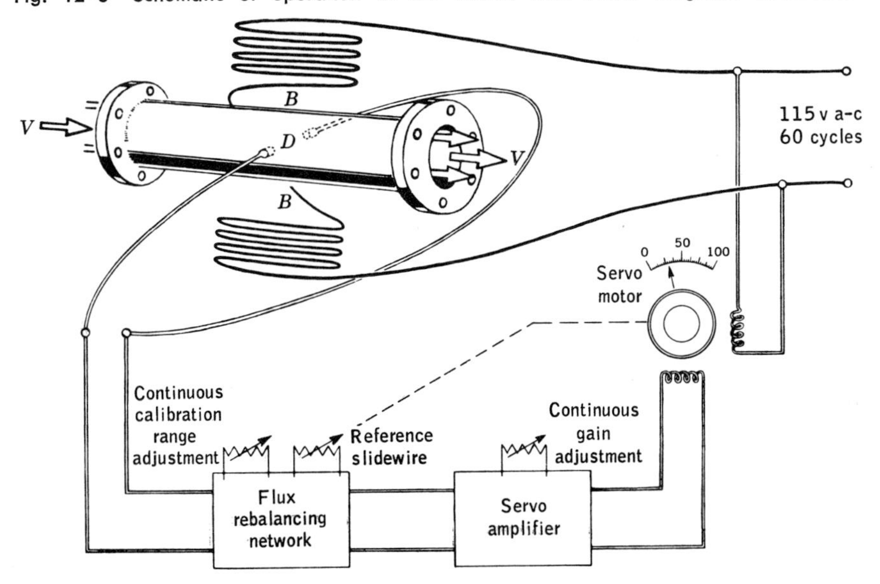

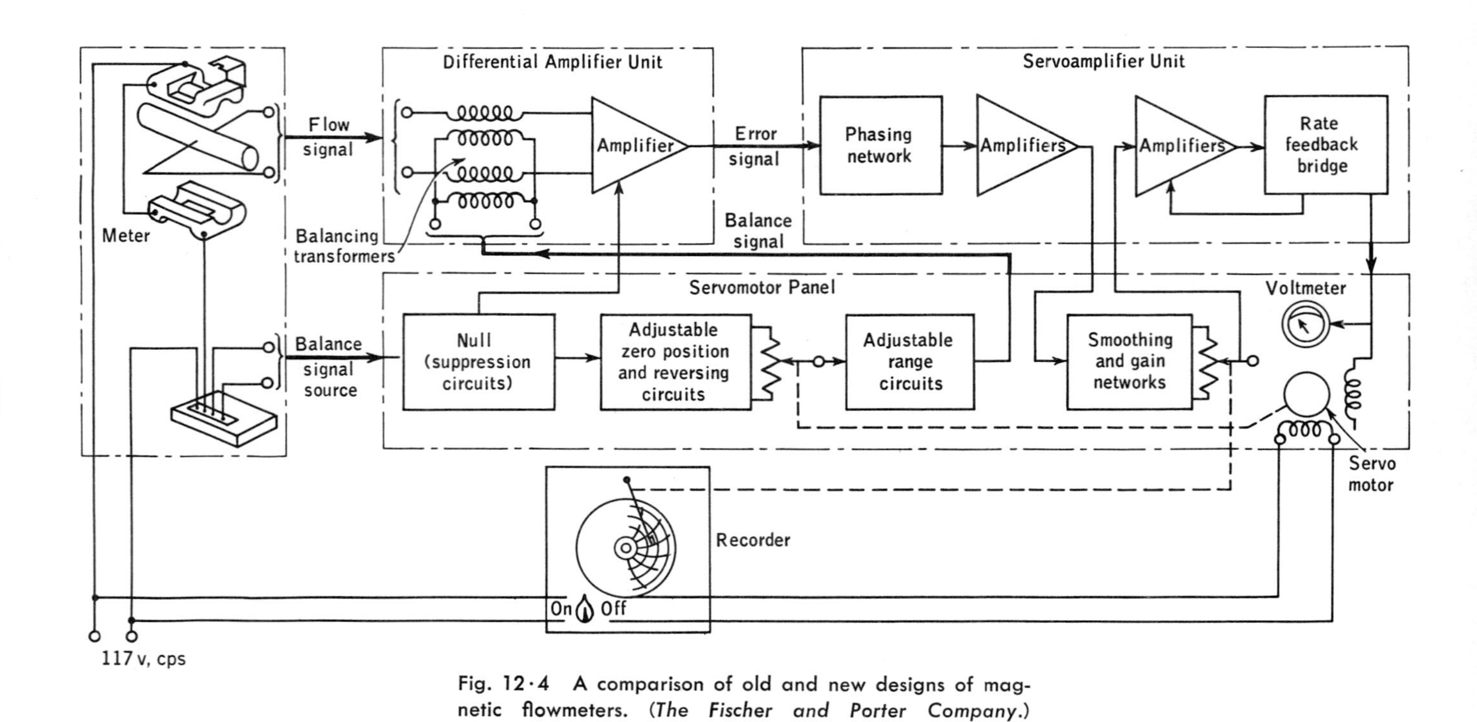

Fig. 12·4 A comparison of old and new designs of magnetic flowmeters. (*The Fischer and Porter Company.*)

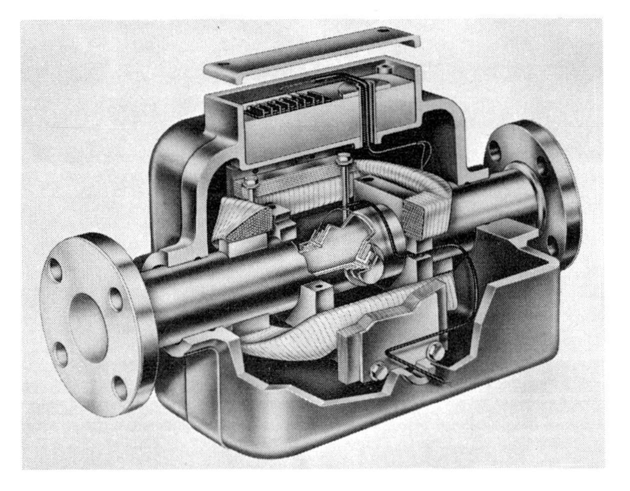

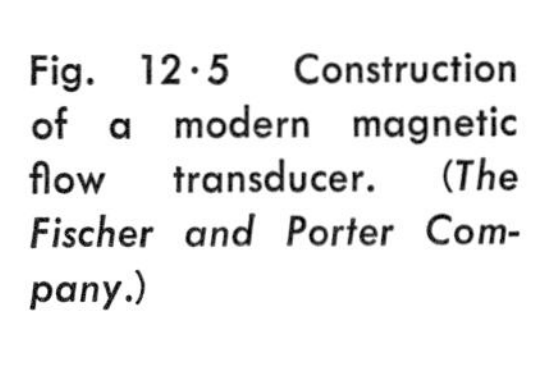

Fig. 12·5 Construction of a modern magnetic flow transducer. (*The Fischer and Porter Company.*)

phase sensitive, so it can determine whether the transformer output is too big or too small and thus move the core in the appropriate direction.

A somewhat different design is shown in Figs. 12·3 to 12·5. The Fischer and Porter design employs the more conventional potentiometer to obtain a measure of the generated emf, rather than some other type of potentiometric device. Because this design imposes no obstruction in the liquid flow path, there will be no pressure drop through the device other than that which would occur in an equivalent length of straight pipe. The

Fig. 12·6 Circuit diagram for a redesigned low-conductivity system. (*The Fischer and Porter Company.*)

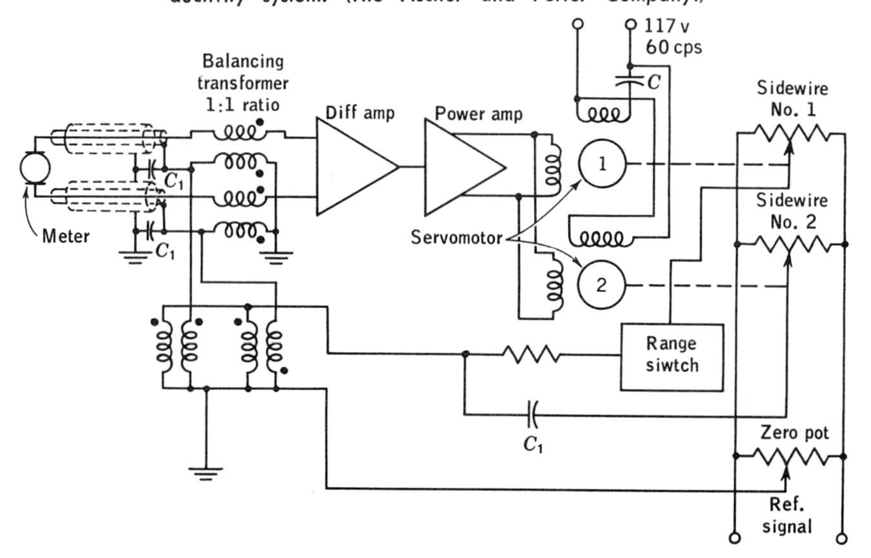

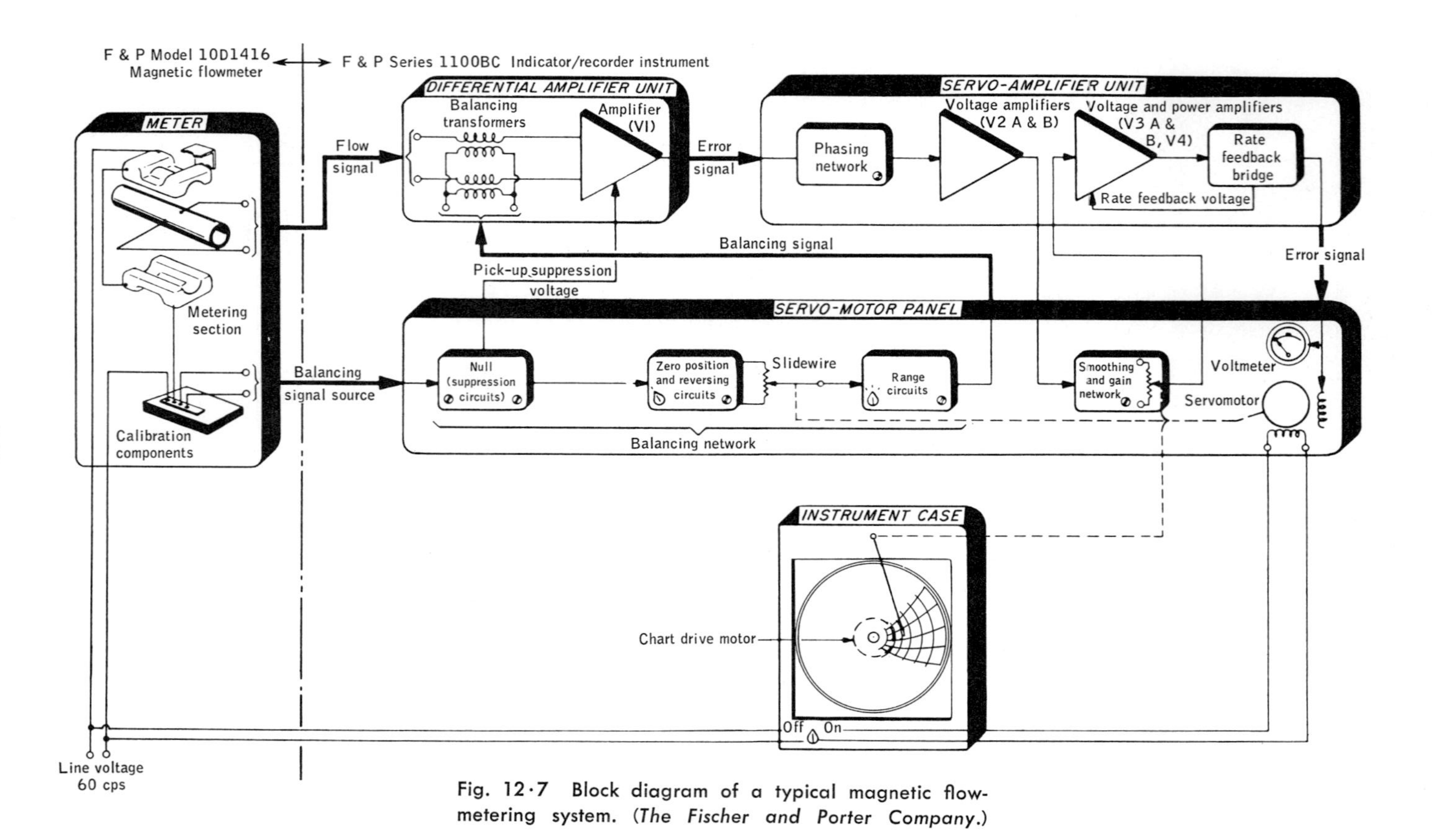

Fig. 12·7 Block diagram of a typical magnetic flow-metering system. (*The Fischer and Porter Company.*)

absence of an obstruction also means that all sorts of slurries and liquids with materials in suspension can be measured with about equal facility. Also, liquids which might flash when going through an orifice do not do so in this device.

In order to operate, some designs specify that the liquids must have a conductivity of at least 20 μmhos, which is the equivalent of 10 parts per million of salt in pure water. (For certain designs for measuring very low flow, the conductivity of the liquid may have to be as high as 100 μmhos.)

OTHER ELECTRIC FLOWMETERS

One of the major problems in connection with liquid flow is to determine the total flow. It is a simpler job to measure the instantaneous rate of flow than it is to measure the total flow over some period of time. The design shown in Figs. 12·8 and 12·9 uses well-established methods to determine both rate of flow and total flow.

It will be noted that the design shown in Fig. 12·8 is quite conventional in so far as the use of an orifice plate and a mercury manometer is concerned. But the mercury manometer is secured to a beam in such a manner that, as increased flow drives the mercury down in one leg and up in the other, the beam will become unbalanced and require a greater restoring force to bring about torque equilibrium. The added downward force on the left-hand side of the beam is applied by a governor system in which, as the motor-integrator speeds up, the balls will assume a position farther from the center of rotation. Through a system of cams and levers, this movement is transferred to a member connected with the left side of the beam. As the motor speed increases—and it will increase above the normal speed of the unit unless the motor circuit is opened—more and more force will be exerted downward by the governor system trying to counterbalance the mercury in the right leg. A point of equilibrium will be reached and exceeded because the motor will still continue to speed up. When this equilibrium point is exceeded, the magnetic switch will be operated, thereby opening the motor circuit. The motor will then decrease in speed until the beam is slightly unbalanced and the left end will move up, at which point the magnetic switch will be closed again and the motor will again speed up. The process will be repeated indefinitely until the flow conditions require that there be more or less mercury in the right-hand leg.

In this manner, the motor will maintain an average speed which is directly proportional to the flow rate. A counter which records motor revolutions will then indicate the total flow in terms of the integrated values of time and flow magnitude. An electrical tachometer which can measure the instantaneous speed will then indicate the flow rate.

Fig. 12·8 A schematic representation of the essential components of an integrating, indicating flowmeter. The rotation of the motor drives a flyball system which requires centripetal forces to maintain the balls in their correct position. This centripetal force is obtained from a member which is attached to the beam. By applying a force to the flyballs, the beam also has a force applied to it. For conditions of equilibrium the torque established on the beam by the flyball system is just equal to that provided by the unbalanced mercury column. (*Leeds and Northrup Co.*)

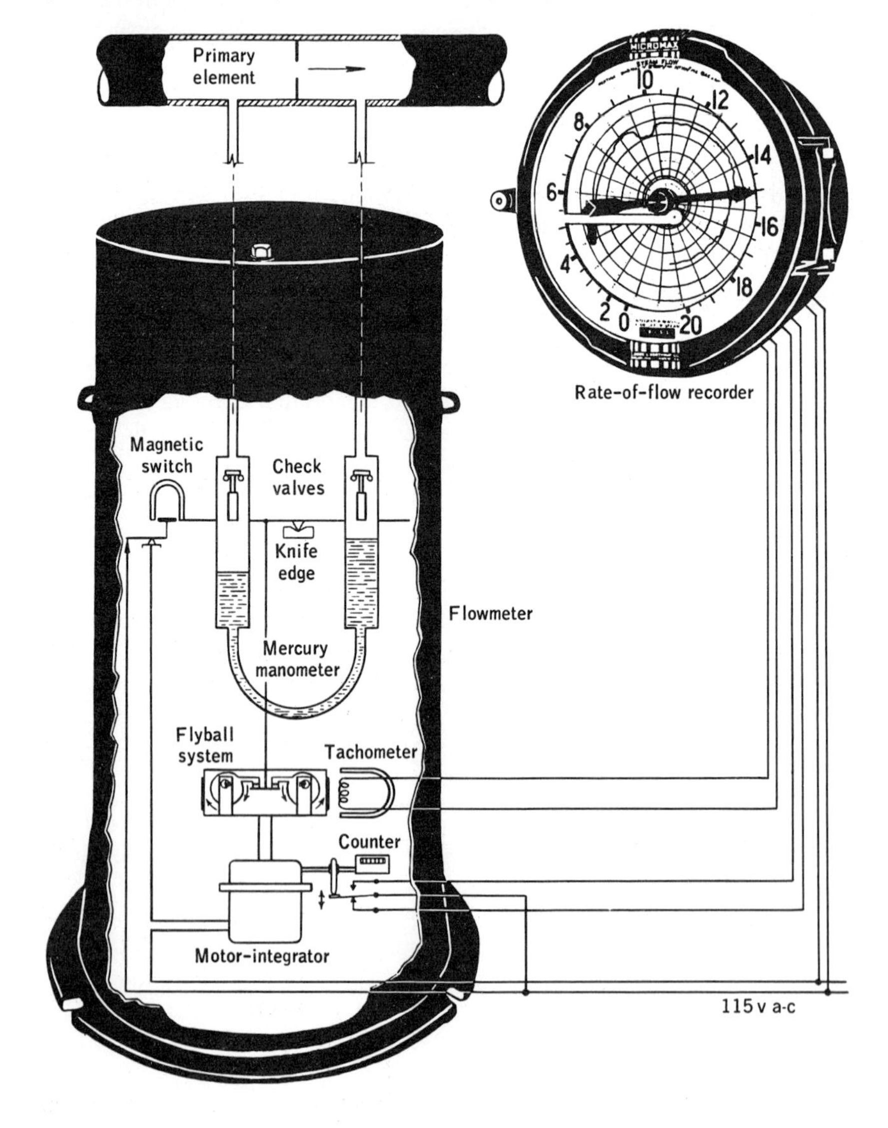

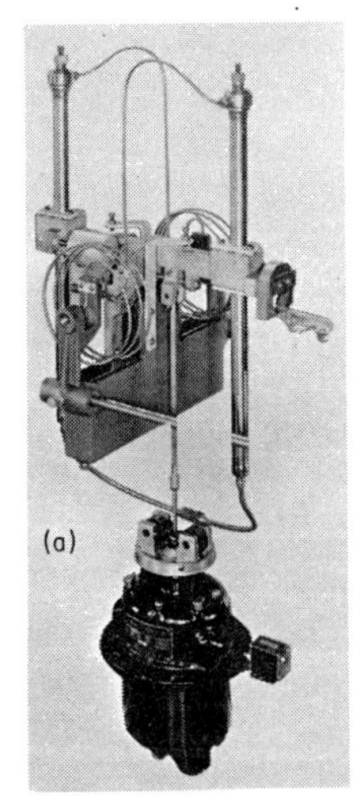

Fig. 12·9(*a*) To integrate flow directly by simply counting flyball revolutions, this design uses this beam-type balance and motor-integrator. (*b*) Front view of assembled meter body with cover removed. (*c*) Rear view of assembled meter body with cover removed. This view shows the motor-integrator and the flyball system which balances the tilting mercury manometer. (*Leeds and Northrup* Co.)

Because the force developed by the governor system is dependent upon the square of the speed and the differential pressure is a function of the square of the flow rate, we may say that the governor speed and the flow rate are directly proportional.

Turbine-type flowmeters. This type of meter is considered in Chap. 11, where the essential mechanical details are discussed. The designs discussed in Chap. 11 indicate flow rate and total flow by using a mechanical gear train or a similar device that depends on the number of turns of the turbine. But a number of limitations are imposed by these mechanical designs. For example, it is necessary to furnish stuffing boxes which are pressure tight but at the same time create only an insignificant frictional drag. To overcome such difficulties, and to simplify certain other problems, certain electrical designs have been developed, as shown in Figs. 12·10 to 12·13. Figure 12·11 shows a design that rotates a magnetic member so that it changes the flux linking a coil. This then induces in the coil an emf the frequency of which is directly related to the rate at which the turbine is turning. It is then necessary to convert the frequency into a measure of flow rate. Each impulse means that a certain volume of liquid has been measured; therefore, if the number of voltage impulses is counted, one has a measure of the total flow.

It might be expected that the speed of the turbine can create an emf which is directly proportional to the speed and that this might be used as a measure of the turbine speed and hence of the rate of flow.

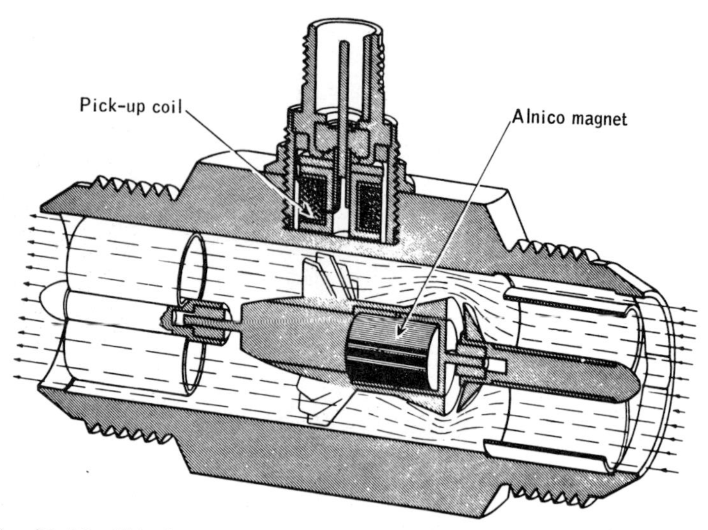

Fig. 12·10 This flowmeter provides a digital output with a frequency directly proportional to flow. (*Potter Aeronautical Corp.*)

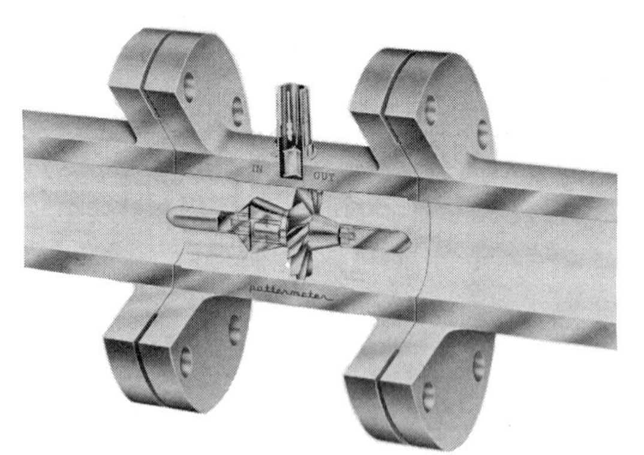

Fig. 12·11 Cutaway view of a flanged meter used in industrial applications. (*Potter Aeronautical Corp.*)

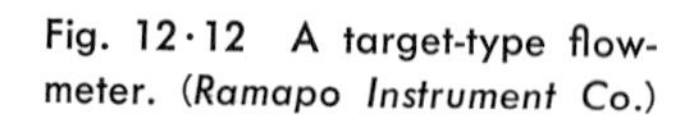

Fig. 12·12 A target-type flowmeter. (*Ramapo Instrument Co.*)

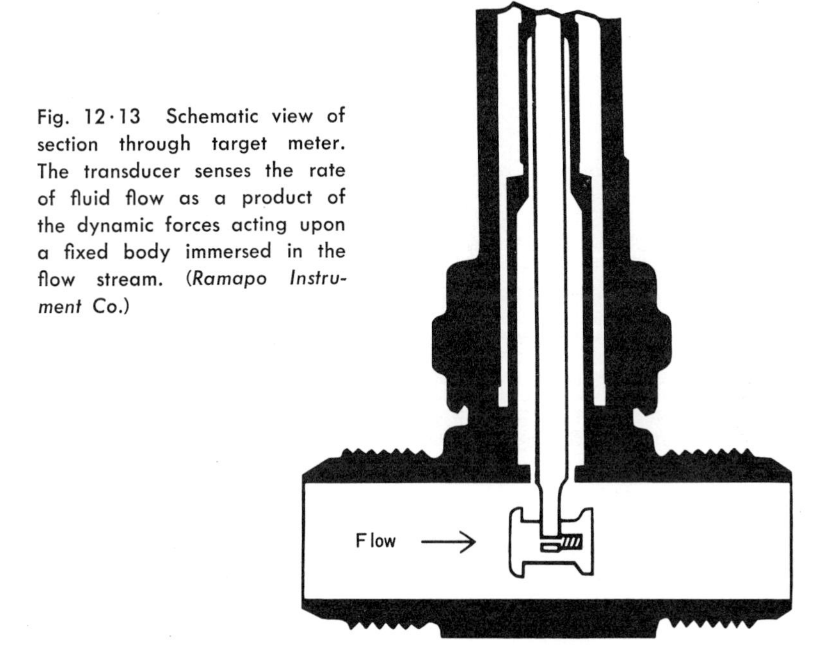

Fig. 12·13 Schematic view of section through target meter. The transducer senses the rate of fluid flow as a product of the dynamic forces acting upon a fixed body immersed in the flow stream. (*Ramapo Instrument* Co.)

Strain-gage flow transducers. An entirely different approach to the measurement of flow is found in the design illustrated in Figs. 12·12 and 12·13. In this design an obstruction is placed in the flow path, but not for the primary purpose of creating pressure differentials. Rather, the obstruction is placed in the flow path in order that the force of the liquid impinging on the obstruction may be used as a measure of the flow rate. The force against the sensing element is determined by

$$\text{Force} = C_d A p \frac{V^2}{2g}$$

where C_d = drag coefficient
A = area of the sensing element
p = fluid density
$\frac{V^2}{2g}$ = velocity head

The force which is developed by the flow is then used to deform a spring element to which are secured bonded strain gages. The gages are connected in a Wheatstone-bridge circuit so that the change in resistance is then a function of flow. Translation of the resistance changes into flow rates in millivolts, Fig. 12·14, then depends upon the ingenuity of the designer. There are many ways of making these conversions.

Variable-area meters. The measurement of flow when the liquids are viscous or tar-like in nature oftentimes requires the use of a meter such

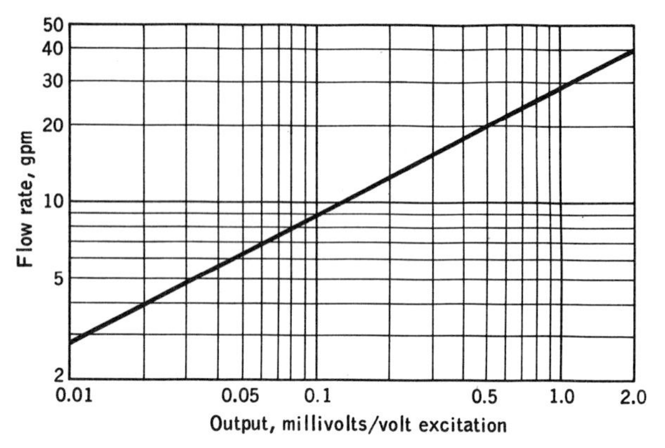

Fig. 12·14 Flow rate versus output, in millivolts. The output is directly proportional to the square of the flow rate.

as is shown in Fig. 12·15. This meter is quite similar to the rotameter in that a constant force must be exerted on the piston. If the pressure differential across the ports of the meter increases, a greater force is available to push the piston up. This movement will then increase the port areas, which will then decrease the pressure differential. In this manner, the piston will take positions which will maintain fixed differential pressure conditions. But the position of the piston, which must be mechanically free to move up or down, is generally communicated

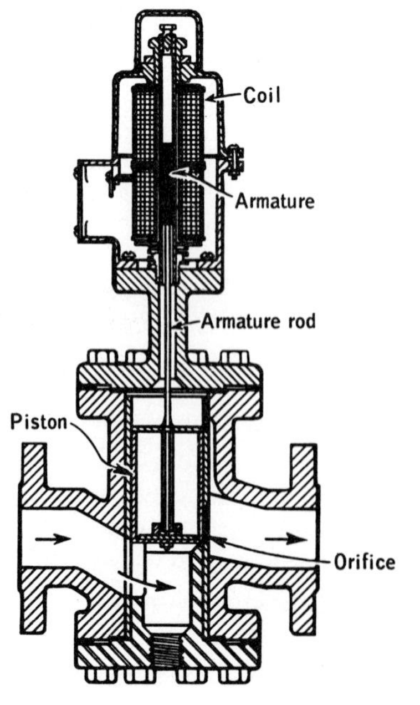

Fig. 12·15 An area type of flowmeter. As the piston moves up or down with changes in the flow rate, it positions the core. In this figure, the core alters the value of an inductor. If a differential transformer were substituted for the inductor, then the output signal would be a variable potential. (*Minneapolis-Honeywell Regulator* Co.)

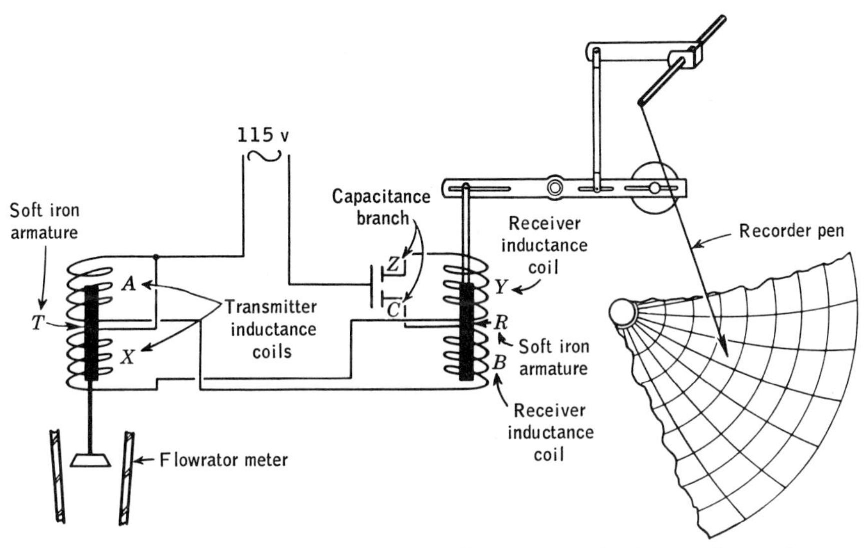

Fig. 12·16 A variable-area meter may employ a tapered tube to obtain changes in area. Capacitors are used in this circuit to tune the system so that the core movements will produce the maximum current changes. The larger the current change, the easier it will be to provide the torque to reposition the receiver core and its connected pen. (*The Fischer and Porter Company.*)

by some electrical means. Positioning the core in an inductance is one means of expressing the piston position.

The position of the float in a design, such as is indicated in Figs. 12·16 and 12·17, may use exactly the same methods as are indicated for the other variable-area meter shown in Fig. 12·15.

Mass flowmeters. In the same sort of way that the variable-area meter just considered employs a combination of mechanical and electrical components to indicate the flow rate, so the design of the mass meter, Figs. 12·18 to 12·21, integrates mechanical and electrical components. The important components and their operation are:

1. An impeller motor which, through a gear train and a magnetic coupling, drives the impeller at a speed which is essentially constant. As the impeller rotates it imparts a rotary motion to the fluid flowing through it. We will assume that the fluid has had virtually no rotary motion before reaching the impeller.

2. A turbine which the rotating fluid tries to rotate. The turbine is prevented from rotating by a linkage with the gyro-integrating system. The force developed in resisting the rotary flow of the fluid is a measure of the rate of flow.

3. When the force required to prevent rotation of the turbine is applied to the gyro-integrating unit, it causes the gyro to precess (or rotate) about its major axis at a rate which is dependent upon the force

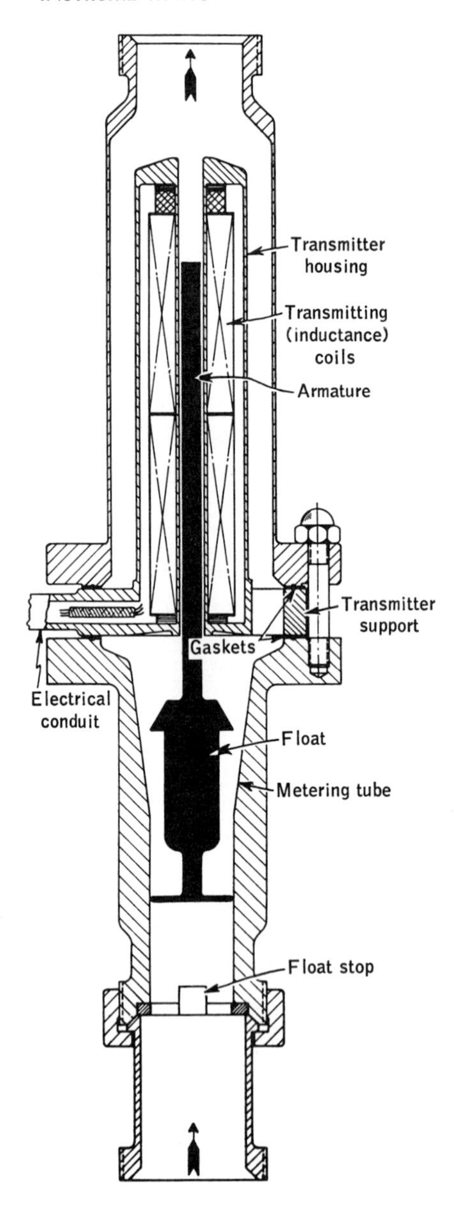

Fig. 12·17 It makes little difference whether the transformer core is above or below the measuring element; the meter is designed both ways. (*The Fischer and Porter Company.*)

applied to the minor axis. By counting the turns which the precession causes, the flow can be totalized.

These meters are termed mass meters rather than volumetric meters because the momentum which is employed here to create the torque in the turbine varies with the mass of the fluid per unit time rather than with the volume per unit time. Suppose, for example, that two fluids were being measured by this meter and that one weighed 64.4 lb/ft^3 and

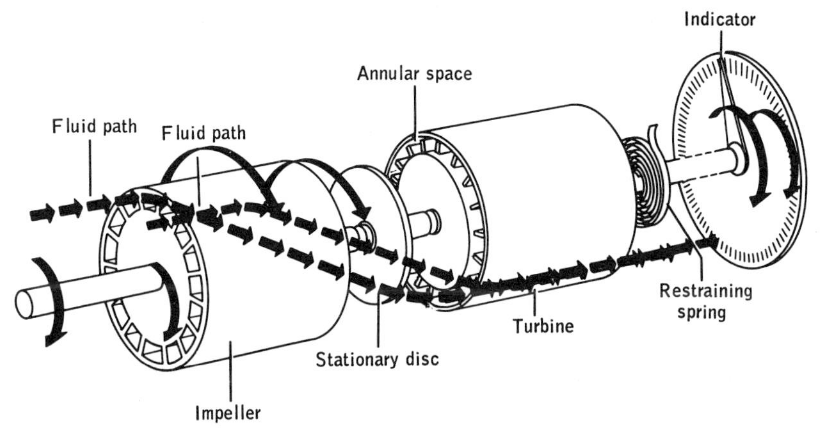

Fig. 12·18 A mass type of flowmeter. When the impeller is driven by a constant-speed motor, it will impart a rotary motion to the fluid. This rotary motion is then removed by the turbine. The force required to return the motion to straight-line flow then becomes a measure of the flow rate. (*General Electric* Co.)

the other weighed 32.2 lb/ft³. Then for *the same rate of flow in feet per second or minute* (this is another way of saying that the same number of cubic feet per second pass through the meter) the heavier liquid, which has a mass of 2 slug/ft³, will register *twice* the flow of the lighter fluid, which has a mass of 1 slug/ft³.

In most commercial installations, the deviation of the electrical frequency from the rated value is of little importance. But when the electrical power is not controlled by interconnections with commercial

Fig. 12·19 Gyro-integrating mass-flowmeter schematic. The axial conservation of momentum flowmeter is combined with a three-axis gyroscope and register to provide a direct and continuous integration of mass fluid flow insensitive to power voltage or frequency variations. (*General Electric* Co.)

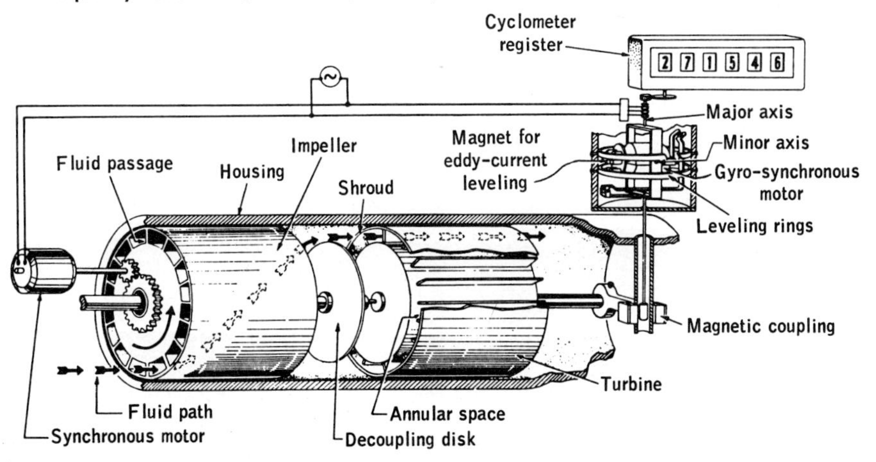

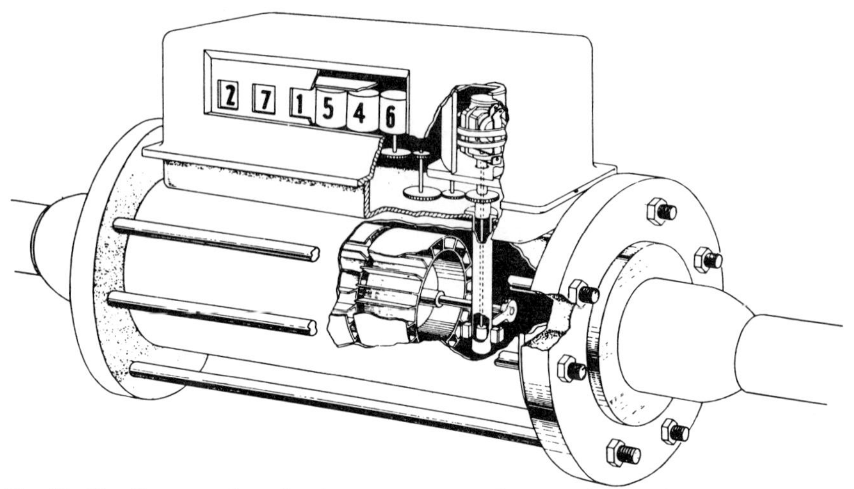

Fig. 12·20 Cutaway view showing transmission of torque from turbine to gyroscope and subsequent gearing to register. (*General Electric* Co.)

sources, the frequency may vary by 5 to 10 per cent. This design is not frequency sensitive, because both the impeller and the gyro motor operate at the same frequency and any changes in frequency will affect both motors to the same degree. This makes the performance independent of the frequency.

Fig. 12·21 The flow rate sensor measures the torque generated by the removal of the angular momentum. (*General Electric* Co.)

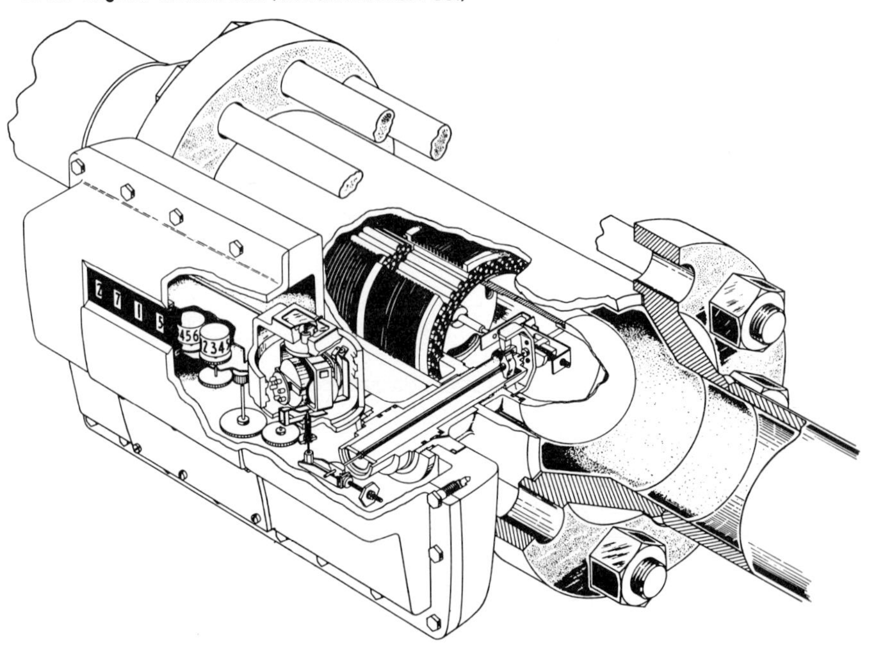

Air and gas flow measurements employing heat. A modification of the principle of measuring the amount of heat required to raise the temperature of flowing fluid by a fixed amount is the system which determines the flow by ascertaining the temperature of a heated member when the input power is kept constant. The temperature of the heated element, for a constant temperature of the flowing fluid, is dependent upon how fast the fluid is moving past the heated member and carrying away heat. If the temperature of the flowing fluid varies, then corrections will have to be made because of the change in the ambient. But if the input power is kept constant, we have the following relationships for a resistor:

$$P = VI$$

$$= \frac{VI}{R} = \frac{V^2}{R}$$

But R is a function of T, so

$$P = \frac{KV^2}{T}$$

However, the temperature is an inverse function of the flow rate F, so

$$P = K'V^2F$$

Rewriting, we have

$$\text{Flow} = \frac{PK''}{V^2}$$

By keeping the power input constant, we have

$$\text{Flow} = \frac{K''}{V^2}$$

This method lends itself to the measurement of air and gases. It might prove useful in measuring the flow of liquids such as gasoline, ether, and alcohol, which have high heat absorption rates. Oils of the nature of lubricating oils have relatively poor heat transfer rates and might not be satisfactory for this type of measurement because the film around the resistor proves to be a very effective insulator of the heated member.

Differential transformer transducers. If the movement of a differential bellows is made to position the core of a differential transformer, then it is possible to convert flow rates into electrical signals of potential. Figure 12·22 shows the significant details of the unit.

Unlike the transducers which produce signals in terms of current, devices which employ potentials as a measure of the variable generally do not feed back or compare the output signal with the input. This type of transducer should be compared with other devices studied in Chap. 9 under Differential Pressure Transmitters. The very small volume, the

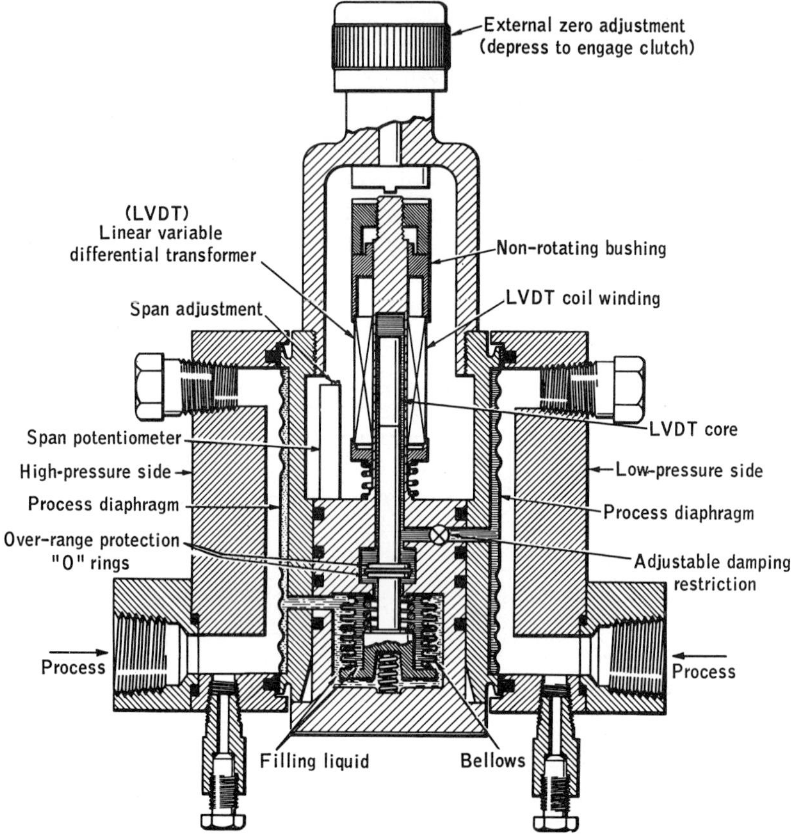

Fig. 12·22 Cutaway view of differential pressure transmitter. Diaphragms are used to keep the process fluid from entering the working parts of the meter. Note the provision for overrange protection. (*The Taylor Instrument Cos.*)

provision against overranging, and the manner of damping out any oscillations should be carefully observed.

In marked contrast to the design illustrated in Fig. 12·22 is the differential pressure–current transducer shown in Figs. 12·23 and 12·24. This design provides for comparison of the electrical output with the pressure input. In the schematic representation, the process pressures are admitted to both sides of the diaphragm capsule and will try to move the capsule in the direction of the smaller pressure. This motion will carry with it the force bar, which is provided with a flexible seal that also acts as a pivot. If we assume that there has been an increase in the differential pressure, then the capsule is carried to the right. This will move the upper portion of the force bar to the left. A suitable connection between force bar and range rod causes the range rod to pivot on an edge of the range wheel. The position of the range wheel provides the necessary

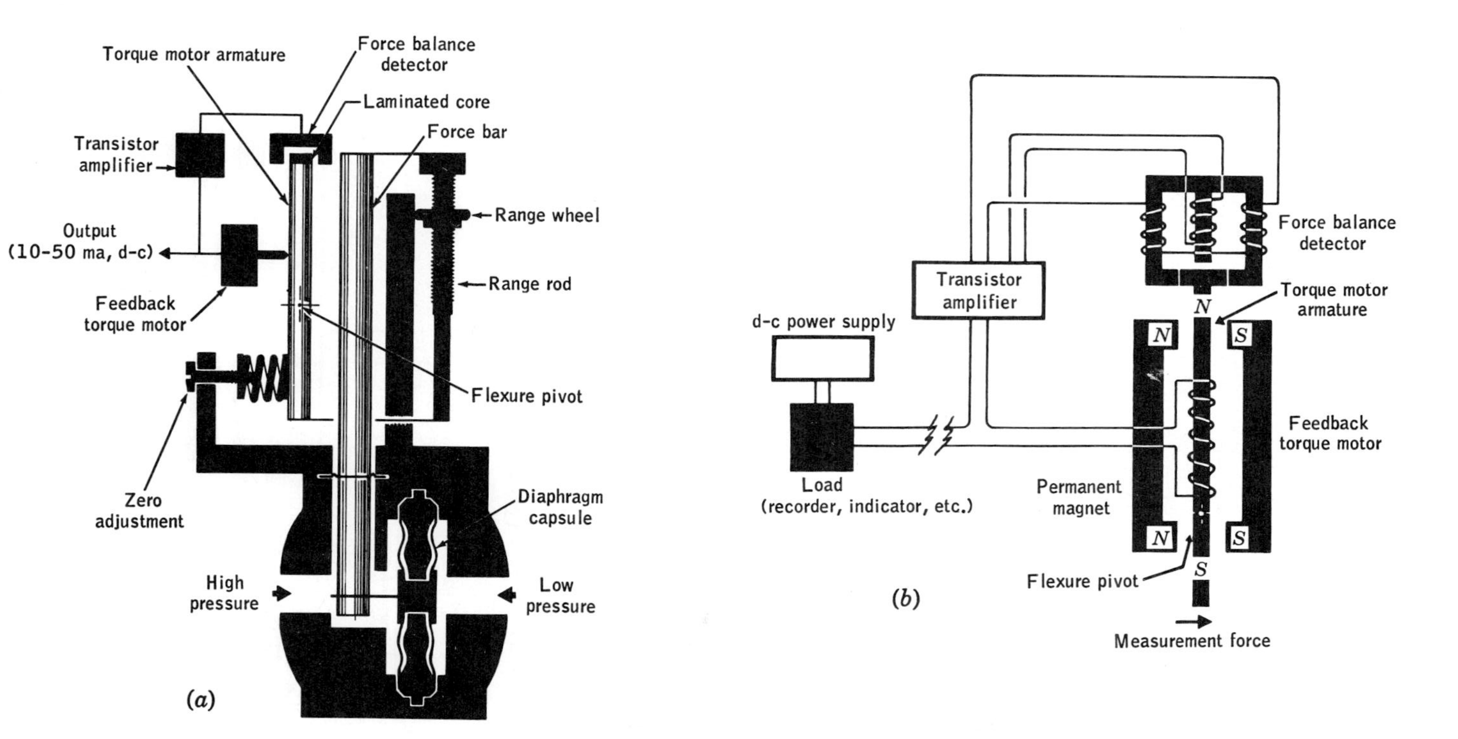

Fig. 12·23 Foxboro electrical differential pressure transmitter. The output current produces a force which balances the torque created by the differential pressure.

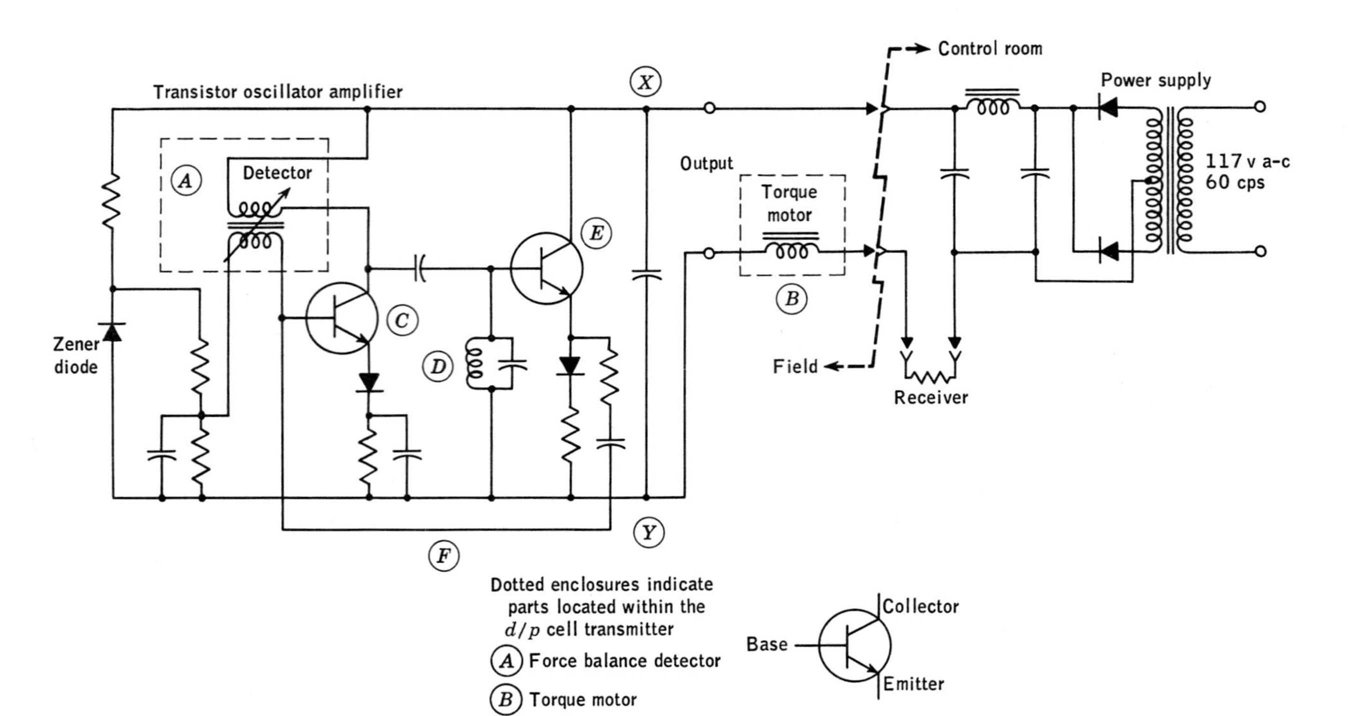

Fig. 12·24 The schematic diagram for the Foxboro electrical differential pressure gage.

mechanical amplification to provide *torque equilibrium* for the various ranges of the device.

Another linkage transmits the movement of the range rod to the feedback torque motor armature and to the differential transformer. The laminated core will be moved to the left, which will increase the output of the differential transformer. Such an increase in signal value will now cause the transistor amplifier to increase its output current. This increase in current to the feedback torque motor will increase the repulsion effects and try to cause the motor to rotate clockwise about its flexure pivot in opposition to the torque created by the increased differential pressure.

MEASURING FLOW OF DRY MATERIALS

The measurement of dry materials is somewhat more direct than the measurement of flow of gases and liquids. But there are the same general divisions of measurements involving volume or weight. The use of screw conveyors which are run at constant speed would furnish flow in terms of volume, and so would a bucket conveyor. Weight can be determined only on the basis of some fixed density or specific gravity. Most measurements undoubtedly involve weight or mass because they are made somewhat more easily. Any type of scale, platform, spring, or some related type, will respond to weight and only incidentally to volume as it is related through density.

Most platform scales are nothing but a grouping of levers so that the moment of the load can be balanced by a known torque created by a known weight at an ascertainable distance. The unknown is identified by the known magnitude required to offset or balance its turning effects.

Whether we use the familiar platform scale in which the weight is established by the positioning of a poise, Fig. 12·25, or whether we use the more modern version, Fig. 12·26, the same fundamental relationships

Fig. 12·25 The weight of the load on the weighing platform may be determined by finding the position of the poise which will provide a torque which just offsets that created by the load on the weighing section.

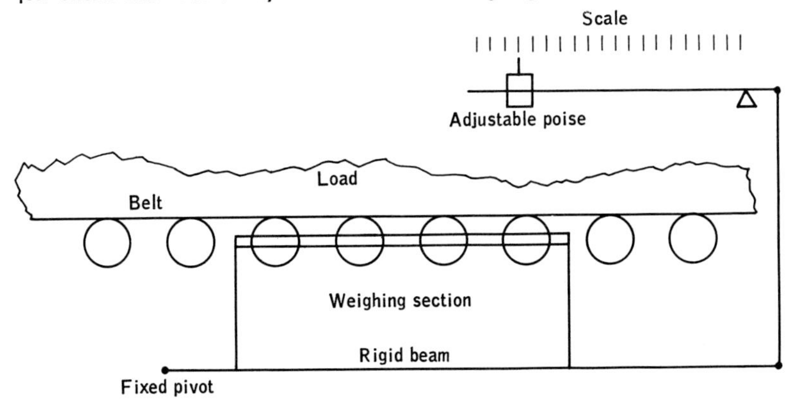

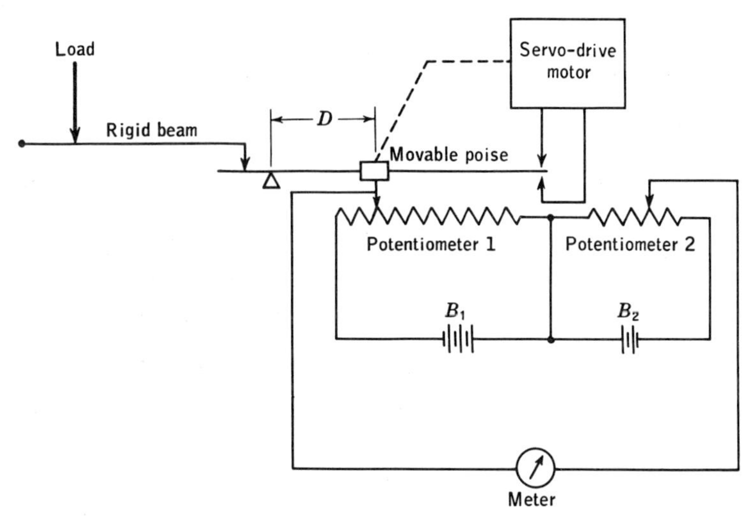

Fig. 12·26 This sketch shows the essential components of an automatic scale with either local or remote indications. By adding battery 2 and its potentiometer, corrections may be made for the dead weight of the weighing mechanism and also for nonstandard conditions.

are involved. About the fulcrum F, the moment of the load is some fixed quantity multiplied by the load. The weight of the poise acting through distance D provides the equal and opposite moment. To obtain remote-reading electrical signals, the lack of balance completes circuits which cause a motor to position the poise, and with it the pointer. The pointer can also position a potentiometer rheostat, so that an electrical signal is obtainable. By adding a second emf and potentiometer rheostat, it is possible to cancel out the effect of the weight of the scale pan or any other tare weight.

Differential transformers can be used to measure weight in terms of

Fig. 12·27 The output of a differential transformer may be used as a measure of a load. The zero compensation voltage not only can correct for the weight of the measuring structure but also can compensate for nonstandard conditions such as excessive moisture or adulterants in excess of some specified amount.

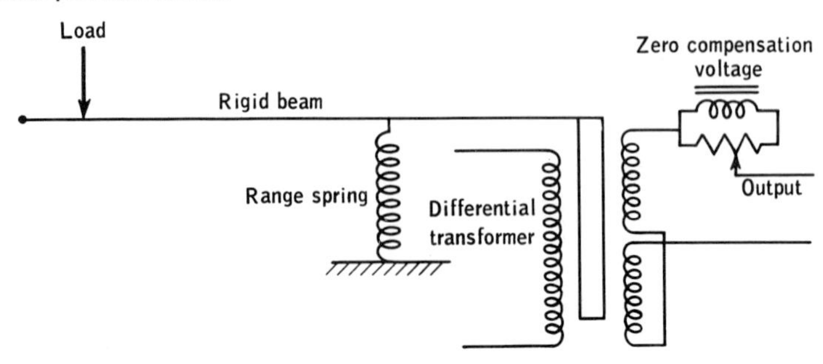

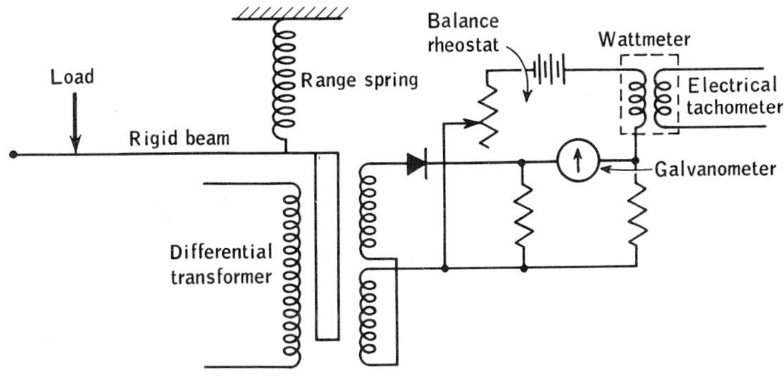

Fig. 12·28 By combining current, which is a measure of the load, with a potential which is directly proportional to the speed of the belt, the wattmeter will provide a reading which is directly proportional to the instantaneous flow rate. By supplying the same two quantities to a watt-hour meter, the total integrated flow may be obtained.

deflection, Fig. 12·27, by weighing the load on a short section of belt. Then, by knowing the rate of travel of the belt (from a tachometer) and the weight per foot, we can quickly compute the rate of flow. Provided we know the time during which this flow occurs, we can determine the total number of pounds or tons. Figure 12·28 shows in schematic form how this might be done. The load on the section being weighed positions the platform by stretching the spring. This moves the core of the differential transformer whose emf is then compared with that developed across a resistor. If the two potentials are not equal, then a servomotor varies the resistance to reestablish the equality. The current then becomes a measure of the weight per foot of load. If the current is passed through a wattmeter whose potential coil is energized by a tachometer driven by the belt, the wattmeter is an indication of that moment's output in tons per hour or pounds per minute. If the same current is caused to pass through a watthour meter and the same tachometer emf is applied to the potential coil, then the watthour meter will indicate the total flow in pounds or tons, and so on.

By applying potentials which add or subtract from the tachometer output, it is possible to correct for nonstandard conditions and come out with an answer in equivalent pounds or tons. The material might be too wet for standard conditions, hence it might weigh 5 per cent, for example, too much. Or it might be that 10 per cent is substandard and should not be counted at its full value.

One of the fundamental relationships is that any body whose speed or direction is changed will require an acceleration. For a given acceleration, the magnitude of the accelerating force is directly proportional to the mass being accelerated. In Fig. 12·29 we find that material is being dumped from one belt to another which is at right angles to the first. This nullifies any effect of the original motion. Now if the second belt is

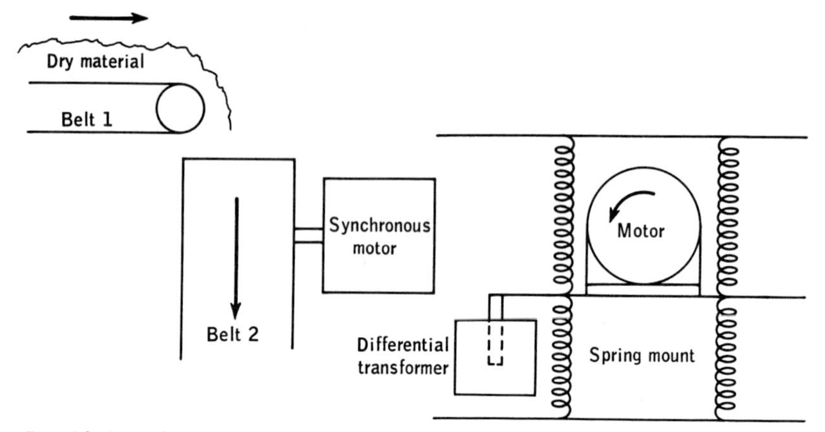

Fig. 12·29 The instantaneous torque required of the motor depends entirely upon how much material is being dumped on belt 2. This torque may be measured in many ways. One method is suggested here.

driven by a constant-speed synchronous motor, all material falling on the second belt will have to be given essentially the same acceleration to bring it up to the speed of the second belt. The torques which are required for these accelerations will cause supporting spring L_2 to be compressed an additional amount beyond that which occurs with no load on the belt. This extra movement can be used to position a differential transformer whose output can be recorded on a strip chart. The area under the curve is then a measure of the total load deposited on the second belt.

Exactly the same principles are employed in Fig. 12·30. However, in this instance the output is pneumatic instead of electrical. Provided that the load is created by the weight of material on a moving belt, the total

Fig. 12·30 When air pressure is a linear function of the variable, this simple integrator may be used.

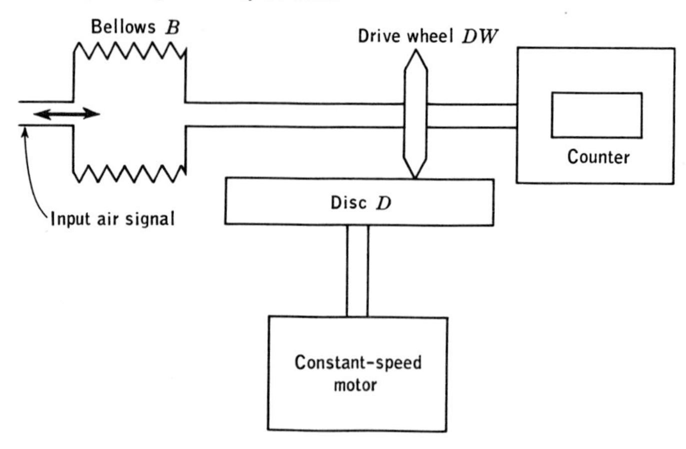

amount of material moving over the weighing device is equal to the product of the rate and the time. The force needed to balance the load becomes a measure of the rate; it will be assumed that the belt is moving at a constant rate. The length of time that the loads are sustained must be integrated if the total flow is to be determined.

When the integrated sum is based on the first power of the variable, the integrator may be a very simple device such as that represented in Fig. 12·30. As bellows B undergoes a linear expansion or change with the input air signal, it positions drive wheel DW. If disk D is driven at a constant speed by a synchronous motor, then the number of revolutions transmitted to the counter will vary directly with the distance that the drive wheel is from the center of the disk. When the air input signal is zero, the drive wheel is at the center of the disk and has zero speed of rotation. The speed of any point on the driven disk D is directly proportional to the distance from the center.

The pneumatic integrator employed in flow measurements furnishes one method of finding the total value of such a series of varying loads, Fig. 12·31. Its principles of operation are quite straightforward. An air signal proportional to the differential pressure from the transmitter is applied to the integrator receiver bellows A. The force exerted by the bellows positions a force bar B with respect to nozzle C. With an increase in the signal pressure, the force bar approaches the nozzle. The resulting back pressure at the relay regulates the flow of air to drive the turbine rotor E. As the rotor revolves, the weight F, which is mounted on a flexure-pivoted bell crank G on top of the rotor, requires a centripetal force to maintain it in position. This force reacts on thrust pin H to balance the torque exerted on the force bar by the bellows.

The turbine rotor is geared directly to the counter J through gearing

Fig. 12·31 A section through a pneumatic integrator. (*The Foxboro Co.*)

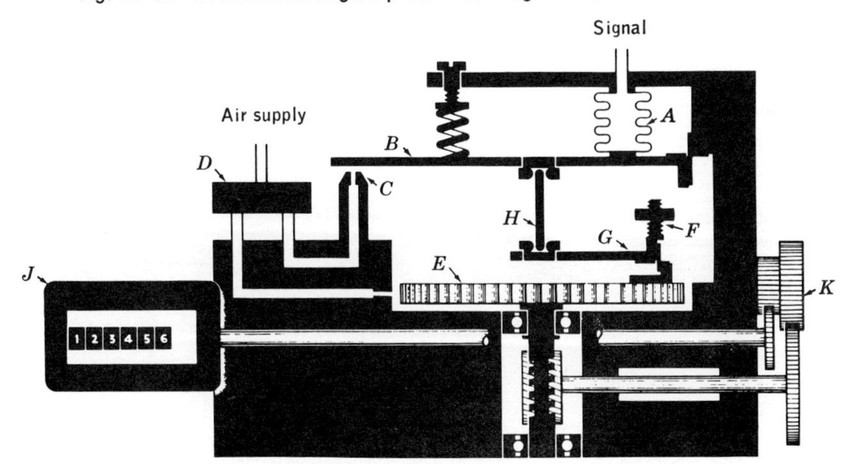

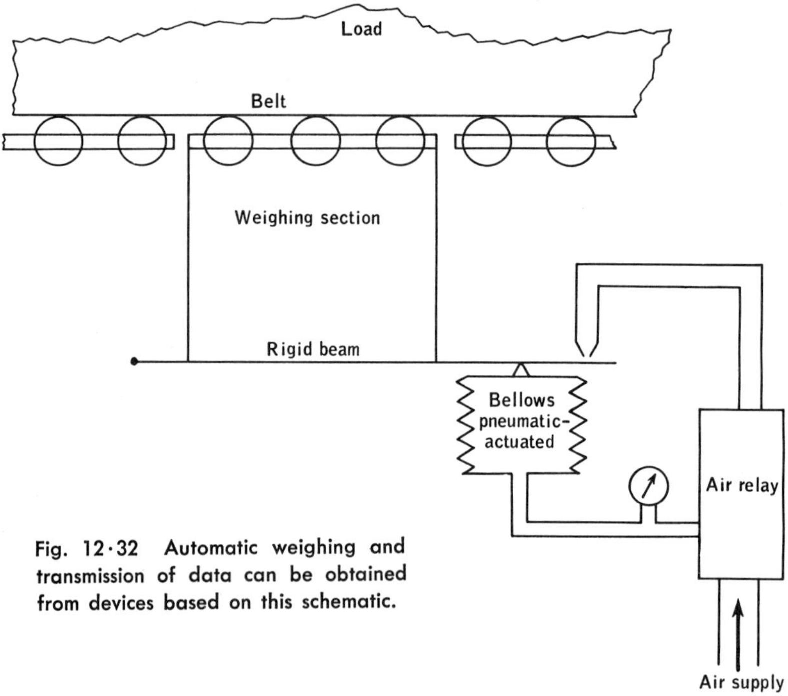

Load
Belt
Weighing section
Rigid beam
Bellows
pneumatic-
actuated
Air relay
Air supply

Fig. 12·32 Automatic weighing and transmission of data can be obtained from devices based on this schematic.

Fig. 12·33 Just as the pressure within a confined bellows or diaphragm system may be used to weigh static loads, so can it be used to measure the weight of material on a given length of belting.

Load
Belt
Weighing section
Rigid beam
Bellows-
liquid-
filled
Pressure
transmitter

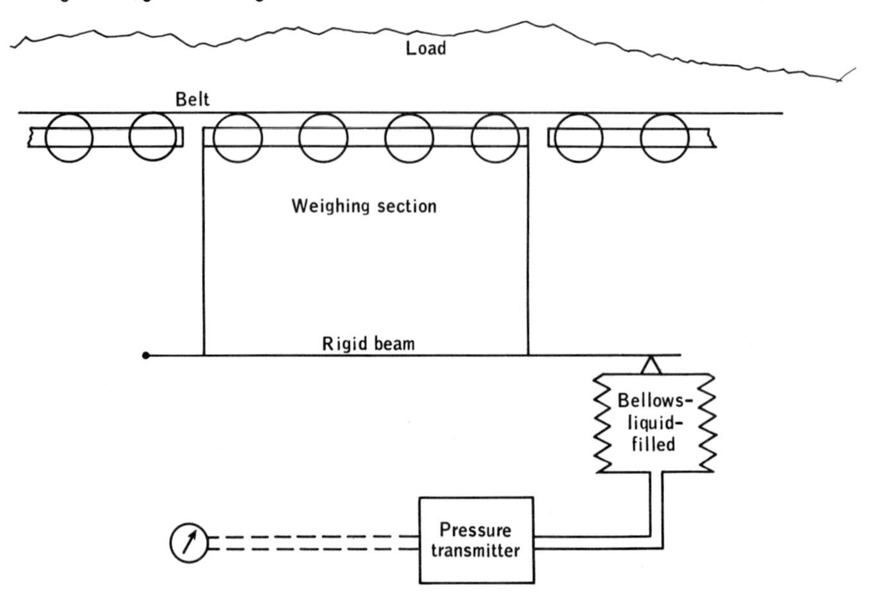

K. Changes in flow continuously produce changes in turbine speed to maintain a continuous balance of torques.

The centripetal force on H is proportional to the square of the turbine speed. The resulting torque balances the signal pressure, which is proportional to the square of the flow. Therefore, turbine speed is directly proportional to flow and the integrator count, which is the sum of the number of revolutions of the turbine rotor, is directly proportional to the total flow.

A slightly different use of pneumatic principles is shown in Fig. 12·32. The air pressure required to balance the torque of the load is then proportional to the load, or one might use the hydraulic pressure created by the load on the short section of conveyor belt.

Still other ways to weigh the load on the test or inspection section is to use the same apparatus previously considered in determining liquid level, Fig. 12·33.

FLOWMETER PRACTICES; LIQUID SERVICE

Review differential pressure instrument practices.

1. The 100-in. differential pressure is the one most commonly used in industry. With this pressure drop, there are not likely to be undesirable effects on the operation of associated equipment. In general, the best results are obtained in the range of 20 to 100 in. of water.

2. Variations caused by temperature, specific gravity, or filling liquid in the lead lines tend to be minimized when the 100-in. differential is used.

3. The selection of the primary device—orifice plate, nozzle, or venturi—should be guided by the following:

a. The orifice plate is the most commonly used primary device. Its characteristics are very well known. It can be readily fabricated if necessary. It is the device normally chosen unless there are other factors.

b. The flow nozzle is recommended instead of the orifice plate when there are moderate amounts of solids in suspension or whenever the following expression gives a value in excess of 140.

$$\frac{W}{D^2\sqrt{wh}}$$

where W = rate of flow, lb/hr
D = internal pipe diameter, in.
w = weight density, lb/ft^3
h = differential pressure, in. of water

c. The pitot tube is recommended only when the fluids are clean, the line is large, and the speed is relatively high.

d. The venturi tube is recommended when:

1) The permanent pressure drop is to be reduced to a minimum.
2) There may be large amounts of suspended solids.

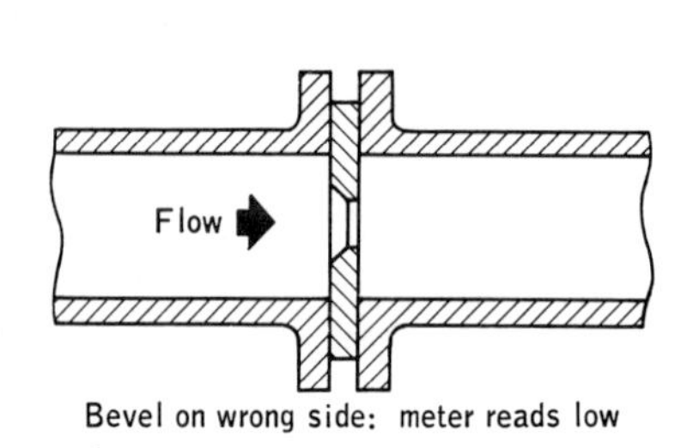

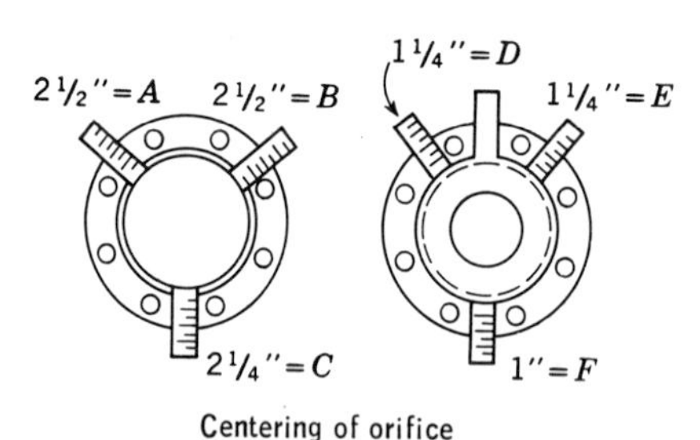

Fig. 12·34 To ensure accurate flow measurement when using the orifice plate, the plate must be centered and installed with the sharp edge on the upstream side. (*Courtesy of Louis Gess, Minneapolis-Honeywell Regulator Co.*)

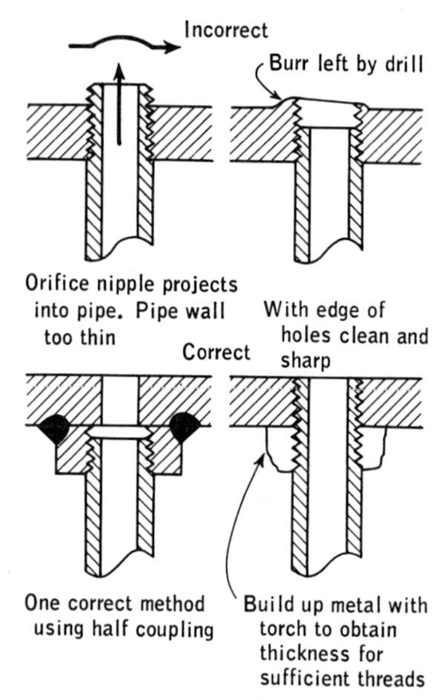

Fig. 12·35 Right and wrong ways to install pipe taps. Any irregularities in the pipe wall or the fittings which will alter the smooth flow of fluid will produce changes in the flow pattern. The type of changes and their effects are unpredictable and may change with the flow rate. (*Courtesy of Louis Gess, Minneapolis-Honeywell Regulator Co.*)

3) The maximum accuracy of measurement is desired when the flow involves highly viscous liquids.

4. The thin-plate concentric orifice can be used with clean, homogenous fluids (liquid, vapor, or gas) whose viscosity does not exceed 300 SSU (seconds Saybolt universal) at 15°C. In general, the Reynolds number should exceed 10,000.

a. The *upstream edge* of the orifice plate *must be sharp and square with the sharp edge always installed upstream.* If the beveled edge is installed on the upstream side, the orifice plate must be removed and reversed inasmuch as there are no suitable correction factors which can be applied. Moreover, the plate must be plane. Some common conditions and their results are shown in Fig. 12·34. Orifice plates are not suited to two-directional flow. Coefficients and operating characteristics which would permit use when the plate is installed incorrectly are not available. If the plate has a small hole to permit the escape of gases, the hole must be at the top of the pipe.

b. The orifice opening must be accurately centered in the pipe. Improper location can cause errors which may be in either direction;

the sense of the error depends upon how the plate is centered with respect to the taps.

c. If the pipe taps are installed with the pipe screwed so far in that it interferes with the smooth flow of fluid, or if burrs are left in the pipe, the flow pattern will not be correct and the accepted flow coefficients will be erroneous. Common errors and how to avoid them are shown in Figs. 12·35 and 12·36.

d. Whenever steam or a similar heated gas or vapor which might condense in the piping is to be metered for flow rate by using sealed chambers as illustrated in Fig. 12·37 (to prevent the flowing fluid from entering the meter), *there must be a continual, free interchange of gas and condensate.* The piping must be insulated to ensure that the steam reaches the reservoir. One-inch pipe is suggested for connecting the supply main with the metering section. This will provide for a free interchange of vapor and condensate.

5. The taps, regardless of the type used, must always be so positioned as to prevent dirt and sediment from clogging the lines and to keep the instrument sealed with fluid. Provision must also be made for permitting gases to return to the line. Steam and liquid measurements require taps in the side of the pipe. Gas taps should be made to the top of the pipe.

a. Vena contracta taps. In order to function properly, these taps are located one pipe diameter upstream and about one-half pipe diameter downstream. The exact position depends upon the value

Fig. 12·36 Burrs and icicles cause errors. Great care must be exercised to ensure that good practices as specified by such groups as the American Society of Mechanical Engineers are strictly adhered to. Long and oftentimes sad experiences have indicated what must be done to ensure good results. (*Courtesy of Louis Gess, Minneapolis-Honeywell Regulator Co.*)

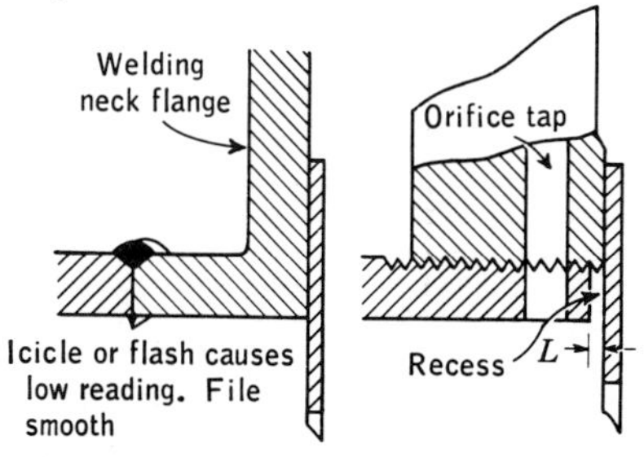

A.S.M.E. orifice coefficients can be used provided "L" does not exceed 1/4" and d/D ratios are less than 0.3 with a 2" line, 0.4 for a 3" and 4" lines and 0.5 for all lines greater than 4"

Fig. 12·37 If water collects in unequal amounts in either leg, the differential pressure at the meter will be incorrect.

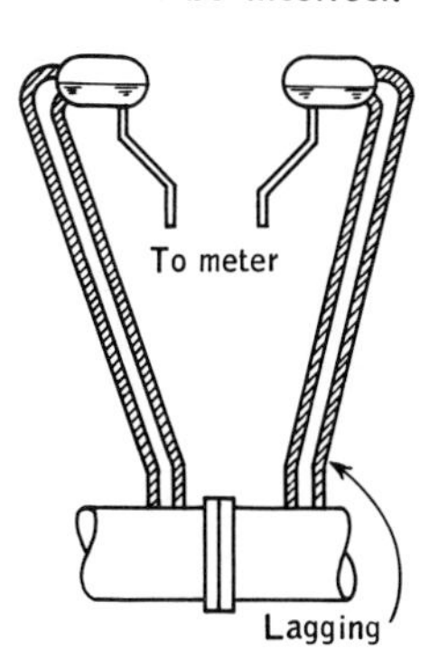

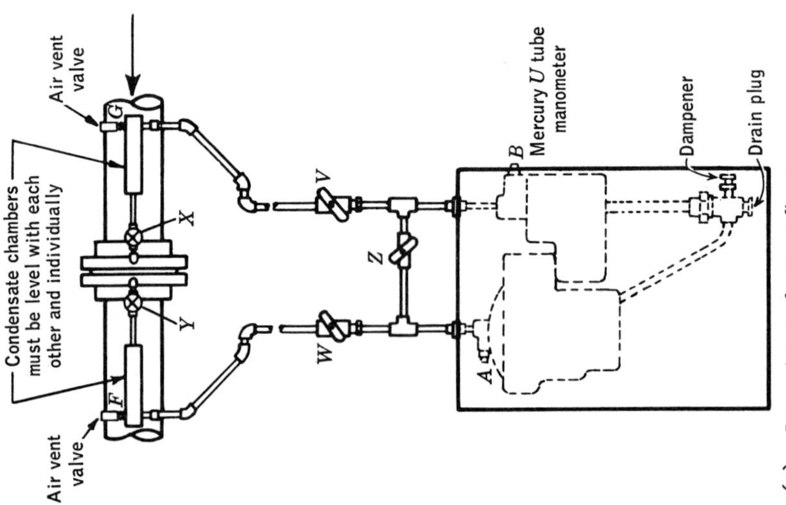

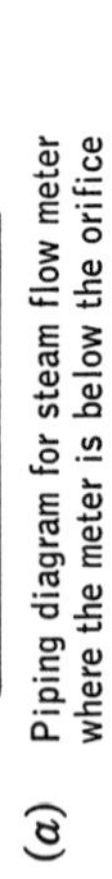
(*a*) Piping diagram for steam flow meter where the meter is below the orifice

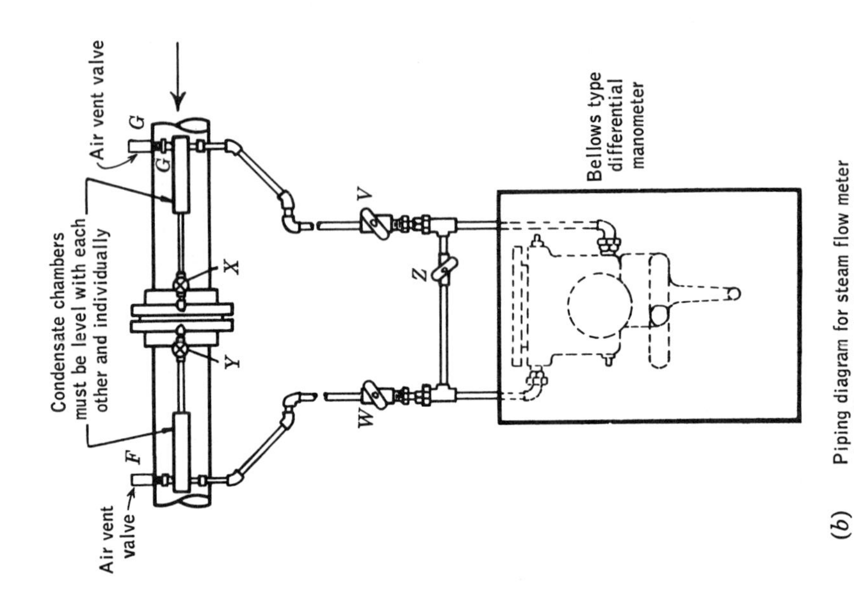

(*b*) Piping diagram for steam flow meter where the meter is below the orifice

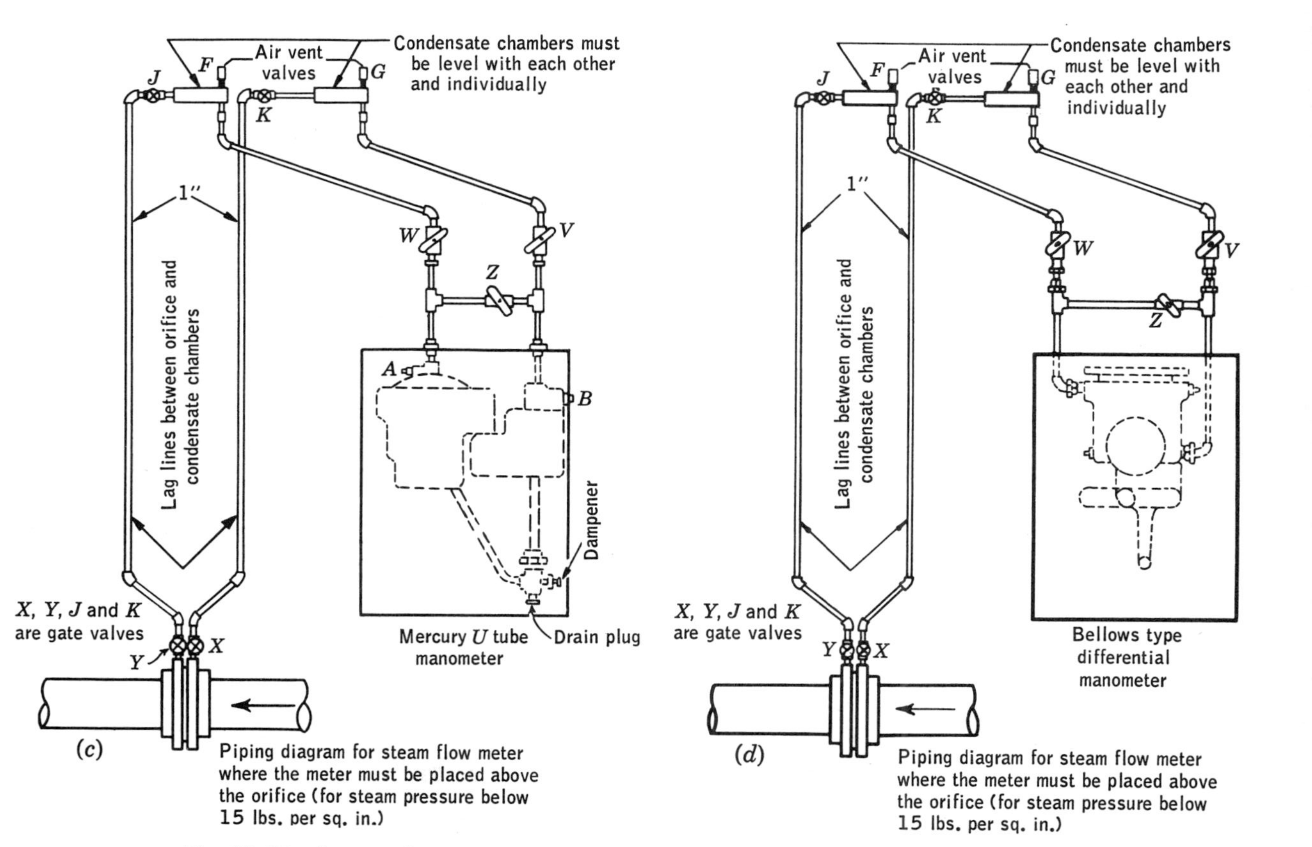

Fig. 12·38 Suggested piping connections for measuring steam flow. (*The Bristol* Co.)

of the orifice ratio, which may be between 0.1 and 0.8. Because of mechanical problems, these taps are not used with pipe smaller than 4 in. in diameter.

b. Flange taps. The correct position of these taps is 1 in. from both the upstream and the downstream faces of the plate. For pipes up to and including 3½ in., the permissible orifice ratio is 0.20 to 0.70. The larger sizes of pipe may employ ratios of 0.10 to 0.75.

c. Pipe taps. In order to obtain the maximum pressure drops, pipe taps are made 2½ pipe diameters upstream and 8 diameters downstream. The best results will be obtained when the orifice ratios are limited to 0.20 to 0.50 for pipe sizes of 2 in. and under and to ratios of 0.20 to 0.60 for pipes between 2 and 3 in.

6. When condensate traps are used, they must both be at the same height. Inequalities in level will cause significant errors. When the two traps are not at the same elevation, a false head will be developed. Install traps as shown in Fig. 12·38. A somewhat different type of error is created, as in Fig. 12·39, when the bent nipple fills with condensate. This may either add or subtract from the pressure differential, depending upon the direction of flow. To prevent such action, the nipple should be adequately insulated so that condensation will not occur in it. The performance is also aided by using 1-in. pipe, with ¾ in. as a minimum. Any such pipe should be well insulated. Should it be desirable to mount the meter and the condensate traps above the steam line, the lines should be 1 in. in diameter and be provided with adequate insulation.

7. *Flow nozzles.* A flow nozzle will handle about 60 per cent more fluid than will an orifice plate for any given pressure drop. This is true because the full area of the flow nozzle will be filled with liquid, whereas the orifice plate will have a constricted stream passing through the orifice. Aside from the increased capacity, the flow nozzle can handle liquids with greater amounts of suspended materials. Also, a shorter amount of straight pipe is required before the nozzle than before the orifice plate.

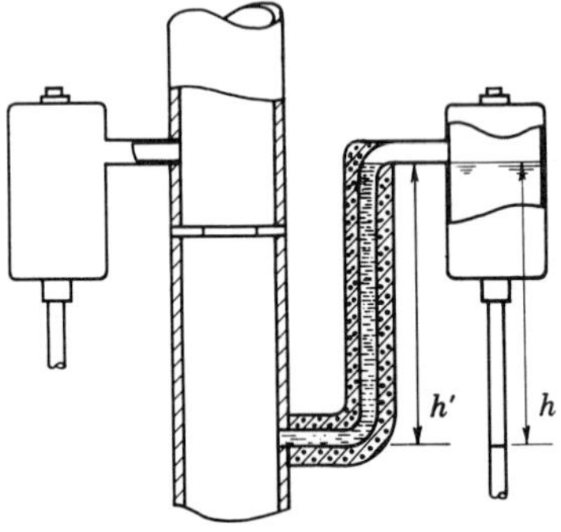

Fig. 12·39 Bent nipples can fill with condensate, causing error, on vertical steam lines. The direction of the flow determines whether the error is plus or minus. Adequate insulation and a pipe at least 1 in. ID should be employed to avoid these potential hazards. (*Courtesy of Louis Gess, Minneapolis-Honeywell Regulator Co.*)

FLOW APPLICATION NOTES

1. *Avoid measuring flow of any liquid when it is near the boiling point.* Even though the liquid is somewhat removed in temperature from its normal boiling point, the turbulence created as the liquid passes through the orifice may cause it to boil, and the work which is done on the liquid at this time also helps to induce boiling. The tendency toward boiling is increased by using a sharp-edged orifice. Venturi meters are less subject to these disturbances than are meters using orifice plates. For example, a liquid which boils at 120° under the usual conditions of pressure but which is being metered while it is at 70°F may well cause trouble by boiling.

SYMPTOMS. The chart of the recording meter indicates many random or noisy measurements when the flow is constant. The general complaint which characterizes this trouble is that the meter is unstable.

REMEDY. The situation can be improved by reducing the differential pressure to a minimum. This may be done by using a larger pipe.

2. *Inaccuracies may be caused by the misalignment of pipe.* A raised section that was improperly positioned when welded to the flanges and rough particles or edges on the inside of a weld may also cause inaccuracies.

SYMPTOMS. The inaccuracy increases as the flow increases.

REMEDY. Use extreme care when welding, fabricating, and assembling pipe which is to be used near the sensing elements of the flowmeter.

3. *Orifice plates, venturi sections, and so on should always be located upstream from any valve or source of a major pressure drop.* Valves may cause gas or vapor to separate from the liquid. Because the mixture is no longer homogeneous, readings are in error.

SYMPTOMS. Meter reads high and shows periodic erratic behavior.

REMEDY. Locate valves and so on downstream.

4. *Improper selection of the primary measuring element can cause improper metering performance.* Orifice plates and so on must be selected on the basis of the material to be measured and the amount of material in solution which may drop out, thereby changing the approach to the plate.

SYMPTOMS. For a known constant value of flow, the meter readings become smaller and smaller.

REMEDY. Remove material from in front of the plate or replace plate with a flow nozzle or venturi if this problem is serious.

5. *Improper arrangement of piping to differential pressure unit can create reading errors or even give an indication of negative flow of steam when the flow is known to be positive.* The indication of negative steam flow can be caused by an inadequate exchange of steam and condensate in the piping to the differential pressure unit. Correct operation of these

instruments requires that there be a constant flow of steam into both sections and of condensate out of both sections. The presence of a burr or dirt in the pipe can impede the flow of condensate back to the pipe. When this happens, the pipe becomes full of water and has the effect of creating a negative head which then might even indicate a negative flow.

REMEDY. (1) Examine carefully differential pressure piping and remove obstruction. (2) Use nothing but gate valves in piping. (3) Insulate pipe to condensate traps; use ¾- to 1-in. pipe to permit free flow of condensate.

6. *Steam flowmeter gives negative readings for positive steam flow.* The preceding operating difficulty has the same symptoms as this one, but the cause and cure are quite different. During a start-up operation, boilers sometimes discharge water into the steam lines. Water also may be present in the lines because of condensation. As this water comes down the pipe, it enters and clogs the differential meter piping.

SYMPTOMS. If a metal object such as a rod or screwdriver is held against the condensate chambers while the ear is held to the other end, and a slurp or gurgle is heard, the indication is that the chamber is full of water.

REMEDY. Install needle valves on the top of each condensate chamber and bleed until live steam comes through whenever this trouble occurs.

7. *Differences in the specific volumes of steam at various measuring points will give values that do not add up to the volume generated at the boiler.*

SYMPTOMS. The sum of the recordings of the various flowmeters do not add up to the flow indicated by the master meter. Errors may be as much as 50 per cent.

REMEDY. If measurements are to be accurate, they should be made only at the point of use. Temperature measurements should also be made at the instrument so that the proper factors may be applied. The assumed temperature may be grossly incorrect. The condensate which may have been accumulated at various points along the line should always be considered in determining the performance of the meters.

8. *Erroneous density values of a gas can cause serious errors.*

SYMPTOMS. When gases of low molecular weight are being measured, the results are erroneous.

REMEDY. The presence of water vapor is generally neglected. A measure of the importance of this term is the fact that dry hydrogen gas has a density of 0.0695, while saturated hydrogen has a density of 0.112.

9. *Inadequate analysis data result in incorrect flow measurements.*

SYMPTOMS. Meter readings are thought to be inaccurate.

REMEDY. The density of the gas being measured must include all the components, not just one or two of them. To the degree that the presence of some of the gases is neglected, the final answers will be in error. To assume a gas composition which is not true can only lead to incorrect values.

10. *Incorrect pressure taps impair performance of flowmeters.*

SYMPTOMS. The meter has momentary periods of insensitivity.

REMEDY. When there is a possibility of the liquid flashing, pipe taps should be used. Their use will accelerate the dissolving again of the gases and vapors which were previously liberated. When the fluids have changing viscosities, flange taps should be used. When gases carry moisture, drip pots should be used to remove the liquid.

11. *Obstructions of any type upstream of the orifice plate may create errors.*

SYMPTOMS. A meter develops an error which increases with the flow rate.

REMEDY. *No obstruction of any type should ever be installed closer than the limits prescribed by the A.G.A., the Transactions of the ASME, July, 1945 (Sprenkle, vol. 67, no. 5), or by the manufacturer of the equipment.* Even though the obstruction might be as small as a thermometer well, it will disturb the flow sufficiently that the error will become appreciable. There should be no exceptions to the rule stated above.

12. *The errors of the gas flowmeter are a function of flow rate.*

SYMPTOMS. The flow rate governs the error in certain gas measurements.

REMEDY. *The valves governing flow rate must always be downstream from the orifice or sensing device.* This tends to keep the pressure at the orifice more nearly constant, and when there may be a vigorous reaction between the chemicals involved, the downstream location of the valve tends to isolate the measuring section from the ensuing pressure drops.

13. *Improved flow valve performance.*

REMEDY. *Always use a valve precisor or positioner* with a flow valve to eliminate dead time and friction. The benefits exceed the disadvantages.

14. *Selection of the correct valve size.*

SYMPTOMS. No controller setting provides stable operation.

REMEDY. It is quite probable that the wrong size of valve has been selected. Proper flow controller operation requires that the valve just be wide open for the maximum flow. *Never install a larger valve to anticipate future needs.* The valve must never be on the "over" side. Experience often indicates that only a bevel-disk valve should be used. Constant-percentage valves, or similar contour valves, are not recommended for flow control by some manufacturers, but are specified by others. There is a marked difference of opinion.

EXCEPTIONS. When flow control is achieved by the use of a pump bypass, the pump characteristics enter the picture and frequently require a valve other than the bevel-disk type.

15. *Flowmeter ranges.* The usual range for differential pressure flowmeters is 30 to 100 per cent of rated flow. Low sensitivity of instruments is conducive to errors. If the differential pressure is too small to ensure

Fig. 12·40 Standards prescribed by the American Gas Association, the American Society of Mechanical Engineers, the Instrument Society of America, and the manufacturers of equipment should be observed strictly. (*American Society of Mechanical Engineers.*)

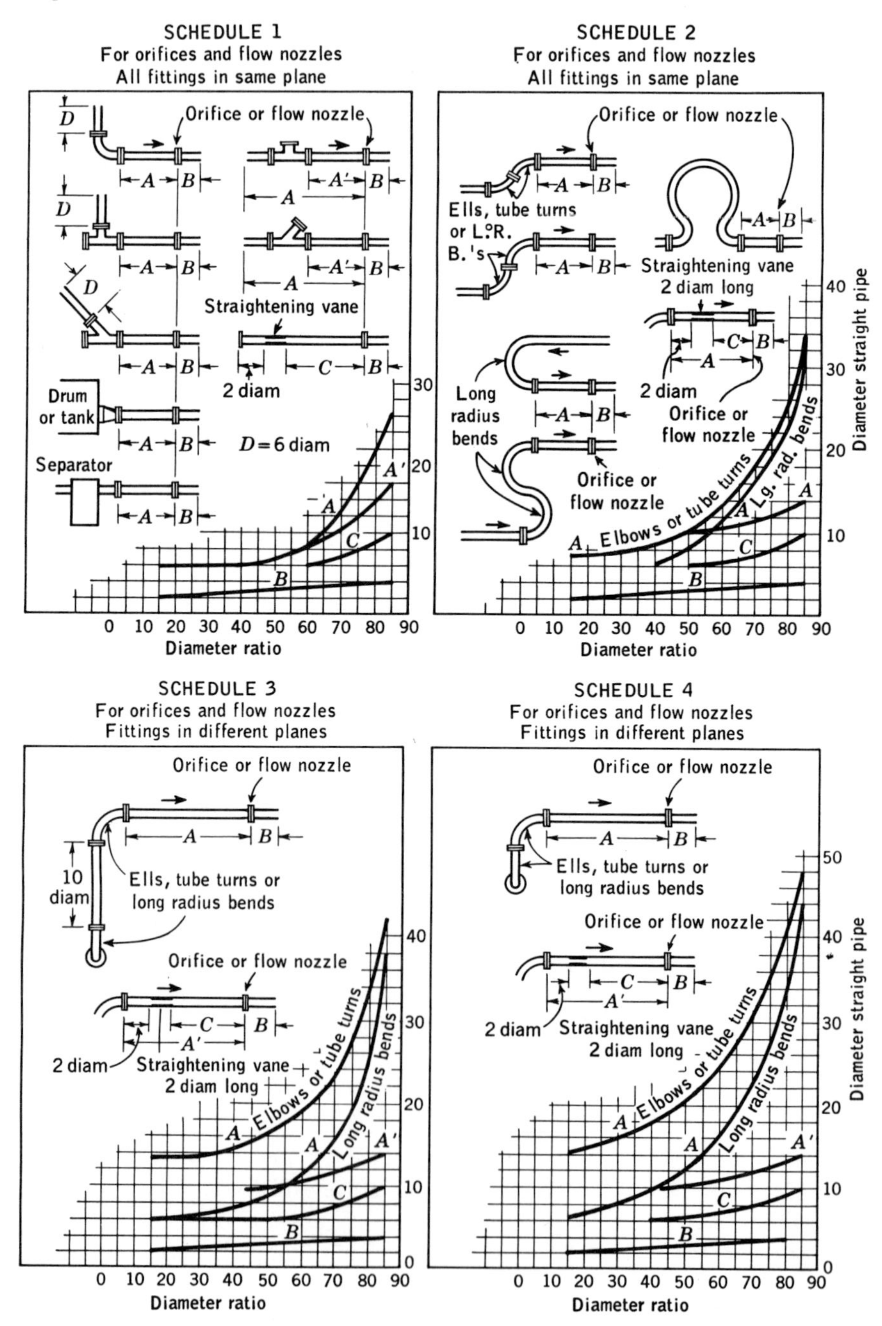

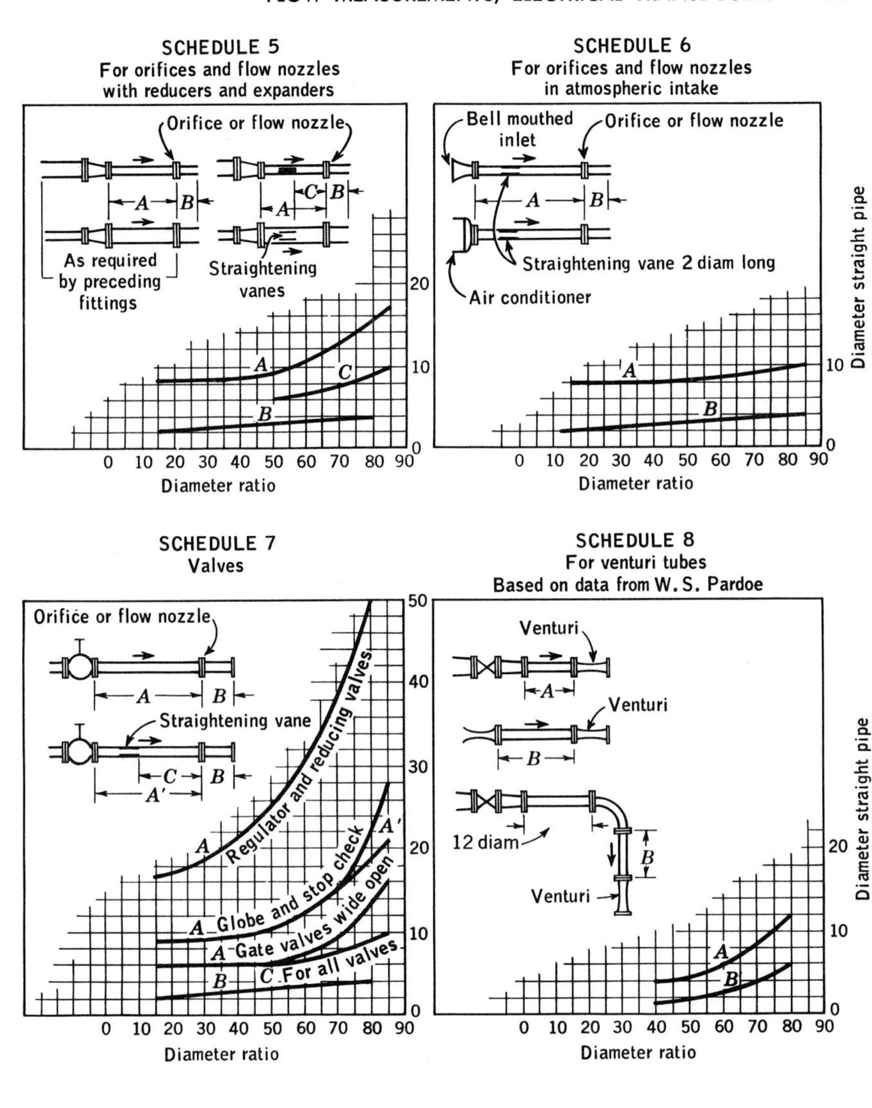

Fittings Allowed on outlet side in place of straight pipe	0 - 50 ratio	50 - 60 ratio	60 - 70 ratio	70 - 80 ratio
	1. Tees 2. 45° ells 3. Gate valves 4. Separators 5. Y-fittings 6. Expansion jts	1. Tees 2. Expansion jts 3. Gate valves 4. Y-fittings 5. Separator (if inlet neck is 1 diam long)	1. Gate valves 2. Y-fittings 3. Separator (if inlet neck is 1 diam long)	1. Gate valve 2. Long radius bend

good performance, alter the primary element or change the instrument in order to bring the readings upscale. Rotameters can measure down to 10 per cent of their maximum flow and retain good accuracy. Turbine meters can do likewise.

16. *Some installation inaccuracies are caused by the use of pipe sizes which are assumed.* Measure metering-pipe internal diameters before assembly; they may be quite different from the nominal or assumed values.

17. *Measuring pulsating flow.* If possible, pulsating flow should not be measured because of the errors which are likely to arise. However, if there is no way to avoid this type of measurement, the bellows-type of differential unit with hydraulic damping furnishes uniform damping. To minimize differences in the flow coefficients in the damping section, one solution is to use a long tapered needle valve such as a Hoke RB281 or Ideal No. 501. Mercury manometers are not generally recommended, because they are too slow in their response. Feedback types of differential pressure or force-balance units are not recommended, because feedback pressures can easily be out of phase with the measured variable.

18. *Selection on the basis of differential pressure.* In the petroleum field, the differential pressures sought are generally 50 to 100 in. of water.

19. *The pulp and paper industries require large orifices because of the large amount of suspended material.*

SUMMARY

1. Flow measurements are among the most important made by the process industries, because cost and selling price often depend on the cost of the materials. Accurate costs are as necessary when apportioning charges as when purchasing materials.

2. With the exception of the direct-volume measurements, flow measurements are all of the inferential type. Almost all properties, qualities, or conditions which are the result of flow conditions are, or have been, used to indicate the flow rate.

3. The measurements divide themselves into two general classes: those which are strictly mechanical in their design and operation and those which use electrical means to obtain an indication. There is also an in-between group which might be said to employ both methods.

4. The positive-displacement pumps provide some of the few examples of a direct measurement; the vast majority of measurements are of the inferred type.

5. The factors associated with measurement of flow through a pipe are the following: (*a*) A reduction in static pressure when the speed is increased. (*b*) A force which is proportional to the acceleration given the fluid. (*c*) An emf which is directly proportional to the rate of flow. (*d*) Buoyant or lifting forces which are proportional to the flow rate and inversely proportional to the area of the flowing liquid. (*e*) The cooling effect created by causing the flowing fluid to carry away heat. The heat required to maintain a fixed temperature becomes a measure of the speed. (*f*) The speed of rotation of a turbine which is driven by the flow.

6. All existing flowmeters depend on one or more of the factors listed in (5).

7. Meters which deliberately insert an obstruction to create a reduction in the

static pressure are commonly called head meters. Most of them are quite sensitive to flow conditions upstream of the meter. There are some notable exceptions.

8. The type of meter selected will be governed by the operating conditions: the kind of fluid, materials in suspension, electrical resistivity, boiling point, piping conditions, allowable pressure drop, and the volume of flow.

9. Many mechanical designs of meters are available. Inasmuch as all perform the same functions, accuracy, cost, and life will be important factors. The pressure drop within the device may also be significant.

10. Weirs are commonly used for the flow-rate determination of liquids in open channels. The geometry of the weir is critical and must be carefully observed.

11. Weir measurements are much less accurate than most other flow measurements. Weighing the liquid passed by a weir is the most accurate check on weir performance.

12. (*a*) The venturi tube provides the smoothest flow pattern and therefore has the smallest energy loss. The flowing liquid completely fills the opening. The venturi tube is the most costly of the head-type primary units. (*b*) The flow nozzle approaches the venturi tube in performance; the liquid fills about 98 per cent of the throat area. (*c*) The contraction of the flow stream as it passes through an orifice plate reduces the net effective area to about 60 per cent of the nominal value. The orifice plate costs the least of the head-type units, is the simplest to construct, and has the greatest pressure loss.

13. The Reynolds number R_D indicates the flow rate, for a given size of pipe and viscosity of fluid, at which the ratio of average speed to maximum speed through the pipe approaches a constant. For lower values of R_D the ratio changes so rapidly as to be of little use. The higher the Reynolds number the more stable the flow coefficients. The type of flow in which the Reynolds number is small is called viscous or laminar. At the higher ratios of flow at which the flow coefficients stabilize, the flow is termed turbulent.

$$\text{Reynolds number} = \frac{K(\text{pounds of material per unit of time})}{(\text{diameter of pipe})(\text{absolute viscosity})}$$

Because turbulent flow will be obtained with certainty at high values of R_D, flow calculations based on this assumption should be accurate. If (1) the flow rate is so low, (2) the pipe diameter so large, or (3) the viscosity so high that the value of R_D is low enough to indicate that turbulent flow will not be attained, then corrections must be made, because the true ratio of average speed to maximum speed will be less than that anticipated for turbulent flow.

14. The magnetic flowmeter creates no pressure drop for metering purposes. It operates by employing the moving fluid as a conductor in a magnetic field.

$$\text{The value of potential} = K(\text{speed})(\text{pipe diam})(\text{strength of magnetic field})$$

15. Turbine-type meters may employ mechanical or electrical read-out systems. The rate of rotation may be measured by a potential or by the frequency, which is directly proportional to speed.

16. The target meter employs the impact force of the moving fluid to bend an elastic member or control the output of a nozzle bellows combination. It can operate where very severe operating conditions restrict the use of many differential pressure meters.

17. Area meters have their greatest use when measuring hot, tarry materials.

18. Flowmeters of the rotameter class have operating spans of about 10:1, while

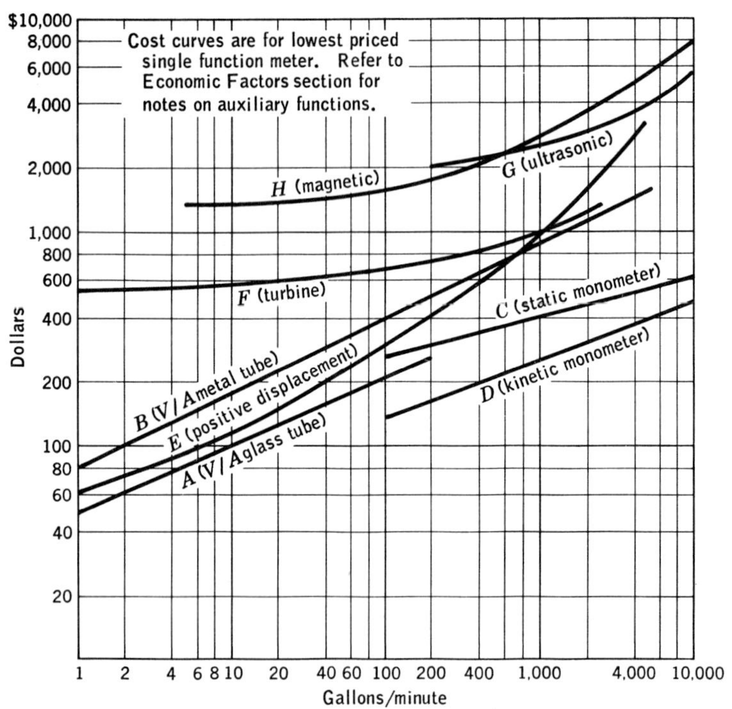

Fig. 12·41 For quick summaries or comparisons, this figure provides most of the pertinent information. (*The Fischer and Porter Company.*)

differential pressure meters generally are limited to about a 3:1 span when using round-chart instruments. Rotameters are relatively independent of upstream piping.

19. The measurement of flow of dry materials employs many of the basic factors involved in other measurements: strain, forces of acceleration, and torque. The rate and total flow can be determined either electrically or mechanically.

20. Table 12·1 and Fig. 12·41 show a comparison of the various types of flow-measuring instruments and some curves for cost approximations.

21. Because many measurements are influenced by it, the piping should conform to values indicated in Fig. 12·40.

22. A cost comparison will be found in Fig. 12·41.

QUESTIONS

12·1 Why should the resistivity of the liquid have any effect upon the operation of a magnetic flowmeter?

12·2 What are the factors which govern the magnitude of the output signal of the magnetic flowmeter?

12·3 What principle is employed when an elbow is used for the primary flow-measuring element? Is the same principle used in any of the other flowmeters which you have studied?

12·4 Discuss the chief differences among the meters classed as head meters.

12·5 Why are head meters so called?

12·6 What are the characteristics of positive-displacement meters? Are there any limitations on their use?

12·7 Why are differential pressure measurements used rather than the difference of two pressure measurements made at the same point? Are these practical reasons or theoretical ones?

12·8 If you had to make an approximate determination of flow, what coefficients would you select for an orifice plate, a flow nozzle, and a venturi tube?

12·9 What practical considerations would limit the use of a glass **U** tube for differential pressure measurements?

12·10 If a differential pressure meter were connected to pipe taps instead of vena contracta taps, what difference would it make?

12·11 What are the factors which determine which type of taps are to be used in making a pressure measurement?

12·12 Could a pressure transmitter be substituted satisfactorily for a differential pressure transmitter?

12·13 Tabulate the advantages and disadvantages of each of the major classes of flowmeters.

12·14 What are some of the factors which might cause a flowmeter to indicate negative flow when it is known that the flow is positive?

12·15 What precautions must be taken when liquids are metered close to their flash point?

12·16 What precautions should be taken when measuring the flow of steam? Of gases?

12·17 Is there any limitation as to angle with the horizontal for pipes containing metering elements? In other words, may the pipe be vertical, horizontal, or anywhere in between?

12·18 Discuss the relative advantages and disadvantages of mechanical and electrical read-out devices.

12·19 What is required to totalize flow? Discuss some of the mechanisms.

12·20 What might be reasons for the elbow being used only to a limited degree for flow-measuring purposes?

12·21 What are the relative advantages of the impact type of flowmeter which employs strain gages? Differential pressure transmitter methods? Magnetic torque-balance methods?

12·22 What is the significance of the terms in the expression for hydraulic drag?

12·23 How do the terms for the flow in a head meter compare with those for the drag meter?

12·24 What is the significance of the Reynolds number?

12·25 Define the Reynolds number in terms of rate of flow, pipe diameter, and liquid viscosity.

12·26 What is the difference between mass and volume meters?

12·27 What factors affect mass meters but do not affect volume meters?

12·28 Classify the various types of meters studied as mass or volume meters.

12·29 Why do some meters require specified lengths of straight pipe ahead of the measuring element?

12·30 Which meters need straight pipe and which are almost independent of it?

12·31 Tabulate the lengths of straight pipe required for five common metering situations.

12·32 What differential pressure ranges are generally preferred? Why should there be any preference?

12·33 What factors make a rotameter preferable for certain flow measurements?

12·34 For what conditions are variable-area meters often preferred?

12·35 If an orifice plate is installed with the sharp edge downstream, what can you say about the reliability of the flow readings?

Table 12·1 Flowmeter

Meter type	Fluid Property Limitations			
	Type of fluid	Pressure limits	Temp limits	Suitability for slurries, etc.
Variable-area float type, glass tube	Suitable for most fluids except alkalis and molten metals	Varies with sizes: ½ in. 300 psi; 1 in. 200 psi; 1½ in. 100 psi; 2 in. and larger 50 psi	−50°F to +400°F	Not recommended
Variable-area float type, metal tube	Suitable for all fluids	150 to 2,500 lb ASA	−300°F to +1600°F	Recommended
Variable-head; static manometer	Suitable for most fluids	To 1,500 psi	−60°F to +400°F	Requires purges of pressure-tap connections
Variable-head; kinetic manometer	Suitable for most fluids	Glass tube to 200 psi; metal tube to 1,500 psi	Glass tube: −50°F to +400°F Metal tube: −300°F to +1600°F	Requires purges of pressure-tap connections
Integrating or volume; positive-displacement type	Suitable for all clean or filtered fluids	150 to 300 lb ASA standard	0°F to +300°F	Not recommended
Integrating or volume; turbine type	Suitable for all clean or filtered liquids	Up to 3,000 psi	−300°F to +700°F	Not recommended
Obstructionless; ultrasonic type	Suitable for most liquids	Up to 1,500 psi	−300°F to +600°F	Recommended
Obstructionless; magnetic type	Suitable for most liquids that are electrical conductors	Up to 600 psi	−200°F to +360°F	Recommended

Selection Guide *

Application Factors				
Range of maximum flows	Individual meter range	Accuracy	Type of scale	Construction materials available
½ cc/min to 200 gpm	10:1	±2% of full scale uncalibrated; ±1% of full scale calibrated	Linear	All metals, plastics, and ceramics (floats and fittings)
0.1 to 5,000 gpm	10:1	±2% of full scale uncalibrated; ±1% of full scale calibrated	Linear	All metals
100 gpm and up	4:1 useful range	±1% of differential measured—additive to accuracy of primary restriction	Square root	Steel and stainless steel (purges or seal pots used on corrosive services)
100 gpm and up	Varies with pressure loss—3:1 to 10:1 available	±1% of differential measured—additive to accuracy of primary restriction	Linear	All metals, plastics, and ceramics
1 to 5,000 gpm	Up to 20:1	0.1 to 0.5% of total volume	Linear	Bronze, steel, and stainless steel
0.25 to 2,500 gpm	Up to 15:1	±0.5% of rate	Linear	Most metals
200 gpm and higher	20:1	±2% of full scale	Linear	Bronze, steel, and stainless steel
5 to 5,000 gpm	20:1	±1% of full scale	Linear	Kel-F- or neoprene-lined stainless steel

* Reproduced by permission of Fischer & Porter Company.

Table 12·1

Meter type	Installation Factors				
	Line sizes available	Normal connections	Special piping considerations	Power requirements for basic measurement	Installation limits
Variable-area float type, glass tube	¼ to 4 in.	Screwed or flanged	None	None	Meter body must be installed in vertical plane
Variable-area float type, metal tube	½ to 12 in.	Flanged	None	None	Meter body must be installed in vertical plane
Variable-head; static manometer	2 in. and up	Screwed or flanged	Straight pipe run required for approx 10 pipe diam upstream, 3 downstream	None	Horizontal or vertical
Variable-head; kinetic manometer	2 in. and up	Screwed or flanged	Straight pipe run required for approx 10 pipe diam upstream, 3 downstream	None	Horizontal or vertical
Integrating or volume; positive-displacement type	½ to 12 in.	Screwed or flanged	None	None	Horizontal recommended
Integrating or volume; turbine type	½ to 6 in.	Screwed or flanged	None	Power supply needed (115 volts, 60 cycles, 350 watts)	Operates in any position
Obstructionless; ultrasonic type	2 in. and up	Flanged	Straight pipe run required for approx 10 pipe diam upstream	Power supply needed (115 volts, 60 cycles, 100 watts)	Horizontal or vertical
Obstructionless; magnetic type	1 to 12 in.	Flanged	None	Power supply needed (115 volts, 60 cycles, 200 watts)	Horizontal or vertical

(Continued)

Economic Factors		
Life expectancy	Pressure loss	Maintenance and servicing
2 to 10 years	Available from ¼ to 30 in. of water; constant throughout meter range	Cleaning of metering tube every 6 months
2 to 10 years	Available from 1 to 200 in. of water; constant throughout meter range	Cleaning of metering tube yearly
2 to 10 years	Available from 1 to 800 in. of water; varies as square of flow change	Check every 6 months
2 to 10 years	Available from 1 to 800 in. of water; varies as square of flow change	Check every 6 months
1 to 3 years	Up to 7 psi; varies with flow	Check of accuracy to determine effects of wear every 3 months
1 to 3 years	Up to 7 psi; varies with flow	Cleaning every 3 months
1 to 5 years	None	Service of electronic equipment every 3 months
1 to 5 years	None	Service of electronic equipment every 3 months

12·36 If the meter fittings were screwed into the flanges so that they extended ⅛ in. into the pipe, would this have any effect upon the meter performance?

12·37 What particular attention must be paid to steam lines going to condensate traps?

12·38 Explain why the position of the metering element should be above the pipe in one case and below it in another when measuring liquids and gases. What are the governing factors?

PROBLEMS

12·1 For flows of 400, 600, 800, and 1,000 gpm, compare in tabular form the differential pressure drops using Figs. 11·4, 11·39, and 11·40. Assume d/D ratios of 0.7 and 10-in. pipes. (These are not strictly comparable situations and should not be regarded as comparisons under equal flow conditions. They are only exercises in using the graphs.)

12·2 Find the ratios of the pressure drops, Figs. 11·4, 11·39, and 11·40, for flow rates of 200, 400, and 600 gpm. Are these differential pressures proportional to the square of the speed, which would also be proportional to the square of the flow rate?

12·3 The piston and its magnet support and guide weigh 1 lb. If the area of the piston is 4 in.^{2}, what must be the difference of P_H and P_L, Fig. 12·15? Assume a flow rate of 10 gpm.

12·4 If the flow rate of Prob. 12·3 is doubled, what will be the difference between P_H and P_L?

12·5 If the flow rate of 10 gpm uncovers but one port and moves the piston up from its at-rest position 1 in., what should be its position and how many ports should be opened for flows of 20 gpm? 30 gpm? Assume that the ports are spaced one per inch of vertical travel.

12·6 If the maximum height of water flowing over a weir is 2 ft, what is the minimum distance from the weir at which the height of the water above the notch should be measured? Why has this value been assumed? What is wrong with measuring the height immediately upstream from the weir?

12·7 What are the values of the quantities under the radical of the flow equation if the height is 2 ft for the triangular and rectangular weirs?

12·8 Determine the flow from a 90° V notch when the head is 15 in., 10 in. Determine the cubic feet per second for each.

12·9 Plot the ratios of the flow, in gallons per minute, for 5-ft, 3-ft, and 1-ft rectangular weirs for heads of 2, 5, and 10 in. of water. Do you think that you could interpolate to find the flow rates for 2- and 4-ft rectangular weirs?

12·10 Determine the inside diameter of the pipe tap to be used with a flow nozzle when used on a 12-in. pipe, a 6-in. pipe, a 3-in. pipe, and a 1-in. pipe.

12·11 The flow rate through the center of a pipe is 50 fpm. What will the probable rate be 3 in. from the center of a 6-in. pipe? A 12-in. pipe? A 10-in. pipe? Assume turbulent flow. If the fluid is water, find the Reynolds number and determine the value of average speed divided by maximum speed, Fig. 11·27.

12·12 If the average rate of flow is obtained about 70 per cent of the way from the center of the pipe toward the inside face of the pipe, at what points should the measurement be made in the pipes of Prob. 12·11?

12·13 What is the average percentage of flow in terms of the maximum if there is a "normal" speed variation across the pipe?

12·14 (*a*) What is the value of the Reynolds number at which the flow of water stabilizes so that the ratio of the average speed to the maximum speed is the same as that stated in Prob. 12·12? (HINT: Use Fig. 11·28.) (*b*) Using Figs. 11·28 to 11·33.

what are the values of Reynolds numbers within which no correction of more than +1 per cent would have to be made for a 2-in., 6-in., 10-in. pipe using D and $\frac{1}{2}D$ taps? Specify d/D. (*c*) Repeat, but use 14-in. flange taps. Specify d/D. (*d*) Repeat, but use 4-in. pipe and corner taps, 14-in. pipe and corner taps.

12·15 Compute the Reynolds number for water at 100°F when the flow is 1,000 lb/min and the pipe diameter is 10 in. Obtain the viscosity from Fig. 11·43. Also check your answer against the nomograph, Fig. 11·43.

12·16 Repeat Prob. 12·15 but use kerosene at 200°F and a 6-in. pipe.

12·17 Repeat Prob. 12·15 but use heavy lube oil at 100 and 400°F. Take pipe sizes of 10 in.

12·18 Repeat Prob. 12·15, but use gas oil at 400°F, with a flow rate of 10,000 lb/hr in a 10-in. pipe.

12·19 If a conductor 10 in. long were moving at the rate of 10 fps in a magnetic field whose strength was 2,000 gauss, what potential would be induced? Relate these values to a magnetic flowmeter.

12·20 Using Fig. 12·8, if the difference of the mercury columns is 10 in. and each column has a cross-sectional area of ½ in.^{2}, what force must the lever arm exert on the flyball system in order to maintain equilibrium? Assume that the distance from the knife edge to the cord is one-half of the distance from the knife edge to the mercury columns.

12·21 Plot frequency versus speed of rotation for a turbine meter having a two-pole rotor. Repeat, but use a four-pole rotor.

12·22 Using Fig. 12·14, what will be the output potential signal when the flow rate is 20 gpm and the bridge excitation is 1 volt? 2 volts? 5 volts?

12·23 For a flow rate of 40 gpm, what will the output signal be when the bridge excitation is 2 volts? (Use Fig. 12·14.)

12·24 Design a span potentiometer for use with the Taylor differential pressure transmitter, Fig. 12·22. The potentiometer should take a negligible current from the differential transformer and should be in series with such a resistor that only 50 mv will be the maximum potentiometer output when the transformer has its maximum value of 200 mv.

12·25 A 6-in. line carrying water has an orifice plate with a d/D ratio of 0.60. What will be the flow when the pressure differential is 100 in. of water? What will be the Reynolds number for these conditions? Will a correction be necessary? Assume D and $\frac{1}{2}D$ taps.

12·26 A 10-in. steam line carrying saturated steam at 400 psia will carry how many pounds per hour when the differential pressure is 100 in. of water and (*a*) a venturi tube is used with d/D equal to 0.50, (*b*) a flow nozzle is used with a d/D equal to 0.50, and (*c*) an orifice plate with a d/D ratio of 0.50?

12·27 Find the value of the Reynolds number for each of the above conditions in Prob. 12·26. Are corrections required? Use Fig. 11·42.

12·28 What d/D ratio should be used in an 8-in. pipe when the fluid is kerosene and the flow is 10,000 lb/hr? The temperature of the fluid is 90°F and the pressure drop is 100 in. of water with an orifice plate.

12·29 For a flow rate of 1,000,000 lb/hr, a differential pressure drop of 200 in. of water, a d/D ratio of 0.70, and a light lubricating oil as the fluid at 200°F, what size of pipe should be selected if (*a*) a venturi tube is used, (*b*) a flow nozzle is used, and (*c*) an orifice plate is used? Specific gravity is 0.85.

13 Liquid-level Measurements

MECHANICAL TRANSDUCERS

The measurement of liquid levels is of great importance in the industrial field, particularly when control may also be a factor. Whether the process be continuous, semicontinuous, or of the batch type, it is necessary to know the level. The level of the materials may affect both the pressure and the rate of flow in and out of a container; hence, the quality may be affected.

Many methods are used by industry to determine level. In general, they fall into the following classifications:

1. Level determination by measuring hydrostatic or pneumatic pressure
2. Level determination by measuring differential pressures
3. Level determination by measuring the movement of a float
4. Level determination by electric strain-gage measurements (see Chap. 10)
5. Level determination by radiation absorption (see Chap. 19)
6. Level determination by heat transfer (see Chap. 23)

The division may also be made on the basis of direct and inferential methods:

Direct Methods

Sight glass (not used in some places because of difficulty in providing automatic reading of level)

Cable and float

Conductivity—for single-position indication, unless more than one unit is used

Heat transfer—for single-position indication unless more than one unit is used

Inferential Methods

Hydrostatic pressure

Float and hydraulic pressure

Pneumatic pressure
Float and pointer (position transmitted through sealed vessel)
Differential pressure gage
Strain gage
Capacitance gage, electrical
Radiation absorption gage (see Radiation Gages, Chap. 19)

In terms of simplicity of construction, the sequence probably would be:

1. Sight glass
2. Cable and float
3. Pressure gage to measure hydrostatic head (open tank or without pressure)
4. Pneumatic pressure (tank without pressure)
5. Float and hydraulic pressure
6. Hydraulic head (closed tank under pressure)
7. Differential pressure (closed tank under pressure)
8. Conductivity, electrical
9. Conductivity, heat
10. Strain gages, electrical
11. Capacitance gages, electrical
12. Radiation absorption gages

With such a wide variety of ways of measuring liquid levels, the selection of the proper device requires the consideration of many factors if the most suitable one is to be selected. The first portion of this chapter is devoted to explaining the theory of the first seven types, the suitability of each type, and the applications where it is superior. The concluding section deals with the five electrical methods.

Another classification might also be made: liquid level measurements fall into the three general categories (1) position measurements, (2) volume determination, and (3) weight determination.

Position measurements. This type of measurement is satisfied when it has been determined that the level is at or below some fixed point. This is essentially a volume measurement without regard for any changes in density or other characteristics because of pressure or temperature. Methods 8, 9, and 12 are of this general classification. Method 11 may be a position measurement, but it may also be a height or volume indication.

Volume determination. If the purpose of the level measurement is to determine the volume of liquid contained in a vessel of uniform cross-sectional area, then a direct measurement of the level height is desired because:

$$V = AH$$

where V = volume of liquid in the vessel or container
A = cross-sectional area of the container
H = height of the liquid, measured in the same units as A and V

In this expression, density and pressure effects are ignored. Should it be desired to include hydrostatic pressure density effects, then

$$V = \frac{AP}{D}$$

where V = volume at the given level
D = weight density of the liquid for the existing pressure conditions
P = pressure caused by hydrostatic head

Weight determination. If the weight of the liquid or material is determined without regard for temperature-density effects, then this is really a determination not of a liquid level, but some equivalent level. The weight or force on the bottom of the container, or on the supporting members, is equal to

$$W = AP$$

where W = weight of the material
A = cross-sectional area
P = pressure caused by the hydrostatic head

When the level must be inferred from the weight,

$$W = AHD$$

where D is the weight density of the material, in pounds per cubic foot, pounds per cubic inch, or other basic units that are the same as those used for measuring the other quantities.

Gage glass. One of the earliest methods used for the continuous indication of liquid level within a tank or vessel employed an open-end manometer, Fig. 13·1, in which the weight of a column of liquid inside

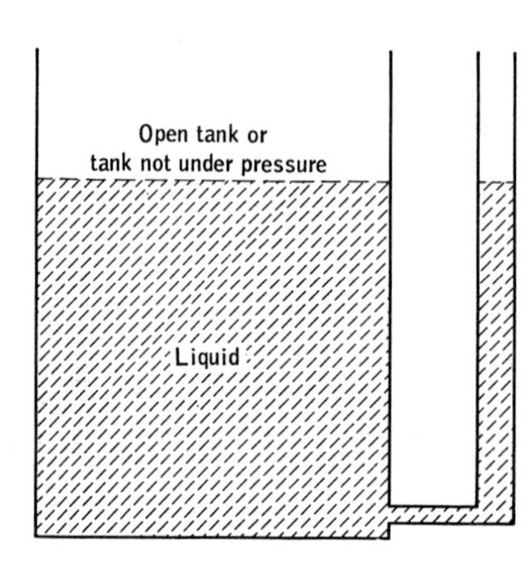

Fig. 13·1 Pressure balance is obtained at any points in the fluid which are at equal distances above or below some reference. If the liquids are subjected to the same external pressure and have the same specific gravity, the level will be the same in both vessels.

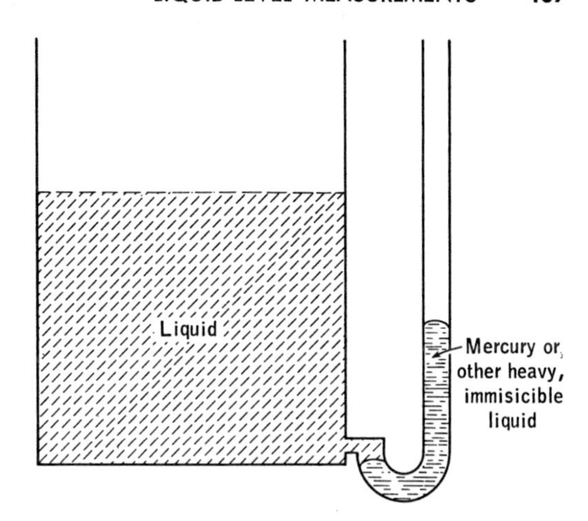

Fig. 13·2 A liquid having a specific gravity greater than that of the liquid in the tank can create a condition of pressure balance with a column which does not rise to the tank level. The tank level must be inferred from a calibration.

the tank was balanced by the weight of a column of the same liquid in the manometer tube. Such a sight tube was not necessarily restricted to the use of the same liquid but could employ any desired liquid by constructing the manometer as shown in Fig. 13·2. By using heavy liquids, such as mercury, the height of the indicating column could be only a small fraction of the height of the liquid in the tank.

When it is desired to measure the liquid level with the liquid under pressure, the manometer is modified as shown in Figs. 13·3 and 13·4. Commercially available designs are shown in Figs. 13·5 and 13·6. Great attention must be given to adequate strength when the pressures may reach as high as 5,000 psi and temperatures may attain 1000 to 1200°F in service. Such temperature and pressure limits probably will not be found in the same installation. The design shown in Fig. 13·6 utilizes special optical properties to create a sharp line of demarcation of the liquid level. This greatly simplifies reading the gage, particularly when one is interested only in a general level, as in boiler operation, rather than an exact reading.

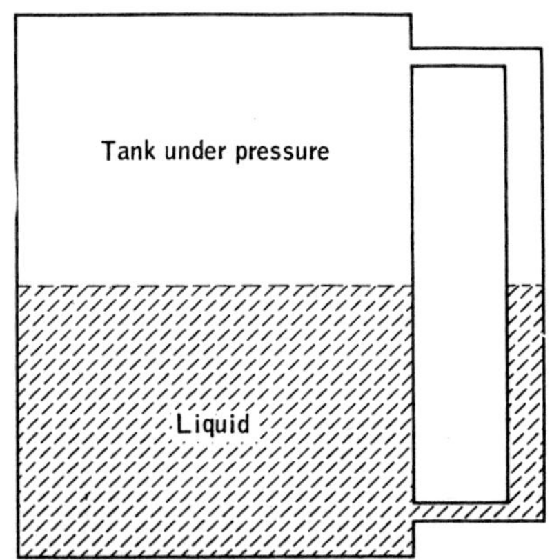

Fig. 13·3 When the liquid in the tank is under some external pressure, the pressure can be nullified by connecting the sight glass as indicated.

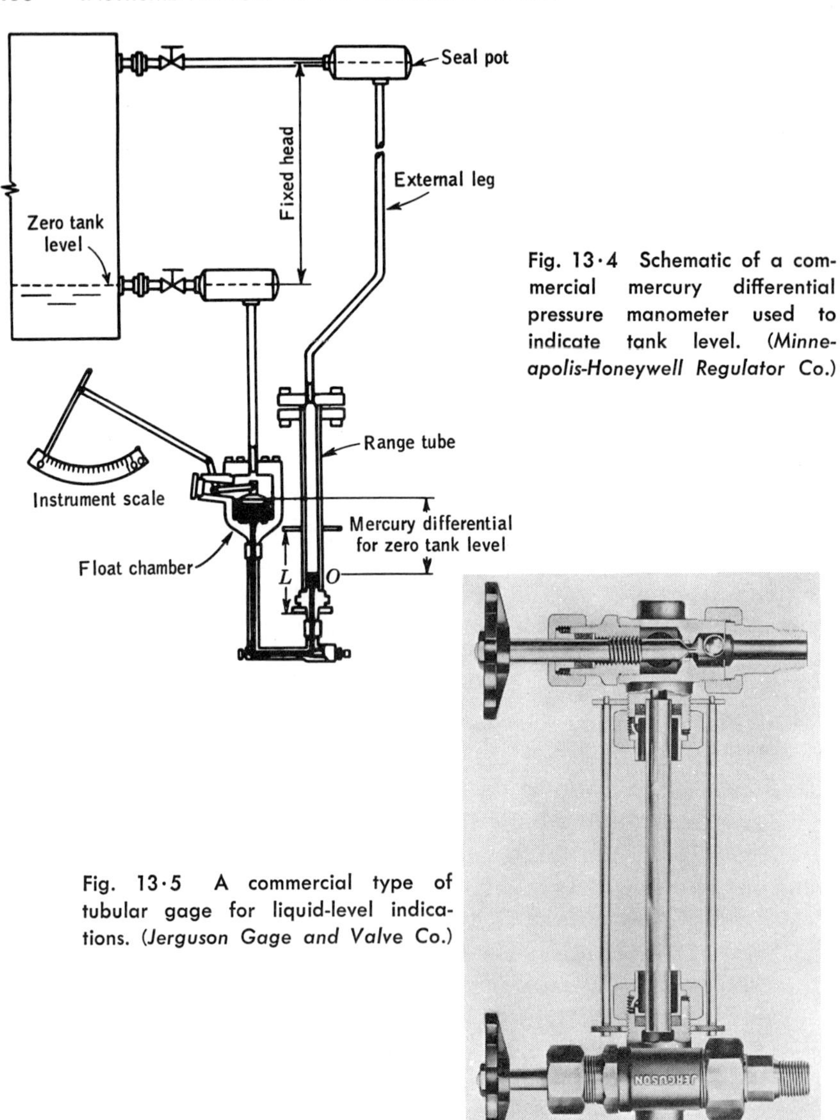

Fig. 13·4 Schematic of a commercial mercury differential pressure manometer used to indicate tank level. (*Minneapolis-Honeywell Regulator Co.*)

Fig. 13·5 A commercial type of tubular gage for liquid-level indications. (*Jerguson Gage and Valve Co.*)

Fig. 13·6 Special optical features have been designed to facilitate reading gages at a distance. One does not depend solely upon the color of the fluid for the indication. (*Jerguson Gage and Valve Co.*)

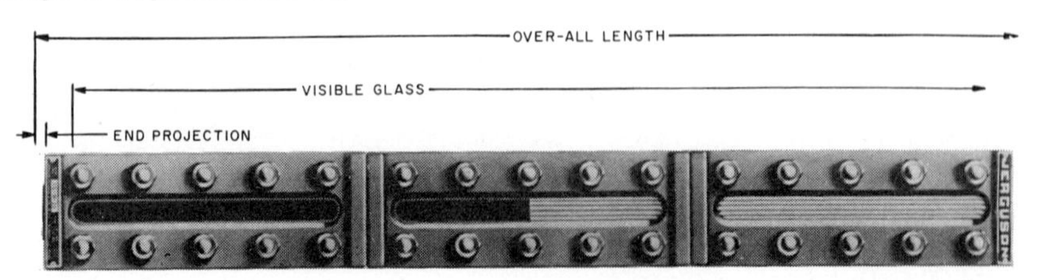

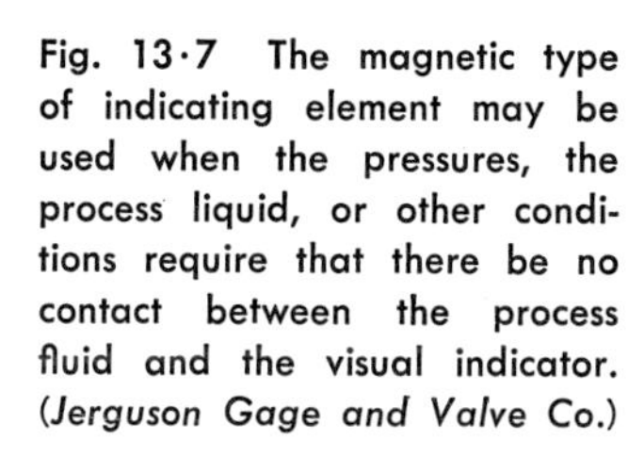

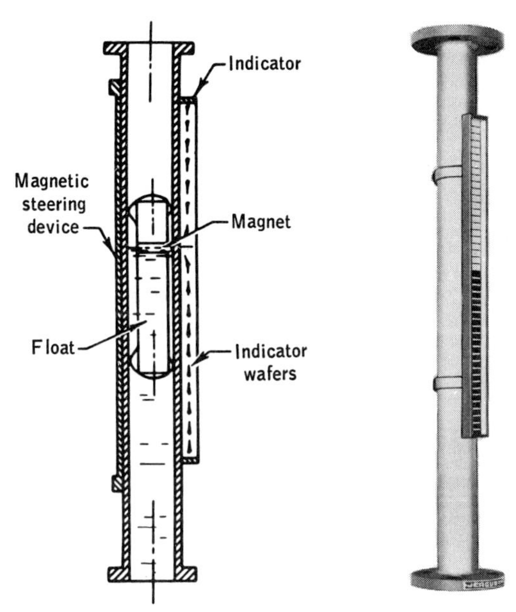

Fig. 13·7 The magnetic type of indicating element may be used when the pressures, the process liquid, or other conditions require that there be no contact between the process fluid and the visual indicator. (*Jerguson Gage and Valve Co.*)

The gage glass has problems that restrict its use to situations in which the liquid will not stain or discolor the tube to such a degree that the level cannot be read or in which the amount of tube cleaning is reasonable. Heavy, viscous liquids or liquids containing material which can fall out of solution and clog the tube cannot be measured satisfactorily by a gage glass. Also, certain materials which will attack the gage glass will have to be measured by other methods. Certain of these objections have been overcome with the designs shown in Figs. 13·7 and 13·8.

The device shown in Fig. 13·7 consists of a special guided float carrying a permanent magnet within a stainless-steel chamber and a scale consisting of a series of magnetized elements which are mounted on the outside of the chamber. The magnetized elements are edge-magnetized so they are attracted to each other, thus forming a continuous scale, one face being painted red and the other silver. Because the floating magnet is stronger than the other magnets, it actuates the indicating elements as it passes each element. By such means a clearly defined level indication is obtained.

The device shown in Fig. 13·8 also uses magnetic techniques. The height of the liquid determines how much of a mercury column is required to balance the hydrostatic pressure. As the level of the mercury changes in response to level changes, a solid float which rests in the mercury also changes position. A magnet which is mechanically coupled to the float and is also within the mercury chamber transmits its position to a pointer which is mechanically balanced and which requires but a small torque to move it. In this manner the vertical movements of the float are converted to rotary movements of the pointer.

Fig. 13·8 Isolation of the indicating portion of the gage from the level-sensing section can be effected by using magnetic means to couple the pointer to the float. (*Jerguson Gage and Valve Co.*)

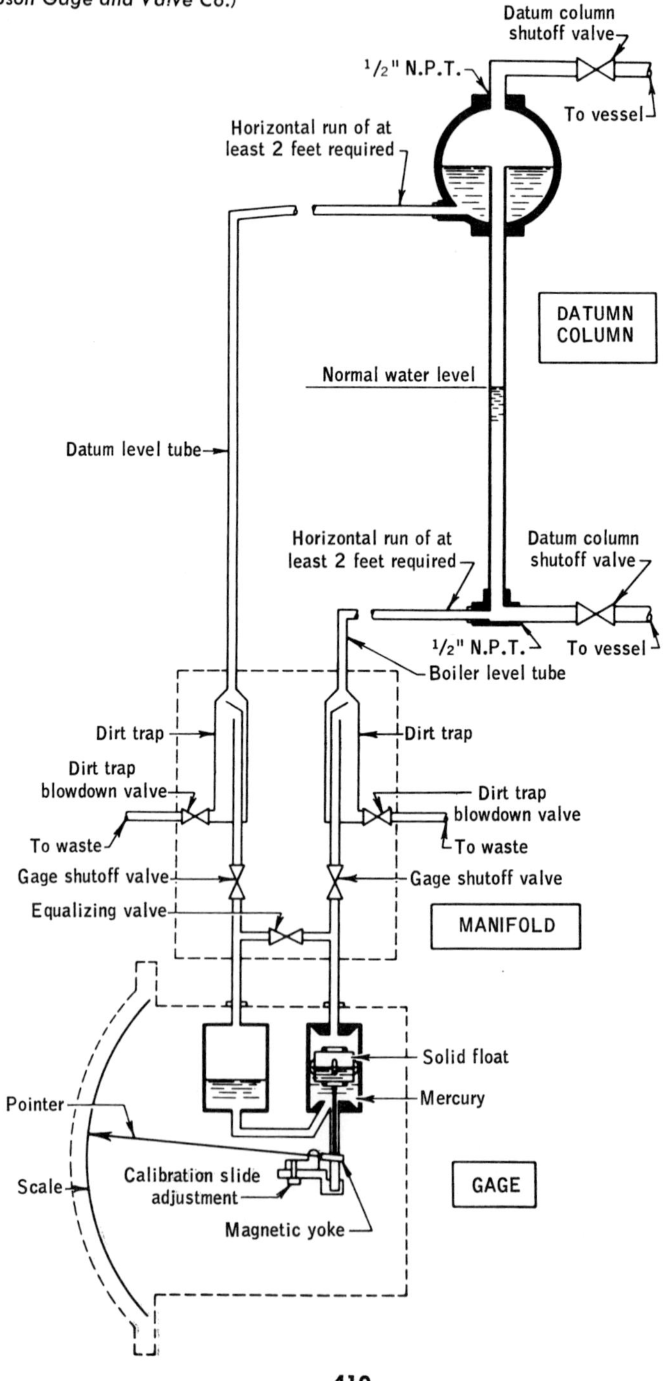

Pressure gage. Inasmuch as a simple manometer can be used to indicate liquid level, a pressure gage which responds to the same pressure range can also be used to indicate level. The indicated readings in pounds per square inch can be converted easily to level or the dial can be recalibrated. Inasmuch as the pressure at the bottom of a tank is created by the weight of the liquid column, the pressure may be expressed in pounds per square inch:

$$P = hd$$

where P = pressure, psi
h = height of the liquid above some reference point, ft
d = pressure of column 1 ft high, psi

EXAMPLE 13·1 What pressure will be indicated by a gage if the liquid level is 10 ft above the gage and the specific gravity is 2?

SOLUTION. If the specific gravity is 2, then each cubic foot of liquid weighs twice as much as a cubic foot of water, or $2(62.5 \text{ lb/ft}^3) = 125 \text{ lb/ft}^3$. But this weight is distributed over 144 in.², so each square inch then has 125/144 lb to support, or the pressure is 0.868 psi. For these assumed conditions,

$$\begin{aligned} P &= (10 \text{ ft})(0.868 \text{ psi/ft of height}) \\ &= 8.68 \text{ psi} \end{aligned}$$

If the tank has a variable pressure applied to the liquid surface, then a simple pressure gage cannot be used and a different gage must be employed.

There are definite precautions to be observed in the use of pressure gages for level indication:

1. The liquid must not corrode the gage.
2. The liquid must not contain solids which will enter the gage and drop out of solution.
3. When (*a*) corrosive liquids, (*b*) thick, viscous liquids, or (*c*) liquids containing solids which can drop out of solution are to be measured, purging systems or capsule seals should be employed, as indicated in Chap. 9.
4. The distance between the tank and the gage must be limited.

Purge or bubbler systems. Inasmuch as a liquid column creates a hydrostatic pressure which is directly proportional to the height of the liquid above the reference point, pneumatic techniques are sometimes used to ascertain the pressure at the reference point. Figure 13·9 shows a very simple type of installation.

By restricting the air available, the pressure at the gage would drop to zero if the end of the pipe were not covered by any liquid at all. If the end of the pipe were completely closed, then the gage would indicate the pressure of the air supply system. Therefore, for any condition between being open and completely closed, there will be some definite pressure associated with each height above the reference.

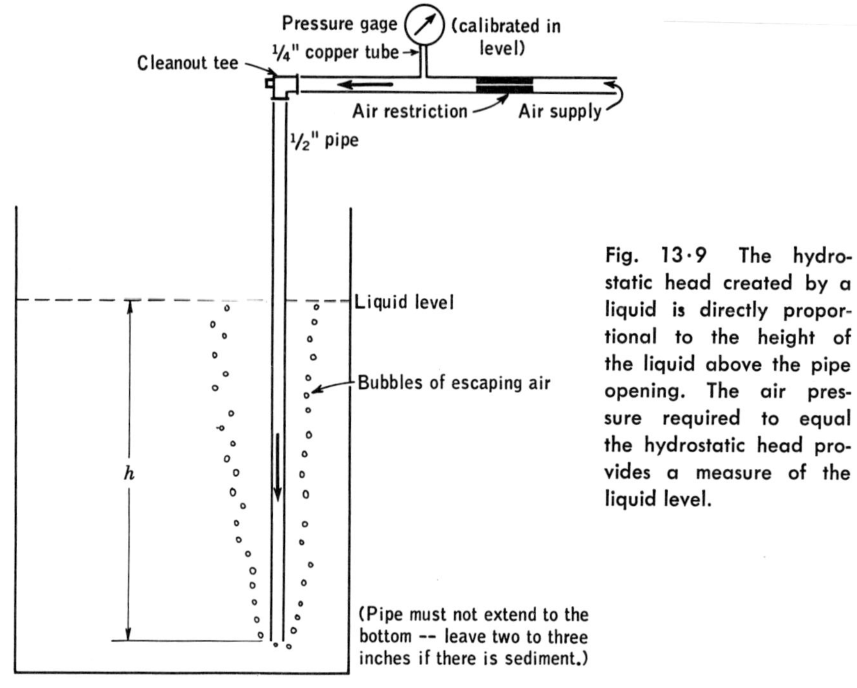

Fig. 13·9 The hydrostatic head created by a liquid is directly proportional to the height of the liquid above the pipe opening. The air pressure required to equal the hydrostatic head provides a measure of the liquid level.

A somewhat more complex system, indicative of a commercial installation, is shown in Fig. 13·10. A rotameter or some other device is used to indicate a positive flow of purging material. Unless one knows that there is an even flow of purging gas or liquid, the indicated pressure may not truly reflect the liquid level. With the air supply shut off for any reason, the gage could not indicate any material increase in level, but would respond to decreases in level. On the other hand, if the liquid were to absorb any of the purging gas, then the pressure would drop for that reason and not because the level had changed. Keeping an even flow of purging material is an important operating factor. One advantage of an air purge system is that the indicating gage may be placed at almost any convenient distance from the tank. Because the purging material will keep corrosive, viscous liquids out of the purge line and will also keep out liquid with suspended solids, the system shown in Fig. 13·10 is useful where the level of such liquids is to be gaged.

The following are the operating considerations:

1. The supply pressure of the purging material must always exceed the static pressure of the liquid by at least 2 psi. If possible, a pressure before the regulator of at least twice the maximum hydrostatic pressure will minimize the variations in purging material as the hydrostatic head varies.

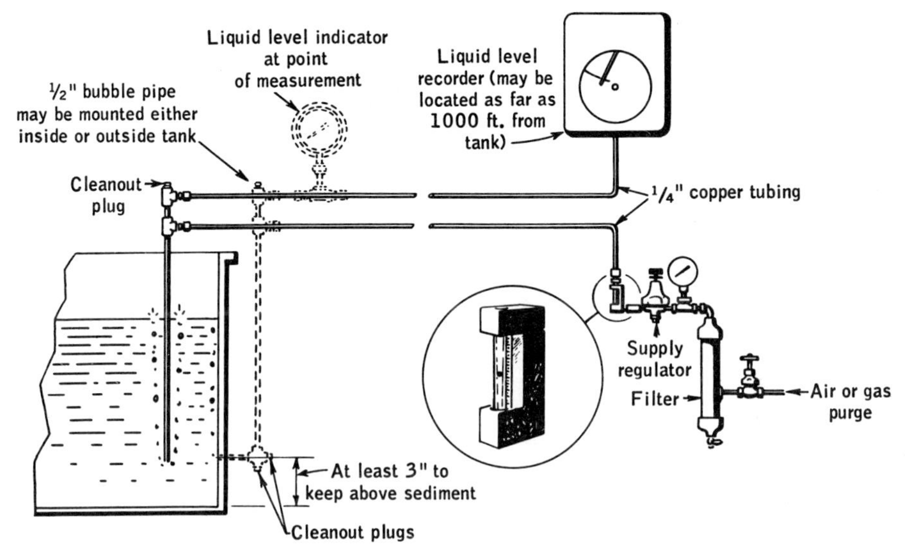

Fig. 13·10 A continuous-purge type of liquid-level measurement. (*The Foxboro Co.*)

2. The maximum pressures for these systems is often limited to 150 psi, or about 300 ft of head of water.

3. The purging material should be turned on before liquid is admitted to the tank.

4. The tank should be drained before the purging material is shut off.

For those situations in which the liquids are of such a nature as to make a purge system desirable but the purging material will react unfavorably with the liquid or for some other reason it is not desired that the purging material mix with the liquid, then a pneumatic force-balance seal may be used. One situation employing such a system is shown in Fig. 13·11. In operation, the force-balance seal functions essentially as shown in Fig. 13·12. The limp diaphragm D is forced by the hydrostatic pressure of the liquid against the orifice O. When the air is no longer permitted to escape at O, the pressure in the line will increase until such a time as it does. This increase in pressure registers on the pressure gage. As soon as the pressure within the purge line exceeds the pressure of the diaphragm, the diaphragm will be forced away from the orifice. This will permit some air to escape, and the diaphragm pressure will again exceed that of the purging fluid. This action will continue with the purging pressure slightly above and then slightly below the hydrostatic pressure created by the height of liquid. The average pressure should be that of the representative liquid level. Whether the diaphragm is made of some limp material or whether it is made of a metal makes little or no difference. In each case, movement of the diaphragm opens or closes the orifice with ensuing slight pressure variations.

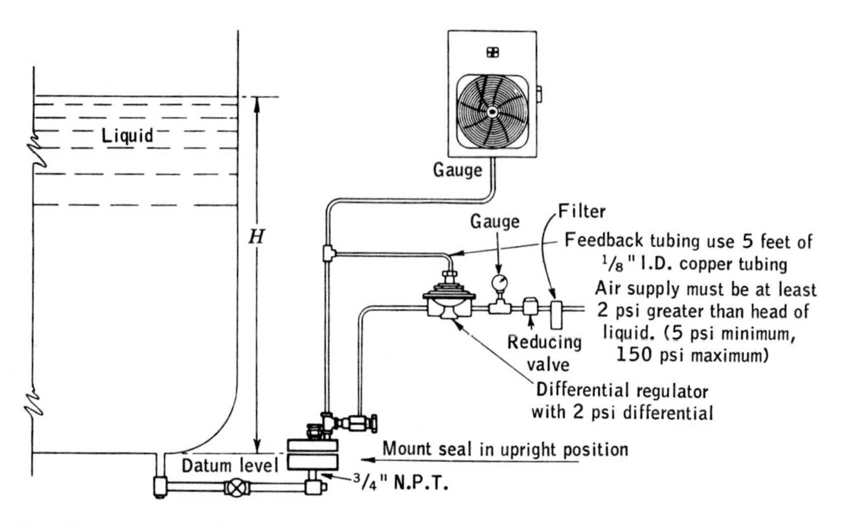

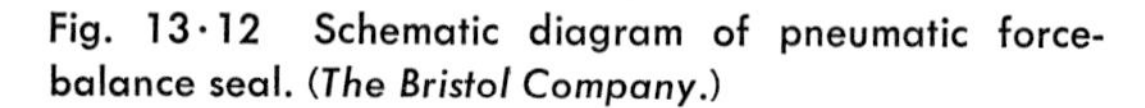
Fig. 13·11 Piping diagram for a liquid-level gage using a pneumatic force-balance seal. (*The Bristol Company.*)

Fig. 13·12 Schematic diagram of pneumatic force-balance seal. (*The Bristol Company.*)

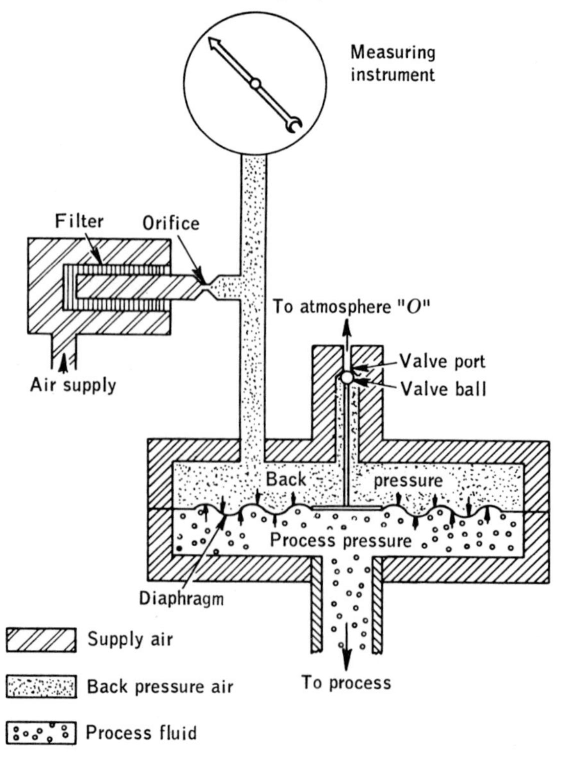

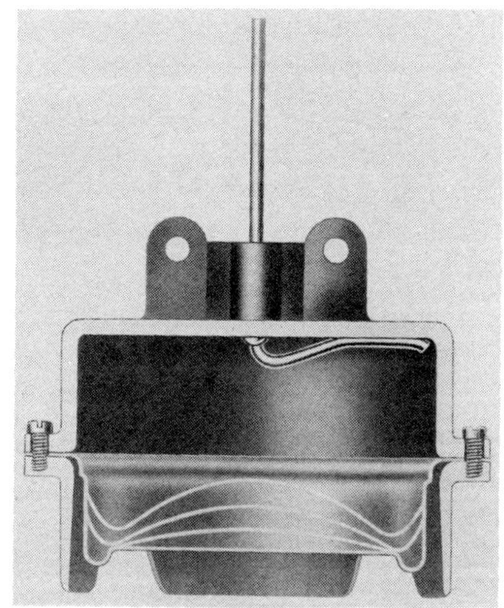

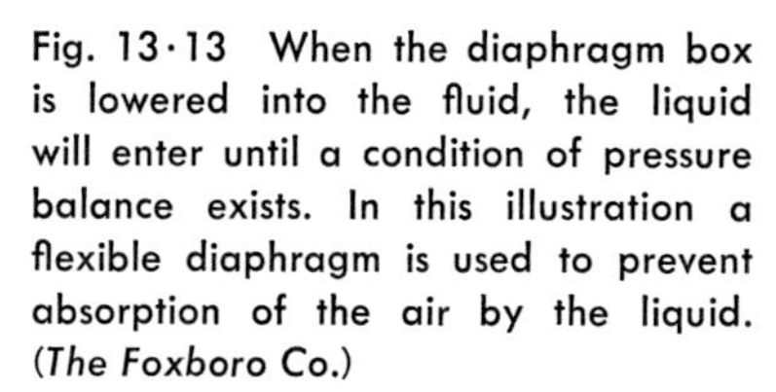
Fig. 13·13 When the diaphragm box is lowered into the fluid, the liquid will enter until a condition of pressure balance exists. In this illustration a flexible diaphragm is used to prevent absorption of the air by the liquid. (*The Foxboro Co.*)

Commercial versions of the diaphragm-box liquid-level indicator will be found in Figs. 13·13 and 13·14. In this design, the pressure created by the head of liquid is balanced by the pneumatic pressure of the gas confined within the box by the flexible diaphragm. If there were no absorption of the gas by the liquid, then we would not need a diaphragm which is impervious to both the gas and the liquid.

Fig. 13·14 One method of using the diaphragm box when it may not be submerged in the liquid. (*The Foxboro Co.*)

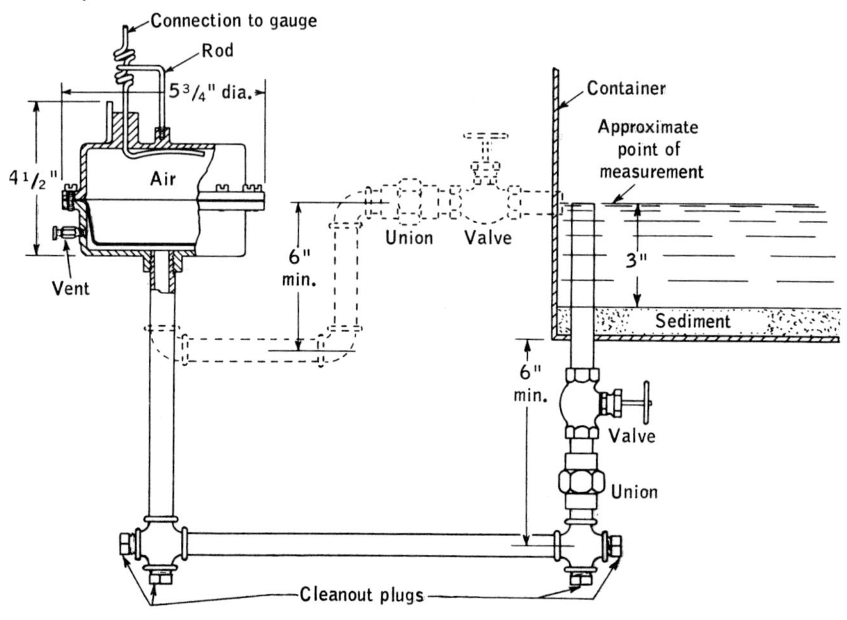

The mechanical simplicity of the device is quite desirable. It also has the advantages of being suited to positioning at any point and of keeping all solids and liquids away from the gage or meter. It is unaffected by turbulence or flow. Also, it may be used to measure levels of reservoirs or tanks where ice may form on the surface.

If the conditions appear to warrant the use of the diaphragm box, then there are several practical considerations which must be observed:

1. Choose box and diaphragm materials which will be unaffected by the liquid.

2. Select the correct size of box and tubing. Figure 13·15 shows how an increase in the size of the tubing limits the transmission distance for any given pressure. Reducing the size of the tubing does, however, seriously increase the response time. The response time for systems using ⅛-in. tubing is about one-sixth that of those using 1/16-in. tubing. The larger size tubing should be chosen wherever possible, although some manufacturers may specify other sizes.

3. *Never* immerse the diaphragm box to an appreciable depth while the union connection to the gage is broken.

4. *Never* open the union connection while the box is under pressure.

5. Be sure that the union connection is absolutely tight.

6. If the temperature of the liquid exceeds that allowable with the diaphragm or if the tank must be kept free of measuring equipment, the installation may be made according to Fig. 13·14.

Fig. 13·15 Maximum tubing length versus range of gage, in feet of water. (*The Bristol Company.*)

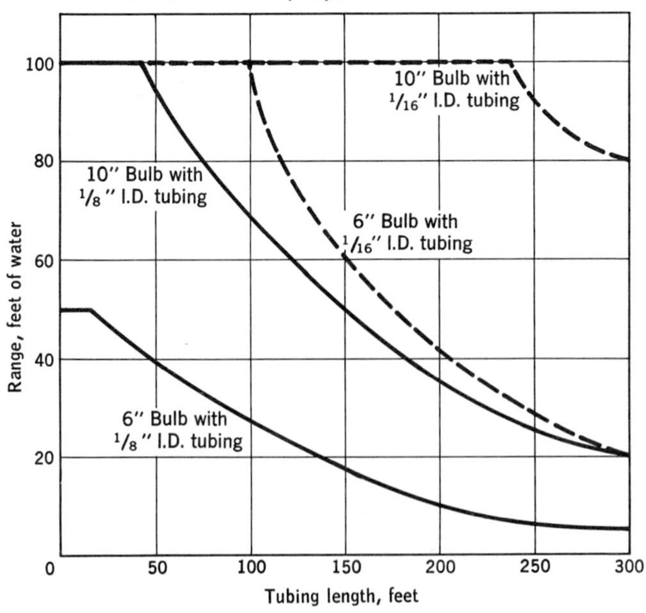

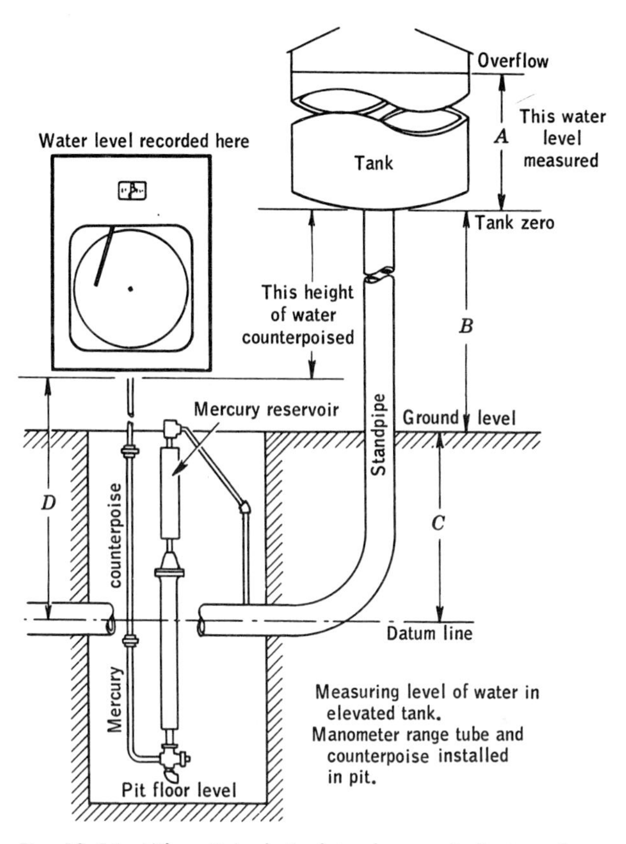

Fig. 13·16 When it is desired to show or indicate only the height of the liquid in the tank or reservoir, a mercury counterpoise in the piping going to the meter can nullify any pressure effects created by the column of liquid in the standpipe. (*The Bristol Company.*)

7. Keep the diaphragm far enough above the bottom that it will never be in sludge or other material: 3 in. above any sediment line or within 1/8 in. of the bottom if there is no sediment.

Manometer range tube and counterpoise. When the liquid level in a tank is to be gaged, it is often desirable to make the instrument respond to tank-level changes only. In Fig. 13·16 it will be noted that a mercury column is used to create a pressure which offsets that created by the liquid in the standpipe. Consequently, the level to which the gage responds is created by the level within the tank.

Float systems. When a float may be used and it is possible to take a reading at the tank, there are many ways in which the position of the float may be communicated to the observer. One of the simplest is to use a gear train and a pointer which are coupled to the float. When there

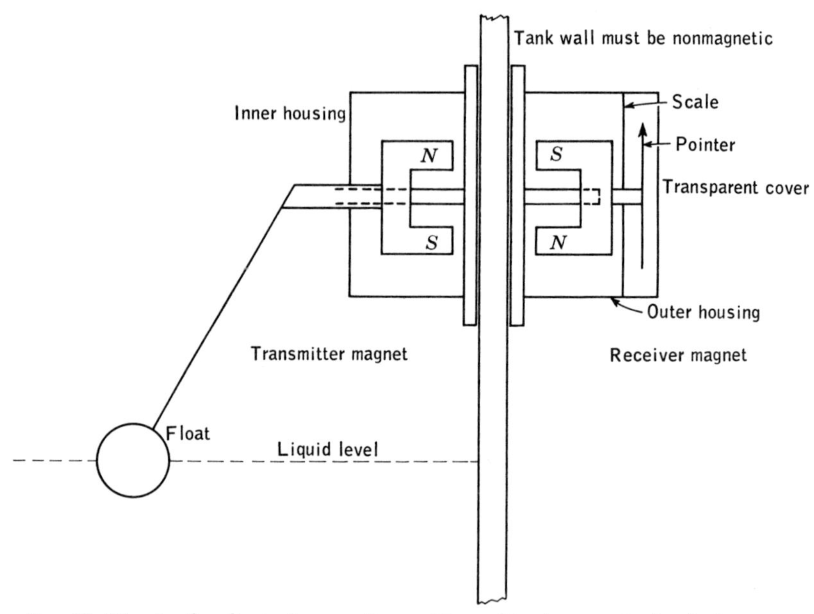

Fig. 13·17 As the float changes its position with changes in level, it rotates the transmitter magnet. The magnetic field causes the receiver magnet to follow the transmitter magnet. Because of the small forces available, the receiver magnet must have a minimum amount of friction.

Fig. 13·18 The movement of the float positions the dials which indicate the level in both feet and inches. The tape which connects the float and the indicating elements is kept taut with a constant-torque spring. (*Vapor Recovery Systems* Co.)

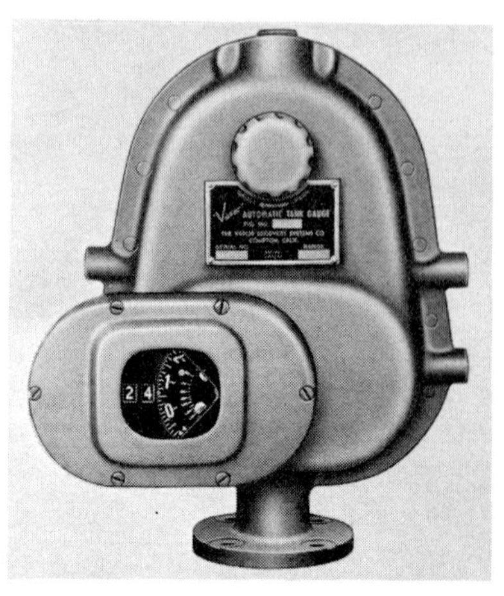

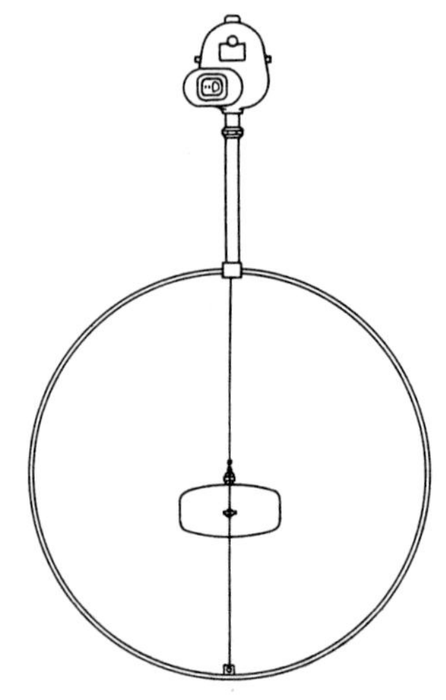

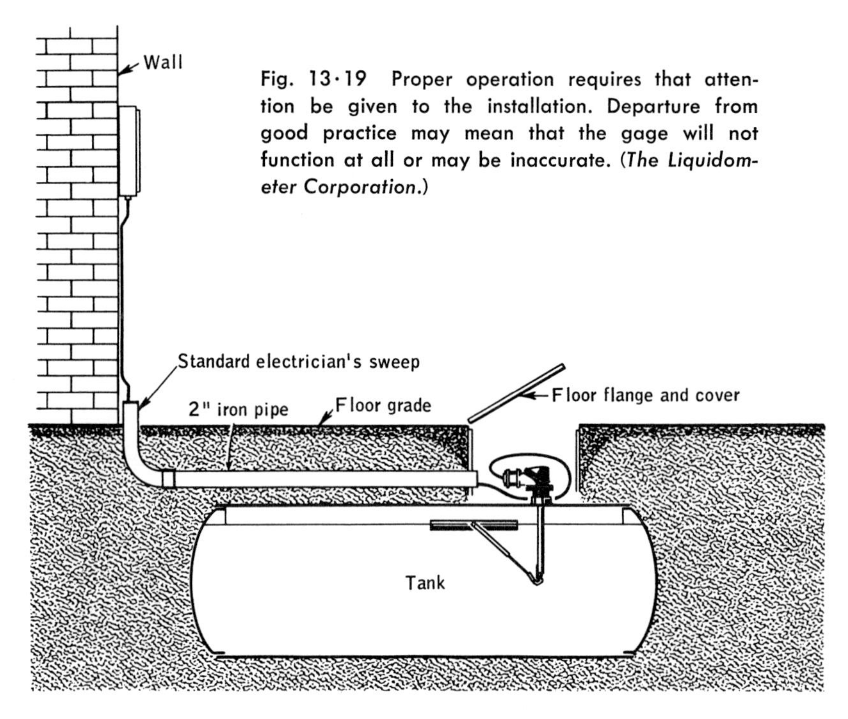

Fig. 13·19 Proper operation requires that attention be given to the installation. Departure from good practice may mean that the gage will not function at all or may be inaccurate. (*The Liquidometer Corporation.*)

is pressure in the tank, the friction of the packing may introduce an error of significance.

For tanks under pressure, particularly those that contain inflammable materials, a magnetic drive is highly desirable because it permits the float position to be indicated without having any openings in the tank, Fig. 13·17. There are two mating magnets which are separated by a nonmagnetic wall; one magnet is attached to the float and the other to the pointer. Moving the driver magnet will cause the pointer magnet to follow. Units of this type have been used on hundreds of gasoline and oil tanks, as well as other tanks. When the pressures and temperatures cause the designer to avoid holes, the float may communicate its position by magnetic action, Fig. 13·17.

A somewhat different use for a float is shown in Fig. 13·18. The position of the float is indicated by the numbers on a tape which can be viewed through a special window provided for the purpose. Quite another use for the float is shown in Figs. 13·19 and 13·20. In this design, the movement of the float actuates a push rod which acts upon either of two bellows. If the liquid level is rising, for example, then the stroke lever tries to compress bellows B, which must then force its fluid into bellows C at the receiver. But bellows C cannot expand without compressing bellows D and returning some of its fluid to bellows A. This is a tight hydraulic system which can function only through the transfer

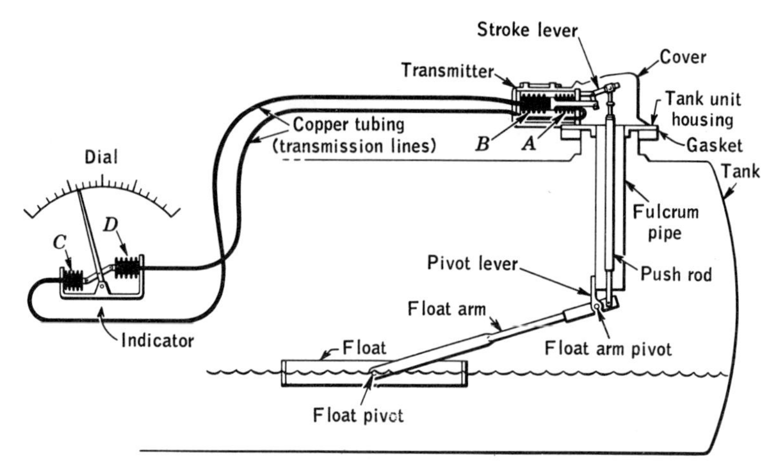

Fig. 13·20 Pictorial diagram of hydraulic gaging system. (*The Liquidometer Corporation.*)

of fluid from one place to another. Its principles will be found in many hydraulic servo systems, the chief difference being the way in which the actuating force is developed.

When the mechanical characteristics of the liquid are such as to preclude any significant movement of gage elements, it may be possible to employ the method suggested in Fig. 13·21. The force which is required to keep the float submerged below its free-floating position is directly proportional to the liquid level, provided that the liquid level is always high enough, or can be so arranged, to cause the object to float freely. The buoyant force may then be measured by the air pressure necessary to maintain a condition of torque equilibrium. The level is then read on a calibrated pressure gage.

Miscellaneous systems. The average level of the contents of a tank may be inferred from the weight of the tank, and this provides a quite different way to determine liquid level. This may be desirable when the tank contents are so viscous that the use of a float is impracticable or the materials are so active chemically that ordinary methods and materials are not adequate.

One of the easiest ways to make such a measurement requires that one-half of the tank be supported on a fixed support and the other half be sustained by the hydraulic pressure created within the filled system. As the load in the tank increases, the tank will press on the flexible bellows with more and more force. Inasmuch as the confined liquid is virtually incompressible, it will resist with a pressure which will create a force equal and opposite to that created by the load. The weight of the tank is known or can be determined, and the pressure gage can therefore be calibrated in terms of the net load instead of the gross load, which would include the tank.

It is obvious that care must be taken to ensure that the tank is not improperly restrained and thus prevented from transmitting to the hydraulic element the calculated proportion of the load. With such a system, there is no reason why the gage cannot be located at some distance from the tank. Attention might have to be given to any difference in elevation of gage and tank unless the unit pressures are high.

The diaphragm type of element measures tension loads, compression loads, or a combination of both. The load is applied through a connection at the center of the diaphragms. If a compression load is applied, the diaphragms flex in; if a tension load is applied, they flex out to reduce the pressure in the system, which is preloaded to ensure that, under all conditions normally encountered, it will be under a positive pressure.

The spool type of gaging element employs the pressures generated by compressing a housing containing mercury. An invar slug within the chamber compensates for changes in pressure created by temperature.

The ring type of element is similar to the diaphragm type but can be used only with compressive loads. One of its chief advantages is that it can permit the loading screw or shaft to pass directly through it. The ring element has been used for many years to measure processing pressure in paper-stock beaters.

Fig. 13·21 Liquid level may be inferred from the buoyant force acting on a partially submerged body.

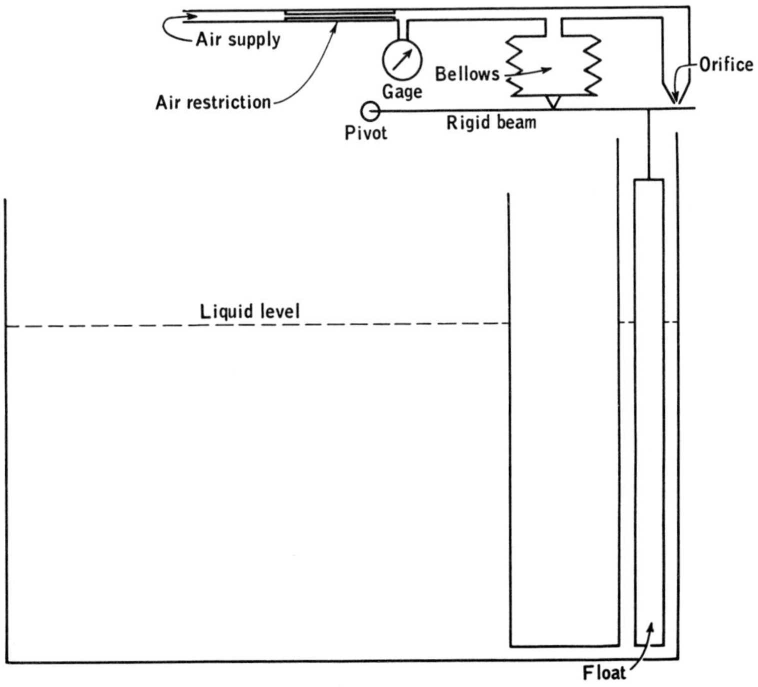

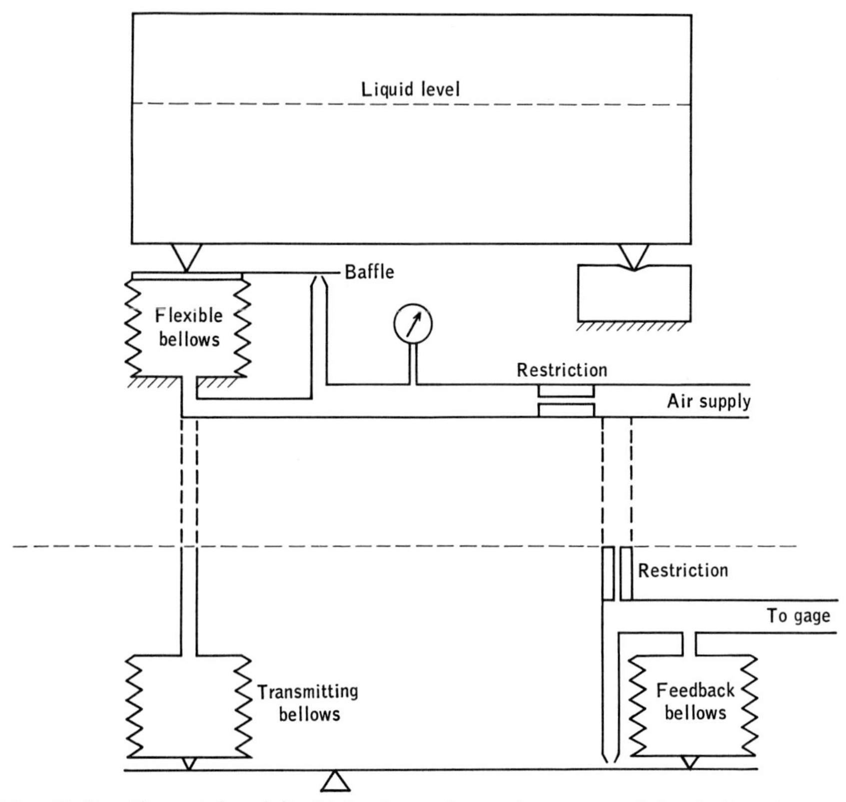

Fig. 13·22 The weight of liquid in the tank can be measured by finding the air pressure required to create an equal and opposite force within the flexible bellows. This pressure, converted into weight or level, may be read locally or transmitted for remote reading.

A somewhat similar installation is shown in Fig. 13·22 except that the resisting force is provided by pneumatic instead of hydraulic means. The position of the tank will control the pressure in the system. One of the chief requirements when the baffle is away from the orifice is that the air be able to bleed out of the orifice faster than it can pass through the restriction. This gives the baffle complete control of the pressure at the orifice. Consequently, when the baffle is brought closer to the orifice by an increase in load, the pressure within the bellows will increase. If the pressure within bellows does not increase fast enough, the baffle will be brought even closer to the orifice, which will increase the pressure still more. The cycle will continue until a condition of force balance has been achieved.

If it is desired to transmit for some significant distance the pressure that corresponds to the load, it may be desirable to add the pneumatic pressure transmitter which has been studied previously. In its simple schematic form, it would be as shown in Fig. 13·22.

ELECTRICAL TRANSDUCERS

There are perhaps a greater number of electrical methods for indicating level of both liquid and dry material than there are mechanical methods. Many of the electrical means will employ some principles which have been studied previously; others will be quite new in their principles.

Almost all of the electrical equipment will be more complex than its mechanical counterpart, but the electrical apparatus is used for situations in which the mechanical designs would be somewhat lacking in the desired performance characteristics. One other factor which is emphasizing the electrical techniques is the great interest in signals which can be used with data logging and computing equipment. As we demand more and more from it, the equipment must of necessity be more complex in its design and more accurate in its performance.

Probe-type level detectors. One of the simplest instruments to understand is represented in schematic form in Fig. 13·23. Only a simple triode tube is required. When the level in the tank is below the level of probe G, there is no complete circuit to the grid of the tube, which then accumulates a charge of electrons. This makes it very negative in effect, and this reduces the tube plate current to a very small amount. Under these conditions, there is too little current in the relay coil to cause it to operate and complete the alarm circuit.

But if the liquid rises so that it completes the circuit between probe G and the other terminal G', then the electrons can be removed from the grid circuit. As the grid is made less negative, the plate current will continue to rise until sufficient current flows in the relay coil to operate it.

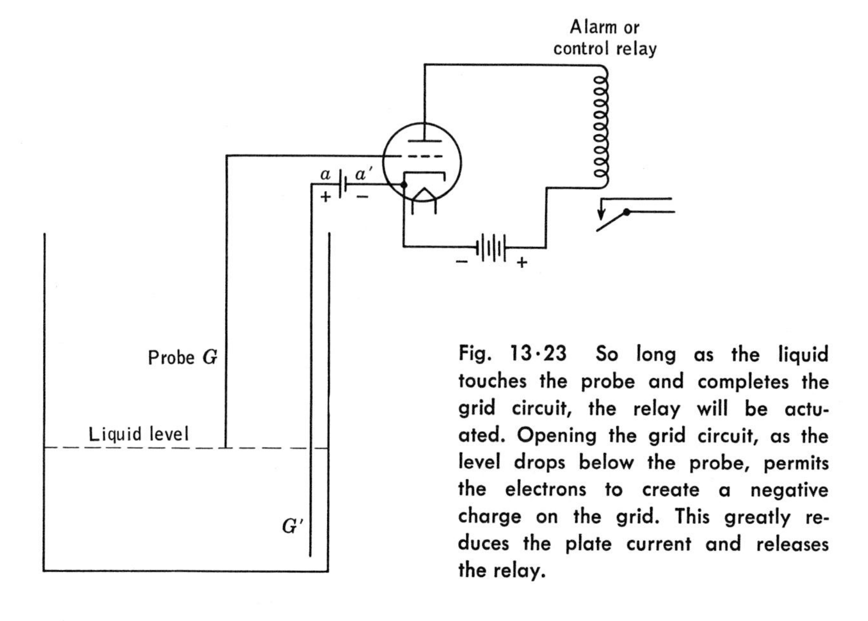

Fig. 13·23 So long as the liquid touches the probe and completes the grid circuit, the relay will be actuated. Opening the grid circuit, as the level drops below the probe, permits the electrons to create a negative charge on the grid. This greatly reduces the plate current and releases the relay.

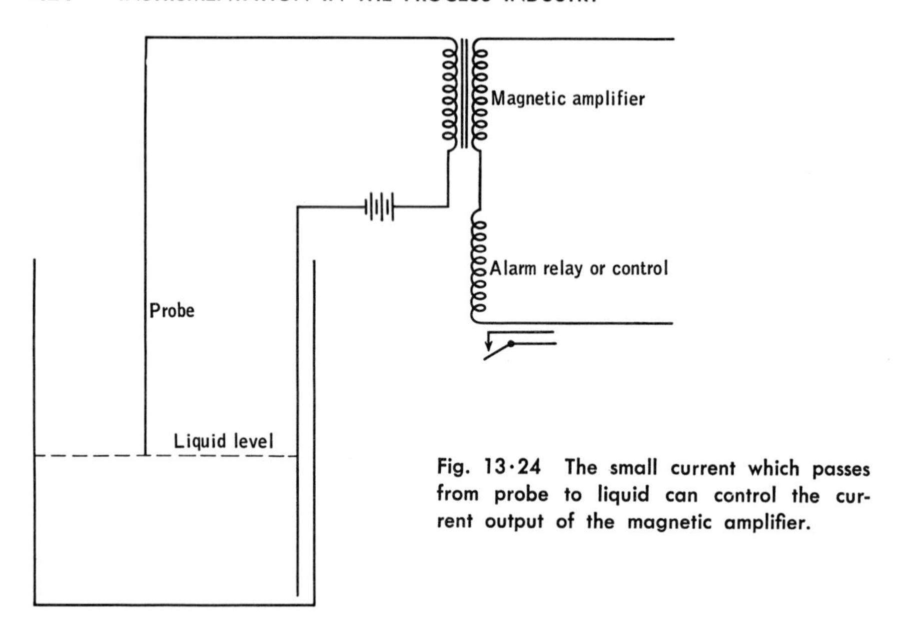

Fig. 13·24 The small current which passes from probe to liquid can control the current output of the magnetic amplifier.

Provided that the liquid is of a very high resistance, then it may be desirable to add a battery such as is shown connected to points A and A'. Ordinary tap water is sufficiently conductive for circuits of this type, and the extra battery normally will not be required. Inasmuch as the transistor can perform many of the tasks which the vacuum tube has performed for years, it might be well to note that the transistor can replace the tube in Fig. 13·23.

Another type of probe uses the conductivity of the liquid to permit enough current to flow to control the associated magnetic amplifier. The magnetic amplifier, in turn, regulates the relay current. This also is an on-off device which can indicate that the liquid level is below the set point or is at or above that point. For situations in which the conductivity of the liquid is very poor, it may be desirable to use a transistor for a current amplifier. About all that is required is to substitute the control winding of the magnetic amplifier, Fig. 13·24, for the relay winding.

A probe type of detector which deliberately introduces a time delay in its response is shown in Fig. 13·25. The probe contains a thermistor which is so placed that any heat generated within it because of the current which passes through it can be readily transferred to the metallic portion of the probe which is in contact with the liquid. So long as this heat can be transferred quickly, the thermistor remains comparatively cool and its resistance remains relatively high. But if the liquid level falls, then the probe housing can no longer dissipate heat at its former rate; so its temperature increases. At the same time, the temperature of the thermistor also increases and the resistance of the thermistor de-

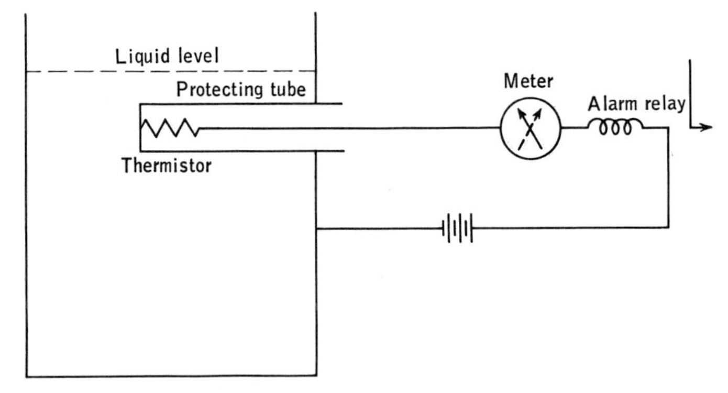

Fig. 13·25 When the liquid level falls below the tube and it can no longer transfer heat from the tube to the liquid, the thermistor temperature will rise. As it does so, the resistance of the thermistor undergoes a large decrease. The increase in current can be read on a meter or indicated by an alarm.

creases. This increases the current still more, which again increases the heating rate until a point of thermal equilibrium is reached. Because of the thermal inertia of the probe, there will always be some delay; it may be of the order of seconds between the time that the liquid drops away from the probe and the moment of relay actuation. If a meter is used in addition to the relay, then an increase in the current would be indicative of an impending alarm or other signal.

If the probe must respond only to changes produced by the temperatures generated within the probe and not to changes in the ambient temperature, then the design shown in Fig. 13·26 may be employed. An outer and an inner tube are joined with a pressure-tight seal at the end which projects into the liquid. A resistance wire which will heat the outer tube is cemented to the inner surface of the outer tube, and the outer tube is secured to a bushing or fitting which can be screwed to the tank. Any

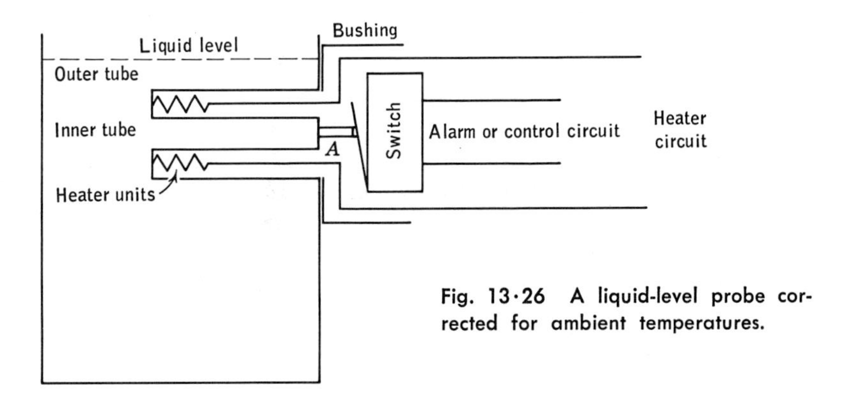

Fig. 13·26 A liquid-level probe corrected for ambient temperatures.

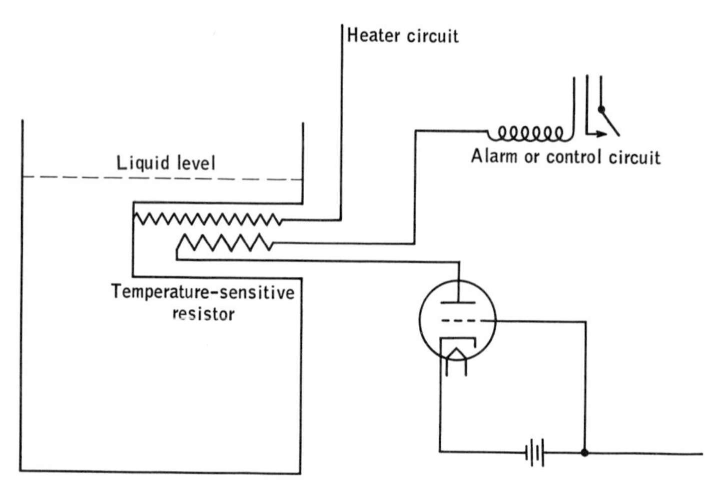

Fig. 13·27 The resistance of the probe drops markedly when the level of the liquid falls below that of the probe.

increase in the ambient temperature will cause the end of the outer tube to move to the left. But the inner tube, which is made of the same material and also has the same length, will have its right-hand end moved an equal distance to the right. In this manner, temperature changes which affect each tube equally will not bring about any significant change in the position of point *A*.

When liquid surrounds the tube, the heat transferred by the heating resistor to the outer tube is effectively dissipated to the surrounding liquid. But if the liquid drops away, then the temperature of the outer tube will rise much faster than will that of the inner tube. This will move point *A* away from the switch actuator and thus permit the switch to function. With switches which require only 0.0003 in. for their actuation, an extremely fast response may be secured. The type of liquid in which the probe is immersed will dictate in part what the speed of response will be. Very volatile fluids such as gasoline or ether will give a faster indication than can be obtained with lubricating oils, which are relatively very poor in their heat transfer characteristics.

Still another probe design is shown in schematic form in Fig. 13·27. This is a very unusual unit because of the characteristics of the sensing material in the probe. As in some of the designs discussed previously, the probe is kept relatively cool by being surrounded by fluid, but when the fluid level drops, the probe can be no longer cooled. The energy furnished by the resistor is then used largely to increase the temperature of the probe and its contents. The thermal sensitive material, in this case, has the general characteristics shown in Fig. 13·28. It will be noted that the resistance remains essentially constant for a significant temperature change, but when a certain critical temperature is reached, there is

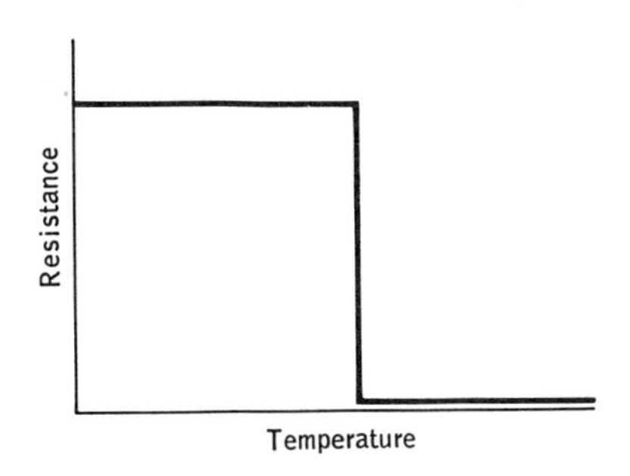

Fig. 13·28 The resistance versus temperature characteristics of the resistor used in the plate circuit. The extremely sharp decrease in the resistance at a critical temperature is characteristic of only a few special materials.

a most abrupt change in the resistance which then approaches some very low value. With the resistance in the plate circuit of the tube suddenly lowered, the tube can conduct sufficient current to actuate the relay. This very unusual temperature-resistance characteristic seems to have been given an extremely limited use in light of its capabilities.

An entirely different means of determining level employs the essential elements of a *fathometer,* or an instrument for determining the depth of water under a boat. Just as sound waves in air are reflected when the transmitting medium changes, so are they reflected when they pass from a liquid of one density to a liquid of another density or when they pass from a liquid to a gas. Any discontinuity in the path will result in a reflection of greater or lesser intensity.

Figure 13·29 shows in diagrammatic form a fathometer but with the transmitter of sound wave on the bottom of the tank instead of on the surface. The waves will now be propagated upward to the surface, where there will be a reflection wave sent to the bottom and there detected by the receiver. The time required for the wave to travel to the surface and back will be a measure of the depth of the liquid; the closer the level is to the transmitter, the sooner will the reflected wave be received at the

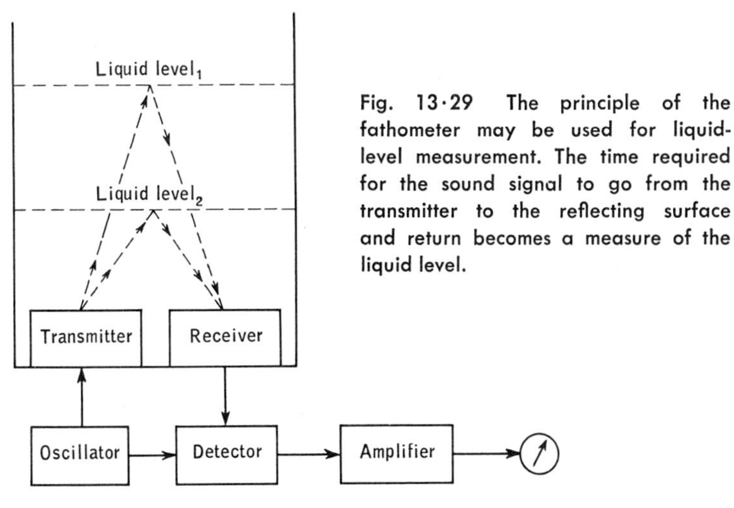

Fig. 13·29 The principle of the fathometer may be used for liquid-level measurement. The time required for the sound signal to go from the transmitter to the reflecting surface and return becomes a measure of the liquid level.

receiver. If the liquid in the tank is composed of two or more materials which do not mix, then there will be reflections from each interface.

Capacitance level gaging. One of the very desirable characteristics of level-gaging apparatus is that there be no moving parts in the tank or vessel. This requirement is met by a system which is based on changes in the capacitance of a probe as the level changes.

It will be recalled that a capacitor consists of at least two electrodes which are separated by an insulator which is termed the dielectric. When a potential is applied to the plates, there will be a removal of electrons from one of the plates and an addition of the same number of electrons to the other plate, assuming that there are two plates.

The number of electrons available for any given potential difference depends upon the spacing between the plates, the area of the plates, and the nature of the material between the plates. Expressed in a mathematical relationship, the capacitance of a two-parallel-plate capacitor, in micromicrofarads, or 10^{-12} farads, may be found from

$$C = 0.225K \frac{A}{d}$$

where C = capacitance, $\mu\mu$f
A = area of the plate, in.^{2}
d = distance between the plates, in.
K = dielectric constant

The characteristics, in terms of capacitance, depend entirely upon the dielectric for any given size of plates and spacing. This means that some materials can furnish far more electrons than can others for any given potential difference. As the spacing between the plates becomes smaller, we create a greater potential pull on the electrons in terms of the volts per inch. This greater electrical pull or stress detaches more electrons from their parent material. In this manner, the available electrons are inversely proportional to the spacing of the electrodes.

Some materials have electrons which can be made readily available, while others share their electrons very reluctantly. Table 13·1 compares the availability of electrons in various materials with those available from air, under identical conditions. It will be noted that water and various aqueous solutions make excellent dielectrics in terms of the large changes in capacitance which result from replacing air with them. It will be noted that one very unusual material, barium titanate, has a dielectric constant of about 5,000, which makes it useful for many applications.

The capacitance, which varies directly with the level of the liquid in the tubes, can be measured in many ways and then related to the height of the liquid. The capacitance of the probe, Fig. 13·30, will be a mini-

Table 13·1 Availability of Electrons in Various Materials Compared with Those Available from Air

Material	*Dielectric Constant*	*Material*	*Dielectric Constant*	*Material*	*Dielectric Constant*
Water	80	Cellulose acetate	3.7–7.5	Cottonseed oil	3.0
Aqueous solutions	50–80	Sand	3–5	Petroleum oils	2–3
Glycerine	47	Nylon	4–5	Asphalt	2.7
Barium titanate	5,000	Rice and cereals	3–5	Benzene	2.3
Methyl alcohol	37	Cellulose	3.9	Liquid chlorine	2.0
Glycol	35–40	Glass	3.7–10	Paper	2.0
Dolomite	8	Sulfur	3.4	Liquid CO_2	1.59
Porcelain	5–7	Turpentine	3.2	Liquid air	1.5
Salt	6	Sugar	3.0	Air and most gases	1

mum when the tubes contain nothing but air and a maximum when the dielectric is the liquid and fills the entire space between the electrodes.

By using some suitable circuit such as the Wheatstone bridge in Fig. 13·31, the liquid level can be determined manually. By using quite conventional means, the output of the bridge can be fed to an amplifier and servo unit which will rebalance the bridge and indicate the level. Particular attention should be paid to the fact that the servomotor is controlled by but a single tube which requires no switching of outputs to obtain the necessary direction of drive. The input signal from the bridge is modulated with the frequency of the power supply which is also supplied to one winding (or phase) of the servomotor. Variations in the bridge balance will shift the incoming phase relationship by either 0 or 180° as determined by the capacitor across the second winding of the servomotor.

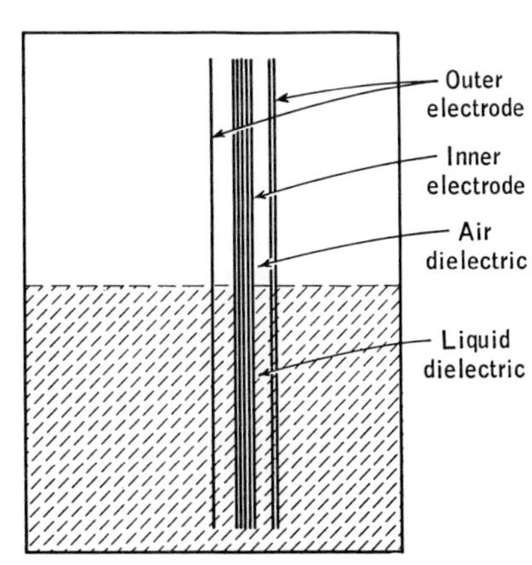

Fig. 13·30 The capacitance of the probe varies with the liquid level.

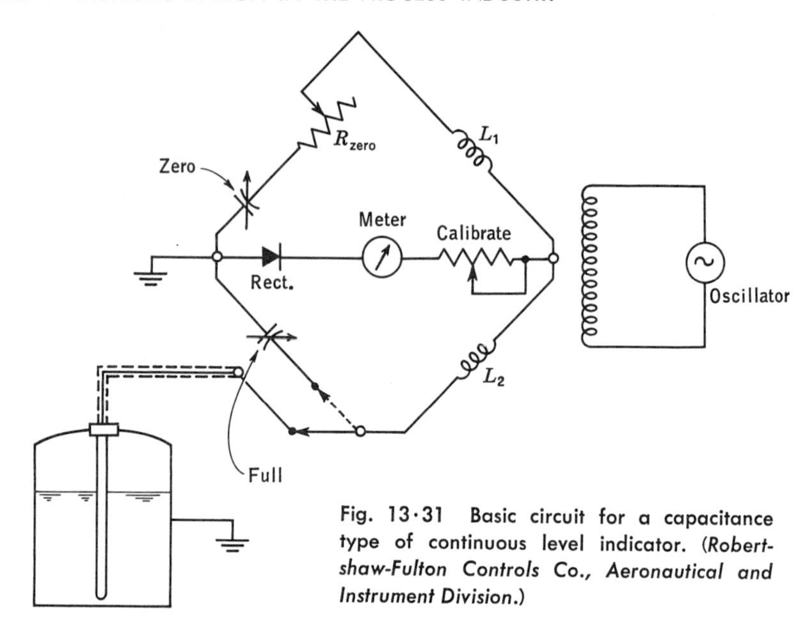

Fig. 13·31 Basic circuit for a capacitance type of continuous level indicator. *(Robertshaw-Fulton Controls Co., Aeronautical and Instrument Division.)*

When the level of solids is to be determined, the approach is essentially the same except that the tank or container becomes one electrode and the intervening material becomes the dielectric. The other electrode is an insulated wire located vertically in the center of the tank, for example.

All of the electrical gaging methods which have been considered so far have required that part of the device be inserted within the chamber where the gaging is to take place. There may be conditions in which it is desirable to keep the gaging equipment away from the process fluid if that is at all possible. Such gaging methods are feasible when using certain materials which are designated as radioisotopes. These materials are treated in much more detail in Chap. 19, and only the fundamentals of their application will be considered here.

In its most elementary form, the use of radioisotopes for level gaging requires some source of energy, indicated by a block in Figs. 13·32 to 13·34, and a detector which is connected to an amplifier and some indicating meter. For all practical purposes, we can assume that the energy given off by the source is constant in value. And so long as only the two tank walls are interposed between the source and the detector units, a definite amount of radiation will be received continuously by the detector. As the liquid level rises from L_1 to L_2, some of the radiated energy will be absorbed by the intervening liquid. When a smaller amount of radiation is received by the detector element, this will be indicated by the meter. But if the level rises to L_3, still more of the radiated energy will be absorbed by the fluid, and the instrument will indicate a lower value of radiation, or a higher level of liquid, depending upon the calibration.

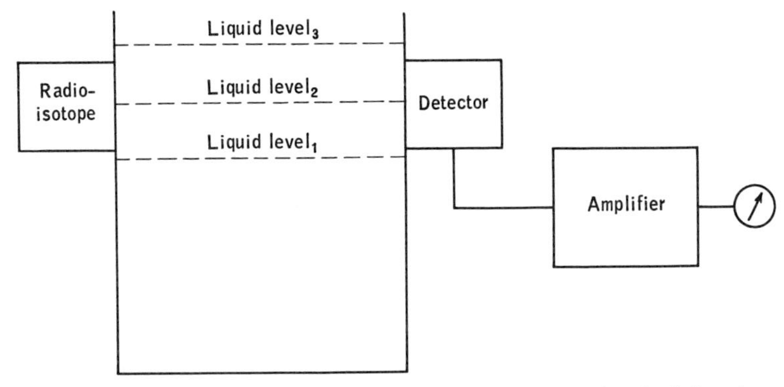

Fig. 13·32 The amount of radiated energy which is absorbed by the intervening liquid is a measure of the liquid level.

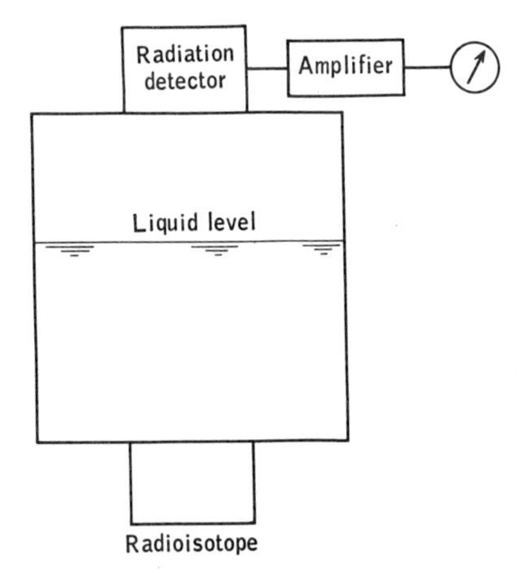

Fig. 13·33 The amount of energy which is absorbed varies directly with the depth of the liquid.

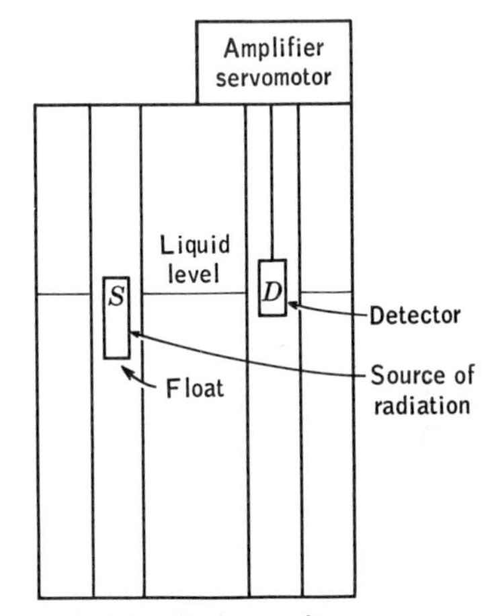

Fig. 13·34 If the radioisotope can be made to move with the liquid level (on a float), a servomechanism can be made to follow the movement. This indicates the level.

Instruments of this general design are used to indicate that the level is within certain limits. When a series of source and detector elements is employed, it is possible to indicate the level within extended ranges.

Radiation types of level gages. Almost all of the discussion so far has been concerned with large containers or tanks. But it may be equally necessary to know the level of the product in containers which are sealed and about ready for shipment. For this type of measurement, the measuring elements must be separate from the container. One means of making such determinations is shown in Fig. 13·35. By keeping the measuring unit close to the filling device, any deviations in level can be quickly detected and corrected.

Fig. 13·35 By measuring the absorption of the radioisotopes, level control may be obtained. (*The Industrial Nucleonics Corp.*)

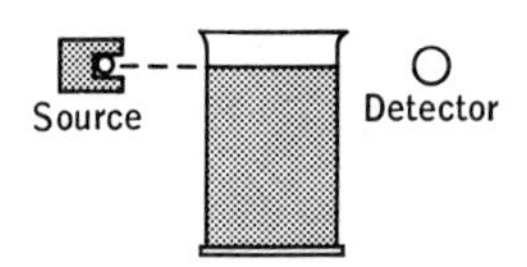

Proper level effectively blocks energy reaching detector. No action is taken by the system.

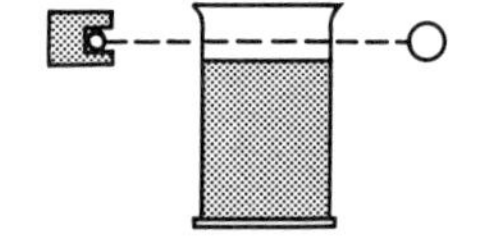

Energy reaching the detector increases sharply with slack-fills. Rejection device is actuated by the Control System.

In a related method of making the same sort of measurement, X rays are absorbed in amounts which are governed by the type and thickness of material doing the absorbing. If a narrow, concentrated beam of radiated energy is transmitted through the can or container, the amount of radiation reaching the detector element will be a measure of the level. By using detector materials which in themselves can liberate energy when subjected to the X rays, very high sensitivities can be obtained. By using greater source intensities, this same basic equipment can also be used to determine the thickness of metal strip and similar rolled materials.

Radioisotope container inspection-rejection system. By using two inspection systems to monitor the liquid level in containers in process, it is possible to control the filling system while maintaining an inspection of the system, Fig. 13·36. If the amount of energy reaching the first monitoring unit is not standard, then the proper correcting signal is applied to the filling unit. A similar unit is used to ensure that all containers which are passed are at or above some definite level.

Containers which are too full require that the filling cycle be altered, but such a container should not be rejected. Moreover, the type of corrective action called for makes it simpler to use one unit for each type of inspection, rather than a single unit of more complicated construction.

Determining level of dry materials. The determination of the level of dry materials is also an industrial problem. However, because it does not require extensive treatment, it is considered in this chapter in order to make the measurement of level that much more inclusive.

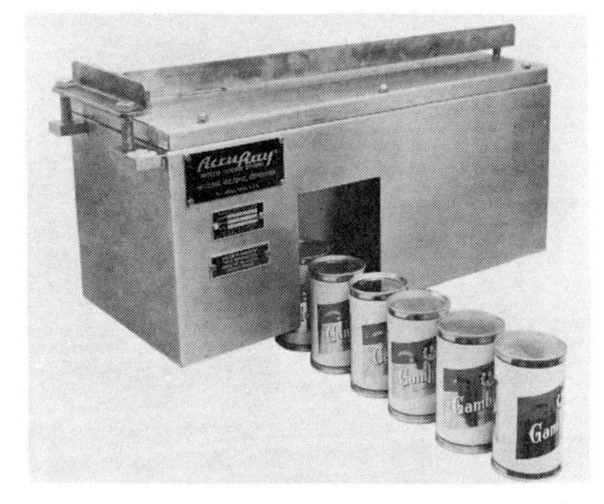

Fig. 13·36 Inspection-rejection systems measure the amount of radiation absorbed by the contents of the container. (*The Industrial Nucleonics Corp.*)

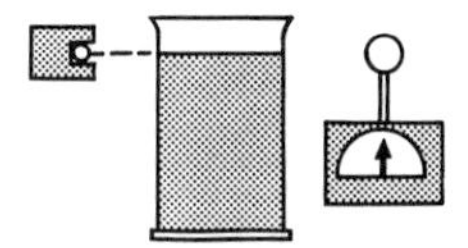

Control circuitry is not triggered when fill height falls within specifications.

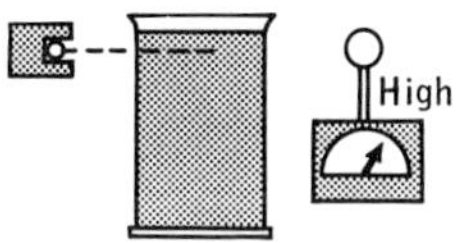

Over-fills reduce energy reaching detector; Control Instrument energizes motor to decrease fill.

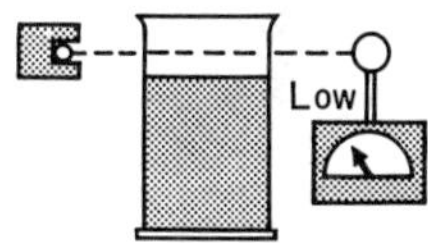

Slack-fills allow increased energy to reach detector; an "increase" signal is fed back to filler.

When the amount or level of dry materials in a bin is to be ascertained, certain difficulties arise because the material is not self-leveling. It may be sufficient or adequate to know that the level is at least up to a certain point. One very simple way of getting the desired indication is to secure a flexible diaphragm to the inside wall of the bin or tank. Then as the level of the material rises, a force will be exerted against the diaphragm and in turn transmitted to the switch against which the diaphragm pushes. This arrangement keeps the switch clean and also free of dust which might be explosive. By mounting a number of these devices around the bin, a series of level indications can be obtained.

Another way to determine the level of dry materials is shown in Fig. 13·37. This method is dependent upon the fact that a motor which is working against a resisting torque will require more current than will a motor which is running with what is essentially no load. For the indicated level L_1, the rotor which is driven by the motor does not touch the material. Consequently, the motor load is small and the current required by the motor is also small. But as the level rises to some value, such as L_2, the rotor blades will encounter more and more material which will have to be moved by the rotor. As the amount of material increases, the torque required to force the rotor through it will increase, and with it the motor current. By reading an ammeter, one can tell whether the level is up to the rotor. And by using a relay which is adjusted for a certain motor current, an alarm can be sounded, or the motor can be shut down when a predetermined level has been reached.

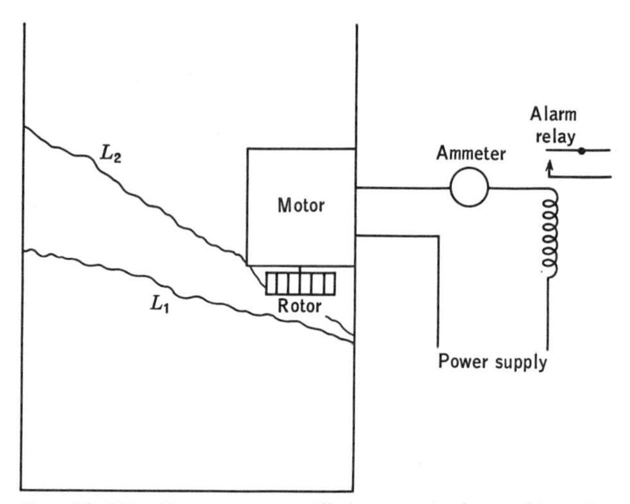

Fig. 13·37 More energy will be required to drive the motor when the rotor is impeded by the presence of material. This will require an increase in the motor current. The relay can be calibrated to operate at some preestablished value of current.

Finding level by weighing. In the preceding section the level of the liquid was inferred from a measurement of the weight of the material and hydraulic and pneumatic signals were used to evaluate the amount of the liquid. By using certain other relationships, the amount of material can also be weighed and the weight can be converted into electrical signals of some desired type.

For example, instead of supporting the left end of the tank on a bellows, as was the tank in the arrangement for weighing the loads, the tank is supported by a metal member which will change dimensions with the weight of the liquid. Cemented to the support material is a strain gage made of fine resistance wire. As the support contracts with the increased load, the resistance of the gage will decrease. With calibration, the change in resistance can be related to the amount of material in the tank. When suitable techniques are used, the point of measurement may be quite remote from the tank.

But if the material being weighed is a solid, a powder or some other dry material and not a liquid, then it will not distribute itself uniformly over the bottom of the bin or tank. Such nonuniform loading requires that the strain gages be applied to each of the supports, three or four as the case may be, Fig. 13·38. Then it becomes necessary to add the respective values of the strain gages in order to know what the total load is.

If the desired type of signal indication is current, then it becomes necessary to alter the system to something essentially like that indicated in Fig. 13·39. We may employ a system previously studied to furnish a

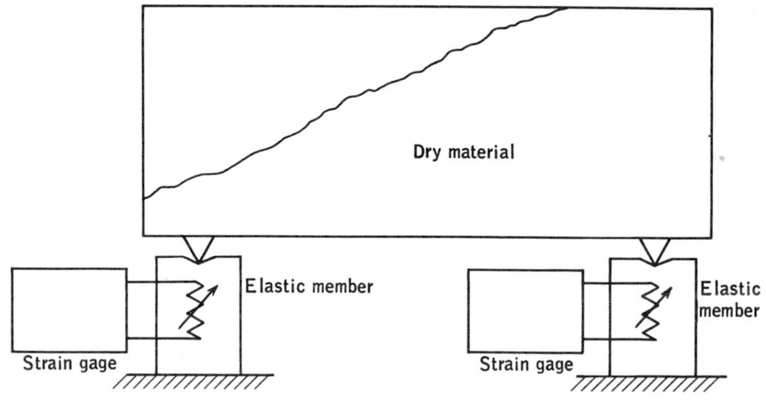

Fig. 13·38 Because most dry materials will not maintain a uniform level in a tank or container, strain gages must be provided at each support.

Fig. 13·39 The pressure created in a filled hydraulic system by the imposition of a load can be converted in any of several ways into electrical signals for transmission.

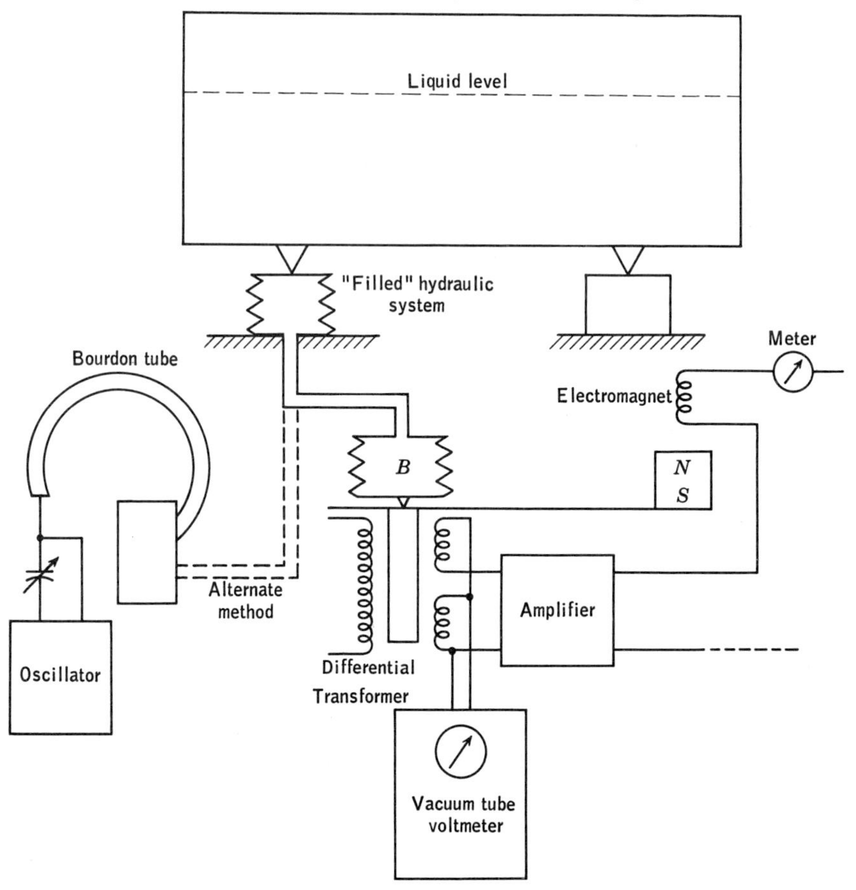

hydraulic pressure which can move the beam, and with it the differential transformer. The output of the transformer is then fed into an amplifier which furnishes an output signal in milliamperes. This output current establishes a magnetic field whose reaction with the permanent magnet creates a force of opposition that drives the beam in the opposite direction. A balance point is reached between the input signal and the torque developed by the output signal current. The value of the current is then a measure of the tank load.

For systems which employ potential measurements, bellows *B*1, Fig. 13·39, may position the core of the differential transformer and produce an emf which is an adequate representation of the tank load. Or one may use the hydraulic pressure to alter the capacitance of a tuned circuit, Fig. 13·40. As the capacitance changes with load, the output frequency of the oscillator may be made to change with it. In this manner, frequency can be made a measure of load.

When telemetering over extended distances is to be provided, the transmission of specific frequencies to indicate tank levels or other quantities simplifies the transmission problem. Inasmuch as the final value depends not on the magnitude of the received signal but only on the pitch or frequency, the intensity or magnitude variations which arise because of repeater-amplifier action have no effect upon the final interpretation. A weak signal of 800 cps has exactly the same meaning as a very strong signal of the same frequency.

Fig. 13·40 The pressure created by the liquid level may be used to alter the capacitance, and thus control or alter frequency or provide some other effect.

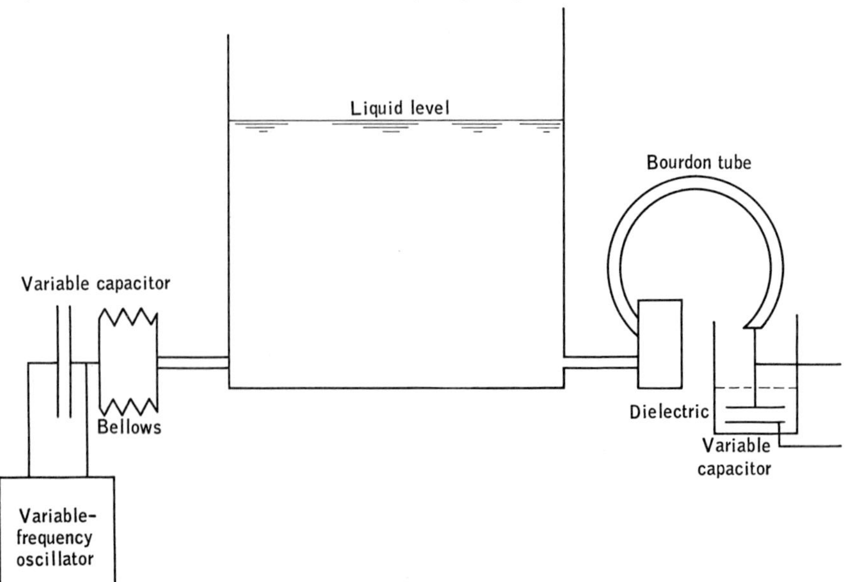

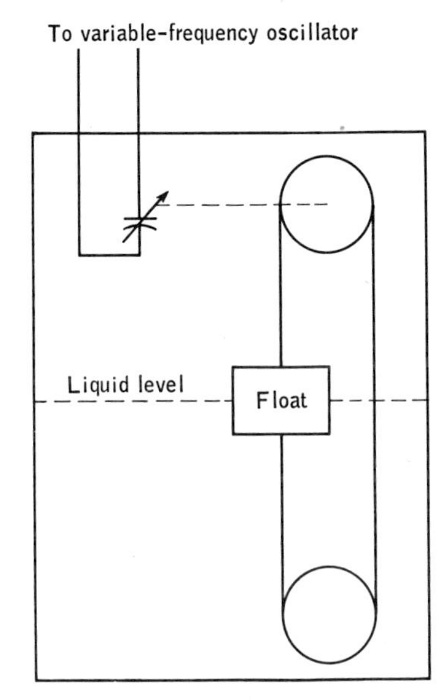

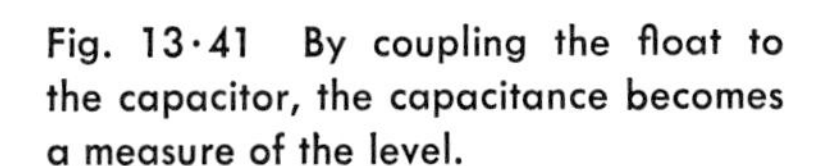
Fig. 13·41 By coupling the float to the capacitor, the capacitance becomes a measure of the level.

Pressure variations caused by level changes will also vary the capacitance. These capacitance changes may be used in many ways to produce the desired signal characteristics. For some industrial measurements the essential elements of Fig. 13·41 are employed. As the float moves up and down, it changes the capacitance of C. Again, how the variations in capacitance are used depends upon the system requirements. Or, a float which moves up and down with liquid level may drive two potentiometers, one of them designed for continuous rotation and geared to the second potentiometer. Each revolution of the potentiometer requires a change in level of 1 ft. The potential transmitted by this unit is then a direct measure of the position in parts of a foot. The second potentiometer then transmits a signal which indicates an integral number of feet. One industrial unit that uses this combination is shown in Figs. 13·42 and 13·43. When it is desirable because of temperature, pressure, or some other installation problem to keep the potentiometers or the indicating units outside the tank, magnetic drives may be used to communicate the position of the float to the unit.

The buoyant effect of liquids may also be used to determine liquid level. Assume that, in Fig. 13·44a, the supporting spring is stretched when there is no liquid in the tank. Then when the liquid level would just cause the float to be completely supported by the liquid, the spring would have contracted to its unloaded length. By attaching the lever arm to any desired part of the spring, any amount of movement may be made to correspond to the complete change in lever. For example, if the weight of the float stretches the spring 1 in., then a point halfway down the spring from the support will move but ½ in., a point a quarter of the

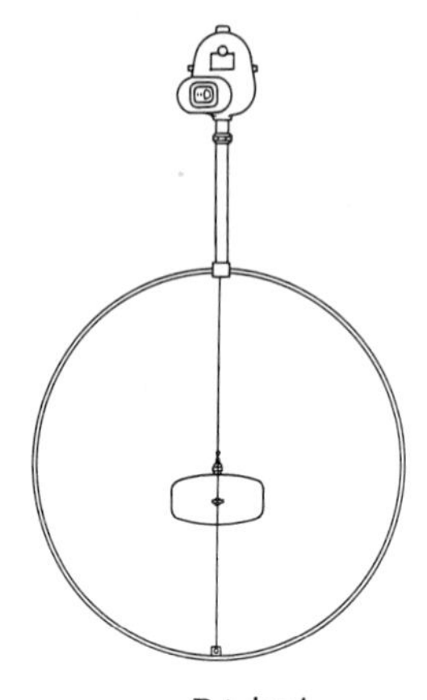

Fig. 13·42 Two potentiometers and associated meters may be used for showing the level in both feet and inches. The two potentiometers are geared together, with the inch potentiometer providing for an indefinite number of complete rotations. (*Vapor Recovery Systems Co.*)

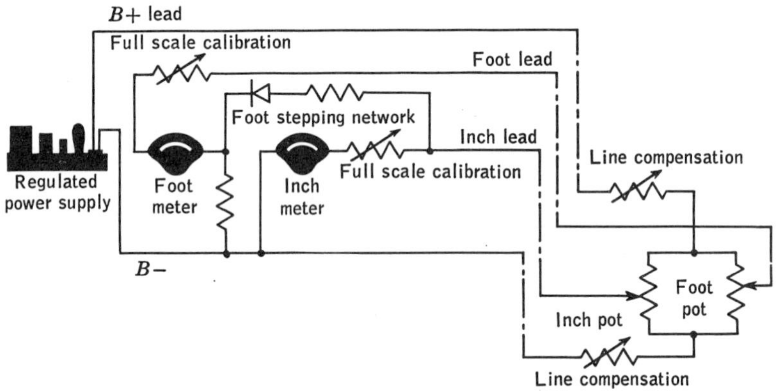

Fig. 13·43 These two meters are connected as shown in Fig. 13·42. To make the readings of significance, temperature measurements may be made at the same time so that temperature-weight density effects can be included. (*Vapor Recovery Systems Co.*)

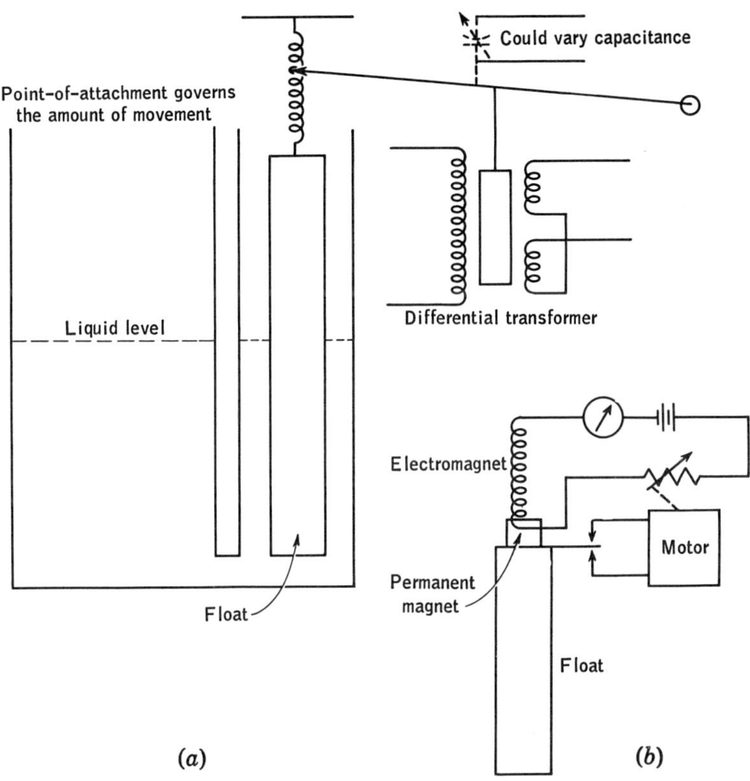

Fig. 13·44 The buoyant force on a partially submerged body is directly proportional to the liquid level, provided the object does not float. The buoyant force may be converted into electrical signals by the means suggested.

way down will move but ¼ in., and so on. Whether the lever moves a capacitor, the core of a differential transformer, or some other device is of little importance.

The liquid level may also be determined by measuring the current required to overcome the buoyant force of the liquid, Fig. 13·44*b*. If the level rises, the float will rise with it and complete a circuit to the motor. The motor, in turn, will reposition the rheostat until such a current passes through the electromagnet that forces of *repulsion* between the electromagnet and the permanent magnet are equal and the float is driven down away from the contact. The current required to accomplish this is then a direct measure of the level. Lever arrangements would be necessary to employ this system commercially.

With all of the many types of transducers available, there are hundreds of ways of converting any given reading into its electrical equivalent. The economics of the situation and the ingenuity of the designer will determine just what will be used in a vast majority of instances.

Experiment 13·1

Subject. Liquid level through conductivity measurement.

Purpose. To show how to construct a simple conductivity type of liquid-level gage, Fig. 13·23.

Theory. Almost all liquids will conduct electricity to some degree. Pure distilled water does not dissociate to form ions and electrons, but water which contains even slight degrees of impurity conducts current to some slight degree. If we can obtain a meter which takes so little current to operate it that the resistance of the liquid is of little consequence, then we should be able to determine the liquid level by finding the point at which conduction begins.

The vacuum tube voltmeter is an instrument of this type. It primarily requires potential for its actuation, and practically no current. When there is a complete break between the electrodes joining the grid and cathode, the resistance of the circuit is essentially infinite. The meter then indicates that no signal is being applied to the grid.

If, however, the liquid rises to such point that both electrodes are in the liquid, then there will be a slight current flow with a major portion of the battery voltage appearing across the grid resistor. This will then cause the meter to give a pronounced movement. Thus, the meter will indicate by a marked deflection or by remaining at zero. This meter and this type of circuit are good only for single-point indication unless more than one circuit is installed.

When the liquid can be prevented from collecting as a drop on the end of the electrode E, very exact level determinations can be indicated.

Apparatus. Vacuum tube voltmeter, battery, jar with tap water, pieces of wire.

Procedure. Connect the apparatus as shown in Fig. 13·23. Fill the jar with tap water to such a level that one of the wires can be easily removed or replaced in the solution. The other wire is to remain in the solution.

Record the amount of separation of movable electrode and the liquid when the meter makes its maximum deflection.

Record how much the electrode can be withdrawn from the solution without causing the meter to deviate materially from its maximum value.

Questions. How closely could the level be determined? Could it be determined within ¼ in. or closer?

Did the formation of drops on the electrode cause difficulty?

In a practical installation using these techniques, what would you have to consider?

Would there be any maintenance or service problems aside from tube replacement?

With what liquids would this method work well?

With what liquids would this method not work well?

SUMMARY

1. Liquid level may be measured in many ways. Unless floats are being used, almost all liquid-level measurements which use mechanical techniques are pressure or differential pressure measurements. Table 13·2 simplifies a comparison of methods.

2. The instruments, the precautions, and the installation procedures are essentially the same as they are for straight pressure measurements.

3. The sight glass is the simplest level indicator, but it does not lend itself readily to automatic reading. It cannot be used with certain viscous liquids, with liquids which are dirty, or with liquids which will attack the glass.

4. Floats cannot be used without much thought to the type of liquid, whether the liquid will cling to the float or react with it chemically, arc of travel, and the possibility of mechanical interference. If there are significant changes in the specific gravity of the liquid, then the float will not always indicate the correct level.

The use of magnetic means for transmitting to the pointer the position of the float eliminates, for many designs, the necessity for stuffing boxes and so on and largely eliminates the possibility of leaks from high-pressure tanks containing inflammable materials.

5. The determination of liquid level by weighing the tank and its contents makes it feasible to determine the level of materials which are too corrosive or too viscous for ordinary instrumentation.

6. The same means which can be used to weigh liquids can be used to weigh solids in bins or on conveyor belts.

7. Many level indications will be correct only so long as the specific gravity is at its normal value. Floats can indicate true height, but differential pressures and weights are expressed in equivalent levels for some given specific gravity or weight density.

8. Electrical means are employed for level determination when the signal must be transmitted long distances or when electrical signals are required for computers or data-logging systems.

9. Electrical systems may be extremely simple in nature when probes or simple conductivity tests are used, or they may be extremely complex. The particular system used will vary with the output signal requirements. Some systems furnish potentials; others furnish a current which is a measure of the variable; and still others convert the variable into frequency values.

10. Changes in electrical conductivity and heat conductivity are the basic elements in two of the simplest level devices which do not employ moving parts. By using several of these devices on a given tank, more than a high-low alarm is made available.

11. The change of capacitance with change in level has been one of the important ways of measuring level when the liquid is clean and nonviscous.

12. The use of radioisotopes has greatly expanded level measurements where the conditions are severe. Ability of a meter to operate without a hole in the tank is of much importance in gaging many types of materials.

13. For gaging the fill of small containers, either metal or fiber, radiation devices possess definite advantages.

14. Many different types of transducers may be used to convert some characteristic of level into a usable signal.

QUESTIONS

13·1 What is the importance of specific gravity or weight density values in the determination of liquid level?

13·2 What is the relationship between pressure and liquid-level measurements?

Table 13·2 Liquid-level Measuring Systems *

Systems	Applications	Components	Principles of operation	Ranges	Limitations	Advantages
Pressure gage	Open vessels	Standard pressure element; standard instrument components	Pressure element measures static pressure	Minimum is 0.1 in. water; maximum is practically unlimited	Distance between instruments and vessel limited; not applicable to tanks under pressure	Simple and inexpensive
Diaphragm box	Open vessels	Flexible diaphragm in bronze box connected to pressure gage	Static pressure transmitted through diaphragm, which seals gage line from liquid	Minimum imposed by diaphragm box is 1 ft water; maximum 50 ft water	Length of tubing to instrument should not be greater than 250 ft; closed-type box system needs purging equipment for corrosive or viscous liquids	Diaphragm keeps all fluids and solids out of line to meter; can be located at any level
Purge or bubbler	Open or closed vessels—depends on instrument used with purge system	Pressure or differential pressure measuring instrument; pressurized air, gas, or clean liquid supply	Purge fluid keeps measured fluid out of system, transmits liquid head pressure	Minimum is 0.1 in. water; maximum is practically unlimited	Limited only by range of instrument used, purge line pressure, and added purge maintenance	Ideal for corrosive fluids, viscous liquids, and liquids with suspended solids; pressure gages with air purge need not be installed at minimum level; mechanical manometers with gas purges do not need dropped range tube; pressure elements may be at any level
Differential pressure converter	Open or closed vessels; pneumatic transmission of variations in level to other instruments and controllers	Chamber with diaphragm; pivoted beam; feedback bellows; pneumatic system	Null balance-producing pneumatic signal proportional to liquid level	Infinitely adjustable from 14 to 200 in. water; for pressures to 1,500 psig, suppression up to 100%	Requires an air supply	Small, light, rugged, easy to maintain and calibrate; ambient temperature range 32 to 225°F; fast response; extremely small displacement, generally eliminating need for seals or purges
Mercury manometer	Open or closed vessels; interface level	**U**-tube meter body, mechanical, electric, and transmission, with indicators, recorders, and controllers	Pressure applied to legs or **U**-tube containing mercury; float transmits motion by giving signal proportional to level	Wide choice of ranges from 2½ to 700 in. water full scale	Must have clean, noncorrosive fluid in meter body	Meters are interchangeable for flow or specific-gravity applications; available for static pressure of vessels up to 4,000 psig

Nonindicating liquid-level controllers	Control of open or closed vessels	Bellows or bells actuating elements with electric or pneumatic control	Pressure due to liquid head actuates measuring element that direcly effects switching action or positions proportioning control transmitting systems	Minimum is 2 in. water; maximum is 160 ft water	Must have clean noncorrosive fluid in measuring system	Inexpensive, small size, easily adjusted for desired level control
Electric strain gage, unbonded	Open or closed vessels	Bellows or diaphragm, armature and frame with stressed wires; electronic recorder or recorder and controller	Pressure due to liquid operates bellows or diaphragm, changes stress and resistance of wires in bridge circuit with electronic instrument	Minimum is 0–0.05 psi; maximum is 0–10,000 psi	Clean fluids must be in contact with pressure elements	Well suited for remote electric transmission to centralized control room; can be extremely fast
Conductivity electrode	Open or closed vessels; control at set point or limit points	Electrode(s); detecting relay; valve or other controlling device	Conducting liquid completes circuit with electrode at predetermined level to initiate control action	Unlimited for high-low control	Liquid must not be chemically injurious to stainless-steel electrodes; not suited for viscous electrically conductive liquids that tend to build up deposit and short out electrodes	Simple and inexpensive; range unlimited for high-low control
Gamma-ray detector	Open or closed vessels	Radioactive source(s); Geiger counter or Ohmart cell(s); amplifier; electronic recorder or recording controller	Geiger counter or Ohmart cell picks up gamma radiation; amount depends on height of liquid; current transmitted to electronic recorder	2 ft per single fixed source; number of sources unlimited with Geiger counter; 15 ft for float-type installation; unlimited with Ohmart cells	Safety precautions may be needed to prevent overexposure of personnel to gamma radiation; gamma radiation might not be recommended for certain liquids	Gamma rays penetrate intervening materials; no connections needed from inside of vessel to outside
Float-actuated meter	Open vessel	Float, cable and counterweight operating pulley and cam	Rise or fall of float rotates pulley; cam actuates pen or pointer of instrument; readings can be transmitted electrically or pneumatically	Minimum is 0–6 in.; maximum is 0–30 in.	Narrow measuring span	Simple in operation; uses standard mechanical liquid-level meter
Change in resistance	Open or closed	Simple probe and relay or meter	Liquid "cools" the probe; low level prevents cooling, causes change in resistance	$\pm 1/16$ in.	Probe when out of liquid must not get dangerously hot	Very simple
Buoyant force or "loss in weight"	Can be designed for either	Float, force, and position-measuring devices	The buoyant force, or the loss in weight is directly proportional to the liquid level	Infinitely adjustable	Requires air or electrical power	Wide range, wide span, simple adjustment

* Used by permission of Minneapolis-Honeywell Regulator Company.

13·3 What is the difference between a sight glass and the **U**-tube manometer?

13·4 What are the objections to the use of floats for liquid-level measurements?

13·5 What precautions in the use of pressure gages must be observed when using such gages to measure level?

13·6 In what respects do bubbler and purge systems used for liquid-level measurements differ from similar systems used for pressure measurements?

13·7 If the biasing battery, Fig. 13·23, were to be reversed, what would be the effect upon the operation of the relay?

13·8 What must be the direction of the control current in the control winding, Fig. 13·23? Will it be affected by reversing the polarity of the battery?

13·9 If the relay in Fig. 13·25 were omitted, what type of instrument would we have? Does that type have any commercial applications today?

13·10 To function as a level detector, what must be the conditions for transfer of heat from the thermistor to the housing, and vice versa?

13·11 If the probe in Fig. 13·26 were to be made of stainless steel with the active elements 4 in. long, how much would point A move for a temperature differential of 20° between the two tubes? 100°? Would either of these temperatures be adequate to actuate a switch requiring a movement of 0.0003 in.?

13·12 If the tank, Fig. 13·29, were to contain two liquids which had specific gravities of 2.5 and 0.5, what intensity signal would you expect to receive from the interface? Would you expect this to mask the true level of the combined liquids?

13·13 From Chap. 19 select a radioisotope which might be effective if the tank, Fig. 13·32, were made of ½-in. steel.

13·14 What alternative methods can you suggest to determine the product level or weight of sealed containers? The use of radiation means, whether X rays or radioisotopes, may be impractical because of the potential damage to photographic products.

13·15 What are some of the reasons why you might prefer X rays rather than radioisotopes for level measurements? In what ways might the isotopes be superior?

13·16 For a level unit such as is shown in Fig. 13·33, what would have to be true of the relative absorption characteristics of the liquid and the gas above it?

13·17 With reference to Question 13·16, would the same conditions have to be true in Fig. 13·34? Assume that variations in the amount of energy received by receiver R will cause the servomotor to move the receiver up or down so that it again receives the correct amount of radiation. If there is only air between source and receiver, the energy received will be too much; if there is too much liquid intervening, then the energy received will be too little.

13·18 Under what conditions might strain gages and hydraulic or pneumatic means be used to measure dry materials with only one active gage instead of one on each support?

13·19 Assume that a mistake was made in connecting up the electromagnet, Fig. 13·36, with the result that the two magnets attracted instead of repelled each other. What would happen to the operation of the transducer? Would the output current be correct? Would there be any effect upon the output of the oscillator, which we will assume is also connected?

PROBLEMS

13·1 What pressure will be created by a column of liquid 20 ft high if the weight density is 125 lb/ft^3?

13·2 If a column of liquid 10 ft high creates a force of 100 lb on an area of 10 in.^{2}, what is the specific gravity of the liquid?

13·3 A float for a liquid-level gage moves through a vertical distance of 4 ft. If a change of specific gravity causes the float to ride 2 in. higher than the designed value, what per cent error will this cause?

13·4 A pressure box contains air at 70°F. What will be the internal pressure, and what will be the per cent of the original volume if the box is submerged in water 100 ft at 40°F?

13·5 A bubbler system is to be used to measure the level of a reservoir. If the span is to be 10 to 30 ft of water, what are the minimum pressures required? What would be the desirable minimum supply pressure?

13·6 If a tank weighing 10 tons with its normal load is to be weighed by supporting it on a piston having an area of 100 in.^{2} at one end, what hydraulic pressures will be set up in the system?

13·7 Draw a calibration graph for the tank in Prob. 13·6. Show the relationship between the indicated pressure and the load in the tank.

13·8 If a maximum air pressure of 100 psi is available, how large a lifting area will be required to sustain the 10-ton load of Prob. 13·6?

13·9 What should be the size of a steel support if it is to bear one-half of a combined load of 20,000 lb and to have a strain of 0.0005 in. when the load goes from 0 to 20,000 lb? If the length of the elastic support is 10 in., what is the change in length?

13·10 If brass replaced steel in Prob. 13·9, because of a corrosion problem, what would be the diameter of the brass necessary to keep the strains and total elongation the same?

13·11 Substitute aluminum for steel in Prob. 13·9 and repeat.

13·12 If the current in a thermistor is 1 amp and the resistance is 6 ohms, what electric power is being converted into heat?

14 Temperature Measurements, Mechanical Transducers

Temperature measurements are among the most common and the most important measurements made in controlling industrial processes. The precise control of temperature is the key factor in many chemical operations. Consequently, it is most important that the various ways of measuring temperature be well mastered, that the advantages and disadvantages of each method be well understood, and that the operating limitations in terms of time of response, temperature range, distance of operation, and compatibility with other control elements be considered for each installation.

Temperature measurements can be made in many ways. The more common and most important are listed below. They have been divided into two general classifications: those which are primarily electrical or electronic in nature and those which are not. The second grouping will be listed first because it was developed first and because there are probably more applications of this type of equipment than of electrical or electronic equipment at the present time.

Nonelectrical Methods

1. Change in volume of a liquid when its temperature is changed
2. Change in pressure of a gas when its temperature is changed
3. Changes in the vapor pressure when the temperature is changed
4. Change in dimensions of a solid when its temperature is changed

Electrical Methods

1. Electromotive forces generated by thermocouples; the temperature is measured by measuring the voltage generated when the common junction of two dissimilar metals is heated.

2. Change in resistance of materials as their temperature is changed.

3. Temperature measurement by comparing the colors of a controllable filament and the object whose temperature is sought.

4. Temperature measurement by ascertaining the energy received by radiation.

Another classification of temperature-sensing devices might be made by using the temperature ranges of the units as the basis for comparison. Figure 14·1 relates the various methods. It will be noted that several methods may be indicated for a given temperature span. However, all will not be equally well suited to any given temperature measurement. Therefore, the selection must be based not only on the range and span but also on such factors as life, speed of response, accuracy, and means of mounting the sensing element. In some situations one will not be able to get within several feet of the object whose temperature is sought.

Fig. 14·1 A wide variety of instruments is required to cover the temperature spans used in industry. There is a greater and greater need for instruments which will measure accurately at ranges far above and below those now commonly used.

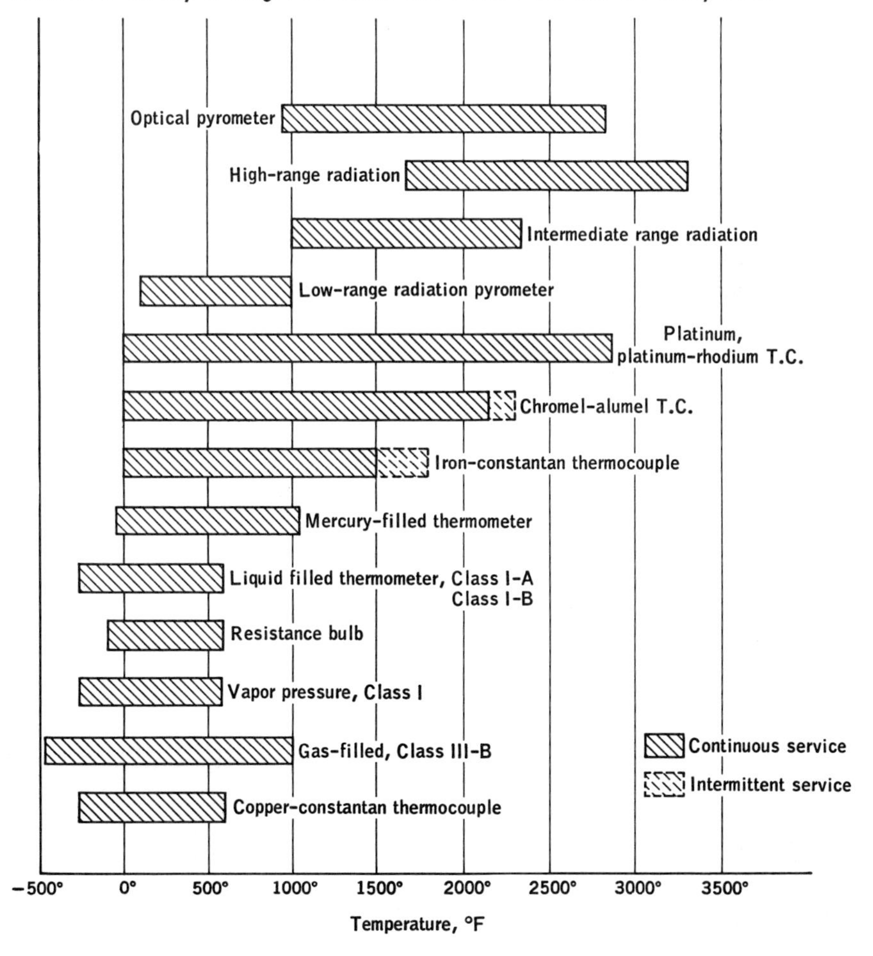

Table 14·1 Temperature Range of Various Manufacturers' Thermometers in the Four Classes

(Temperatures are expressed in degrees Fahrenheit)

Manufacturer	Class I	Class II	Class III	Class IV
Bristol	−125 to +500	−50 to +600	−125 to +1000	
Fischer and Porter	−300 to +500	0 to +600	−400 to +1000	−38 to +1000
Foxboro	−300 to +600	−300 to +560	−450 to +1000	
Minneapolis-Honeywell	. . . .	−40 to +600	−125 to +800	−40 to +1000
Taylor	−125 to +500	0 to +600	−350 to +1000	−70 to +1200

In this section we will consider the various factors which govern the suitability of each type of mechanical transducer. Much attention will have to be given to the suitable location of the sensing element to ensure that we are measuring what we desire to measure. Electrical units will be treated in Chap. 15. The subject summary, questions, and problems will also be found in Chap. 15.

Classes of thermometers. In order to avoid confusion when specifying thermometers, the manufacturers of American instruments use the following classifications:

Class I Liquid-filled thermometers, not including mercury-filled thermometers

Class II Vapor-pressure thermometers

Class III Gas thermometers

Class IV Mercury-filled thermometers

The Society of Manufacturers of Scientific Apparatus designates mercury-filled thermometers as Class V.

Table 14·1 indicates the temperature ranges which are available in the various classes.

Tables 14·1 and 14·5 relate many characteristics of the various classes of thermometers. They will be found useful as guides to performance.

LIQUID-FILLED THERMOMETERS

The use of volume changes to indicate temperature changes is probably one of the oldest methods employed for temperature determination, yet it is as important today for temperature measurements in the general range of −50 to +1000°F as any other method available. In many instances, it offers definite advantages. It is probably the one used in thermometers of our common acquaintance. The height of the column of liquid indicates the temperature by measuring the volume of that liquid at that temperature. This is, or by accurate calibration can be made, an accurate way to express temperature. Table 14·2*b* lists several liquids and their coefficients of cubical expansion. Theoretically, any of the liquids listed could be used in a thermometer. However, practice has limited use to only a few—alcohol, the hydrocarbons, and mercury being the most common. When the liquid is allowed to expand into a glass tube so that its height may be read, two limitations are experienced: a person must determine the temperature by observation, and the upper limit cannot be greatly above the boiling point of the materials because of the pressures that would be created. For these reasons, this type of thermometer is limited to on-the-spot reading and to the lower temperature ranges.

When a bellows is used, it is quite likely that the stresses employed

Table 14·2*a* Coefficients of Cubical Expansion

Material	Temp, °C	Coefficient $\times 10^{-4}$
Iron	0–100	0.355
Paraffin	20	5.88
Platinum	0–100	0.265
Porcelain	0–100	0.108
Quartz	50–60	0.353
Silver	0–100	0.583
Sulfur	13–50	2.23
Tin	0–100	0.689
Zinc	0–100	0.893

Table 14·2*b* Cubical Expansion of Liquids *

Material	Temp range, °C	$\alpha \times 10^3$	$\beta \times 10^6$
Amyl alcohol	−15–80	0.900	0.657
Benzene	11–81	1.176	1.278
Carbon tetrachloride	0–76	1.184	0.899
Ether	−15–38	1.513	2.36
Glycerine	....	0.485	0.489
Mercury	0–100	0.182	0.0078
Water	0–33	−0.064	8.505

* $V_t = V_0(1 + \alpha t + \beta t^2 \cdots kt^n)$.

The values given above for α and β are too large by the indicated factor. In use the values of α must be divided by 10^3 and the values of β by 10^6.

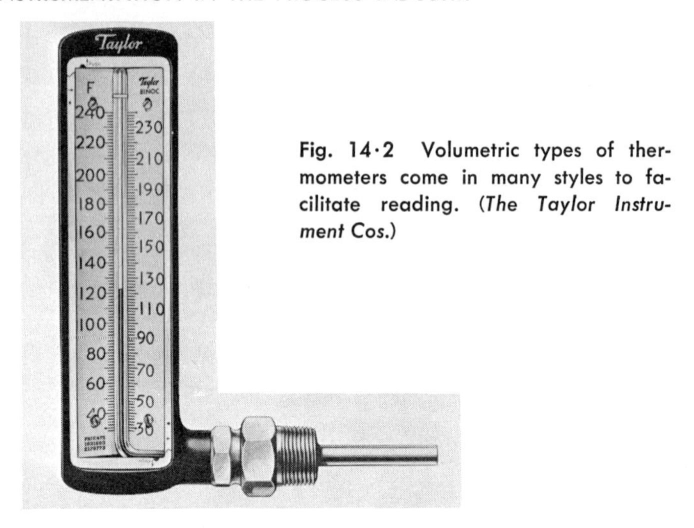

Fig. 14·2 Volumetric types of thermometers come in many styles to facilitate reading. (*The Taylor Instrument Cos.*)

to expand the bellows will be quite small, only a few pounds per square inch. But in the Bourdon spring, the pressures may rise as high as many thousands of pounds per square inch. It might be pointed out at this point that the Bourdon spring is used extensively in pressure gages of medium and high range; the bellows type of gage is more often used to measure low values of pressure in the region of a few ounces or pounds per square inch.

Strictly speaking, with the bellows method of temperature measurement we are doing nothing but measuring the pressure established in a closed system when a liquid attempts to expand or contract as it is heated or cooled. To simplify our use of this information, we calibrate the dial in terms of the temperature which created the pressure instead of the pressure created at that temperature.

The Bourdon spring type of gage was apparently invented in France in 1849. Over the intervening time there has been only one material change in it. Improved materials, better heat-treating procedures, and superior fabricating techniques are now combined to produce gages of high accuracy, long life, and superior performance. The various pressure ranges, the materials with which the spring will come in contact, the number of operating cycles, and the forces which must be developed will dictate which of the many materials available will be selected. Additional information will be found in Chap. 9.

Industrial versions of thermometers which employ volumetric changes of the fluid are many. Some are permanently installed, while others are portable. Requirements for easy reading have created designs such as those shown in Fig. 14·2. When only occasional temperature checking by an operator is required, this type of thermometer is quite satisfactory. It will be noted that it cannot be used for temperatures in excess of 1200°F.

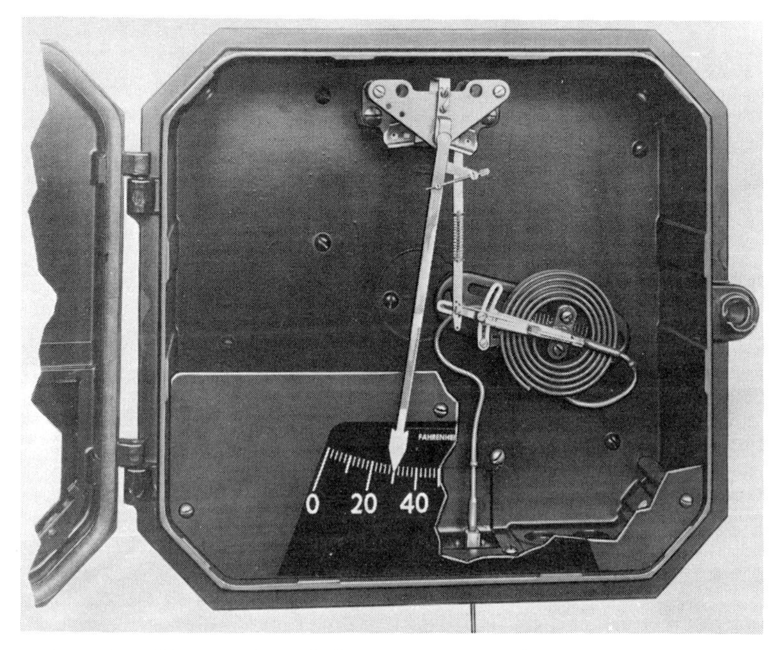

Fig. 14·3 The use of a spiral pressure-sensing element in a temperature device of the filled-system type generally eliminates any need for the gear train which must be used if a Bourdon spring is used. (*The Taylor Instrument Cos.*)

In industrial work, it is often advantageous to employ liquids listed in Table 14·1*b*, or other liquids, in a closed system of the general type shown in Figs. 14·3 to 14·5 to obtain limited degrees of remote reading, as well as greater ranges. When the liquid expands in a closed system, it has to do so by changing the volume of the system which holds it. Even though mercury be enclosed in a steel bulb and a steel capillary tube to connect the bulb with the bellows or Bourdon spring, the pressures developed by the mercury will be great enough to cause the steel to expand. It will be recalled that, in Chap. 3, elastic materials are defined as those materials which will return to their original dimensions when the dis-

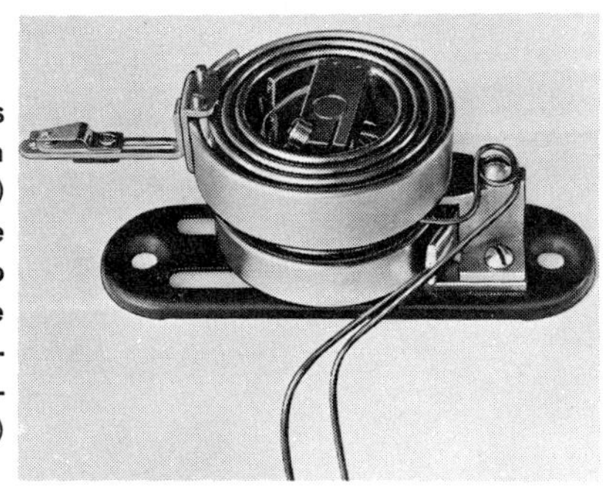

Fig. 14·4 Typical Bourdon spring assemblies of the spiral type. By mounting two Bourdon springs together, it becomes possible to (1) measure two temperatures with the same instrument, (2) obtain the difference of two temperatures by causing one spring to rotate opposite to the other, and (3) provide complete instrument compensation for any ambient condition. (*The Taylor Instrument Cos.*)

torting forces are removed and whose change in dimension is proportional to the force. The materials used for the bellows and the Bourdon spring are selected especially for their ability to change their dimensions at a rate which is almost exactly proportional to the pressures applied to them.

One of the advantages of the closed pressure system is that the indicating portion can be located at a distance from the sensing bulb. Certain limitations are, however, imposed on the distance. Because the capillary tubing has an internal diameter of approximately 0.020 in., there is time lag in transmitting pressure variations to the spring element. When the capillary is long, this lag may be of significance in certain operations, particularly if gas or vapor pressure is the actuating force.

Perhaps of even greater importance is the requirement for some type of temperature compensation when the capillary tubing has a length of more than approximately 20 ft. Temperature compensation, which may take several forms, is required for high accuracy when the volumes of the capillary and the spring are an appreciable percentage of the volume of the pressure system. If the spring element and the capillary are at

Fig. 14·5 Class I liquid-filled thermometer systems employ bimetallic compensation to provide for temperature changes within the case. Such compensation is generally adequate when the capillary tubing has a length of less than 10 ft. (*The Bristol Company*.)

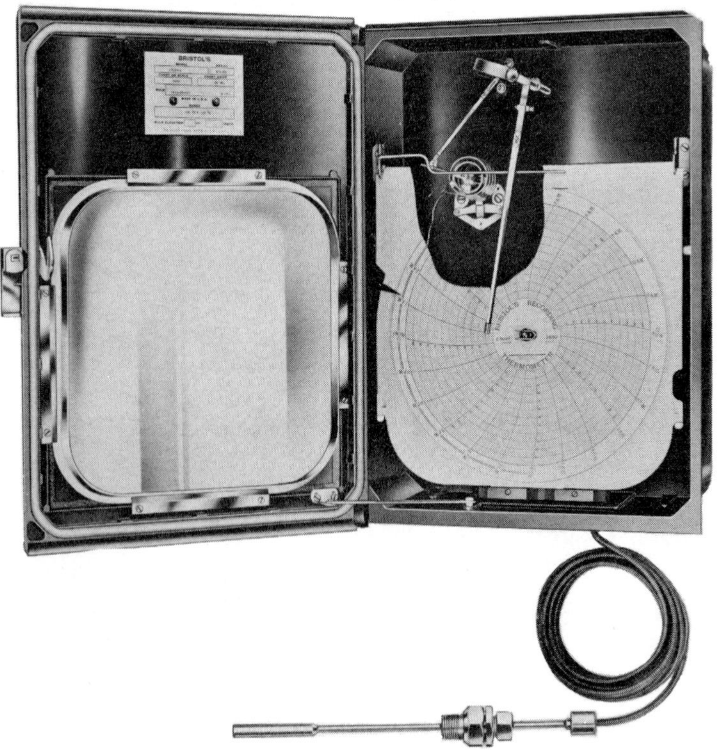

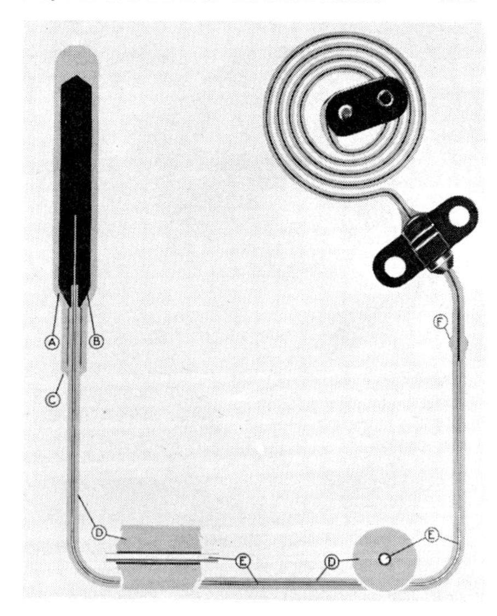

Fig. 14·6 In one design, temperature compensation for mercury-actuated systems is obtained by inserting a special alloy wire into the capillary. Any change in the volume of the system is offset by a corresponding change in the volume of the wire. (*The Taylor Instrument Cos.*)

different temperatures than those used in calibration, then their volumes will be different. The pressures developed by the confined liquid will then not be the same as the calibrating values, so an error will exist.

1. To take care of temperature changes in the case, a piece of bimetal can be employed to move the pen or pointer an amount which just compensates for the changes in the volume of the Bourdon spring and the capillary within the case, Fig. 14·5.

2. To compensate for volume changes of the capillary and of the mercury filling it, Taylor Instrument Companies employ a wire whose coefficient of expansion is different from that of the tubing. The difference in the expansions of wire and tubing leave the volume of the tubing changed by enough to correct for changes in volume, Fig. 14·6.

3. If the volume of the bulb is made large with respect to the volume of the tubing and spring, ambient temperature errors are minimized.

4. A duplicate spring and tubing system is used to give complete compensation. Figure 14·7 shows the way in which the extra system is used.

Fig. 14·7 When capillary lengths exceed 10 ft, complete compensation may be required for liquid-filled systems. Two systems are required. With the exception of the bulb, the two systems are identical and work in opposition to each other. (*The Bristol Company.*)

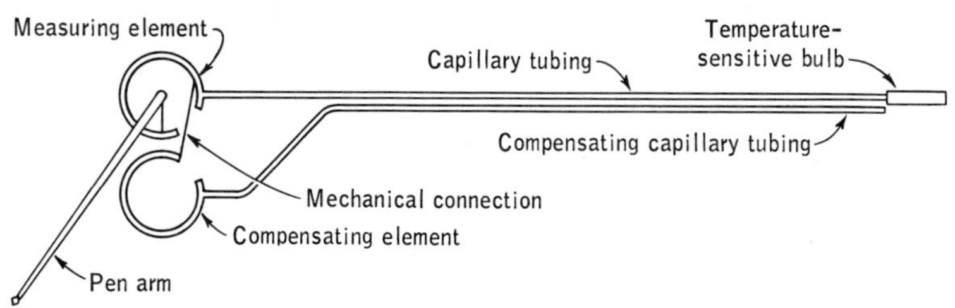

Except for the sensing bulb, the second system is the same as the first. All of the factors which cause the system to deviate from the true value are corrected by the second unit, which is so connected that it works in opposition to the main unit to position pen or pointer. Since this system can sense only changes in the capillary and spring, it can in no way affect the temperatures reported by the bulb. Occasionally the second system is not a duplicate of the first but does insert into the system the proper correction.

The bulb. The rate of response of a thermometer is dependent upon the material of which the bulb is made, the thickness of the walls of the bulb, and the volume of material that undergoes the temperature change. Because of the chemical activity of mercury, steel or stainless steel is used for the bulb and the capillary. Pressures of approximately 2,000 psi require relatively thick walls. For uncompensated units, the bulb usually has a diameter of about $\frac{1}{2}$ in. and a length which is between $3\frac{1}{2}$ and $5\frac{1}{2}$ in.

Operating range and span. By employing eutectic mixtures of mercury amalgam for the lower temperatures and mercury under high pressure to raise the boiling point, the mercury thermometer using the closed pressure system can operate between -70 and $+1200$°F. More usual limits are from -40 to $+1000$°F. Temperature spans as low as 100° can be obtained over the entire range.

When the use of mercury is undesirable because a broken instrument could poison food or fog film, organic liquids are sometimes used. This makes possible lower minimum temperature ranges to -300°F but limits the upper values to about 600°F.

GAS THERMOMETERS

In Chap. 3 it was pointed out that the change in pressure of a confined gas is directly proportional to the absolute pressure (Charles's law). Stated mathematically, this becomes:

$$\frac{T_1}{T_2} = \frac{P_{1\ \mathrm{ab}}}{P_{2\ \mathrm{ab}}}$$

Although no known gas obeys Charles's law perfectly, there is so little deviation that the gas thermometer becomes one of the most accurate methods for measuring temperature within its operating range, which extends from approximately -450 to $+1000$°F.

Although any of the gases listed in Table 14·3 could be used in commercial gas thermometers, nitrogen and helium are usually employed in practice. Both of these gases are chemically inert, have good values for their coefficients of expansion, and have low specific heats.

Table 14·3 Thermometer Gases

Gas	Critical temp, °C	Specific heat at constant pressure	Viscosity, cgs units $\times 10^{-6}$	Coef of expansion, constant pressure
Air	−140	0.237	170	0.0037
Carbon dioxide	31.1	0.203	139	0.0037
Helium	−267	1.25	195	0.0037
Hydrogen	−235	3.40	97	0.0037
Nitrogen	−146	0.24	163	0.0037
Oxygen	−118	0.218	212	0.0037

Despite the fact that hydrogen has a low viscosity, it has a specific heat which is many times that of the other gases listed. It has one other characteristic which rules out its use, particularly at the higher temperatures: it can diffuse through the walls of the system. The rate may be low, but it can happen. Of equal concern about hydrogen when it is present in a gas whose temperature is being measured is its ability to infuse *in* through the sensing bulb and create within the bulb an additional pressure which is not a function of temperature. Helium also possesses this characteristic, but to a much smaller degree. For the higher temperatures, therefore, nitrogen is preferred. One manufacturer makes it possible to use a quick visual inspection for decrease in the pressure created by the pressure-creating medium. Before filling, a straight line is drawn across successive turns of the helix. When filled, and at the operating range, the lines should be displaced. If they are aligned, then leakage has occurred and the instrument must be repaired.

Construction. The gas thermometer has a construction which is almost identical with that of the liquid-filled system. The only major difference is the use of a confined gas instead of a liquid to generate a pressure which is proportional to the temperature. In practice, we find the same problem of temperature compensation for case and capillary that was discussed when the liquid-filled thermometers were considered. The solutions are practically the same for both problems: keep the volume of the bulb large compared with that of the rest of the system; employ a bimetallic strip as part of the indicating linkage for case temperature compensation; or use a second system which is a duplicate of the first but works in opposition to it so far as the ambient temperature is concerned.

Inasmuch as the capillary and the Bourdon spring are essentially the same in both systems, those portions of the system will not be discussed again. Attention should be paid to the bulb, which is different in a number of respects.

The bulb is much larger than that of the liquid-filled system. Its length may vary between 5 and 20 in.; its diameter is approximately 1 in. The volume is likely to be between 50 and 100 cm^3. If an averaging bulb is to be used to sense the temperature over an extended space, its internal diameter may vary between $\frac{1}{16}$ and $\frac{1}{2}$ in. The length of an averaging bulb may be as great as 200 ft. For good performance, the volume of the bulb should be at least 8 times that of the rest of the system.

Temperature range and span. The customary pressures used when the gas thermometers are filled at room temperature vary between 150 and 500 psi. These values are dictated by the necessity for using pressures such that the pressure changes which occur with small temperature changes will be great enough to overcome the friction of the device and hence register correctly. Whereas the mercury-actuated thermometer can have a span of 100°F or less, the span of the gas thermometer depends upon the range of the instrument.

$$\frac{\text{Maximum absolute temperature}}{\text{Minimum absolute temperature}} = \text{more than } 1.3$$

$$\frac{\text{Maximum temperature, °F} + 460}{\text{Minimum temperature, °F} + 460} = \text{more than } 1.3$$

VAPOR-PRESSURE THERMOMETERS

Because the vapor pressure of a liquid is a function of temperature, it is very often used in commercial temperature-measuring devices. Its high speed of response, lower cost, ease of repair, and nonlinear scale (the scale can be large where the thermometer is normally used) are reasons for its extensive use.

This system has all of the major components of the liquid-filled and the gas-filled systems for temperature measurements. However, it is a hybrid in that its sensing bulb contains a liquid but is not full of it and the pressure transmitted to the spring is created by a vapor. However, it is normal for a liquid to fill the capillary and transmit the pressure to the pressure-responsive element.

All liquids with free surfaces have molecules leaving and returning to the surface at all times; the number leaving depends upon the temperature. These molecules develop a pressure which is indicative of the temperature. If enough leave so that the pressure that they develop is equal to the atmospheric pressure, we say that the liquid is boiling. If we reduce the pressure above the liquid, the boiling will take place at a lower temperature because it is easier for the molecules to escape; if the surrounding pressure is increased above atmospheric, boiling will take place at a higher temperature. In a confined space these escaping molecules will increase the surface pressure and automatically raise the boiling point. So long as liquid is present in the bulb there will always be a perfectly definite relationship between the temperature and the pressure developed by the escaped molecules for any given liquid.

The process might be compared with that of a steam boiler. As the boiler pressure increases, the temperature of the water is raised. For a given boiler pressure, there is a fixed temperature at which there can be both water and steam in the boiler. Figure 14·8 shows the vapor pressures developed by the more common liquids that could be used in vapor-pressure thermometers. In selecting the liquid for a particular temperature range, attention must be paid to the following:

1. The boiling point of the liquid must be below the lowest temperature to be measured; otherwise, the vapor pressure cannot develop sufficient power to move the meter or pen.
2. The liquid must be inert with respect to the bulb, capillary, and pressure element.
3. The chemical must be commercially available in pure form.

Fig. 14·8 Vapor pressure–temperature relationships of common charging liquids. (*Robertshaw-Fulton Controls Co.*)

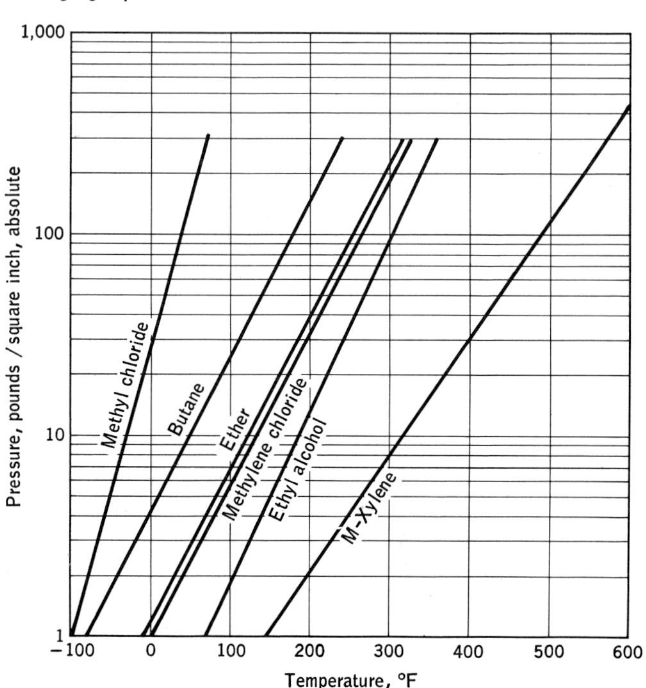

4. The highest temperature to be measured must be well below the critical pressure for the liquid.

Five of the more common liquids used for thermometer fills are ethyl alcohol, ether, methyl chloride, sulfur dioxide, and toluene.

Whereas the mercury-filled thermometer has such high pressures that the pressures established by the columns of mercury in the capillary are of no consequence and the gas-filled thermometers have a fill whose weight is insignificant, thermometers of the vapor-pressure type are generally designed for definite bulb positions with respect to the pressure element. If the vapor pressure is 20 psi when the bulb is moved 6 ft below its designed position, this might decrease the pressure at the gage by 10 per cent. If the bulb is moved to a position 6 ft above the gage, the pressure can be increased by 10 per cent. Consequently, it is important that any differences in position of bulb and indicating element be considered. This is most important if the instrument is not used under the conditions for which it was designed.

Ambient effects. Because of the possibility of having the free surface at some point other than the surface of the liquid in the bulb, vapor-pressure thermometers are often not reliable when measuring temperatures near the ambient. The use of a second liquid, of much higher boiling point, in the capillary tubing will ensure that the evaporating surface is in the bulb. For temperatures which will always be above or always be below those of the capillary and case, vapor-pressure units may be satisfactory.

Overtemperature protection. In both the liquid-filled and gas types of thermometers, the pressure is essentially a linear function of temperature, so the parts must be designed to take above-normal pressures that could be encountered when the thermometer is subjected to high temperatures. In the vapor-pressure type, overtemperature protection can be built in by limiting the amount of liquid in the bulb. After all of the liquid has been evaporated, increases in temperature will cause only small increases in the pressure of the system.

Bulb design. The construction of the bulb plays a large part in the speed of response of all vapor-pressure thermometers. The larger the bulb and the thicker the bulb walls, the more sluggish is the response. In the design of vapor-pressure thermometers, the bulb can be made very small, relatively. It need have a volume only slightly more than that of the capillary and pressure-sensitive element together. This small size together with relatively thin bulb walls give this type of thermometer its higher speed of response.

If the bulb is made of a long capillary tube, the vapor-pressure thermometer can be used to monitor temperatures at all points along the capillary and will respond to the highest one. It does not average the temperature, but indicates the highest value.

Table 14·4 Comparison of Filled Thermal Systems *

	Class I,† liquid	Class II,†‡ vapor pressure	Class III,† gas	Class IV,†§ mercury
Scale	Even	Uneven unless mechanically corrected; increases with temperature	Even graduation	Even graduation
Range	−300 to +1200°F	−300 to +600°F	−350 to +1000°F	−70 to +1200°F
Minimum span	50°F	60°F	150°F	100°F
Bulb size	Medium	Relatively small; controlled partly by length of tubing	Relatively large; should be at least 8 times volume of tubing and gage	Depends on desired temperature span
Maximum tubing lengths	200 ft	200 ft	175 ft	150–200 ft
Response time		Next to gas thermometer in response time	Fastest response	

* All of these values may not be obtainable from any one manufacturer.
† Class I (A: fully compensated for case and ambient temperature)
II (B: case compensated)
‡ Class II. A suffixed D indicates dual fill.
§ Class V, Scientific Apparatus Manufacturer's Association.

Tables 14·1 and 14·5 should be considered as guides in the selection of suitable instruments.

BIMETAL THERMOMETERS

There are many temperature readings which are not of such a nature that they must be monitored continually. Nor are they indicators of temperature which will be used for control purposes. They are guides only as to satisfactory operation. For temperature determinations of this sort, the bimetal thermometer is widely used. It is easy to read, will cover a temperature range up to 1000°F, and will take much abuse in times of vibration. It is also easier to read than is the conventional fluid-column thermometer. Newer designs make it possible to position the face to improve ease of reading.

For its actuation, the bimetal thermometer depends upon the difference in the expansion ratios of two unlike metals. By coiling the metal into a spiral or a helix, Fig. 14·9*b*, it is possible to obtain sufficient ro-

Table 14·5 Typical Thermometer Characteristics *

Points of comparison	Class IA Liquid expansion, fully compensated	Class IB Liquid expansion, case compensated	Class II Vapor pressure, uniform scale	Class II Vapor pressure, nonuniform scale	Class IIIB Gas pressure, case compensated
Bourdon-spring type	Five-turn helical	Five-turn helical	Five-turn helical	Five-turn helical	Five-turn helical
Bourdon material	Beryllium copper	Beryllium copper	Beryllium copper	Beryllium copper	Beryllium copper
Tubing lengths	Up to 150 ft	Up to 40 ft	Up to 150 ft	Up to 150 ft	Up to 200 ft
Typical sensitive bulb size	3/8 by 3 in.	3/8 by 3 in.	9/16 by 4 in.	9/16 by 4 in.	7/8 by 6 in.
Filling fluid	Hydrocarbon	Hydrocarbon	Volatile liquid	Volatile liquid	Nitrogen
Temperature limits	−300 to +500°F	−300 to +500°F	−50 to 450°F	−50 to 450°F	−400 to +1000°F
Type scale	Uniform	Uniform	Uniform	Incr'g increments	Uniform
Minimum span	40°F	40°F	100°F	100°F	200°F
Maximum span	500°F	500°F	150°F	150°F	1000°F
Case compensation	Complete	Complete	None required	None required	Complete
Tubing compensation	Complete	Partial	None required	None required	Partial
Elevation error	None	None	Can correct	Can correct	None
Gross ambient error	None	None	Yes	Yes	None
Barometric error	None	None	Very slight	Very slight	Negligible
Overrange available	Yes	Yes	Small	Small	Yes
Underrange available	Yes	Yes	Yes	Yes	Yes
Torque (output at pen spindle)	12 to 20 gr-cm/%	25 to 30 gr-cm/%	12 to 20 gr-cm/%	12 to 20 gr-cm/%	10 to 20 gr-cm/%
Linearity (max deviation from)	0.5% span	0.5% span	0.5% span	10 to 20%	0.5% span
Sensitiveness	0.2% span	0.2% span	0.2% span	0.2% span	0.2% span
Hysteresis	0.25% span	0.25% span	0.4% span	0.4% span	0.4% span
Repeatability	0.4% span	0.4% span	0.4% span	0.4% span	0.4% span
Accuracy (factory calibration)	0.5% span	0.5% span	0.5% span	0.5% span	0.5% span
Guaranteed accuracy	1% span	1% span	1% span	1% span	1% span
63% response:					
Bare bulb, agitated water	4 to 8 sec	4 to 8 sec	2 to 5 sec	2 to 5 sec	5 to 10 sec
Bare bulb, agitated air	20 to 40 sec	20 to 40 sec	10 to 25 sec	10 to 25 sec	25 to 50 sec
Well, agitated water	12 to 24 sec	12 to 24 sec	6 to 15 sec	6 to 15 sec	15 to 30 sec
Well, agitated air	60 to 120 sec	60 to 120 sec	30 to 75 sec	30 to 75 sec	75 to 150 sec
Capillary tubing materials	Copper or 321 stainless steel				
Protective armor materials	Bronze, 302 stainless steel, lead, or polyvinyl chloride–covered bronze				
Sensitive bulbs	Copper, lead, Everdur, or 316 stainless steel				
Bushings	304 stainless steel, brass, or 316 stainless steel				
Wells	304 stainless steel, hardware bronze, 316 stainless steel, Hastelloys, nickel, Monel, glass, tantalum, lead, or silver				

* This tabular comparison of the various classes of thermometers indicates what may be expected from high-grade instruments. Reproduced by permission from Fischer and Porter catalogue 12A-10.

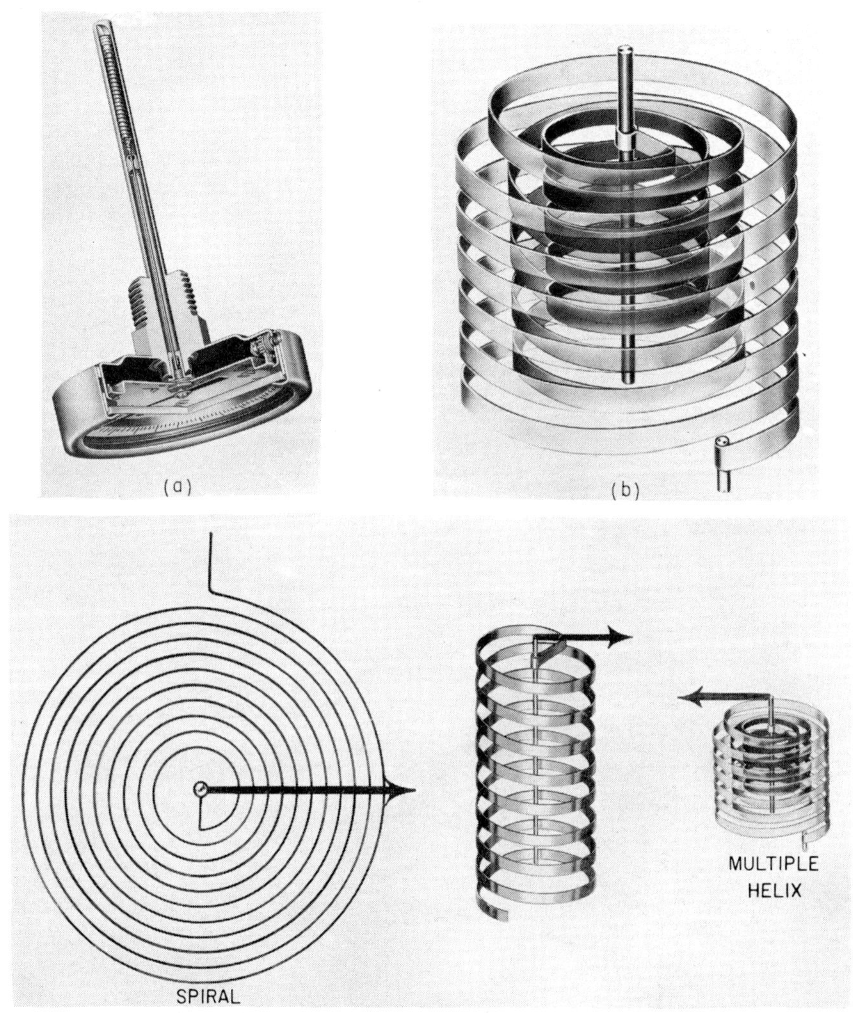

Fig. 14·9 This bimetallic thermometer employs a double helical member, which is also shown. The multiple helix occupies less space than some other designs of bimetal elements. (*Weston Electrical Instrument* Co.)

tation of the free end so that an angular rotation of about 270° is obtainable without the use of gears or linkages. Both single and double helixes are employed. The double helix is shown in Fig. 14·9*a*, *b*, and *c*. Attention is drawn to the scale because it illustrates a little-known fact about most metals: they have quite uneven expansion rates. This causes the scale factor to change several times where the span is high, Fig. 14·10.

It will be noted in Fig. 14·11 that different bimetals have deflection rates that the novice would never anticipate. Some even go from a negative to a positive deflection in some portions of their span.

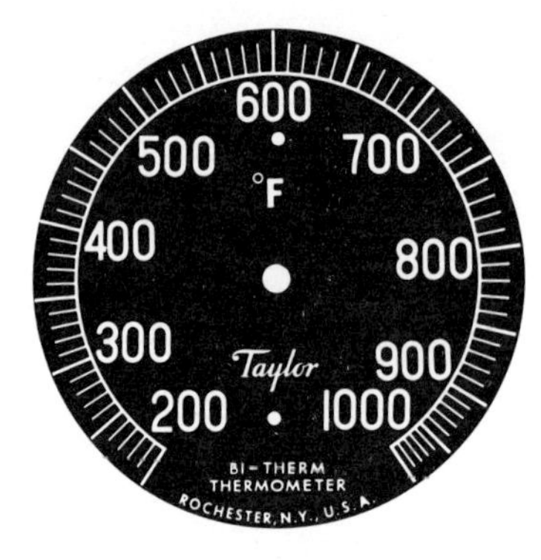

Fig. 14·10 The expansion rates of metals are not constant over wide ranges. This may not be evident to the eye, but take a pair of dividers and determine the chord lengths for 100° intervals. In particular, note the rapid change above 900°. (*The Taylor Instrument Cos.*)

Fig. 14·11 Expansion coefficients are constant over very limited temperature spans. Such variability must be anticipated in design. (*H. A. Wilson Division, Englehard Industries.*)

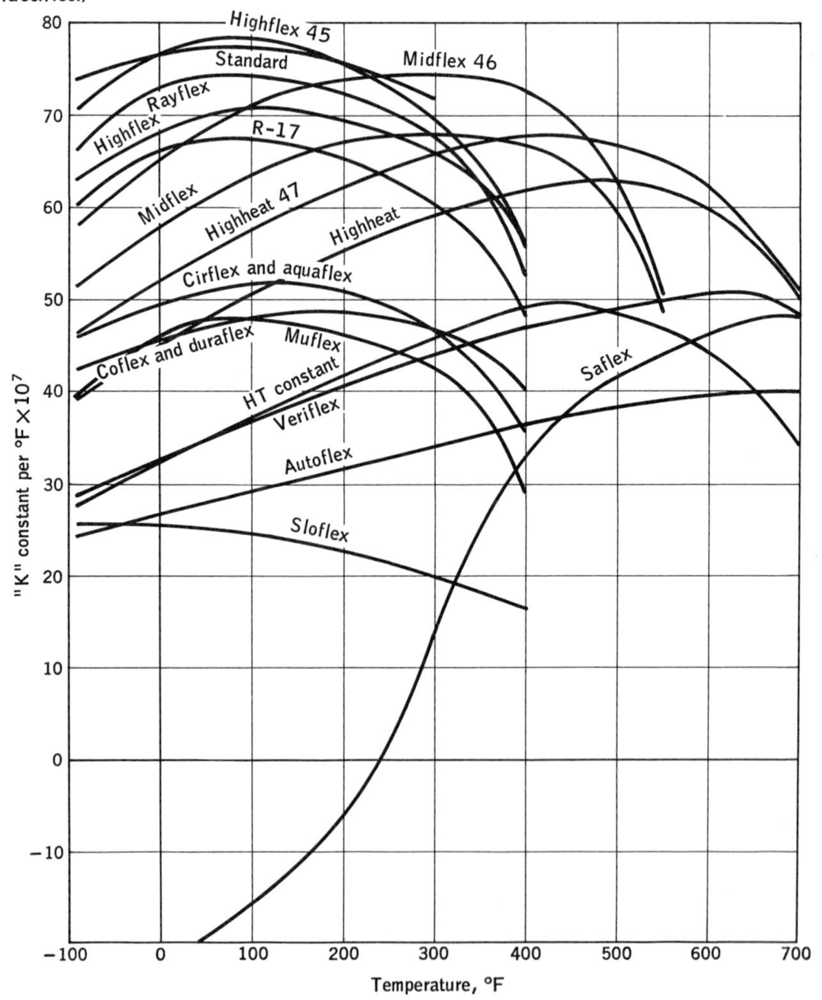

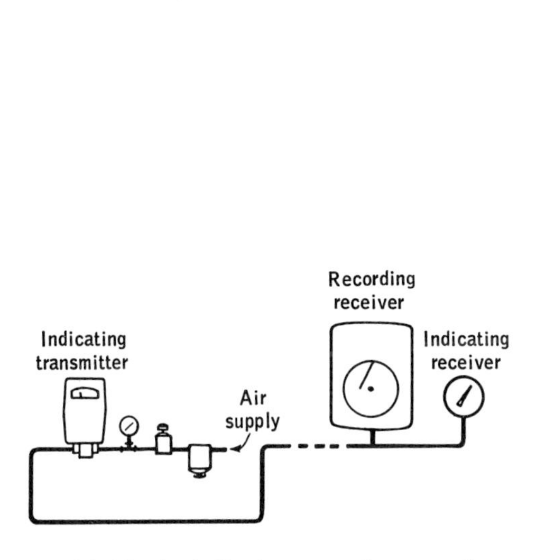

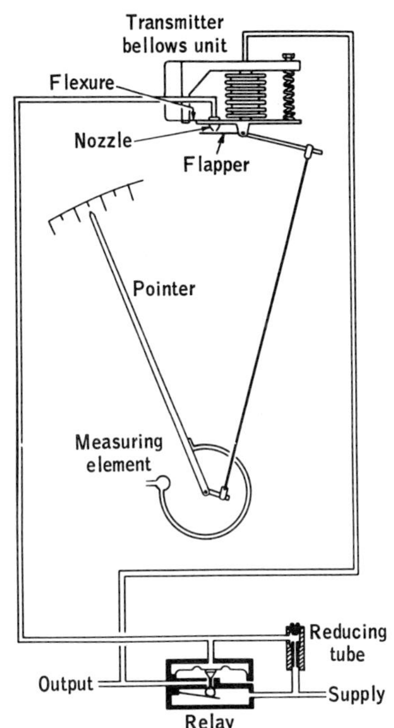

Fig. 14·12 An indicating type of pneumatic temperature transmitter converts pen or pointer position into an analog pressure which can then be transmitted far beyond the limitations of capillary tubing. (*The Foxboro Co.*)

PNEUMATIC TEMPERATURE TRANSMITTERS

Any of the four general classifications of pressure-creating thermometers can be employed with pneumatic-pressure transmitters if it is found that the signal must be sent materially more than 200 ft or if it is desired that there be multiple points at which the temperature may be monitored visually. Inasmuch as so many temperature measurements are nothing but pressure measurements, the pneumatic pressure and temperature transmitters should be essentially alike, differing only in relatively unimportant details.

Figure 14·12 shows a very simple device in which the pointer is moved directly by the Bourdon tube. Also connected with the tube is a member which controls the position of the flapper. If we assume that the temperature is rising, then any increase in temperature will move the flapper closer to the nozzle. This will increase the pressure at the orifice, which in turn will increase the pressure output of the pneumatic relay. The increased output of the relay goes two ways: to the remote gage or gages and to the transmitter feedback bellows which causes the flapper to be away from the nozzle until a new condition of equilibrium has been established.

A different way of accomplishing the same result is shown in Fig. 14·13. In this design, the pressure created by the liquid in the temperature-

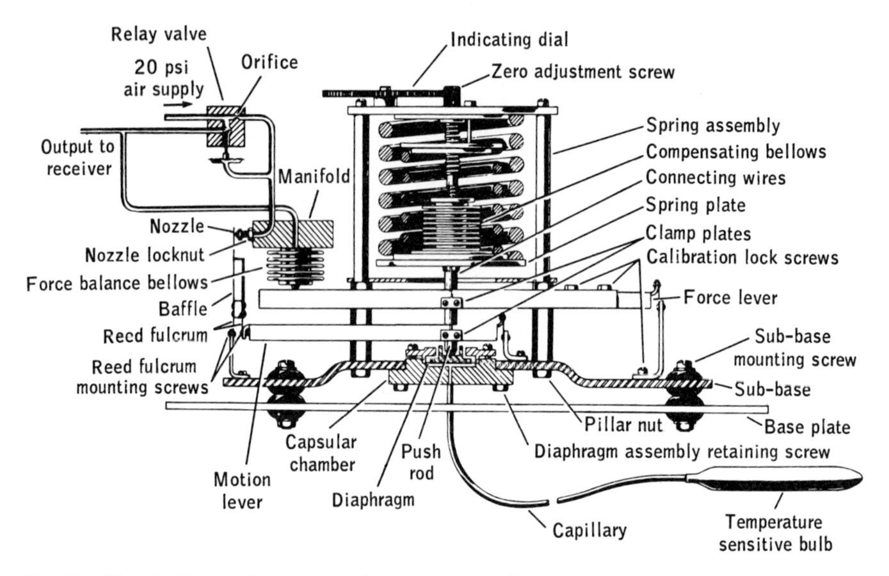

Fig. 14·13 A design for a nonindicating type of pneumatic temperature transmitter. Except for the way in which the pressure variations are created, this is essentially a pressure transmitter. Particular attention should be given to the baffle-reed arrangement which provides a high amplification of the displacement of the motion lever. (*The Taylor Instrument Cos.*)

Fig. 14·14 Another design of a pneumatic temperature transmitter. The position and appearance of the actual components will be found in Fig. 14·12*b*. (*The Foxboro Co.*)

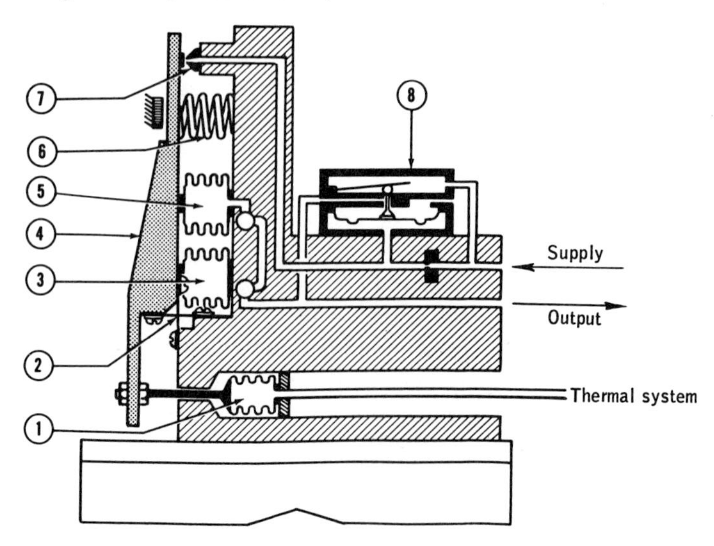

sensing bulb tries to push the diaphragm up, and with it the motion lever. Any movement of the motion lever will move the baffle with respect to the nozzle. With the construction shown, the movement of the baffle in the vicinity of the nozzle will be 40 to 50 times as great as that of the motion lever. Inasmuch as the full range of the device can be obtained with a baffle movement of about 0.001 in., the motion lever need not move much more than about 0.000020 in. to effect full control.

As the baffle moves away from the nozzle, the pressure in the tubing to the nozzle decreases. Because the relay with which the tubing is connected is of the reverse-acting type, this decrease of pressure at the nozzle will create an increase in the output pressure of the relay. This increase in pressure output is then fed into the force-balance bellows, which now acts to oppose the original diaphragm movement. As it acts in opposition to the original movement, the motion lever brings the baffle back toward the nozzle until a condition of equal torques is reestablished, but at a very slightly different spot from the one at which equilibrium existed prior to the temperature increase.

Still another version of a temperature transmitter is shown in Fig. 14·14. Essentially the same operations are carried out, but with different components. For purposes of this text, ignore the use of the capacity tank which is shown in the cutaway model.

15 Temperature Measurements, Electrical Transducers

It will be noted in Fig. 14·1 that the filled temperature systems cover the range of −450 to +1200°F. (These are extreme limits and are not the values likely to be available from any one supplier of equipment.) These systems also have the following characteristics:

1. They are mechanically simple.
2. They are limited in their distance of transmission of response, unless used with auxiliary transmitters, to about 200 ft.
3. They employ sensing elements which have appreciable thermal mass.
4. They are simple to use with pneumatic control systems.

Because of (2) and (3) and the large number of instances when temperatures over 1000 to 1200°F must be not only measured but controlled, certain electrical systems have been developed, Fig. 15·1. The electrical principles of operation on which the different methods are based are quite varied:

1. The use of transducers to convert from mechanical to electrical signals.
2. Change of resistance with change of temperature, range −100 to +600°F.
3. Creation of an emf which is proportional to the difference in temperature between the heated junction and the cold junction of two dissimilar wires.
4. Color or brightness, which is a function of the temperature.
5. The radiation of energy, which is a function of temperature. The value of the radiated energy may be sensed by one or more thermocouples.

TRANSDUCERS

The resistance thermometer. There are three general classes of materials whose resistance-temperature characteristics must be recognized:

1. Almost all pure metals undergo a change of resistance which is directly proportional to the temperature change. Figure 15·2 indicates that nickel does have some peculiar temperature-resistance relationships.

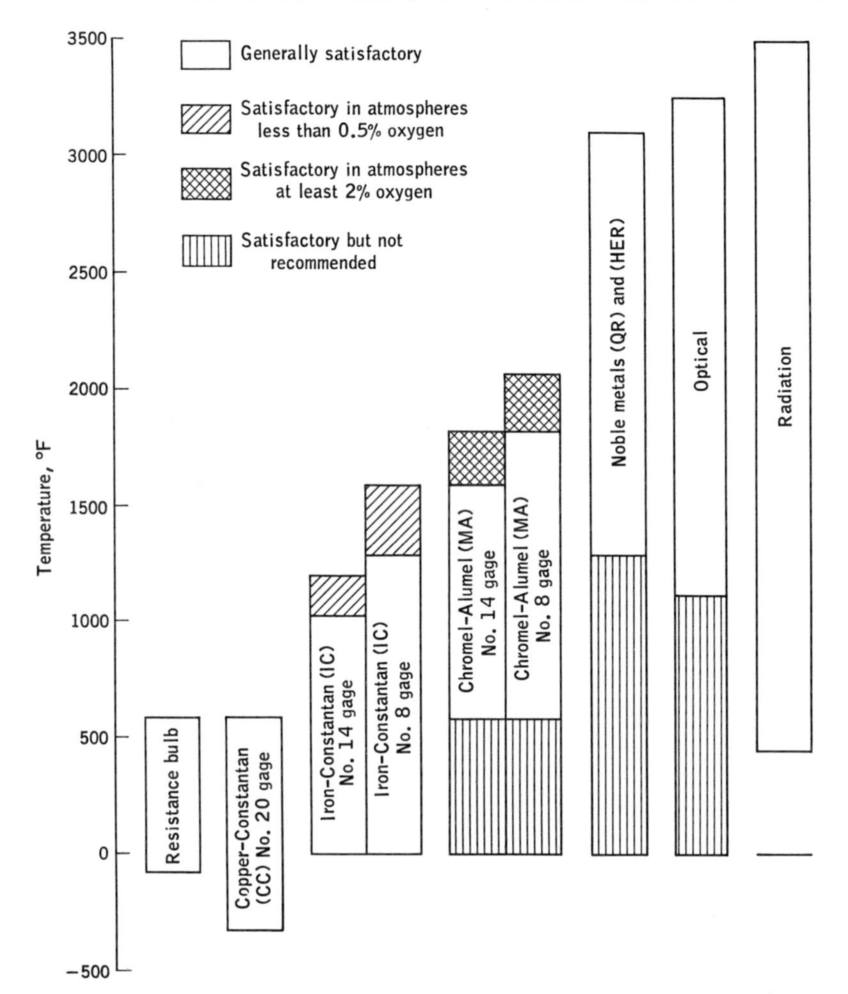

Fig. 15·1 The various electrical measuring systems for temperature determination are restricted, in general, to the indicated spans and ranges. Resistance measurements are more generally employed within the span of −100 to +600°F. Electrical-optical methods can go to even higher temperatures than are shown. (*"Fundamentals of Instrumentation,"* *Minneapolis-Honeywell Regulator Co.*)

2. Certain alloys have almost no change of resistance with temperature. They are used in instruments when any temperature effects must be reduced to a minimum.

3. The oxides of certain materials undergo large reductions in resistance as the temperature is increased. Their rate of change of resistance with temperature is not constant, as is that of pure metals to a great degree, but is variable and negative in effect. This general class of materials is called *thermistors,* Fig. 15·3.

At the present time, the greatest number of industrial measurements

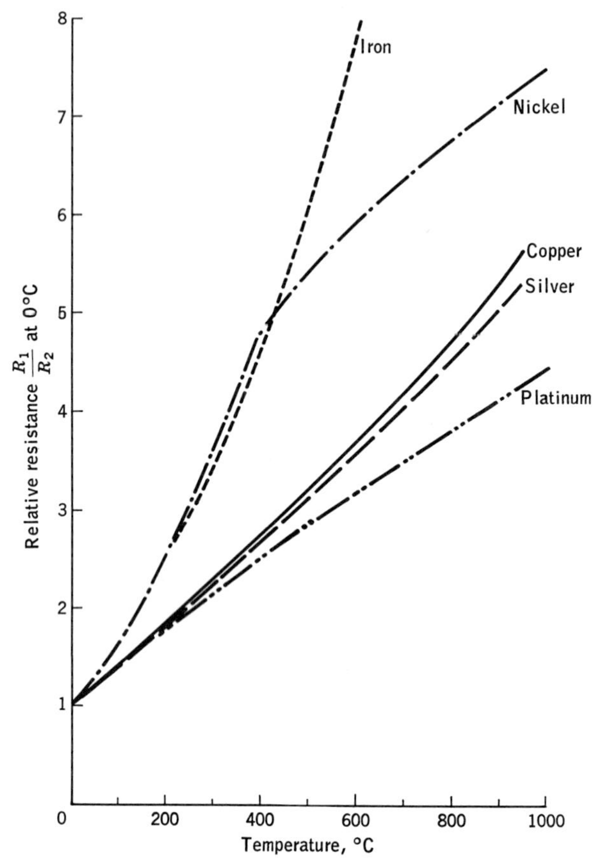

Fig. 15·2 Relative resistance versus temperature for some common metals.

employ the pure metals. The third type of material is of increasing importance as its particular characteristics become better known. Not only are thermistors important in temperature determination, but their resistance characteristics are suitable for use in anemometers, liquid-level units, and low-pressure gages.

Of the metals in the first group, copper, nickel, and platinum are most often used. The physical shapes of these temperature-resistance sensing elements may be many, but all are based on a minimum thermal mass and a maximum rate of heat transfer to or from the active element, which is the winding. In fact, this method is so accurate, it is one of the standard methods used within the temperature span of −297 to 1167°F.

In the typical resistance bulb, the temperature-responsive wire is wound around a high-conductivity solid silver core. When the tip of the core is in contact with the bottom of the well, heat is transferred rapidly by

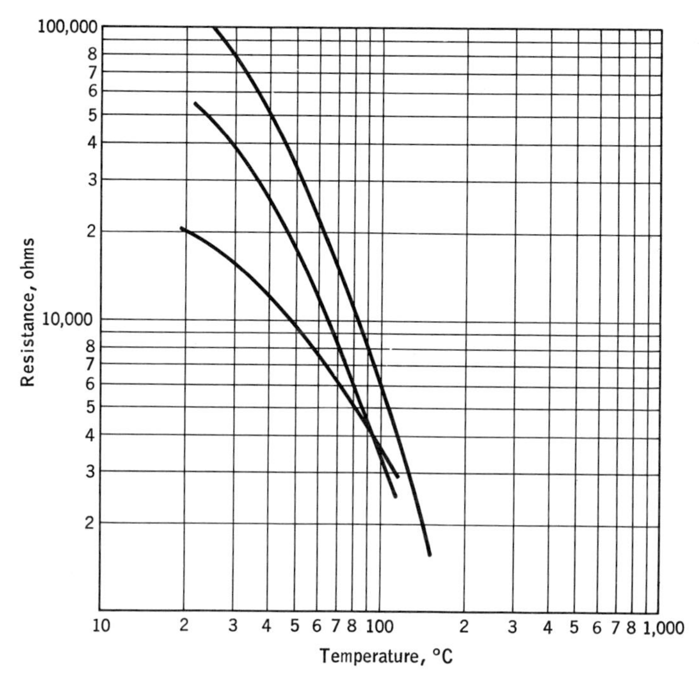

Fig. 15·3 Typical resistance-temperature relationships for thermistors. (*Bell Telephone Laboratories.*)

metal-to-metal contact to the winding. Sustained high conductivity rates are assured through use of a spring which forces the end of the bulb against the bottom of the well. Response is fast, and measurement accuracy is not affected significantly by ambient temperature changes.

Resistance bulbs are a practical means for measuring the temperatures of stationary, moving, or roll surfaces. They are also widely used for chemical and laboratory measurement applications. As has been previously discussed, the rate of response of the resistance unit will be decreased by the use of thick-walled thermometer protective tubes and heavy assemblies.

The use of thermistors is more common at the lower temperature ranges than at the high ones. Their large resistance changes, which are 5 to 10 times those for metals having the same temperature changes, make it easier to measure smaller increments of temperature. Figure 15·3 shows some typical resistance-temperature values for commercial units.

It cannot be emphasized too much that *all resistance readings reflect the temperature of the sensing element and not the temperature of the medium transferring heat to or from the element. Everything possible must be done to make the element accurately reflect the temperature of the surrounding medium.*

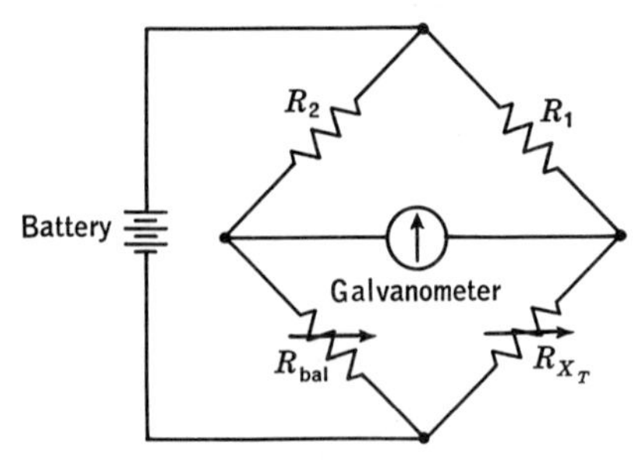

Fig. 15·4 A simplified Wheatstone bridge. This type of bridge is generally used for the measurement of resistance when direct-current potentials are employed.

Because accuracy of measurement is generally of great importance, many simple ways of measuring resistance, and consequently the inferred temperature, cannot be considered. One of the most accurate methods uses the Wheatstone bridge shown schematically in Fig. 15·4. In Chap. 5 the theory of this instrument is discussed. It is basically simple, sturdy, and accurate; it may be portable or permanently installed. The relationships for this bridge are:

$$\frac{R_2}{R_1} = \frac{R_{\text{bal}}}{R_{X_T}} \tag{15·1}$$

This equation is true only for the condition of balance when the meter current is zero.

Rewriting the equation,

$$R_{X_T} = \frac{R_1}{R_2} R_{\text{bal}} \tag{15·2}$$

The relationship between the resistance and the temperature is obtained from

$$R_{X_T} = R_{20}(1 + \alpha t) \tag{15·3}$$

where α is the average temperature-resistance coefficient and t is the difference between the reference temperature and the actual temperature.

If we rewrite Eq. (15·3) to solve for t, then

$$R_{X_T} = R_{20} + R_{20}\alpha t \tag{15·4}$$

$$R_{X_T} - R_{20} = R_{20}\alpha t \tag{15·5}$$

$$\frac{R_{X_T} - R_{20}}{R_{20}\alpha} = t = \text{temperature change from } 20°\text{C} \tag{15·6}$$

if that is the reference temperature for which the coefficient of resistance was computed.

Wheatstone bridges may be so calibrated that they indicate the temperature which created the resistance rather than the resistance which resulted from the temperature. Wheatstone-bridge designs for industrial use may be improved by:

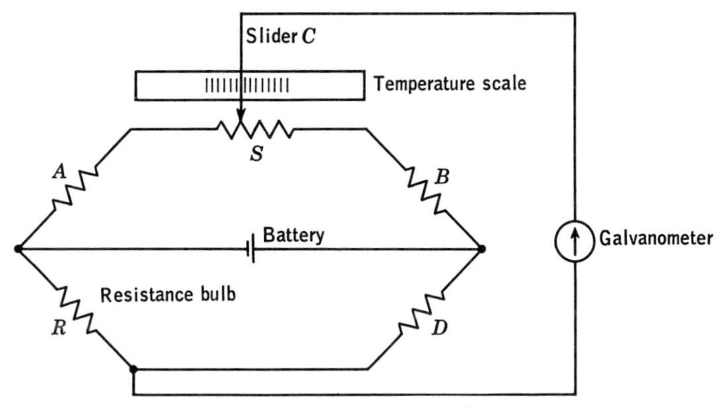

Fig. 15·5 A modified Wheatstone bridge. The simple Wheatstone bridge may be improved for commercial use by having no variable resistors in the bridge. The bridge is balanced by changing the position of the movable contact on resistor S.

1. Eliminating variable resistances from current-carrying circuits. The presence of dirt and wear at the resistor may change the resistance of the parts, thus changing the calibration.

2. Correcting for the change in resistance of the leads to the thermometer bulb.

3. Connecting the galvanometer to the slide wire. At the time of bridge balance, the galvanometer carries no current, so a change in the resistance of this circuit will have little or no effect. Figures 15·5 and 15·6 show how some of these modifications can be made.

In Fig. 15·5

$$R = D \frac{A/S + f}{B/S + 1 - f}$$

where f is the fractional part of the slide-wire resistance S in the A arm of the bridge. The simultaneous positioning of sliders S_1 and S_2, Fig. 15·6, which are mechanically coupled together, will balance the bridge. Any contact resistance is now in a noncritical part of the circuit. The position of the sliders at balance is then read on a calibrated temperature scale.

In all resistance-type thermometer measurements, the power supplied to the sensing resistor must not change appreciably the temperature of the sensing resistor, or the indicated temperature will be that of the resistor and not that of the process liquid or gas where the measurement is being made. If greater sensitivity is necessary, it is frequently better to pulse the current through the sensing element. If 2 volts applied to the bridge does not furnish adequate sensitivity but with this voltage the resistor accurately reflects the temperature of its surroundings, then ap-

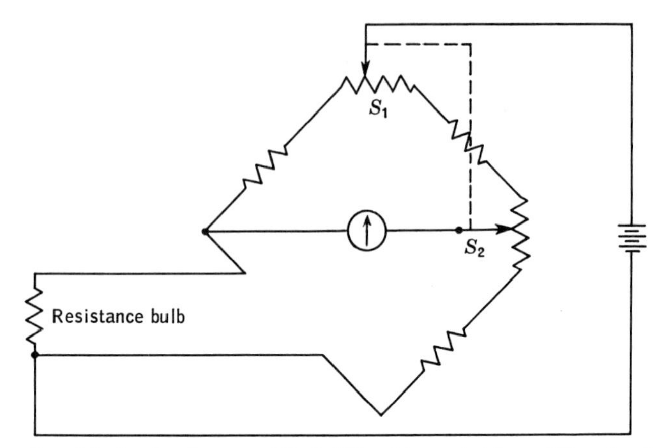

Fig. 15·6 Another modified Wheatstone bridge. This circuit eliminates the effects of lead resistance. Both sliders are mechanically coupled and move together.

plying 8 volts to the bridge for 4 sec each minute will increase the sensitivity by a factor of 4. But the heat added will be the same as in the original condition. For processes in which the temperature changes are slow, such intermittent measurements might be quite acceptable. For systems with low thermal inertia in which temperature changes might be rapid, such a system would not be satisfactory.

If only a temperature reading is desired, but without the high accuracy of the Wheatstone bridge, the circuit shown in Fig. 15·7 may be used. Whereas the former circuits do not require a constant-voltage source, this type of instrument requires that the voltage be constant or that the

Fig. 15·7 High-resistance indicating pyrometer. If a high-resistance meter is used, any unbalance will cause the meter to deflect an amount proportional to the change in temperature. The temperature may then be read on a calibrated scale. (*The West Instrument Corporation.*)

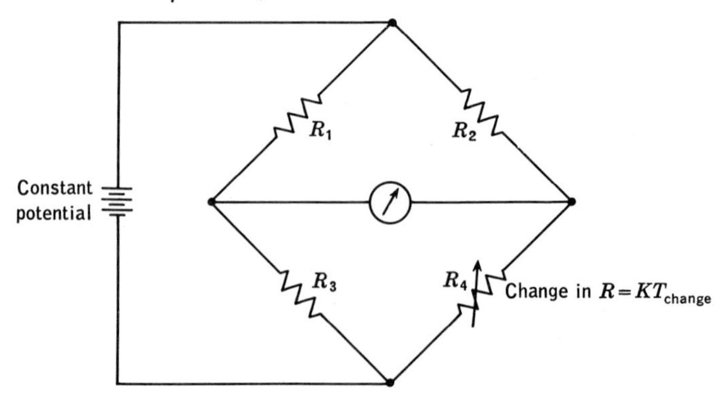

bridge be standardized at intervals. The unbalance current of the bridge causes the galvanometer (or meter) deflection. Except for one position, which normally would be near the center position on the scale, there will always be an unbalance current in the bridge elements. The direction of the unbalance current causes the appropriate meter deflection.

Capacitance bridges. Another version of the conventional bridge is illustrated in Fig. 15·8*a*. In this design, two of the resistors have been replaced by capacitors. A capacitor of fixed value is used as a standard of reference, while the other capacitor is variable. The variable capacitor has the same function as the balancing resistor in Fig. 15·4.

When switch S is closed on contact 1, both capacitors are charged; when switch S is closed on contact 2, both capacitors are discharged. The quantity of electricity which can be stored in each capacitor is expressed by

$$Q = CE = \int i\,dt$$

But

$$Q_1 = C_1E_1 \qquad \text{and} \qquad Q_2 = C_2E_2$$

If C_1 has one-half the capacitance of C_{std}, it will take only one-half the charge of C_{std}, and the current going to C_1 to charge it should at any instant be one-half of that charging C_{std}. To produce equal voltages across the resistors, which will be indicated by the galvanometer having zero deflection during the charging and discharging cycles, R_{X_T} will have to have a value twice that of R.

$$i_{C_1} = \tfrac{1}{2} i_{C_{\text{std}}}$$

But

$$(i_{C_1})(R_{X_T}) = (i_{C_{\text{std}}})(R)$$

$$(i_{C_1})(R_{X_T}) = (2i_{C_1})(R)$$

Therefore

$$R_{X_T} = 2R$$

The capacitance bridge of Fig. 15·8*a* functions perfectly well for teaching purposes, or where manual manipulation is permissible. But when an automatic indication of temperature is desired with the capacitance bridge, the circuit sketched in Fig. 15·8*b* may be employed. A commercial design using the same basic components is shown in Fig. 15·9. It can measure temperature spans as small as 5°F. The measuring circuit consists of a modified Wheatstone bridge energized from a 1,000-cps alternating-current supply at points 2 and 4. A change in the value of the resistance bulb appears as an unbalanced bridge voltage across points 1 and 3. This potential is fed to a voltage amplifier and then to a power amplifier for driving the self-balancing system. By using 1,000-cps excita-

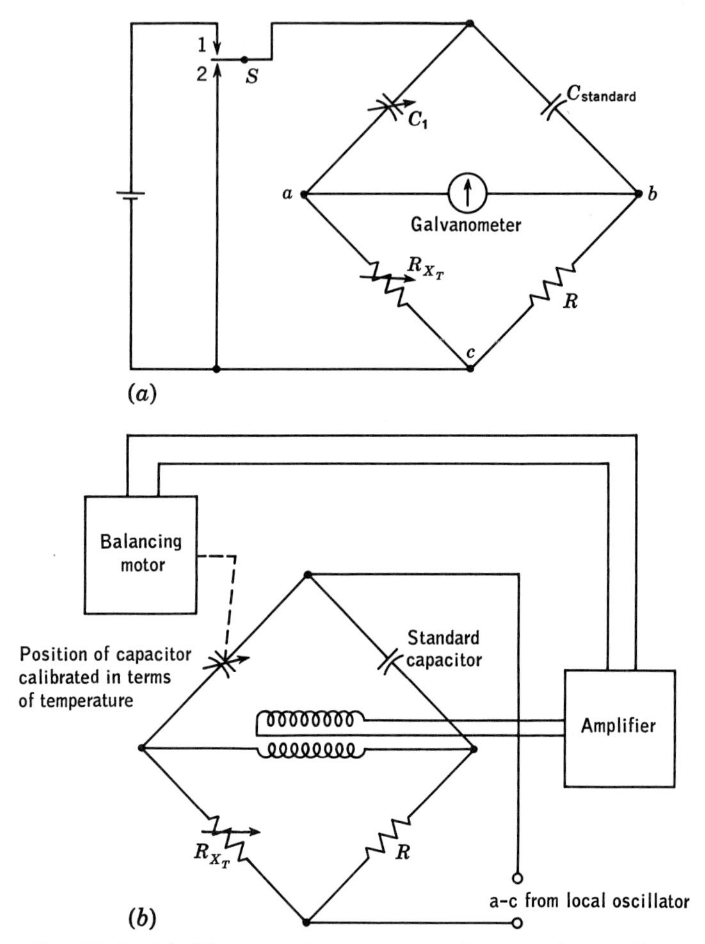

Fig. 15·8 (a) When a pulsating source of potential is used, capacitors may be substituted for resistors in a conventional Wheatstone bridge circuit. (b) The position of the capacitor used to balance the bridge may be calibrated in terms of temperature in the same way that a change in temperature may be inferred from a change in resistance.

tion voltage, filters can be used to eliminate any 60-cycle pickup voltages which can interfere with the accuracy of measurement.

It will be noted that in the commercial design there is no *reference capacitor* in the sense that its value is fixed. In this design, the capacitors in both legs of the bridge change; one increases while the other decreases.

Whereas most instruments use two-phase motors for reversal and speed-control purposes, this design employs two coils which work in opposition to each other. When the bridge is in balance, the two supply wires assume an equal negative potential relative to the common return wire. Equal

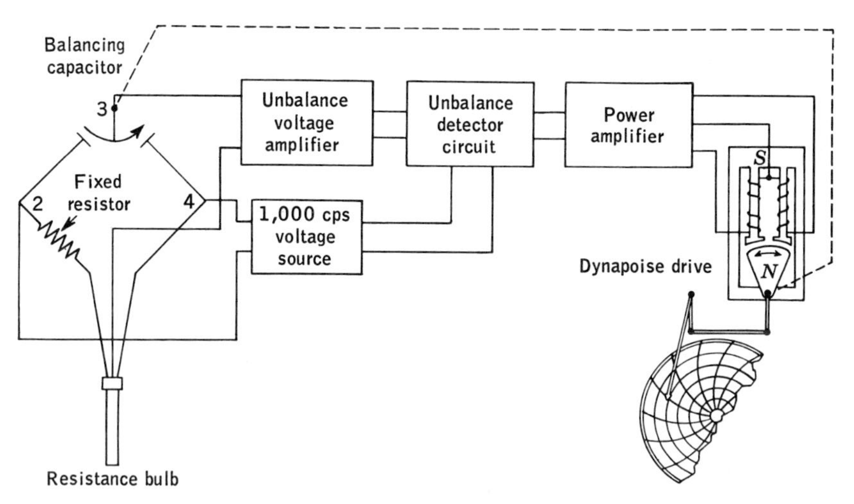

Fig. 15·9 Simplified diagram of an automatic temperature recorder. Employing a variable capacitor for balancing the bridge avoids the use of movable contacts on resistors. (*The Foxboro Co.*)

current thus flows through the two sets of coils. With equal current through each coil, the electromagnetic effects of the coils cancel because the coils are connected in opposition. The net effect is that the magnetic attraction between the projecting south-seeking poles and the north-seeking rotor is eliminated. It would be the same if the coils were removed or the current reduced to zero in each coil. When the bridge is unbalanced, one supply wire assumes a more negative potential, relative to the common return wire, than the other. Therefore, more current flows through one set of coils than through the other. The resultant effect is that one of the south-seeking poles, depending upon the direction of bridge unbalance, is reduced in its magnetic intensity while the other one is increased in strength. As a consequence, the north pole moves. This rotates the balancing capacitors and reestablishes a condition of bridge balance. At balance, the coil currents are again equal, and their net magnetic effect is zero. Consequently, they no longer exert any force on the rotor pole, which remains in the position where the balance condition obtains.

The capacitive bridge represents a different approach to the problem of eliminating moving potentiometer contacts or other current-carrying contacts which might be sources of trouble and require maintenance.

Most of the discussion has been concerned with resistance-temperature relations which can be made basically linear over most of their operating range. But the metals which give us such simple relationships have a relatively small change of resistance with temperature. A class of materials which is becoming of greater importance is generally classed as *thermistors.* The more common semiconducting materials used in their construction are oxides of cobalt, copper, iron, magnesium, manganese, nickel, tin, titanium, uranium, and zinc. Thermistors have two characteristics which set them completely apart from the metals:

1. Their temperature-resistance coefficient is about 10 times that for the common metals.
2. Their resistance decreases with temperature, whereas the resistance of most metals increases. Thermistors are said to have a "negative" temperature-resistance coefficient; metals have positive coefficients.
3. Their temperature-resistance coefficient is not constant, but is an inverse function of temperature.

The materials are stable over a period of years and can be used for smaller temperature changes than are feasible with metallic resistance bulbs. They may be shaped in the form of beads, disks, washers, rods, and flakes.

TEMPERATURE DIFFERENCE MEASUREMENTS

When there are requirements for determining the differences between temperatures, the resistance-temperature method has certain advantages.

Fig. 15·10 The temperature of the leads and any difference in the lead lengths may create significant errors in temperature indications. (*The Foxboro Co.*)

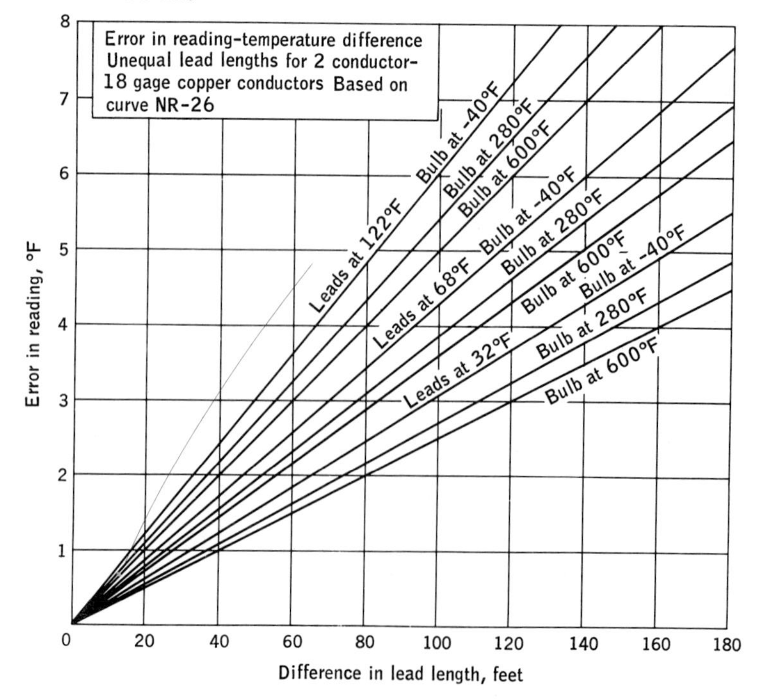

The linear resistance-temperature relationship is important when the span is only a few degrees. In order to obtain a difference temperature indication, it is necessary only to substitute a second resistance bulb for the standard or reference resistor in the usual bridge connections. When both bulbs are at the same temperature and hence have the same resistance, the bridge indicates zero degrees difference. It makes little or no difference what this temperature is, provided that the bulbs are within their temperature ratings. Now, if one increases by 5° while the other decreases by 5°, the net result is about the same as though one had changed by 10° while the other had remained constant.

So long as the two bulbs have leads of the same length and the leads have the same temperatures, no difficulty exists. But if the leads are of unequal length, then corrections are in order. Figure 15·10 shows how the error varies both with the difference between the bulb and the lead temperatures and with the length of the leads. This set of relationships is offered only to show the magnitude of the errors which may be expected with resistance bulbs having similar values of resistance. Differences in the ambient temperatures affecting the two sets of leads are shown in Fig. 15·11. The values indicated are lead lengths of 500 ft; for any other distance the values will be proportional parts of those shown.

Fig. 15·11 The greater the difference between the bulb temperature and the ambient temperature, the smaller is the error caused by resistance changes in the leads. (*The Foxboro Co.*)

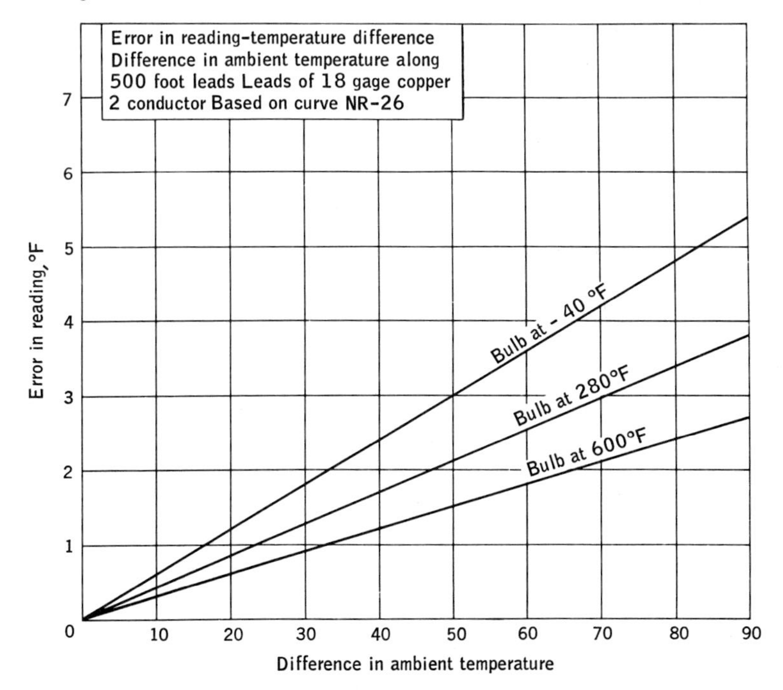

THERMOCOUPLES

For both high- and low-temperature measurements, thermocouples are most important. In 1821, Seebeck discovered that when two dissimilar metal wires are twisted together and heated, a sensitive meter connected to the other end of the pair will indicate a voltage (often called an electromotive force, and abbreviated to emf) which is very nearly directly proportional to the *difference* in temperature between the heated, or *hot*, junction and the other end, which is called the *cold* junction. The theory of exactly how this emf is generated is not well understood even today. Much progress is being made as more complete studies of solid-state physics are extended. The study of semiconductor devices is extremely important as better thermocouple materials are sought. We are also seeking to understand the phenomena occurring at the junction because:

1. Applying heat to the junction produces an electric current, at least it can if the circuit is complete.
2. Sending a current from an external source through the junction results in a cooling effect which may someday be used in refrigerators.

For a relatively complete theory of the various effects noted by Seebeck, Thomson, and Peltier, a standard college physics text should be consulted.

Figure 15·12 shows the response of the more common thermocouple

Fig. 15·12 Electromotive force versus temperature, with respect to platinum. The available potential is equal to the *algebraic difference* of the potentials created by each material with respect to platinum. The greatest output for a thermocouple pair will be obtained when one material is positive and the other is negative with respect to platinum.

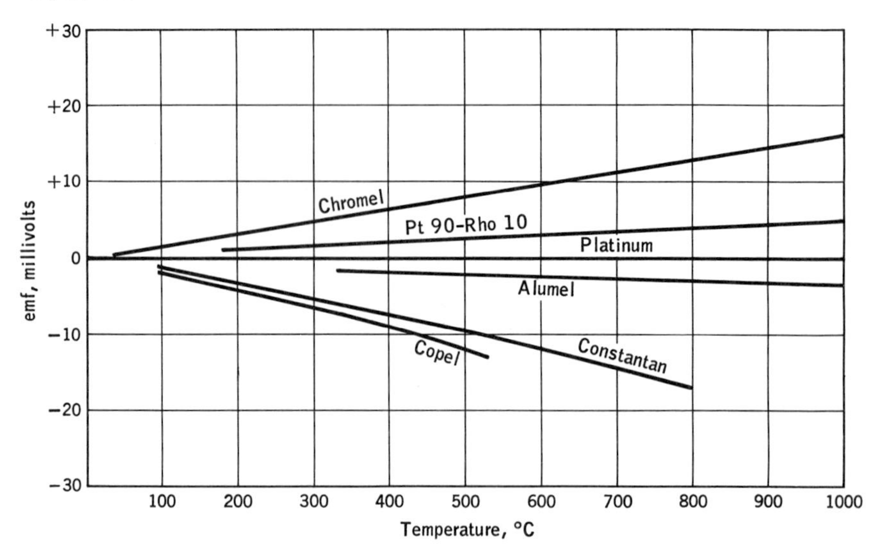

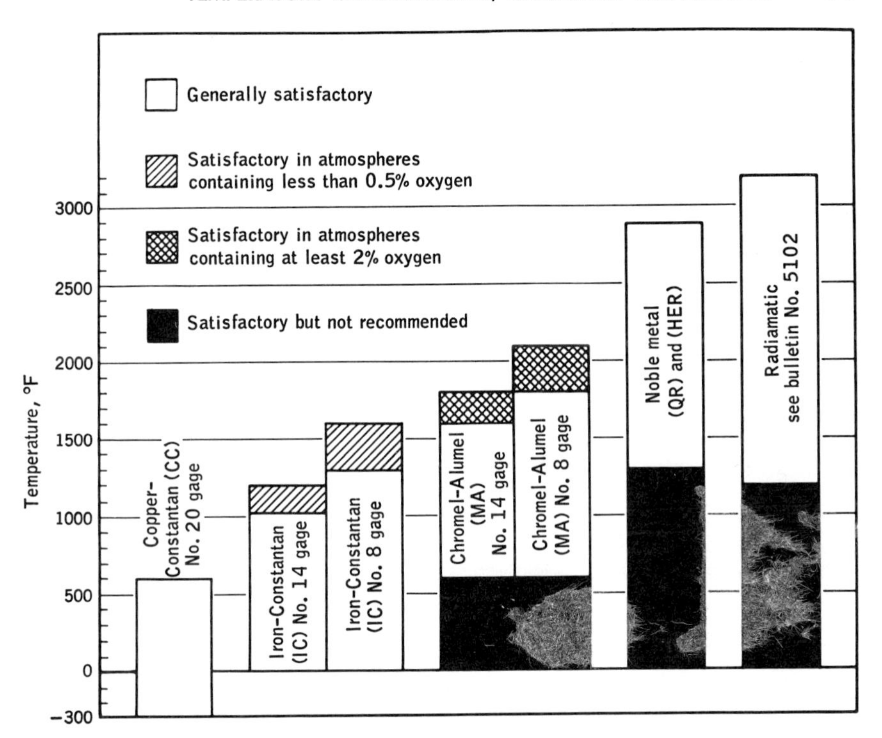

Fig. 15·13 Suggested thermocouple combinations, wire sizes, and temperature spans for the usual commercial application. No single combination of wires will accurately sense the entire temperature range. Each combination has been carefully selected for extreme accuracy in a sector of the range. (*Minneapolis-Honeywell Regulator* Co.)

materials when pure platinum is used for one of the thermocouple wires. It will be noted that these materials are both positive and negative with respect to platinum. In selecting a pair of thermocouple wires, one wire should be positive and one negative in order that the pair will create the maximum emf. The larger the value of the potential produced, the greater will be the sensitivity of the system and the more accurately can the temperature be ascertained. Commercial couples are represented in Figs. 15·13 to 15·15. Standard thermocouple potentials are listed in Table 15·1.

Through experience based on operating life, chemical stability, uniform quality, and reproducibility of couples, the commercial thermocouples are generally limited to those listed in Table 15·2 and Fig. 15·13. It will be noted that thermocouple wire size has a bearing on the maximum temperature for continuous operation, except in the case of platinum and its alloys, where cost is a major factor. Longer life generally is obtained at the expense of high speed of response. The errors to be expected, under normal conditions, are listed in Table 15·3.

Table 15·1 Thermocouple

(Reference

For conversion to 32°F reference junction, add following millivolt values.

T/C	ISA symbol	Millivolt value	T/C	ISA symbol	Millivolt value
C/C	T	+0.94	C/A	K	+0.95
I/C	J	+1.22	P10	S	+0.136
I/C	Y	+1.26	P13	R	+0.135
CR/C	E *	+1.43	P30	P30 *	+0.006

* Thermoelectric symbol.

−320 to 0°F					
°F	C/C T	I/C J	I/C Y	CR/C E	C/A K
−320	.....	−9.01	.....	−10.03	.....
−300	−6.23	−8.74	−9.13	−9.73	−6.46
−275	−5.97	−8.36	.....	......	−6.20
−250	−5.69	−7.93	−8.28	−8.87	−5.91
−225	−5.39	−7.47	.....	......	−5.59
−200	−5.06	−6.98	−7.27	−7.83	−5.24
−175	−4.70	−6.45	.....	......	−4.87
−150	−4.32	−5.90	−6.13	−6.66	−4.47
−125	−3.92	−5.31	.....	......	−4.05
−100	−3.50	−4.71	−4.89	−5.37	−3.60
−75	−3.06	−4.08	.....	......	−3.13
−50	−2.60	−3.44	−3.57	−3.97	−2.65
−25	−2.12	−2.78	.....	......	−2.15
0	−1.61	−2.11	−2.18	−2.45	−1.63

0 to 400°F								
°F	C/C T	I/C J	I/C Y	CR/C E	C/A K	P10 S	P13 R	P30
0	−1.61	−2.11	−2.18	−2.45	−1.63			
25	−1.09	−1.42	.....	.....	−1.10			
50	−0.56	−0.72	−0.74	−0.84	−0.55	−0.080	−0.080	−0.004
75	0.00	0.00	.....	.....	0.00	0.000	0.000	
100	0.57	0.72	0.74	0.84	0.57	0.085	0.085	0.007
125	1.16	1.45	.....	.....	1.14	0.173	0.173	
150	1.77	2.19	2.24	2.61	1.71	0.265	0.265	0.024
175	2.39	2.94	.....	.....	2.29	0.360	0.361	
200	3.02	3.69	3.77	4.44	2.87	0.459	0.461	0.045
225	3.67	4.44	.....	.....	3.45	0.560	0.565	
250	4.34	5.20	5.31	6.33	4.02	0.664	0.672	0.069
275	5.01	5.96	.....	.....	4.58	0.771	0.782	
300	5.70	6.72	6.86	8.28	5.14	0.881	0.895	0.095
325	6.41	7.49	.....	.....	5.70	0.992	1.009	
350	7.12	8.26	8.42	10.28	6.25	1.106	1.126	0.126
375	7.84	9.03	.....	.....	6.80	1.221	1.246	
400	8.58	9.81	9.98	12.32	7.36	1.338	1.369	0.163

1200 to 1600°F							
°F	I/C J	I/C Y	CR/C E	C/A K	P10 S	P13 R	P30
1200	34.79	34.91	47.61	26.03	5.590	5.990	2.092
1225	35.64	.....	.....	26.62	5.735	6.150	.....
1250	36.49	36.60	49.84	27.20	5.879	6.311	2.275
1275	37.35	.....	.....	27.79	6.025	6.474	.....
1300	38.21	38.32	52.07	28.37	6.171	6.638	2.467
1325	39.09	.....	.....	28.96	6.317	6.802	.....
1350	39.97	40.08	54.28	29.54	6.465	6.968	2.662
1375	40.85	.....	.....	30.12	6.613	7.134	.....
1400	41.74	41.86	56.49	30.70	6.761	7.301	2.861
1425	42.63	.....	.....	31.27	6.910	7.470	.....
1450	43.53	43.66	58.68	31.85	7.060	7.639	3.064
1475	44.42	.....	.....	32.42	7.211	7.810	.....
1500	45.31	45.48	60.87	32.98	7.362	7.981	3.275
1525	46.20	.....	.....	33.55	7.514	8.152	.....
1550	47.09	47.30	63.04	34.12	7.667	8.325	3.492
1575	47.96	.....	.....	34.68	7.820	8.499	.....
1600	48.83	49.13	65.20	35.24	7.974	8.674	3.718

1600 to 2000°F						
°F	I/C Y	CR/C E	C/A K	P10 S	P13 R	P30
1600	49.13	65.20	35.24	7.974	8.674	3.718
1625	.....	.....	35.81	8.129	8.849	
1650	50.94	67.33	36.36	8.284	9.026	3.948
1675	.....	.....	36.92	8.440	9.203	
1700	52.61	69.47	37.48	8.596	9.381	4.181
1725	.....	.....	38.02	8.754	9.559	
1750	54.25	71.59	38.58	8.912	9.739	4.422
1775	.....	.....	39.12	9.070	9.921	
1800	55.90	73.69	39.67	9.229	10.102	4.669
1825	.....	.....	40.21	9.389	10.285	
1850	.....	.....	40.75	9.550	10.468	4.924
1875	.....	.....	41.29	9.711	10.651	
1900	.....	.....	41.83	9.873	10.838	5.184
1925	.....	.....	42.36	10.035	11.025	
1950	.....	.....	42.90	10.198	11.213	5.449
1975	.....	.....	43.43	10.362	11.402	
2000	.....	.....	43.96	10.526	11.591	5.719

Temperature Millivolt Tables

junction 75°F)

400 to 800°F								
°F	C/C T	I/C J	I/C Y	CR/C E	C/A K	P10 S	P13 R	P30
400	8.58	9.81	9.98	12.32	7.36	1.338	1.369	0.163
425	9.33	10.58	.....	.....	7.92	1.457	1.493	.....
450	10.09	11.35	11.54	14.41	8.48	1.576	1.619	0.215
475	10.85	12.12	.....	.....	9.05	1.697	1.747	.....
500	11.63	12.90	13.09	16.52	9.62	1.820	1.877	0.287
525	12.42	13.66	.....	.....	10.18	1.944	2.009	.....
550	13.21	14.43	14.63	18.66	10.76	2.069	2.142	0.373
575	14.02	15.20	.....	.....	11.33	2.195	2.277	.....
600	14.83	15.96	16.17	20.82	11.91	2.322	2.412	0.465
625	15.65	16.73	.....	.....	12.49	2.451	2.550	.....
650	16.48	17.50	17.69	23.01	13.07	2.580	2.688	0.563
675	17.31	18.27	.....	.....	13.65	2.710	2.828	.....
700	18.16	19.04	19.22	25.22	14.23	2.841	2.968	0.668
725	19.01	19.80	.....	.....	14.81	2.972	3.110	.....
750	19.86	20.57	20.75	27.43	15.40	3.104	3.252	0.779
775	.....	21.33	.....	.....	15.99	3.237	3.396	.....
800	.....	22.10	22.27	29.66	16.58	3.370	3.542	0.895

800 to 1200°F							
°F	I/C J	I/C Y	CR/C E	C/A K	P10 S	P13 R	P30
800	22.10	22.27	29.66	16.58	3.370	3.542	0.895
825	22.87	.....	.....	17.16	3.504	3.688	
850	23.63	23.80	31.89	17.75	3.639	3.835	1.019
875	24.40	.....	.....	18.34	3.774	3.981	
900	25.18	25.33	34.14	18.94	3.910	4.129	1.149
925	25.95	.....	.....	19.53	4.046	4.278	
950	26.73	26.88	36.39	20.12	4.183	4.428	1.290
975	27.52	.....	.....	20.71	4.321	4.580	
1000	28.30	28.44	38.63	21.31	4.460	4.733	1.437
1025	29.10	.....	.....	21.90	4.599	4.887	
1050	29.90	30.02	40.88	22.49	4.738	5.041	1.590
1075	30.70	.....	.....	23.08	4.879	5.197	
1100	31.50	31.62	43.13	23.68	5.020	5.353	1.750
1125	32.32	.....	.....	24.27	5.162	5.510	
1150	33.14	33.25	45.38	24.86	5.304	5.670	1.908
1175	33.96	.....	.....	25.45	5.447	5.829	
1200	34.79	34.91	47.61	26.03	5.590	5.990	2.092

2000 to 2400°F				
°F	C/A K	P10 S	P13 R	P30
2000	43.96	10.526	11.591	5.719
2025	44.48	10.691	11.781	.....
2050	45.01	10.856	11.970	5.993
2075	45.53	11.021	12.161	.....
2100	46.05	11.187	12.353	6.274
2125	46.57	11.353	12.544	.....
2150	47.08	11.519	12.736	6.561
2175	47.59	11.686	12.928	.....
2200	48.10	11.853	13.120	6.851
2225	48.60	12.019	13.313	.....
2250	49.11	12.186	13.506	7.146
2275	49.61	12.354	13.698	.....
2300	50.10	12.521	13.892	7.446
2325	50.59	12.688	14.084	.....
2350	51.08	12.855	14.277	7.749
2375	51.57	13.022	14.471	.....
2400	52.06	13.189	14.663	8.054

2400 to 2800°F				
°F	C/A K	P10 S	P13 R	P30
2400	52.06	13.189	14.663	8.054
2425	52.54	13.355	14.856	......
2450	53.02	13.522	15.049	8.361
2475	53.49	13.689	15.242	......
2500	53.97	13.855	15.433	8.671
2525	.....	14.021	15.626	......
2550	.....	14.188	15.819	8.981
2575	.....	14.354	16.012	......
2600	.....	14.520	16.205	9.296
2625	.....	14.686	16.397	......
2650	.....	14.852	16.590	9.609
2675	.....	15.017	16.783	......
2700	.....	15.183	16.975	9.923
2725	.....	15.348	17.166	......
2750	.....	15.513	17.358	10.237
2775	.....	15.678	17.550	......
2800	.....	15.843	17.740	10.550

2800 to 3270°F			
°F	P10 S	P13 R	P30
2800	15.843	17.740	10.550
2825	16.008	17.930	
2850	16.172	18.120	10.864
2875	16.337	18.311	
2900	16.501	18.501	11.178
2925	16.665	18.691	
2950	16.829	18.881	11.492
2975	16.992	19.070	
3000	17.156	19.259	11.806
3025	17.319	19.448	
3050	17.482	19.638	12.120
3075	17.644	19.827	
3100	17.807	20.015	12.434
3125	17.969	......	
3170	18.260	......	12.873
3220	......	......	13.187
3270	......	......	13.499

Table 15·1 (Continued)

Abbreviations used in table	ISA symbol	Conductor polarity	
		Positive	Negative
I/C J	J	Iron	Constantan
I/C Y	Y	Iron	Constantan
C/A K	K	Chromel	Alumel
C/C T	T	Copper	Constantan
CR/C E	E*	Chromel	Constantan
P10 S	S	Platinum, 10% rhodium	Platinum
P13 R	R	Platinum, 13% rhodium	Platinum
P30	P30*	Platinum, 30% rhodium	Platinum, 6% rhodium

* Thermoelectric symbol.
SOURCE: Data furnished by Thermo-Electric Co., Inc.

Protection of the thermocouple. Protecting the thermocouple for mechanical reasons, or to ensure long life in its operating environment, generally involves the use of open- or closed-end metal or refractory protecting tubes or wells (see Fig. 15·16). To obtain the maximum rate of

Fig. 15·14 To have a valuable industrial thermocouple, a rise in temperature must always produce a substantial linear rise in emf. (*Minneapolis-Honeywell Regulator Co.*)

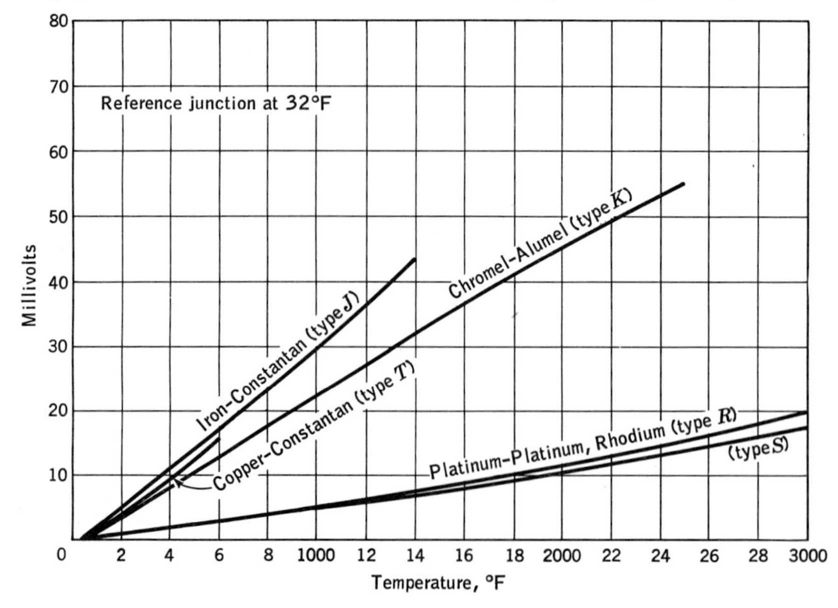

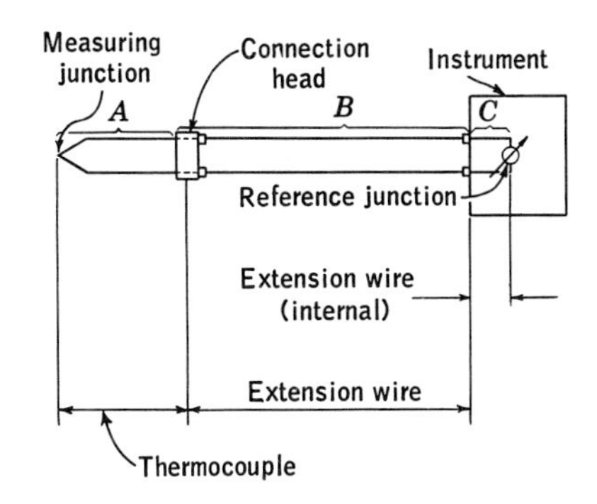

Fig. 15·15 A thermocouple is more than just the sensing element in the process. It extends from the process to the inside of the instrument. (*Minneapolis-Honeywell Regulator* Co.)

response where the atmosphere will not unduly affect the couple, the open-end tube will furnish adequate protection. Conditions that will adversely affect the couple require the use of closed-end tubes. When precious metals (platinum and its alloys) are used, the protecting tube must not only be chemically inert but vacuum tight as well so as to prevent contamination of the couple. *Thermocouples using precious metals are never used without vacuum-tight protecting tubes of a refractory type when the temperature exceeds 1000°F.*

The ceramic tubes, in turn, also require protection against thermal shocks caused by rapid temperature changes. It must be recognized that the heavier and more complex the shielding tubes, the greater will be

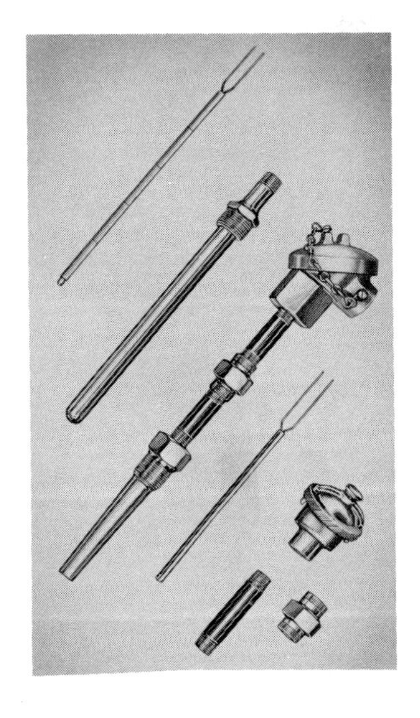

Fig. 15·16 Components of an industrial thermocouple. The industrial thermocouple consists of many more parts than just the element. (*Thermo Electric Co., Inc.*)

Table 15·2 Summarized Data for Selection of Thermocouple Element *

Type of element combination	Positive wire	Negative wire	Wire size, Awg	Max temp for continuous exposure (protected), °F	Atmospheric conditions for which thermocouple is best suited	Remarks
Copper-constantan (T)	Copper	Constantan	14 20 24 28	700 500 400 400	Oxidizing or reducing; upper limit set by oxidation of copper; stands up well against moisture	Commonly used for relatively low temperatures down to −300°F; good reproducibility and stability
Chromel-constantan	Chromel	Constantan	8 14 20	1300 1000 800	Oxidizing or neutral	Generates higher emf per degree Fahrenheit than others; particularly suitable for short spans down to −400°F; popular for special purposes (needle type, rolled strip, etc.)

Iron-constantan (J)	Iron	Constantan	8	1600	Reducing or neutral; iron deteriorates rapidly at high temperatures if oxidizing	Most popular for general purposes and inexpensive; available in concentric or tube form
			14	1200		
			20	900		
			24	700		
			28	700		
Chromel-alumel (K)	Chromel	Alumel	8	2300	Oxidizing or neutral; susceptible to attack by carbon, sulfur, and cyanide fumes	Satisfactory for high temperatures, although continuous operation at 1475°F is critical
			14	2000		
			20	1800		
			24	1600		
			28	1600		
Calido	242 Alloy	33 Alloy	8	2300	Oxidizing or reducing	Has wider application and is more stable than chromel-alumel
			14	2000		
			20	1800		
			24	1600		
			28	1600		
Platinum-rhodium–platinum (S)	10% rhodium 90% platinum	Pure platinum	24	2700	Requires protection in all atmospheres with impervious ceramic tubes	Has highest precision if properly protected
Platinum-rhodium–platinum (R)	13% rhodium 87% platinum	Pure platinum	24	2700	Requires protection in all atmospheres with impervious ceramic tubes	Has higher output emf than type S
5-20 platinum rhodium	20% rhodium 80% platinum	5% rhodium 95% platinum	24	3200	Requires protection in all atmospheres with vacuum-tested ceramic tubes	More stable than platinum-rhodium S and R at very high temperatures

* By Anthony Turner, Taylor Instrument Companies.

Table 15·3 Limits of Error of Thermocouples

(Standard wire size)

Type	Temp range, °F	Limits of error
Copper-constantan	−150 to −75	±2%
	−75 to +200	±1½°F
	200 to 700	±¾%
Chromel-constantan	−320 to −150	±2%
	−150 to +400	±3°F
	400 to 1300	±¾%
Iron-constantan	0 to 530	±4°F
	530 to 1600	±¾%
Chromel-alumel	0 to 530	±4°F
	530 to 2300	±¾%
242-33 alloy	0 to 530	±4°F
	530 to 2300	±¾%
Platinum–platinum-rhodium	0 to 1000	±5°F
	1000 to 2700	±½%
5-20 platinum rhodium	0 to 2000	±10°F
	2000 to 3200	±½%

NOTE: When special forms of thermocouples fabricated from finer wire or strip material are supplied, the accuracy and reproducibility may deviate from the tables, although repeatability is assured.

the time lag of the unit in responding to temperature changes. In fact, if there are small cyclical changes, a large thermal mass will show very little response to them.

Connecting heads. It is customary to use a connector or terminal block to provide good electrical contact between the thermocouple and lead wires to the measuring instrument whenever it is considered impracticable or uneconomical to run the thermocouple wires directly to the instrument. In most industrial installations this terminal block is supplied inside a weatherproof connecting head which is firmly attached to the protecting tube. The terminal block is readily removable from the head so that thermocouples can be readily replaced without disturbing the installation of the protecting tube or process.

Thermocouple extension wires. In view of the fact that the emf developed by the thermocouple is dependent upon the temperature difference of the measuring and reference junctions, it is necessary to extend the thermocouple to the compensating coil of the automatic reference junction compensator within the measuring instruments by means of extension or "compensating" wires. The connection of compensating cable

Table 15·4 Limits of Error of Extension Wires

Type of thermocouple	Type of extension wire	Temp range, °F	Limit of error, °F
Copper-constantan	Copper-constantan	−75 to +200	±1½
Chromel-constantan	Chromel-constantan	0 to 400	±4
Iron-constantan	Iron-constantan	0 to 400	±4
Chromel-alumel	Chromel-alumel	0 to 400	±6
242-33 alloy	242-33 alloy	0 to 400	±6
Platinum-rhodium–platinum	Copper-alloy	75 to 400	±12
5-20 platinum rhodium	Twin copper	75 to 200	±12

to the thermocouple element is made where the ambient temperature is never likely to exceed 200°F. Although these flexible leads have thermoelectric properties similar to the actual thermocouple, they are not necessarily of the same materials but are such that the characteristics are identical over the ranges as listed in Tables 15·4 and 15·5. By using extension wire of the same material as that of the thermocouple element, maximum accuracy is assured, although this approach is not economically possible with the rare-metal thermocouples. The 5-20 combination, however, has an output that is so low over the first 200°F that extension wires are not justified unless extreme accuracy is required.

Extension-wire insulation. Extension wires are available in duplex form with four specific types of insulation and protective covering as detailed

Table 15·5 Basic Data on Thermocouple Extension Wire

Type of thermocouple and symbol	Extension wire combination and symbol	Color code of insulation			Recommended limits of error, °F		
		Over-all	Positive	Negative	Standard	Special	Temp range, °F
Iron-constantan (J)	Iron-constantan (JX)	Black	White	Red	±4	±2	0 to 400
Copper-constantan (T)	Copper-constantan (TX)	Blue	Blue	Red	±1.5	±0.75	−75 to +200
Chromel-alumel (K)	Chromel-alumel (KX)	Yellow	Yellow	Red	±6	...	0 to 400
	Copper-constantan (VX)	Red	Brown	Red	±6 *	...	75 to 200
	Iron-alloy (WX)	White	Green	Red	±6 *	...	75 to 400
Platinum–platinum-rhodium (R or S)	Copper-alloy (SX)	Green	Black	Red	±12 *	...	75 to 400

* Applicable for a reference junction temperature of 75°F only.

Note: Data for this table were extracted from the ISA Tentative Recommended Practices on Thermocouple 19. Limits of error given are those *recommended* by ISA and may not reflect those for wire currently available from suppliers.

in Table 15·6. They are grouped into general service, high-temperature, moisture-resistant, and the more expensive metal-clad corrosion-resistant types.

Notes on installation. It should be realized that a thermocouple element can only indicate the temperature experienced by its measuring junction. For this reason, it is essential so to install the thermocouple that the hot junction is not in the path of flames or streams of gases hotter than the general space whose temperature is to be measured. Flame impingement on the protecting tube will also materially shorten the tube life.

When fluid temperatures are to be measured, stagnant areas which are not representative of the true temperature should be avoided, and every care should be taken to locate the thermocouple in a turbulent zone of the hot medium to reduce transfer lag and improve controllability of the process.

When the interior temperatures of continuous hearth heat-treatment furnaces and ovens vary widely from point to point, it is often necessary to use portable or traveling thermocouples to determine the temperature distribution in order to select the best thermocouple locations for optimum control.

The protecting tube should be immersed sufficiently to prevent erroneous readings due to heat transfer along the tube as well as along the conductors, a general rule being that the depth of immersion in a liquid should be as great as conditions permit and should never be less than 10 times the outside diameter of the protecting tube. However, special designs of thermocouple assemblies permit shallower immersions, and recommendations based on previous experience will generally be taken into consideration.

When the temperature to be measured is relatively high for the type of protecting tube, the thermocouple should be installed vertically if possible to prevent sagging of the tube. Protecting tubes installed in electrically heated furnaces operating at temperatures above 1500°F should be grounded.

The over-all length of the thermocouple assembly should be such that the assembly extends beyond the outer casing of the furnace or vessel so that the temperature of the connecting head can remain relatively low in order to avoid errors introduced by exceeding the specified operating temperatures of extension wires.

It is essential that the correct type of extension wires be used for the specified type of thermocouple, and color coding has been introduced as recommended by ISA to avoid confusion. Whenever practicable, it is advised that conduit be used with adequate grounding to prevent interference from power lines. While it is permissible to run other extension wires in the same conduit, it is essential that no other electrical transmission wires be included.

Table 15·6 Basic Characteristics of Pyrometer Cable Insulations

Material	Insulating resistance	Temp range, °F	Resistance to chemicals *	Abrasion resistance	Moisture absorption resistance	Resistance to aging	Color coding	Use—prim or sec insulation	Cost
Fibers:									
Cotton	†	To 200	‡	Fair	Very poor	Fair	Unlimited	Prim and sec	Low
Asbestos	†	To 1200	‡	Fair	Very poor	Excellent	Limited	Prim and sec	Moderate
Standard fiber glass	†	To 900	‡	Fair	Very poor	Excellent	Limited	Prim and sec	Low
Tempered fiber glass	†	To 1200	‡	Fair	Very poor	Excellent	Not available	Prim and sec	Low
High-temp fiber glass	†	To 2000	‡	Very poor	Very poor	Excellent	Not available	Prim and sec	Very high
Rubbers:									
Natural rubber	Good	−50 to 167	Very good	Very good	Excellent	Poor	Limited	Primary	Moderate
Neoprene	Fair	−50 to +225	Very good	Excellent	Good	Good	Limited	Secondary	Moderate
Buna-N	Poor	32 to 275	Very good	Very good	Very good	Good	Not available	Secondary	Fairly high
Buna-S	Good	−65 to +140	Good	Very good	Good	Good	Not available	Secondary	Moderate
Silicone rubber	Good	−50 to +500	Very good	Very poor	Very good	Excellent	Very limited	Primary	Fairly high
Plastics:									
Polyvinyl chloride	Good	−50 to +220	Very good	Very good	Very good	Excellent	Unlimited	Prim and sec	Very low
Nylon	Good	−50 to 250	Very good	Excellent	Poor	Fair	Limited	Secondary	Moderate
Polyethylene	Excellent	−94 to 175	Very good	Fair	Excellent	Good	Fair	Prim and sec	Moderate
Teflon	Excellent	−190 to +482	Excellent	Good	Excellent	Excellent	Very limited	Primary	High
Oxide:									
Megopak	Excellent	1400 to 2000, depending on diameter	Excellent	Excellent	Not affected	Unlimited	Not used	Both (complete assembly with sheath)	Very high
Enamel:									
Electrical enamel	Good	To 225	Not used	Poor	Good	Unlimited	Not available	Only primary	Very low

* Not all types of insulation are equally resistant to the same chemicals.
† Insulation resistance good in perfectly dry atmospheres.
‡ Not used for this purpose.

Fig. 15·17 Thermocouples and wires. A magnified cross section shows construction of Megopak thermocouple wire. The pure magnesium oxide insulation provides high electrical resistance and long life at high temperatures. (*Minneapolis-Honeywell Regulator Co.*)

Insulation of thermocouples. Electrical insulation between the two conductors is essential except at the measuring junction, and it is preferable that each conductor be insulated from ground. Although the electric stresses between the two conductors are always relatively low, the insulation must be good enough to remain unimpaired by any high temperatures experienced by the thermocouple in service. Therefore, twin-hole refractory insulators are generally used, and they are essential for operating at temperatures above 1000°F. The insulators are carefully selected so that all volatile impurities have been removed. A more compact but expendable thermocouple has the metal-clad swaged magnesium oxide insulation, Fig. 15·17.

For temperatures below 1000°F and where design necessitates, satisfactory insulation can be obtained by utilizing and combining the various qualities of fire, heat, chemical, abrasion, and moisture resistance of asbestos, glass, silicone, rubber, and Teflon.

Maintenance of thermocouples. Whenever possible, a visual inspection of the protecting tube should be made periodically without actually disturbing the installation. Tubes showing signs of deterioration should be replaced before the thermocouple becomes contaminated. Data on protecting tubes and wells will be found in Table 15·7. When replacing protecting tubes, all those made of refractory materials should be preheated to reduce the thermal shock if the plant is up to temperature, and care should be taken to replace the thermocouple so that the depth of insertion remains unchanged. It is further suggested that the accuracy of the thermocouple be checked periodically at a rate determined by experience, especially on replacement of protecting tubes.

Contamination of the thermocouple can cause an increase or decrease in output emf, so it is practically essential to check the thermocouple in its installed location. This is conveniently done by placing a check thermocouple alongside the service couple, or in its place, and comparing the readings. Checking by the replacement method is possible only when the temperature is likely to remain stable, and in certain installations where all practical means of testing are ruled out, many users have established an estimated life cycle of replacement to assure optimum operation of plant.

Table 15·7 Summarized Data on Protecting Tubes and Wells *

Type	Material composition	Recommended upper temp limit, °F		Remarks
		Oxidizing atm	Reducing atm	
Wrought iron	Wrought iron	1200	1300	Inexpensive material suitable for general applications except corrosive atmospheres
Cast iron	Cast iron	1300	1600	Useful in chemical industry for concentrated sulfuric acid and caustic solutions
Calorized carbon steel	Mild steel coated with aluminum oxide	1700	1700	Increased corrosion resistance and suitable for molten-metal temperatures to 1700°F
Type 304 stainless steel	70% iron, 18% chromium, 8% nickel	1800	1800	Corrosion resistant; used for protecting tubes in wet-process applications such as steam lines and chemical solutions
Type 446 chrome iron	70% iron, 27% chromium	1900	1700	General-purpose tube particularly suitable for strongly sulfidizing conditions
Nickel	Nickel	2000	2000	Principally used for hot caustic and cyanide baths and sulfur-free atmospheres
Inconel	79% nickel, 13% chromium, 6% iron	2200	2200	Better mechanical strength than chrome-iron and preferred to nickel where atmosphere is slightly sulfur-bearing
Silica	Fused silica	2400	2000	Excellent resistance to thermal shock and unaffected by most acids regardless of concentration or temperature
Corundum	High aluminum content	3000	3000	Good resistance to thermal shock and highly resistant to destructive action of fluxes, corrosive slags, and gases
Carbofrax	Silicon carbide	2900	3100	Generally used as outer protecting tube where rapid temperature fluctuations and mechanical shock are prevalent

Table 15·7 (Continued)

Type	Material composition	Recommended upper temp limit, °F		Remarks
		Oxidizing atm	Reducing atm	
Metal-ceramic	Chromium and aluminum oxide	3000	3100	High thermal conductivity and superior strength at high temperatures; used for brass and bronze melts
Recrystallized alumina	Pure alumina	3500	3500	Vacuum-tested for porosity, has high thermal conductivity, good electrical resistivity at high temperatures; is inert in hydrogen and carbonaceous atmospheres
Steel	Carbon steel	400	500	Useful for low-temperature applications except in corrosive atmospheres; commonly used for fabricated wells
Monel †	67% nickel, 30% copper, iron, manganese	700	800	Suitable for wide uses in chemical-process applications; is strong, tough, and highly resistant to most types of corrosion
Nickel †	Nickel	900	900	More expensive than monel but withstands higher temperatures; has good thermal conductivity
Type 304 stainless steel †	70% iron, 18% chromium, 8% nickel	1100	1100	Corrosion resistant; used in wet-process applications such as steam lines and chemical solutions
Type 316 stainless steel †	65% iron, 18% chromium, 12% nickel, molybdenum	1500	1500	Better strength than 304 for high-pressure wells

* Permission of Anthony Turner of Taylor Instrument Companies.
† Pounds per square inch rating depends on construction.
NOTE: It is customary to use protecting wells in preference to tubes where operating pressures are likely to exceed 50 psi.

Thermocouple certification. Thermocouples can be supplied at extra cost with a National Bureau of Standards certificate of calibration or a certificate of test based upon precision comparison against NBS certified couples. It is requested that no base-metal thermocouple be less than 3 ft in length and that rare-metal thermocouples have a minimum of length of 2 ft for certification.

TEMPERATURE-INDICATING AND RECORDING INSTRUMENTS FOR USE WITH THERMOCOUPLES

The simplest type of temperature measurement that can be made with a thermocouple is shown in Fig. 15·18. The indicating instrument is a sensitive millivoltmeter which will take only a small current for its full-scale deflection. Under these conditions, the deflection of the meter will be almost directly proportional to the electromotive force of the couple, and hence to the difference in temperatures between the cold junction and the hot junction. *It must always be remembered that the emf of any couple depends upon the difference in temperatures of its junctions, and not on the temperature itself.* An instrument of this sort will not indicate the correct temperatures if the temperature of its cold junction differs from its calibrated value. To compensate for changes in the cold-junction temperature, some meters are provided with bimetallic elements within the meter housing which will provide the necessary mechanical correction so that the indicated value is correct. There are three limitations in the use of such an instrument:

1. The resistance of the series circuit must be kept to a minimum and have some definite value if the calibration is to be correct.
2. The current taken by the measuring circuit must be a minimum.
3. The mechanical power that can be developed is extremely limited.

Because of the first limitation, the distance that the meter can be from the thermocouple is to some degree limited; because of the second limitation, a very sensitive meter must be employed. The meter will be unable to do more than indicate; recording and controlling will generally be beyond its capabilities, although controlling is possible. In order to over-

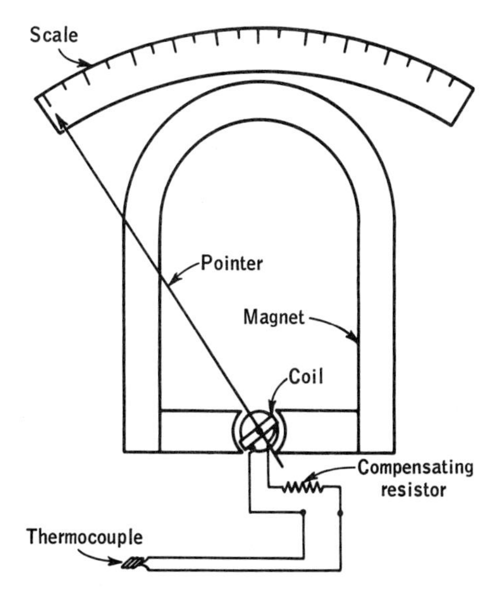

Fig. 15·18 Simple deflection type of thermocouple pyrometer.

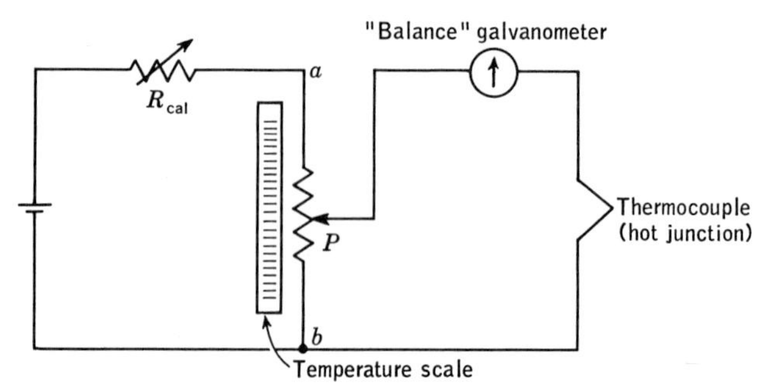

Fig. 15·19 Elementary potentiometer circuit for determining temperature.

come these deficiencies, a series of circuits employing potentiometric techniques has been developed. One of the simplest is shown in Fig. 15·19. Before attempting to analyze the performance of the circuit shown in Fig. 15·19, the theory of the potentiometer should be reviewed in Chap. 7. The galvanometer will indicate a balanced condition when the emf of the thermocouple is exactly equal in magnitude to the potential between points P and b; it is assumed that the thermocouple and the potentiometer are connected in series opposition, so that the two sets of emf's will tend to cancel each other. Only the net voltage remaining will send current through the meter circuit. Inasmuch as there is a very definite relationship between the millivolts developed by the couple and the difference in temperature of the two junctions of the couple, the scale may be calibrated in millivolts, temperature difference, or both. The true temperature will then be obtained by adding the cold-junction temperature to the indicated difference value. If the cold junction is to be kept at some definite value, then the scale can also be made to read the true hot-junction temperature.

The simple circuit shown in Fig. 15·19 is highly desirable, but the temperature indications can be true only when the current through the potentiometer is precisely equal to the calibrating value. Unless the battery is some very unusual type whose emf does not vary with temperature, age, and current flow, one never knows the current, and thus the millivolts developed across the resistor winding ab. Ordinary current-measuring devices will not give the desired accuracy either.

To overcome some of the deficiencies just mentioned, the circuit shown in Fig. 15·20 is employed by many temperature indicators, recorders, and controllers. When the switch is closed on the contacts T and C, the circuit is almost identical with that of Fig. 15·19. But when the switch is moved to the opposite position, a new circuit is established: a standard cell is placed in series opposition with the emf existing between points X and b. If these two voltages are not almost exactly equal, the galva-

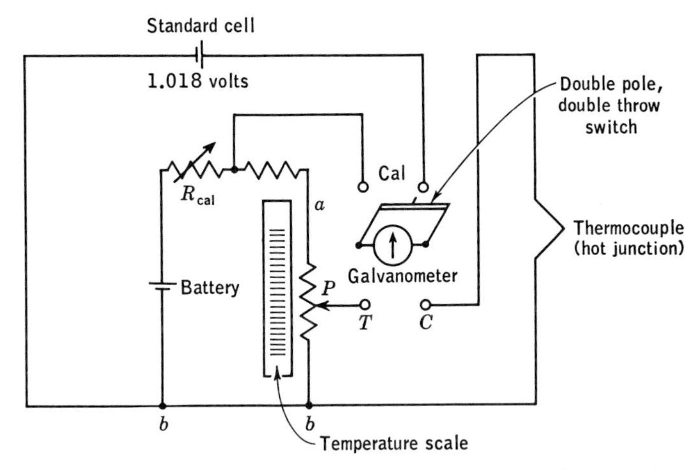

Fig. 15·20 Basic potentiometer circuit for temperature determination.

nometer will indicate the unbalance. Adjusting the resistor R_{cal} can reestablish the equality of these voltages.

The standard cell. The standard cell, Fig. 15·21, is a special battery whose terminal voltage remains almost constant for long periods of time. There are two types of Weston or cadmium cells used for standard voltage references.

1. The saturated Weston cell. The emf of this cell varies by only a few microvolts over a period of 10 years or more. It is, however, temperature sensitive, having a coefficient of 40 μv/°C. At 20°C it has a value of 1.0186 volts. This is the standard cell used by the standardizing laboratories.

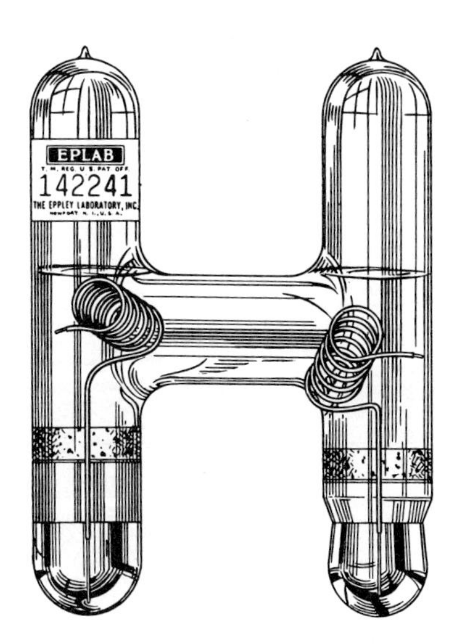

Fig. 15·21 Typical standard cells. (*The Eppley Laboratory.*)

2. The unsaturated Weston cell. This cell differs from (1) in that it has a negligible temperature coefficient, but it does have a yearly change of about $-100\ \mu v$. This is the cell in common use by industry. Its nominal value is 1.019 volts.

Both of these cells develop the indicated emf's when there is no current flow; they are not designed to furnish more than a few microamperes at any time. To take more than this current from them will invalidate their use as voltage standards. In certain circuits that will be developed later, these cells are used for working cells instead of reference cells, but in each instance, the current taken from them is on the order of a few microamperes. The unsaturated cell is not reliable at temperatures lower than 32°F or above 140°F. If these limits are exceeded, the cell should be given a week under normal conditions to regain its usual value before being tested.

In practice, the standard cell may be used in several ways. Figure 15·20 suggests a very common one. The switch may be moved to the calibrate position manually whenever it is considered desirable to check the potential developed by the potentiometer. When the calibration is complete, the switch is moved to the opposite side. By using a clock-actuated switch with automatic self-balancing, the calibrating procedure may be made automatic in nature. During the few seconds required for the calibration check, the process is not being monitored.

Fig. 15·22 Schematic diagram of continuous standardizing circuit. By connecting the standard cell in opposition to the output of the regulated-voltage transformer, the standard cell is able to supplement low outputs and reduce deviations caused by high transformer voltage. Under these conditions the standard cell furnishes continuous stabilization. (*The Bristol Company.*)

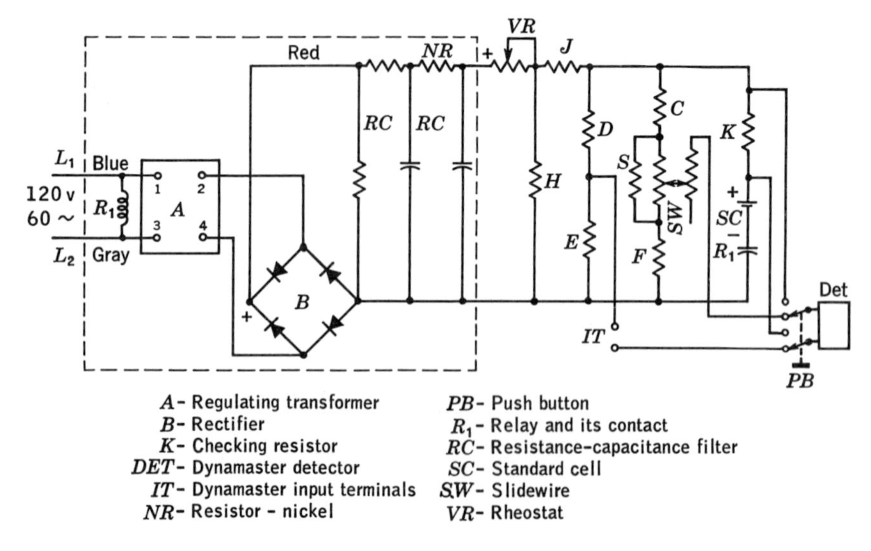

When interruptions in the measurement are undesirable, or for some other reason it is not desired to rely upon occasional calibration, a circuit such as is employed by the Bristol Company, Fig. 15·22, suggests itself. This circuit provides for a constant-voltage transformer to deliver the correct potential to the device. By individually tuning each transformer, the optimum constancy of a-c potential is delivered to the rectifiers and the voltage-dividing network. For the correct input voltage and frequency, the output of this unit should be almost exactly equal to the potential of a standard cell. If this condition exists, then connecting a standard cell across these terminals in series opposition will not cause any current to flow through the cell. If, however, the rectifier output should drop slightly, the standard cell will try to maintain the potential at its proper value by delivering a slight current of its own. On the other hand, if the potential is high, then current will go through the standard cell in the opposite direction.

The action of the standard cell in this circuit is almost exactly that of the storage battery in your car with respect to the generator: if the voltage of the generator is exactly equal to that of the storage battery, no current will flow from one to the other, even though current is being delivered by the generator to other circuits. If the terminal voltage of the generator should now decrease slightly, the battery will deliver current to the circuit and try to maintain the original voltage. If on the other hand, the voltage of the generator rises above that of the battery, then current will go through the battery to charge it. But under all conditions, the battery is trying to maintain a voltage which is essentially constant. If a voltage regulator is not used with a car generator, then the storage battery is about the only method of keeping the voltage somewhere close to the rating of the battery.

A relay connected to the input power supply permits the standard cell to remain in the circuit only while commercial power is available. If the standard cell were to try to deliver the potentiometer current, there would be too much of a drain on the cell, approximately 60 μa. Under normal conditions, a 1 per cent change in the d-c bridge supply voltage requires a cell current of 0.6 μa. It should be realized that the potentiometer circuit is one of high resistance (about 17,000 ohms) and is not of the order of resistance suggested by some of the studies cited in earlier chapters.

Another method of using the standard cell continuously is shown in Fig. 15·23. This design by the Foxboro Company employs a totally different approach to the problem. The circuit will be discussed at another point.

Cold-junction compensation. The temperatures indicated by the pointers in Figs. 15·19 and 15·20 can be true only when the cold-junction

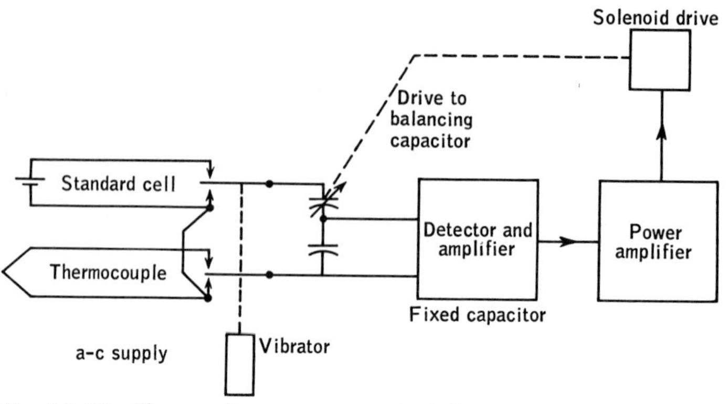

Fig. 15·23 The small currents required by certain capacitor-type bridges permit the use of the standard cell as a working cell. (*The Foxboro Co.*)

temperature is equal to the calibrated value. Any variations in the cold-junction temperature can be corrected automatically if an additional circuit is added to that shown in Fig. 15·20. One method is indicated in Fig. 15·24, where two additional resistors are shown in series. One has its resistance variable with temperature, while the other is unaffected by temperature changes. This then creates a variable-resistance series circuit. If the resistor R_{temp} gets colder, its resistance decreases, so the circuit resistance decreases. This produces an increased current through the fixed resistance $R_{bb'}$. This increased current will increase the potential from b' to b by enough to cancel the increased emf of the couple. When the cold junction is above its design temperature, the action is reversed.

Fig. 15·24 Schematic of a potentiometer used for the measurement of temperature.

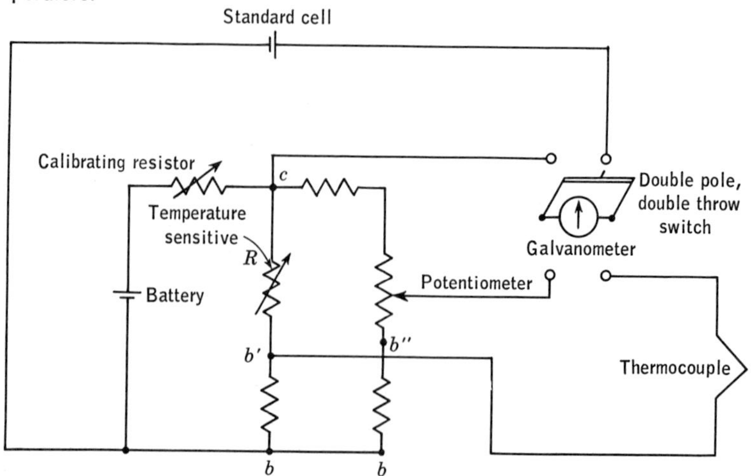

Without cold-junction temperature compensation, when the cold junction is below the design temperature, the indicated temperature will be too high. When the cold-junction temperature is above the design value, the indicated reading will be lower than the true value.

Regardless of the accuracy of the calibration of the thermocouples, the accuracy of any reading will be in part dependent upon maintaining the cold junction at its proper temperature or correcting the readings to allow for any deviation.

Automatic potentiometers. The implication of the remarks to this point is that the positioning of the potentiometer to obtain the balance position is entirely a task for a person. For purposes of simplifying the circuits for study, this has been assumed. But in actual practice, automatic indication is, of course, the usual thing. The methods employed to rebalance the galvanometer are many. All of them cannot be studied here, but some of the most representative ones will be considered. The same basic potentiometric devices are used not only for temperature measurements but also for any measurements of potential within the range of the instrument. The measurements may include pH, oxidation-reduction, gas analysis, and radiation, among others. Consequently, it is imperative that the principles of operation be *thoroughly learned,* not just memorized.

Inasmuch as the galvanometer has been used to indicate conditions of unbalance, it would appear to be only reasonable to use it to reestablish the desired conditions of balance. For many years, the Leeds and Northrup Company manufactured a potentiometric device which employed a galvanometer and a clutch of unusual design, Fig. 15·25. When there was perfect balance of the potentiometer galvanometer, it was

Fig. 15·25 The position of the galvanometer pointer governs the amount and direction of the correction applied to the potentiometer. (*Leeds and Northrup Co.*)

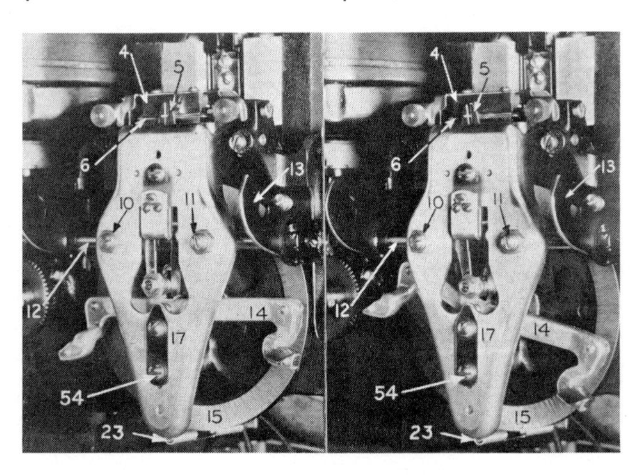

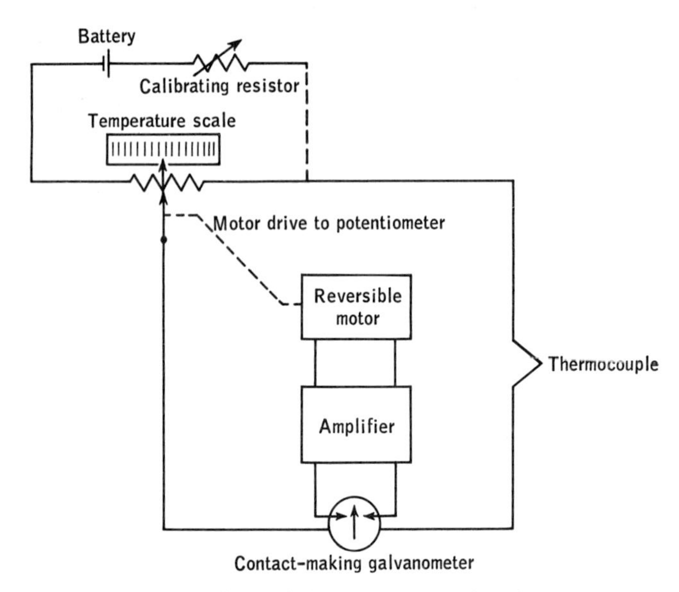

Fig. 15·26 A mechanical clutch may be replaced by a contact-making galvanometer and an amplifier.

centered between two arms that were connected to a clutch which could drive the potentiometer in the proper direction to secure balance. At intervals of 2 or 3 sec, a cam would cause the galvanometer pointer to be clamped if it were away from the center position. At the same time, a second cam was being rotated to furnish the driving power for the balancing mechanism. With the galvanometer pointer in the center, there was no mechanical interference to hinder the movements. But when the galvanometer was away from center, it created a degree of interference which produced motion of the clutch proportional to the amount the galvanometer was away from center. For small deviations from the center, a correction could be made with one or two movements of the clutch. But if there were a very large degree of unbalance, then eight or ten attempts might be made before perfect balance was secured. It should not be inferred that this potentiometer was slow in its operation; only a few seconds were required to move the pointer or pen from one end of the scale to the other.

This use of the deviation of the pointer to affect a return to the balanced circuit is worthy of enough study to master the performance. Perhaps the next application of the galvanometer that suggests itself is to use the galvanometer deflection to actuate certain circuit components. The circuit shown in Fig. 15·26 has been used by the Bristol Company. Such a circuit must employ only very low voltages at the galvanometer contacts. Voltages of the order of 100 volts will create electrostatic forces of such magnitude that the galvanometer cannot open its contacts even when the bridge is balanced. In fact, the current may have reversed itself to an appreciable degree before the force developed by the galvanometer can overcome the electrostatic effects. When this happens, the galvanometer will immediately call for a correction in the opposite direction; the same sort of thing happens again, and the device will continue this overcorrection, first in one direction and then in the other. Consequently, with any attempt to use the galvanometer pointer to complete some circuit, make sure that it is a low-voltage circuit of just a few volts.

To ensure that the galvanometer is subjected to no physical load, the galvanometer has been used to direct light to a photocell, Fig. 15·27*a* and *b*. In the circuit shown, it is assumed that when the bridge is balanced there will be a definite amount of light reflected to the photocell. Under these conditions, the photocell will have a definite current and in conjunction with resistor *ab* will determine the current flowing in its own circuit. The magnitude of this current in the photocell will govern the voltage between points *a* and *b* inasmuch as $V = IR$. If the voltage *ab* can be so adjusted that it has the same value as that of battery 1,

Fig. 15·27 (*a*) Photoelectric devices may also be used to obtain automatic balancing of potentiometers. (*b*) A single photocell may be used to provide two-directional positioning of the potentiometer.

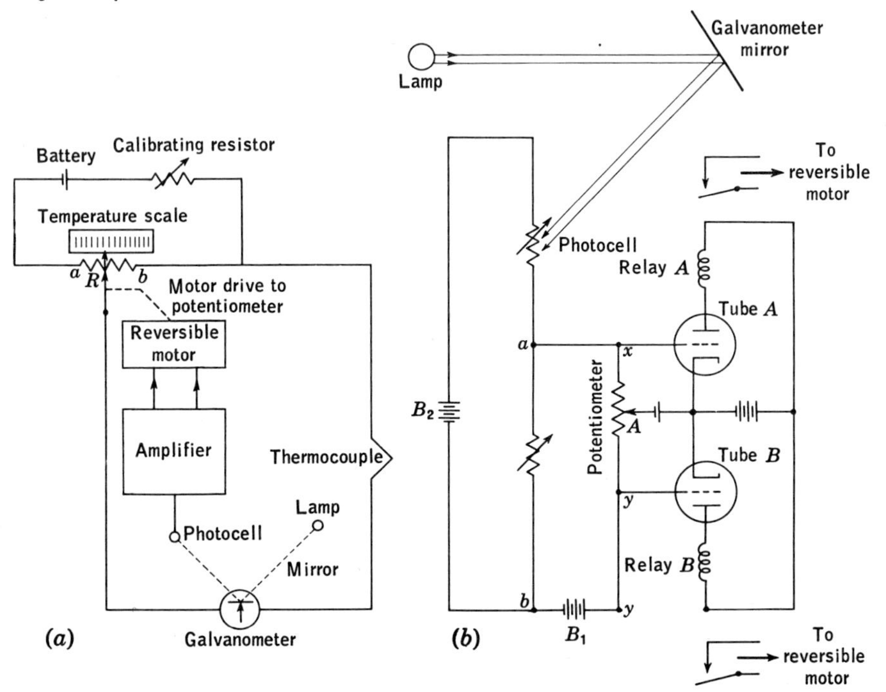

(varying the value of R_{ab} will help to do this) and the two potentials are connected in series opposition, no current will flow, because the voltages are equal in magnitude and opposite in direction. Thus, for some fixed amount of light falling on the photocell, there will be no current through potentiometer A. Accordingly, each of the tubes has the same potential applied to its grid, and this grid potential is not sufficient to cause the plate current to cause the relay to function.

If the light striking the photocell is now reduced because of some movement of the galvanometer (caused by a change of temperature at the thermocouple junction) the resistance of the photocell will change and with it will change the current through resistance ab. This change in current will in turn establish a different potential between points a and b. If we assume that the current in the photocell circuit decreases, the potential from a to b will decrease, and the potential of battery 1 will then exceed that across the terminals a and b. This inequality will cause a current to flow through potentiometer A, thereby raising the voltage of y and lowering that at x with respect to the tube cathode circuit. The positive increase in potential at y will cause the tube current in tube B to increase. If the variation in the light striking the photocell is adequate, the change in current in relay B will cause the relay to close its contacts to the balancing motor. The motor action will then be in such a direction as to try to reestablish the balanced condition.

If the amount of light striking the photocell increases, the action will be similar except that the potential between a and b will exceed that of battery 1, and current flow will be in such a direction as to increase the potential of point x and decrease that of point y.

When it is desirable to employ tubes which are totally inactive when no rebalancing action is required, cold-cathode tubes may be used in a circuit similar to that shown in Fig. 15·28. Such tubes should have a useful life several times that of the more usual hot-cathode tube.

A voltage-divider circuit similar to that in Fig. 15·27b is used. When the photocell has little or no light falling on it, its resistance is very high, and the voltages between points a, a', and b are correspondingly low.

Under normal conditions of bridge balance, the pointer, which carries a very lightweight opaque metal vane, is positioned over each photocell as shown in the sketch. Any unbalance of the circuit will cause less light to fall on one cell and more to fall on the other. The one receiving the greater amount of light will then have its resistance fall, with a consequent lowering of the potential across its terminals and an increase in the potential at the terminals of the variable resistor in series with it. Inasmuch as the cells used (Clairex) have a resistance range of about 1,000,000 to 1, the potential of point a for all practical purposes goes

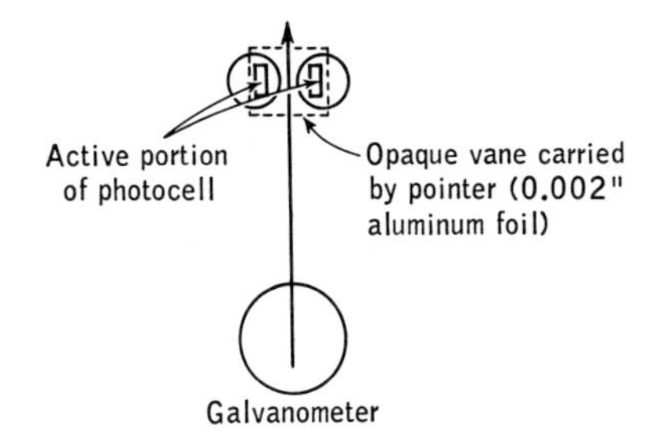

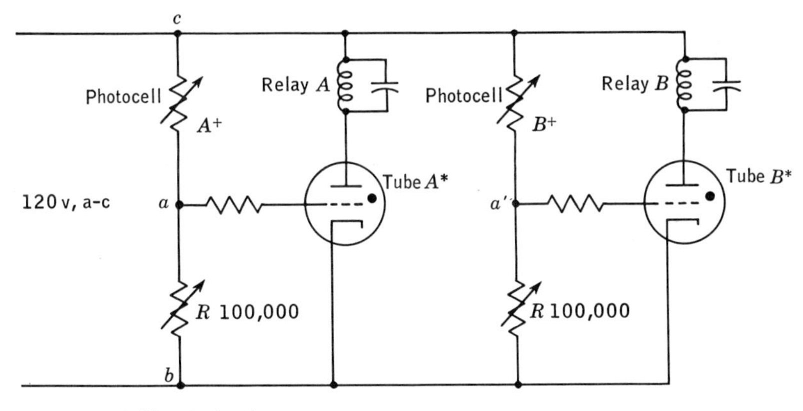

* Cold-cathode tube
+ Similar to "Clairex"

Fig. 15·28 By employing cold-cathode tubes, the tubes are energized only when corrective action is required to rebalance the bridge. This provides for extremely long operating life.

from that of b to that of c for a galvanometer pointer movement of about 0.030 in. For example, as the light on photocell A increases, that on photocell B decreases; the potential applied to the grid of tube A increases immediately to the firing point, while that of tube B is held inoperative. It might be noted at this point that a gas tube such as is indicated in the diagram will be either on or off; it cannot exist in a partially conducting condition. Some temperature indications using the circuits shown in Fig. 15·28 have indicated variations of less than $\frac{1}{10}$°F.

A somewhat different approach to the problem of balancing the bridge is shown in Fig. 15·29. This represents a fairly standard method of dealing with small direct-current error signals by (1) converting them to alternating-current signals which (2) can then be amplified, and then (3) sending them to balancing motors to reestablish the balanced-bridge condition. The *converter* is the heart of such a system. In theory and practice, it functions in a manner almost identical with that of the vi-

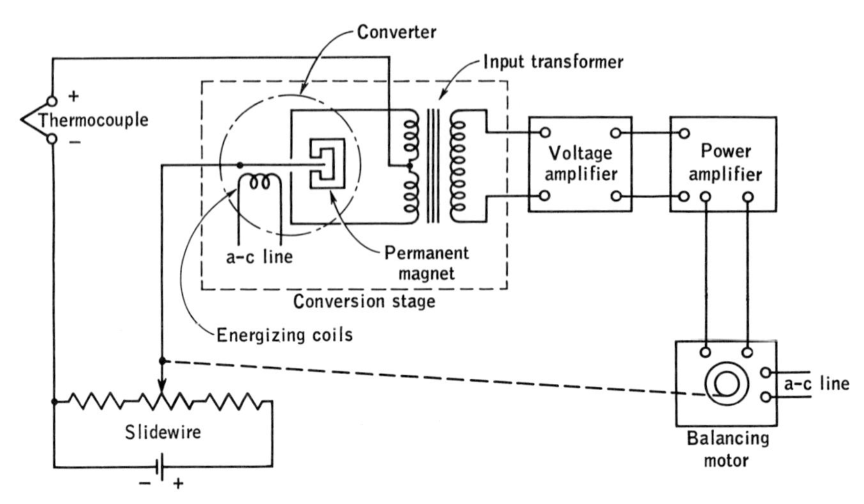

Fig. 15·29 Basic circuit when error emf's are converted to alternating-current signals for amplification and use. (*Minneapolis-Honeywell Regulator* Co.)

brator in the ordinary automobile radio. In the car radio, the vibrator interrupts the battery current, thereby converting it from a steady value to an intermittent one. This conversion is made only so that a transformer can be used to increase the voltage to some value required by the tubes. It will be recalled that in Chap. 8 it is pointed out that constant currents cannot be used with transformers; only those undergoing changes, and preferably those with sine wave inputs, can be transformed.

In instrumentation work, the reason for using the converter is almost identical: alternating currents are needed to obtain the desired amplifier action. Amplifiers using alternating-current signals are much more stable in their performance than are those using direct-current inputs. The instantaneous polarity of the error signal indicates the direction of the correction.

The physical construction of the two types of converter differs primarily in detail: one is a fairly inexpensive device for handling appreciable amounts of current, while the other is designed for long life while handling currents of the order of microamperes. When power is applied to the driving coil, the magnetic armature vibrates in synchronism with the frequency of the voltage applied to the coil. The alternating current in the coil magnetizes the armature first in one direction and then in the other, with the result that the free end which carries the contacts is attracted first toward one pole and then toward the opposite pole. The amount of movement of the armature may be of the order of 0.001 or 0.002 in. In this way, by limiting the travel, the armature can follow alternating currents of 60 to 400 cps.

Another factor that should be appreciated is that the polarity of the error signal coming from the bridge determines whether the pulse going into the transformer is of one polarity or the opposite polarity. At the output of the amplifier, the error pulse, suitably amplified, is fed into a motor. The polarity of the amplified signal with respect to the instantaneous line potential will determine the direction of rotation of the motor.

A different way of arriving at the same answer is indicated in Fig. 15·30. In this design a *magnetic motor* is used to change the resistance of the bridge error circuit. Again, this device is required because the error signal is direct current but alternating current is desired for the amplifiers and the balancing motor drive. Inasmuch as a steady direct-current error signal would not pass through the transformers, a variable resistance is used to vary the current flowing in the error circuit. The variable resistance is obtained by using what is, in essence, a telephone transmitter through which the direct-current error signal passes but the resistance of which is caused to vary by applying a mechanical force to the diaphragm. When force is applied, the carbon granules in the transmitter provide better paths for the current and their resistance decreases. But when the pressure is removed, the carbon granules develop greater resistance. If one could sing a 60-cycle note (a little more than two octaves below middle C) with sufficient intensity, he could do the same

Fig. 15·30 Any current resulting from an unbalance of the potentiometer and the thermocouple emf's is converted into pulsating direct current which can be amplified and used to control the potentiometer-drive motor.

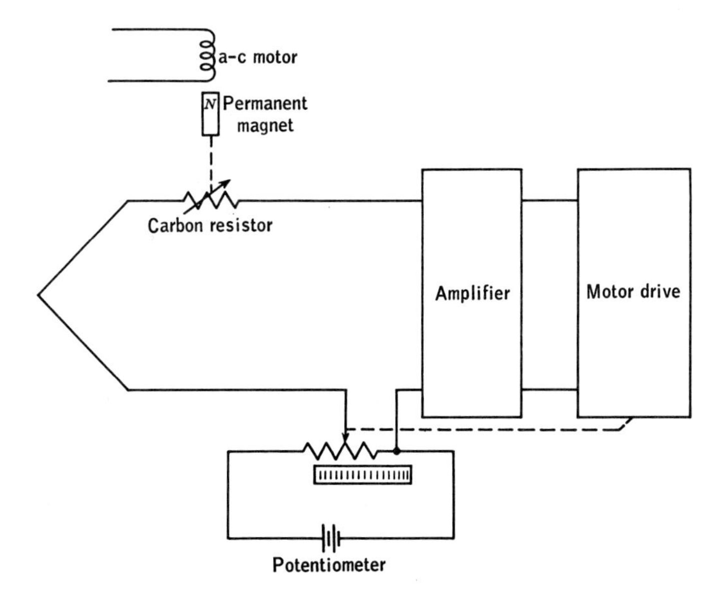

thing to the current in the transmitter that the magnetic motor does; in effect the motor simulates the voice or some instrument. The magnetic motor is really only a loudspeaker in which the movement of the voice coil applies force to the transmitter instead of moving the cone as happens normally.

Quite a different way to correct for bridge unbalance is seen in Fig. 15·31. The plate circuit of an oscillating tube carries a much smaller current when the tube is oscillating freely and delivering little power to any external elements. If a small metal piece is now interposed between the turns of the oscillator, the vane will absorb energy from the magnetic circuit set up by the coil. This extra energy is indicated by an increase in the plate current. In fact, the demand of the plate circuit may be such that the circuit no longer will oscillate, and the plate then goes to a still higher value which may be many times the value when the tube is freely oscillating. This change in oscillator plate current with the approach of a metallic vane is the basic principle employed both by Wheelco Instruments and by the Microsen system of Manning, Maxwell and Moore, Fig. 15·31. It is the principle used by certain elevator control circuits to effect leveling and speed control.

The circuits exemplified by the designs of Wheelco; Manning, Maxwell and Moore; Taylor; Swartwout; and others employ an entirely different type of potentiometer. Whereas all previous potentiometers have used movable contacts to vary the potential provided for balancing purposes, these circuits control the current through a resistor, and hence control the emf developed across the resistor terminals. Any problems caused by inertia of the balancing motor, rolling or moving contacts, and wear of the potentiometer element are eliminated by these designs. It will be noted that each circuit has its own way of accomplishing the balancing operation. (The effect of wear in the potentiometer is a debatable point in the industry.)

The basic circuit of the Microsen potentiometer, Fig. 15·31, is a bridge circuit, but an output current of 1 to 5 ma will always be required to effect the potentiometric action. In this design, we also have both torque-balance and emf-balance systems working simultaneously.

It will be noted that a series circuit is formed by the thermocouple, the input coil, and a resistor in the output of the bridge circuit. The input coil may be thought of as a conventional winding on a milliammeter or microammeter; the pointer of the meter is then the metal vane whose position near the grid oscillator coil determines the value of the tube current. Hence, the unbalance current is read on a milliammeter which may be located nearby or far away, inasmuch as the external circuit may have a wide range of resistances.

If we assume an increase in the thermocouple temperature, the emf

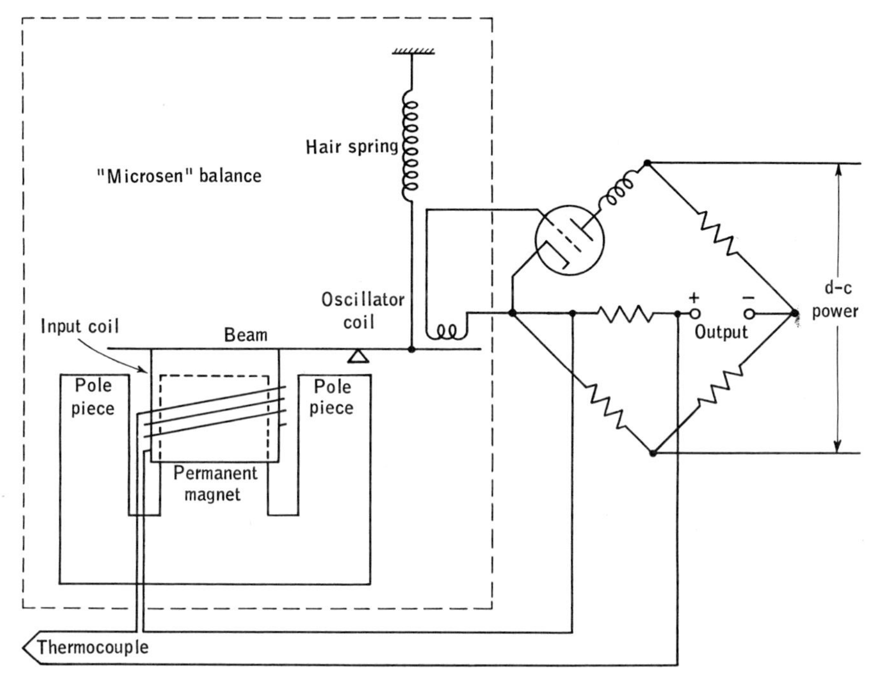

Fig. 15·31 Basic circuits for the Microsen potentiometer.

of the couple will increase, thus causing more current to pass through the series circuit. This extra current will cause the coil to attract the armature a bit more, bringing the other end of the metallic armature closer to the oscillator coil. This increases the plate current. An increase in plate current will cause a greater voltage unbalance across the mid-points of the bridge and send more current through the output circuit.

To reach the output circuit, this current must flow through the resistor in the series circuit that we previously noted. An increase in this voltage will be such that it will oppose the increased thermocouple voltage. This reduces the current in the input coil circuit, which in turn requires a readjustment of the armature position. The final torque balance that is produced causes the armature to take up a slightly different position from what it had before the assumed temperature increase. There will be a very slight current, microamperes or less, in the series circuit at the time of balance; it can never be exactly zero when operating. A stable mechanical system cannot have more than one set of conditions for equilibrium in one physical position.

The extremely high amplification factor of the system is the ratio of the change in output current to the change in position of the vane. This very high performance factor makes it possible for the torque balance to be obtained with almost imperceptible armature movement.

Because the unit requires *current* to balance the system, we are not particularly concerned with the resistance of the output circuit, provided it does not exceed approximately 3,000 ohms. The voltage output of the bridge is of no concern, it is only the current output that is employed to balance the thermocouple voltage, and it is current which produces the mechanical balance. In this one instrument are employed many basic situations:

1. Torque balance.
2. Potential balance (almost).
3. Change of oscillator plate current when energy is transferred to some other circuit.
4. The elimination of a standard voltage cell; the current and its magnetic effects are used as the standard of reference.
5. The accuracy of the system is largely dependent upon the value of the series resistor which is in the balancing circuit.
6. The bridge is designed to operate in an unbalanced condition.
7. The current to the output circuit never reverses polarity, but varies between predetermined limits.

An entirely different approach to the design of a potentiometer to balance the emf generated by a thermocouple, or any other emf of com-

Fig. 15·32 Basic circuit for the potentiometer section of Taylor Instrument Companies Transet potentiometer transmitter.

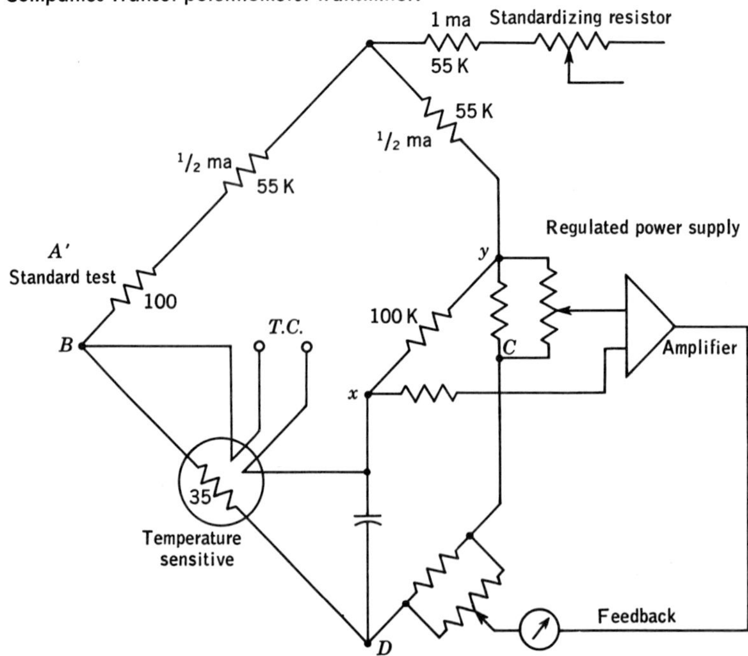

parable value, is that of the Taylor Instrument Companies No. 700T temperature transmitter. Whereas the Microsen system uses the thermocouple emf as generated and requires the movement of a vane to produce the current which is a measure of the emf, the Taylor unit introduces the thermocouple emf into what in most bridges would be the zero-balance circuit. The bridge is then balanced by causing the voltage developed in one of the bridge arms, CD, to change until points C and B are once again at almost the same potential, Fig. 15·32.

An analysis of Fig. 15·32 should not be attempted until after the bridge circuits in Chap. 7 have been completely mastered. Once these are understood, the commercial version will become apparent.

It must be emphasized that the device shown in Fig. 15·32 is truly a potentiometer, because voltages, or potentials, are balanced out. A potentiometer need not involve a movable wire slider or mechanically adjustable parts.

The use of a regulated power supply is absolutely necessary because the bridge must operate under standard conditions of current if the correct voltages are to be developed at each of the various resistors. By using two regulator tubes, a very stable output is produced. Therefore, in this circuit no standard cell is employed. By measuring the potential developed between points A' and B, it can be determined when the unit is correctly adjusted. The voltage at $A'B$ is designed for 50 mv, or 0.050 volts.

$$\begin{aligned}\text{Voltage} &= (\text{current})(\text{resistance}) \\ &= (0.0005\text{ amp})(100\text{ volts/amp}) \\ &= 0.050\text{ volts} = 50\text{ mv}\end{aligned}$$

If the voltage—this must be read with a very accurate potentiometer—is not 50 mv, then the standard adjusting resistor is varied until the correct value is obtained.

There are two other resistors that should be mentioned.

1. Resistor BD is temperature sensitive. With a constant current through it, its voltage is directly proportional to the resistance. By choosing the proper value of resistance, the changing voltages can provide reference-junction compensation. The reference junction of the couple is embedded close to the resistor to ensure that both are at the same temperature.

2. The use of resistor XY causes the full output of the instrument (5 ma) to be reached in the event of a thermocouple burnout. This will cause the instrument to indicate the maximum temperature and should shut off the furnace or unit.

The value of the current required to produce bridge balance then becomes a measure of the temperature as sensed by the thermocouple. Suitable calibrations are, of course, necessary to permit direct temperature evaluation.

RADIATION AND OPTICAL PYROMETRY

Radiation and optical pyrometry instruments have the highest ranges of any of the temperature-measuring devices; they are from +125°F to as high as 10,000°F for certain models. Whereas all other temperature-sensitive devices have required attachment to the furnace or object whose temperature is sought, radiation and optical instruments require no physical connection with the object, but depend upon:

1. The matching of the brightness of a calibrated lamp filament with that of the heated object, the optical method.
2. The measurement of the energy radiated by the object, the radiation method.

The optical method obviously cannot be used for temperatures under 1400°F, approximately, because the eye is too insensitive to wavelengths characteristic of this temperature and lower, but it can be used at the higher temperatures with high degrees of accuracy.

The radiation pyrometer will, however, respond to the extremely long heat waves of objects at temperatures of about +125°F as well as to the much shorter wavelengths characteristic of the higher temperatures. Most devices of this type concentrate the radiated energy onto delicate thermocouples of very small mass. The concentration of this energy can raise the thermocouple temperature by amounts which ensure high degrees of accuracy when the instrument is used correctly.

Both types of instrument are very fast in their response, compared with other temperature-sensing devices, because of the small mass to be heated in each type. Not only is the response fast, but the accuracies are excellent. Speed has not been obtained at the expense of precision. The outstanding characteristics of the two types are that they can be used to:

1. Measure temperatures above those of the thermocouple and resistance thermometer.
2. Measure temperatures where the atmosphere or conditions prevent the satisfactory operation of other types.
3. Measure the temperature of moving objects.
4. Measure the temperature of objects which are not easily accessible.
5. Measure the temperature when the objects would be damaged by attaching thermocouple leads or other sensing units.
6. Measure the average temperatures of extended surfaces.

The optical pyrometer. The optical pyrometer employs principles which all of us have used at one time or another: the color of an object is an indication of its temperature, and the brightness of a hot object is also a measure of its temperature. Both of these phenomena are related; as the color becomes brighter more blue light and less red light is likely to be present. In commercial instruments the reference filament in the pyrometer is made to match the brightness created by the radiated energy from the unknown temperature source. When the controlled filament is indistinguishable, in terms of brightness, from the background of the hot object, then the two temperatures are assumed to be equal. Variations in commercial design center around the manner in which the filament is made to match the brightness of the object:

Filament too dark. As seen through the telescope, the straight horizontal filament appears to be superimposed on the hot object, such as the stream of molten metal. The filament, in this instance, appears colder (darker) than the metal. Filament brightness must be increased by causing more current to pass through the filament.

Filament at correct brightness. When the current has been adjusted to the correct value, the filament disappears; it has matched and merged with the object being measured. This match is made with great precision owing to the rapid change in brightness of the filament with changes in temperature and to the sensitivity of the eye to small differences in brightness.

Filament too bright. When the filament is too hot, it appears brighter than the object being measured. The current through the filament must be reduced to provide the correct merging of filament and object.

In one method a very precisely made lamp is used in which the filament temperature is a known function of the current through the filament. When the filament has been made to match the brightness of the hot object by controlling the filament rheostat, the current can be measured. At the time of manufacture, a calibration which indicates the filament temperatures for various filament currents is made. Therefore, if the filament current can be measured, the lamp temperature may be inferred. In the circuit shown, Fig. 15·33, a potentiometer is used to measure the potential created by the filament current passing through a carefully

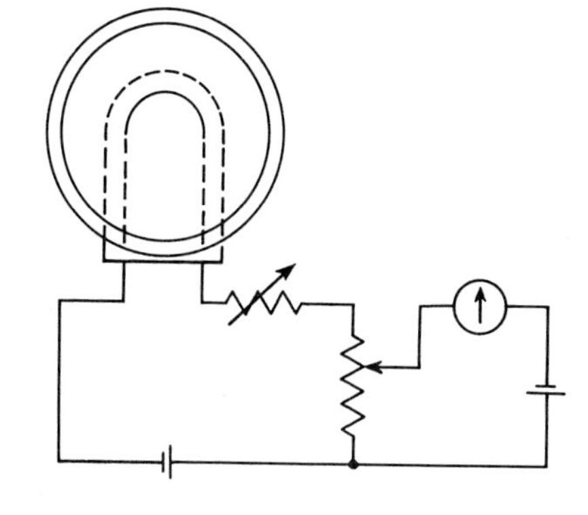

Fig. 15·33 Basic circuit used with optical pyrometers.

standardized resistor. At the time that the potentiometer is balanced and ready for reading, it will take no current from the filament circuit; it also provides a far higher degree of accuracy of current measurement than could be obtained with conventional milliammeters or microammeters.

In the circuit of a commercial instrument a unique feature is current measurement by the potentiometer without the usual standardization of current in the potentiometer circuit. Instead, the measurement is made by varying the value of a resistor in the lamp circuit that is bridged by a standard cell until a balance is obtained. Actually the calibration is not in terms of filament current but is in equivalent *blackbody temperature*. Four flashlight batteries supply the filament current. This current, passing through a rheostat, is manually adjusted until the brightness of the filament matches the brightness of the image of the object sighted upon. The battery rheostat contact and the potentiometer contact are associated through a friction clutch in such a way that when the current is adjusted, the potentiometer is carried along to maintain an approximate balance. As a result, when the standard cell is put into the circuit, only a minor adjustment is needed to bring the galvanometer to zero. A switch keeps the standard-cell circuit open except when the position of the contact is being adjusted to obtain balance.

A drum scale attached to the potentiometer contact reads directly in temperature to three, or by estimation to four, significant figures on the various scales provided. Resistor G provides the necessary suppression for establishing the instrument range.

The theory of such a device is quite simple, but to obtain accurate results in practice requires that the basic design assumptions be realized:

1. The relationship of current-temperature must be either maintained or determined by careful calibration.
2. Emitted energy is not absorbed before reaching the pyrometer.
3. The hot body is radiating as a true blackbody, or suitable corrections must be made, Fig. 15·34.

Another method uses different means to obtain essentially the same end. If a predetermined value of current passes through a filament, the temperature of the filament should always be the same. If the filament is caused to be hotter than any object whose temperature is to be measured, then the brightness of the filament can be made equal to that of the hot object by absorbing some of the luminous energy radiated by the filament. This is accomplished by rotating an optical wedge to an intensity-matching position. The position of the wedge is then interpreted in terms of the temperature.

Because the life of the lamp filament is seriously reduced and the filament calibration is changed by high temperature, it is desirable to keep the filament temperature under about 2300°F. When temperatures

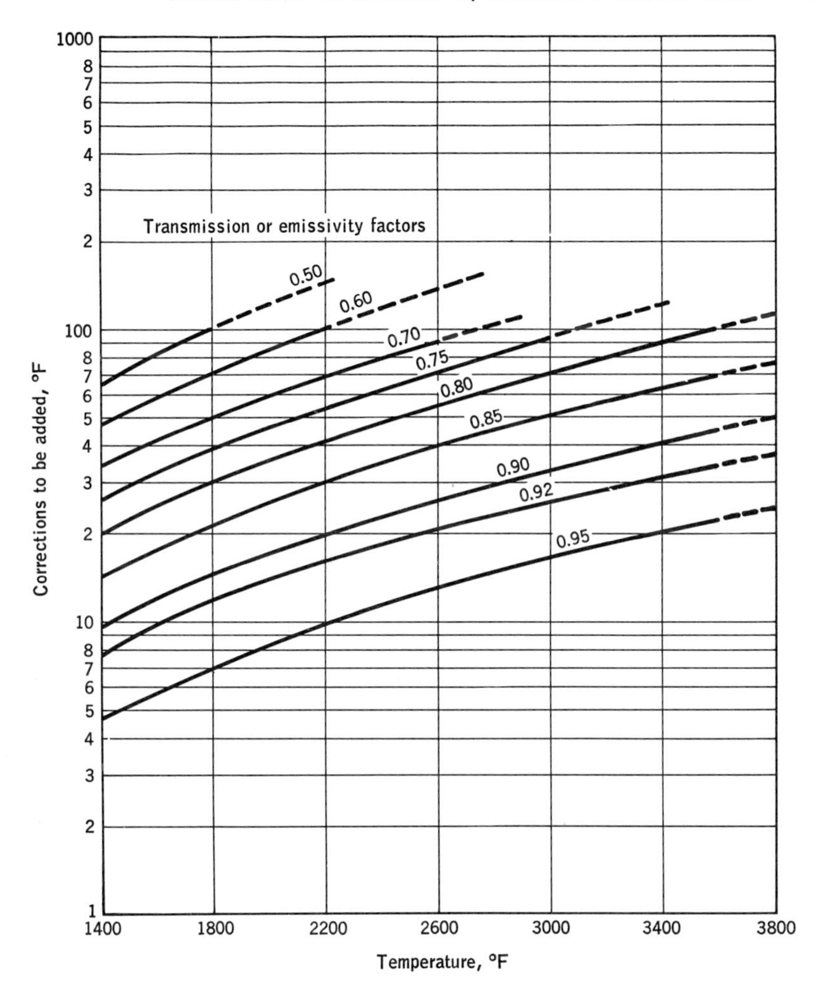

Fig. 15·34 Corrections for optical pyrometers to be added because of low transmission and emissivity. (*By permission of John Wiley & Sons, Inc.*)

higher than that are to be measured, an optical wedge is employed to reduce the brightness of the image which must be matched. In this way, higher temperatures can be measured without impairing the life and calibration of the filament.

It has been stated previously that the surface of the heated object is a factor in determining the brightness. When precise values of temperature are required, corrections should be made as indicated in Fig. 15·34. The emissivity of common materials will be found listed in Table 15·8.

One of the major shortcomings of the optical pyrometer is that it requires manual adjustment and reading. When a skilled operator is used,

Table 15·8 Total and Spectral Emissivity of Open Surfaces *

Material	Type of surface	Emissivity: Total e_t	Emissivity: Spectral e_n, $n = 0.65$
Alumel	Clean	...	0.37
Aluminum	Clean	0.03	
	Oxidized	0.19	0.30
Brass	Clean	0.04	
	Oxidized	0.61	
Carbon	....	0.08	0.90
Chromel	Oxidized	0.82	0.87
Chromium	Clean	0.08	0.34
	Oxidized	...	0.70
Constantan	Clean	...	0.35
	Oxidized	...	0.84
Copper	Clean	0.02	0.10
	Clean, molten	0.15	0.15
	Oxidized	0.60	0.70
	Calorized	0.26	
	Calorized, oxidized	0.19	
Firebrick	....	0.75	
Glass	....	0.94	
Iron	Clean	0.05	0.35
	Clean, molten	...	0.37
	Rusted	0.65	
	Oxidized	0.84	0.70
Iron, cast	Clean	0.21	0.37
	Clean, molten	0.29	0.40
	Oxidized	0.78	0.70
	Strongly oxidized	0.95	
Iron, wrought	Dull, oxidized	0.94	
Lead	Clean	0.05	
	Oxidized	0.63	
Monel	Clean	...	0.37
	Oxidized	0.43	
Nichrome	Clean	...	0.35
	Oxidized	0.87	0.90
Nickel	Clean	0.12	0.36
	Oxidized	0.85	0.90
Platinum	Clean	0.15	0.30
Porcelain	....	...	0.40
Silica brick	....	0.85	
Silver	Clean	0.04	0.07
Steel	Clean	0.08	0.35
	Clean, molten	0.28	0.37
	Oxidized	0.79	0.80
	Calorized, oxidized	0.57	
	Rough plate	0.97	
Steel (18·8)	Oxidized	0.97	0.85
Tungsten	300°K	0.032	0.472
	1000°K	0.114	0.458
	1500	0.192	0.448
	2000	0.260	0.438
Tungsten	2500	0.303	0.428
	3000	0.334	0.418
	3500	0.351	0.408

* Data from The American Institute of Physics, "Temperature, Its Measurement and Control in Science and Industry," 1941. By permission of Reinhold Publishing Corporation.

NOTE: Emissivity of oxidized surfaces may vary by 30 per cent and depends on roughness of surface. Values are given at operating temperatures, wherever possible.

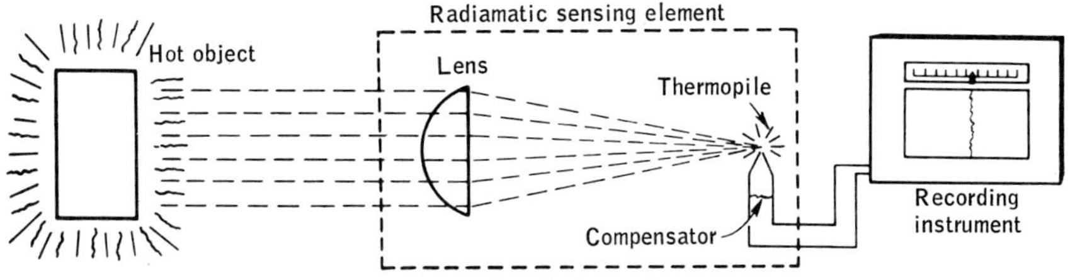

Fig. 15·35 Principle of the radiation pyrometer. (*Minneapolis-Honeywell Regulator* Co.)

accuracies of ±6°F can be attained for temperatures of 2000°F. Some of the literature refers to measurements made with a precision of 0.2°C at the temperature of melting gold, which is 1063°C, or 1945°F.

For installations where fumes or smoke prevent a clean, unobstructed path to the hot object, a refractory tube closed at one end and purged with clean air, may be used. Of course, the tube should have a temperature which is representative of the objects whose temperature is sought.

The radiation pyrometer. Whereas the optical pyrometer requires an operator to match brightness to ascertain the temperature, the radiation pyrometer can be made automatic in its operation because temperature is inferred from the *total* heat energy which is caused to impinge on a thermocouple or series of thermocouples which comprise a thermopile. The essential elements of such a pyrometer are shown in Figs. 15·35 and

Fig. 15·36 If the radiated energy is caused to heat a resistor, it can cause an unbalance in the bridge circuit. Any system for inferring temperature from resistance change may be used.

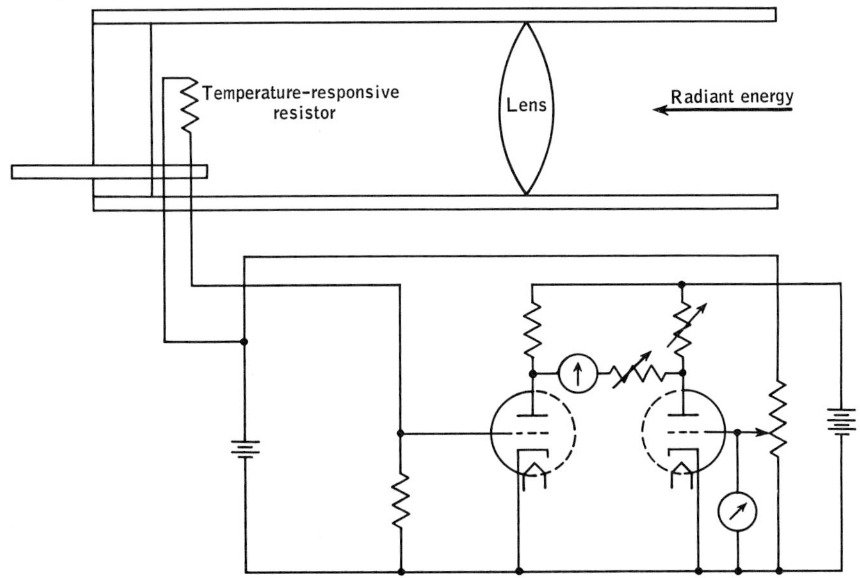

15·36. The radiated energy is converted into an electromotive force by the thermopile. This potential may then be measured in any one of a number of ways.

The tremendous changes in the amounts and kinds of radiant energy, in terms of the wavelengths of the radiated energy, are shown in Fig. 15·37. It should also be noted that the human eye can perceive but an extremely small portion of the radiated energy and thus is quite unsatisfactory for detecting the amount of energy radiated. This is particularly true for temperatures under 2000°F. It should also be observed that different lens systems have different radiation limits, and the use of

Fig. 15·37 Distribution of energy in the spectrum. The amount of energy radiated and the wavelengths vary with the temperature of the emitting surface. (*Minneapolis-Honeywell Regulator Co.*)

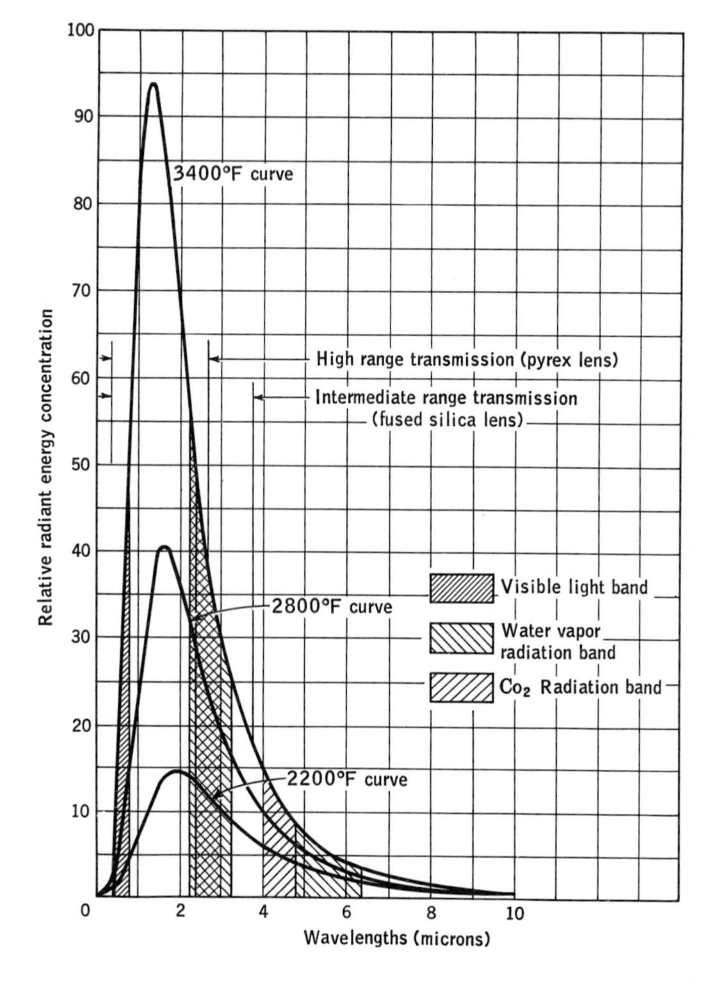

a particular lens system for the wrong wavelengths could result in very large errors.

The development of satisfactory radiation instruments was delayed until about 1879, when Stefan formulated his law which states that the

$$\text{Energy radiated by a blackbody} = KT_K^4$$

where K is a constant and T_K is the absolute temperature measured on the Kelvin scale.

In 1884, the theoretical relationship was stated by Boltzmann. The value of K in the above expression is now called the Stefan-Boltzmann constant and has a value of $5.6699(10^{-12})$ watts/(cm^2) (°K^4). Inasmuch as all bodies are continually receiving and radiating energy, the net amount radiated per second is expressed by

$$U = \sigma A(T_h^4 - T_l^4)$$

where U is expressed in watt-seconds, σ is the Stefan-Boltzmann constant, T_h is the higher temperature on the Kelvin scale, and T_l is the lower temperature on the Kelvin scale.

The above expression was derived for the ideal blackbody. One simple definition of a blackbody is that it is a body which will absorb all and reflect none of the energy which falls on it. In practice, we cannot always attain such an ideal surface. To correct for this deficiency, another factor must be applied. In practice, therefore,

$$U = e_t\sigma A(T_h^4 - T_l^4) \qquad e_t = \frac{W_1}{W_2}$$

where W_1 = actual radiation per unit area
W_2 = theoretical radiation per unit area of a blackbody

and e_t is always less than 1.

The general arrangements of the components of a radiation pyrometer are shown in Fig. 15·38*a*, while Fig. 15·38*b* illustrates the manner in which the tiny thermocouples are arranged to form a thermopile. In order to increase the speed of response, the mass of the junctions is reduced to a minimum consistent with stability, long life, and resistance to shock. By keeping to a minimum the amounts of material which must be heated by the received radiation, many instruments can attain 98 to 99 per cent of their final temperature readings in 1 to 2 sec. Increased sensitivity can be obtained by using thermocouples which are housed in evacuated bulbs, but the response time is greatly increased. In a vacuum, the thermocouple can attain temperature equilibrium, aside from heat losses along the leads, only by radiation, whereas the thermocouple or thermopile in air can reach equilibrium temperature faster because it loses heat by radiation, convection, and conduction. If the thermocouple or thermopile has to attain a higher temperature in one case than in the other,

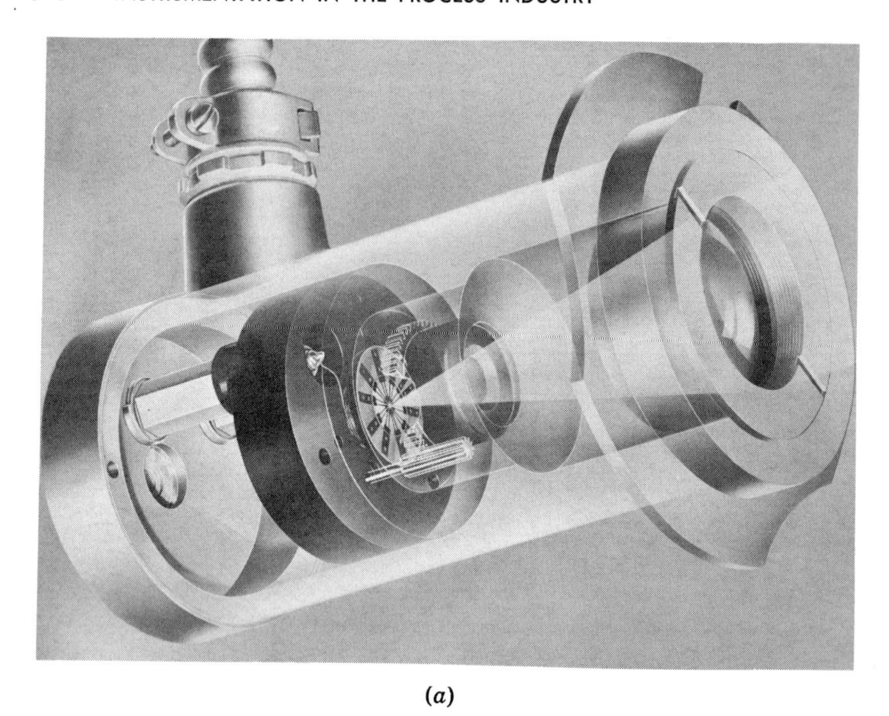

(*a*)

Fig. 15·38 (*a*) Phantom view of Radiamatic head. To increase the sensitivity its thermocouples have their junctions at the focal point. (*b*) Cross section of low-range Radiamatic head. (*Minneapolis-Honeywell Regulator* Co.)

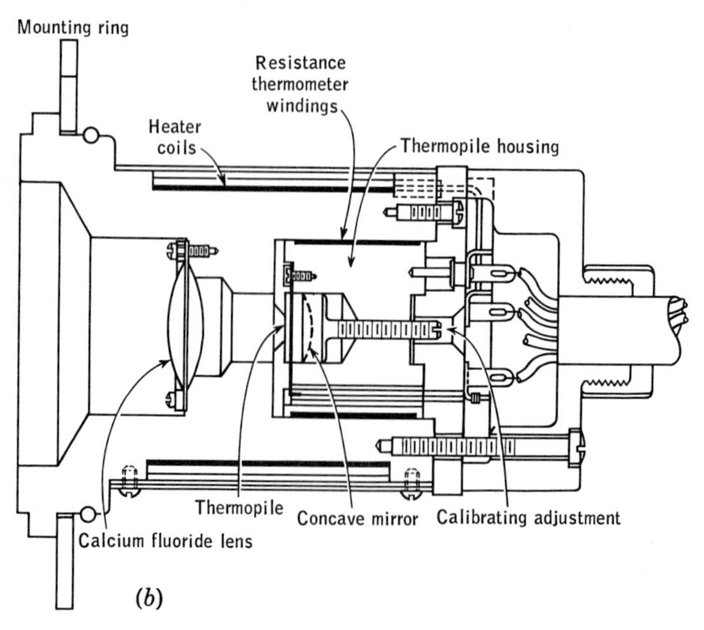

(*b*)

more heat must be delivered to it to attain the higher temperature. Obtaining additional heat requires extra time for response.

A somewhat different design of a radiation pyrometer is illustrated by Fig. 15·39. In this model, the energy is concentrated on the thermopile by using two concave mirrors instead of a lens system. The radiant energy which passes through window *A* falls on mirror *B*, which concentrates the energy on the small opening at *C*. The energy which passes through this opening is further concentrated by the action of mirror *D*, which in turn concentrates the radiant energy on the thermopile at *E*. This particular design makes it possible to position the pyrometer so that very specific areas may be monitored.

Considerations in the use of the radiation pyrometer. Although the following material is reprinted here from "Temperature Measurements with Rayotubes" by Paul H. Dike, with permission of the Leeds and Northrup Company, the comments are applicable to all types of radiation pyrometers. The word "Rayotube" is a registered trade-mark name for a "radiation pyrometer."

Fig. 15·39 The use of this particular type of optical system makes it relatively easy to measure the radiation from a specific portion of the heated body. Since the image of the radiating source, on surface G, can be clearly seen through lens F, the detector is readily positioned to measure a specific area by observing that the image of the area covers the opening C. The pyrometer is shown full size. (*Leeds and Northrup Co.*)

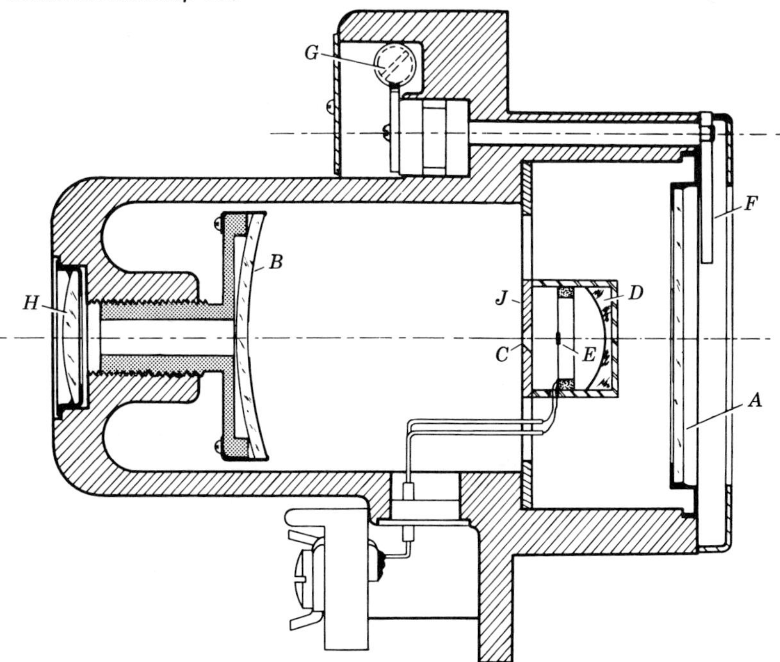

Errors due to uncertain or variable emissivity of the object under observation are inevitably present when sighting on a nonblackbody. It is only when sighting into a closed-end tube or other deep cavity in the heated object that emissivity errors may be neglected.

Errors resulting from absorption of radiation by carbon dioxide, water vapor, or other apparently transparent gases which have absorption bands in the infrared are often troublesome. These are apt to be overlooked since they are invisible to the eye. Smoke and fumes are less likely to be disregarded because they are visible, but are no more serious as a source of error. Closed-end tubes may fill with gases or fumes, giving erroneous temperature readings. It is sometimes necessary to flush closed-end tubes continuously with a gentle stream of air to remove absorbing gases. When sighting on a hot body in a furnace, the furnace gases absorb some of the radiation, or, if they are hotter than the surface sighted on, they radiate into the Rayotube, giving too high a reading. To avoid this, an open-end tube may extend through the furnace wall, nearly to the hot surface, and a gentle stream of air or of a neutral gas forced through it to maintain a nonabsorbing path for the radiation.

It is sometimes desired to measure the temperature of a body inside a furnace which is sealed to maintain an atmosphere of hydrogen or other gas. Windows provided for this purpose introduce large errors, whether they are of glass or of quartz. Such errors must be determined experimentally if possible and corrected. Closely related to these errors are those that result from deposits of foreign materials on windows, lenses, or mirrors. These errors are entirely controllable by the user of the instrument, being eliminated by cleaning at suitable intervals.

Error caused by fluctuations in ambient temperatures are sometimes encountered. In the newer Rayotubes, these errors are in general not appreciable, but with the older Rayotubes such effects are observed. A water-cooled mounting sometimes serves to stabilize the temperature of the Rayotube which, in any case, should be kept below 300°F. Sudden large changes in ambient temperature should be avoided with the older Rayotubes.

When sighting into a closed-end tube, it is important that the bottom of the tube is focused on the diaphragm in the Rayotube, and that its image is centered on the aperture in the diaphragm. Otherwise the temperature reading may be affected by variations in temperature along the walls of the tube. The Rayotubes to be used with closed-end tubes are focused in the factory for the particular length of tube with which they are to be used, and should be used with no other length of tube without refocusing, which is best done in the factory. The centering of the image of the bottom of the tube on the aperture in the diaphragm is accomplished with adjusting screws on the mounting which supports the Rayotube on the end of the closed-end tube.

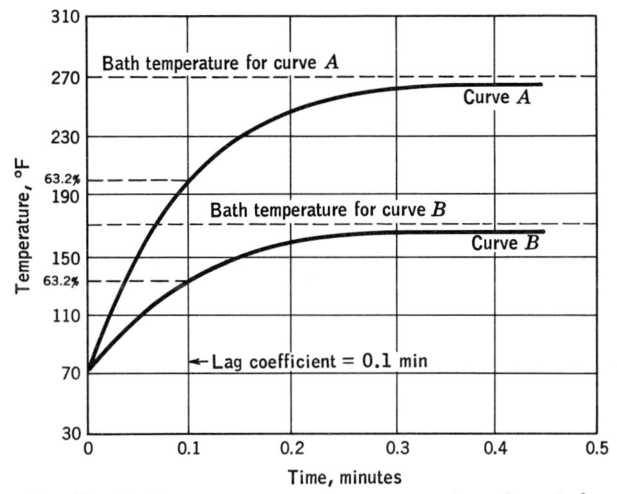

Fig. 15·40 Temperature response versus time. Speed-of-response curves show that the thermometer bulb attains a given percentage of the total change in a given time, irrespective of the magnitude of the change. (*Minneapolis-Honeywell Regulator Co.*)

GENERAL TEMPERATURE-MEASURING PRACTICES

Even though the correct instrument has been selected for a given installation, this does not guarantee that the device will function satisfactorily, because there are many other factors which combine to govern the over-all performance.

Speed of response. *It always takes time for a thermometer to respond to any temperature change.* No matter what bulb, fill, or length of capillary may be involved, time will be required for the thermometer to respond. Observation and experiment have shown that the thermometer readings, as they approach the final value, change at a slower and slower rate, Fig. 15·40. Because of the long time required to reach the final value, several other "times" have been selected because less time is required to reach them. One of the most important of these is the time required to reach 63.2 per cent of the final temperature. *The time required to reach 63.2 per cent of a step change is known as the time constant.* It will be found that regardless of whether the temperature has made a change of 10 or 100°, the same time will be required for the same percentage of change, 6.3 or 63°, provided all other conditions remain constant.

Time constants may be expressed in seconds, minutes, hours or any other unit of time. The time to reach any percentage of the final value is readily obtained from Fig. 15·41.

The effect of the well. Because of the pressures which exist where the thermometer is installed, because of the chemicals which may react with

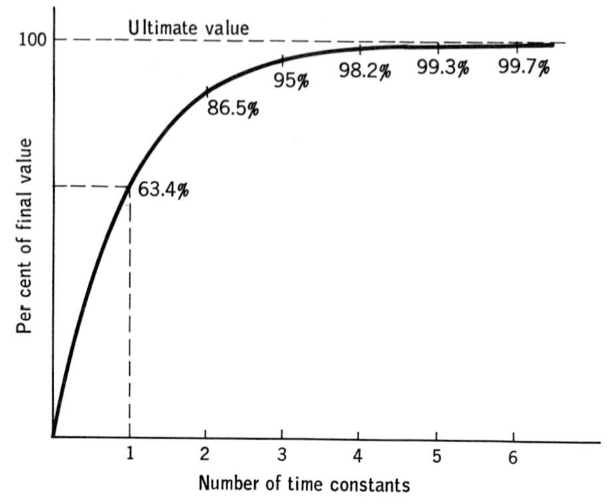

Fig. 15·41 Per cent of final value versus the number of time constants. Time constants may be expressed in seconds, minutes, hours, or any other suitable units of time.

the bulb, or because of the desirability of replacing the thermometer without shutting down the system, thermometers often are installed within a protective tube known as a well. The material of which the well is made, the distance between the bulb and the inside wall of the well, and the way in which the well is installed will determine the total time constant for the sensing unit. Figure 15·42 indicates how the well should protrude into the liquid the maximum amount and thus prevent any cooling (or heating) of the bulb by liquid which is not at the temperature of the rest of the body. Insulation to prevent the loss (or gain) of heat along the bulb and well will help to give more nearly correct readings. How serious the delay imposed by the well may be is indicated in Fig. 15·43.

It cannot be emphasized too strongly that anything that impedes the transfer of heat to the sensing element will greatly, and perhaps seriously, delay the response of the device. For identical conditions of agitation of the water, enclosing a sensing element in a stainless-steel protect-

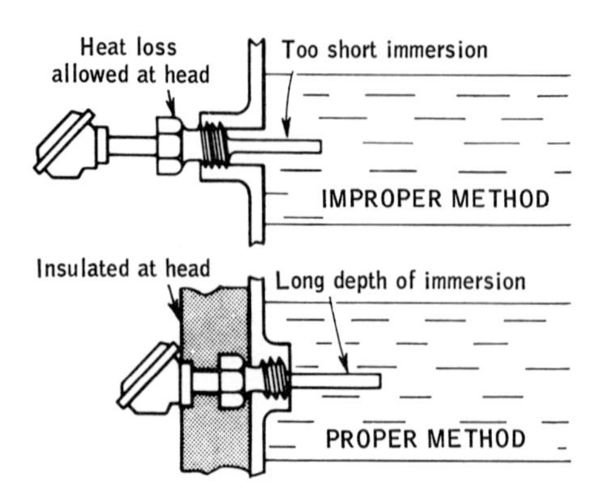

Fig. 15·42 Insufficient immersion of the thermocouple or thermometer well increases the measuring lag as heat flows along the well toward the outside wall instead of being transferred to the thermocouple or bulb. (*Minneapolis-Honeywell Regulator Co.*)

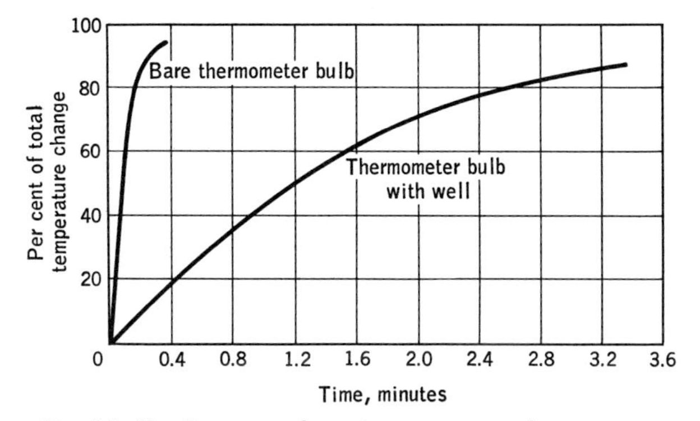

Fig. 15·43 Per cent of total temperature change versus time. The addition of a protecting well to a thermometer bulb increases the lag coefficient from 0.10 to 1.6 min, an increase of about 16 times. (*Minneapolis-Honeywell Regulator* Co.)

ing well may delay the response by a factor of 8 or 10, Fig. 15·44. The amount of heat which must be transferred will naturally control the time required to effect the transfer. For this reason, it is most important that thermal sensing units have a minimum amount of thermal capacity.

By maintaining a rapid movement of the fluid over or around the sensing element, the maximum temperature differential will be maintained. This will increase the speed of response up to a point. The response time in a still fluid may be reduced to one-quarter, or less, of this time by providing for agitation of the fluid. Figure 15·45 gives some idea of the decreased response time when a high rate of flow exists. It will be noted, however, that little is gained by increasing the speed over 60 fpm.

Because of the low thermal capacity of air, it is more difficult, in terms of time, to effect a given temperature change of a sensing element in air than would be the case if the element were in some liquid. Figure 15·46

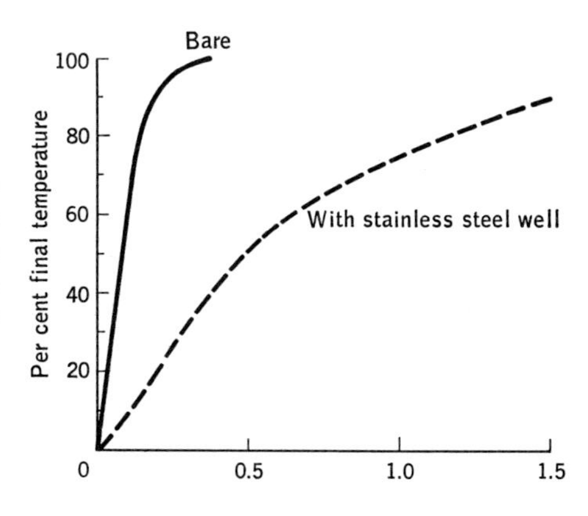

Fig. 15·44 Per cent temperature change versus time. It will be noted that the bare bulb has reached its final temperature when the device in the stainless-steel well has reached only 40 per cent of its total change.

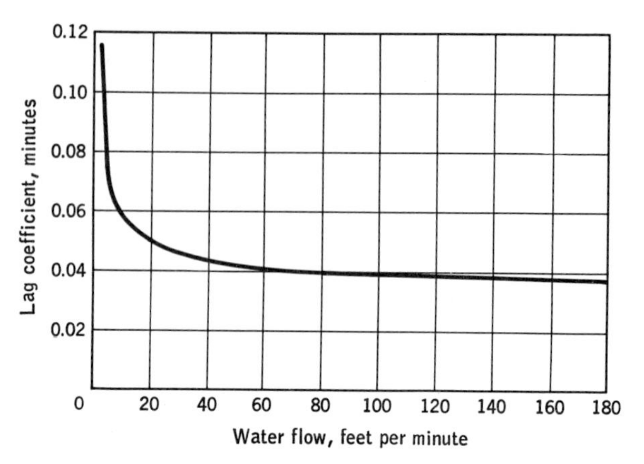

Fig. 15·45 Lag coefficient, minutes, versus water flow, in feet per minute. The lag coefficient of a thermometer bulb immersed in water decreases rapidly as the speed of the water flowing past the bulb increases. Note that little is gained by having speeds in excess of 60 fpm. (*Minneapolis-Honeywell Regulator* Co.)

indicates how significant is the time of response in moving air compared with the response in still air. The increased thermal capacity of the sensing element, caused by the use of the well, has in the example plotted, increased the response time by a factor of 3.

When the temperature is likely to change at a fairly rapid rate, the temperature-sensing device must be of such type and design that it can accurately reflect the temperature of the medium and that any time lags shall be as small as possible.

Even though the instrument be perfectly accurate, it can only measure the temperature of its bulb. Delays in getting the heat to the bulb, or conditions which cool the bulb, will be reflected in long time lags as well as erroneous readings. Unless pressure, corrosion, or continuous operation require the use of it, a well should never be used with an automatic controller.

Location of the bulb. The sensing bulb tends to become covered with a film of air, gas, or liquid, and that is equally true of a well. This film has an insulating effect which reduces the speed of response of the system. Therefore, the bulb must be located where there is a rapid flow in order to provide for a rapid transfer of heat. Figure 15·45 shows how the rapidity of liquid flow influences the lag. The performance of the same thermometer will be materially different in moving air from what it is in moving liquids because of the greater rate of heat transfer in a liquid. Therefore, one must expect inferior performance when the medium surrounding the bulb can transfer heat at a low rate. Liquids and solids

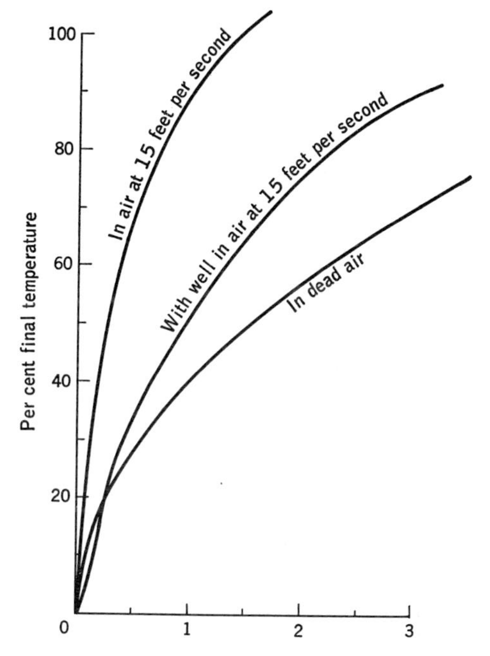

Fig. 15·46 Per cent final temperature versus time. The response of all thermometers depending primarily upon conduction or convection for the transfer of heat will be greatly speeded up by a rapid movement of the heated material.

generally will provide greater rates of transfer than will gases, and rapidly moving gases and liquids will provide better performance than will those whose movements are only a few feet per minute.

Errors. The static error of an instrument is the amount that the reading deviates from the true value. Large static errors in instruments are to be avoided. However, in automatic process control, a large static error may not be too important. Provided that the error remains constant, the settings can be adjusted to compensate for the error.

The accuracy of an instrument is generally expressed as the ratio of the static error to the span. The span of an instrument is the number of units it measures, whereas the range of the device states the maximum value of the variable which it will measure or for which it is calibrated. Thus, a thermometer which has a scale reading from 400 to 1400° has a span of $1400° - 400° = 1000°$, while its range is 1400°. If the maximum static error of this instrument is 10°, its accuracy, in per cent, is $10/(1400 - 400) \times 100 = 1$ per cent. This must be understood to mean that the readings at any point on the scale are correct to within 1 per cent of the span, or 10°. As indicated, a reading of 500°, for example, means $500° \pm 10°$.

The *reproducibility* of an instrument is a measure of the closeness with which the same values of the variable are indicated. One might say that it is a measure of the repeatability of the instrument.

An instrument which has no static error, or one which can be ignored,

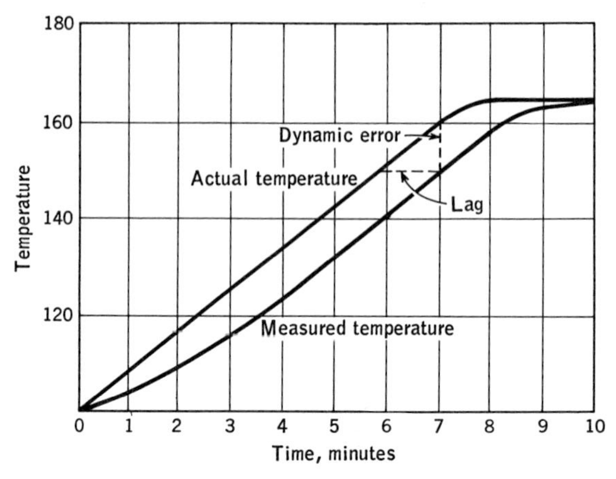

Fig. 15·47 Temperature response versus time. Temperature-measuring instruments with no static error develop dynamic errors when the measured temperature changes rapidly. (*Minneapolis-Honeywell Regulator Co.*)

will have a dynamic error which depends upon all the factors which cause an instrument to give a slow or delayed response. The "lag" of a thermometer is measured by the additional time required for the instrument to indicate the true value which previously existed. In Fig. 15·47 it will be noted that for the sample temperature selected, an extra 1¼ min was required for the instrument to catch up.

Whatever can be done to speed up the rate of heat transfer between the bulb and the surrounding medium, be it liquid, gas, or solid, will tend to reduce the lag and improve the performance of the instrument. Whether the device is a controller or an indicating device is not the important question.

Because the sensing bulb lags behind the surrounding variable, it cannot give a true indication of the temperature of the variable. The difference between the true value and the indicated value we call the *dynamic error*. If the true temperature is varying at a high rate, the dynamic error becomes greater. Figure 15·48 indicates the dynamic error when the temperature change shown in Fig. 15·44 takes place in one-half the time and in one-quarter the time.

Dynamic error and lag are present only when the temperature is changing. When there is no change in the temperature, only the static errors should be present. It must always be remembered that the dynamic error is independent of the static error, and both are present at the same time when the variable is changing. Even an instrument with no static error will have a dynamic error because energy must be transferred to or from the bulb, and such transfers always require some time. Pressure-measuring devices may be quite free of this error.

Almost all instruments also possess a characteristic which is called the dead zone. This is the largest amount by which the variable can change without creating a changed reading or position, or indication. The dead zone effectively delays the response of the device and is another factor in the lag of an instrument.

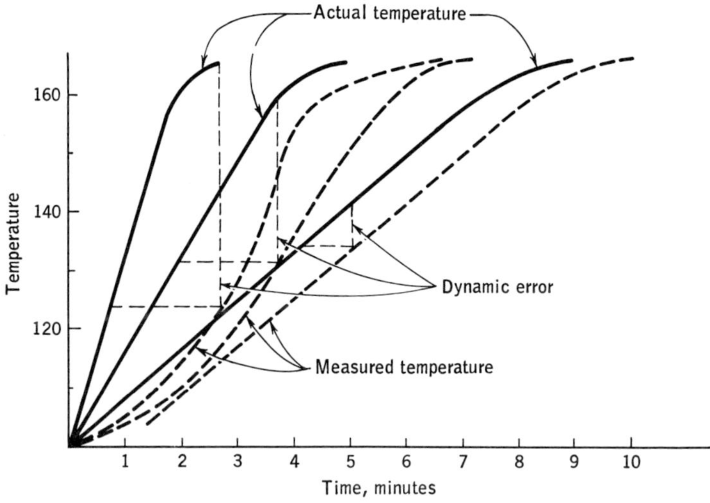

Fig. 15·48 Temperature versus time. For any given instrument, the greater the rate of change of temperature, the greater will be the dynamic error. The magnitude of the deviation from true values varies with the temperature rate of change.

SUGGESTIONS TO IMPROVE SPEED OF RESPONSE, ACCURACY, AND LIFE OF TEMPERATURE-MEASURING INSTALLATIONS

1. If there is any opportunity to make a choice, a small bulb will give a faster response than will a large bulb, irrespective of the type of bulb.

2. Always install the sensing bulb where there is a fast movement of gas or liquid. This will reduce the effect of the film which surrounds the bulb.

3. Too little thought is given to the type of response sought, just what temperatures are to be obtained, and for what purposes. The type of response desired will dictate where the sensing bulb or thermocouple should be installed.

4. In almost all instances, the sensing unit should be located as close as possible to the place where the temperature is desired. Locating it at a more remote point, even though there is no error in the temperature steady state, will make the time lag of the system greater by the time required for the liquid or gas to move from one point to another. Time lags are a major problem in automatic control.

5. When steam and water or other liquids are to be mixed directly, the sensing element must be far enough from the point of mixing to ensure that the temperature measured is based on complete mixing.

6. A well should not be used with automatic controllers if it is possible to get along without it.

7. If a well must be installed because of the pressure, type of liquid

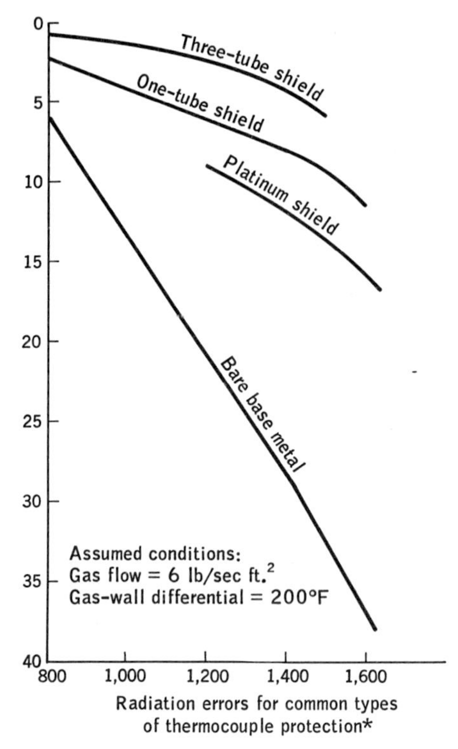

Fig. 15·49 Radiation errors for common types of thermocouple protection. As the temperature increases, the errors also increase. (*By permission of Reinhold Publishing Co., from Mullikin and Osborne, "Accuracy Tests for High Velocity Thermocouples,"* 1941.)

or gas, and so on, or because the bulb must be removed without shutting down the system, attention must be paid to cleaning the well—it can become covered with film, grease, and solids which have come out of solution. All of these greatly decrease its rate of heat transfer.

8. Where the temperature of a gas or air is quite likely to be different from that of the walls of the container, the sensing element will be influenced by the wall temperature. Therefore, the unit should be protected against false sensing by placing a shield between the temperature-sensing element and the wall (see Fig. 15·49).

9. If it is possible to arrange it, better results will be obtained if the heat can be transferred to or from the sensing element by a liquid rather than a gas.

10. Long, averaging capillary bulbs give the best results with air or gas.

11. To increase the response speed of sensing bulbs in a well, make sure that the bulb bottoms in the well. Special fins attached to the bulb may be used to speed the response. Aluminum powder, mercury, or oil may be used to speed the heat transfer.

12. If the temperature is measured at X_1, Fig. 15·50, to control steam flow, the response will be sluggish. The temperature of X_1 will have to be higher than X_3 and will change as the heat exchanger pipes become coated with sludge, etc. For a given value of temperature at X_1, the

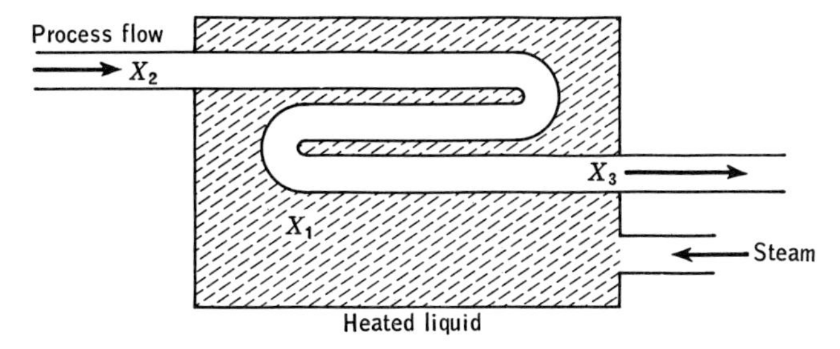

Fig. 15·50 The temperature-sensing unit should be placed where the measured temperature truly represents the temperature of the material being processed. It should also be located as close to the heat exchanger as possible in order to detect any temperature deviations as soon as possible.

product temperature will vary with the rate of flow of the liquid and the temperature of the incoming liquid. If the temperature is measured at X_2, the rate of flow will not affect the steam delivered to the exchanger. Even if the flow is zero, the heat called for will be based on the temperature of X_2. For the best control, the temperature should be measured at X_3, Fig. 15·50.

13. If high-pressure gas, such as hydrogen, is used in a process, it may penetrate metallic bulbs or wells. To avoid penetration of the bulb when a well is used, drill a small hole to let the hydrogen escape from the well. At low pressures, its infusion rate into the bulb will be negligible. Its effect is greatest with filled systems.

14. Mercury and other liquid-filled thermometers rated at 1200°F probably will employ materials different from those ordinarily used at 600°F. The nonlinear expansion rates, and other physical differences when the volume of the liquid is small, create calibration and scale problems.

15. When measuring the temperature of condensate, the use of a tray as shown in Fig. 15·51 will give much better results than will be obtained by letting the liquid dribble over the bulb.

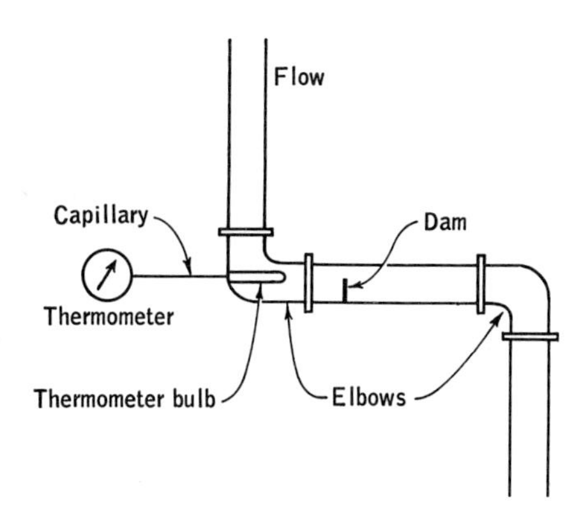

Fig. 15·51 When fluid is passing down through a pipe but does not fill it, a dam will provide for a liquid-to-thermometer heat transfer.

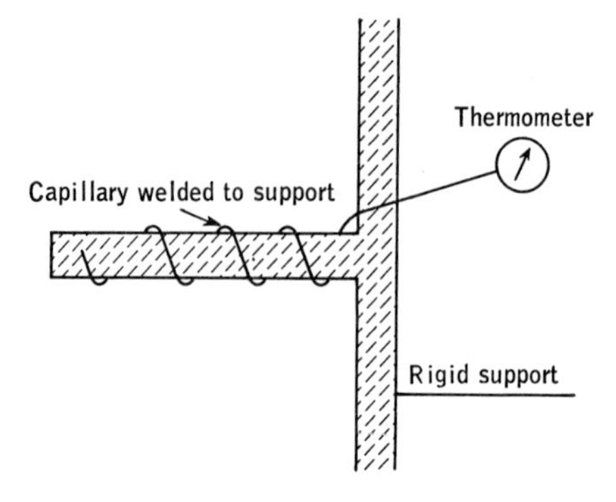

Fig. 15·52 A long capillary which is securely fastened to a strong support can furnish average temperatures which have significance. By having so much capillary in contact with the material, there cannot be appreciable temperature differences which are not measured.

16. To measure the average temperature of a batch being mixed, a capillary tube welded to a support may be satisfactory (see Fig. 15·52).

17. When it is desired to determine any hot-spot temperature in a mix or process, or along a grinder pulverizing wood, and so on, a vapor-pressure thermometer should be used; it will indicate the hottest, and not the average, temperature. Any point of localized heating caused by improper grinding conditions will be detected, Fig. 15·53.

18. Temperature readings in large pipes may be erroneous because the liquid does not at all times cover the bulb to a proper depth. When there is a variable level in a pipe, locate the bulb so that it will at all times have the necessary amount of surface in contact with moving liquid.

19. On stills or evaporators, make sure that the wells are long enough to penetrate sufficiently for the bulb and well to be immersed in the process fluid for the required distance. On a new installation, sluggish performance would suggest a check to see whether the wells are long enough; they may not have been installed according to the design.

20. When the boiling point is used as a measure of the concentration of a product, it is suggested that the installation be made according to the sketch shown in Fig. 15·54.

Fig. 15·53 A vapor-pressure thermometer can be used to provide hot-spot temperatures. In such processes as grinding, a hot-spot temperature may be far more significant than an average value, which could completely mask an improper operating condition.

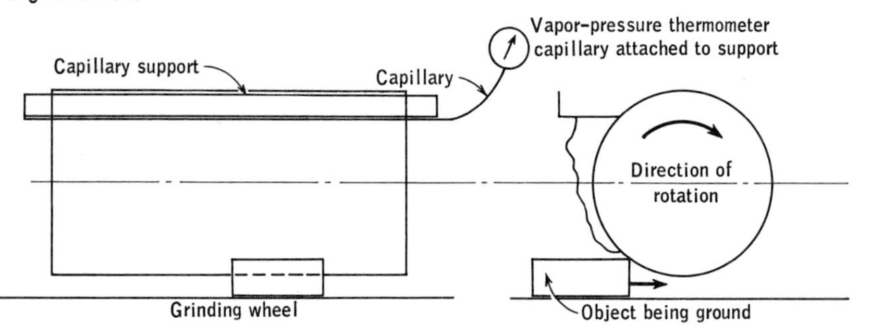

21. If an oil or mercury is used to increase the rate of heat transfer from the well to the bulb, make sure that the maximum temperature is below the boiling point of the liquid and that the minimum temperature is above the freezing point. The liquid should not react with either the bulb or the well. The well must tilt down to make it possible for it to hold the fluid prior to insertion of the bulk.

22. When a temperature bulb and fitting are screwed directly into a pipe, the following conditions should exist:

a. The pipe pressure is low.
b. The temperature is relatively low.
c. The measured liquids and gases are not dangerous.
d. The liquids and gases are not corrosive.
e. It will not be necessary to repair or replace the bulb while there is pressure on the system.

23. A fitting should be welded to the pipe to receive the bulb or well when:

a. The pressure is high.
b. The temperature is high.
c. The liquids are corrosive or dangerous.

24. For automatic operation, or when fast response time is very important, bulb sockets or wells should be avoided if at all possible. *The bulb must be located where change of process conditions produces significant bulb temperature changes.*

25. Use a 2-in. or larger pipe to help get a uniform radius when bend-

Fig. 15·54 By comparing the temperature of the processed fluid or material with that of the thermometer under the same pressure conditions, but heated only by steam, the concentration may be inferred.

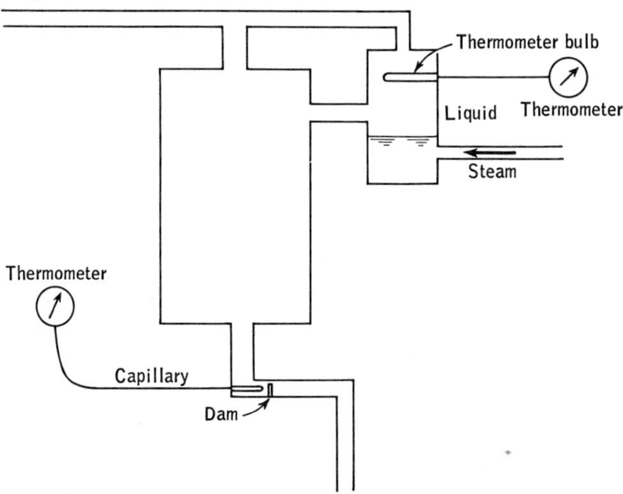

ing the neck tubing on bulb thermometers. In no case should the radius be less than 1 in.

26. When measuring steam or other high-temperature gases and fluids, the sensitive element of the gages should always be filled with condensate to protect the sensitive element from the high temperatures which may damage the instrument or change its calibration.

27. When installing capillary tubing, there are certain basic rules to observe:

a. *Handle thermometer tubing and bulb with care. Do not twist, kink, or stress tubing. Do not bend it unduly. Never cut tubing or disconnect it from the bulb or case. Do not use the tubing to support the instrument.*

b. Uncoil the tubing carefully to avoid kinking it.

c. Use at least a 3-in. radius for all bends.

d. Protect tubing from mechanical damage by crushing, abrasion, strain, or movement.

e. Support tubing every 2 ft if attached to wall or pipe.

f. Tubing should *never* be attached to heated pipe.

g. Class I and Class III systems must be installed to avoid extremes of temperature caused by hot or cold pipes, ovens, chimneys, air ducts, open windows, or direct sunlight.

h. Coil any excess tubing in an 8- to 12-in. coil near the instrument and tape in place. *Do not cut off the excess tubing.*

28. Bulb locations. Because tank, oven, and furnace temperature may vary widely from point to point, the following should be considered:

a. The temperature that is to be measured—maximum, minimum, input, output, and so on. A test may be necessary to determine the locations of these temperatures.

b. Bulbs should be kept away from heating and cooling coils, extreme top and bottom of ovens, rooms, etc., unless these are the desired points.

c. What will provide for the fastest heat transfer to the bulb?

d. Avoid dead spaces, corners, and locations where sludge or sediment can insulate the bulb.

e. How to locate the bulb near the paddle or blower.

f. How to locate the bulbs in pipes or ducts at right angles to the flow.

g. How to avoid radiation errors by (1) placing where unshielded high-temperature radiation cannot be sensed by the bulb, (2) insulating from "cold" walls.

29. Do not subject thermometers to temperatures over or under their design ranges when they are installed, in storage, or in shipment.

30. Make sure that Class II systems are employed with the head position used during calibration; otherwise, a correction may be necessary.

31. Special bulbs and conditions:

a. Averaging bulbs are made for maximum response and to measure average conditions rather than those at a point. However, if a Class II system is used, the temperature will be the maximum and not the average.

b. With the long bulbs which may be employed to average, do not bend sharply or kink. Follow the instructions for handling the capillary tubing.

c. When measuring the temperatures of plating tanks:

1) Keep the bulb up out of the sludge but not in a direct line between the anode and the cathode.

2) Insulate the bulb and tubing from tank or electrode.

3) Insulate instrument case and all metallic components so that there is no path to ground. Conduits and air piping must be insulated by plastic bushings.

32. Pyrometer extension wires must always be of the type specified. To use copper wire or extension wires for other types of couples will create unpredictable errors. Solder all connections carefully and tape.

33. Always use a separate conduit for thermocouple lead wires. Do not run parallel or close (within 12 in.) to a-c power lines. Ground all conduit.

34. When installing thermocouples, the general principles stated in item 8 should be observed. In addition:

a. Where the temperatures exceed 1500°F, the couple and the tube should be suspended vertically through the crown or roof of the oven or furnace. Otherwise, for temperatures in excess of 1500°, bending is likely.

b. Good operating experience indicates that the amount of tube inserted in the furnace should conform to the following: (1) 8 to 10 tube diameters insertion into the furnace or below surface if the thermocouple does not make physical contact with the protecting tube, Fig. 15·55*a*, (2) 6 tube diameters of tube insertion into furnace or below surface of liquid if the thermocouple touches the end of the tube, Fig. 15·55*b*, (3) 4 tube diameters tube insertion into furnace or below liquid surface if the thermocouple is welded to the tube, Fig. 15·55*c*.

c. Once a thermocouple has been installed, do not remove or disturb it. To check, use a second couple so that the couples touch. In general, it will be better to use a new couple than to try to reinstall an old one.

d. Couples used in liquids or molten metals should be so mounted that they may be removed for inspection. No metallic protecting tube will permanently protect against the action of molten aluminum and its alloys. Painting the tubes daily with a thick white-

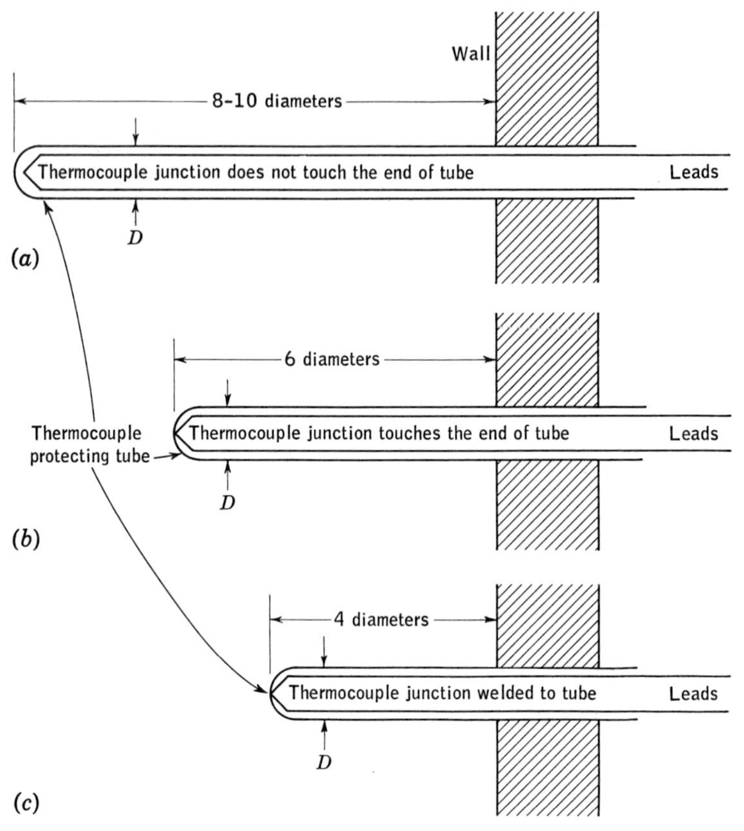

Fig. 15·55 The way in which the thermocouple is placed in its protecting tube will determine the length of tube required.

wash solution or a wash of sodium silicate or silicon carbide should extend the tube life.

e. Distortion of the couple junction by rolling, twisting, or hammering will alter the calibration. The output of the couple will also be affected by excessive vibration and the frequency of heating and cooling. *Always handle the junction carefully.*

f. To ensure the highest thermocouple accuracy:

1) Obtain the correct location.

2) Insert the protecting tube or couple the correct distance, depending upon the method of transferring heat from the tube to the couple.

3) Use the correct couple for the temperature range and atmospheric conditions.

4) Always protect platinum–rhodium-platinum thermocouples with a gastight ceramic tube to prevent contamination by hydrogen, carbon, and furnace atmospheres.

The accuracy of a thermocouple generally deteriorates with age; a thermocouple should be checked every week to every 3 months, depending upon the severity of the service. Iron-constantan couples change very little during their life. *Chromel-alumel thermocouples undergo a continued change.*

NOTES ON TEMPERATURE MEASUREMENTS

1. If the thermometer does not read the true temperature of the variable to be measured and if a calibration check reveals that the instrument itself has no serious error, then one common source of error may be the location of the sensing element. If the sensing bulb is not completely in the path of moving liquid or is located in a pocket, Fig. 15·56*a*, so that the adjacent liquid insulates the bulb from the main stream, the response will be poor. The bulb must be located where the flow of liquid or gas is greatest, Fig. 15·56*b*. Reduced rates of movement will result in a higher film coefficient which delays the response.

If a tee must be used to insert the bulb, the first method, Fig. 15·56*a*,

Fig. 15·56 (*a*) The sensing element should be so located as to promote the fastest movement of fluid over the greatest amount of bulb or well. (*b*) This arrangement is always to be avoided, if possible. The flowing fluid can come in close contact with but a small portion of the bulb. The portion of the bulb which is surrounded by stagnant fluid will not have a temperature which is that of the moving fluid. This arrangement causes large temperature errors and very sluggish response.

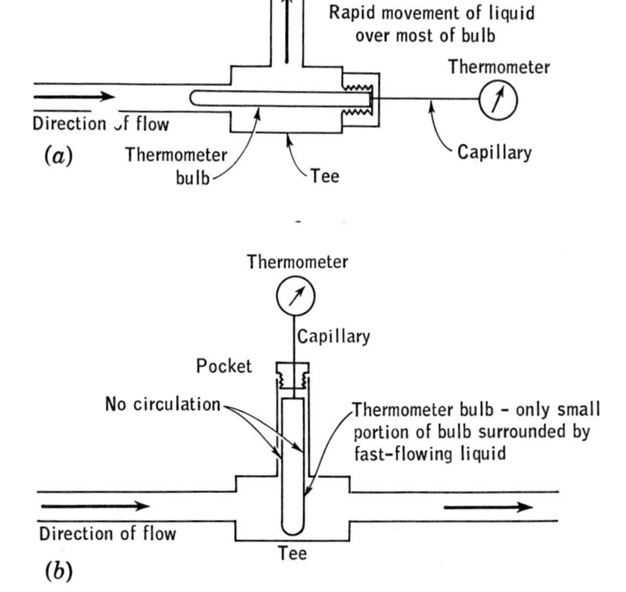

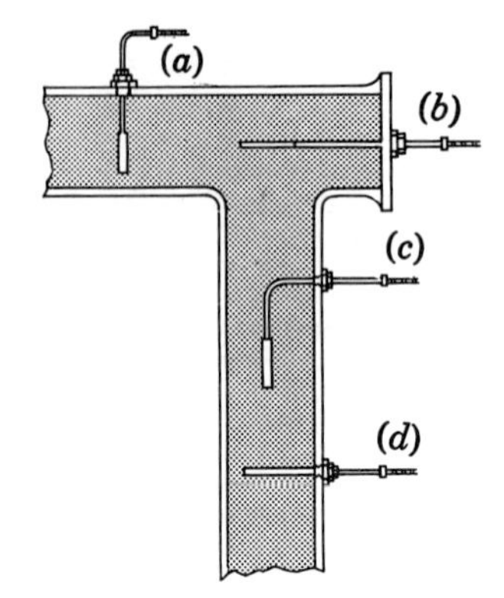

Fig. 15·57 Sensitive-bulb installation in pipes. (*a*) Plain bulb; union-connected, adjustable gland; bendable extension with plain bushing. (*b*) Plain bulb; union-connected, fixed gland; rigid extension with plain bushing. (*c*) Plain bulb; union-connected, fixed gland; bendable extension with plain bushing. (*d*) Plain bulb; union-connected, adjustable gland; bendable extension with plain well. (*The Fischer and Porter Company.*)

will give vastly better performance than will the second where the bulb is well removed from the main flow.

2. If the process temperature continues to rise slowly, but the instrument indicates that the correct value is being monitored, entrained air or gas may have formed around the thermometer bulb. This film of air or gas then acts as an insulator. In order for the bulb to reach the set temperature, the surrounding medium must differ from it by enough to transfer the energy. As the film becomes thicker, the temperature differential must also increase. The formation of any scale, film, or coating on the bulb will also require an increase in temperature differential.

3. When measuring the temperature of mixed liquids, the sensing bulb should be set as close as possible to the point of mixing in order to eliminate dead time. However, the mixing of steam and water requires that the sensing unit be about forty pipe diameters downstream; otherwise, mixing will not be complete. Location of the sensing bulb too far downstream will result in erratic controller behavior.

4. Provided that it does not conflict with other considerations, the temperature-sensing bulb should always be located at the point of highest velocity, be it in liquid or gas, Figs. 15·57 to 15·59.

5. Temperature-measuring problems in kettles are serious. Good agitation is required at all times. The lack of an adequate speed of agitation is one of the major difficulties, particularly as the viscosity increases. The bulb must be completely covered at all times, and it should be at least 6 in. from the source of heat so as not to be directly affected by the heat source. If it is directly affected, erratic behavior is likely to result.

6. Mercury thermometers are not used where photographic emulsion and similar products are made. Organic-liquid fills should be used instead.

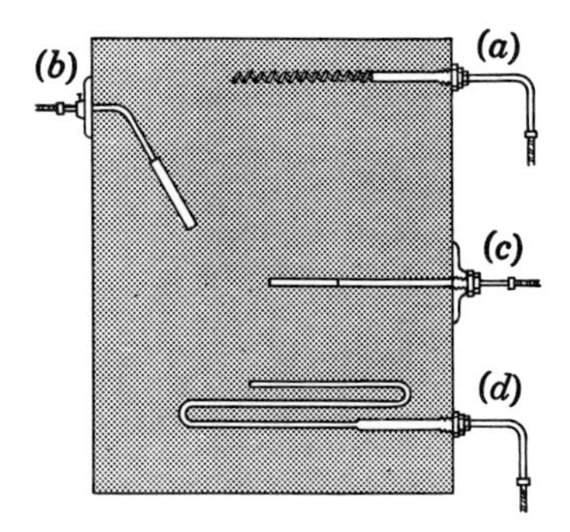

Fig. 15·58 Sensitive-bulb installation in ducts and flues. (*a*) Capillary bulb; union-connected, adjustable gland; bendable extension with plain bushing. (*b*) Plain bulb; bendable extension with adjustable split-ring flange. (*c*) Plain bulb; union-connected, fixed gland; rigid extension with plain bushing and threaded flange. (*d*) Averaging bulb; union-connected, adjustable gland; bendable extension with plain bushing. (*The Fischer and Porter Company.*)

7. The use of toxic thermometer fills should be avoided in food processing.

8. If the controller or thermometer response is sluggish or shows a slow deviation, the cause is probably sludge on the bulb. The remedy is to clean the bulb. If thermometer wells are welded into place, consideration of the manner by which they are to be cleaned must be given in the original construction.

9. Valves used for temperature control should be sized for the maximum flow. If only the normal load is known, choose a valve which will give 130 to 140 per cent of *normal* load (30 to 40 per cent more than normal) or 100 per cent of the maximum known flow.

10. Many thermometer wells are used where they are not needed. A well always slows down the rate of response. Unless the situation absolutely requires a well, don't install it.

11. Crystallizers are likely to be sources of trouble because of the poor heat conductivity of the material. An averaging bulb should be installed on some firm support and carried well into the tank.

12. Where the temperatures of viscous materials are to be determined, use a long capillary sensing element, adequately supported, wound around the inside of the container.

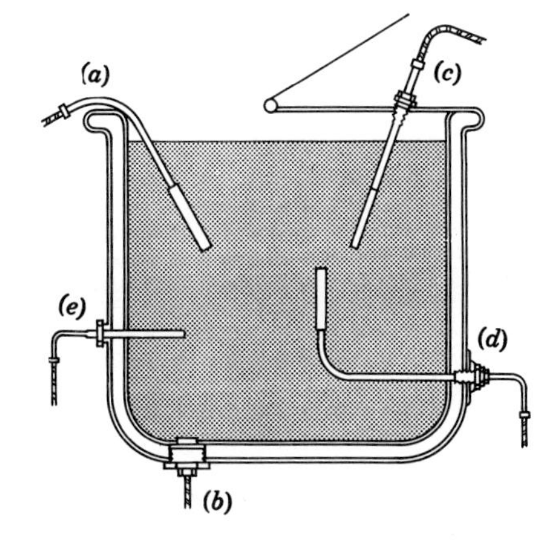

Fig. 15·59 Sensitive-bulb installation in kettles, tanks, and vats. (*a*) Plain bulb with bendable extension and flush bulb with bushing. (*b*) Plain bulb; union-connected, fixed gland; rigid extension with plain bushing. (*c*) Plain bulb; union-connected, adjustable gland; bendable extension with plain bushing and threaded flange. (*d*) Sanitary bulb, dairy-type fitting. (*The Fischer and Porter Company.*)

13. Selection of the right bulb size will improve the response.

14. Shielding the bulb to prevent breakage and so on may give the bulb such a long time constant that it no longer responds to cyclical changes. Placing the sensing element where the air, gas, or liquid has impeded flow will result in incorrect readings.

SUMMARY

1. High degrees of accuracy are possible with all common types of commercial temperature-measuring devices. The choice of any one depends upon the range, the span, the speed of response, the distance the signal has to be transmitted, compatibility with other system components, servicing, and costs. The results are summarized in Table 15·9.

2. Except for certain pyrometers which match color, all systems require the transfer of heat to and from the sensitive element. The transfer of heat always requires time. This prevents the instantaneous response of the sensing element to the changed condition. The greater the amount of heat required, or the greater the thermal resistance, the greater the lag of the device.

3. Because of the time required for the indicated value to get to within ½ or 1 per cent of true value, it is customary to use special names to indicate unique values: the time required for an instrument to register 63.2 per cent of a step change of temperature is called the time constant of the unit. It will change with different conditions. Increasing the rate of flow will decrease it; trying to measure the same temperature change in a gas, as contrasted to a liquid, will give a much greater time constant.

4. Liquid-filled and gas-filled thermometers have linear scales. Systems employing vapor pressure have nonlinear scales unless special cams or mechanical elements are used to correct the nonlinear pressure changes to linear pointer movements. If the nonlinear response provides a large pointer travel for the temperatures generally measured, this may facilitate reading the instrument.

5. All filled temperature systems are primarily pressure-sensing systems.

6. To get high accuracy, every one of the components must be as linear as possible or have its nonlinearity compensated for in the design.

7. For the pure metals, the change of resistance with temperature tends to be linear. The changes generally are about 0.4 per cent per degree centigrade of the resistance at 20°C. The resistance type of pyrometer, when properly calibrated and linearized, is such that it is one of our standard temperature-measuring devices. Resistance increases with temperature for the metals.

8. The extremely large nonlinear temperature-resistance coefficients of certain oxides provide one of the best means for measuring small temperature changes within the operating range of the oxides.

9. The temperature-resistance characteristics of thermistors causes them to be employed in still other fields: gas flow, gas analysis, vacuum pressure measurements, and certain types of liquid-level gages. Thermistors can also be used to tell if liquid particles are being carried along by gases and vapors.

10. Platinum is the best of the metals for accurate resistance thermometry. Its useful temperature range is −258 to 900°C. Platinum does not oxidize, but it is contaminated by carbon monoxide and other reducing atmospheres. In general, it should always be used in a sealed housing.

11. Nickel has a nonlinear temperature-resistance coefficient, but it is the metal most widely used for resistance bulbs. The nonlinearity is corrected at the factory

Table 15·9 Industrial Temperature Measurements

Type	Basic element	Commercial accuracy	Max span, °F	Min span, °F	Temp range, °F	Response time	Distance of transmission, ft	Basic math. expression
Liquid filled:								
Class IA	Elastic member deformed by expanding fluid	±½ to 1%	300	25	−125 to +600	Long (slowest of the filled systems)	·200	$T = kP$ $= k'\Delta$ volume
Class IB			300 to 500					
Class VA			1000 (1200)	40	−38 to 1000 (1200)	Intermediate	200	
Class VB			600		−65 (Microalloy®)		25	
Vapor pressure	Elastic member deformed by pressure	±½ to 1%	40	300	−40 to 600	Next to gas thermometer in speed	200	$P = kT_{ab}$
Gas filled	Elastic member deformed by expanding fluid	±½ to 1%	1000	100	−400 to 1000 (1500)	Fastest of the filled systems	200	$P = kT_{ab}$
Bimetallic	Differential expansion of two different metals	±1%						
Thermocouples	Chromel-Alumel (K)				−300 to 2000	Depends upon construction and mass; small bare welded junction fastest—slowest in tubes and wells	Almost any distance	$e = K - (T_{hot} - T_{cold})$ not linear except for short spans
	Copper-constantan (T)				−40 to 2200			
	Iron-constantan (J,Y)				−200 to 650			
	Platinum–10% rhodium-platinum (S)				32 to 2650			
	Platinum–13% rhodium-platinum (R)				32 to 2650			
Radiation	Thermocouple				100 to 700	About 2 sec	Limited to direct field of view which will fill viewing area	$T = kT_{ab}^4$
					200 to 3200			
					1800 to 3200			
					1800 to 7000			
Optical	Brightness of filament made to match that of hot object				1000 to 10,000	Essentially zero		$T = k$ (filament current) nonlinear

by connecting a different material in series (or parallel) with it. Its useful range is about −150 to +300°C.

12. Copper of high purity is quite consistent and predictable; also, it is simple to calibrate. Its useful range is −200 to +120°C.

13. For high accuracy, a bridge circuit should be used for resistance determination. It may be of either the resistance or the capacitance type. Constancy of bridge voltage is required only to maintain sensitivity.

14. If a deflection type of indication is employed in a bridge circuit, calibration is not independent of the bridge voltage.

15. Thermocouples do not respond to temperature, but they do respond to *temperature difference*. The active and the reference junctions cannot have the same temperature if an emf is to be generated.

16. Thermocouples must be selected on the basis of the temperatures to be measured and the atmosphere surrounding the couple.

17. The amount of material included in the active junction will affect the response rate. For a fast response, the material should be minimal in amount and so positioned that it picks up heat quickly. It should not be located in a flame unless the flame temperature is desired.

18. The use of protecting tubes may be absolutely necessary, as with platinum couples, but the protection will seriously retard thermocouple response. Any indication of temperature will be that of the couple temperature, which may differ appreciably from the ambient, particularly where the temperature is changing or where the thermocouple or thermocouple housing may be radiating to some cold portion of its surroundings.

19. When a potentiometer is balanced, it takes no current from the circuit being measured; therefore, the resistance of this circuit approaches infinity. The use of instruments which take current to actuate them may result in serious errors either in analysis of a condition or in making a calibration.

20. The use of electronic potentiometer circuits makes possible very close balancing. There will almost always be some very minute current when the potentiometer is balanced, because the amplifiers do not have an infinite gain or amplification factor. Otherwise, certain types of transducers could not tell whether the lack of current is caused by a broken connection, zero emf, or a perfect balance.

21. To obtain accurate temperature readings with the minimum dynamic error, the following must be taken into consideration. (*a*) All temperature-sensing devices should be so located that there is a very rapid flow of gas or liquid over the bulb, or the bulb is exposed to radiant energy. (*b*) The sensing element must not gain or lose heat to the adjacent material. (*c*) The thermal resistance of the bulb or well must be kept to a minimum if good response is desired. If the bulb accumulates an insulating film, its thermal resistance will increase. Thought must be given to ways to keep clean both bulbs and wells. (*d*) Liquids and solids can transfer heat much more rapidly than gases can.

QUESTIONS

15·1 What is the difference between the meanings of *range* and *span?*

15·2 What does it actually mean to say that an instrument has a stated accuracy of 1 per cent?

15·3 Why is it undesirable to use readings low on the scale if readings of high accuracy are desired?

15·4 To get readings of high accuracy, what part of the scale should be used?

15·5 Can you suggest a reason why gas thermometers often employ wider spans than do liquid and vapor-pressure systems?

15·6 What is the meaning of the term "time constant"? What is its significance in industrial measurements?

15·7 Why is the time constant used instead of the time required to reach the final temperature?

15·8 Does the time constant of an instrument change from liquid to liquid, or does it depend only upon the instrument?

15·9 What would a frothy, bubble-filled liquid be likely to do to the time constant of a thermometer?

15·10 If you were designing a thermometer, what could you do to increase the speed of response? Do not limit your answers to practical suggestions.

15·11 What practical considerations prevent your taking advantage of the suggestions made in Question 15·10?

15·12 If costs were the same and chemical reactions were of no importance, what common material would make the best bulb? Which one the worst?

15·13 Would you use a gas thermometer for extended periods of time when the pressure of hydrogen in the reactor exceeds that of the gas in the bulb? The temperature is 900°F. Justify your answer.

15·14 If in Question 15·13 helium is used for extended periods at 1000°F, can you expect the calibration to stay reasonably constant?

15·15 What are some of the factors that determine instrument choices in temperature measurements? Which do you think are the most important ones?

15·16 In what temperature measurements and under what conditions is ambient temperature compensation required?

15·17 What types of instruments require consideration of the difference in "head" between the instrument and the bulb, either for calibration or accurate use?

15·18 Are there times when nonlinear scales are desirable? What temperature-measuring system gives this type of indication?

15·19 What are the general classes of filled thermal systems?

15·20 What is an averaging bulb? Where is it used?

15·21 To get the fastest response, where should the sensing bulb be located when measuring the temperature of a liquid leaving a heat exchanger?

15·22 When finding the temperature of water which is being heated by live steam, do you use the answer to Question 15·21? Should these two questions require different answers?

15·23 What is meant by the dynamic error of an instrument? Is it present in an accurately calibrated system? What is the effect of one error upon the other?

15·24 Is the dynamic error always the same in value? What factors determine its size?

15·25 What are some of the factors that cause electrical systems to be preferred over mechanical systems for certain applications?

15·26 What are the temperature ranges ordinarily employed with electrical resistance gages, thermocouples, and electro-optical systems?

15·27 Are the factors that govern the time constants of filled systems of the same general type as those found in electrical systems?

15·28 Why does one need to limit the electrical power input to a resistance bulb? What can be done to increase sensitivity with the same heat input?

15·29 With the resistance method of measuring temperature, what is the effect of a poor joint or connection (high resistance) upon the apparent temperature? If the indicated temperature were to go in the opposite direction from the true value, what could have happened?

15·30 What might happen when thermocouple and resistance bulb lead wires are placed near power wires or run in the same conduit with annunciators, telephone, or control wires?

15·31 What are the cardinal rules for installing leads to electric pyrometers?

15·32 Does a variation in the voltage of a bridge system employing a resistance bulb have the same effect as a variation in the emf applied to a self-balancing potentiometer pyrometer? Why?

15·33 What is affected by the voltage applied to a resistance bridge?

15·34 What is the purpose of the standard cell in thermocouple measurements?

15·35 Is a standard cell required for all electrical pyrometers? What are the exceptions?

15·36 What two instruments described dispense with the usual type of standardizing?

15·37 Why do so many thermocouple balancing systems use a chopper? Compared with other parts of the system, would you be justified in checking its performance reasonably often? In cases of trouble, would you consider this a likely source of trouble?

15·38 Why are platinum resistance bulbs and platinum thermocouples protected against contamination? What sort of protection must be furnished at the high temperatures?

15·39 What factors govern the sizes of wires used for thermocouples? Are your answers true for all couples?

15·40 Why are compensating lead lines used with thermocouples?

15·41 Does it make any difference how the lead wire extensions are connected to a thermocouple?

15·42 Why should electrical connections in instruments be soldered?

15·43 Why should the resistance of a millivoltmeter circuit be kept to a minimum?

15·44 What is meant by thermocouple burnout protection? How is it often provided?

15·45 If several thermocouples were installed in a furnace to provide average temperature, how would you recommend that they be connected for best results?

15·46 Can anything be done to speed up the response of a thermocouple in a well?

15·47 How much current can safely be taken from a standard cell without appreciably changing the terminal voltage of the cell?

15·48 How does the two-phase motor used in electronic balancing systems differ from an ordinary reversible a-c motor? In case of trouble, could you replace one by the other, assuming the same line voltage and mechanical dimensions?

15·49 Some potentiometer systems use a potentiometer and movable contact; others use a fixed resistance and a variable current to effect a similar end result. What advantages can you think of for each system? Are these factors real or theoretical?

15·50 What is meant by cold-junction compensation? Of what practical importance is it? How may it be obtained?

15·51 What is the result of improper immersion of the sensing bulb?

15·52 What happens if the sensing bulb is surrounded by trapped liquid or air?

15·53 Would a given thermometer have the same time constant in a liquid and in a gas?

15·54 As a liquid becomes more viscous in processing, what should be done to the rate of agitation to minimize temperature errors?

15·55 If a sensing bulb is located within about six inches of the heated walls of a kettle, how accurate may the indicated readings be? What may they indicate?

15·56 What different materials are used to fill thermometers?

15·57 What factors are important in selecting a particular type of thermometer?

15·58 What determines where the thermometer bulb should be placed?

15·59 If stainless steel were substituted for copper of equal bulb thickness, what would happen to the response rate of a thermometer?

15·60 What are the primary standards for temperature measurement?

15·61 What are the standards that you might use to calibrate an instrument?

15·62 What are the steps involved in calibrating a thermometer? Explain each one.

PROBLEMS

15·1 A gas thermometer is calibrated for the interval of 300 to 1000°F. What is the span? What is the range?

15·2 For a 1 per cent accuracy meter used with an instrument whose span is 1000°F, what is the probable error at any point on the scale? If the pointer indicates a reading of 100 units, what could be the true values in degrees? What is the percentage error at these readings?

15·3 For the same instrument as in Prob. 15·2, what are the possible values, error, and percentage error when the pointer indicates 900?

15·4 Plot the following data and determine the time constant for the thermometer when plunged into a liquid at 400°F. The temperature of the thermometer originally was 70°F.

Indicated temp, °F	Time, sec	Indicated temp, °F	Time, sec
129.5	2	336	15
179	4	359	20
218	6	384	30
254	8	394	40
280	10	398	50

15·5 From your plot of Prob. 15·4, how long does it take to reach 90 per cent of the final value? 95, 98, and 99 per cent?

15·6 Draw a very simple schematic diagram, using the correct symbols, to measure the input and output temperature to a heat exchanger. Indicate the important distances.

15·7 An electrical resistance bulb is made of copper wire. Its resistance at 20°C (68°F) is 100 ohms. Determine its value at −100°C and at +150°C.

15·8 Work Prob. 15·7, but use the average coefficient for nickel from 0 to 400°C, Fig. 15·2.

15·9 At the time of balancing a Wheatstone bridge, $R_1 = 100$ ohms, $R_2 = 200$ ohms, and $R_3 = 300$ ohms. What is the value of R_X? Assume that at 20°C the resistance of the copper coil was half of its resistance when the bridge was balanced. What was its hot temperature? Use Fig. 15·7.

15·10 A 10.00-ohm copper resistor at 68°F is to be used to indicate bearing temperatures. What resistance should not be exceeded if the maximum bearing temperature should not exceed 300°F?

15·11 A temperature alarm unit with a time constant of 2 min is subjected to a sudden 50° rise in temperature because of a fire. If an increase of 30° is required to actuate the alarm, what will be the delay in signaling the sudden temperature increase?

15·12 A resistance bulb is at −40°F. If the difference in lead length is 100 ft, what error is introduced by having the ambient temperature of the leads 68°? 32°?

15·13 A remote power station needed to measure the temperature of the steam delivered to a turbine, but no thermometer was available. Because it was an emergency, it was decided to make a thermometer by screwing a short piece of pipe, closed at one end and connected at the other end to a pressure gage with a 0–500 scale, into the steam line. The short pipe had previously been partially filled with water. Using conventional steam tables, make a calibration chart for this improvised thermometer; that is, show the relationship between pressure and temperature.

15·14 Compare the standard table values with the computed ones if a copper-constantan couple is used at −200, 0, 200, and 400°C. Assume the reference cold junction is a 0°C. Explain any difference.

15·15 If the time constant of a thermometer for a given set of conditions is 20 sec, what is the maximum cyclic deviation from the set point which will be indicated if the cyclic disturbance takes place every minute and the cyclic deviation is 10°? Review Chap. 6.

15·16 How often would the cyclic disturbance of Prob. 15·15 have to occur if the indicated maximum deviation were half of the true value?

15·17 How much more effective is water at 400 fps than air at 400 fps in effecting the response of a ½-in.-diameter bar bulb industrial thermometer?

15·18 Rework Prob. 15·17 but with the air at 4,000 fpm. Water speed is still 400 fpm.

15·19 Approximately what is the time constant for the same ½-in. bulb as in Prob. 15·17 when the air velocity is only 40 fpm?

15·20 If the air in a duct were to undergo a cyclic variation of 20° every 3 min, what variation would be indicated by the ½-in. bulb of Prob. 15·17 for an air speed of 100 fpm? For 1,000 fpm?

15·21 What conclusions might you draw from the preceding six problems as to the reliability of certain measurements?

15·22 To increase the accuracy of the indications, what should you try to do with every installation in which temperature measurements are required?

16 Potentiometric Devices

It is quite probable that a majority of the people in industry who are concerned with the operation of the vast systems of instrumentation would say that their chief interest in potentiometers is created by a requirement for measuring the output of a thermocouple in order to check its calibration. The chances are that there are many occasions when resistance measurements are made, and a Wheatstone bridge is employed for the resistance determination. That too involves the use of a sort of potentiometer in that one circuit is adjusted until the potentials of two points are made equal with respect to some third point.

It seems appropriate, therefore, that we widen our concepts of what potentiometers are and what they do. Perhaps we should say that any measurement in which we seek to obtain an equality of potentials is a potentiometric measurement. What creates the potentials, what the measurement is for, whether it is made automatically or manually should not enter into our notion of what a potentiometric instrument is.

Historically, potentiometric measurements have been reserved for direct-current potentials that had to be measured very accurately when the current was essentially zero. With little or no current drain in the circuit, the electromotive force and the terminal potential are essentially one and the same thing. It might well be repeated that the outstanding characteristics of potentiometer measurements are two:

1. The unknown emf is determined by finding the known potential which is just equal to it.

2. The current taken by the measuring circuit at the time of balance is zero, or essentially so.

There has been no other common way for determining potentials with great accuracy, to better than 0.1 per cent or so. Such high degrees of precision require a standard of reference, that is, the standard cell whose emf is extremely stable over long periods of time.

It is only to be expected that these characteristics, modified in form, will be present in newer developments.

COMMON TYPES OF POTENTIOMETERS

In reviewing the distinguishing characteristics of the well-recognized standard potentiometer, we find the main elements of Fig. 16·1.

1. Some source of potential to be measured. A thermocouple is shown because it is so often used, but the source can be any direct-current potential within the range of the instrument.

2. A very linear potentiometer rheostat so that the position of the contact is an accurate representation of the potential at that point P with respect to point b, for example.

3. A working battery in series with a variable resistor so that the potential from a to c can be adjusted to a very definite value, namely, that of the standard cell.

4. A sensitive galvanometer to indicate when the emf of the source and that provided by the potentiometer rheostat are equal.

5. A *standard cell* whose emf shall be practically invariable and with which the potentials of the operating circuit can be compared and corrected to its value in the case of differences, or a Zener reference diode.

6. Figure 16·1 indicates that a second galvanometer is used for adjusting the value of potential ac to that of the standard cell. This has been done for the sake of circuit clarity. In practice, only one galvanometer is used, and it is switched from one circuit to the other.

The accuracy of such a potentiometer depends upon the accuracy of the original calibration and the degree to which the present device

Fig. 16·1 The conventional type of potentiometer uses these components to measure an unknown potential by comparing it with some known value which has been obtained by standardization.

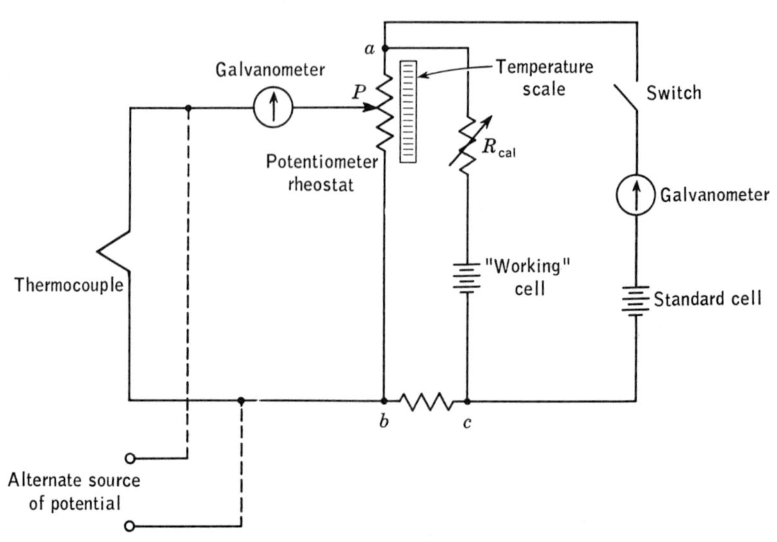

resembled the original when it was calibrated. We have become accustomed to expressing position not as position, but as a potential which would exist if all of the original conditions still prevailed. A change in resistance between points *b* and *c* would alter the calibration even though the potentiometer rheostat were in no way altered. (It must always be recalled that no potentiometer rheostat will provide the calibrated value of emf if it is delivering current—unless it was calibrated for some known value of load current. But this would be most unusual.) The extra current being delivered to the load circuit would create an additional potential drop through that portion of the potentiometer rheostat through which it flowed, thus destroying the calibration.

Automatic potentiometers. Assuming that the potentiometer has been accurately calibrated, the measurement of the input potential is dependent primarily upon positioning the movable contact until the galvanometer indicates zero current. At this time, the position of the contact with respect to the scale is noted. However, the desire for an automatic determination and recording of the potential (often in terms of temperature or some other variable) has prompted modifications of the basic elements.

Perhaps the first thing which comes to mind when automatic operation is desired is to use the galvanometer position to initiate and control the positioning of the potentiometer contact. The physical movement of the galvanometer pointer and the creation of "mechanical interference" might suggest itself first. It is perfectly possible to create an extremely satisfactory instrument this way; an excellent one based on such a design principle is discussed in detail in Chap. 15. Despite the mechanical and electrical excellence of such a design, Fig. 15·25, other designs have completely superseded it. The first of these is represented in Fig. 16·2. Many of today's instruments of the potentiometric type

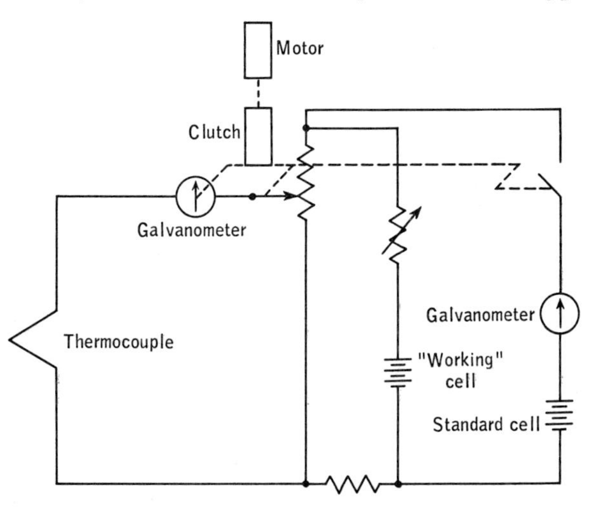

Fig. 16·2 An automatic potentiometer can be made if any inequality of measured and comparison voltages can be made to initiate action to restore the equality. The well-known Micromax potentiometer employs the principles indicated here.

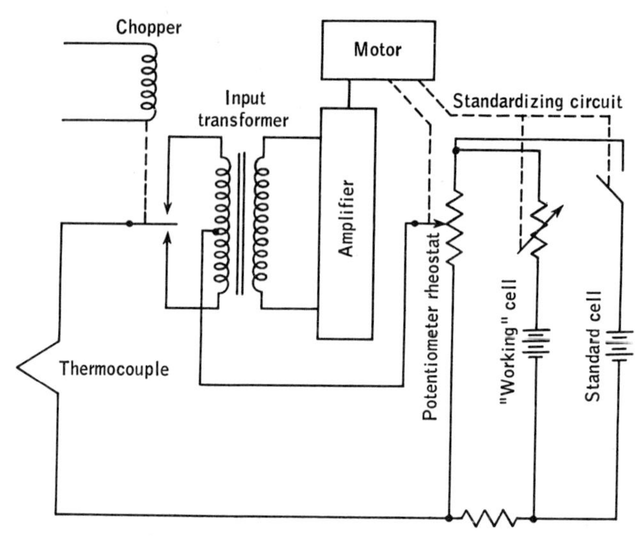

Fig. 16·3 This type of potentiometer performs essentially the same functions as does its mechanically operated counterpart. The potentiometer rheostat and the standardizing circuits are similar in both designs.

employ its basic principles: The difference in potential between the thermocouple (or other potential) and the calibrated balancing potentiometer will be converted by a chopper into an alternating current which can be amplified by conventional means. The output current is adequate to operate a two-phase alternating-current motor. The amplifier is also phase sensitive, so that it supplies current of such a magnitude and polarity that the motor drives the potentiometer in the direction to reduce the error signal to zero, or very close to it. The motor then ceases to run until some other difference potential develops between the input and comparison potential circuits. Figure 16·3 shows one commercial design, while Fig. 16·4 shows another design in which calibration has been eliminated because the standard cell is employed for the working battery. It might be added, somewhat parenthetically, that the standard cell may be used for other than occasional calibration purposes only if the current is limited to a few microamperes; taking more current than this from the cell will invalidate the constant-potential rating of the cell. For example, the cell may be used to charge a small capacitor, because the charging current is but a few microamperes. By using a source of this type for the constant potential, any change in charging current is caused by changes in capacitance of the variable capacitor.

In Fig. 16·5 capacitor C_v is adjusted until its charging current creates the same potential from B to O as the current charging the fixed capacitor creates from a to O. This operation is that of a potentiometer. Any inequality of the potentials is amplified and used to adjust the variable

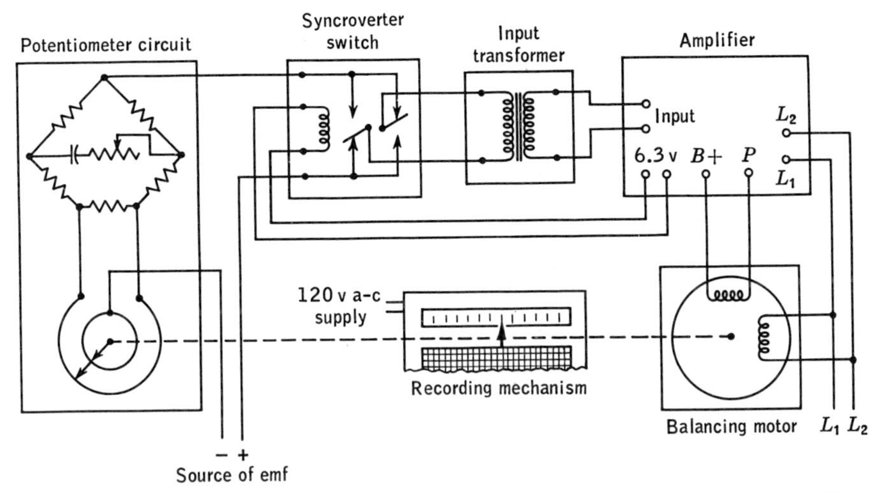

Fig. 16·4 Simplified schematic wiring diagram of an automatic potentiometer which can be made to provide continuous standardization. (*The Bristol Company.*)

capacitor. The position of the variable capacitor is then used as a measure of the input emf, which is also representative of the temperature.

Electronic potentiometers. Although the potentiometers previously discussed have required the flow of electrons for their functioning, they

Fig. 16·5 The charging current required by a small capacitor is so small that it may be supplied by the standard cell without significantly impairing the performance of the cell. (*From a Foxboro Co. design*)

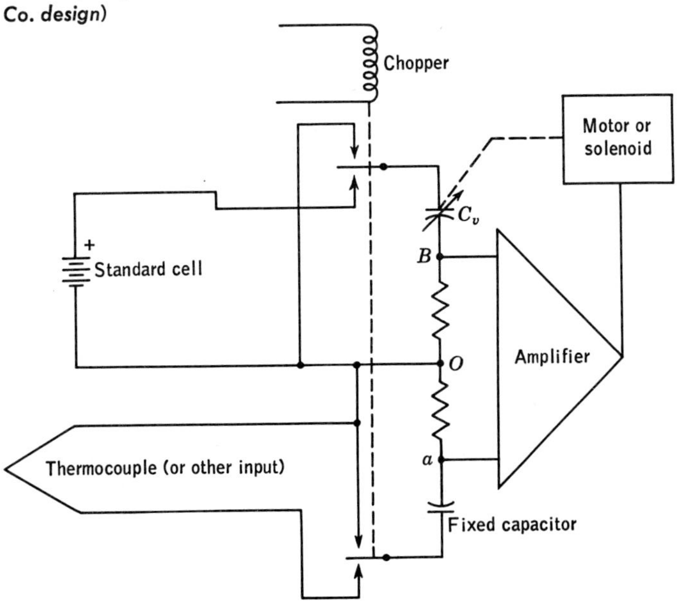

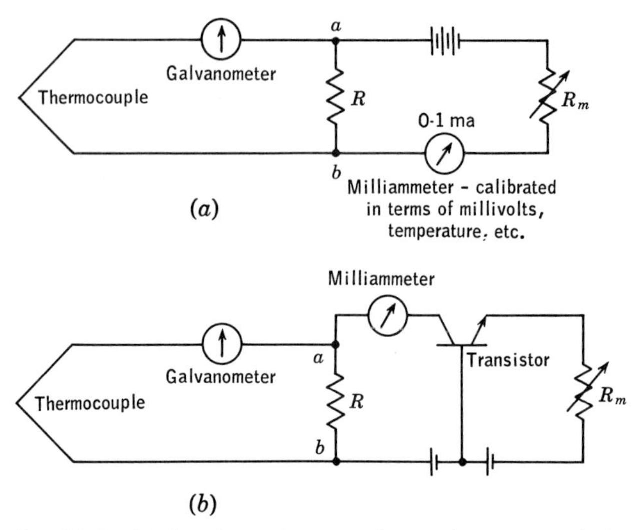

Fig. 16·6 (*a*) An electronic type of potentiometer, employing no moving parts in the balancing circuit, can be made from the simple elements shown. The value of resistor R_m is adjusted until the galvanometer deflection is zero. The milliammeter then furnishes either the millivolts or temperature, according to the calibration. (*b*) The sensitivity of electronic potentiometers of this design can be increased by using a transistor to control the current through resistor R.

might be termed mechanical potentiometers because the act of balancing requires the physical movement or rotation of some part or parts. Some of the designs which we will consider under electronic potentiometers may require some movement, but many of them will not. So we will distinguish between the two general classes in the main way in which potential equality is achieved.

We all know that the potential developed across the terminals of a resistor is directly related both to the value of the current and to the value of the resistance. Since this potential, for a fixed value of resistance, varies directly with the current, the current may be used as a measure of the potential.

Inasmuch as

$$V = \text{current} \times \text{resistance} = IR$$

it might occur to us to use the circuit shown in Fig. 16·6*a*. By manually controlling resistor R_m until the galvanometer has no deflection, we can balance the potentiometer. If we know the value of resistor R, and we will, and the value of the current which is indicated by the milliammeter, then the product of these two quantities will give us the potential difference of terminals a and b, and that is what we were seeking. By employing a transistor, a smaller change in current in the controlling circuit can bring about a large change in the current in resistor R. With small currents to control, the resistor might be light in weight and have little frictional drag, Fig. 16·6b.

It should be recognized at this point that commercial ammeters or milliammeters are not the precisely calibrated devices that potentiometers are. And any determination of the potentials will be no better than the accuracy with which the value of R and the calibration of the meter have been determined.

Some simple automatic potentiometers. To make a simple automatic potentiometer might cause us to follow the previous development: use the position of the galvanometer to initiate the corrective action. This is indicated in Figs. 16·7 and 16·8.

One characteristic of many automatic potentiometers which employ electrical balancing means, in contrast to mechanical or position balancing, is that during normal operation there must always be a difference

Fig. 16·7 A self-balancing type of potentiometric device can be made by causing the galvanometer to control the amount of light falling on a photocell. If a 0–1 milliammeter is employed for the galvanometer, approximately 10 μa will give enough pointer movement to control certain types of cells which have narrow sensitive areas.

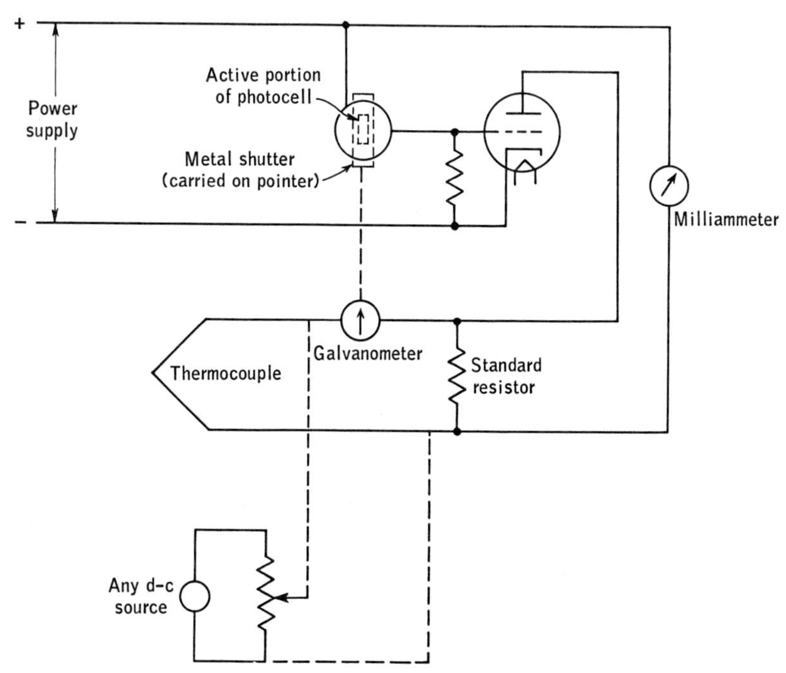

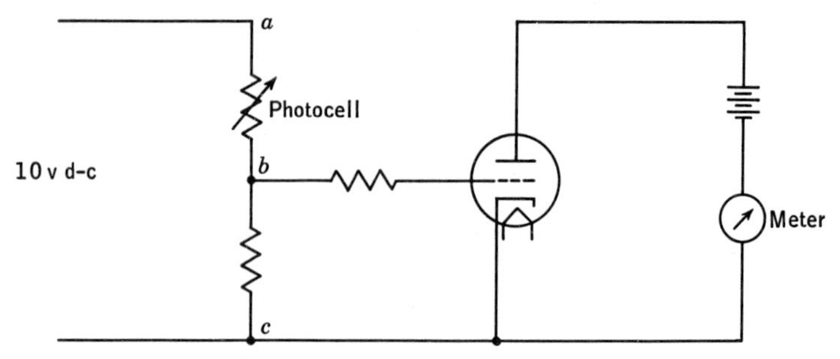

Fig. 16·8 When the photocell is in a voltage-divider network, Fig. 16·7, it can greatly change the grid potential of the tube. When the photocell is dark, its resistance is high, which causes the potential of the grid to approach zero. But when the cell is illuminated, the grid potential approaches that of the source.

between the input signal and the balancing potential. This difference may be of the order of microvolts and may be insignificant so far as the accuracy of the device is concerned, but the amplifier output current which is used to obtain the potential balance will drop toward zero if the input signal is zero. These designs, then, must always be characterized by *high-gain amplifiers,* so that any errors or differences in potential may be made inconsequential, but at the same time they must be extremely stable. However, changes in the amplification factors, provided that they are not excessive, will have no significant effect upon the performance. It is the current which provides the balancing potential. Therefore, any

Fig. 16·9 This automatic electronic potentiometer employs photocells to control the position of the shutter, which in turn governs the amount of current through the balancing resistor R_2.

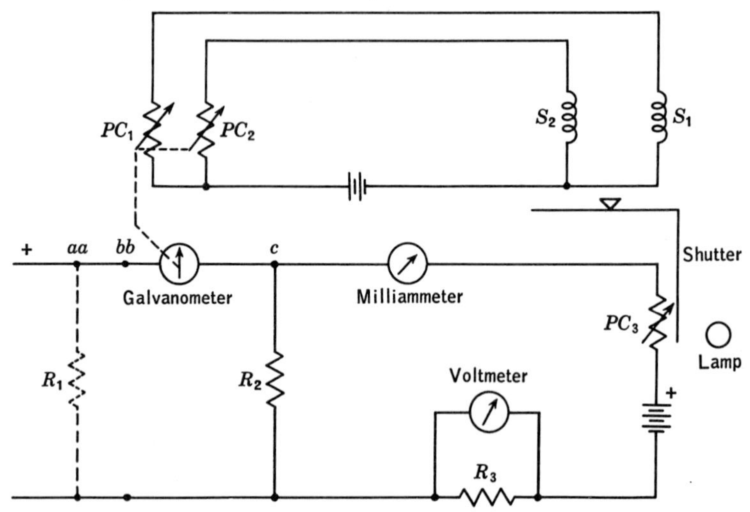

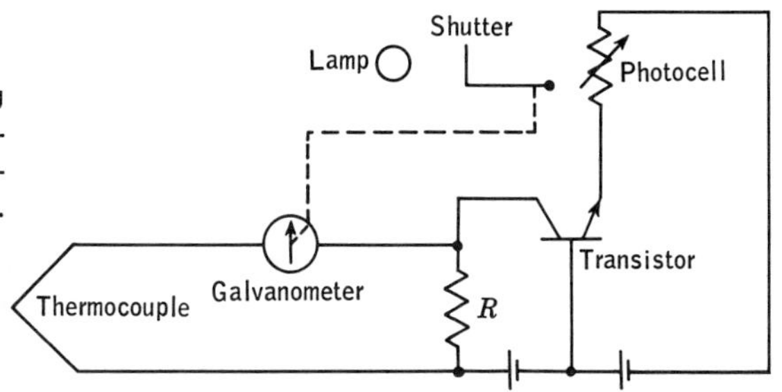

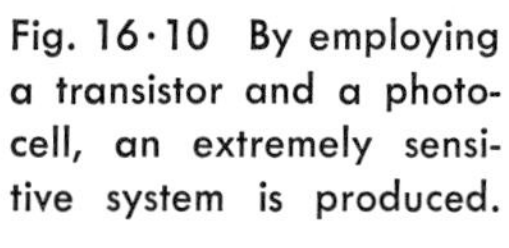
Fig. 16·10 By employing a transistor and a photocell, an extremely sensitive system is produced.

change in amplification merely means that the error signal will change with it.

One electronic type of automatic potentiometer which does not require an error signal is shown in Fig. 16·9. Currents and potentials can easily be converted into other currents or potentials for use with other components. The analog values may be either larger or smaller than the input values.

The basic principles of this circuit are few and simple: The galvanometer needle carries an opaque vane which controls the light falling on photocells PC_1 and PC_2. When the galvanometer current is zero, the cells are equally illuminated and the currents in coils S_1 and S_2 are equal but create opposite torques. The shutter controlling the light which falls on photocell PC_3 remains in position because it has no restoring torque of its own. But any deviation in the position of the galvanometer will cause the two coil currents to become unequal. The unequal torques thus produced change the position of the shutter, and hence the current through PC_3, until the galvanometer is again returned to its zero position.

Because a transistor is a current-operated device and has current-amplifying capabilities of 100 to 200 times in some types, a slight movement of the galvanometer, Fig. 16·10, can be made to produce a relatively large change in the transistor current which passes through resistor R in the balancing circuit. The amplifying action, or the gain, of such a device can be improved by adapting the high-voltage, low-current characteristics of the photocell to the low-voltage, high-current requirements of the transistor.

The step-down transformer, Fig. 16·11, will provide a larger current signal to the transistor than can be obtained directly from the photocell. If the transformer had a step-down ratio of 20, then the current delivered to the transistor should be about 20 times that obtainable without the transformer. This will greatly improve the response of the system.

A commercial design, Fig. 16·12, employs a chopper to convert the direct-current error signal to an alternating-current potential which can be readily amplified and then demodulated to provide an output direct current.

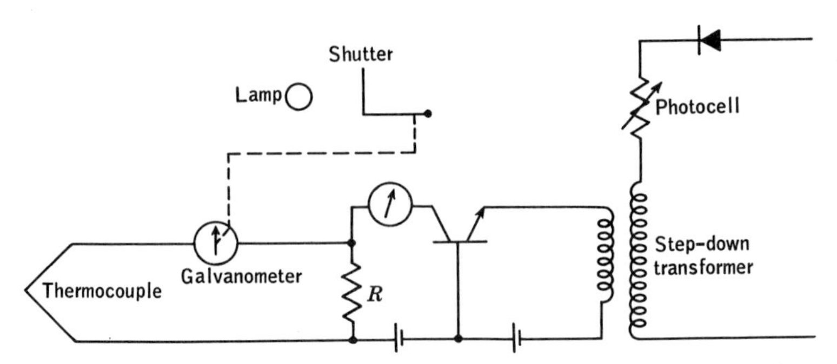

Fig. 16·11 For the best performance of systems, steps should be taken to provide the correct impedance matches. Employing a high-voltage, low-current device to control a device which needs a relatively large current at a low potential will not provide the best performance.

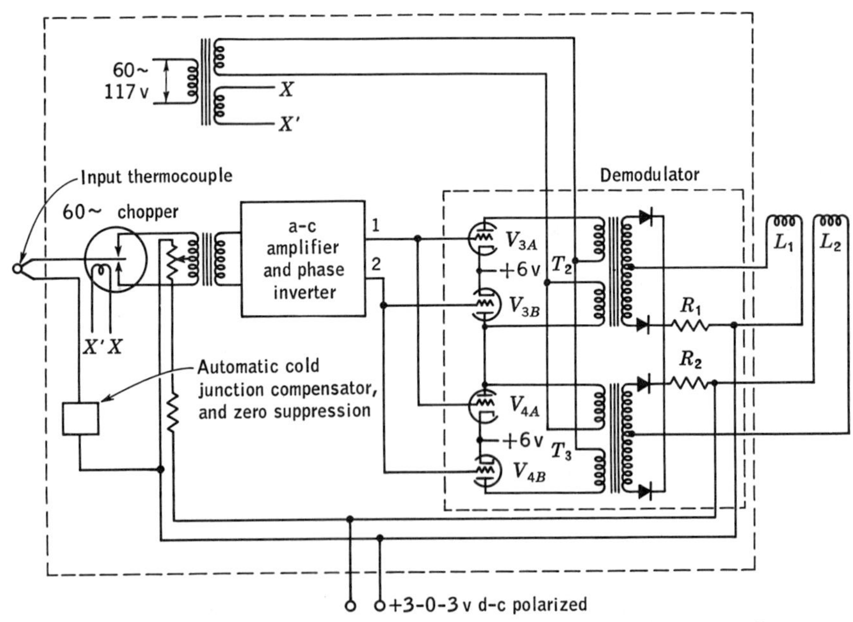

Fig. 16·12 The schematic circuit of a commercial design which uses a *chopper* to convert the error signal to alternative current but which causes a flow of current through a resistor to balance the input emf. (*Hagan Chemicals and Controls, Inc.*)

Oscillator-type electronic potentiometers. A marked departure from other designs is represented in Fig. 16·13. In this instance the input emf is also connected in series opposition with the emf developed across bridge resistor R span. Any inequality between the value of the input signal and that developed across the terminals of resistor R will cause a current to flow through the magnet motor. Assuming that the current-carrying

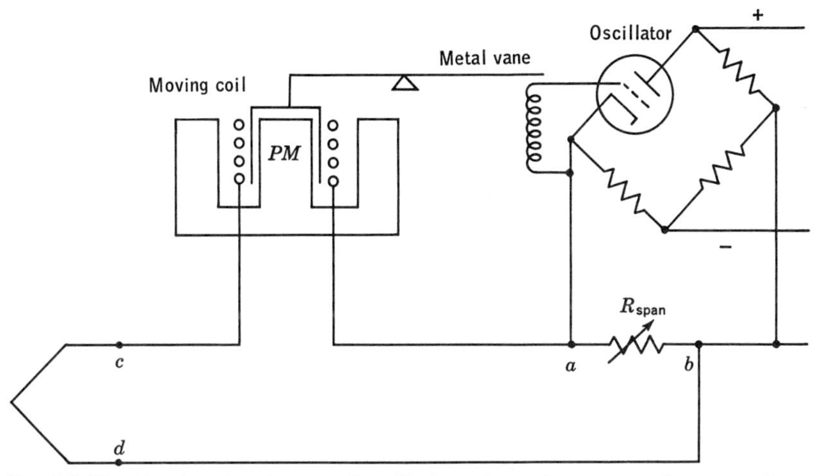

Fig. 16·13 Varying the degree of oscillation of a tube can produce very large changes in the tube current. An almost imperceptible change of position of the vane can cause changes in current that will reestablish equality of potentials *af* and *cd*. If the moving-coil assembly were perfectly counterbalanced and had no friction, then at the time of potential balance there would be nothing to change the position of the vane until the input potential changed.

coil, with its associated assembly, is perfectly counterbalanced, then the force set up by any current in the current coil will try to displace the coil in a direction that depends on the direction of the current. (It will be recalled that whenever there is a current there will be an associated magnetic field. The field around the conductor is made to react with the field produced by the permanent magnet. The reaction of these two magnetic fields produces the force required to adjust the position of the coil.) However, any movement of the coil will alter the position of the metal vane with respect to the oscillator coil. The particular virtue of using an oscillator and a controlling vane of the type indicated is that a very small physical movement of the vane—of the order of a thousandth of an inch—can cause a change of many milliamperes in the tube current. The degree of oscillation can control the apparent resistance of the tube far more than can a small change in the input grid potential; therefore, this type of circuit functions as an extremely high gain amplifier. Very small input variations, in the form of vane movement, create large changes in the output currents.

Provided that the beam is so balanced that it will maintain any position with no signal—in other words, that it has zero restoring torque—perfect balance between input and balancing circuits can be achieved. This type of circuit need not require an error signal in order to have an appropriate output potential. However, if there is a slight restoring torque, then an error current will be required to make the coil maintain an equilibrium position.

It should be noted at this time that if some physical means is used to control the position of the vane, then an output potential which is proportional to the displacement is produced. This phase of the subject is covered more completely in Chap. 10 under Electrical Transducers.

Bridge-type potentiometric devices. In accordance with our previous definition of what constitutes a potentiometric type of instrument, we will have to include the conventional Wheatstone bridge and many of its derivations. With the exceptions of bridge circuits which employ a difference current, as indicated by the position of a meter when read against a calibrated scale, bridge circuits seek to obtain the equality of two or more potentials, this equality being denoted by the zero-current position of the indicating element.

The theory of the Wheatstone bridge has been covered previously and will not be reviewed here except to note that:

1. The bridge is balanced only when the galvanometer current is zero.
2. The resistance and potential relationships which are assumed are true only for the condition of zero current in the galvanometer circuit. Provided that the error-signal current in the indicating meter or circuit is very small, the assumed relationships may not be seriously in error.

Assume that the bridge in Fig. 16·14 is balanced and that resistance R_2 is created by a given temperature—it might just as well be created by strain, by the apparent resistance of a vacuum tube, in fact, by anything whose resistance can be made to change. Using conventional techniques (switches S_1 and S_2 are open), we might control the value of R_3 in order to rebalance the bridge by bringing the galvanometer current to zero.

Fig. 16·14 With this simple circuit, changes in resistance or the value of the resistance itself can be converted into electrical equivalents which are measured by the value of the current required to balance the bridge.

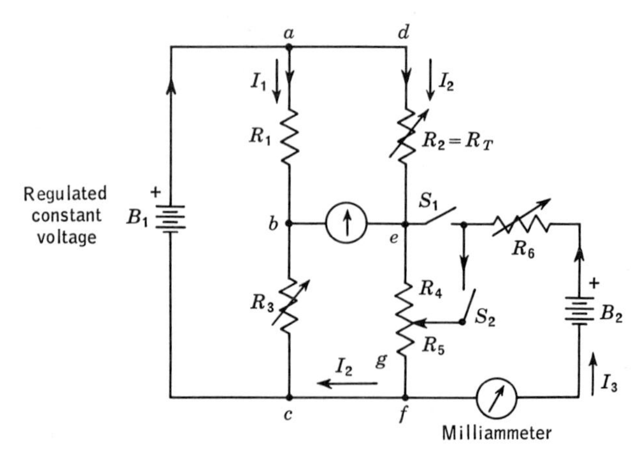

For an increase in the value of R_2, the value of R_3 would have been decreased in this process in order that there be the same ratio between R_1 and R_3 as between R_2 and R_4 plus R_5. The bridge could also have been balanced by increasing the value of R_1, leaving R_3 and R_4 plus R_5 unchanged. An increase in R_2 with R_4 plus R_5 unchanged would increase the potential between points d and e and decrease it between points e and f. In fact, *all bridge balancing using resistive or other passive elements requires such adjustments of the values that the indicating galvanometer will be connected to points at the same potential.*

We have previously considered that the potential existing across the terminals of a pure resistor is dependent only upon the values of the current and the resistance. Consequently, any means which can be used to control the current through a given resistor can be used to control the terminal potential. Let us assume that, in Fig. 16·14, R_2 has undergone an increase for some reason or other. So R_2 and R_4 plus R_5 will divide the potential from d to f in the ratio of their resistances. Consequently, as the resistances of R_2 increases, the potential between points d and e will change in the ratio of $R_2/(R_4 + R_5)$. Recall that an increase in the potential from d to e will decrease the potential from e to f.

The values of the three other resistors in the bridge have not been changed; so the indicating galvanometer will indicate the lack-of-balance condition by moving from the zero position. To restore the bridge to balance, we must increase the potential between e and f. This can be accomplished by closing switch S_1. If we now cause some additional current from battery B_2 to flow through resistors $R_4 + R_5$, then their terminal potential will be the sum of the potentials caused by each current passing through the two resistors in series. Consequently, by controlling the value of current I_3, we can *increase* the potential between e and f by any desired amount, the value of the increase being the product of $(R_4 + R_5)$ and I_3. Because we will assume that the value of $R_4 + R_5$ is constant, the increase in potential difference between e and f will be linearly and directly proportional to the value of current I_3. By means of a suitable calibration, any change in the value of R_2 can be measured by noting the value of current I_3 required to bring the bridge back into a condition of balance.

It should not be inferred from this that all of resistors $R_3 + R_4$ must be used. Any desired portion of the sum may be employed. By closing switch S_2 instead of switch S_1, I_3 must be increased to compensate for passing through only R_5. The chief requirements are to adjust the changes of potential required to the available or desirable changes in current I_3.

AUTOMATIC BRIDGE BALANCING FOR RESISTANCE MEASUREMENTS

If we continue with our previous lines of thought, we will soon want to obtain automatic operation instead of intermittent manual manipu-

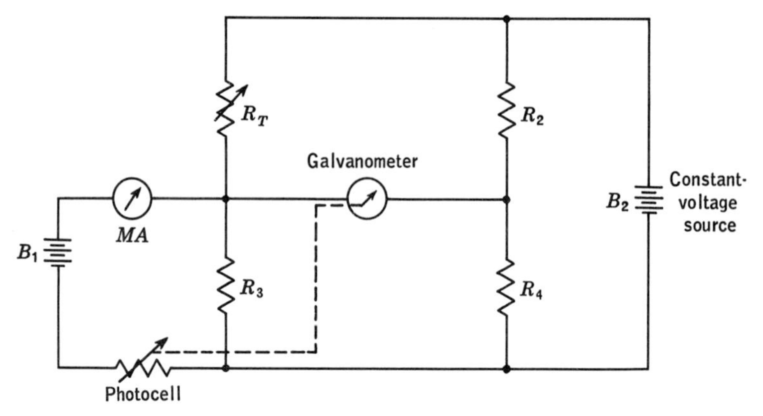

Fig. 16·15 A photocell which is controlled by a galvanometer can be used to balance a bridge automatically. The size of the error signal, for a stable bridge condition, is controlled by the sensitivity of both the galvanometer and the photocell.

lation. Consequently, let us employ the methods sketched in Figs. 16·15 and 16·16. We have met these methods previously and are quite well acquainted with their principles.

CAUTION: The performance of such bridges, either manual or automatic, is based on two conditions which must always be realized:

1. With a *constant* potential source B_1, all parts of the bridge always have the designed values of currents I_1 and I_2. It is only with a constant value of I_1 that any change in resistance will provide a predictable change in its terminal potential.

Fig. 16·16 Resistance values or resistance changes may be converted into analogs in many ways. This figure shows the use of photocells to control the balancing current through resistor R_4. As indicated, the magnitude of R_3 can be represented by either a voltage or a current.

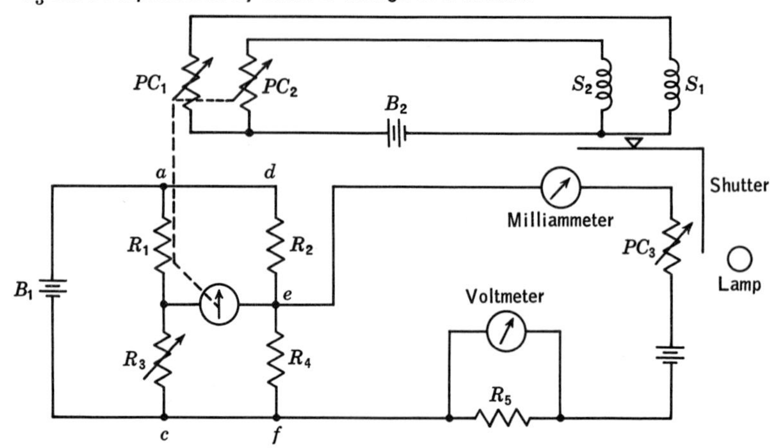

2. Any change in the value of R_1 must be so small that there is no significant change in the value of I_1. For example, if R_3 is 1,000 times as much as R_1, then any reasonable change in the value of R_1 will change the current through them by a factor of less than 0.1 per cent.

3. Devices of this type require that:
   a. The value of the resistance remain fixed while the current is allowed to vary.
   b. The current remain constant in value while the resistance changes with conditions.

Bridge balancing to obtain values of potential. We have noted in our preceding discussions that when we sought to balance a bridge we were really bringing two points to the same potential. It might occur to us, then, that if these two points were already at the same potential, we could deliberately insert some potential which would unbalance the bridge. Then we could reestablish bridge-balance conditions by using the principles which we have just reviewed.

One way of doing such a thing is indicated in Fig. 16·17, which it will be noted, bears a very close similarity to Fig. 16·14. Let us assume that, with the values chosen for R_1, R_2, R_3, and R_4, the bridge is balanced when terminals *a* and *b* are connected with a wire of low resistance. If we now remove the wire and apply some potential—it may come from a thermocouple, it may be the potential across the terminals of a resistor, or it may be the rectified output of a transformer—the bridge will be immediately unbalanced. This will place terminal *b* at a higher potential than terminal *c*. This will be indicated by a deflection of the galva-

Fig. 16·17 A voltage may be easily converted into an analog of current by finding what current through R_5, from battery *B*, will bring the galvanometer current to zero. This situation is almost identical with that seen previously in Fig. 16·14.

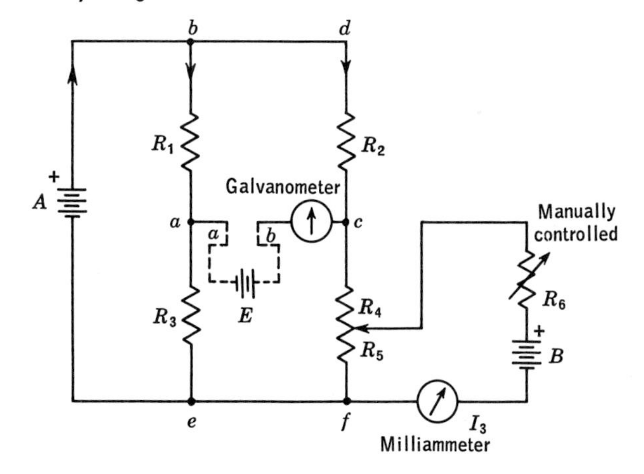

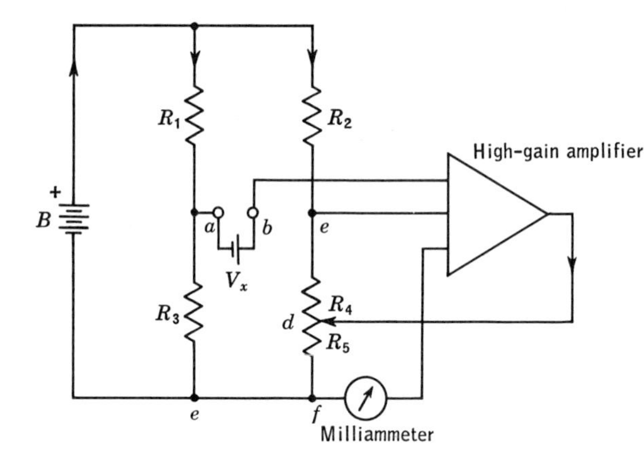

Fig. 16·18 Schematic representation of a transducer for converting a potential to an equivalent measured in milliamperes or amperes.

nometer. By controlling the value of R_6, we can increase current I_3 to regain bridge balance. The product of ΔI_3 (the increase required of I_3) and R_5 gives the value which will restore terminals c and b to the same potential. This is the value of emf E.

Automatic potential bridges. By supplying the difference in potential to a high-gain amplifier, Fig. 16·18, the amplifier will create an output current which will raise the potential of point d to create equality of potentials at b and e. The potential from e to f minus V_x must equal V_{ae}, or $V_{ef} > V_{ae}$ by V_x. A commercial version of this arrangement is shown in Fig. 16·19.

It should be noted that *electronic bridges for the measurement of*

Fig. 16·19 A commercial counterpart of Fig. 16·18. The value of the current used to represent the thermocouple emf, temperature, or any other desired value is governed by the design of the particular system. Certain manufacturers have standardized on current values of 10 to 50 ma, while others have chosen 1 to 5 ma. The ratio in most instances is 1 to 5. This is the same as the ratio used in pneumatic devices, 3 to 15 psi. (*The Foxboro Co.*)

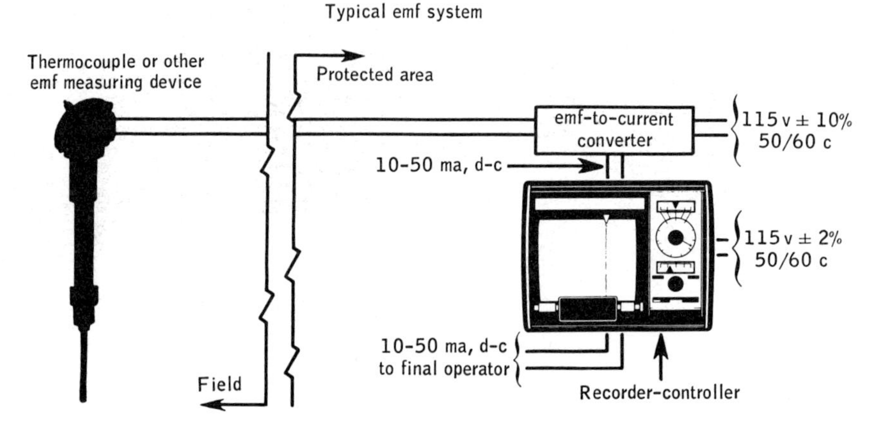

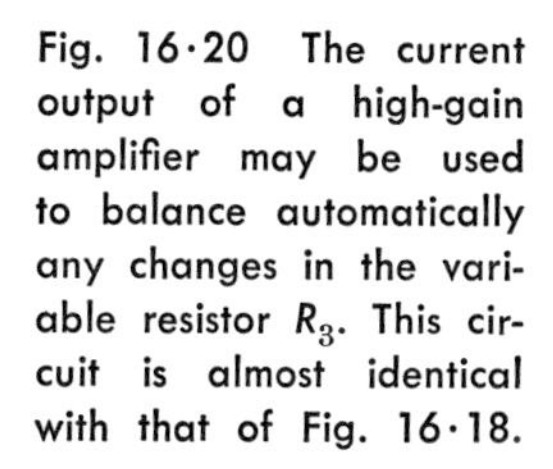

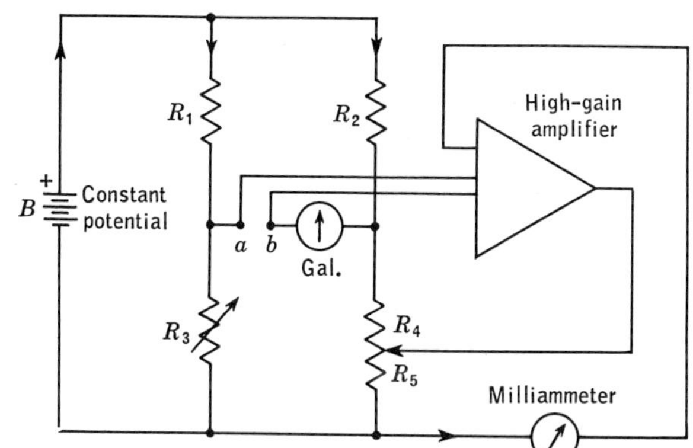

Fig. 16·20 The current output of a high-gain amplifier may be used to balance automatically any changes in the variable resistor R_3. This circuit is almost identical with that of Fig. 16·18.

potential do not require other than a reasonably constant potential source. The incremental value of the current through the balancing resistor R_5, for a fixed value of resistance, need furnish an emf just equal to unknown potential being measured. Changes in the value of battery B change the potentials of points a and e by equal amounts.

Changes in the resistance, or the value of the resistance itself, Fig. 16·20, may be determined automatically when the output current of the amplifier is used to bring points a and b to essentially the same potential. There must always be a slight difference in the potentials of a and b in order that the signal input to the amplifier create an adequate current through R_5. If the two terminals a and b were to have the same potential, then the input to the amplifier would be zero and the output current would then have the same value, theoretically. However, the difference in potential at the time of balance is generally insignificant.

With such a type of bridge for resistance measurements, there must be an extremely stable supply voltage. Moreover, any change in the value of R_3 must make an insignificant change in the current through R_1 and R_3. Consequently, R_1 must have a large value in comparison to R_3.

Figures 16·21 and 16·22 are for related types of circuits.

The emf-to-current converter, Fig. 16·19, has its input emf signal opposed by a high negative voltage feedback from the output of the primary amplifier to ensure stable amplification. The feedback also causes a high input impedance to make insignificant any variations in lead-wire resistance, Fig. 16·23.

The final amplifier acts as an isolation component, since there is no feedback between it and the primary amplifier. Either input lead or either output lead may be grounded. The final stage also amplifies the signal to some standard, such as 10 to 50 ma, for actuation of receiving devices. Current feedback around the final amplifier provides a regulated d-c output which will not change with normal variations either in line or load resistance.

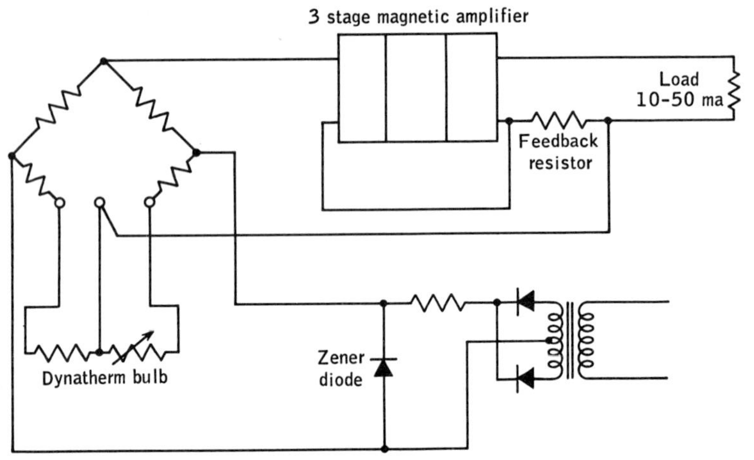

Fig. 16·21 A commercial version of Fig. 16·20. Because this bridge measures resistance changes, it must be supplied with a constant potential which, in this design, is obtained by using a Zener diode. (*The Foxboro Co.*)

The remarks made about resistance-type bridges apply equally to capacitive ones. If capacitors are used, Fig. 16·24, there are problems of charging and then discharging the capacitors when direct current is used. If rectified a-c is used periodically to charge the capacitors, the current to charge each capacitor will have to pass through the resistors in its own leg. If I_1 creates the same potential drop in R_x as I_2 does in R_2 and R_3, then the galvanometer will not deflect, because it is

Fig. 16·22 Magnetic amplifiers may be used to convert the error-signal current to one adequate for bridge balancing. (*The Foxboro Co.*)

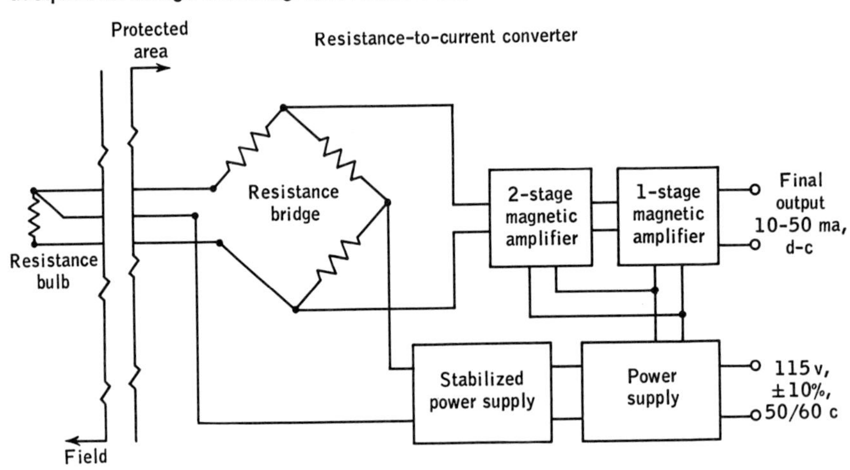

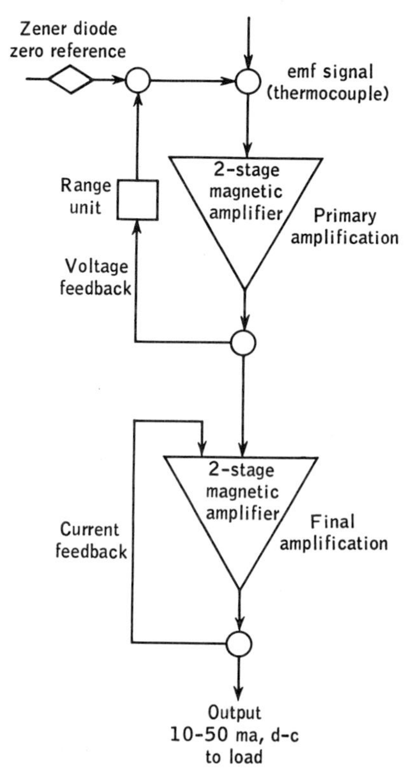

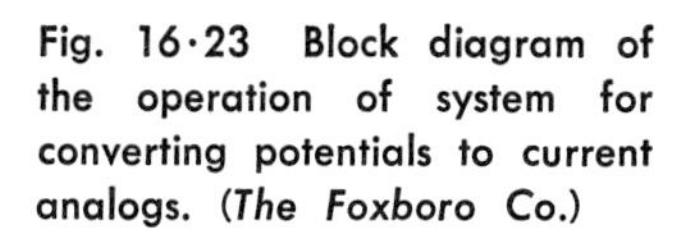

Fig. 16·23 Block diagram of the operation of system for converting potentials to current analogs. (*The Foxboro Co.*)

connected to points at the same potential. But if R_x were to increase for any reason, then the potential across its terminals would exceed that from c to b. The potential of point b with respect to point c can be raised by adjusting current I_3 so that it creates the additional emf across the terminals of R_3.

Fig. 16·24 By supplying a pulsating direct current, the conventional bridge can be converted to automatic balancing.

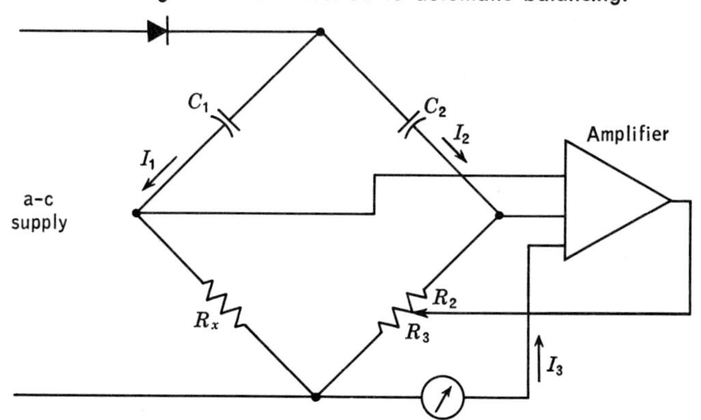

POTENTIOMETERS WHICH USE DIFFERENTIAL TRANSFORMERS

Because the linear differential transformer provides an output potential which is directly proportional to its displacement, many modern instruments and measurements are built around it. For example, Figs. 16·25 and 16·26 represent any situation in which some condition can position the core and in this manner produce a potential which is a measure of the input. If we connect an identical transformer in series, opposing, with the first, then any net potential which is developed by the pair indicates that the receiver-transformer needs to be repositioned until potential equality is again achieved. By feeding any difference signal into an amplifier of suitable design, an output current will be obtained which will drive the positioning motor to the correct position. The position of the receiver core is used as a measure of the input quantity when transmitter and receiver transformers are creating equal potentials.

If it is desired to use direct currents to measure the output of the dif-

Fig. 16·25 Any variable which could position the core of the transmitter can be measured if the core position of the receiver can create conditions of potential equality. Transmitter and receiver must be connected to supply lines of the same frequency and phase. A specific-gravity condition is used only for illustration.

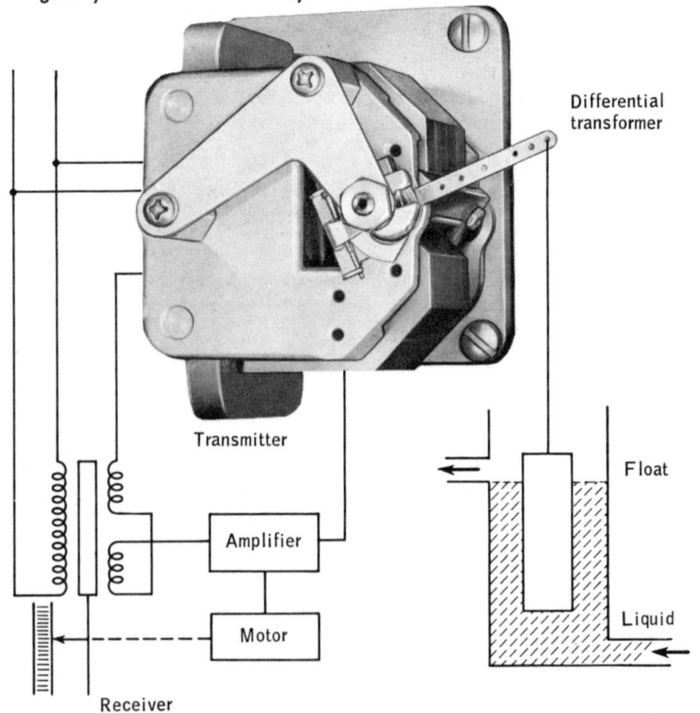

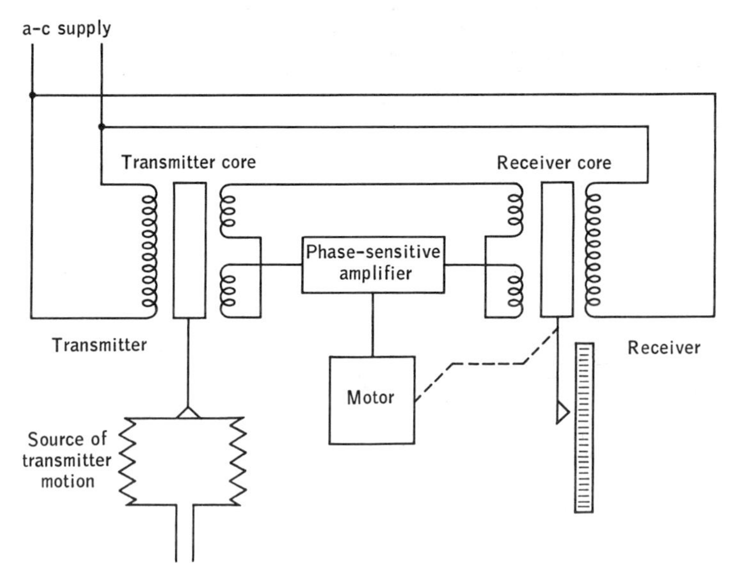

Fig. 16·26 Pressure or differential pressure measurements lend themselves easily to potentiometric methods.

ferential transformer, then a circuit such as Fig. 16·27 might suggest itself. The output of the transformer is converted to direct-current potential and used to charge the capacitor. This will impose virtually no load on the transformer. A word of caution is perhaps in order: for small potentials in the low-millivolt range the rectifier may not function, or it will be quite nonlinear. This will require that the transformer output be in the range in which such troubles can be successfully avoided.

Potentiometric circuits may be simple or complex, a-c or d-c, automatic or manual in their operation. In marked contrast to earlier ideas that they are reserved for the measurement of thermocouple outputs, potentiometric measurements can be employed regardless of the source of the transmitting signal. Motion, position, tachometer outputs, or thermocouple potentials all may be measured by this method.

SUMMARY

1. A potentiometric device is one which depends for its operation on the equality of two or more potentials. It may employ direct or alternating current. It may employ either manual or automatic manipulation to secure equality of the potentials.

2. Potentiometric devices generally require some standard of reference, which may be a standard cell, a potential derived from the Zener voltage of a rectifier, and so on. Measuring circuits which employ current generally do not employ reference standards.

3. Potentiometric devices generally take from the circuits being compared little or no current when the measurement is made. Certain types of self-balancing

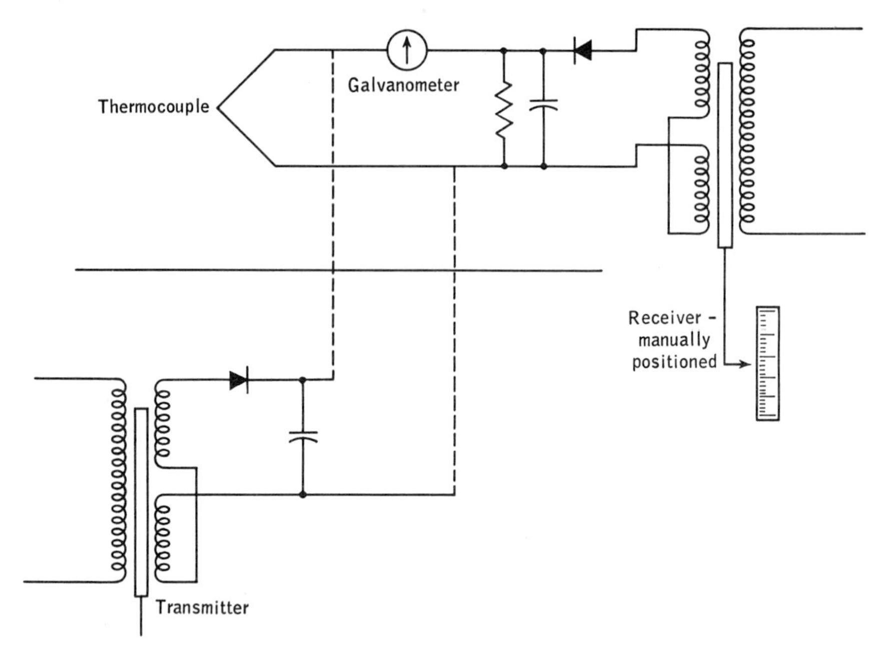

Fig. 16·27 The rectified output of a differential transformer might be employed to balance the output of a thermocouple or any other direct-current input.

devices do require a small current for operation, but the resistance or impedance will be very high at the balance point. This ensures that the current which exists at that time will have little or no effect on the potentials of the circuits.

4. All potentiometric devices require a comparison of potentials, that is, a decision as to which is larger and what correction is required. The correction is an increase or decrease of potential to restore a condition of equality. The restoration of potential equality may require manual, mechanical, or electronic means.

5. In potentiometric devices of this sort any current resulting from potential inequality is the basis for the correcting action.

6. The standard cell is extremely limited in its ability to deliver current yet maintain its correct potential. For this reason, the cell should at no time deliver current in excess of a few microamperes.

7. Whereas potentiometers can measure to a fraction of a millivolt, devices which infer the potential from a direct measure of the current passing through a resistor do not have comparable degrees of accuracy. Accuracies of 0.5 per cent are about the best that one can hope for. Inasmuch as this percentage is based on the maximum indicated value, the error may be quite significant at the lower end of the scale.

8. Electronic potentiometers tend to be faster than their mechanical counterparts because they have no moving parts. However, their amplifiers may be complex and their read-out devices may be more complicated than those of a mechanical design.

9. In general, potentiometric devices which are of the mechanical design do not have any potential inequality, or current flow, at the time that the circuit is balanced, whereas electronic types often require some small current for the operation of their amplifier systems.

10. Automatic electronic bridges for the measurement of potential generally do not require some fixed reference potential for their operation.

11. The measurement of resistance with electronically balanced bridges requires the use of a very carefully regulated source in order that all variations in potential across the measured section will be caused by variations in resistance and not by a variation in bridge current resulting from a changed circuit potential.

12. Potentiometric devices are no longer limited to the measurement of temperature and a few similar tasks. Today they are used to indicate pressure, position, speed, motion, or any other quality or property which may be represented by a potential analog.

QUESTIONS

16·1 In what major respect do potentiometric measurements differ from other electrical measurements? Of what importance is this difference?

16·2 Why do certain potentiometric devices use both working cells and standard cells? Is it always necessary to use both? Why?

16·3 What important precautions should be taken when using standard cells?

16·4 In what important ways do electronic and mechanical types of potentiometer differ? In what ways are they alike?

16·5 Why should the use of an electronic bridge for the measurement of resistance require a very stable source of voltage?

16·6 Does the measurement of potential by the use of an electronic bridge require the use of an extremely stable bridge potential? Justify your answer.

16·7 In what important ways do Questions 16·5 and 16·6 require different answers?

16·8 When using an electronic bridge for the measurement of resistance, or resistance change, how large a change may be measured? Give your answer in terms of the fixed resistor in the same leg. Assume a Wheatstone bridge arrangement.

16·9 What parts of Fig. 16·2 might be said to be counterparts of Fig. 16·3?

16·10 In the event of power failure, the standard cell shown in Fig. 16·4*b* supplies the current to the bridge. Must this be a high- or low-resistance bridge? Justify your answer.

16·11 Figure 16·5 represents a commercial potentiometer. Compare and contrast it with Fig. 16·1. With Fig. 16·2.

16·12 What circuit elements have to be known if the current is to be a measure of the thermocouple potential, Fig. 16·6*a*? Within what limits are their accuracies generally expressed? W. uld this furnish the same degree of accuracy as is obtainable with the conventional potentiometer which uses the standard cell for standardizing purposes?

16·13 What is the advantage derived from using a photocell rather than a variable resistor to control the tube current, Fig. 16·8?

16·14 In what way are Figs. 16·7 and 16·8 alike? Unlike? Explain fully.

16·15 What is the purpose of the chopper, Fig. 16·12? Why must it be used?

16·16 Compare the performance of the device shown in Fig. 16·9 with that of the device shown in Fig. 16·13.

16·17 When a system like that in Fig. 16·26 was installed, it was decided to save on the initial wiring by energizing the receiver differential transformer from a different source. What might happen under such conditions? What sort of measuring system would it be?

PROBLEMS

16·1 What is the minimum value of the calibrating resistor, Fig. 16·1, if the maximum value of the working cell may be assumed as 1.6 volts? The resistance of the potentiometer may be assumed as 100 ohms. The value of resistor R_{bc} is 918 ohms. $V_{ab} = 0.100$ volt.

16·2 In Fig. 16·2 the drop across the potentiometer is to be 10 mv. Determine the potential across the fixed resistor and the minimum value of the calibrating resistor. The standard cell is the usual Weston cell. Working cell current is 0.001 amp.

16·3 If R_{BO} is twice that of R_{Oa}, what must be the ratio of the capacitor currents at the time of balance, using Fig. 16·5?

16·4 If the maximum thermocouple potential is 30 mv, determine the value of the variable resistor, Fig. 16·6*a*, when the battery has an emf of 2 volts and R has a value of 6 ohms.

16·5 The photocell, Fig. 16·8, has a dark resistance of 1,000,000 ohms. What should be the value of R_{bc} so that the potential applied to the grid of the tube shall be less than 2 volts?

16·6 If the photocell in Prob. 16·5 had a light resistance of 100,000 ohms, what would be the potential applied to the grid of the tube if $R_{bc} = 250{,}000$ ohms?

16·7 In Fig. 16·11 the step-down transformer has a ratio of 20 to 1. If the maximum current through the photocell is 20 ma, what should be the emitter current? What should be the current through resistor R if the transistor has a current amplification factor of 30?

16·8 Using Fig. 16·14, determine the additional current which will have to flow through resistor R_5 to balance the bridge when voltage *bc* is increased by 20 mv. $R_5 = 10$ ohms.

16·9 In Fig. 16·17 the value of E is 50 mv. What current through resistor R_5 will be required to balance the bridge if $R_5 = 10$ ohms? Battery $A = 2$ volts, and battery $B = 6$ volts. Before inserting E, the bridge was balanced with $I_3 = 1$ ma.

16·10 Determine the value of resistor R_6, Prob. 16·9, when the bridge is balanced.

17 Indicating and Registering Equipment

There are many ways of presenting the values of the variables which can be measured by our various types of measuring and indicating instruments. There are times when a simple mechanical indication is quite adequate. At other times, we want a record of the performance. For still other operations, we are interested only when the process moves away from the set point at which the process is designed to function best. As a result of this, there are many different ways of presenting the results: circular charts, rectangular-strip charts, off-limit lights and alarms, and digital or analog presentation. All require consideration. The final answer depends, generally, upon what is to be done with either the reading or the record. We may also classify the devices as self-acting and servo-acting units. An ordinary pressure gage is a good example of a self-acting device, and an automatic recording potentiometer is illustrative of servo-controlled units.

Certain instruments may be actuated by more than one sensing element. This requires selective identification of the points on the record in order that each variable may be viewed properly with respect to its neighbors. The use of printing wheels, different-color inks, and other coding techniques will be considered.

For purposes of this discussion, the following phases will be considered:

Pointer indication: gages and indicators, self-acting devices
Charts: circular and strip
Servo-actuated indicators and recorders
Multiple indications on one chart
Data logging and indicating

RECORDING DEVICES

Pointer indications. Whether one considers a simple pressure gage or a large indicating meter, the pointer is moved directly by the pressure-sensitive element. In almost every instance, there will have been some

Fig. 17·1 An elastic spring, often of the Bourdon type, with some means for actuating the pointer, constitutes an almost standard way of presenting values of the variables which can create pressure changes within a system.

type of mechanical amplification to make readily apparent the small movement of the primary element. In most cases this increase in movement will be obtained from a gear train or from a lever system.

When the force which is available is very small, a lightweight pointer will generally be moved over the face of the dial. Figure 17·1 shows the conventional gear train and linkage required for a Bourdon tube gage in which the movement of the Bourdon tube is amplified in order to move a large pointer for use with conventional indicating-recording equipment. It should be pointed out here that in almost all pneumatic controllers the active element moves the pointer only and the position of the pointer or pen actuates the controller mechanism. But when there is sufficient torque and the bearings are adequate, it may be more desirable to rotate the dial, having the pointer at the top. In this manner, the deviation of the dial from the 12 o'clock position is readily apparent, even from appreciable distances.

Almost all of the indicating instruments are of such a nature that the measuring portion of the device is incapable of exerting any appreciable force through any significant distance. Where Bourdon springs are concerned, so much of the energy delivered to the spring is used up in distorting the spring that there is none left over to do anything else. Consequently, any gage or meter using similar elements is incapable of exerting any real force to move a pointer, pen, or alarm elements. In a similar manner, millivoltmeters and galvanometers are incapable of pressing with adequate force to mark the value of the variable. If a pen is used, then there need be but little force between the pen and the paper, and graphical representations sometimes can be obtained. For similar reasons, certain types of electrical recordings (hot wires and high voltage, for example) are employed because the registration is then not dependent upon the force between paper and marker.

Chart types. There are two general classes of charts used in industry today. One is the round or circular chart which contains the record for the past day or week, as the case may be, and the other is the strip chart which gives a rectangular presentation and is more useful when all parts of the scale must be given equal emphasis.

The circular chart can be obtained with hundreds of variations of the scales and quantities. There are some very distinct advantages to circular charts for some purposes:

1. They permit an easy comparison of present operations with past performance during the previous few hours or days. Variations within any given unit of time are easily spotted and evaluated.
2. Being small and flat, they are easy to handle, store, and use for reference.
3. With fairly simple mechanical means, as many as four pens or indicators may be used.
4. It is very easy, mechanically, to convert the motion of many gage elements to adequate pointer movement.

The use of the circular chart also has a number of liabilities:

1. The chart must be changed at fairly frequent intervals.
2. For low values of the variable, there may be too much crowding so that reading the chart may be more difficult.
3. The nonlinear area relationships make the task of integrating the area under the chart somewhat more difficult than if rectangular charts are used.
4. For the common types of flowmeters, the minimum flow is limited to about 0.3 of the maximum flow.
5. Variations in chart speed are rather impractical.

The principal parts of the circular chart–recorder assembly are shown in Fig. 17·2.

Fig. 17·2 A circular-chart type of recorder for temperature or pressure. (*The Taylor Instrument Cos.*)

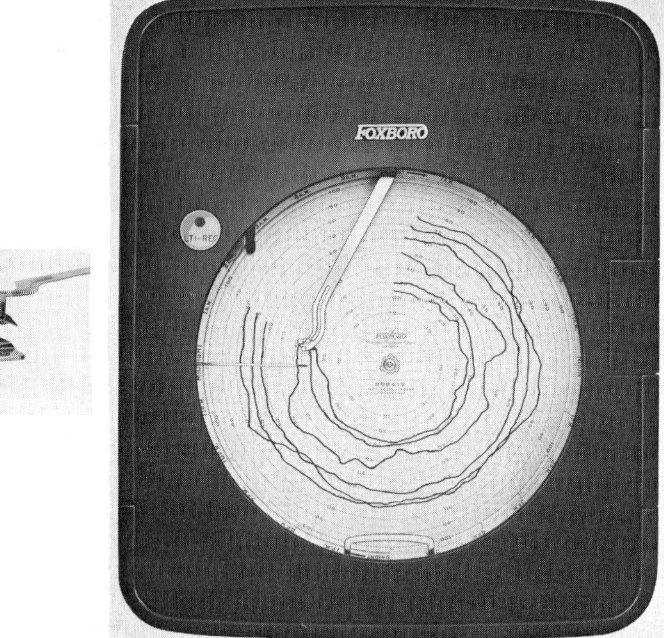

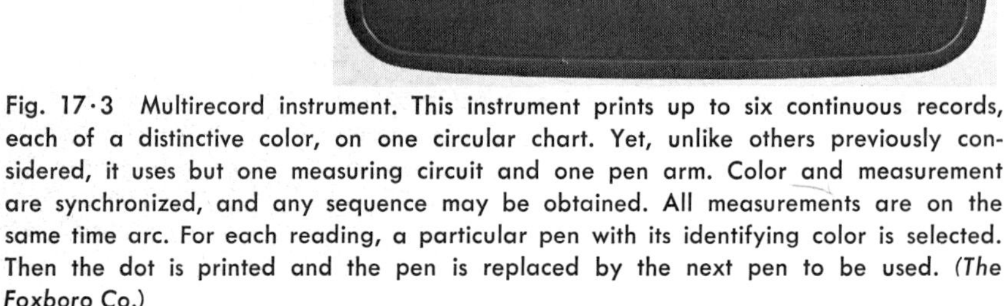

Fig. 17·3 Multirecord instrument. This instrument prints up to six continuous records, each of a distinctive color, on one circular chart. Yet, unlike others previously considered, it uses but one measuring circuit and one pen arm. Color and measurement are synchronized, and any sequence may be obtained. All measurements are on the same time arc. For each reading, a particular pen with its identifying color is selected. Then the dot is printed and the pen is replaced by the next pen to be used. *(The Foxboro Co.)*

Circular-chart instruments. In order to increase the versatility of the circular-chart instrument by presenting the values of more than one variable, Fig. 17·2, several means have been used. The most obvious one is to use several pens. This requires, of course, that there be as many primary measuring elements as there are pens. In order that the pens do not interfere with each other, they must all be of slightly different length. Use of a color wheel permits one pen to present the values of several variables. Figure 17·3 shows one design. Detachable pens are carried on a rotary holder and are so indexed that as the pen is connected to each indicating variable in turn, the holder will give up the pen, with the distinctive ink previously used, and pick up a new pen whose ink is indicative of the new variable. In this way, five or six variables can be recorded intermittently with but one pen holder. Such a design can monitor all of the variables in a fractional part of a minute, so the record becomes a series of short broken lines. For processes in which any variations are most likely to occur at a slow rate, this is quite adequate. Unless the pen-pointer mechanism is servo-operated, none of these devices can use any appreciable amount of force for positioning. Such a condition is in marked contrast to the situations in which servo elements are used and in which pneumatic or electrical power is available.

Strip-chart indicators and recorders. During the past few years, very little has been done to miniaturize the circular-chart instruments, but in the design of strip charts there has been a marked trend toward the smaller type of instrument, one using a chart 3 to 4 in. wide (Fig. 17·4) instead of 10 to 12 in. wide, which had formerly been more or less standard. Much of this has come about because of the design of panels to simulate the flow of the products through the process, with the instruments and associated controls located at the appropriate point on the panel. If several dozens of instruments were to be employed, it became physically impossible to provide the space which would be required for standard instruments. Of necessity then, the smaller units have been presented.

More or less parenthetically, it might be added that there are some very pertinent facts about instruments and charts which must always be kept in mind if one is not to make serious errors:

1. Unless especially calibrated, or otherwise indicated, the percentage

Fig. 17·4 The miniature strip-chart recorder-indicator can provide more than one pen. Whereas all strip-chart recorders formerly were restricted to electrical signals only, the newer designs, when suitably modified, will function with pneumatic input signals. (*The Taylor Instrument Cos.*)

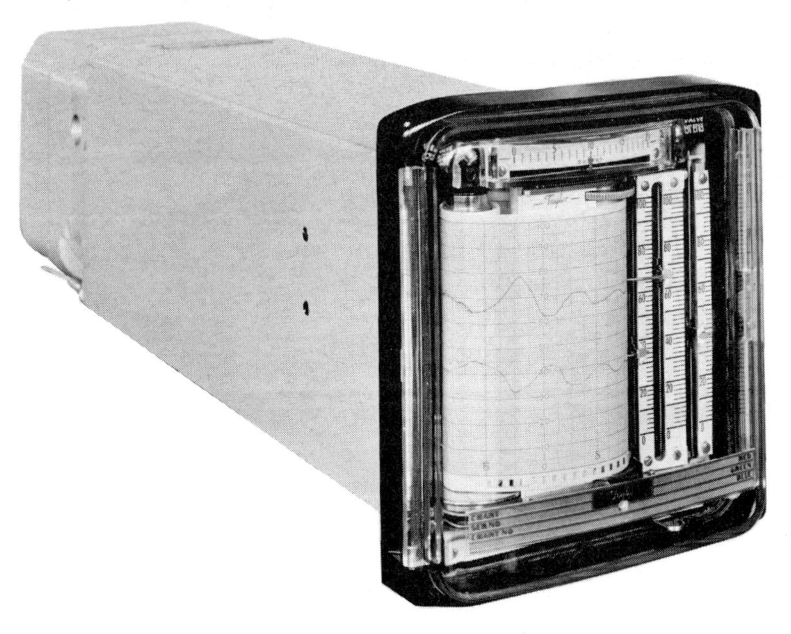

errors which apply are with reference to the maximum value indicated. Thus a ½ per cent instrument which has an indicated maximum value of 200 units will have an uncertainty of ± 1 unit at any point on the scale. When the instrument indicates 10 per cent full scale or 10 units, then the error is still ± 1 unit, which is 10 per cent of that value. Strictly speaking, it should be read as 10 ± 1. We get into bad habits when we neglect to indicate this uncertainty of measurement.

2. Trying to read a chart to very high degrees of accuracy may be misleading. If, for example, we use great care and read a 200° thermometer which has a 1 per cent accuracy and end up with a reading of 136.4°, we are indicating that we "know" the answer to the nearest one-tenth of a degree. Actually, we know nothing of the sort. We know that answer no closer than the inherent accuracy of the thermometer, and not to the accuracy of our reading and estimation of fractional spaces. If we are perfectly honest about that reading, we will have to write it $136.4° \pm 2°$. There is quite a difference!

3. Trying to read any chart or instrument with degrees of accuracy far beyond the accuracies of the device may lead to serious errors. This does not mean that one should be careless or pay little heed to the measurements which he is making. Rather it means this: know what you are measuring and within what limits you can believe what you are reading.

Until the development of the small strip chart, almost all strip charts were designed for use with electrical servomechanism systems such as automatic potentiometers. Except for those used for self-acting instruments, which use curvilinear charts, almost all of the charts have the same construction and, consequently, the same advantages and disadvantages which are summarized here. For the advantages, we may list the following:

1. Conversion of data is easier when rectangular coordinates are used.

2. Many more separate variables can be recorded on a strip chart than can be placed on a circular chart.

3. The rate of movement of the chart can easily be changed to spread out the trace of the variable being observed.

4. Relatively large amounts of paper can be inserted at one time in the form of a roll.

5. Since the chart is in roll form, it is almost impossible to mix up sequences.

6. By using special reading devices, it is possible to scan a large amount of recorded data quickly and select maximum and minimum points, and so on.

The use of the strip chart also has certain undesirable features:

1. Observing behavior several days or hours back is not as easy as picking out one circular chart which covers the desired period of time.

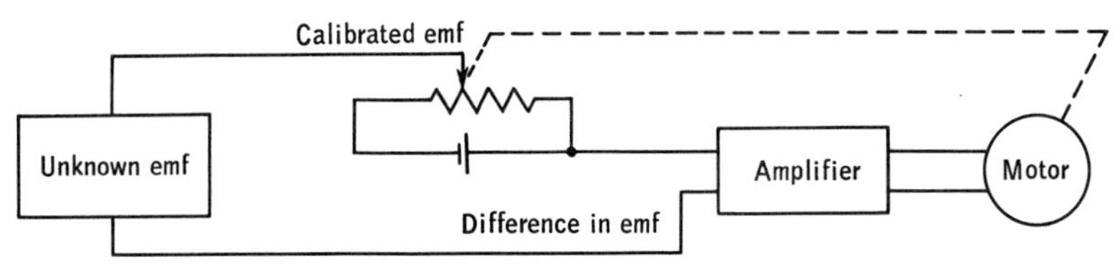

Fig. 17·5 Almost all potentiometric devices, except those using purely electronic means, operate essentially as indicated by this schematic. The unknown potential is compared with some known value. Any difference is amplified and then used to actuate the drive motor to reduce the differential. In some instances the potentiometer may be in the form of a bridge, but bridges also require an equality of potential for their balance. (*Brown Instrument Division, Minneapolis-Honeywell Regulator* Co.)

2. The mechanism is considerably more complicated than is required to drive a circular chart. Provision for inspecting the balancing drive while the unit is operating requires special gear drives and introduces mechanical complications.

Almost all rectangular-strip charts are moved by servo units; the variable merely controls a balancing function, the position of balance being indicated by the position of the pointer. In this way there is no real limitation on the power required to actuate the printing or positioning mechanisms. With an adequate amount of power available, frictional drag between pen or marker and the paper is of little importance. Similarly, there can be very high speeds of movement of the pen mechanism. Overshoot can be prevented by the use of tachometer circuits which reflect into the measuring circuit the speed of the balancing unit. Abilities such as these make it feasible for the pen to travel from one side of the chart to the other in a second or less and impose no load on the primary measuring circuit. The basic balancing circuits are shown in Figs. 17·5 to 17·9.

The use of solenoid motors which are damped by a very viscous silicone oil also prevents any overshoot. But whether we use a two-phase reversible motor to position the pointer and balance the potentiometer or whether we use a solenoid-type motor, the end result is the same.

The action of the balancing circuit shown in Figs. 17·6 and 17·8 is as follows. The thermocouple input potential is compared with the potential developed between points a and P on the potentiometer. Any discrepancy will cause a current to pass through the windings of the step-up transformer. By using a chopper to interrupt this current flow, the steady direct current is converted into a series of pulsations which are applied alternately to the transformer input terminals. If the two potentials created by the thermocouple and the potentiometer have a net sum of zero, then the input to the transformer is also zero, and there is no net action by the drive motor which positions the potentiometer.

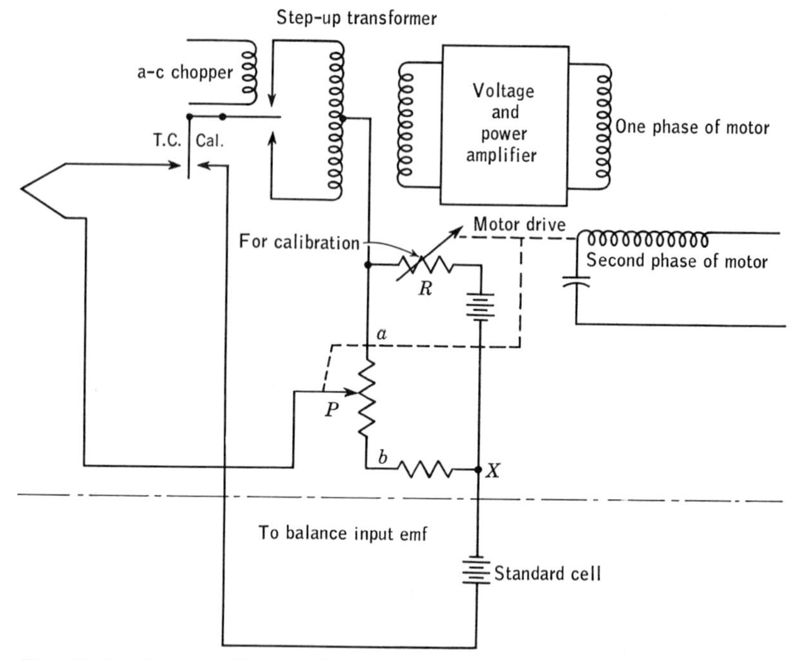

Fig. 17·6 Any significant difference between the emf of the thermocouple and that provided between terminals *a* and *P* of the potentiometer is amplified. The output current causes the drive motor to move the potentiometer in such a direction as to minimize the difference between the two potentials. Manual or automatic standardization can be provided.

In order to assure correct performance, provision must be made for calibration. In the example shown, the potential between points *a* and *b* will be correct when it is equal to the potential developed by the standard cell with which it is compared. When the instrument is being calibrated, the potential of the standard cell is substituted for the thermocouple emf. If the sum of the two potentials which are connected in series opposition does not equal zero, the drive motor will alter the value of resistor *R* until the zero condition does exist.

For our analysis let us assume that, at the moment, the thermocouple emf exceeds the emf at the potentiometer terminals *P* and *a*. This will produce current pulsations in the primary of the transformer which are essentially as shown in Fig. 17·7*b*. (If the emf picked off by the potentiometer exceeded the emf of the thermocouple, then the pulsations at each corresponding instant of time would be about as shown in Fig. 17·7*c*.) These signals pass through the amplifier and are applied to the output tubes which are represented in the schematic of Fig. 17·8. If there is no error signal, then equal signals should be applied to the two grids. In turn, the tubes, which conduct current only when the plates are positive, pass equal currents. Equal currents in the winding marked phase *A*

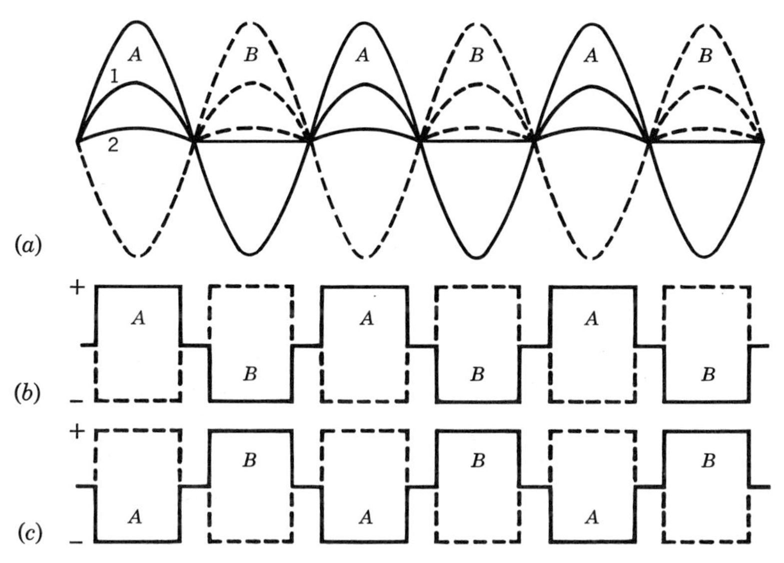

Fig. 17·7 (*a*) Representation of signals applied to the grids of the output drive tubes when potentiometer is not balanced. (*b*) If the emf of the thermocouple exceeds that of the potentiometer, tube *A* will have its grid potential increased, while the grid potential of tube *B* will be decreased (as in *a*). Reversed conditions are represented in (*c*).

Fig. 17·8 Schematic representation of the output stage of an amplifier. As the grid signal changes because of changes in the input signal, the potential applied to the grid of the output tubes will produce a net torque to bring about corrective action, unless both tubes are receiving equal grid signals. Then the net torque will be zero.

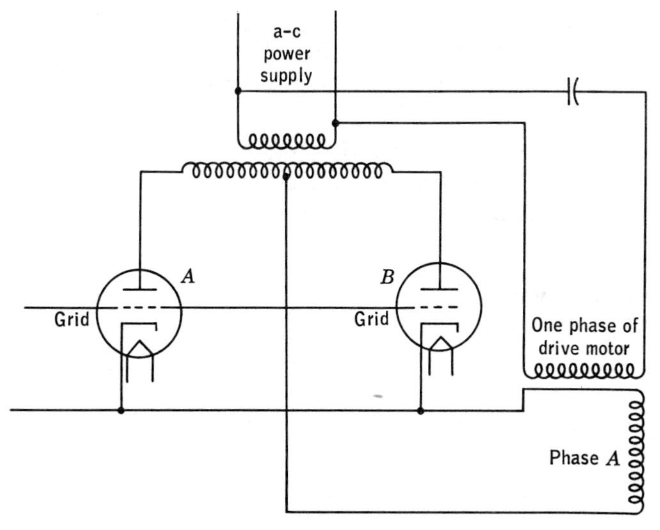

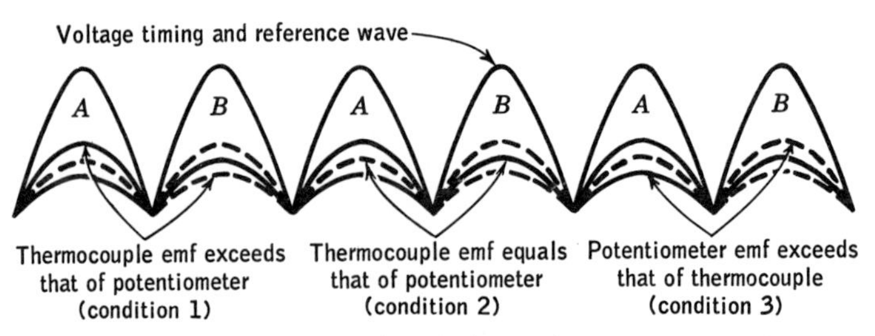

Fig. 17·9 Relationship of grid and plate voltages in automatic potentiometer drive. When tube *A* has a large plate current, tube *B* will have a small one, and vice versa. Any inequality of plate currents will produce a net torque to provide motor rotation. When the current in tube *B* exceeds that of tube *A*, the net torque will provide rotation in an opposite direction. When both tube currents are equal, the net torque will be zero.

create alternately equal torques for clockwise and counterclockwise rotation. The result is that no net motion results. Only when the net value of the torque is not zero can any significant amount of rotation take place.

Let us assume in Fig. 17·7*a* that the maximum positive error signal provides a current value in the output tube indicated by curve 1 and that the maximum negative signal provides a value of current represented by curve 2. But when tube *A* is receiving a maximum positive signal, tube *B* is receiving a negative maximum and its plate current is at the value 2. As the drive motor positions the potentiometer so that it more closely matches the value of the thermocouple emf, the maximum value of current in tube *A* will decrease while that in tube *B* will increase until, at the time of balance of the two potentials, tubes *A* and *B* will each be passing currents of equal value.

Should the potentiometer emf exceed the thermocouple emf, then for the assumed conditions tube *A* would receive a large negative signal while tube *B* would receive a large positive signal. The large current in tube *B* would then provide a torque greater than that provided by the current from tube *A*, and the net torque would then be opposite to that obtained when the thermocouple input emf exceeds the potentiometer emf. This would result in the potentiometer being so driven that it provided smaller potentials, Fig. 17·9.

It must be realized that any system such as this cannot provide *perfect* balance. *An error is always required before any signal is fed into the amplifier. This is a basic feature of all such designs. With very high gain amplifiers this error can be reduced to an inconsequential amount.*

A very practical problem in such apparatus is to control the speed of the motor to such a degree that the motor does not drive the rheostat too far and thus overcorrect for the original error. If this happened, then the signals applied to the grids of the tubes would be approximately as shown in Fig. 17·7*c*.

The motor which drives the balancing motor also positions the pen, pointer, or marker which is used. It should have such mechanical and electrical characteristics that it does not drive the balancing element beyond the balance condition and thus create conditions of overcorrection which might result in indefinite periods of hunting. To prevent overcorrecting because of motor inertia, a solenoid type of motor may be employed. An example of a mechanical solution to the problem is found in the Leeds and Northrup Micromax units, Fig. 17·10. These units are no longer being manufactured, but there are so many of them around and their principles are so fundamental that it is well to consider them. All later designs of balancing recorder-indicators are simply using somewhat different methods and components to solve the same problem.

Fig. 17·10 Typical schematic circuit for a potentiometer using mechanical balancing means. (*Leeds and Northrup* Co.)

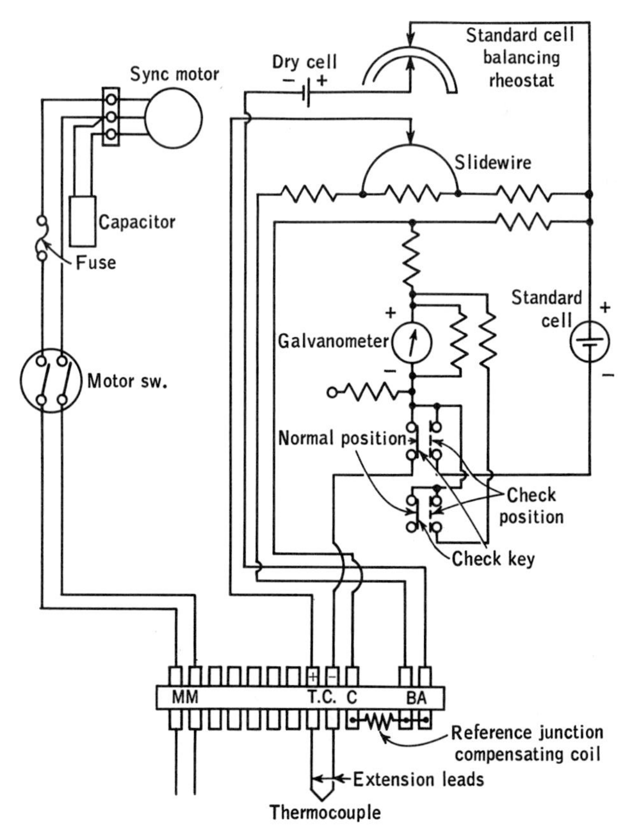

The Leeds and Northrup Micromax potentiometer. This particular strip-chart type of potentiometer recorder has been selected for analysis not because it is the only one which possesses the characteristics described, but because it is probably better known than any other which might be considered. It should be well understood that our chief interest is in the design and how later models of potentiometers built by most manufacturers deviate from this earlier concept of what the potentiometer-recorder should be.

The basic circuit, Fig. 17·10, is not noticeably different from the circuits of many similar devices built at the same time. As in all such circuits, any difference between the thermocouple and the potentiometer emf's creates an unbalance current which causes a galvanometer or meter to deflect an amount which is proportional to the unbalance current. Now as the galvanometer moves away from its neutral or zero-current position there is a periodic sensing of its position. Actually, every two or three seconds, regardless of the degree of balance or unbalance, the position of the galvanometer pointer is determined.

A synchronized clamping plate periodically holds the galvanometer firmly in the position given it by the unbalanced current. Two fingers then find the pointer, and as they do so they position the clutch, part 14, Fig. 17·11. Two heart-shaped cams, synchronized with the movement of the clamping plate and the positioning of the clutch, then rotate the clutch in such a direction as to reduce the unbalance current. It should be understood at this point that another cam lifts the clutch free from its connection with the driven disk, part 15, while the clutch is being positioned. The clutch is then returned so that it presses against the driven disk, part 15. As the heart-shaped cams return the clutch to the normal, neutral position, the driven disk is rotated, and as it rotates it moves with it the potentiometer and the indicator. The correction provided by the clutch repositioning depends entirely upon the amount of unbalance current, unless the correction to be made exceeds the maximum designed value. For major corrections, during start-up operations, for example, five or six balancing attempts might be required.

We should recall the following:

1. The direction of the unbalance current (determined by whether the thermocouple or other emf is more or less than that of the potentiometer) dictates which way the galvanometer deflects, and consequently the direction in which the potentiometer is moved.

2. The amount of rotation of the potentiometer, and movement of the pointer, during each correction period is directly proportional to the error between the two balancing potentials.

3. The driving motor runs at constant speed in one direction, but two-directional corrections are available.

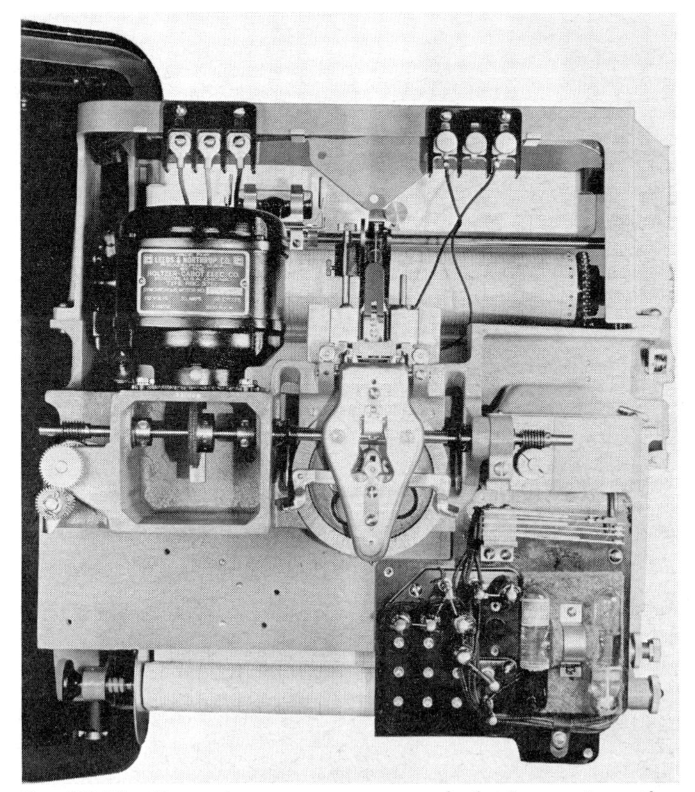

Fig. 17·11 The galvanometer, cams, and clutches perform the same functions that in later designs are carried out by choppers, amplifiers, and phase-sensitive motors. (*Leeds and Northrup* Co.)

More recent designs obtain the same potential balancing by using components which:

1. Use a synchronous vibrator, or chopper, to interrupt the unbalanced direct current so that it may be amplified. General experience is that amplifiers built for alternating-current signal inputs are more stable and satisfactory than are those designed for direct-current inputs.
2. Amplify the error signal so that there will be sufficient power to drive a motor which will reposition the potentiometer and the pointer or other indicator.
3. Determine the "phase" or the direction in which the correction must be made by the correcting drive motor.

Practically every potentiometer-recorder of recent design will have to provide, in one way or another, the three steps outlined here.

Automatic periodic standardization is provided by substituting the standard-cell circuit for the thermocouple input and rebalancing. Newer

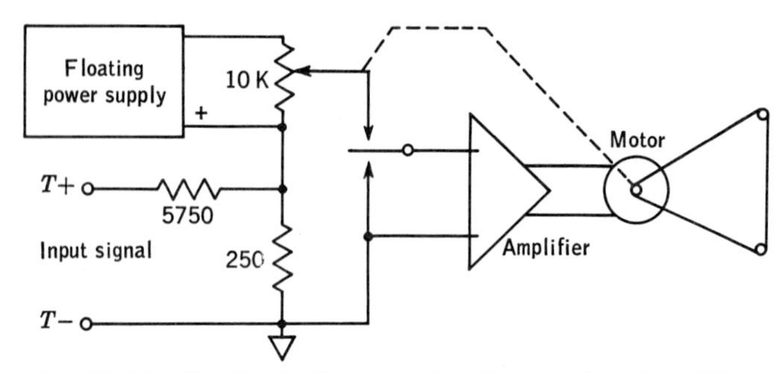

Fig. 17·12 Simplified schematic of a d-c recorder. Any difference between the input signal and the signal obtained from the potentiometer within the recorder will transmit a signal to the amplifier. The output of the amplifier will then drive the motor in such a direction as to equalize the two potentials. The movement of the potentiometer is synchronized with the position of the pointer. (*The Taylor Instrument Cos.*)

designs employ Zener diodes to furnish a constant potential which acts as the reference standard.

The circuit elements used in Fig. 17·6 were vacuum tubes, in part. If transistors are substituted for tubes, then we have circuits much like those shown in Figs. 17·12 and 17·13. It will be noted that they have all of the basic elements of the other circuit.

The output power is transmitted through an output transformer instead of being used directly; this might be expected inasmuch as the transistors are essentially high-current, low-voltage devices, whereas the motors probably were originally designed for high-voltage, low-current tubes. Using standard-voltage motors with a transformer would present less of a problem than trying to obtain nonstandard motors.

The miniature chart recorders and the associated indicating mechanisms are shown in Fig. 17·14.

A somewhat different approach to the problem is shown in Fig. 17·15. In this design, the motor used is essentially a D'Arsonval meter movement. The moving coil is located in the field created by the permanent magnet. As the current in the coil varies, a torque which is directly proportional to the current is produced. As a result, the coil changes the length of the calibrating spring until a condition of torque balance is again established. As the movable coil changes its position, it also repositions the pen with which it is mechanically linked.

Strip-chart multiple recorders. It is just as desirable to have multiple indications with strip-chart devices as it is with circular-chart units. However, it is more difficult to have more than two variables being registered continually. Strip-chart instruments generally are limited to having one indicator above and one below the instrument scale. In order to pro-

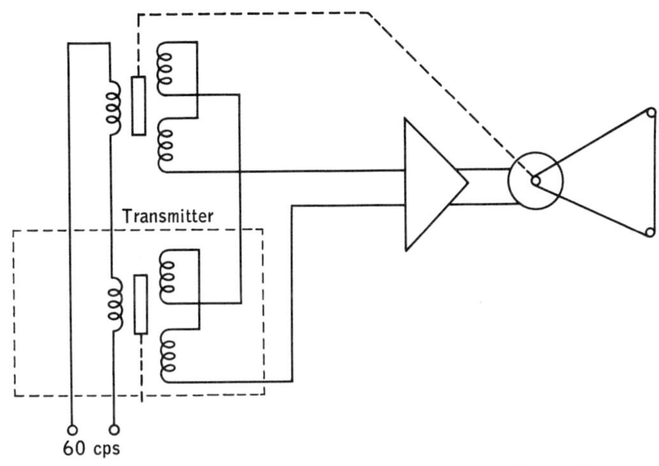

Fig. 17·13 Simplified schematic of an a-c recorder. When an alternating-current potential is supplied to the recorder, it must be balanced by another alternating-current potential, which in this case comes from a differential transformer whose position is linked with that of the pointer or pen. The motor will position the recorder's differential until the two potentials are equal. (*The Taylor Instrument* Cos.)

vide for multiple registrations of variables, there is synchronized switching of the variable to the potentiometer circuit, with rotation of a printing wheel which carries numbers or some other means for identifying the variable. When the potentiometer has achieved balance, the marking

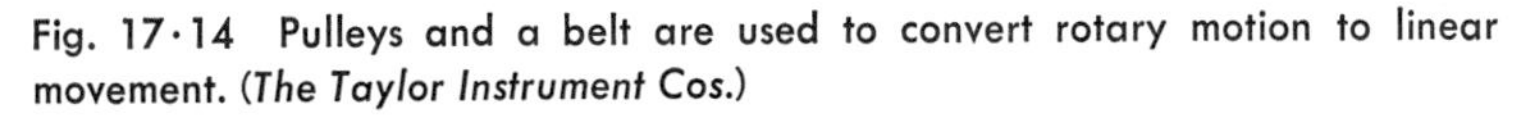

Fig. 17·14 Pulleys and a belt are used to convert rotary motion to linear movement. (*The Taylor Instrument* Cos.)

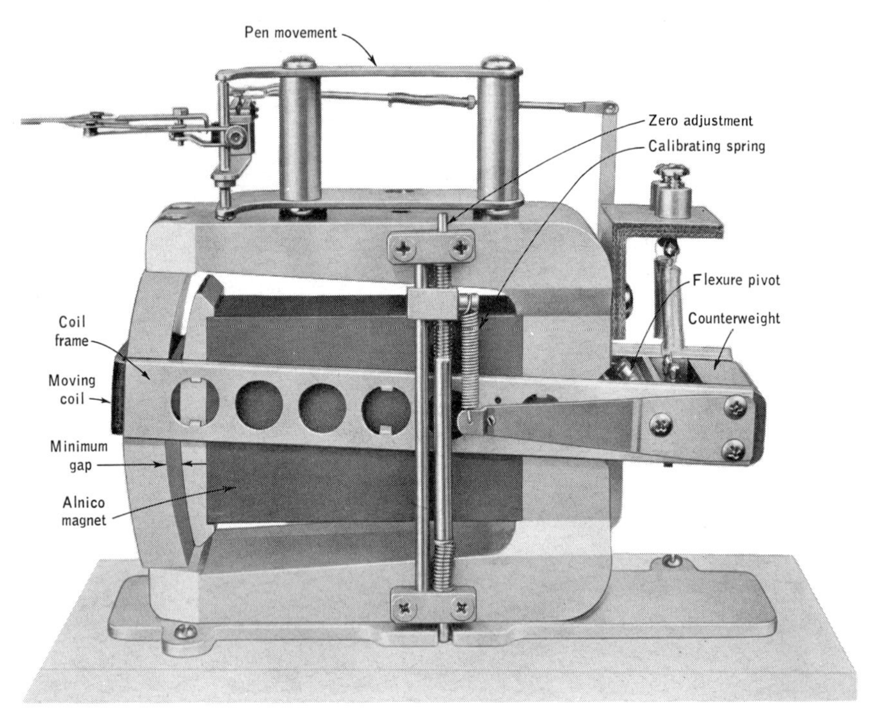

Fig. 17·15 A very heavy duty type of meter movement may also be used to position the pen. (*The Foxboro Co.*)

wheel with its symbol properly positioned for printing is struck against a ribbon which makes an impression on the chart in exactly the same manner in which a typewriter works. This type of indication means that the variables are monitored at a predetermined rate and in definite sequences. It might be said that this intermittent type of evaluation and printing is the earliest of the scanning systems.

When only a few variables are to be logged, it may be practicable to use a series of colored ribbons instead of one ribbon with a positioned wheel. Such systems synchronize the color of the ribbon with the variable.

Pneumatic strip charts. It should not be presumed that the miniature strip-chart recorders can be operated only by electrical means. It is possible to convert pneumatic or other pressure signals so that they too will operate the strip chart. Previously the applications of pneumatic and other pressure signals had been chiefly in the field of the circular-chart type of device. One design by Taylor Instrument Companies is shown in Fig. 17·16. Some designs use a bellows or Bourdon tube and a long pen arm to approximate straight-line movement. In the device under study, the indicating and recording mechanism is driven in a vertical direction while the chart moves horizontally.

One of the major problems in applying pneumatic techniques to graphical presentations of a rectangular nature is to get adequate movement of the chief element. For example, there must be enough force and motion so that the gear trains and other mechanical devices can be rugged and durable. In this design we find that a special diaphragm is used to permit a long stroke for the piston. This provides sealing over a wide range of travel; yet the diaphragm is limp and exerts virtually no restraining force on the operation of the unit.

With certain reservations, it might be said that this is nothing but a pressure transmitter similar to those discussed in Chap. 9. The chief difference is that the conditions of torque balance are produced by different kinds of motion. In the pressure transmitter, there is a beam which is made to pivot about a point but there is very little movement; in this design the motion is great enough to produce rotation of the clutch pulley and to drive the pointer.

As in the pneumatic transmitters previously considered, the input signal, regardless of the source of the pressure, positions the baffle with respect to the nozzle. This then controls the pressure of the signal applied to the diaphragm which pushes on the top of the piston. If we assume that the pressure being measured is increasing, then the baffle will be moved closer to the nozzle and an increased pressure will be applied to the top of the piston. This will compress the main spring and consequently decrease the pull on the flexible strip B, which transmits to the shaft the movement of the piston. Also connected to this shaft, which carries the clutch pulley, is a second flexible strip A, which is wound around the shaft. Movement of the shaft is thus transmitted through the flexible strip to the range spring, which is in turn connected to the baffle. The connections of the components are so made that as the piston is driven down by increasing pressure, the rotation of the shaft will apply a greater force to the range spring to move the baffle away from the nozzle. As a result, a balance point is reached when the pressure on the piston just balances the upward force of the main spring. At this point the movement of the shaft ceases. There is nothing new in the principles used. This device merely uses a modification of the mechanical elements to achieve the same sort of pneumatic feedback necessary for conditions of torque balance.

Graphic panels. As industrial processing has become more and more complex with increased numbers of steps, it has become more and more difficult to train personnel. The function of each instrument and its place in the process has to be learned. The interrelationships of the various steps are also important, because a variation in one place may mean a severe deviation some other place.

By concentrating and arranging the instruments on a panel of the type shown in Fig. 17·17, the function of each instrument becomes apparent and supervision becomes better. The time required to train operators is reduced, and the possibility of becoming confused as to the significance of some deviation is reduced. As useful as such a graphic panel is, there are conditions which have initiated still further changes:

1. Whenever the process is changed, either through adding some new step or through eliminating a step, it is rather difficult to modify the panel.

2. Panels of this type have become so large as to be unwieldy and too expensive.

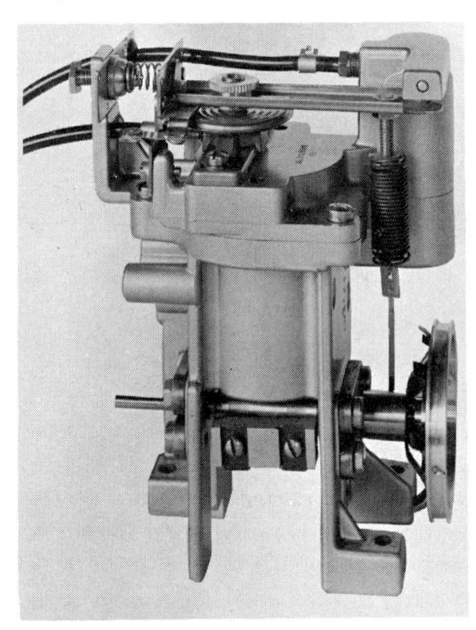

Fig. 17·16 A strip-chart indicating-recorder which accepts pneumatic inputs is a relatively new device. There is mechanical feedback from the output shaft which ensures that the pointer position accurately represents the value of the input signal. Details of the recorder are shown below and on page 587. (*The Taylor Instrument Cos.*)

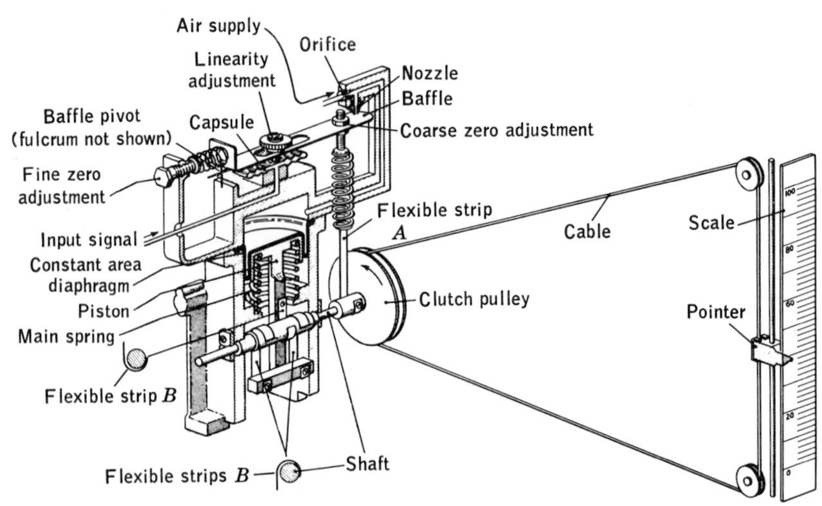

3. If one relies on visual observation to note deviations from the set point, the preoccupation of the operator with some other task may delay recognition of the deviation.

4. Significant and coordinated data are difficult to obtain from so many diverse instruments; it might be necessary to reduce the data from 10 to 30 charts, for example, to produce a composite result of the operations. This would be both costly and time consuming; moreover, it would furnish the data hours or days after the event, and perhaps too late to be of any assistance in determining how the process could have been adjusted or improved.

One of the major factors in the recent development of the miniature indicators and recorders has been the development of the graphic panel. At the same time, there have been concentrations of instruments in common control points which involved such numbers of instruments that the value of the space required was prohibitive. The desire for the use of more indicating and controlling devices, the concentration of greater numbers of units at some common center, and the desire to improve the performance of the process operator by providing representations of what is happening at all points in the process made it imperative that miniature representations be employed.

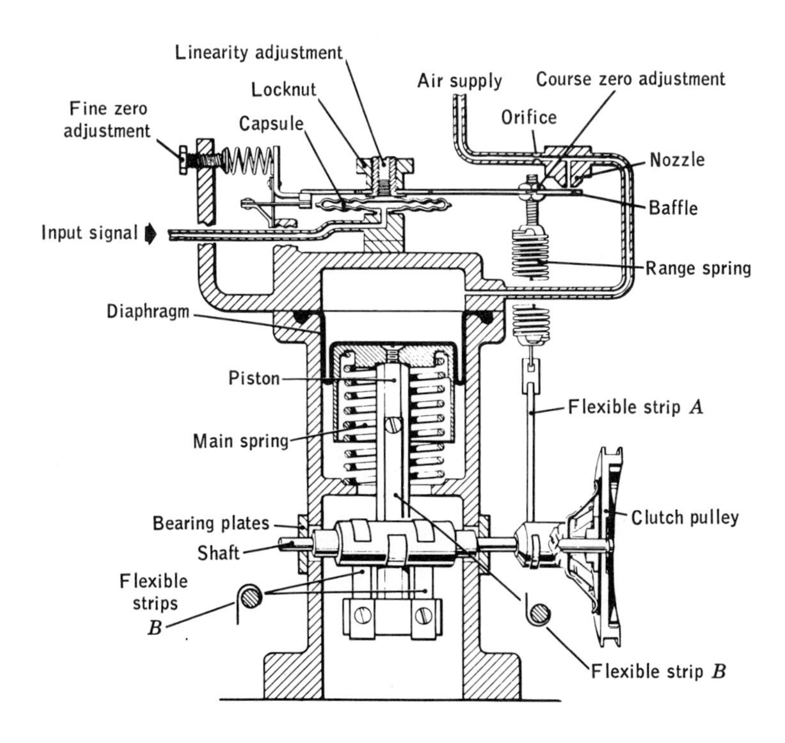

Fig. 17·17 (*a*) A graphic panel has the instruments mounted in a flow diagram. (*b*) A modern control center. (*Minneapolis-Honeywell Regulator* Co.)

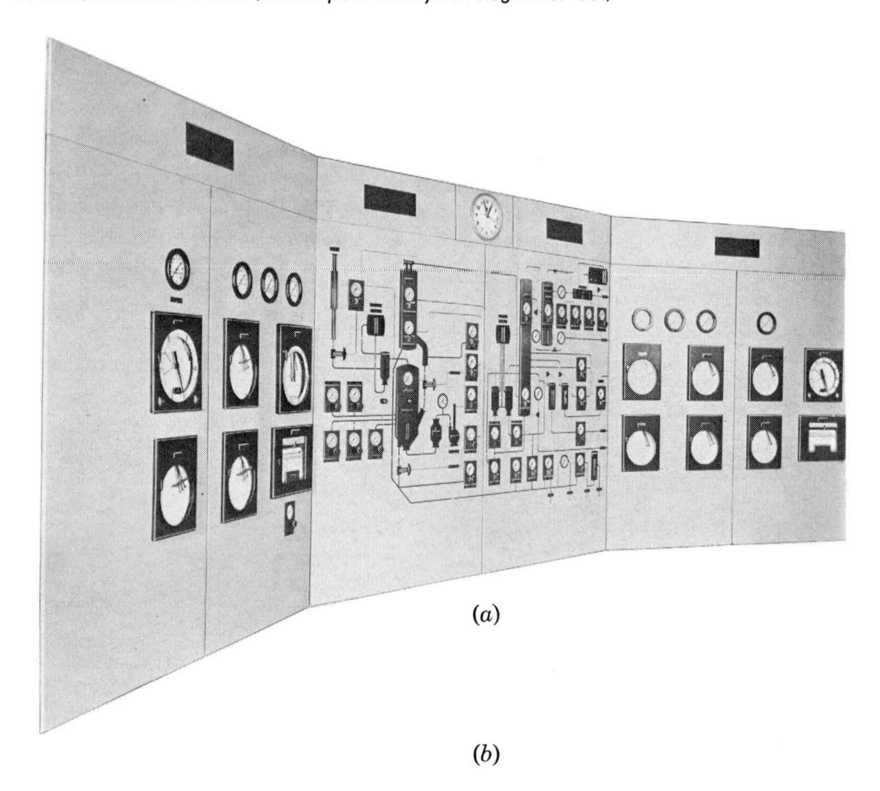

(*a*)

(*b*)

In any such system, there are somewhat conflicting points of view. From the operator's viewpoint, the graphic panel system should be a true representation of the process in terms of importance, sequence, relative position, and function. Management's point of view probably is that aiding the operator is well enough and most desirable, but there is information that may be required for other departments to simplify internal cost accounting, charges, and so on. There are also certain phases or steps of a process which take precedence over any other phases. For example, in a particular process, it may be imperative that the pressure at some particular point be maintained without regard to emergencies at any other point, or it may be that the flow of cooling water is more important than anything else. Any instrument system, then, will have to reflect these conditions.

As a consequence of these conflicting requirements, there are several systems for presenting data. The *full-graphic panel* was the first of the general design, Fig. 17·17. Now there are the scanning and data-logging systems which employ still more sophisticated means for collecting data, recording it, and sounding alarms when the variables exceed predetermined limits, Fig. 17·18.

Fig. 17·18 A center that concentrates all control information for more efficient and economical process operation. A complete off-normal condition would follow this sequence:

1. Audible alarm sounds. Off-normal signal lights appear at graphic station and on control panel. Visual read-out of time, value, and station appears and is printed on auxiliary column printer.
2. Supervisor presses acknowledgement button. Audible alarm is silenced. Lights remain on, and printer can continue to print off-normal condition each time station is scanned.
3. Signal lights draw attention of supervisor to instrument or jack plug associated with the off-normal variable. Recorder station may be plugged in to record trend.
4. On return to normal, signal lights are extinguished. Printer can be set to print confirmation of return to normal conditions. (*The Taylor Instrument Cos.*)

It will be noted in Fig. 17·17 that this device provides an easy way for the operator to associate each instrument with its function and just where it fits into the operating sequence. This is in marked contrast to older designs for which the operator must know precisely what the function of each instrument is. Until he is fully trained and conversant with the operating system, lack of familiarity could produce undesirable results.

Because process changes are made from time to time, there are some who feel that the full-graphic panel does not lend itself well to alterations in the process sequence or modifications in the product itself which require different instrumentation. This situation has produced the second round in the development of the graphic panel, Fig. 17·18. With systems of this sort, it is possible to scan or observe the value of each process variable as frequently as one to five times each second. So long as the values are within acceptable limits, the scanning continues. In some installations, provision is made for a typewriter to record at specified intervals the value of each variable. In this way, a record is kept automatically. But if the process should deviate from the desired set point, then the following normally would occur:

Audible and visual alarms would be actuated.

The value of the variable would be indicated, and the station would be identified.

The off-limit values would be printed in a special column or indicated by some special color.

The recorder can be connected to indicate the deviations of that particular process so that trends can immediately and easily be noted.

Such control points are actuated by any type of variable measurement: temperature, pressure, level, or flow. Whether the signal is electronic or pneumatic or hydraulic is of little significance.

Alarms such as flashing lights, and other means for showing that the limits for the variables are being exceeded, are now being grouped at the proper point on the panel, but the indicating and recording units are being grouped on both sides of the panel. In this way, any changes in the process or other conditions require only the addition or deletion of painted flow lines and the substitution of one set of alarm units for another. Inasmuch as all the indicating and recording units are located at one end of the panel, there will be no extensive changes in the piping or wiring.

Certain instrumentation groups feel that while a process is functioning properly there is little or no need to accumulate voluminous amounts of data. When the process begins to deviate is the time to record the performance in order to determine the cause, if possible, of the upset. Recording will indicate trends and tell whether the corrective steps have been effective. A graphical chart is much easier to evaluate quickly than is a column of figures.

DATA LOGGING

Certain of the data-logging centers are designed to record the performance of every variable at stated intervals, perhaps every fifteen minutes or half hour. If during this time one of the variables deviates beyond the acceptable limits, the divergence will be found as the unit makes its periodic check of conditions. Some designs provide for a scanning rate of 5 stations per second; others check at the rate of 30 points per minute; and so on. Consequently, all parts of the process are kept under constant surveillance, but until there is an undesirable deviation, values will be recorded only at predetermined times.

Scanning instruments and systems. Although the operation of the multipoint recorder or indicator, either automatic or manual, constitutes a form of scanning, this is not what is suggested by today's use of the term. Instead, fairly high speed automatic monitoring of each of the variables is presumed. Along with automatic alarms, there may be means for recording, at will, any deviations from the set point, and perhaps even for the conversion of the value of the variable into digital equivalents.

Our study might well start with a very simple system shown in Fig. 17·19. By means of a manually operated switch, one may determine the

Fig. 17·19 By using high resistance values in the sensing element and in the bridge, it is feasible to cause bridge unbalances of the order of 5 volts. An absolutely dependable voltage source is needed if the deflection is to be a true measure of the temperature of the sensing element. Manual selection of the element is quite adequate if only a few readings are required at infrequent intervals.

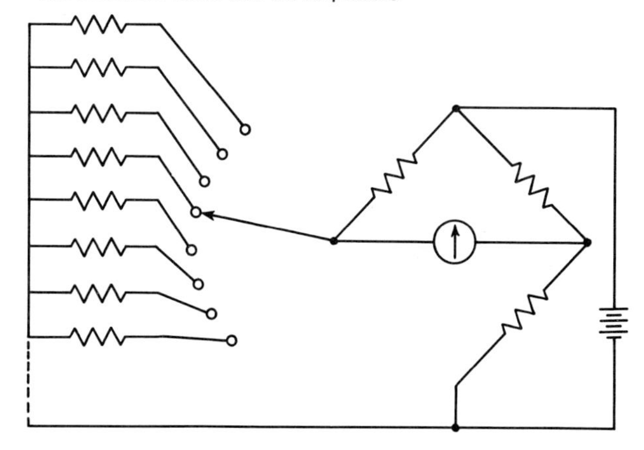

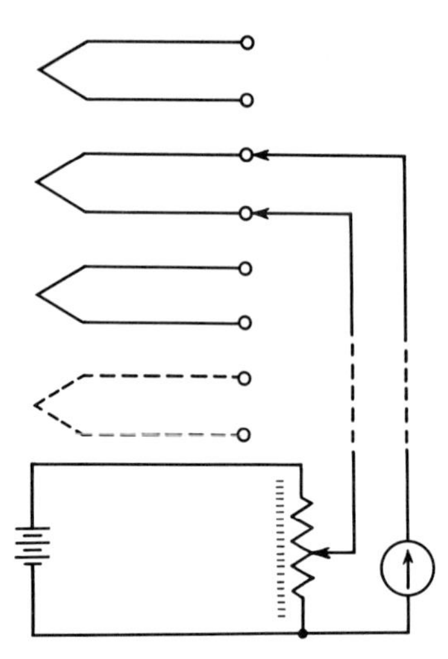

Fig. 17·20 Thermocouple outputs can be scanned manually when only a few values are to be determined. The potentiometer is adjusted manually until the galvanometer reads zero, at which time the thermocouple and potentiometer emf's are equal.

temperature of any one of 20 points, using resistance thermometers. Full-scale output of the bridge is about 5 volts, so any variables which can be converted into resistance equivalents could be measured in a similar manner. In a similar manner, one may select any one of a number of temperatures based on thermocouple outputs, Fig. 17·20. It should be noted that almost all scanning techniques used today depend upon the use of electrical analogs or equivalents.

Inasmuch as such elementary types of scanning are used primarily for monitoring or surveillance purposes, the next logical step might be for the manual operation to be replaced by automatic mechanical or electrical evaluation of each variable as it is scanned or observed. This would require that electrical values be established for every point. Then if the observed value was within the limits, the switch would advance automatically to the next point, etc. Should a variable be outside the established limits, then this would actuate an alarm of suitable type. Such a plan is shown in Fig. 17·21.

If the source of the signals is a group of thermocouples, as is indicated, then there are certain potentials within which each thermocouple output should remain if it is at the set point. The potential which each circuit should provide is then obtained by positioning the contact at the appropriate place on the potentiometer. In the example which we are studying, the thermocouples have been connected in order of their emf. This need not be so, and it is done here because it simplifies the schematic connections.

The stepper relay, or other suitable relay, is so designed that it can complete the circuit to each thermocouple in a series of discrete movements or steps. If we assume that the relay is completing the circuits for thermocouple 3 and its rated potentiometer output, then there will be a circuit to the comparison and alarm circuits. So long as the two emf's are almost balanced, or within the tolerance limits, the stepper relay will advance to the next thermocouple, 4 in the illustration. But if the comparison circuit shows that the output of thermocouple 3 is too high or too low, then appropriate circuit-indicating and alarm circuits must be energized.

Through the use of certain auxiliary timing equipment (not shown) the rate at which the stepper relay searches over each pair of circuits can be controlled. Because of the electrical noise created during the switching operations, the rate of monitoring is held to a rate of perhaps one thermocouple every two or three seconds. The voltages induced by the switching operations make the measurement of small voltages rather difficult. With higher signal voltages, the electrical noise becomes fairly insignificant and the switching rate can be increased.

Fig. 17·21 A very simple type of scanning system which can be made to search over the thermocouple circuits at will or according to some prearranged schedule. These potentials which are outside acceptable limits will then cause alarms to be set off or some corrective action to be taken.

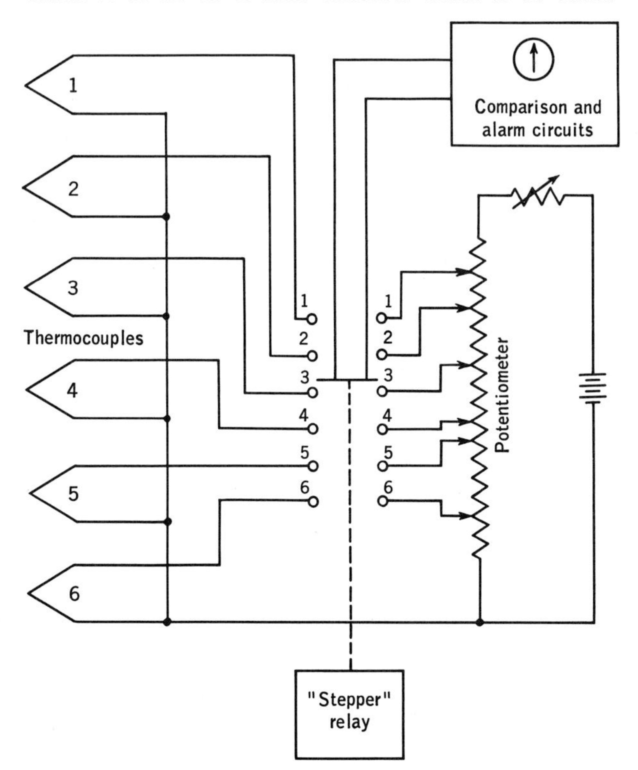

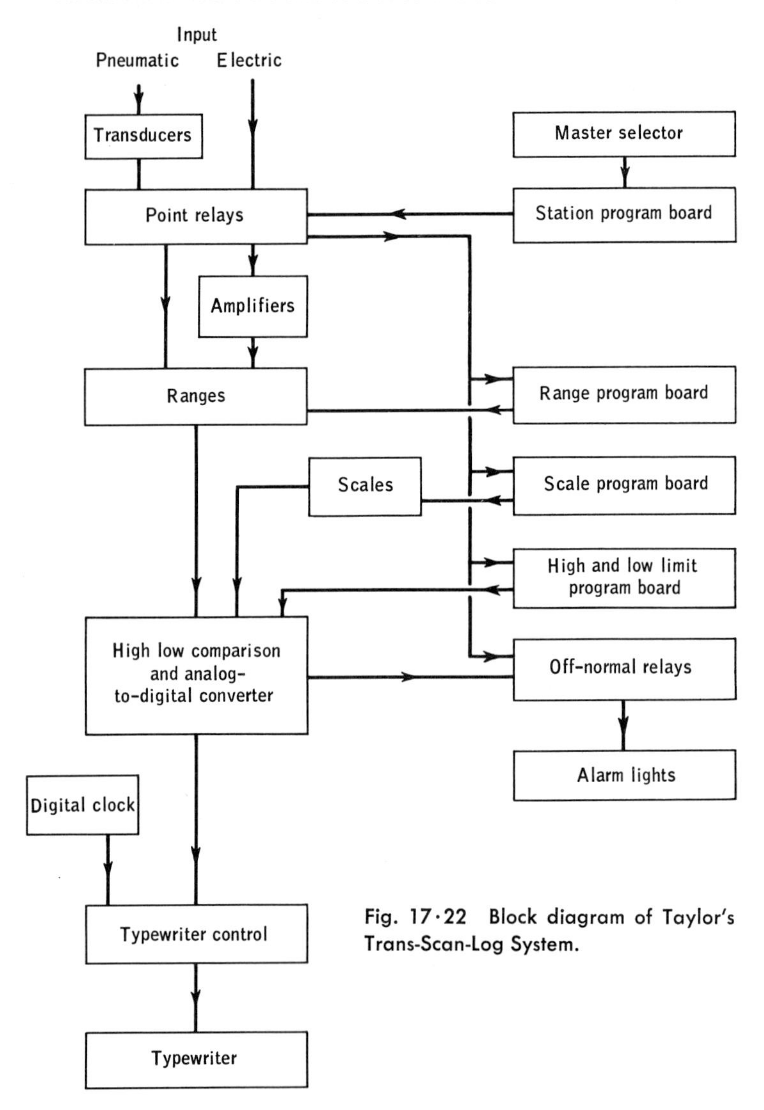

Fig. 17·22 Block diagram of Taylor's Trans-Scan-Log System.

There are many ways of scanning and logging data while controlling the process. There are the manual methods, which require that the operator record the values of the variables, and there are the automatic methods, which carry on a constant surveillance of all points in the process to verify that all functions are proceeding within their permissible limits. With either manual or automatic scanning, the purpose is to determine at the earliest moment:

1. When the process is departing from its desired values
2. What can be done to bring the process back to its correct condition

All automatic scanning and logging processes will have about the same major components as are listed here for the Taylor Trans-Scan-Log System. Some systems will use other means for comparing present values with the desired set-point values; others will use different techniques for converting from analog to digital values. But in all events, every system will perform most of the functions which are listed and discussed here. The other systems which are also considered will bear out this similarity. Our chief interests then are in the principles used to solve this problem, which is a complex one. It should not be expected that it has a simple solution.

Object of the system. The purpose of the system is to:

1. Enhance process operation by presenting, at some central point, visual and recorded values of all the variables involved in the process.
2. Signal by suitable alarms that excessive deviations exist.
3. Indicate and record (if desired) when variables deviate from tolerance limits.
4. Furnish a visual representation, if desired, of the behavior of certain process variables.

Taylor TSL system. Using Fig. 17·22 let us consider what has to be done to convert an incoming signal into suitable components in order that the system may provide us with alarms, digital read-out, and typed records of the performance.

Incoming signals. The input signals to any such system may consist of three distinct groups:

1. Small electrical signals which are direct currents with a potential of 1 volt or less. (The potentials created by thermocouples, bridges, or similar types of equipment.)
2. Alternating-current signals from any of the various transducers.
3. Pneumatic signals which must be converted into electrical signals before they can be amplified, Fig. 17·23.

Point relays. Each incoming electrical signal or the output of one of the various transducers is connected to a separate relay which can be actuated as desired. By connecting the point relays to the rest of the circuit in a definite sequence, each variable can be evaluated in the desired sequence, and only one signal is being evaluated and judged at any one instant. At all other times, the point relay is not operated and there is no connection between its particular variable and the rest of the system.

The master selector. There must be a control and sequencing control to present in the desired sequence the various points being monitored. In addition, there should be a visual indication of the point being compared at any one moment. Provision must be made for selecting any desired point on demand. These three functions are performed by the master selector. There may be a stepper relay to select the various points, or there may be any one of a number of relay selectors.

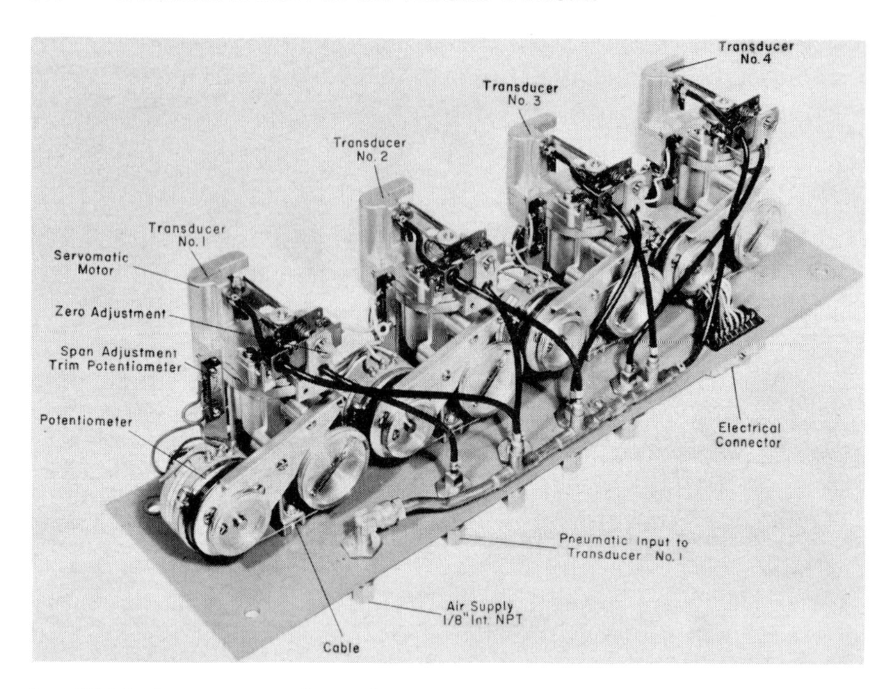

Fig. 17·23 Pneumatic to electrical conversion may employ units such as this. (*The Taylor Instrument Cos.*)

Station program board. The station program board is used to simplify programming or changes in future programs. Whereas the connections and lines to each variable are quite fixed, the use of such a board makes it possible to interchange electrical leads without disturbing any of the physical components. It is at this point that provision is made for skipping certain points—for certain of the less important points, only an occasional reading is desired. We will assume that the values of the various points are being recorded only once each quarter hour, half hour, and so on. In this instance, the value of this particular point will be determined only during the period of regular logging. Its value can, however, be measured at the will of the operator. Not only are some points evaluated only occasionally, but there are other points for which no upper or lower limits are set. Such a system does not require that all points be accorded the same treatment, but each one is programmed in accordance with the information desired.

The range unit. Because all measured spans do not start at zero, it is necessary that the proper voltage be established with correct upper and lower limits. When the values of the variable drop below zero, it is necessary to divide the span into two parts with provision made for reversing the polarity of that section which is measured below zero. This

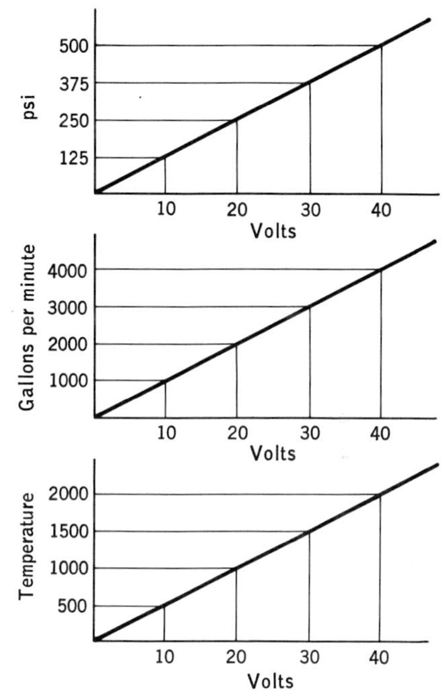

Fig. 17·24 For analog to digital conversion the variable may be converted to a voltage.

reversal is necessary because the amplifier section in the analog to digital converter will accept signals of only one polarity.

Analog to digital converter, Fig. 17·24. In this unit, the various electrical inputs are converted to numerical, digital equivalents. For example, if we consider only the signals which have been converted into pneumatic signals for transmission, they have a uniform span of 3 to 15 psi. They have these values whether the variable has a maximum value of 10 units, 100 units, or 500 units. It is at this point that we correlate 12 psi, for example, with 300 units, 500 units or some other value. It must be clearly understood that electrical signals of equal voltage values do not mean equal numerical values for the process variables being monitored and controlled, Fig. 17·25. At this point, each variable is connected to the proper dividing network which will translate the 40 volts, for example, into 2500°F, 5,000 gps, or 100 psi. The number of dividing networks must provide for each group receiving numerical conversion.

Let us assume that 40 volts will always be the maximum value to be applied to the potential-conversion networks. If we are measuring temperatures which have maximum values of 2000°, then the thousands potential will be divided into two equal parts of 20 volts each. But if a maximum pressure of 800 lb is to be converted into digits—let us assume for purposes of this discussion that the equipment is set up to provide an answer accurate to four figures—then we have to divide the 40 volts into eight equal parts.

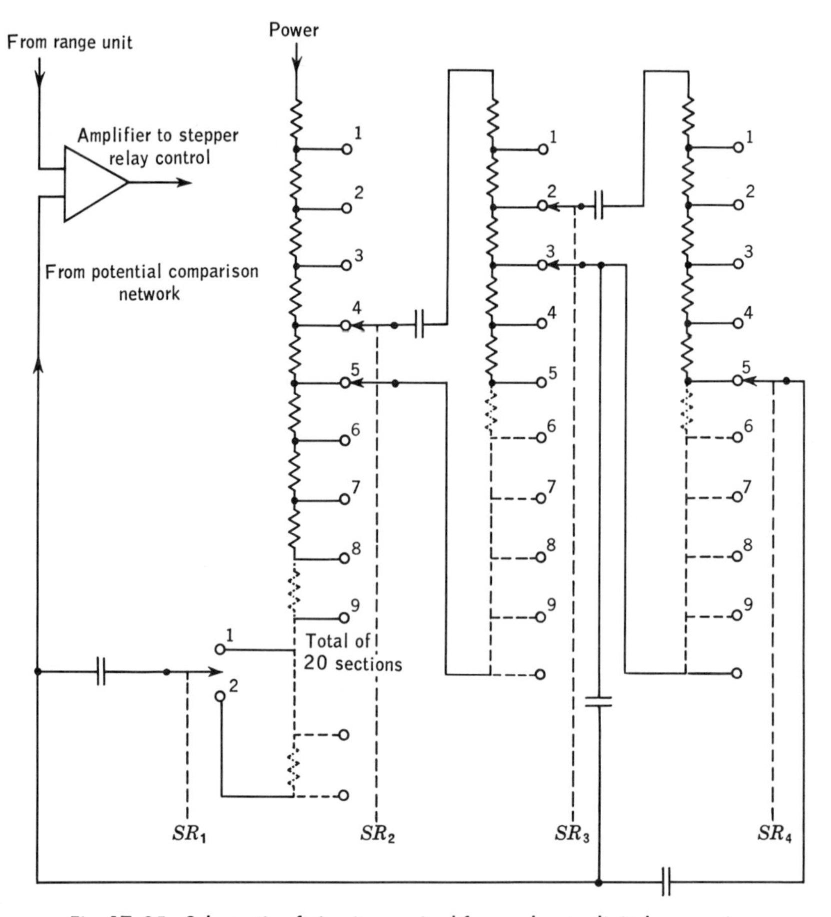

Fig. 17·25 Schematic of circuits required for analog to digital conversion.

For the sake of illustrating the circuit provided in Fig. 17·25, a potential of 24.24 volts is to be digitized for the condition that 40 volts is equivalent to 2,000 units. Stepper relay SR 1 will be actuated first, moving to point 1, and the input will be fed into the comparison amplifier. Because point 1 is the mid-point of the network, it will feed half of 40 volts, or 20 volts, to the amplifier. Because the incoming signal is greater than the comparison value, the stepper relay receives another impulse which takes it to position 2. Here the network potential is greater than the incoming signal, so stepper relay 1 will stop because it has determined that the first digit will be a 1 with the incoming signal having a value greater than one-half of 2,000 units. When this relay is deenergized, it will establish the path for stepper relay 2, which moves in successive steps from position 1 toward 10 until it finds a position where the potential which goes to the comparison amplifier is greater than the 24.24 volts provided by the signal. It comes to a stop with its pair of contacts bridging terminals 4 and 5, so that we know that the signal has a value of 24. When the signals to the comparison amplifier reverse their polarity because the comparison unit provides a signal in excess of that required to balance the incoming signal, an impulse starts stepper relay 3 on its way. In this instance, we find that it stopped when the pair of contacts bridge points 2 and 3, so the value of the signal is now noted at 24.2+. The sequence then repeats itself with stepper relay 4 stopping on position 5, which means that the value is between 4 and 5. But the device cannot be accurate to better than 1 unit. So with the circuits which we have considered here, there will be uncertainty of 1 in the last digit.

In a similar manner, any set of values may be converted into digits. The positions of the stepper relays when they stop provide the means for actuating the electric typewriter which will record the digital values. At the same time, circuits which will furnish the lighted digits for the visual read-out are completed.

It should be pointed out that if a signal of 40 volts is to be split up into 10,000 units (9,999), then the potentials applied to each successive network must be chosen in a somewhat different manner if an accuracy of one digit is to be obtained.

When only a series of pneumatic signals is to be logged and digitized, then a somewhat simpler system may be established. Instead of using individual pressure transducers to provide the electrical analog, Fig. 17·26 shows how the individual pressures may be sampled in sequence,

Fig. 17·26 The master selector determines the sequence and the time for the cycle. To change the sequence or the time does not require any change in the piping; it requires only a modification of the electrical circuits in the selector.

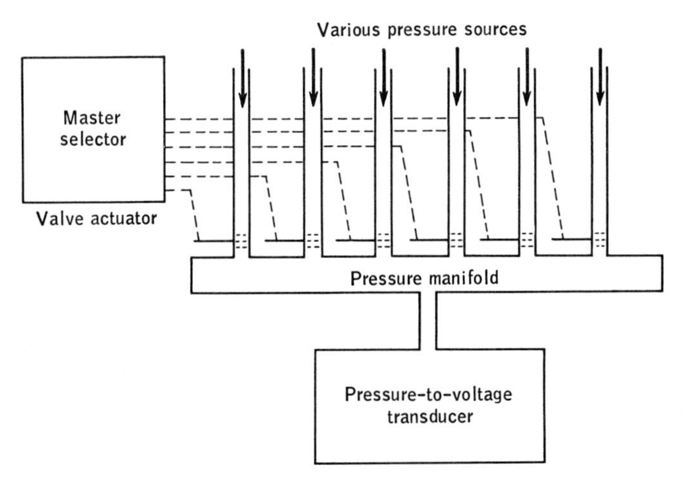

with each successive pressure being transmitted to the pressure transducer which will provide the proper electrical values to be evaluated and compared in the rest of the circuit.

OTHER SCANNING-LOGGING SYSTEMS

A different approach to the problem of presenting temperature, flow, level, and pressure measurements in digital form, either as lighted numerals or as typed values on a log sheet, is employed by Hagan Chemicals and Controls and illustrated in schematic form in Figs. 17·27 to 17·29. It will be observed that all of the steps previously studied are repeated in this design. What is of most interest and value is the difference in the means used to obtain the same end. Once again we are primarily concerned with the principles involved and what we can learn from them.

Hagan Kybernetes system. The various steps may be followed in about the following sequence (see Figs. 17·27 and 17·28):

1. Electrical inputs from thermocouples, strain gages, and so on
2. Inputs from pressure, level, flow, and so on which must be transduced into suitable voltage analogs
3. Point selector switch
4. Amplifier

Fig. 17·27 Most scanning systems require that the input signals be converted into electrical equivalents because this simplifies the selection and sequencing operations. (*Hagan Chemicals and Controls, Inc.*)

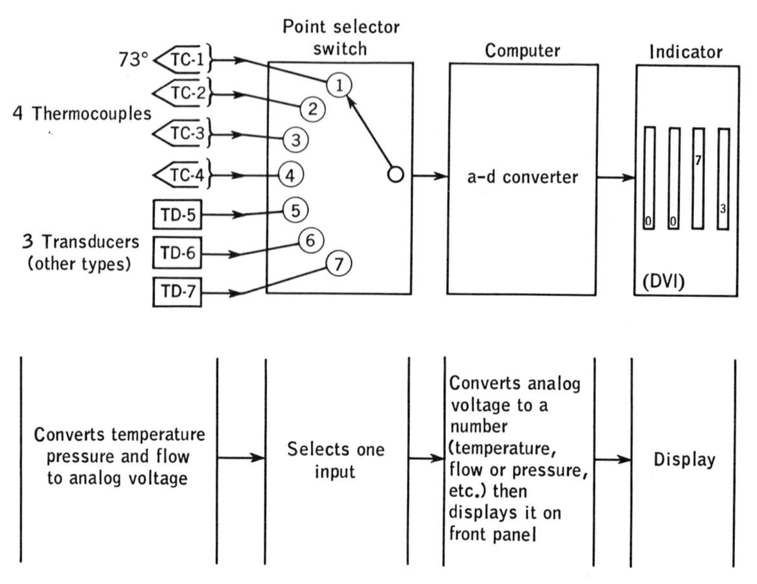

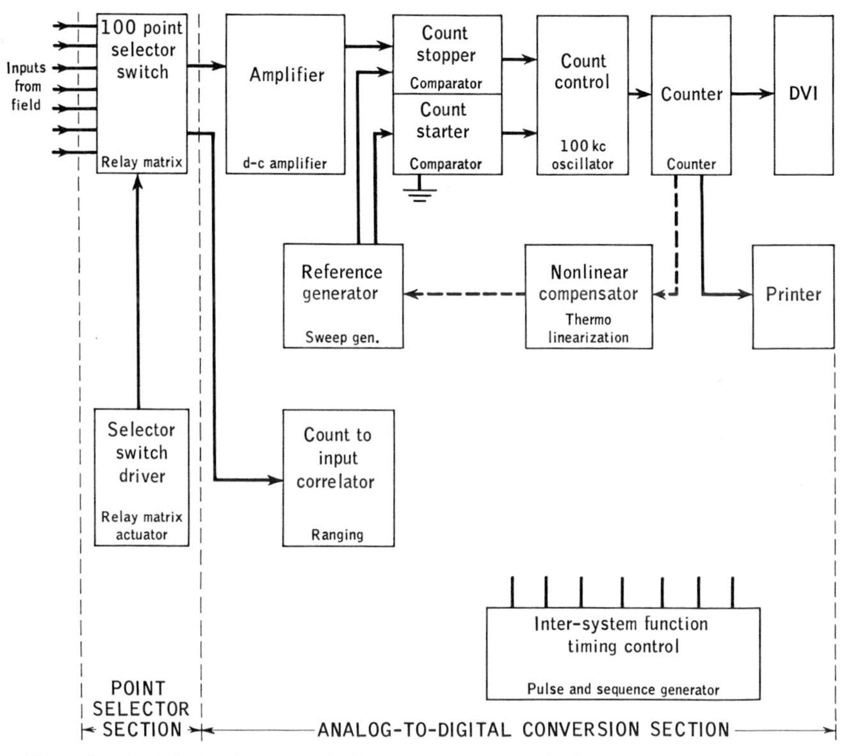

Fig. 17·28 Block diagram of the components and the sequences as a signal completes its passage. (*Hagan Chemicals and Controls, Inc.*)

5. Counter section
6. Alarm
7. Printer-visual indication

The various elements required for each one of these sequences will now be examined. The electrical inputs may be from any suitable source and will in almost every instance be relatively small in value, varying from a few microvolts to a small fraction of a volt. The maximum value of the variable signal is not of too great importance inasmuch as stable high-gain amplifiers are available to boost the variable by a factor of 100 or 1,000 to bring the value to a suitable value for comparison.

Point selection switch. There are many ways of connecting a large number of input circuits to a single amplifier; one may use the stepper switch which we have previously considered, others may use crossbar switches, electronic sampling devices, or high-speed mercury-jet switches. Still others use a matrix of relays.

A small relay matrix is shown in Fig. 17·30, where only nine relays are involved. It will be noted that by closing only two switches, one in

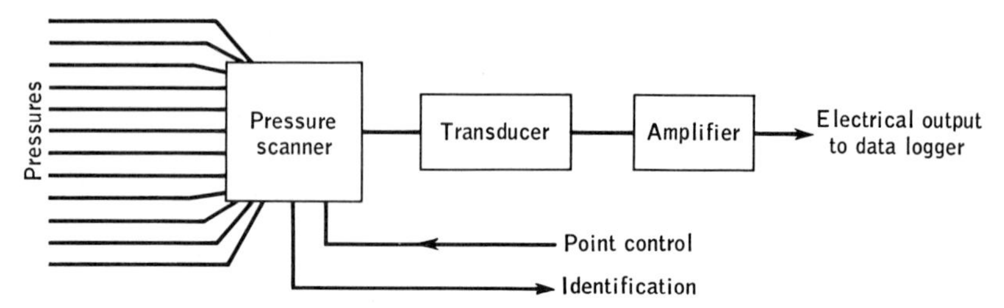

Fig. 17·29 Pressure-scanning system. When only one type of variable, such as pressure, is to be measured, there need be but one type of transducer. This greatly simplifies the selection and amplifier sections because of the similarity of the output voltages. All signals are likely to be volts or millivolts, not mixtures of these values. (*Hagan Chemicals and Controls, Inc.*)

the tens row and one in the units row, that we may select any one of the nine relays indicated. If we were to extend the number to 10 rows of 10 relays each, then closing any two relays could again select any one of the 100 relays in the matrix, Fig. 17·31. Consequently, our programming system must be set up to actuate each tens group at the proper time, then each successive digit in sequence, and return to the starting point when all the points have been evaluated. The sequence is then begun again.

Fig. 17·30 A selection matrix requires the operation of a particular relay to obtain specific operation or selection. When 100 relays are arranged in such a matrix, it requires the operation of but two switches, or other relays, to select any one of the 100. One would be the appropriate tens switch, while the other would be the desired units switch. (*Hagan Chemicals and Controls, Inc.*)

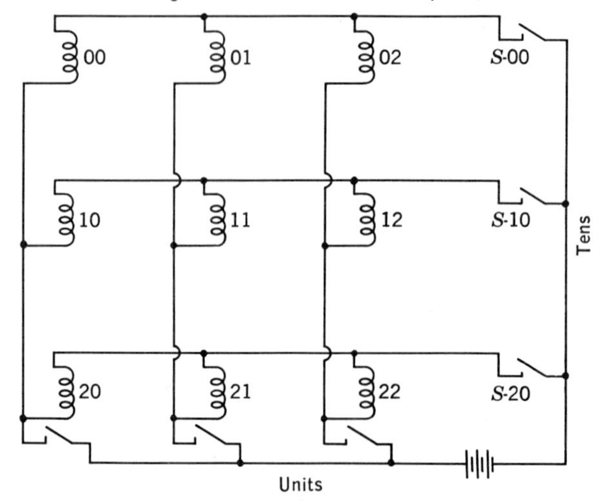

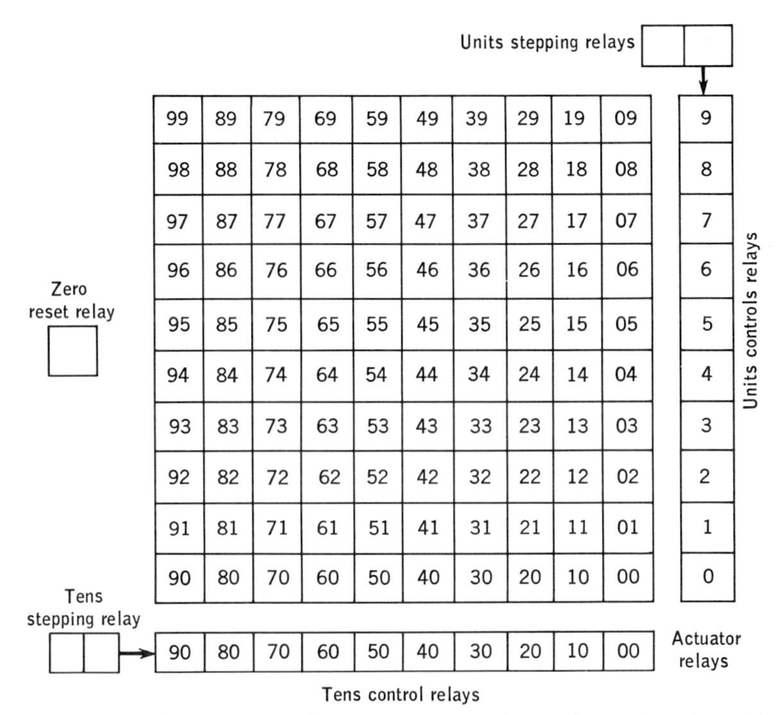

Fig. 17·31 The operation of a tens stepper relay and a units relay will select any of the possible 100 relays in the matrix. (*Hagan Chemicals and Controls, Inc.*)

The significance of the zero count must be understood. It will be noted in Fig. 17·31 that "1" does not have the lowest count. Rather it is the one where there is "zero tens count" and no "zero units count." Such an arrangement is in marked contrast to many other systems with which we are familiar. On a dial telephone, the "zero" really represents a "ten." In this system, it represents a logical condition which permits us to select any one of a group of 10 digits or points in exactly the same way.

A study of Fig. 17·32 will show some of the features basic to any similar method of selecting points. When contact C completes the circuit on the even side, relay 0 will become energized through the chain of contacts which are normally closed when the relays are not energized. When it operates, it closes the circuit for relay 1 and opens its own circuit, which will deenergize it. When contact C closes the odd circuit, relay 1 will become energized through the path prepared by relay 0. In such a manner, each of the relays will become energized in turn, with each preparing a path for the next relay. By giving these relays slight operating delays when current is applied and giving them similar delays in dropping out when their circuits are opened, each relay can be made

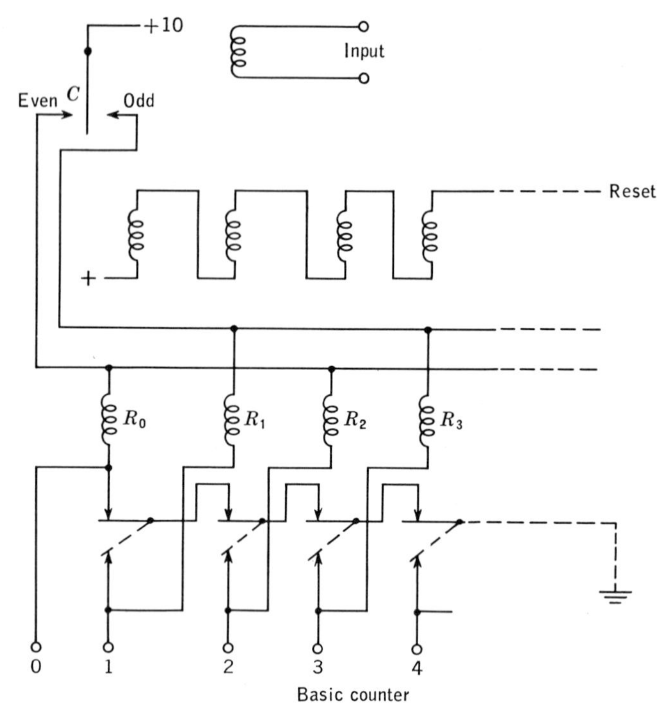

Fig. 17·32 Relay counters such as this require that the pulse advance the system but one step and that each step prepare the way for the next relay to be operated.

to take specific times of total operation. Periods of time of a small fraction of a second to several seconds are possible. Inasmuch as five points can be scanned each second, the time delays are quite nominal.

A slightly more complex network is shown in Fig. 17·33. If the battery is applied to the matrix, with all of the relays deenergized and normally closed contacts as shown, relay 0,0 will pick up first, followed in succession by each successive relay in its column. When relay 0,9 is reached, it will open the circuit to relay 0,8 and at the same time, complete the circuit to relay 1,0, which will then operate and initiate the operation of each relay from 1,0 to 1,9. This sequence will then be repeated until every relay has connected its variable with the input amplifier. As soon as the last relay 9,9 has operated, it sets up the circuit for the first relay 0,0 to initiate the performance all over again.

If we follow the simplified circuit shown in Fig. 17·34, we should find almost the same situation. If we start with the zero count, there will be a circuit available for only relay 0,0. As soon as relay 0 has opened the circuit, its own coil forces the coil current to pass through relay 0,0 as well; the two coils are then in series. Relay 0 has in the meantime pre-

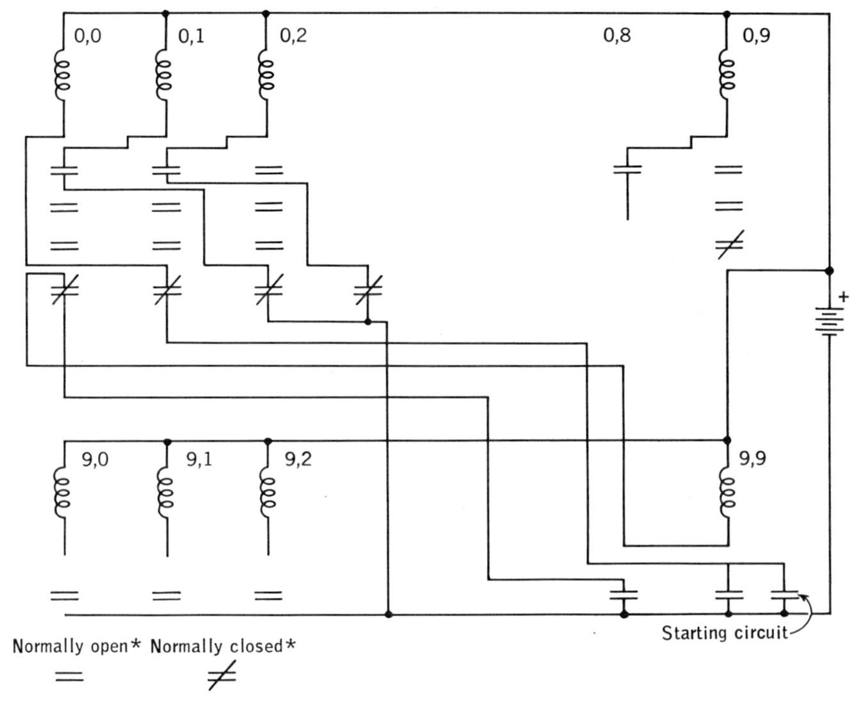

Fig. 17·33 Partial schematic of relay system for automatically searching through a series of 100 positions. Each position prepares the electrical path for the succeeding relay. Once started, this sequence would continue indefinitely.

pared a circuit for relay 1, which then repeats when the units drive completes the circuit for the first odd digit.

Amplifier. One very serious problem often occurs when small signals are to receive a large amplification: electrical noise and interference bring into the circuit many signals which are quite undesirable and which may mask the signal which is coming from the variable. Particular trouble is likely to arise when a thermocouple is grounded for the sake of increasing its thermal response. The ground of the thermocouple and the ground of the amplifier are not at the same potential, so this difference signal, regardless of what causes it—and there may also be large stray magnetic fields, leakage currents, and other situations which may create spurious electrical responses—may create serious difficulty. This particular design provides complete electrical isolation between the variable and the amplifier by using a capacitor to sample the size of the signal. As soon as the capacitor is charged, it is connected to the amplifier input where it now creates a signal output proportional to the potential impressed on the capacitor. Such a means completely eliminates spurious signals caused by having grounds at different potentials.

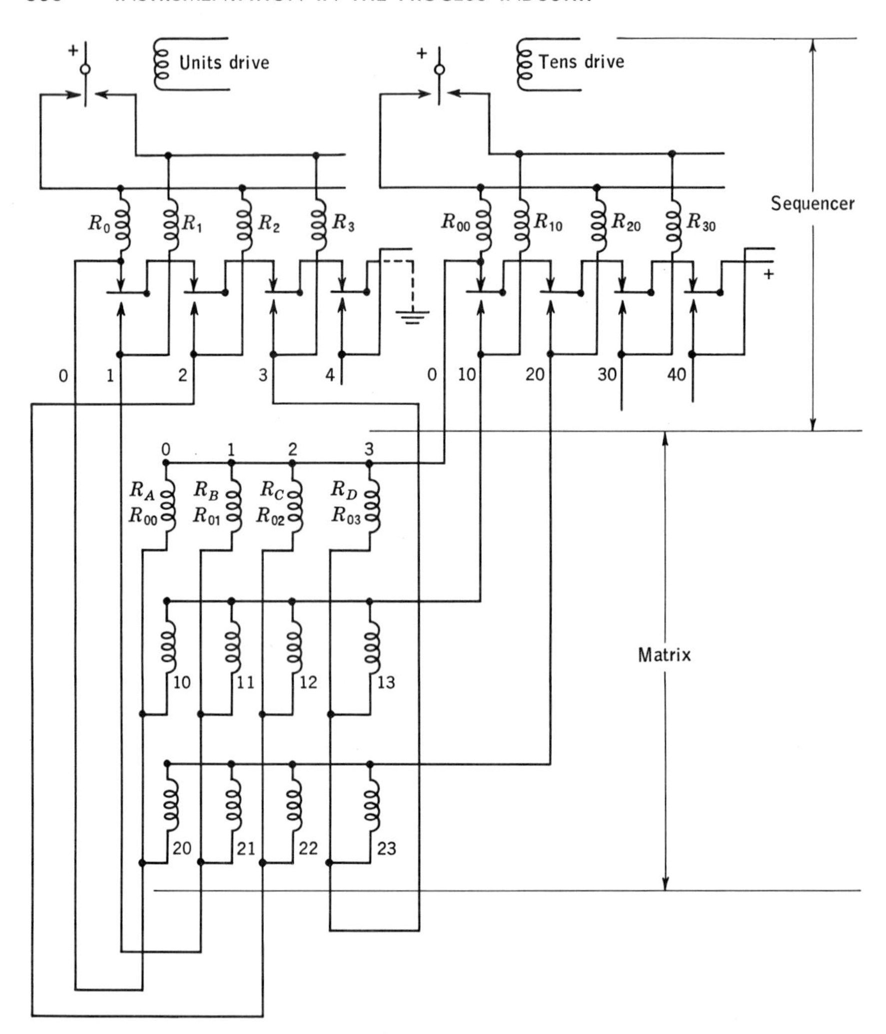

Fig. 17·34 Principles of matrix sequencing. (*Hagan Chemicals and Controls, Inc.*)

Aside from the desire to provide electrical isolation, the chief reason for the amplifier is to increase the signal of the variable to such a value that when it is compared with another signal which is increasing with time and which is associated with the digitizing equipment, the amplifiers which have to judge whether or not the two signals have equal values will not have to evaluate signals in the low millivolt range, but rather will be able to discriminate between differences of 200 to 300 mv, for example.

Counter. In order to convert the electrical analog voltages into digital equivalents, we have to weigh or measure the analog, in this case an electrical potential, in order to determine just how much it really represents. One of the easiest ways of doing this is to add to the analog voltage another voltage of opposite polarity until their sum is zero. This point can be determined relatively easily inasmuch as it involves a reversal of current rather than a measurement of just how much the current or potential may be. There are then at least two ways that we might try to accomplish this:

1. In terms of mechanical balance, we can measure 20 qt, for example, by putting on the other side of our balance a series of pint or quart containers until the two sides balance. In this manner, we can find how large one quantity is by comparing it with a series of known quantities.

We do the same thing whenever we weigh an unknown on a balance or platform scale by counting how many known ounces or pound weights are required to balance the unknown. Or we position a counterweight on an arm to provide a variable moment which is adjusted until it just equals the moment of the load. *We can measure an unknown by comparing its effects with those created by a known quantity.*

Now we might also do it this way: If we add 10 pints/min and it takes 4 min to reach the balance point, then we also have a measure of how large the unknown quantity is. Thus, if we know the rate and the time, we can determine the amount.

2. We might, using the same example as above, be pouring liquid into the balancing side continually at the rate of 5 qt/min. Thus, if we can measure the time accurately and the filling rate remains constant, then we can ascertain correctly the amount of the unknown by noting the balancing time.

The first method is used by some firms, but the second method is the one which is used in the system under study. Figure 17·35 shows, in schematic form, the sawtooth oscillator which by means of its type of feedback through the capacitor furnishes a comparison potential which increases linearly with respect to time, as is indicated. This potential, which starts from zero, is then compared with the electrical analog. The oscillator generates very accurately timed pulses which directly actuate the digital counter; each cycle of the oscillator may correspond to an increase of 1 mv by the sawtooth oscillator. When the emf generated by the sawtooth oscillator "just" exceeds the analog signal, the *stop comparator* will cause the counts going to the gate and the counters to cease. Consequently, if the count stops after 756 cycles, then the value of the analog signal is 756 mv—based on our assumption that the comparison sawtooth voltage is increasing at the rate of 1 mv/cycle. Obviously, longer or shorter counts would then indicate larger or smaller signals to be digitized.

The above discussion is based on the assumption that all relationships

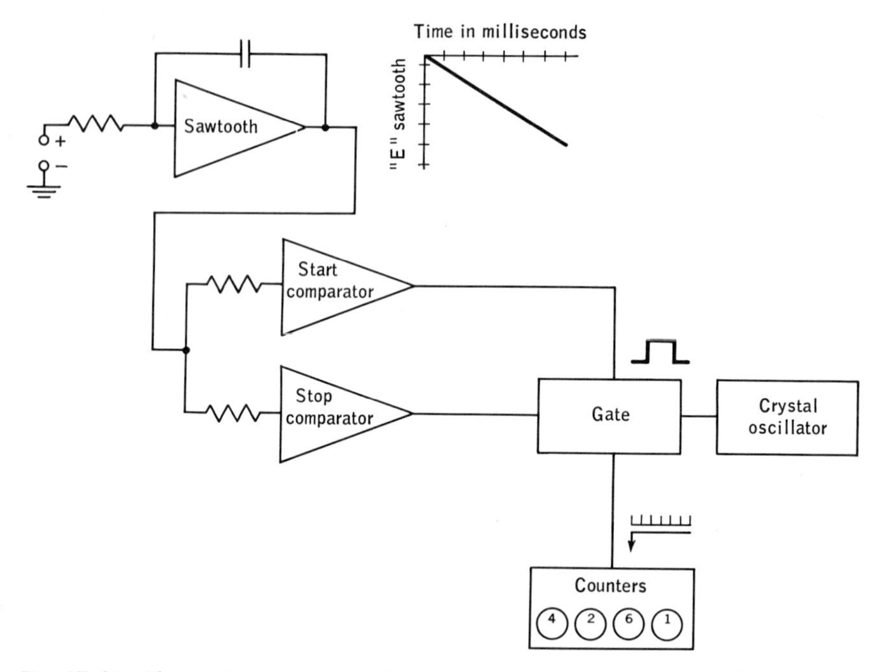

Fig. 17·35 The application of a potential to the sawtooth oscillator will provide a potential which increases at a fixed rate of so many volts per second. Counting the number of cycles created by the crystal oscillator between the time that the sawtooth oscillator output voltage was initiated and the time when its value is equal to that of the variable becomes a measure of the variable. For example, if the output of the sawtooth oscillator is increasing at the rate of 10 volts/sec while the crystal oscillator provides 10,000 cps, and if the stop comparator halts operations after 2,000 counts, it must have taken ⅕ sec to accumulate 2,000 cycles. Therefore, the value of the variable must be the value of the sawtooth output after ⅕ sec or 2 volts.

are linear. But if what we want to know is the temperature and not the number of millivolts created by the thermocouple, then we have a more difficult conversion to make, inasmuch as the potential–temperature difference relationship is far from linear and differs among the more common thermocouples. Figure 17·36 illustrates some of the relationships which might be involved. When there are nonlinear relationships, it is customary to use a series of straight lines to approximate the exact relationship. Thus, in Fig. 17·37 five straight lines can be used to approximate the true values (see also Table 17.1).

By connecting a linearization unit between the counters and the sawtooth generator, the position of the counters can be made to change the rate of the increase per cycle of the comparison emf, Fig. 17·38. If the potential is increasing at a greater rate, then it will take fewer cycles for the sum of the two potentials to become zero; conversely, if the rate

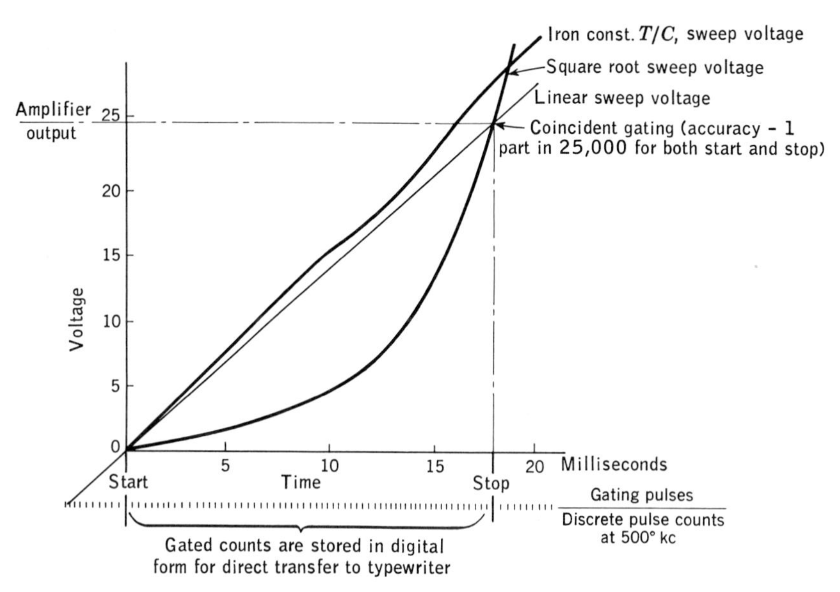

Fig. 17·36 The output voltage of the oscillator can be made to match almost any variable likely to be encountered in process-control work. *(Hagan Chemicals and Controls, Inc.)*

of increase is made smaller, it will take longer for the sum of the potentials to become zero and the counter will run for a longer period of time and thus will show a greater number of units. In this manner, it is possible to convert very nonlinear relationships into numerical equivalents. The

Table 17·1 Relationships of Temperatures to Readings

(Stop pulse time based on linearized, that is, corrected, sawtooth)

Temp, °F	$E_{T/C}$, mv	D-C amp, volts	Stop time, msec	Total count (stop time × 100)
0	0	0	0	0
100	0.3	0.3	1	100
200	0.8	0.8	2	200
300	1.4	1.4	3	300
400	2.0	2.0	4	400
500	2.7	2.7	5	500
600	3.5	3.5	6	600
700	4.5	4.5	7	700

Total count matches temperatures at all points.

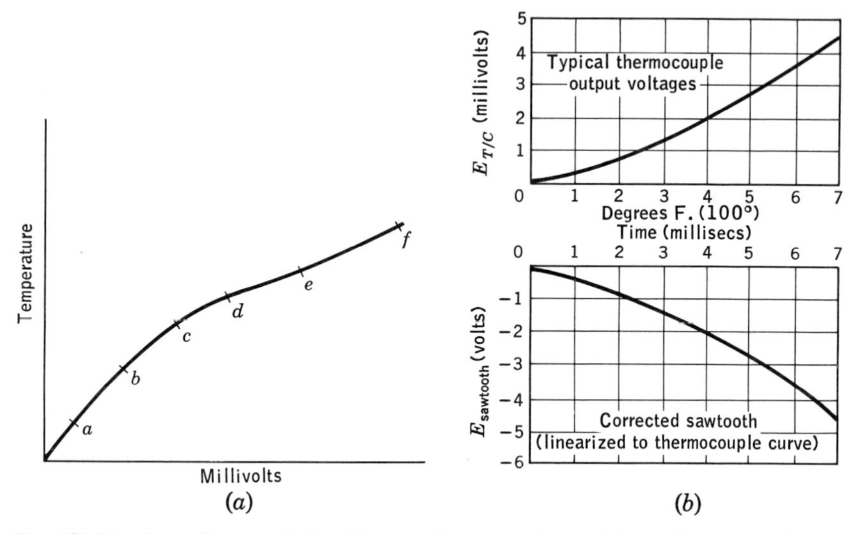

Fig. 17·37 A nonlinear relationship may be approximated by taking a number of tangents to the curve. In this example, five straight lengths have been used to replace the original curve. For practical reasons, it is desirable to employ the smallest number of straight-line sections because each one requires a different voltage gradient. However, the greater the number of sections, the more nearly will the approximation approach the true value of the curve.

Fig. 17·38 By connecting the linearization unit between the counters and the sawtooth oscillator, the slope of the output wave is changed so that it will approximate the nonlinear relationship which exists between emf and temperature, for example. (*Hagan Chemicals and Controls, Inc.*)

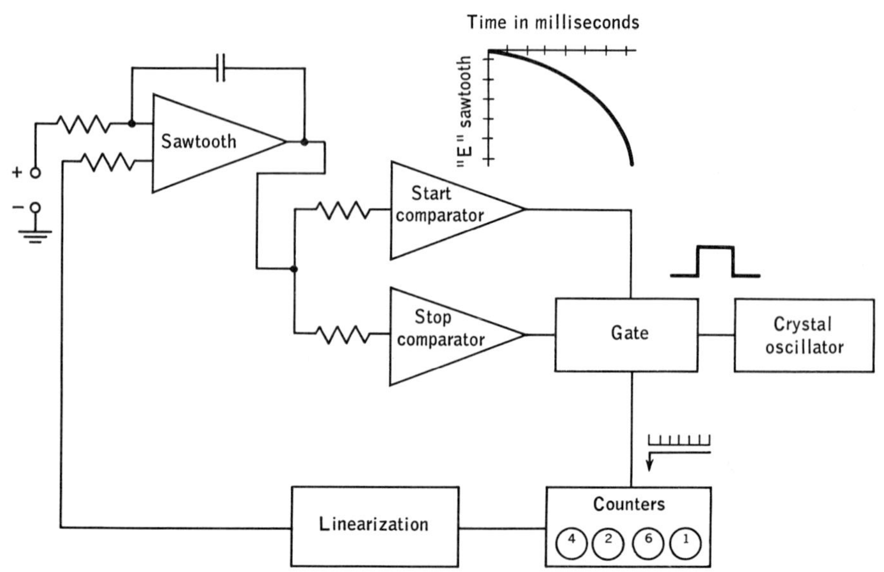

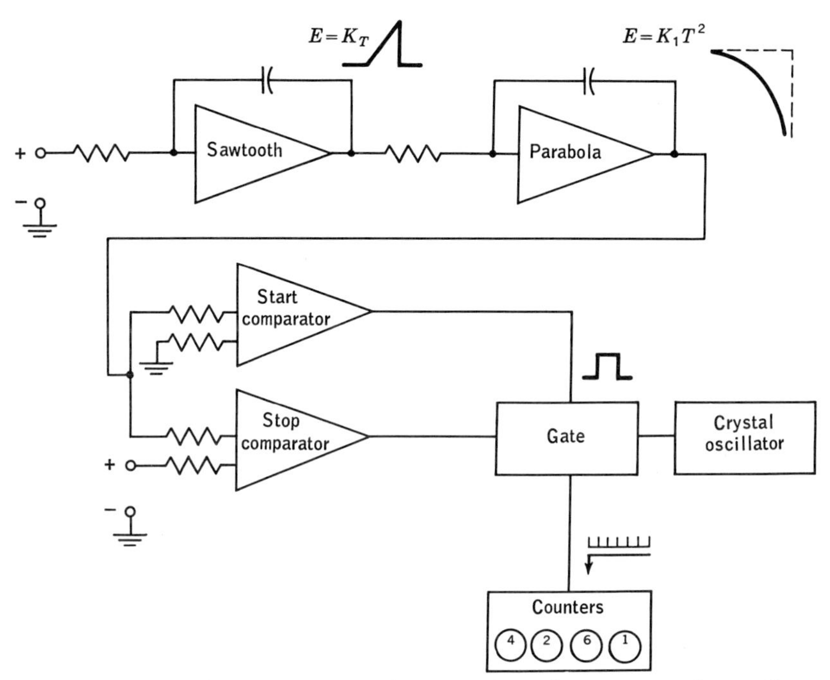

Fig. 17·39 By feeding the output of the sawtooth oscillator into another oscillator, a nonlinear output with respect to time is obtained. This will vary with the value of T^2. The comparison circuits are of the same nature as those considered previously. *(Hagan Chemicals and Controls, Inc.)*

arrangement shown in Fig. 17·39 generates a potential which varies with the square of the time.

Alarm section. Because the chief purpose of a scanning system is to determine whether the many process variables are within acceptable limits, the value of each variable when determined is compared with the high- and low-limit values. If it is within tolerances, then the unit continues its scanning activities. But if the variable is outside tolerances, then suitable alarms are actuated. If the system is also logging the results, they will be noted in an appropriate color to indicate their deviation from normal.

Printing and visual indication. A system that is capable of converting five variables per second into digits for purposes of comparison is far faster than typewriters can follow and faster than the eye can accept information. Consequently, any logging operation must be at a lower rate, probably with logging occurring only at definite intervals, unless it is desired to monitor one particular quantity. Although the end results of both systems are the same, it will be noted that the principles of evaluation, comparison, and digital equivalents have been applied in quite different ways.

SUMMARY

1. When circular charts are used, almost invariably the pen or pointer will be actuated by responses made directly to changes in the value of the variable. As a general rule, the charts are designed to run at but one speed.

2. The use of several pens or pointers is quite practicable, or a single pen can be made to write with a different color ink for each variable.

3. Circular charts have definite advantages in use, inasmuch as a complete record for some specified period is available on a single small piece of paper. Relatively long term trends are easily observed. Storing or removing from storage one particular day's operation is simple.

4. Rectangular charts have the advantage that, regardless of the position on the chart, each interval of time may be presented with equal length of line and equal emphasis. This type of presentation makes it much easier to determine the area under the curve.

5. Rectangular presentations require the use of servomechanisms to position the pointer or pen. With more than adequate power available, there are no real limitations on the weight of the pen, pressure between pen and paper, or length of the pointer.

6. Despite the use of servomotors, mechanical complications generally prevent the use of more than two pointers or pens. When more than two variables are to be presented on a single strip chart, the printing wheel or the use of colored ribbons is resorted to. With certain designs, as many as 40 different sets of values may be recorded. Such a device represented the first logging system. However, generally no provision was made for indicating or signaling when each variable had departed from its set point by predetermined amounts.

7. The graphic panel became feasible only after the small rectangular recorder became available. Such a development was required because of the desirability of locating as many as possible of the process-measuring and controlling elements where they could be monitored by one or two operators. This made it possible for the operator to observe the entire process instead of isolated reactions. Corrective actions could then be made in light of the entire process.

8. The use of the graphic panel simplifies the training of operators. The panel emphasizes the relationship of each component in the production of the process material, which makes it easier to see why some measurements are of far greater significance than others. Relatively unimportant measurements can be monitored visually, but the important ones can be recorded and provided with alarms.

9. When it is felt that the accumulation of vast numbers of records which indicate that conditions are normal is of little value, it is possible to establish systems which provide for recording the variable at will—when it deviates from the set point, for example. With 50 or 100 instruments on a panel board, the operator is unable to concentrate on the most important units unless the system can relieve him of some of the routine observation, leaving him free to concentrate on those variables which either have deviated or may deviate too far from normal.

The most recent logging systems do this very thing; they maintain surveillance over all the variables, logging the values only at predetermined periods unless it is desired to do otherwise and giving alarms only when undesired deviations have occurred.

10. The logging system makes it feasible to present, as it happens, a complete record of the factors which were involved and just what the results were of a given correction. Logging of this type makes it unnecessary to accumulate voluminous records; only that which is significant is presented. Perhaps of as great importance as any factor is the immediate availability of the data.

QUESTIONS

17·1 Why are servomechanisms not used to actuate the pen or pointer in the usual circular-chart indicator or controller?

17·2 With the usual type of circular-chart unit, how much of a mechanical load can be imposed by the indicating element on the responsive element?

17·3 What is likely to be the result if one reads a chart extremely precisely, with no thought of the inherent accuracy of the instrument?

17·4 Discuss the advantages and disadvantages of the circular-chart type of presentation.

17·5 Why are indicating and recording potentiometers servo-operated?

17·6 What are the desirable characteristics of a servo drive?

17·7 What are the relative advantages and disadvantages of the rectilinear presentation obtainable with strip charts?

17·8 Why is it easier to determine only that one electrical analog quantity exceeds another than to determine the amount by which one exceeds the other? Is your answer consistent with the two commercial data-logging designs?

17·9 Could you replace the stepper relay, Fig. 17·21, with a motor-driven contactor?

17·10 How might you provide electrical isolation between the sensing elements and the logging system other than by charging a capacitor and transferring its charge?

17·11 Might it be possible to construct a pneumatic analog system instead of an electrical one? What problems might arise? What might be the advantages and disadvantages of such a step?

17·12 With a pneumatic analog, how might you convert from pressure to digits? Could you digitize nonlinear relationships? (HINT: Would either commercial principle work with pneumatic elements?)

17·13 Could you use a scanner system run by a constant-speed motor? What would be its advantages and disadvantages, if your answer is affirmative?

17·14 Would your answer to Question 17·13 be different if the readings were to be obtained once each 10 sec? Once each second?

17·15 What are the chief requirements of any scanning-logging system?

17·16 In what way may measuring analog values be easier than converting to some numerical equivalent of the variable?

17·17 List the steps in the two data-logging systems which are alike.

17·18 What steps are found in one data-logging system but not in the other? Why?

17·19 What factors might the manufacturers have been considering when one selected a relay system and the other selected stepper relays to perform the same functions?

17·20 What is your evaluation of each of the factors in Question 17·19?

17·21 How would you evaluate a continuous record furnished by a strip chart and a continuous scanning system which furnished only off-normal records?

17·22 If all conditions are normal, what are the advantages and disadvantages of a continuous record?

17·23 Of what use are instrumentation records to departments other than the operating department and the instrument department?

17·24 If normal pressure pulsations frequently actuate alarms, would you be justified in installing snubbers? What may happen when resistance-capacitance devices are used to suppress pressure peaks?

17·25 Discuss the reasons for the various output standards of current and potential.

17·26 What values or ratios do the standards of Question 17·25 have in common, in most cases? Does this aid in analog to digital conversion?

17·27 Study three of the small strip-chart recorders. Classify the units as force-balance or torque-balance units. How do they compare? In what ways are they markedly different?

17·28 For high-speed scanning, Fig. 17·29, what would have to be true of the volume of the A-2 pressure transducer? Are the remarks true for Fig. 17·26?

17·29 How faithfully must output, using the fundamentals of the design of Fig. 17·16, reflect input?

17·30 If a pressure-voltage transducer can be made from the same basic unit, Question 17·29, by positioning a potentiometer rheostat, what factors will govern the dependability of the output potential?

17·31 Is the design, Question 17·29, based on torque, motion, or force balance? Explain the reasons for your answer.

17·32 If the design, Question 17·29, is too slow to use with data logging, what might you change to increase the speed of response? (HINT: Change or add only one component.)

PROBLEMS

17·1 Assume the following, Fig. 17·21:

	Nominal Temp, °F	*Permissible Deviation*	*Millivolt Values*
Thermocouple 1—Copper-constantan	400	±10°	
Thermocouple 2—Copper-constantan	300		
Thermocouple 3—Chromel-alumel	2000	±15°	
Thermocouple 4—Chromel-alumel	1500		
Thermocouple 5—Platinum–platinum-rhodium	2500	±25°	
Thermocouple 6—Platinum–platinum-rhodium	3000	±35°	

Determine from the tables what millivolt values would have to be selected for alarm purposes.

17·2 Using the values given on Fig. 17·24 what would be the electrical equivalents of 100 psi? 200 psi? 400 psi?

17·3 Write an equation which will permit you to determine quickly and accurately the value of the analog voltage for any given pressure.

17·4 Determine the electrical potential equivalents of the flow rate, Fig. 17·24. Take the flow rate at every 250 gpm.

17·5 Write the equation which relates the flow rate and its electrical analog.

17·6 Using 200° spans, determine the electrical analogs for the ten spans between 0 and 2000°, Fig. 17·24.

17·7 Write the equation which relates temperature and potential.

17·8 If the ± limits for Prob. 17·2 are ½ per cent, what will be the potentials between which the variable must remain?

17·9 With limits of ±1 per cent, determine the potentials within which the variable would have to stay, Prob. 17·4.

17·10 Repeat Prob. 17·9, but use the conditions of Prob. 17·6.

18 Analysis

In addition to knowing the degree of acidity or conductivity of a liquid, there are countless times when it is highly desirable to determine the composition of a substance in order to know whether or not the process is producing the proper product. Not only are certain physical properties desired, but oftentimes correct chemical composition is necessary as well if the desired end results are to be achieved.

In general, however, the number of determinations which can be made will be quite small. But the techniques which are used will be improved and enhanced in order to increase their scope and to widen the employment of automatic analysis techniques which will facilitate the manufacture of products of ever-higher quality.

There are many ways of determining the constituents of the process product, because there are many types of information desired. For certain conditions, it may be enough to know the percentage of oxygen or hydrogen present; under other conditions, it may be necessary to determine the presence and the percentages of much less common materials. In order to make these measurements, it is necessary to find some response which is associated with the desired measurement. This chapter will be concerned chiefly with the various phenomena which can be used to indicate the presence or absence of certain constituents. Some gases are quite magnetic, while others are not. Some materials will absorb certain radiations, but others are quite unaffected.

Carbon dioxide has certain characteristics which make possible measurements of its presence and also of the percentage present. This gas has a thermal conductivity which differs from the conductivities of air and other common gases. It can also be absorbed by certain chemicals. Because of its density, which is greater than that of air, certain other techniques can be used to ascertain the amount present. Carbon monoxide, the gas which is chemically closest to carbon dioxide, requires radically different methods of measurement. One of the most common is a measure of the heat created when the carbon monoxide is converted into carbon dioxide.

Certain other techniques use properties or characteristics which are not peculiar to the material in the sense that other materials do not respond the same way. Analysis by the absorption of ultraviolet energy and infrared rays depends upon the material and the special absorption spectra. In a similar manner, mass spectrometry does not depend upon properties which are not possessed by any other material.

Karp, in *Control Engineering,* May, 1957, lists the following fundamental methods for determining the analysis of materials:

- Radiant energy properties:
  - Absorption spectrometry—visible, ultraviolet, infrared, and X ray
  - Refraction and dispersion
  - Emission spectrometry
  - Nuclear magnetic resonance
  - Electron paramagnetic resonance
  - Ionizing radiation
- Electrical properties:
  - Dielectric constant
  - Magnetic susceptibility
  - Conductimetry
  - Electrolysis
  - Electrolytic conductivity
  - Electromotive force (pH)
  - Voltammetry
  - Polarography
- Mechanical properties:
  - Viscometry
  - Specific gravity
  - Velocity of sound
  - Absorption of sound
- Thermal properties:
  - Thermal conductivity
  - Transition temperatures
  - Heats of reaction
- Chromatography
- Mass spectrometry

THERMAL-CONDUCTIVITY GAS ANALYZERS

One of the outstanding characteristics of gases is their vast difference in heat conductivity. This has made the measurement of thermal conductivity one of the most widely used techniques for certain types of analysis, particularly when the gas being analyzed has but two constituents or can be considered as such because of behavior. The basic instrument is similar to the Pirani gage with the exception that the reference unit is not a vacuum. Figure 18·1 includes the familiar bridge circuit. This circuit makes the operation of the device much more independent of the current in the active path of the bridge than would be the case if but a single cell were used.

It should be obvious that the type of unit in which the heat transfer takes place is important to the design of the instrument. When there is a close control over the flow rate so that it is essentially constant, the simple design shown in Fig. 18·2*a* is quite adequate. The current flows through the resistor and heats it. This heat is then transferred to the walls of the container and to the ambient gas. Inasmuch as the reference unit has the same construction, the heat transferred to the walls should be about the same in each instance and therefore should balance out in the bridge. But any inequality in the amount of heat transferred to the

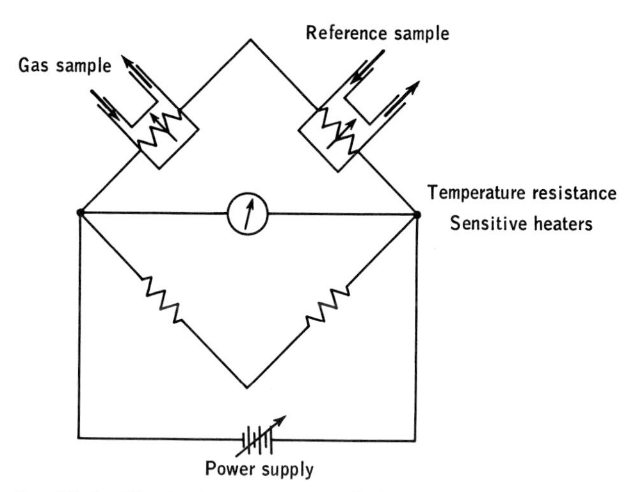

Fig. 18·1 The basic components of the Pirani pressure gage can be used for composition analysis when the reference cell is not evacuated but contains the gas with which the sample is being compared.

gas will be reflected in higher or lower temperatures of the heater, and consequently the resistance of the heater. The heater wire must not be one of the commercial heaters whose resistance-temperature coefficient is very small.

Fig. 18·2 Measuring cells of different designs can minimize false readings caused by flow-rate variations. The cooling effect of the gas must be caused by differences in the composition of the gases, not by variations in the flow rate.

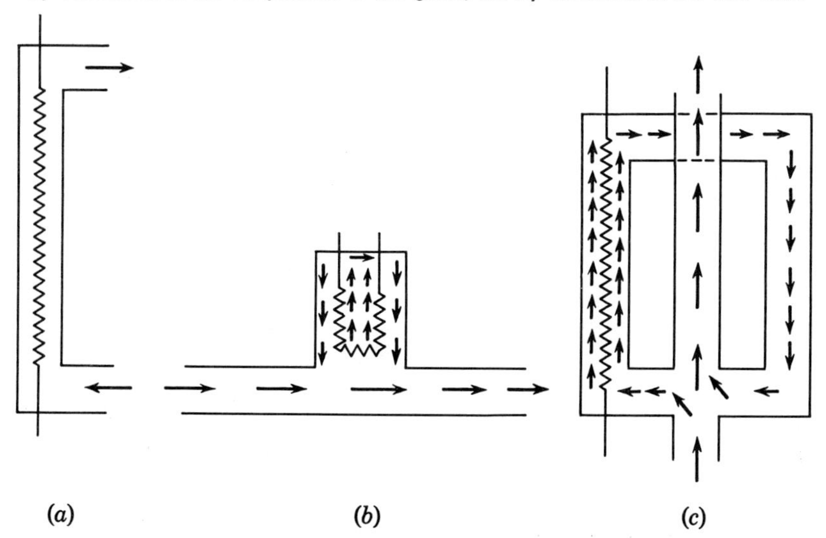

When there is little control over the flow rate, then the conductivity cells must be so designed as to minimize the effect of flow rate. In Fig. 18·2*b*, the main body of the sample flows through the pipe without passing through the heater. Only that part of the sample which is heated will produce convection currents and thereby produce a circulation around the heater. This will be of a very gentle nature which is essentially dependent only upon the amount of heat being transferred to the gas and is independent of the flow rate.

A somewhat different design is indicated in Fig. 18·2*c*. The sample has to be introduced at the bottom of the conductivity cell in order that the heat transferred to the sample gas can create a convection current of the gas. As the gas is heated by absorbing heat from the filament or heater, it rises, pushing the gas ahead of it back into the main sample stream. Inasmuch as the sample to the heater is removed from and returned to the line at the same point (so far as the pressure is concerned) the flow of gas through the heated portion should be a function only of the gas and the amount of heat transferred, and it should be virtually independent of the rate at which the sample is supplied to the cell.

It must be understood that the success of any measurement of this type is dependent upon:

1. Proper conditioning of the sample by drying, adding water vapor to the point of saturation, or purifying the sample to ensure that the desired values will be measured
2. Maintaining the proper temperature and pressure conditions
3. Avoiding contamination of the sample as it passes through the apparatus
4. Clean, filtered samples, free from solids, tars, and so on

Inasmuch as each gas will establish convection currents that depend upon the specific heat, density, and viscosity of the gas and the dimensions of the convection chamber, there will be certain optimum dimensions for each gas convection chamber for each specific gas. By selecting the convection chamber size, one can minimize the response of one gas which is not to be measured and optimize the response of the desired one. Figure 18·3*a* illustrates one method of increasing the selectivity of a gas analyzer by using two conduction cells of different dimensions. One of the cells is small enough so that most of the heat transfer is by conduction through the gas; the other cell is designed to provide for optimum heat transfer by convection. By using such a pair of cells for both the gas being analyzed and the reference gas, the sensitivity of the unit is greatly increased as well as made more selective. The way in which the pair of cells are incorporated in the usual bridge circuit is illustrated in Fig. 18·3*b*.

Erratic gas flow and turbulence which might interfere with the convection currents are minimized by the use of porous metal disks through

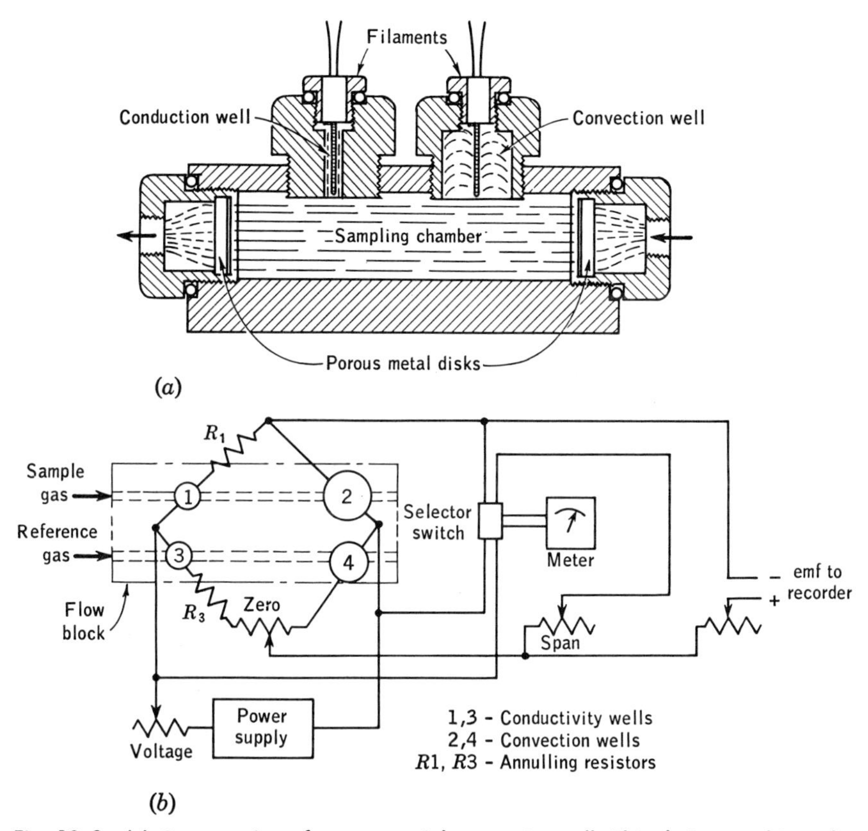

Fig. 18·3 (*a*) Cross section of a commercial convection cell. This design combines the advantages of thermal convection and standard conductivity for the selective analysis of one gas in a complex mixture of gases. (*b*) Higher degrees of selectivity are possible when both conduction and convection units are combined in a single bridge. It will be noted that both the sample and the reference gases pass first through a conduction well and then into a convection well. (*Mine Safety Appliances Co.*)

which the gas must pass. The flow block also has been designed to eliminate turbulence of the sample. Turbulence interferes with the natural convection currents in the large wells and tends to upset thermal conductance in the small wells. Large porous stainless-steel disks in each end of the flow block reduce turbulence and also act as heat exchangers and flashback arresters.

By combining the principle that each gas has its optimum thermal convection loss in a well of different diameter with the principles of thermal conductivity, well diameters are determined for the four filaments. By proper pairing of these wells as opposing arms of a Wheatstone bridge, the effects of background gases are annulled and the cell becomes specific for the gas of interest. Additional accuracy is obtained by incorporating annulling resistors with the bridge arms as required.

OXYGEN ANALYZERS

The correct determination of oxygen content is important in many widely separated industrial operations. A few of the chief ones are outlined here, together with examples and the reasons for making the measurement.

1. One of the most important, and one which readily comes to mind, is the determination of the oxygen content of flue gas from boiler plants or process heaters. If too much oxygen is present in the flue gas, then some heat is being used to increase the temperature of air, which performs no useful function. If there is not enough oxygen to provide for complete combustion, then a certain amount of the carbon will not be completely burned to carbon dioxide because of the incomplete combustion and a proportionate amount of heat will not be available.

2. Certain types of process control require that preceding products left in the equipment be removed or burned if the process is to continue. For example, the burning of the carbon which adheres to the catalyst in certain petroleum refinery operations should be carried out in such a manner that the catalyst is not damaged by an excess of oxygen, but a deficiency of oxygen will not permit the desirable amount of regeneration of the catalyst to take place.

3. The problem of product control is oftentimes simplified, particularly when the product consists of mixed gases (for example, propane with air to provide fuel) and the ratio of oxygen to the other constituents can be maintained within close limits.

4. The large number of installations which employ inflammable or explosive materials in their processes need some means of determining when dangerous concentrations of oxygen are present. Furnaces which are full of pure hydrogen can be used safely for annealing, but if the amount of oxygen in a furnace approached a critical value, the furnace might be destroyed.

Some characteristics of oxygen. A characteristic of oxygen which is most useful in oxygen analyzers is one with which most of us have little personal experience. The ability of oxygen to support combustion or to promote the oxidation which we note as rusting is well known to all. But the fact that oxygen is quite strongly magnetic when cold and that it will respond to a magnetic stress is generally not well understood. In instrumentation we have to employ such unique properties because there are only a few responses which one gas or material will have and another will not have. While oxygen is capable of responding to a magnetic field, other gases are quite unresponsive, or may be considered so for practical reasons.

Figure 18·4 shows the relative responses of some of the more common industrial gases. Fortunately, the oxides of nitrogen very rarely occur

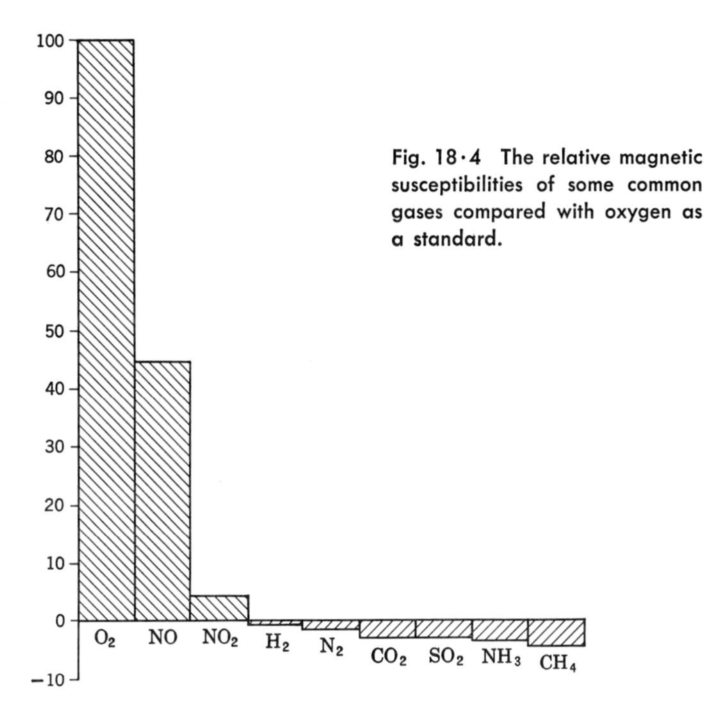

Fig. 18·4 The relative magnetic susceptibilities of some common gases compared with oxygen as a standard.

in significant amounts in industrial gas analysis. For this reason, the magnetic activity of nitrous oxide and nitrogen dioxide will seldom invalidate the performance of an oxygen analyzer which employs the magnetic responses of gases.

The oxygen analyzer differs from the Pirani gage chiefly in the way in which the magnetic field created by a permanent magnet attracts the oxygen and creates an atmosphere of oxygen immediately around the resistor, Fig. 18·5. As the oxygen absorbs heat, it loses its magnetic characteristic and is forced away from the hot resistor by the pressure of the oxygen which has not been heated and which is still attracted by the magnetic field. In effect, this establishes a magnetic draft. In the reference unit there is no magnetic gas, so all cooling is largely by conduction and by convection currents created in the gas sample. Because the heated oxygen becomes nonmagnetic, it is rapidly displaced by the cooler oxygen, which will absorb relatively large amounts of heat. The chief effect of the oxygen characteristic, then, is to promote a far faster movement of cooling gas over the hot resistor. If there is no oxygen or other magnetic gas, then both sets of resistors will cool at the same rate, Fig. 18·4*b*. Any difference in their cooling is then attributed to the presence of some magnetic gas. By removing the permanent magnet, or by moving it out of the way, the two cells should respond in the same way to gas of the same composition. In this way, the device can easily be zoered.

Fig. 18·5 The proper application of a magnetic field can cause a great increase in the movement of oxygen past the hot filament. Note the difference in the convection currents when a magnetic draft is employed, as shown to the left in (*a*). (*The Hays Company*.)

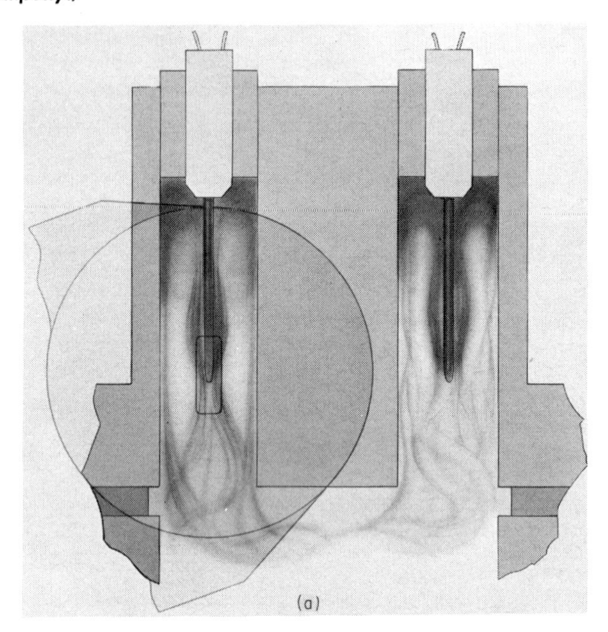

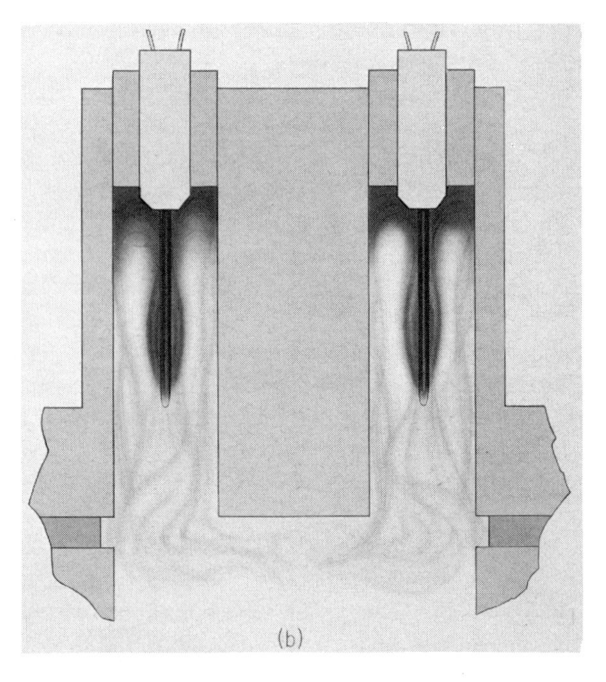

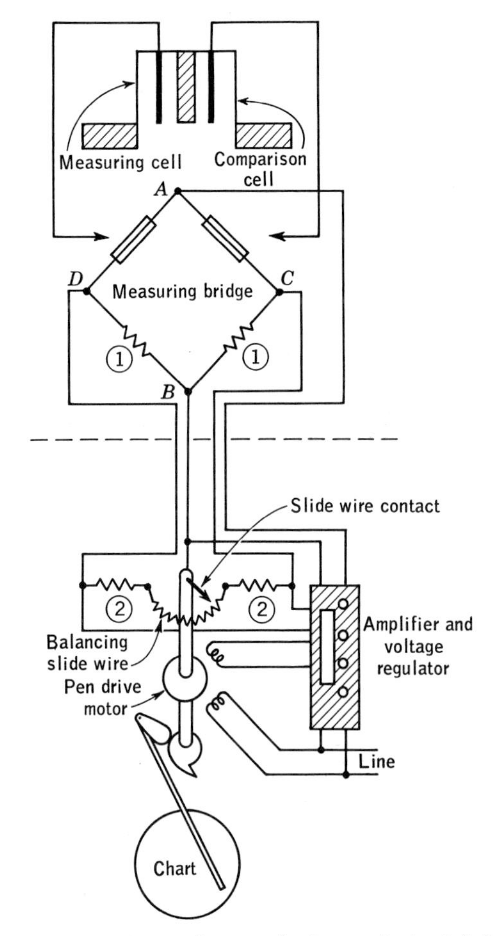

Fig. 18·6 A self-balancing bridge converts differences in resistance into suitable pen or pointer movement over a calibrated scale which shows the percentage of oxygen content. (*The Hays Company.*)

Figure 18·6 shows how this measurement can be made in an industrial unit.

A similar design which is modified to compensate for pressure changes is shown in Fig. 18·7. In all essentials it is the same type of unit just considered. In order that variations in the pressure might be eliminated from the percentage analysis, two measuring bridges are used; one is used to measure the thermal effects of the sample gas, and the other one is used to measure the thermal effects of air which is kept under the same pressure as the sample gas. Although the illustration shows two magnets, only one magnet is used. This prevents any variation in the magnetic field from being introduced into the answer.

Studies have shown that the thermomagnetic effects are theoretically proportional to the square of the static pressure of the gas. By making the calibration dependent upon the ratio of the bridge outputs when both gases are at the same pressure, this effect is virtually eliminated.

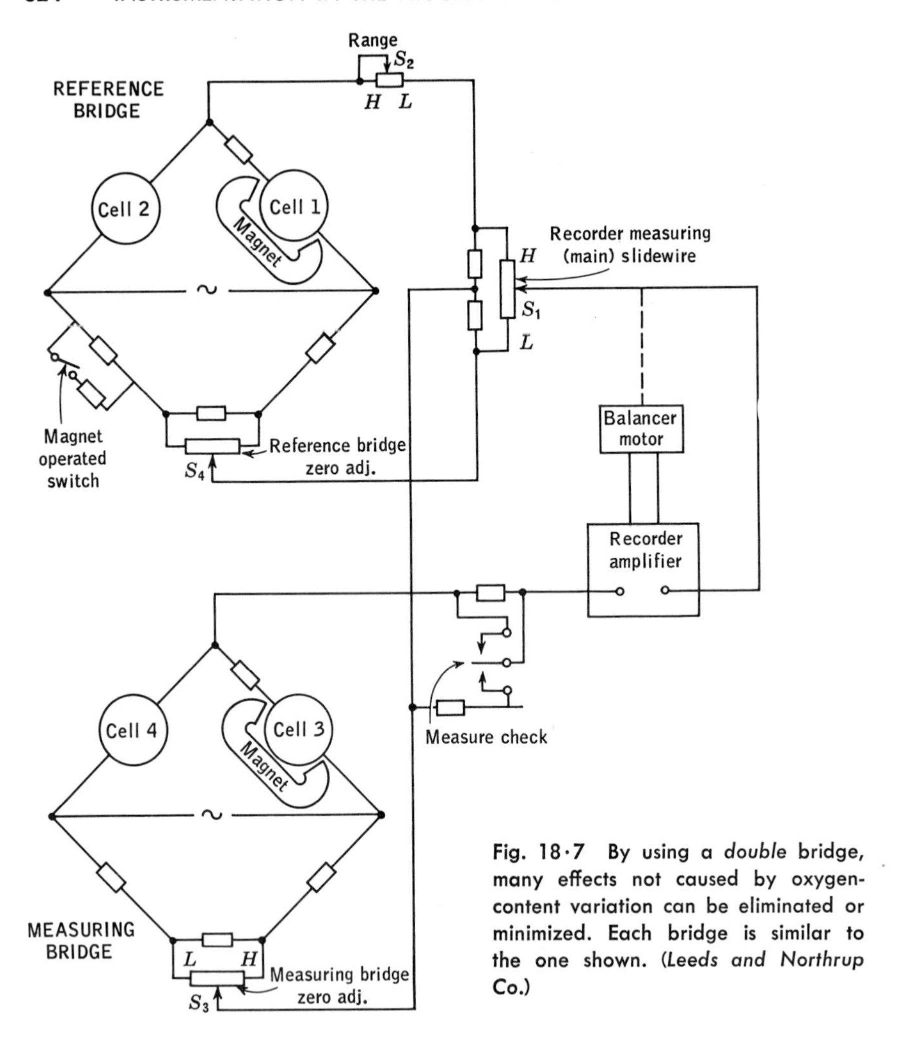

Fig. 18·7 By using a *double* bridge, many effects not caused by oxygen-content variation can be eliminated or minimized. Each bridge is similar to the one shown. (*Leeds and Northrup Co.*)

A radically different way of determining the percentage of oxygen in a gas is shown in Fig. 18·8. This design uses the chemical affinity of hydrogen for oxygen as its means of measuring the amount of oxygen present. In many ways, it resembles the conventional dry cell used in flashlights, because it has a metal and a carbon electrode. The chief difference is in the electrolyte, which in the case of the dry cell is a moist paste containing, among other things, sal ammoniac and manganese dioxide which acts to depolarize the cell. The depolarization is necessary because hydrogen gas gathers on the carbon electrode and, as it covers more and more of the electrode surface, effectively insulates the electrode from the rest of the circuit. The amount of gas is directly proportional to the current between the electrodes.

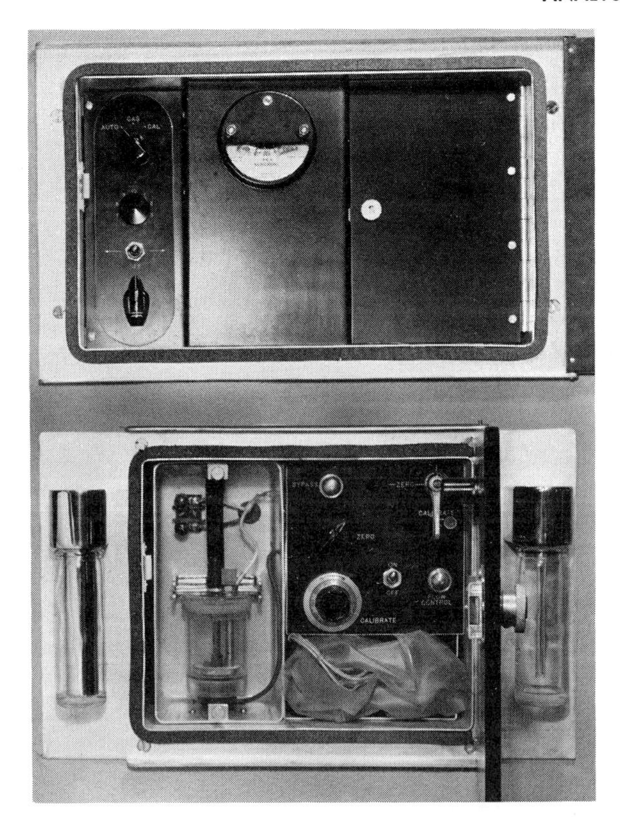

Fig. 18·8 This oxygen analyzer employs a primary cell having a metal electrode and a porous carbon electrode in an electrolyte. As the sample gas passes through the porous carbon it tends to depolarize the cell and thus reduce its internal resistance. The resulting cell current is then a measure of the oxygen in the sample. (*Mine Safety Appliances* Co.)

Exactly the same performance goes on here in the detector cell, hydrogen ions being furnished by the "oxylite" electrolyte whenever current flow occurs. The carbon electrode soon becomes covered with the hydrogen gas bubbles, which increase the resistance of the circuit and reduce the current flow. But in this instance, the gas which is being sampled passes through the porous carbon, and the oxygen which is in the sample has a chance to diffuse through the walls to the surface of the electrode. When the oxygen combines with the hydrogen, the hydrogen gas disappears and the resistance of the circuit is reduced, with a resulting increase in the current. Calibration can be carried on automatically at periodic intervals by sending through the sample cell a gas whose oxygen content has been very carefully established.

The principle used here has been known for many years. However, in this case, the polarizing effect of the hydrogen gas, and its control by the amount of oxygen admitted to the system, converts into an asset what is a liability in most processes.

MASS SPECTROMETRY

In materials made up of many constituents, analysis may be quite difficult if one is trying to determine the presence and the relative amounts of several components. Such a measurement would require the use of principles or behaviors which are characteristic of many materials, and not just those characteristic of one or two. The mass spectrometer employs some of the most basic responses:

1. The acceleration of a particle is directly proportional to the force applied and inversely proportional to its mass, or $A = K(F/M)$.

2. Ions in a uniform magnetic field possess a radius of curvature

$$r = \frac{144\sqrt{mV/e}}{B} \tag{18·1}$$

where r = radius, cm
m = mass, atomic-weight units
V = accelerating potential, volts
B = magnetic field strength, gauss
e = number of electronic charges

3. Materials that have equal kinetic energies but different masses will have unequal speeds and will necessarily take different times to traverse the same distance.

4. If a radio-frequency field is applied to particles that have different speeds, then energy can be imparted to some in far greater measure than it can be imparted to others. Consequently, there is a selective accelerating process.

5. The angular velocity of a charged particle moving in a magnetic field is dependent upon the following relationship:

$$f = \frac{1}{2\pi}\frac{eB}{mc} \tag{18·2}$$

where f = frequency, cps
e = charge of the electron, electrostatic units
m = mass of the particle
B = magnetic field strength, gauss
c = speed of light, cm/sec

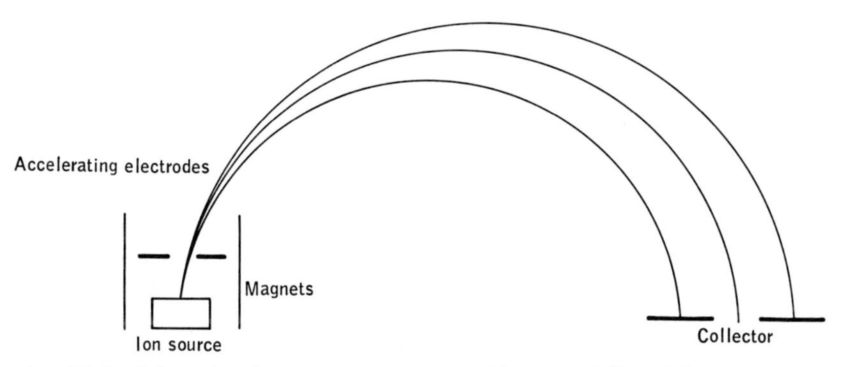

Fig. 18·9 Schematic of a mass spectrometer. The path followed by an ion depends upon the accelerating voltage, the strength of the magnetic field, and the mass of the ion. Adjusting the voltage permits any particular ion to be gathered at the collector.

Magnetic analyzer. For analyzing fluid and gaseous samples, the magnetic type of analyzer is particularly well suited. It is first necessary to convert the sample into ions by bombarding it with a stream of electrons. The application of a negatively charged field will then accelerate these particles. But if at the same time they are located in a magnetic field, then they will follow a curved path with a radius which is proportional to the square root of the mass of the ion, according to expression 18·1. Since each particle of different mass will have a characteristic radius, it is possible to sort out each ion according to its mass. The value of the current delivered by the ions will also be a measure of the degree of concentration of a particular ion. Instead of having a series of collector plates which extend for a long distance, as implied in Fig. 18·9, it is possible to retain only a small collector but bring all of the various ions to it by varying the voltage. The particular ion being analyzed is then a function of the applied potential and can be so designated by calibration.

All of the "mass" analyzers have five important components:

1. The power supply
2. The sample system
3. The high-vacuum system
4. The ionizing and collecting system
5. The amplifying and recording section

The sampling system must provide, at suitable rates, samples that are representative of the material being measured. The control system must provide the accelerating voltages and means for controlling the strength of the magnetic field, the applied voltage, or whatever variables it is desired to manipulate. A high-vacuum system is necessary in order to remove the air and other gases which would invalidate the results. The sample should be capable of providing vapor pressures of at least 0.1 mm Hg. Any currents created by the ions arriving at the collector plate will be extremely small in value, probably of the order of 10^{-10} amp. In order to use conventional recording or indicating equipment, it is necessary to use suitable high-gain amplifiers to raise the current values to usable values.

OPTICAL ANALYZERS

Optical techniques are among the most important of the techniques used for continuous automatic analysis of materials. Some optical techniques employ color, others use changes in the index of refraction of the material as a measure of certain concentrations, and still others are based on the selective absorption of certain wavelengths. In the latter technique, the amount of absorption and the particular wavelength are dependent upon the constituent and the concentration for a given set of base conditions. Each of the methods will be considered in this section.

The following material on the characteristics of ultraviolet and infrared absorptions is reprinted here with the permission of the Perkin-Elmer Corporation.

Since the middle 1940's, instrumental analysis has made some of its greatest advances in the field of absorption spectroscopy. This technique involves passing radiation of different wavelengths through a substance and measuring the radiation absorbed or transmitted to determine the material's identity, concentration, or molecular structure. With this new technique, for the first time chemists have had the means to identify and measure organic and inorganic chemicals with previously unheard-of speed. As a result, the discovery of new compounds has been accelerated and a new degree of control over chemical processes has been achieved.

The radiation ranges used in absorption spectroscopy are divisions of the electromagnetic spectrum, which comprises all electromagnetic radiations arranged in order from the longest to the shortest wavelengths, Fig. 18·10. These light-energy waves are measured in the wavelength unit, the micron, which is one-thousandth of a millimeter and is represented by the Greek letter μ. Thus, the ultraviolet region extends from 0.2 to 0.4 μ; visible, from 0.4 to about 0.7 μ to about 200 μ; and the infrared (IR) region, the most useful in analysis, from 2.5 to 15 μ. Another subdivision of IR is the near-infrared, extending from 1 to 3 μ. This division of the electromagnetic spectrum is based on the methods for producing and measuring radiations of the wavelengths represented by these various regions and on their interaction with matter. There are no sharp boundaries between these regions. Consequently, problems involving absorption

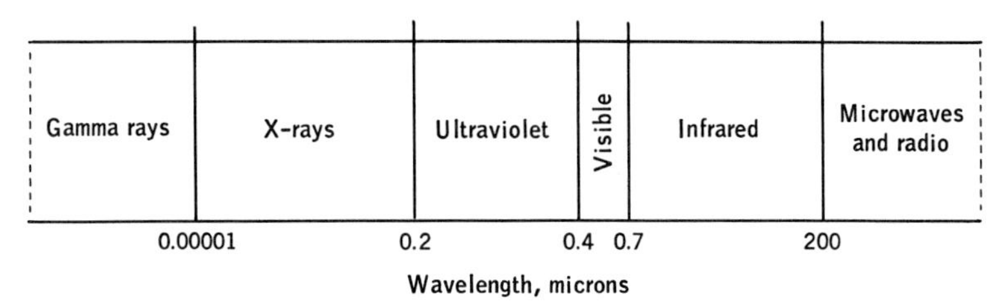

Fig. 18·10 The effect of irradiating chemical compounds with portions of the electromagnetic spectrum—especially those portions on each side of the visible portion—is the basis for identifying and measuring chemicals by absorption spectroscopy.

of radiation in the transition zone between adjacent regions can often be solved by instruments designed for either of the regions.

The value of absorption spectroscopy in chemical analysis lies in the fact that matter absorbs radiation very selectively with respect to wavelength. This relationship can be understood by considering the effect of different kinds of radiation on the molecular structure of compounds. All molecules are made up of atoms connected by electromagnetic bonds. These atoms are in continuous motion, vibrating and rotating about their centers of gravity. Also contributing to the electromagnetic bonding forces holding a molecule together are valence electrons shared by the atoms comprising the molecule.

The atoms of each molecule vibrate at a definite frequency with respect to each other. This frequency is specific to molecular structure. If radiation of a given frequency, or wavelength, strikes a compound which has the same vibration frequency, the radiated energy is absorbed by the molecule, thereby increasing its natural vibration. However, if radiation of some other frequency strikes the compound, it passes through the molecule without change in radiant energy. Since the molecular vibrational frequencies of substances have the same range of values as the frequency of infrared radiation, IR radiation can be used to characterize molecular structure.

The electromagnetic energy involved in electron bonding is of a much higher order than infrared's resonant vibration with the movements of atoms. It corresponds to the shorter, higher-frequency electromagnetic radiations in the ultraviolet (UV) and visible range. Consequently, when molecules absorb UV and visible radiations, they undergo discrete energy shifts. Numbers of electrons participating in valence bonds may change, and so may the number of electrons in atomic shells. Since the electronic energy levels of a molecule are specific to it, the absorption frequencies causing such electronic transitions can be used to characterize the molecule.

Applying these absorption phenomena to instrumental analysis involves

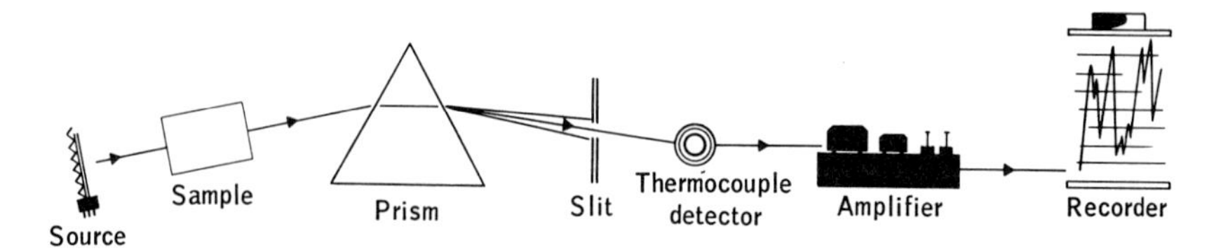

Fig. 18·11 One basic design of absorption spectrometer employs the components indicated here. (*The Perkin-Elmer Corp.*)

irradiating a sample with all frequencies in a given region—in successive wavelengths—and recording those which are absorbed. Figure 18·11 shows schematically how this is done. In operation, the instrument automatically passes the dispersed radiation across the slit, and the detector absorbs each radiation frequency in turn. The energy absorbed by the thermocouple detector is a measure of how much energy was *not* absorbed by the sample. By feeding the output of the thermocouple into an amplifier and a recorder, a continuous record can be made of the absorption at each frequency. The instrument determines not only what compound is present but also how much is present, inasmuch as the total amount of energy absorbed is a measure of the number of molecules present. Each analysis is recorded as a graph of the energy transmitted through the sample versus the frequency or wavelength. This is the spectrum—IR, UV, or visible—of the sample, and it is a fundamental property of the sample which can be used to characterize it and determine its concentration.

What happens when molecules are irradiated with energy of different

Fig. 18·12 The higher energy of the ultraviolet and visible regions produces electronic transitions in the outer orbitals of the atoms. Such shifts appear as absorption bands in the ultraviolet and visible regions. When molecules are excited only by the lower-energy infrared radiation, only vibrational and rotational modes occur; the electronic transition states are not affected. (*The Perkin-Elmer Corp.*)

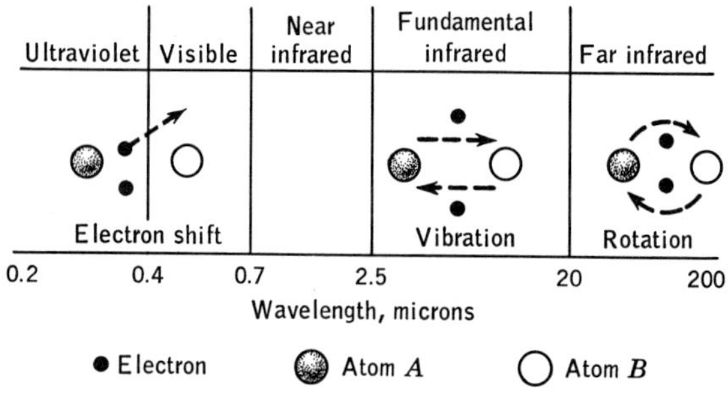

Ultraviolet | Visible | Near infrared | Fundamental infrared | Far infrared

Electron shift | Vibration | Rotation

0.2 0.4 0.7 2.5 20 200

Wavelength, microns

● Electron ◉ Atom *A* ○ Atom *B*

wavelengths is represented in Fig. 18·12. The higher energy of the UV and the visible regions produces electronic transitions in the outer orbitals of the atoms. Such electronic shifts appear as absorption bands in the ultraviolet and visible regions. At the same time, this energy increases atomic vibrations and rotations associated with the lower-energy infrared regions. When molecules are excited only by lower-energy infrared radiation, only vibrational and rotational modes occur; the electronic states are not affected.

Both IR and UV (UV and visible are generally grouped together) are widely used in chemical analysis. However, each gives information that recommends it for particular problems. Since the IR spectrum of an organic compound is its most characteristic physical property, it is generally selected when specificity is the prime requisite. IR can distinguish small differences in chemical structure, which show up as large differences in the IR spectrum, to make the identification of complex mixtures simple.

The near-infrared region is valuable because quartz optics, which permit the use of water, are used in such instruments. Near-IR instruments can be used to measure water, and, by operating at lower concentrations of material, to distinguish between such groups as NH and OH, which overlap in the conventional IR region.

UV is not as specific as IR, but it is a more sensitive technique. Moreover, aqueous samples can be analyzed by UV; this is not possible with IR because water absorbs in the IR region and also because the materials generally used for sample cell windows in IR instruments are water-soluble. UV is generally quantitatively superior to IR in the determinations of complex organic structures, especially aromatic and heterocyclic compounds and those with conjugated double bonds. Absorbing compounds fall into classes the members of which have in common a specific configuration of atoms—called a chromophore—and similar absorption patterns. This characteristic provides a rapid means of determining structural features of a molecule.

Because of their wide application, analysis instruments using absorption techniques will be considered first. Analyzers using the invisible portions of the spectrum, but on each side of the visible portion, differ to a marked degree because the absorption characteristics of materials are quite different in the two different spectra spans. They may be summarized as follows:

1. The absorption of ultraviolet light by a material occurs in a relatively broad pattern, while absorption of infrared energy is likely to be concentrated in two or three very narrow bands. Materials have far more distinctive absorption patterns for infrared than they have for ultraviolet.

2. When there are several process gases, for example, in a product being

analyzed, variations in the absorption can be caused by variations in the concentration as well as by the several gases present. The narrow absorption bands in the infrared spectrum make equal multiple absorptions much less likely.

3. From the viewpoint of sensitivity, more energy is normally available in the absorbed ultraviolet beam than is available in the infrared beam because of the much narrower absorption bands. Consequently, ultraviolet analyzers are likely to be much more sensitive to small changes than are the infrared units.

The principles used in ultraviolet analyzers are somewhat easier to understand than are those employed by infrared analyzers, so we shall begin our study of optical-photoelectric analyzers by considering the ultraviolet design first.

The components required for an optical analyzer are few. There must be some source of radiant energy. But because this energy will be made up of a wide band of frequencies, there generally will be some sort of filter to reduce or limit the spectra to those which can be used most advantageously. A wavelength which has almost identical absorptions in two or more different materials would be a poor one to use in an analyzer intended to measure one of those materials. The sample cell may be one of the critical parts of the analyzer. Through it will flow the fluid being analyzed, and it must be kept clean so that the maximum amount of absorption is caused by the fluid and the minimum amount by the sample cell itself. Finally, there is a photocell, which must be sensitive to the analyzing frequency. The current which flows in the galvanometer circuit will then produce meter positions which are functions of the material and the concentrations.

Such a simple circuit has two shortcomings:

1. Even with regulated power supplies, there may be significant changes in the output current caused by variations in the lamp output only.

2. Variations in the potential of the battery in the galvanometer circuit will indicate concentration changes when there may be none.

By using comparison circuits which employ common light sources and energy (battery) sources, both of these objections can be overcome. Figure 18·13 includes the major components of such an instrument. A source of ultraviolet radiation is suitably filtered and directed to a semitransparent mirror which transmits about half of it and reflects the other half to the comparison portion of the equipment. The radiation which passes through the mirror passes through the sample cell, where some of it is absorbed. The rest then strikes a diffusing medium of some kind which redirects the energy to one of the photoelectric cells, where its relative energy will be determined by comparison with the portion of the beam which did not pass through the sample cell. If we assume that the two photocells and amplifier sections are essentially equal, then unequal

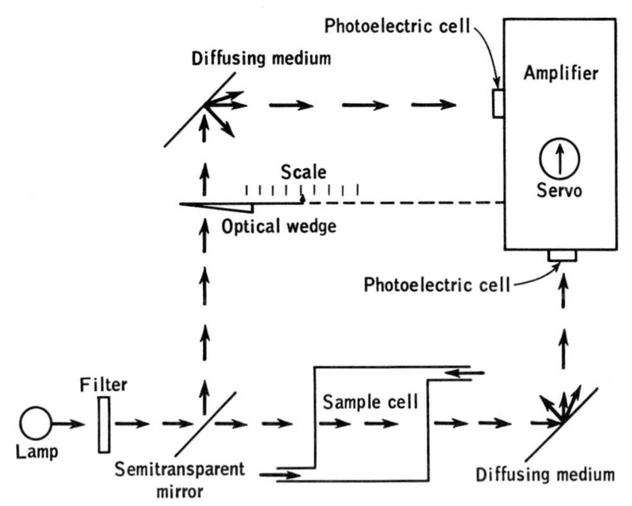

Fig. 18·13 When automatic operation of an industrial ultraviolet analyzer is desired, a servo-controlled optical wedge may be inserted in the reference beam. When the two ultraviolet beams reaching the photoelectric cells are equal, the bridge is balanced. The position of the optical wedge is then a measure of the absorption of the sample.

amounts of energy will be delivered to each input section of the amplifier and their net outputs will not cancel. This will then produce an output which can be used so to position an optical wedge, or a similar device, that the beams striking the photocells shall be equal. The position of the wedge, or its relative absorption, will then be a measure of the absorption going on in the sample cell. Through suitable calibration, the concentration may then be inferred.

The source of the ultraviolet, the filters, and the optical transmitting portions of the device all have an important bearing on the performance.

Absorption spectrometry; infrared. Most materials when exposed to various wavelengths of electromagnetic radiation, whether in the visible end of the spectrum or in the nonvisible portion, will absorb almost all the energy available at certain frequencies and absorb very little at other frequencies. Because the absorption characteristics of molecules are dependent upon the material, it is possible to determine the composition by the characteristic absorptions. These are peculiar to the molecule in terms of the frequency at which the absorption occurs and the amount of absorption. Figure 18·14 indicates the type of "fingerprint" which it is possible to get for each material which responds to infrared energy. It should be noted that the only difference between Figs. 18·14*a* and 18·14*b* is one atom of oxygen, while the difference between Figs. 18·14*b* and 18·14*c* is brought about by the substitution of one atom of chlorine for one of hydrogen.

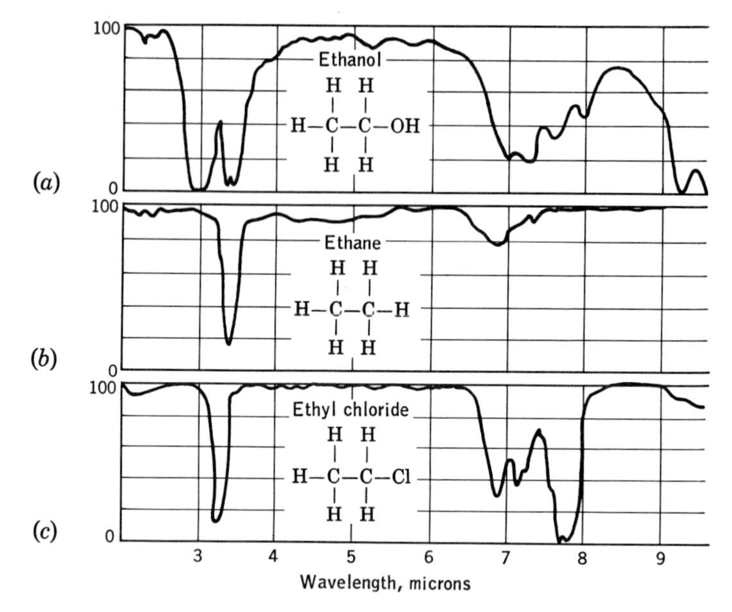

Fig. 18·14 Infrared spectra of ethanol, ethane, and ethyl chloride. Despite the close chemical relationships, each has its own characteristic absorption pattern.

Identification is then dependent upon matching the frequency and the relative amount of absorption with that of some known substance. Other identifications are made by the use of comparison units and sensitizing techniques.

It is perhaps unfortunate that the common gases such as hydrogen, oxygen, nitrogen, and chlorine do not absorb infrared energy. The rare gases such as argon, neon, xenon, and krypton are also nonresponsive to this type of analysis. Table 18·1 lists certain gases which do absorb significant amounts of infrared energy.

If a gas does absorb infrared radiation, the absorption will produce heat, which will in turn produce a pressure variation in the confined gas depending on the amount of energy absorbed. Pressure variations are employed in certain transducers to furnish electrical signals which can be used in one way or another to indicate the percentage of the gas under study. Figure 18·15 is a schematic representation of one commercial infrared analyzer. It will be noted that there are two sources of infrared radiation. Current flowing through a resistor furnishes the required radiant energy which should be produced in equal amounts by each source. The beams then pass through cells which contain gases and are so directed that they both pass into the detector unit which contains especially selected gas or gases.

Table 18·1 Sensitivities of Some Infrared Analyzers

Gases	Ranges, per cent		
Carbon dioxide	0–0.2	0–10	0–100
Carbon monoxide	0–0.2	0–1	0–10
Methane	0–1.0	0–20	0–50
Water vapor	0–0.5	0–2.0	0–3.0
Ammonia	0–0.3	0–1.0	0–5.0
Methylene chloride	0–100	Lower explosive limit	
Ethane	0–2.0	0–4.0	0–100
Cyclohexane	0–3	0–5	0–7.5
Sulfur dioxide	0–0.5	0–5	0–20
Ethylene	0–10	0–20	0–50

Placed between the infrared sources and the gas cells is a disk which is driven by a motor. The purpose of the disk is to interrupt the passage of energy into the gas cells. By having only half of the disk opaque, for example, energy can be delivered to one cell while it is prevented from reaching the other cell. By such a means, the energy which is received at the detector cell comes first from one cell and then from the other, but never from both at the same time. Each cell also receives and transmits energy for equal periods of time.

Let us assume that each cell contains some gas such as oxygen or hydrogen, neither of which absorbs infrared radiation. Let us put carbon

Fig. 18·15 If equal amounts of infrared energy are transmitted to both sample and comparison cells, then the energy arriving at the detector will be the same when comparison and sample gases are equal in composition. By rotating a disk, half of which is opaque, between the infrared source and the gas sample, only one cell receives energy at any one time. Differences in the amount of energy arriving at the detector cell can then actuate the detector and amplifier units. (*Mine Safety Appliances Co.*)

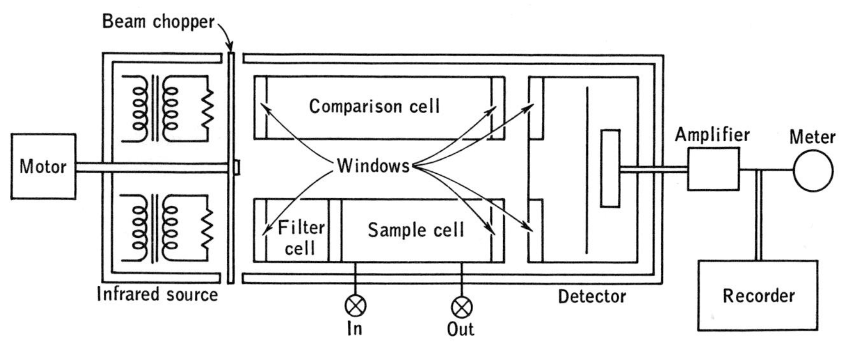

dioxide into the detector cell because it does absorb infrared radiation. If we now permit the unit to operate, energy will pass alternately through one cell and then through the other. It will be received at the detector cell with little of its infrared energy absorbed by the intervening gases. But when the infrared energy arrives at the detector cell, it will be absorbed by the carbon dioxide present. Absorbing this energy will raise the temperature of the detector gas, and consequently the pressure within the cell. This increase in pressure will then cause a thin membrane, which forms one wall of the detector cell, to be deflected outward each time a pulse of energy is received. If this membrane carries a metal plate which is one plate of a capacitor, then the capacitance will be dependent upon the spacing of the plates, and hence on the cell pressure.

A small hole permits slow pressure changes to equalize the pressures on the two sides of the diaphragm or membrane. This tends to make the device insensitive to temperature and pressure changes in the ambient. The associated amplifier is designed to operate only on pulsed differences in the capacitance, and it does not function when the detector cells are heated equally by energy from the reference cell and the sample cell.

But if we now want to employ this analyzer in a conventional manner, we will introduce into the sample chamber a gas which is not the same as the one in the comparison cell. Let us assume that the new gas does contain carbon dioxide, along with several other gases. Under these conditions, as the radiant energy passes first through one cell and then the other, there will be unequal absorptions:

1. The comparison cell will not absorb much of the radiated infrared energy but will let it pass on to the detector where it will cause temperature, pressure, and membrane changes, as it did previously.

2. But the sample cell which now contains carbon dioxide will absorb much of the infrared energy and thus cannot let all of it pass on to the detector cell. As a result, when the energy from this beam arrives at the detector cell, it will have less energy which can be absorbed by the detector cell gas. Of necessity, this will produce a smaller amount of heating, lower pressures, and a smaller movement of the membrane.

Consequently, the movement of the membrane and necessarily the capacitance of the capacitor, will be synchronized with the rotation of the beam interrupter, and the magnitude of the capacitance variations will be a measure of the absorption of the gas in the sample cell. If the sample cell absorbs only a small amount of radiated energy, then the heating effect in the detector cell will be almost the same as it was when the energy passed through the reference unit.

In practice, the gases or products which we want to analyze will not have just one absorbing component. When we want to evaluate just one of the components of a sample, we can place in the detector definite percentages of all of the constituents which will absorb energy. If we also place in the reference cell similar percentage samples of all gases except for the one gas whose analysis we are seeking, then we find the following taking place:

When the infrared energy is passing through the reference cell, the component gases will absorb some of the infrared energy. Energy which is not absorbed will be passed along to the detector unit, which will produce its customary heating and pressure changes. But this unit does not contain the one gas whose presence is sought, and, consequently, it cannot absorb this particular amount of energy.

The sample cell contains not only all of the gases in the reference unit but the additional subject gas as well. Consequently, it can absorb more energy than the reference unit can.

The greater absorption of the sample unit will necessarily produce a smaller amount of heating of the detector with its accompanying pressure changes.

If there is none of the gas present, both sample and reference cell absorb equally, so there will be no periodic moving of the membrane and the capacitor plate. As the percentage of the analyzed gas increases, there will be an increasing difference between the amounts of absorption by reference and sampling units and a greater variation in the capacitance, which may be used to control or modulate the frequency of some device.

The value of the gas being measured may be determined by controlling the amount of energy delivered to the reference cell in order to make the energy at the detector the same for both sample and reference units. Or it may be that an absorbing material is placed in the sample beam to accomplish the same purpose. Still other devices employ a meter which shows a deviation, and the deviation is a measure of the concentration.

A different solution to the problem of analyzing gases by measuring their infrared absorption is shown in Figs. 18·16*a* and 18·17. The source of radiation is a heated wire sealed in nonabsorbing argon gas, Fig. 18·16*a*. The radiant energy is directed and reflected through the sample cell, which has gold-plated walls to ensure maximum transmission of the radiated energy. The path of the energy is through the filter cell, the sensitizing cell, and then to two thermocouples which are connected in series opposition in order that they may measure temperature differences and not the temperature itself.

The parts played by the energy source and the sample cell are quite obvious, but the functions of the filter cell and the two sensitizing cells may not be so clear. However, a complete understanding is necessary if one is to know just how the infrared analyzer can separate the effects of one gas from those of many others which may be in the same sample.

Let us consider the six situations shown in Fig. 18·17. The infrared

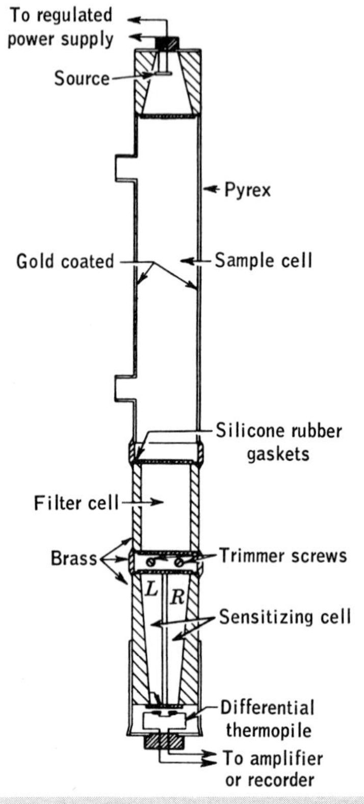

(*a*)

Fig. 18·16 (*a*) This type of infrared analyzer uses a differential thermopile to measure the difference in energy absorptions. (*b*) The components required for a thermopile type of analyzer. (*Leeds and Northrup Co.*)

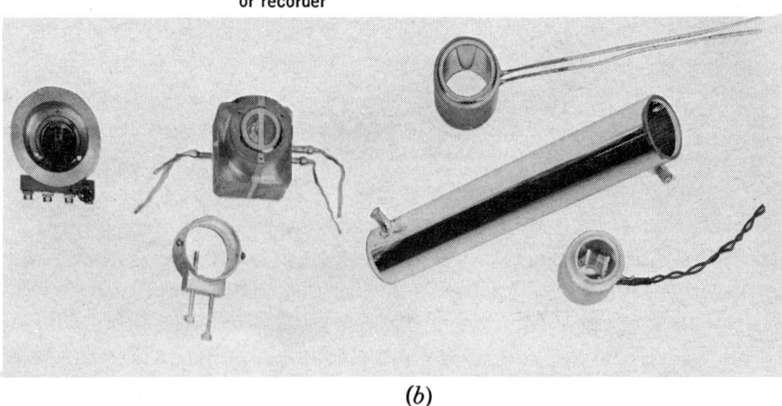

(*b*)

source is at the top of the analyzer. If we consider Fig. 18·17*a*, where there is a nonabsorbing gas such as nitrogen in the sensitizing cell and the same gas is passing through the sample cell, we find that essentially none of the infrared energy is absorbed as it passes from the source to the thermopiles. With no absorption of the energy before it reaches the thermopiles, the thermopiles should be heated to a maximum. Moreover, each one should be heated to the same temperature, so their temperature difference should be zero.

But if we now put enough methane, Fig. 18·17*a*, in the right-hand sensitizing cell to absorb all of the infrared energy coming through that path, then far more energy will reach the left-hand thermopile than will reach the right-hand member. This then should produce a very significant temperature difference, Fig. 18·17*b*. However, we may use this difference as our standard and cause it to appear as a zero on the scale instead of some maximum value; any departures from this maximum difference will then appear as upscale readings.

If we now cause a small percentage of methane to pass through the sample cell along with the nonabsorbing gas, then some energy will be absorbed by the methane and thus will be prevented from reaching the thermopiles. It will be recalled that the methane in the right-hand sensitizing cell has removed the maximum amount of energy. But there will be less energy arriving at the left-hand cell, and consequently, less energy can arrive at its associated thermopile. This smaller temperature difference now appears on the scale, Fig. 18·17*c*, and it is a measure of the amount of methane passing through the sample cell.

If another infrared-absorbing gas such as carbon monoxide, which has no absorption in the spectrum range where methane absorbs, is present in the sample, then it will absorb equal amounts of energy from the energy arriving at the thermopiles, and it should reduce the temperature of each one by essentially the same amount. The difference in the temperatures of the thermopiles should, under these conditions, remain constant, Fig. 18·17*d*.

But if still another gas which absorbs some energy in that part of the spectrum which is characteristic of methane is present, then we will find that it will make a greater decrease in energy going to the left-hand thermopile than it will in that going to the right, because the methane in the sensitizing cell already removes *all* of the energy in its part of the spectrum. (This is based on the assumption that the absorption characteristics of the two gases are not exactly alike, and it is most unlikely that two gases with identical absorption characteristics could be obtained.) The unequal absorption in the two paths will now provide a new temperature difference which will be erroneously interpreted as a variation in the amount of methane present, Fig. 18·17*e*.

However, by introducing into the filter cell in adequate concentration the gas or gases whose absorption band overlaps the absorption band of the methane, it is possible to reduce by equal amounts the energy being delivered to the sensitizing cells. With equal reduction to the inputs to the cells, the effect of the overlapping bands is nullified. Any differential absorption should be then a function of the amount of the

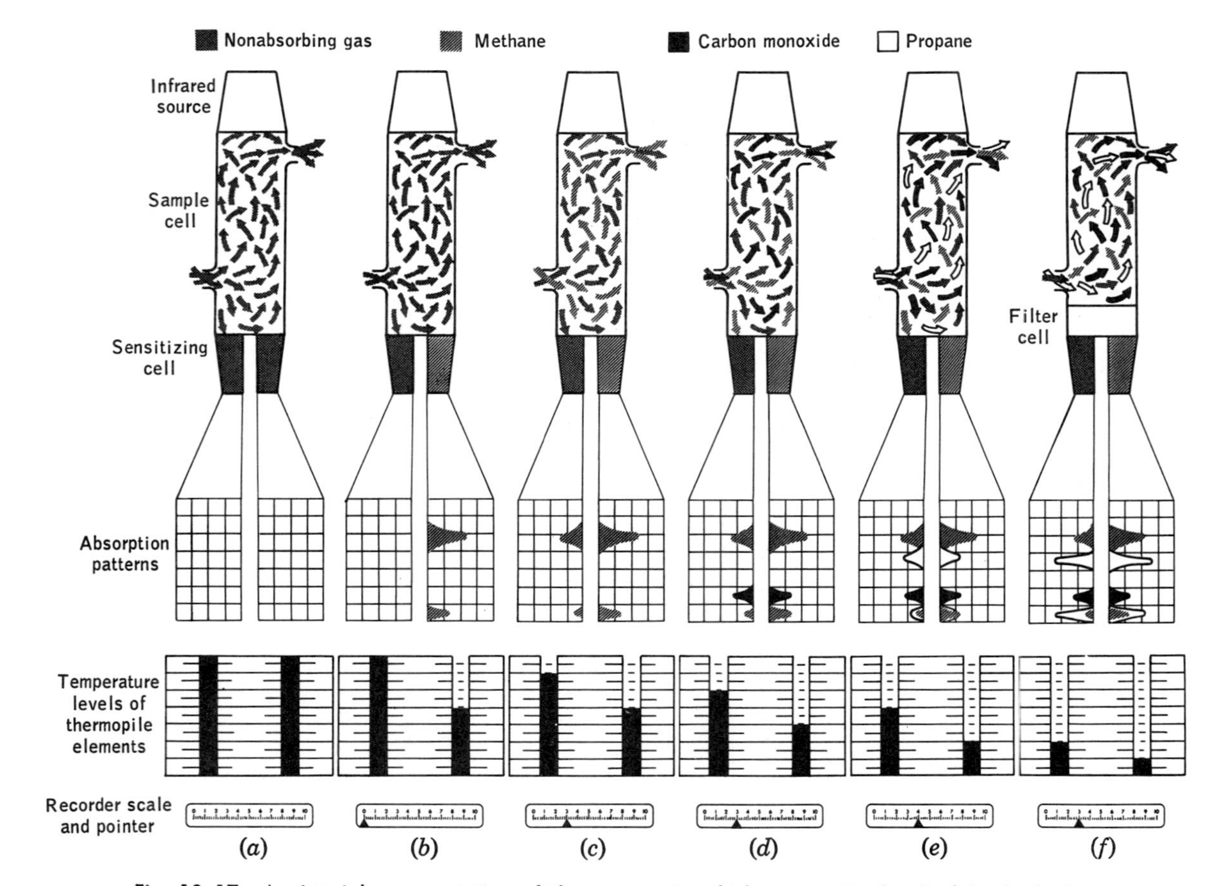

Fig. 18·17 A pictorial representation of the manner in which energy is absorbed by both the sample gas and the sensitizing cells. (*Leeds and Northrup* Co.)

methane present. Absorbing this additional energy from each path should reduce the total energy reaching the thermopiles, Fig. 18·17*f*, but their difference remains the same.

Nondispersive infrared analyzers. The dispersive type of infrared analyzer which has been previously considered uses a fairly wide range of spectra, but the nondispersive type uses but two narrow frequency bands which have been especially selected for the material whose presence is to be measured. Figure 18·18 shows how the choice of a particular frequency will make it possible to distinguish between two spectra which quite closely resemble each other. At about 6.6 μ there is a marked difference in their transmission characteristics. Thus, the choice of this wavelength would make it possible to distinguish between the two types of gases. Inasmuch as the nondispersive type of analyzer uses two different wavelengths, one for reference and one for a measurement of absorption, it is necessary to know the transmission characteristics of the subject gas in order to pick one which undergoes virtually no absorption.

The principles involved in the design of such an instrument will be apparent if we consider each of the components in the order in which the infrared energy proceeds through the equipment, Fig. 18·19*a*.

1. The source. The infrared energy is created by a platinum-rhodium heater which is maintained at a constant temperature of 1000°C. The design is such as to make it radiate as a blackbody.

2. In order to simplify the amplification problems and to provide two signals, the beam is chopped at the rate of about 13 cps. The source is also shielded in such a way that two beams of different wavelengths are

Fig. 18·18 Absorption patterns of four gases. (*Leeds and Northrup Co.*)

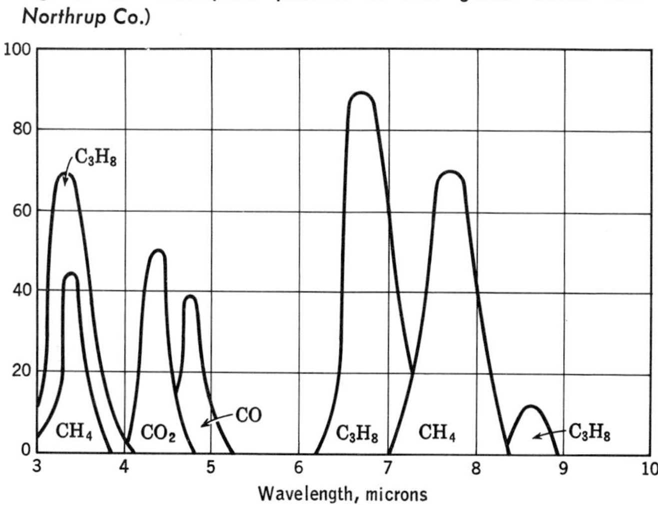

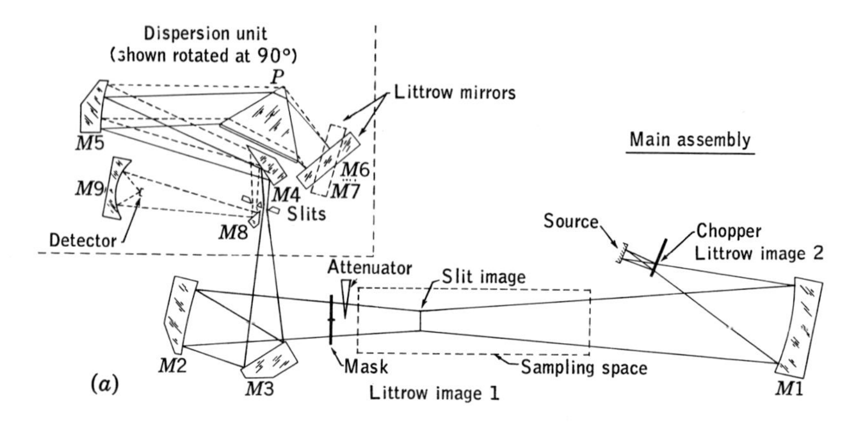

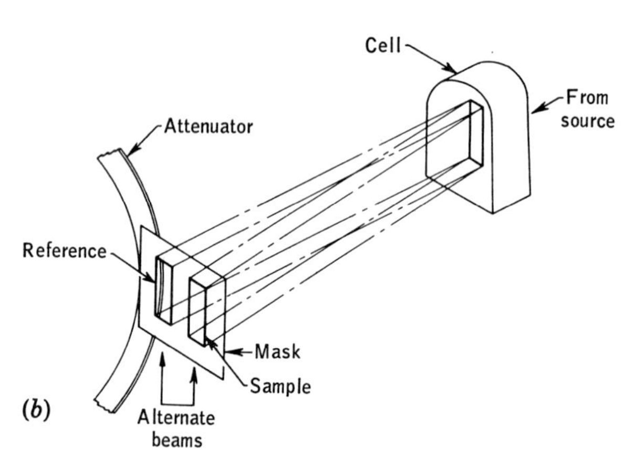

Fig. 18·19 (*a*) The nondispersive type of infrared analyzer, illustrated by the Bichromator analyzer, employs two narrow frequency bands which have been especially chosen to obtain the maximum response. (*b*) Relationship of sample and reference wavelengths at sample cell and mask. (*The Perkin-Elmer Corp.*)

created, and these are transmitted alternately through the rest of the system.

3. Sampling space. Both beams of infrared traverse the sample space where the material being monitored is continually changing to keep up with the process.

4. The attenuator and mask. The two beams which had passed through the sample cell have different optical paths, as is seen in Fig. 18·19*b*. Such a separation makes it possible to insert an attenuator in one path without affecting the other. The amount of attenuation required to equalize the energies of the two light beams is a measure of the amount of the absorbing medium. As in so many other instruments, the attenuator is

positioned to balance the energies delivered to the thermocouple and thus create an output of zero frequency which will not actuate the servomechanisms.

5. The mirror system. The two beams of radiant energy which pass through the sample space and mask will strike mirror M_2 at different angles. This will require that they pass at different angles through the slits, then to mirror M_5, through prism P to the Littrow mirrors, M_6 and M_7. These last mirrors then reflect the beams back through the prism to mirrors M_5 and M_6 and from there to mirrors M_8 and M_9, where the beams are finally concentrated on the detector, which is a thermocouple.

6. The amplifier. The thermal input to the thermocouple is always small in magnitude; in this design it has a frequency of 26 cps. By tuning the amplifier so that it is responsive to this frequency primarily, noise and other troubles which are common with high-gain amplifiers are somewhat reduced. The output of the amplifier is then converted into a direct-current signal of reversible polarity which is then fed into an amplifier of a type previously studied. This final amplifier then actuates the two-phase motor which positions the attenuator.

APPLICATIONS

Furnace atmosphere control. Measurement and control of furnace atmospheres. Carbon dioxide, carbon monoxide, methane, or water vapor content of the atmosphere can be continuously monitored to control oxidizing or carburizing.

Humidity measurement and control. Relative humidity or dew point of the atmosphere or process streams can be quickly and accurately measured.

Combustion control. Measurement of carbon dioxide, carbon monoxide, or unburned hydrocarbons in flue gas to determine efficiency of combustion can be accomplished very simply.

Toxic and combustible gases. Can be used to monitor areas for hazards caused by many toxic and combustible gases and provide immediate warning if the concentration exceeds a safe limit.

Logging of oil or gas wells. Continuous measurement of the methane content is usually desired when drilling an oil or gas well.

Chemical and petrochemical process streams. Many analysis problems encountered in chemical or petrochemical process streams can now be handled.

Reproduced by permission of Mine Safety Appliances.

GAS CHROMATOGRAPHY

Color has been used for many years as a measure of product quality, but color as a measure of chemical composition has also been used for a long time, dating back to about 1910. In the intervening time, the process which still uses the same name no longer uses color as its measure of composition. Consequently, the name is somewhat misleading. Actually, the present processes considered under the heading of chromatography rely largely upon the principles discussed in the section on analysis by thermal conduction.

Principles. A gaseous product may be composed of many different gases, with the result that analysis is made difficult unless one uses a device like the mass spectrograph in which the constituents sort themselves in terms of their mass number. In fact, almost any complex mixture could be analyzed if there were some simple way of making the components so arrange themselves that only one of the components need be considered at any one time.

We might say that the gas chromatograph is such a device in which a column creates a selective absorption of the components which in effect sorts out the components and permits them to enter the gas stream in a fairly regular and predictable succession. Those constituents which are most readily absorbed will take a relatively long time to pass through the column, while those for which the absorption is slight will pass through with only a slight delay. An inert gas is used as a carrier to take the sample through the column, and it is also used as a reference gas when the amount of the constituents is determined. Figure 18·20 shows in simple form the general arrangement which is required when the *elution system* (washing with a gas) is employed. Two other systems of vapor-phase chromatography, *frontal* and *displacement,* are employed, but they are not as suitable as the elution system when employed for industrial monitoring of gas and vapor streams.

The elution system carries the sample, which is a fixed volume, to the column of absorbing material where the carrier gas leaves the sample. As the carrier gas then continues to flow, it carries along with it, in succession, the various constituents as they are released by the column. But

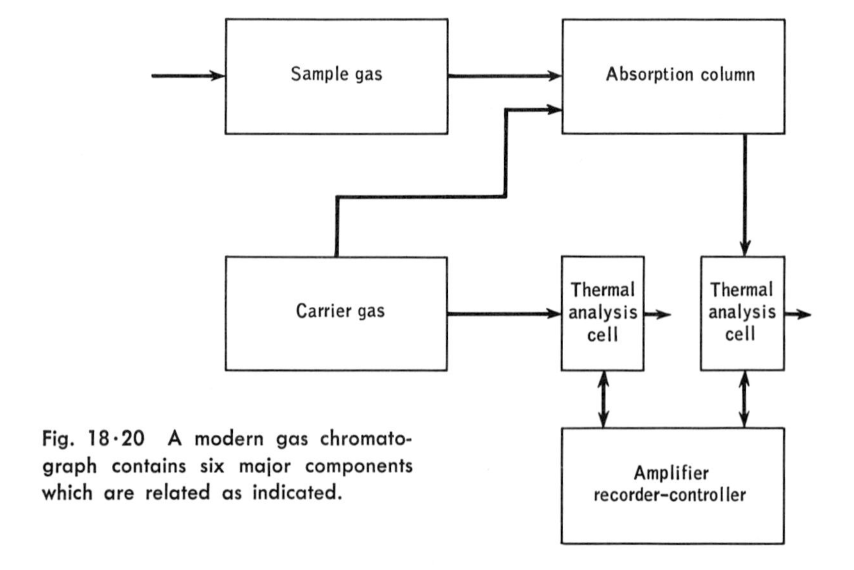

Fig. 18·20 A modern gas chromatograph contains six major components which are related as indicated.

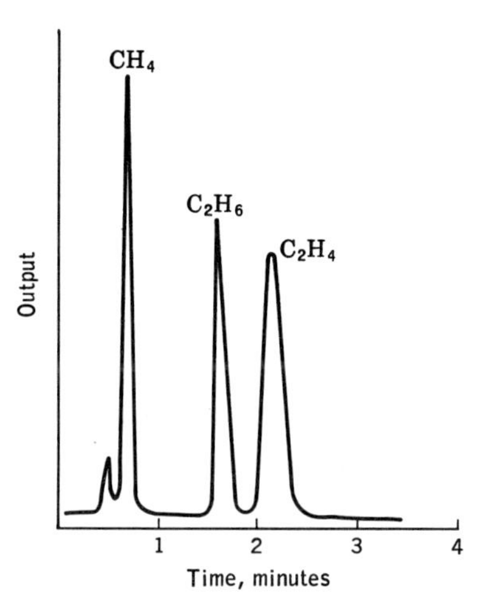

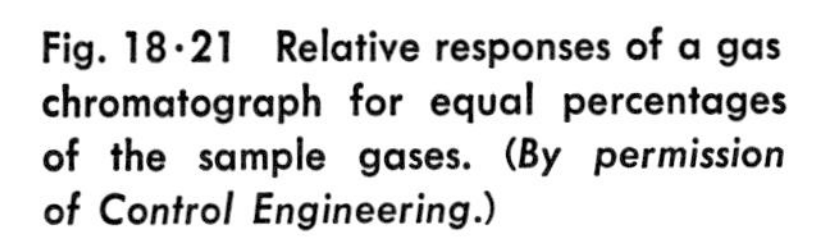
Fig. 18·21 Relative responses of a gas chromatograph for equal percentages of the sample gases. (*By permission of Control Engineering.*)

as it carries them away from the column to the thermal cell where the relative amount of carrier and constituent is determined, it cleans the column and prepares it for the next sample which is to pass through. This process eliminates contamination.

The successful operation of the system is dependent upon the use of such materials in the column that the constituents of the sample will be absorbed and then released to the carrier or stripper gas one at a time. If there were two components of the sample which would be stripped from the column at the same time, then the process would not function for those two materials. Fortunately, in the usual case materials are released with sufficient time between releases so that there is a well-defined change in the response. Such performance is indicated in Fig. 18·21.

The gas used for the carrier plays an important part in the performance of the unit. Helium and hydrogen provide the highest degrees of sensitivity, and helium has the advantage of being inert. Clean, dry air and nitrogen may also be used, but the sensitivity will be somewhat reduced.

REFRACTOMETERS

Many continuous-stream analyzing methods are satisfactory when used with gases but are not designed for use with liquids or solids. Ultraviolet and infrared methods are generally limited to gaseous samples. Refractometers, on the other hand, are designed for use with liquid samples. However, the refractometer suffers from the same liability as the other analyzers which have been considered in that it must be used with binary systems in which only one of the constituents is varying. But if the sample consists of several components which have no significant changes in composition or reaction while only one other com-

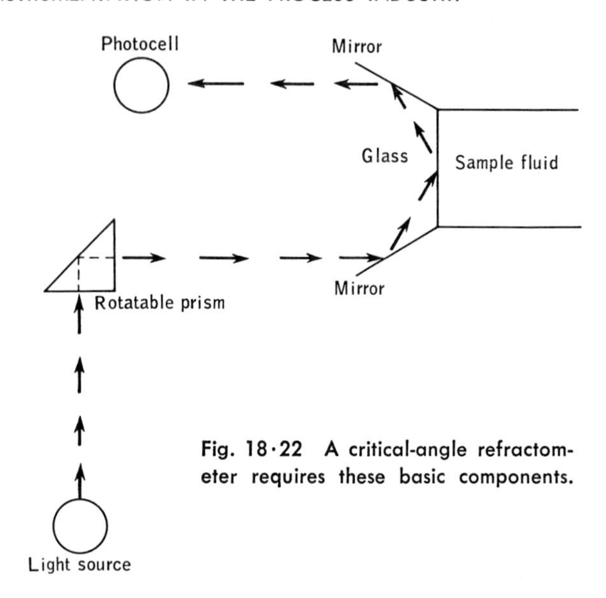

Fig. 18·22 A critical-angle refractometer requires these basic components.

ponent varies, then the system is for all practical purposes a binary system also.

When the sample is turbid or carries much suspended material, the critical-angle instrument is employed. But when a clear liquid is available, the differential refractometer is preferred. Both systems are based on the fact that the index of refraction of a liquid depends upon the composition and concentration of the liquid. Temperature, pressure, and other variables are assumed to remain constant.

Critical-angle refractometer. For any given liquid under specified conditions there is a definite angle at which rays of light will be transmitted through the liquid and not reflected from the surface. Small changes in the concentration or composition of the liquid will bring about large changes in the amount of light which is either transmitted or reflected, provided that the angle at which light impinges on the liquid is close to the critical angle. Figure 18·22 shows in the very simplest schematic the components of such an analyzer. Light enters the prism, which is rotatable, and then is transmitted to a mirror which directs the beam to a glass which separates the sample from the air. If the beam is not transmitted through the liquid, it will be reflected from the surface of the liquid to the second mirror, and then directed to the photocell. Rotating the prism and proper positioning of the photocell will then bring about large changes in the photocell output as energy reaching the cell is varied between the two conditions of transmission and reflection. By means of calibration, the angle at which transmission occurs can be interpreted in terms of concentration.

There are two precautions that must be taken with instruments of this design:

1. The glass should be kept at the temperature of the liquid by suitable shielding. Keeping both liquid and glass at a fixed temperature may be difficult or impossible, so index-of-refraction readings should be corrected for temperature effects.

2. The critical angle is dependent upon the composition of the liquid at the inner face of the glass. It should therefore be representative of the entire batch or sample. The windows must be kept clean by any one of a number of methods.

Du Pont differential refractometer. When clean, clear liquids are available for analysis, it may be desirable to use the differential type of refractometer. The main elements are shown in Fig. 18·23. There is the customary light source, plus some means for obtaining parallel rays which are then directed onto the sample cell. If the cell were filled with a uniform liquid, the rays then would pass straight through to the reflecting prism (if we assume that the walls of the comparison cell have negligible effect). However, if the sample cell fluid and the fluid in the comparison cell are not the same, with respect to their refractive index, then the beam of light will be partially refracted as it passes through the sample cell. This will change the angle at which the rays pass through

Fig. 18·23 The Du Pont differential refractometer requires the adjustment of the position of the comparison cell until equal amounts of light reach each photocell.

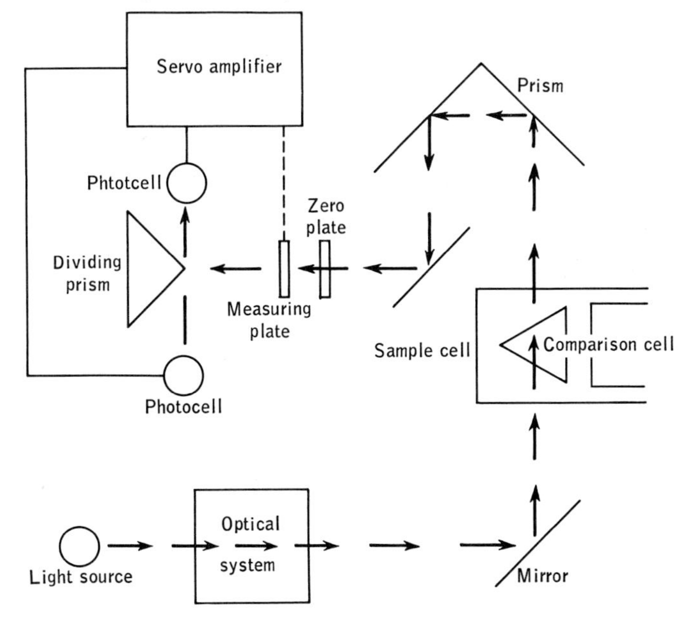

the prism and the rest of the optical system. If there were no diffraction of the light beam, then the light striking the dividing prism would be divided into two equal parts, and each photocell would be affected equally. But when the rays are refracted because of differences between the fluid being analyzed and the comparison cell, then unequal amounts of light will be directed to each photocell. The output of the photocells is then amplified, and the unbalanced output is made to drive the measuring plate in such a direction as to reestablish the condition of having each photocell receive equal amounts of energy. The amount of movement of the measuring plate necessary to bring the amplifier output back to zero is a measure of the index of refraction of the fluid, and hence of its composition.

GAS ANALYSIS BY GRAVIMETRIC MEANS

Perhaps one of the simplest ways to determine some of the characteristics of a material is to weigh the material and then employ some of the basic relationships in terms of kinetic energy, momentum, accelerating torques, and so on. From the data in Table 18·2, it is apparent that there is quite a wide variation in the specific gravities of the common gases. It must be assumed that many which are not listed will also show such values that there is a significant difference between them and their neighbors in the tabulation.

If definite volumes of gases are given specified angular accelerations, then the torque required will be a linear function of the mass and consequently of the specific gravity of the gas. Similarly, the torque required to decelerate or stop the movement of such gases will be a linear function of their density. These principles are employed in the analyzer shown in Fig. 18·24.

Two fans, or impellers, are driven at the same speed by means of a constant-speed motor. Through one fan is forced air, while through the other is forced the gas whose specific gravity, and in turn an indication of its analysis, is desired. Equal volumes of gas are driven through each unit in any given time. As the gas leaves each impeller, it strikes an impulse wheel, which is essentially nothing but a fan blade so positioned that the air or gas striking it will create a torque.

In actual practice these impulse wheels are located on separate shafts and are arranged to turn in opposite directions. The two impulse wheels are then coupled together so that their torques will oppose each other. Any difference in their torques is then used to provide rotation against a spring. When air, for example, is passed through each impeller at the same rate, the torque developed at each impulse wheel should be the same, and the net torque developed is zero. But if a gas such as hydro-

Table 18·2 Specific Gravity of Gases and Vapors

Name	Sp gr, air = 1	Sp gr, oxygen = 1	Name	Sp gr, air = 1	Sp gr, oxygen = 1
Acetylene	0.9073	0.8208	Helium	0.1384	0.1239
Air	1	0.9047	Hydrogen	0.0695	0.0629
Ammonia	0.5963	0.5395	Nitrogen	0.9672	0.8751
Argon	1.3796	1.2482	Oxygen	1.1053	1.0000
Carbon monoxide	0.9671	0.8750	Propane	1.562	1.414
Chlorine	2.486	2.249	Sulfur dioxide	2.2638	2.0482

gen is passing through one impulse wheel while the much heavier air is passing through the other, then the torques developed will not be equal and the torque equality must be realized by the addition of a resisting torque from the spring which restrains the pointer.

Specific gravity measurements can be used to express the percentage of a gas present, hydrogen for example. The measurement of the amount of dissociation, by knowing the specific gravity, is of use during nitriding operations. From the standpoint of safety, purging operations must be carried out until nonexplosive conditions are obtained. Too little purging is unsafe, but excessive periods increase costs and accomplish nothing.

Fig. 18·24 The gravimetric type of instrument operates on a simple mechanical principle: equal volumes of two gases, moving at equal speeds, will develop torques proportional to the density of the gases when each gas undergoes the same acceleration or deceleration. (*The Permutit Co., Ranarex Division.*)

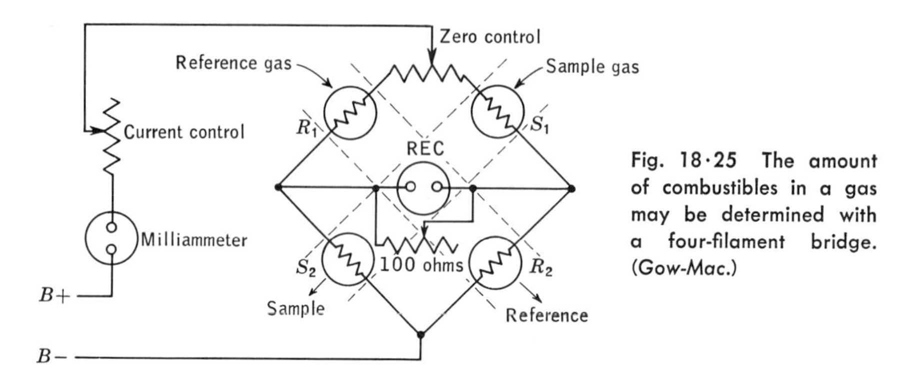

Fig. 18·25 The amount of combustibles in a gas may be determined with a four-filament bridge. (Gow-Mac.)

COMBUSTIBLE-GAS ANALYZERS

There are some situations in which it is desirable to know the amount of combustible gases available in the gas samples, in contrast to the conditions in which it is desirable to know the percentage of specific gases.

Most analyzers of this type depend upon the catalytic action of platinum to oxidize the combustibles. Inasmuch as heat is furnished in such an oxidation process, the temperature of the combustion zone will be higher than some reference point where no heat-producing reaction is taking place. In many ways this is exactly the same sort of condition which is employed in the gas thermal-conductivity analyzers, the difference in the temperature of the two zones being representative of the amount of combustibles present.

When the resistance of the filaments is used as a measure of the temperature, Fig. 18·25, errors may occur with time as the filament evaporates. Thus, the resistance will increase despite the fact that the temperature of the filament has not changed. It has been determined by some that with a fixed supply voltage the filament temperature tends to remain constant despite its decreased cross-sectional area and its increased resistance. Inasmuch as the temperature tends to be independent of the filament resistance, it appears desirable to measure the temperature and not the resistance. This is done by attaching thermocouple leads directly to the filaments, Fig. 18·26. In this manner, any exothermic reactions will increase the temperature of the active-cell filament while the reference unit remains unchanged. The two thermocouples which are connected in series opposition will produce emf's which are then a measure of the amount of combustibles in the gas sample.

SUMMARY

1. When a material is being analyzed for its composition and percentage of constituents, it is necessary to use properties or qualities not possessed by other materials or components being analyzed at the same time.

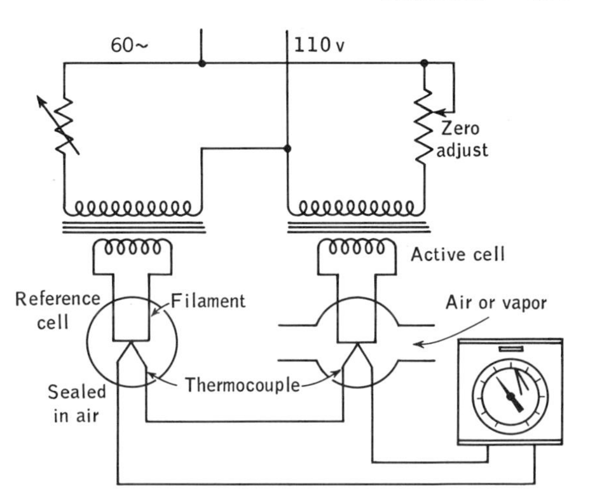

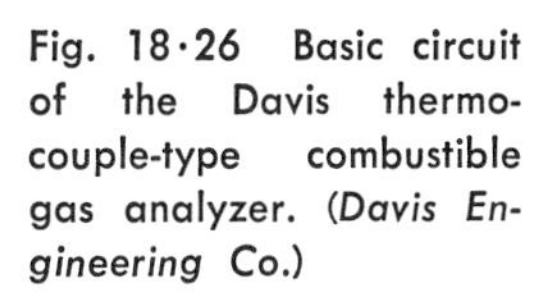
Fig. 18·26 Basic circuit of the Davis thermocouple-type combustible gas analyzer. (*Davis Engineering* Co.)

2. In general, the material being measured by the methods under study must be a binary. It must act as though it is composed of but two parts, one of which responds to the methods employed and the other of which does not. Even though a compound is composed of several components, if only one is affected by the method, then the material is considered to be a binary.

3. The rate at which heat can be dissipated by a filament is dependent upon the gas which carries away the heat by convection currents, for fixed temperature and pressure conditions. Not only do the different gases possess different cooling properties, but the size of the chamber in which the cooling takes place has certain optimum dimensions for each of the various gases. Selectivity in measuring the gas can then be enhanced by selecting the correct critical dimensions.

4. The analysis of gases for oxygen is one of the most important gas analyses. Most analyzers employ one of two properties possessed by oxygen: (*a*) When cold, oxygen is paramagnetic and will respond to magnetic fields. When the analysis is being made with a thermal conduction instrument, the use of a magnetic field creates a magnetic draft which increases the cooling effect as the cool oxygen is attracted to replace the nonmagnetic heated oxygen which is displaced. (*b*) Oxygen combines readily with hydrogen. When the hydrogen exists in an electrolyte in the form of a film of gas covering an electrode, it effectively insulates the electrode. As a direct consequence, the current in the system will be small. But if oxygen can combine with the hydrogen, the insulating film is reduced in effectiveness. For a constant-potential system, the flow of current then depends upon the amount of oxygen.

5. Mass spectroscopy employs some of the most fundamental characteristics of materials: (*a*) For a given accelerating force, the acceleration is inversely proportional to the mass. (*b*) Ions of various materials have different weights and hence are accelerated at different rates. (*c*) These ions follow a curved path when they are caused to pass through a magnetic field. For a given voltage and magnetic field, the radius depends upon the mass. (*d*) For practical design, varying the applied voltage can cause all of the ions to be counted at some common point. Composition and voltage are therefore related.

6. Optical analysis may employ many different phenomena and a fairly wide portion of the electromagnetic spectrum ranging from the ultraviolet through the infrared. In general, methods which employ the ultraviolet will have a greater

sensitivity, because of the greater amount of ultraviolet energy, than will methods employing the infrared energies. But those employing infrared techniques may be more selective, because the absorption bands in the infrared regions are much sharper than the corresponding bands used in ultraviolet spectroscopy. When there are sharp narrow absorption bands of frequencies, absorption by more than one constituent is minimized.

7. The analysis for the identity of the material is quite different from that used when the percentage of a given constituent is to be determined. By measuring the absorption throughout a band of frequencies, an absorption pattern is obtained. By comparing the pattern with other patterns, the identity can be determined when a known material having the same absorption spectra is located. Wide-band absorption spectra are unique and are characteristic of but one compound.

When the percentage composition of some constituent is desired, the method is to cause the material to act as though it were a binary. By different sensitizing techniques, all constituents except the desired one can be made to give no significant response. All variations are then caused by the material under study.

8. Infrared analysis may use either of two general methods: the dispersive and the nondispersive. The dispersive system does not employ an optical system other than a polished reflecting surface which causes the energy to be reflected from surface to surface until it finally reaches the absorbing cell. The nondispersive system, using the more usual optical methods, reflects and refracts the beams as it causes them to actuate the detector-amplifier system.

9. Gas chromatography is a batch type of analysis, in contrast to the other methods which have been considered. A sample is sent to an absorption column; the different constituents are then removed in a definite sequence by a carrier gas. The time of release of the material indicates its composition. The carrier gas thus carries but one variable through the analyzer at any one time, and all systems which are being analyzed are binaries.

10. The refractive index of a material is proportional to (*a*) the composition, and (*b*) the concentration of the material. By measuring the index of refraction, the degree of concentration can be determined. When dirty or turbid materials are to be analyzed, the critical angle is more generally employed. The use of the differential refractometer is restricted to use with clear liquids.

11. Because gases have materially different densities, it is possible to make certain analyses by measuring the forces of deceleration involved when different gases undergo the same change in speed. This same sort of measurement makes it feasible to determine by mechanical means the degree of dissociation of certain gases used for heat-treating operations.

12. An analysis for nonspecifics classified as combustibles uses the principle of the thermal conductivity cell. By using platinum heaters which catalyze the combustion of combustible gases, the presence of such gases will increase the temperature of the filament as the gases oxidize. In the comparison unit, no oxidation can take place. The difference in temperature of the filaments is then a function of the amount of oxidation taking place.

QUESTIONS

18·1 Of what significance is the analysis of materials while they are in process? What industries might be most likely to make great use of instruments for the purpose? Which might find little use for analysis instruments?

18·2 What are the basic assumptions made in the design of continuous analyzers?

18·3 When materials have conflicting or complementary characteristics, what steps can be taken to separate their individual effects?

18·4 What is the essential difference between the Pirani pressure gage and a thermal conductivity gage?

18·5 For any given temperature change, which undergoes the greater resistance change, a platinum filament or a thermistor? What are two outstanding differences between these two types of components used in thermal conductivity bridges?

18·6 Is the size (in terms of volume) or the dimension of more significance in the design and operation of thermal conductivity bridges?

18·7 Which common gases are paramagnetic? Diamagnetic?

18·8 Which gas having strong magnetic characteristics is of the most significance in industrial operations?

18·9 Is oxygen ever magnetic? Nonmagnetic? Are your answers of any significance in the design of analysis instruments?

18·10 What property of oxygen is of use in oxygen analyzers using current-conductivity principles?

18·11 What are the significant characteristics of mass spectrometers? What are the basic principles upon which their design is based?

18·12 What is the significance of each component shown in Figs. 18·3*b* and 18·5?

18·13 What is the requirement for each part shown in the schematic Fig. 18·9*a*?

18·14 Could we say that the ultraviolet analyzer, Fig. 18·13, is a sort of optical bridge? How would you compare it with a Wheatstone bridge?

18·15 Of what practical significance is the fact that different compounds have quite different electromagnetic absorption patterns?

18·16 What are the important differences between the two dispersive types of infrared analyzers which were discussed in detail? Are the basic principles of the one applicable to the other?

18·17 What are some of the gases which have pronounced absorptions in the ultraviolet portion of the spectrum? The infrared portion?

18·18 What means are used to eliminate the effects of gases which absorb in the same portion of the spectrum as does the gas being analyzed?

18·19 From the viewpoint of analysis, what is the most significant frequency shown in Fig. 18·14?

18·20 In what significant ways do nondispersive infrared analyzers differ from the dispersive types? Which has the more involved optical system?

18·21 Would you be correct in saying that each of the analyzers in Question 18·20 produces the same final product at the detector?

18·22 In what way does an analyzer using the principles of gas chromatography differ from one using portions of the visible or nearvisible spectrum?

18·23 What are the basic principles upon which gas chromatography is based?

18·24 What is the significance of each component indicated in Fig. 18·20?

18·25 What is the method used in chromatographic processes to separate the components of the samples?

18·26 Which of the gases listed in Table 18·2 have the lowest specific gravities? The highest? List the first five in order of lightness and the five heaviest in order of their weight.

18·27 What are the essential components of a critical-angle refracting analyzer?

18·28 In what ways does a differential refractometer differ from one of the critical-angle type?

18·29 What are the operating conditions which might dictate which of the refractometers of Question 18·28 would be employed?

18·30 Do all gases respond to the principles employed in the design of an analyzer such as is shown in Fig. 18·24? How fast would such a device be in its operation?

18·31 What would happen to the performance of the bridge shown in Fig. 18·25 if the polarity of the battery were reversed? If you were to reverse the connections to the milliammeter at the same time, what would you be able to say about the operation of the device?

18·32 If one can measure the resistance of filaments and thus infer the temperature, Fig. 18·26, as easily as one can get the filament temperatures, why should one bother about applying thermocouples to find the filament temperatures?

18·33 What would happen to the performance of the analyzer, Fig. 18·26, if the leads to one of the thermocouples were reversed when the unit was repaired in the instrument shop? What can you predict about the performance of the unit? If you reversed the polarity of the a-c supply, what would be the effect upon the operation? If a chromel-alumel couple were replaced by a copper-constantan couple during the overhaul period, what do you feel reasonably sure would happen?

19 Industrial Measurements with Radioisotopes

In Chap. 15, in the discussion of optical and radiation pyrometers, it was noted that energy is radiated into space. In general, the higher the temperature, the greater the amount of energy radiated.

In industrial process control there is another type of radiation which is becoming of greater and greater use, namely, that produced by the spontaneous decomposition of materials into matter having a lower energy level. The differences in energy levels occur because energy has been created at the expense of matter. The radiant energy so produced may be so weak that it can be stopped by a thin sheet of paper, or it may be strong enough to penetrate an inch or more of steel. Because this radiated energy is absorbed to some degree by all materials, the amount of absorption can be used to measure such items as thickness, specific gravity, level and interfaces, amount of corrosion, and so on. Before learning some of the techniques employed, we need to know more about the origin and characteristics of the different radiations.

An isotope is defined as one of two or more species of an element having the same number of protons and electrons, but differing in the number of neutrons contained in the nucleus. There may be any number of isotopes of the same element; uranium may have as many as fourteen, hydrogen may have three.

Some certain arrangements of the atom seem to be preferred by nature, because there are no natural changes from these conditions. Other arrangements are not stable; a change from the unstable condition may be initiated by having a neutron penetrate the nucleus of the atom. When the particles of the atom rearrange themselves, energy is radiated.

A radioisotope may be defined as a form of an element which is unstable and which exhibits the property of radioactivity. About 900 radioactive isotopes and 270 stable ones have been studied.

Table 19·1 Half-life Periods

Isotope	*Half-life*	*Isotope*	*Half-life*
Uranium 238	$4.51(10^9)$ years	Calcium 45	163 days
Potassium 40	$1.3(10^9)$ years	Iron 59	45 days
Uranium 235	$7.1(10^8)$ years	Phosphorus 32	14.3 days
Plutonium 239	$2.43(10^4)$ years	Radon 222	3.825 days
Radium 226	$1.6(10^3)$ years	Iron 52	7.8 hr
Strontium 90	25 years	Nitrogen 16	7.4 sec
Hydrogen 3	12.5 years	Polonium 212	$3(10^{-7})$ sec
Cobalt 60	5.8 years		

RADIOACTIVITY

As unstable materials disintegrate to their more stable states, they emit neutrons, charged particles, electromagnetic waves, or all three. Such a process is known as radioactive decay, and the particles or emanations produced are known as *radiation*. As this radiated energy collides with other materials, it is sometimes able to knock loose electrons. This produces *ionization*, and it is the basic principle upon which most measuring and control instruments depend if they employ radioactive isotopes.

Half-life. Radioisotopes change into the more stable states at a perfectly predictable rate. We cannot tell what particular atoms are most unstable, and thus most likely to radiate next. But we do know that half of the atoms will have had this change within a certain period of time. This period is called the *half-life*. Table 19·1 lists some of the more common materials having this type of decomposition and their half-life periods. Table 19·2 contains a list of radioisotopes and their radiations and uses.

If the half-life is 1 day, then at the end of each successive 24-hr period, the radioactivity left will be but one-half of what it was at the beginning of that 24-hr day. Thus, the radioactivity will decrease and

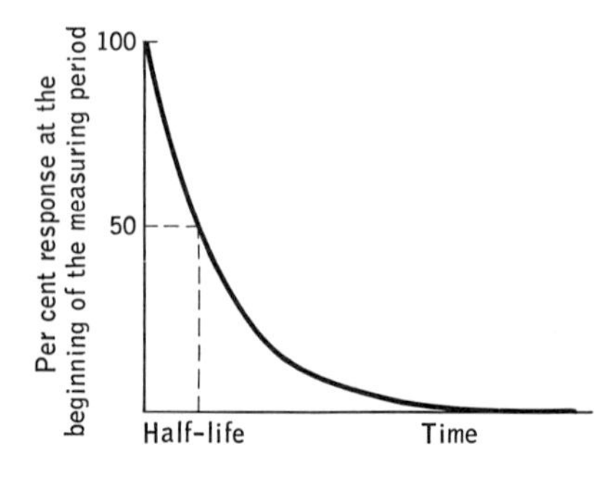

Fig. 19·1 Radioisotope materials undergo a predictable loss in the percentage of the material present at that time. By definition, the half-life is the time required for the number of disintegrations per unit time to decrease to 50 per cent of the value at the beginning of the period. Because one never can get to zero by taking half of a half, and so on, there is no time when the activity becomes zero. Hence the requirement for using some other unit of time, the half-life.

Table 19·2 Radioisotopes Used as Measuring and Control Gages

Isotope	Half-life	Useful radiation	Energy, mev	Used in	To measure and control
Thallium 204	4.1 years	Beta only	0.8	Thin-sheet gages	Fine paper, newsprint, tissue, plastic film
Strontium 89–90	25 years	Beta only	2.3	Thin-sheet gages, reflection gages, coating gages	Board; metal foil; rubber; linoleum; calendering operations, chiefly paper and rubber
Cerium 144	290 days	Beta Gamma	3.0 0.085	Thin-sheet gages	Intermediate grades of steel sheet
Ruthenium 106	1 year	Beta Gamma	3.5 0.7	Thin-sheet gages	Top range of beta gages, up to 0.070-in. steel; glass and slag; wood, fiber
Iridium 192	75 days	Gamma	0.35	Corrosion gages	Plate, pipe- and tank-wall thickness; also used for industrial radiography
Cesium 134	2.3 years	Gamma	0.7	Some types of level gages	Liquid level in tanks
Cesium 137	33 years	Beta Gamma	0.5 0.6	Some varieties of thickness gages	Finest grades of paper
Cobalt 60	5.3 years	Gamma	1.2	*a.* Density gages *b.* Heavy-sheet gages	*a.* Any fluid material *b.* Up to 1-in.-steel sheet, hot or cold; heavy grades of rubber (tires)

SOURCE: Stewart and Leighton, *Control Engineering*, March, 1955.

decrease without end, but it will never get to zero. This situation is represented in graphical form in Fig. 19·1.

It will be noted in Table 19·1 that these half-life periods vary from fractions of microseconds to billions of years. For some purposes, we desire materials which have very short half-life periods. But for most industrial applications, the half-life should be a period of months or years to ensure adequate radiation, constancy of performance, and a minimum of recalibration or adjustment.

TYPES OF RADIATION

For many years, we have known that there are four types of radiations which have quite different penetrating capabilities, Table 19·3: (1) alpha particles, (2) beta particles, (3) gamma rays, and (4) neutrons.

Alpha particles are composed of two protons and two neutrons released from the nucleus. Basically, they are unstable helium atoms stripped of their electrons. To become stable, they must regain two electrons. These particles have the following notable characteristics:

1. They are positively charged.
2. They are the heaviest subatomic particles.
3. They possess very little penetrating power; nearly all can be stopped by thin sheets of paper.

Table 19·3 Penetrating Power of Different Radiations

Ray	*Penetration*
Alpha	Stopped by thin foils
Beta	Stopped by heavier foils and sheets; stopped by ¼-in. steel
Gamma	Not completely stopped by any material, their intensity decreases exponentially with distance or thickness of the material; appreciable radiation through 12 in. of steel
Beta, 0.8 mev	Completely stopped by 0.015-in. steel
Beta, 2.3 mev	Completely stopped by 0.045-in. steel
Gamma, 0.3 mev	Intensity reduced to 1/250 original value by 4-in. steel
1.2 mev	Intensity reduced to 1⁄16 original value by 4-in. steel

Beta particles are quite different:

1. They are negatively charged.
2. They have very small mass.
3. They travel only a few feet in air, and they are stopped by about an inch of wood or one-quarter inch of aluminum.

Gamma rays are entirely different from either the alpha or the beta particles. Consequently, they have quite different uses in industry.

1. They are waves of energy, not particles.
2. They have no electric charge.
3. They travel at the speed of light.
4. They can travel relatively great distances.
5. They are extremely difficult to stop. They can be stopped mainly by thick sections of lead or concrete. They pass easily through most solid materials.

Fig. 19·2 Some of the common radioisotopes and the relative energies of their radiations. (*Tracerlabs, Inc.*)

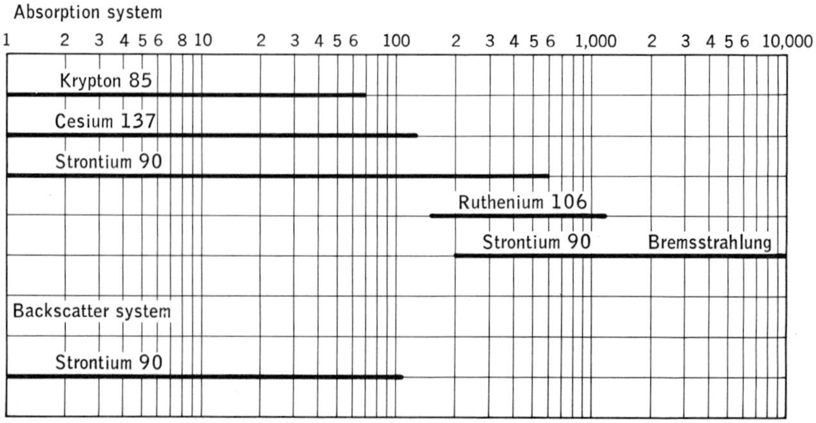

It will be noted that the alpha particle is so easily stopped that its industrial uses probably will be few, whereas the beta particles and gamma rays possess enough energy to be useful in many gaging instruments. Gaging, analyzing, and process control are the applications of most interest in this study. At the present time, the *neutron* has but limited use in industrial measurements.

THICKNESS GAGING

Accurate thickness gaging of rapidly moving materials presents many problems if one uses the conventional techniques of weighing or using micrometers or other gages. Ideally, the measurement should be made while the item is being produced and at such a time that corrections can be applied to ensure a product of correct quality. It is also desirable that the gaging be nondestructive and continuous, without physical contact with the material. Radioisotopes possess characteristics that permit many of these ideals to be realized, Fig. 19·2.

Theory of operation. By placing a source of radiation on one side of a piece of material to be gaged and a radiation detector on the other, a thickness gage can be made. Or the radiation may pass through the material to a reflector, then back through the material to enter an

Fig. 19·3 The amount of energy absorbed by the material being gaged is used as a measure of the thickness. Because the amount of energy reaching the ionization gage is so small, amplification is required before meters and recording-controlling devices may be employed. (*Tracerlabs, Inc.*)

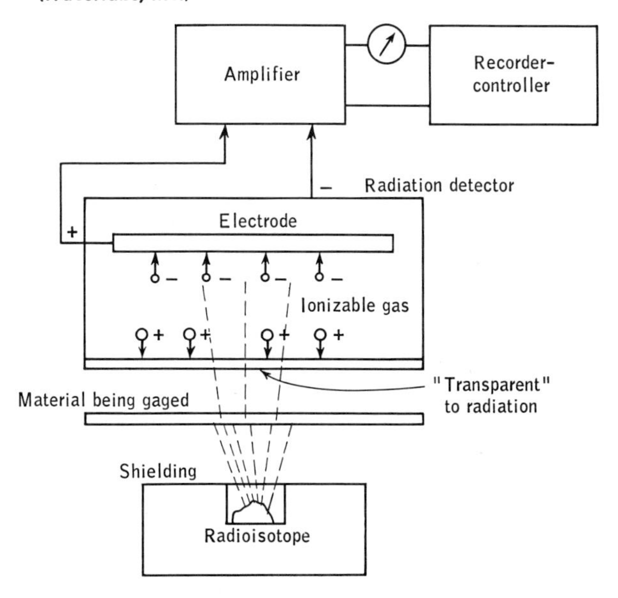

Fig. 19·4 Radioisotopes may be used for the continuous thickness gaging of very thin materials whose strength will not permit the use of gages which make physical contact. (*Tracerlabs, Inc.*)

ionization chamber, Figs. 19·3 to 19·6. As the beta rays collide with the molecules of the gas in the ionization chamber, the gas becomes ionized. If a positively charged electrode is placed in the chamber, many of the electrons liberated by the collisions will reach the electrode. The current created by the movement of the electrons to the electrode is extremely small, being of the order of one-billionth of an ampere, or 10^{-9} amp. If the material has a greater density—in this case we are interested in the weight per square centimeter or per square inch or per square foot for a given thickness—then fewer beta particles will reach the ionization chamber because more of them have been absorbed by the intervening material being gaged, Fig. 19·7.

It should be understood that these gages do not measure thickness directly. What they do measure is an absorption based on the weight per unit area. *We infer the thickness from the weight per unit thickness.* A piece of rubber ¼ in. thick and one ½ in. thick should absorb the same amount of energy if each has the same weight per square inch, or other unit area. Thus, *this is a thickness gage only if the specific gravity remains constant.* Stated another way, the gage measures the *amount* of material per square centimeter or per square inch. Only so long as the density remains constant will our inferred thickness be correct. Unless there is a linear relationship between thickness and weight (for the range being considered), the inferred thickness will be incorrect unless the error can be precorrected by calibration. Random variations in density will create errors in the indicated thickness.

One other consideration in connection with the basic theory is that the materials or constituents have the same average atomic number. The atomic number is the number of orbital electrons of the atoms. If the number is below about 10, no error seems to result because of the constituent materials. According to the Tracerlab, Inc., paper, plastic, unloaded rubber, rubber with small percentages of loadings, wet paper, paper with less than 10 per cent titanium dioxide filler, paper with clay filler, etc., all measure on a beta gage in proportion to their weight per unit area; and a gage calibrated for use with one will be correct when reading another. In cases where a beta gage is to be used to measure

Fig. 19·5 Multiple gages may be required for the control of thickness coatings applied to materials of variable thickness and density. (*Minneapolis-Honeywell Regulator* Co.)

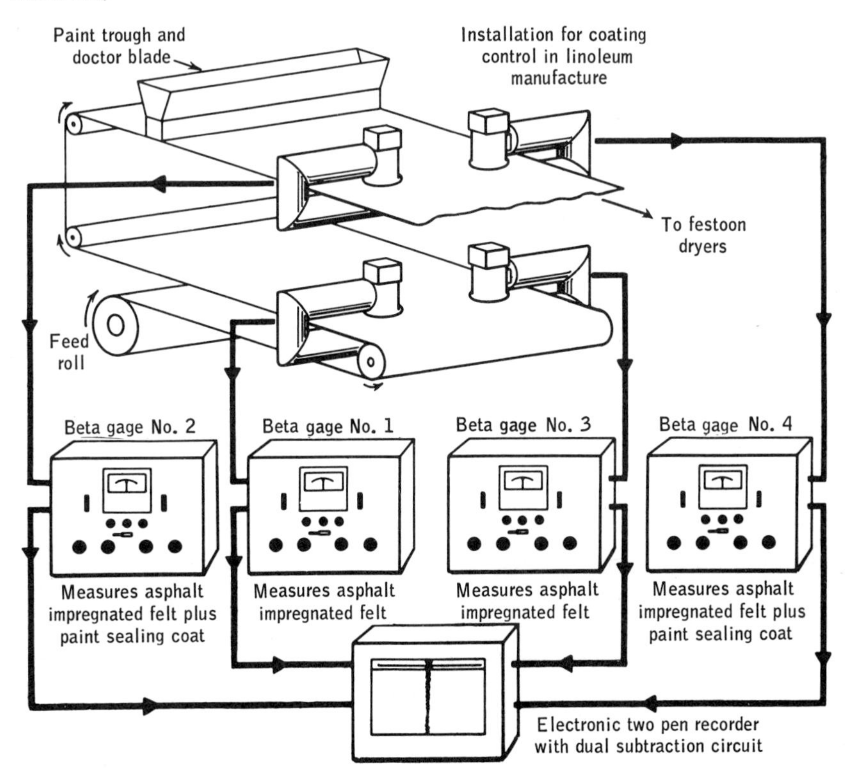

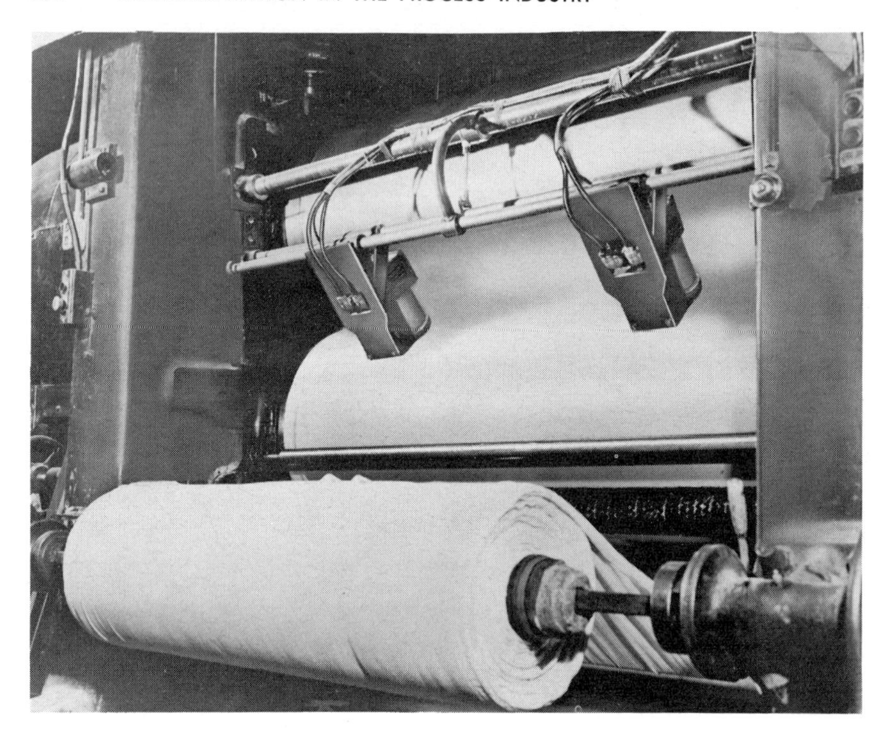

Fig. 19·6 The backscattering principle. (*Tracerlabs, Inc.*)

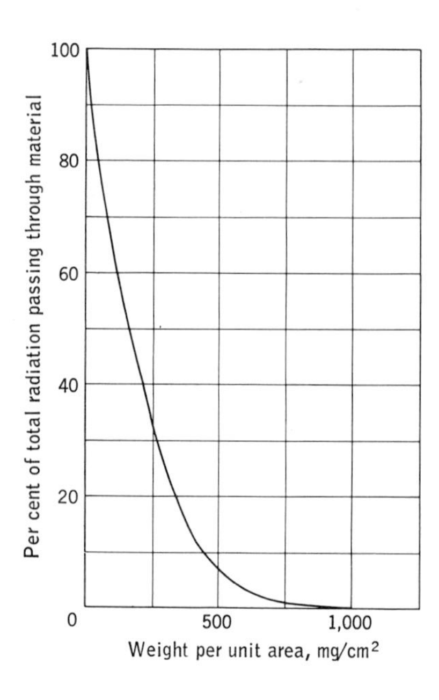

Fig. 19·7 The percentage of radiation passing through the sample decreases very rapidly with increases in the weight density. (*Tracerlabs, Inc.*)

materials of grossly different average atomic numbers, it will be necessary to reset the gage to a calibration corresponding to each material.

Backscatter gage. When it is impracticable to install the source on one side of the material and the ionization chamber on the other, a different principle is used. Just as light will be reflected back toward the source when it falls on certain types of surfaces, so will radiation be reflected by certain materials. In practice, the reflecting material can be aluminum, stainless or chrome-plated steel, or iron. Normally, only thin materials can be gaged by this method because of the double absorption.

Basic circuits. A simplified schematic of an ionization gage circuit is shown in Figs. 19·8 and 19·9. Figure 19·9 may be said to consist primarily of two parts: the portion of the circuit to the left of the dividing line *YY* is the voltage-generating and balancing circuit, while the portion to the right is primarily a bridge circuit in which the electrometer tube is the variable resistor. The batteries *B3c* and *B3d* are two other legs of the bridge, while *R*3 constitutes the fourth member. Any variations in the resistance of tube CK5886AX will unbalance the bridge. The degree of unbalance will be indicated by the meter *M*1.

The voltage-generating portion of the circuit may be said to start with the ionization chamber. A positive potential is applied to the anode to attract any electrons which have been liberated by radiation striking gas molecules in the chamber. They create an extremely small current, about $2(10^{-9})$ amp, which is then caused to flow through a very high resistance element, about 5 billion ohms. However, to cause this small current to pass through this very high resistance will require a potential difference of 10 volts. This potential can be nullified by the proper setting of the potentiometer in the circuit containing battery *B3b*. Now if some material is placed between the source of radiated energy and the ionization chamber, less energy will reach the chamber and the current through the ionization chamber will decrease. This will of neces-

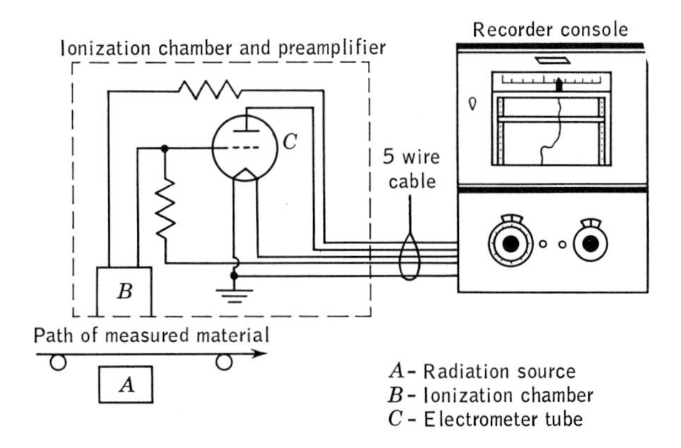

Fig. 19·8 A simple schematic drawing showing the relative locations of the various components of the measuring system. (*Minneapolis-Honeywell Regulator Co.*)

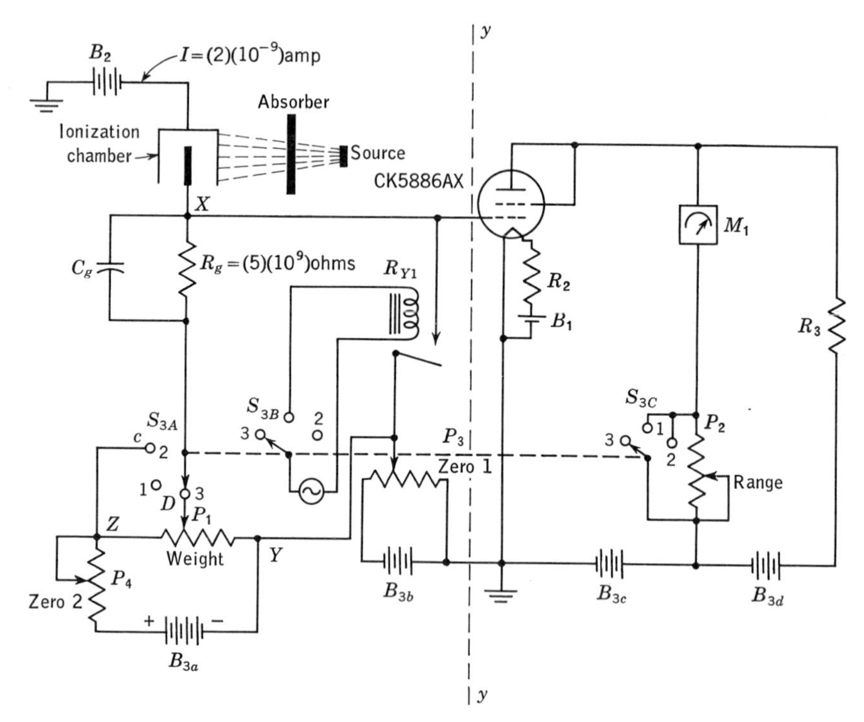

Fig. 19·9 A simplified schematic of a beta gage. (*Tracerlabs, Inc.*)

sity decrease the potential appearing at the terminals of resistor *Rg*. The potential applied to the grid of the electrometer tube is not now balanced to zero by the previous adjustment; and the increased grid potential will decrease the tube resistance. With the bridge unbalanced, the meter will now deflect. The amount of the deflection will be a measure of the amount of energy absorbed by the material, and thus the thickness of the material may be inferred by calibration. In practice, it is desirable to introduce another voltage into the circuit so that the meter will merely indicate deviations from the desired set point. This is accomplished by the weight potentiometer. This can insert into the circuit a potential which just compensates for the reduced voltage across *Rg* when the material of the correct weight is being gaged. With the circuit again balanced in this manner, any deviations in thickness will again create an unbalance in the potentials applied to the grid of the tube. By suitable calibrations, any deviation of the meter from its zero position can be interpreted in terms of deviations of weight per unit of area of the material being gaged.

This unbalance current in the meter circuit can be taken to amplifiers which can by various techniques sense the direction of the deviation and then apply corrective action to the valves, speed of operation, or any of the other factors which may be necessary to return the product to the desired set point.

The behavior of the meter-balancing circuit in Fig. 19·9 may be more easily understood if the two parts which comprise it are considered without reference to the rest of the circuit. In Fig. 19·10*a* we find two circuits having different potentials and different resistances but equal currents. The potentials have such polarities that, in the meter portion of the circuits, the two currents are flowing in opposite directions.

If we combine the two circuits, Fig. 19·10*b*, so that a single meter measures both currents, then the net flow of current through the meter should be zero, and it will remain at zero so long as the voltages and resistances remain fixed in value. But if the resistance of one resistor, the 100-ohm one, for example, decreases, then that resistor will pass more current through the meter in one direction than the other circuit is passing in the reverse direction. Under these conditions, the meter must deflect.

In many circuits, vacuum tubes function as variable resistors. That is the way in which the electrometer tube functions here: as a variable resistor whose value will determine whether or not the net meter current is zero.

Fig. 19·10 Development of the bridge circuit used in Fig. 19·9.

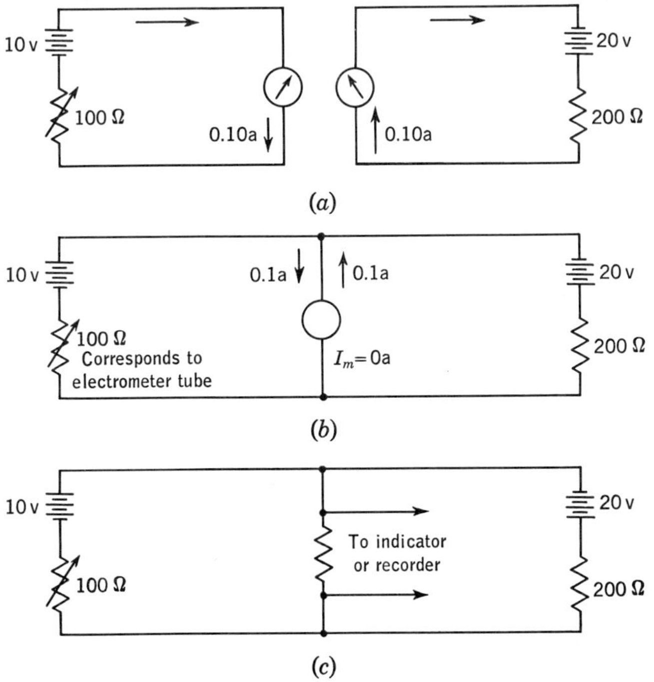

For those installations in which it is desired to have a recorder or a controller connected into the circuit, the meter may be replaced by a resistor, Fig. 19·10*c*. Any unbalance current which flows through the resistor will be accompanied by a potential across the terminals of the resistor. The value and polarity of this potential will indicate any deviations from the set point or standard value.

It will be noted in Fig. 19·7 that the amount of material directly affects the absorption of the radiation. However, doubling the thickness, for a given density of material, does not double the absorption.

For effective measurements high degrees of sensitivity are desirable. Thus, weights per unit area which vary between 0 and 250 mg/cm^2 can be measured relatively easily. But when the weight per unit area exceeds 300 to 400 mg/cm^2, there will not be large changes in the amount of radiation transmitted for small changes in thickness.

It must be repeated that gaging for thickness cannot be made with this type of equipment unless the weight density per unit volume is constant.

Detector gage circuits. Should it be desired to have the detector-amplifier section deal only with a series of uniform pulses in order to simplify certain types of presentation, then it may be necessary to clip each pulse to some uniform height Fig. 19·11. This can be done quite readily with certain types of biased rectifier circuits.

Fig. 19·11 Some systems use a clipping circuit to make all pulses equal in magnitude. Consequently, the meter current is a measure of the total number of pulses received per unit of time.

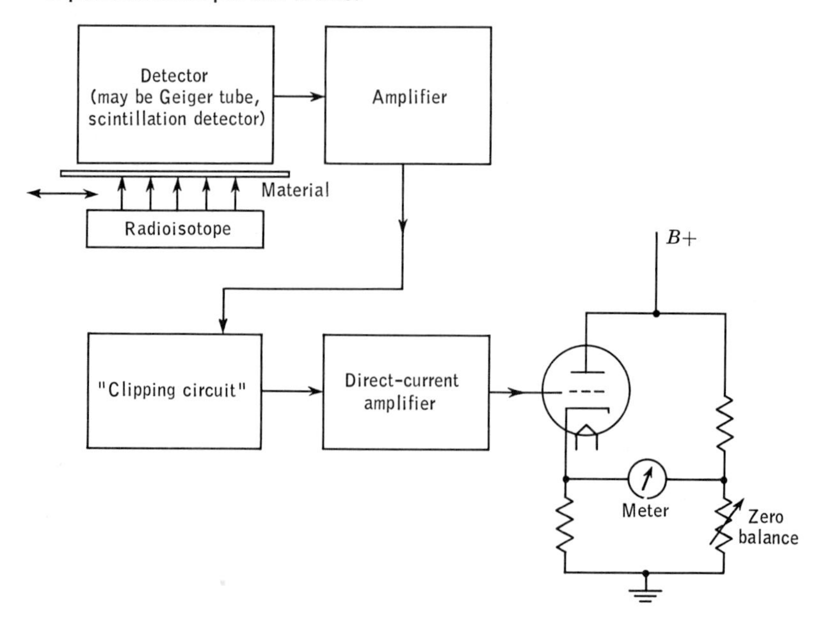

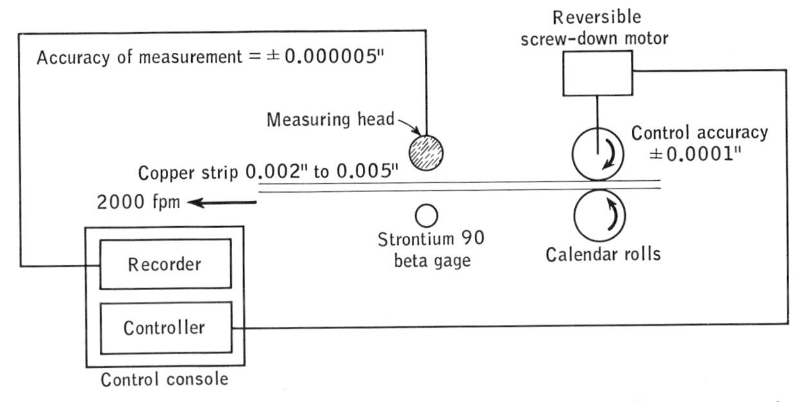

Fig. 19·12 Beta gage instrumentation on a copper-strip mill. (*Courtesy of Stewart and Leighton, from Control Engineering.*)

Figures 19·12 and 19·13 show some of the typical ways in which radiation absorption measurements may be used to control the thickness of rapidly moving materials. By making measurements at particular points in the process, any deviations are detected almost as soon as they occur. When corrections can be made almost as soon as a deviation appears, dead time is minimized and the amount of spoiled product is greatly reduced. If one were to wait until the product had been dried (in the case of paper) or the rolling had been finished (in the case of

Fig. 19·13 The thickness of paper can be controlled very accurately by controlling the concentration of stock delivered to the head box. Strontium-90 beta-ray equipment is used to measure the mass per unit area of paper. (*Courtesy of Stewart and Leighton, from Control Engineering.*)

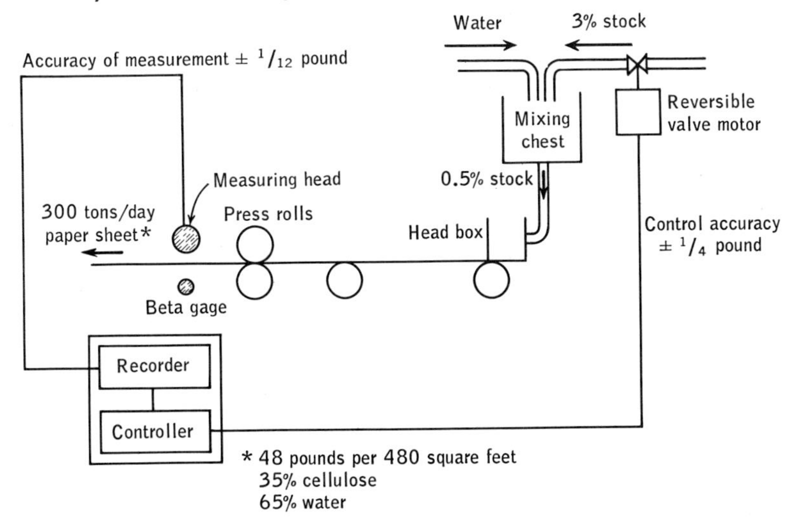

metals), there might be hundreds of feet of off-standard product turned out before there was any indication that corrections should be made in the process.

Inasmuch as all ionization equipment operates in essentially the same way, any situation in which the product or material can be introduced or used so as to absorb some of the radiated energy before it reaches the ionization cell may be able to use these techniques. Some of the ways in which radiation measurements are being utilized, in addition to the measurement of the densities of moving materials, will now be discussed below.

LIQUID-LEVEL GAGES

If we place a source of gamma radiation on one side of a tank and an ionization unit on the other side of the tank (both on the outside), then with nothing but the two tank walls and air between the two units there will be a certain amount of absorption of the emitted rays, Fig. 19·14*a*. But if the tank is now filled so that liquids or solids are interposed between the source and the ionization counter, there will be a greater amount of absorption which may be indicated in any one of a number of ways, Fig. 19·14. By suitably screening the amount of radiation that reaches the ionization chamber, the level can be measured within narrow limits or broad limits, Fig. 19·14*b*. It is possible to gage many materials which are extremely corrosive, or which may be under high pressures, or which may be so viscous that most other techniques are quite unsatisfactory and it would be desirable to have no gage within the vessel, Fig. 19·14*c*.

Fig. 19·14 The liquid level determines the amount of radiation reaching the cell stack. The total output is equal to the sum of the currents produced by each receiver unit. It is also proportional to the level of the liquid. The response may be linear over a wide range. The radiation and detector units may be arranged to provide for a variety of conditions. (*The Ohmart Corporation.*)

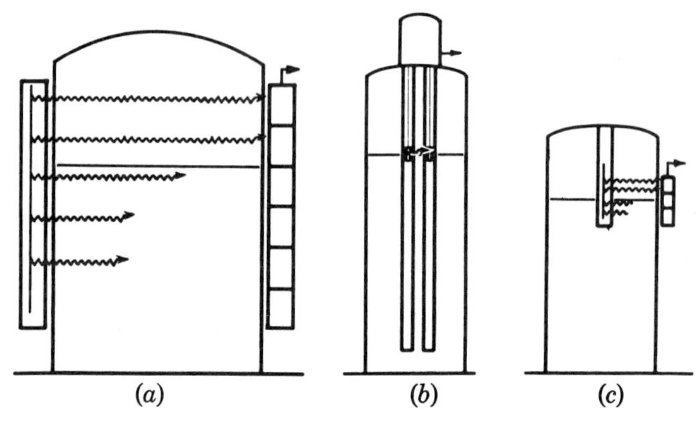

DENSITY GAGES

When the term *density* is employed in industrial measurements, it may mean the weight per unit volume or it may mean *relative density* or specific gravity. It is quite unlikely that it is used as the physicist uses it to indicate the slugs per cubic foot. In this discussion, we will consider that it means the weight per unit volume. Under these conditions, specific gravity, percentage of solids, and degrees of concentration may all be intimately related. Consequently, what is said about one type of measurement probably will apply to the others in this general group.

It has been stated previously that thickness measurements can be made only on the assumption that the weight per unit area is a constant value. Any variations in density would be indicated as thickness variations. For these measurements, thickness is fixed either by the constant diameter of a pipe or by a fixed distance between radiation source and detector. But if the *amount* of material between these points varies because of changes in density, degrees of concentration, and so on, then there will be a change in the amount of radiated energy which is absorbed, and thus the detecting unit will reflect the changed conditions.

Measurements of this type are essentially independent of many factors which make other methods of measurement difficult or impossible. Inasmuch as there need be no physical contact with the liquid, there is no concern about the liquid being corrosive, erosive, or abrasive. In a similar manner, the measurement is independent of the viscosity, the flow rate, and the pressure of the system, Figs. 19·15 and 19·16.

RADIOISOTOPES AS TRACERS

The uses of radioisotopes as tracers are quite apart from the use of radioisotopes for thickness gaging, level measurement and control, and density and specific-gravity determination. For this general class of use, the intensity of the source may be greatly reduced and the half-life may be very short because the effect being measured has only a short life. Another factor is that the long life of the isotope might interfere with subsequent measurements or result in undesirable concentrations of the active materials.

The uses that can be made of radioisotopes for tracers are limited primarily by the imagination and the economics of the situation. For the measurements previously considered, there is no contamination of the material by the isotope. But for tracer methods, there is a direct mixing of the isotope with the work or material in process. Consequently, this method is limited to those situations in which no problems will be raised by this direct mixing.

The type of detector or radiation counter used for tracer work is quite different in theory and design from that used with ionization tubes. A

Fig. 19·15 Many difficult specific-gravity or weight-density determinations become practicable when radioisotope energy absorption is used as the basis of the measurement. (*Industrial Nucleonics, Inc.*)

very simple schematic is shown in Fig. 19·17. The isotope produces current pulses in the detector, and these pulses vary directly in number with the intensity of the beta or gamma rays. There are a series of spurts of energy which are then amplified. In order to make the count

Fig. 19·16 (*a*) Specific gravities or constancy of concentration can be monitored without physical contact with the material being measured. (*b*) If the change in density is quite small, the signal voltage will tend to be a linear measure of the density.

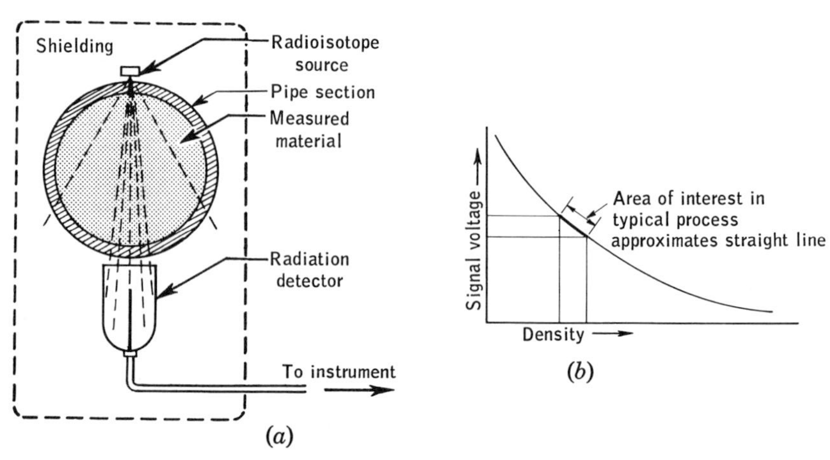

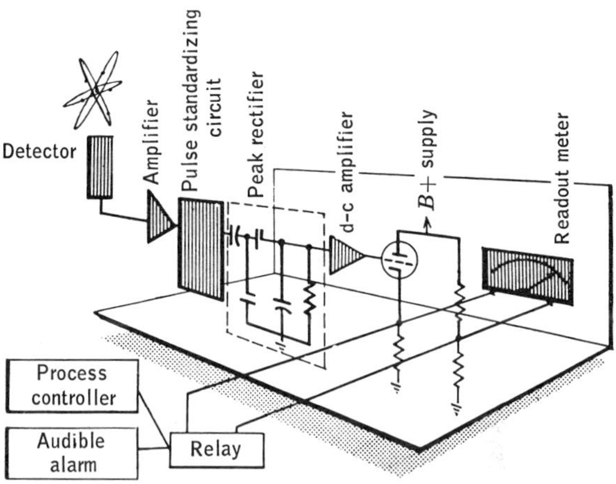

Fig. 19·17 The radiation ratemeter produces a current directly proportional to the number of pulses received. (*Tracerlabs, Inc.*)

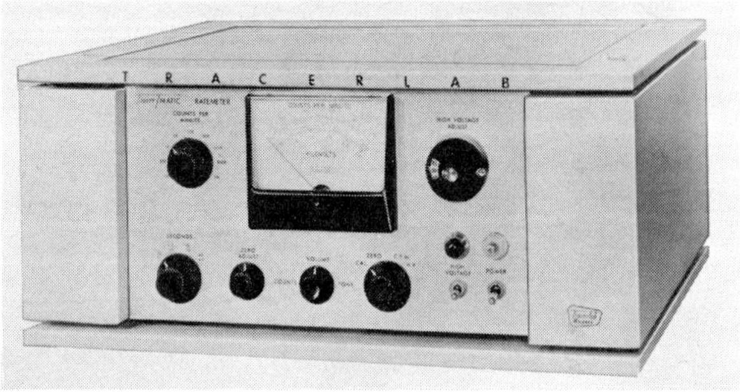

meaningful, these amplified signals are then clipped or cut down to some predetermined size so that each pulse leaving the clipping circuit will have exactly the same magnitude; in this manner each pulse can be given equal weight in the count. The pulses can then be fed into a resistive-capacitive network in which the potential developed will be a measure of the number of pulses being received per second, the pulses per second being a measure of the amount or concentration of the radioisotope.

By feeding this signal into the grid of a suitable tube used as a vacuum tube voltmeter, the magnitude of the signal can be read on a calibrated meter. By substituting a resistor for the meter and measuring the potential developed, the concentration may be recorded by a potentiometer or other device. Suitable alarms can be sounded or corrective action can be taken to bring the concentration to the desired point.

For certain applications in which very high speed of response is required, the scintillation counter may be needed. In this instance, the radiation pulse is converted into a pulse of light in the tube, which in turn uses a photomultiplier to change the light pulse into another electrical signal.

The more common groupings for such tracer techniques might be:

1. Determination of position
2. Completion of mixing
3. Detection of leaks
4. Flow rates
5. Functioning of components
6. Contamination

The first group is intended to show the position of a particular point in some material being processed; it might be a splice in a roll of paper, a joint in a strip of nonferrous material, or a joint in a piece of wire which is covered by insulation. Because of the speed of operation or because processing operations have covered the joint, visual detection may be impractical or costly. The application of a suitable radioisotope makes feasible high-speed detection of the imperfection even though it be completely obscured. Figure 19·18 illustrates how this might be done.

The time required to complete mixing processes depends upon many factors: the amount of material, the viscosity, the rate of mixing, and the type of agitator. In order to ensure adequate mixing of the components, enough time probably is allowed to take care of the worst set of conditions, and this time may be too great for the majority of occasions. But to be sure, extra time is allowed. A method similar to that shown in Fig. 19·19 indicates how the completeness of mixing can be determined. When the mixing is complete, there will be no further change in the isotope concentration. By using such a count, one is no longer dependent upon time as a measure of the situation.

A fairly typical manner of detecting leaks is shown in Fig. 19·20 and needs no further explanation.

The determination of flow rates might involve the method shown in Fig. 19·21, in which the tracer material is injected in fixed amounts and

Fig. 19·18 The exact location of material in process is found by radioisotope instrumentation. Wire splices, made radioactive, are spotted by detectors as the wire travels at high speed. (*Tracerlabs, Inc.*)

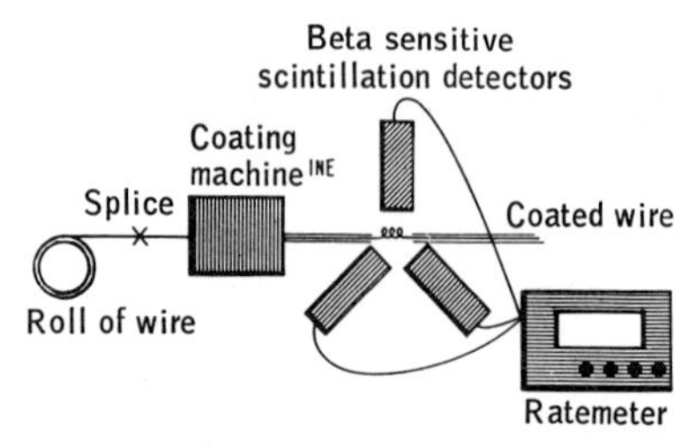

Fig. 19·19 Radioisotopes may be used to measure the degree of complete mixing in a surge tank. As the mixture becomes more and more homogeneous with prolonged mixing, the ratemeters approach some constant value. (*Tracerlabs, Inc.*)

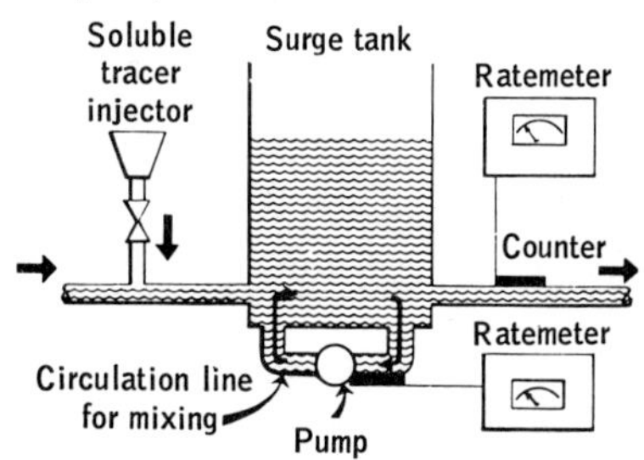

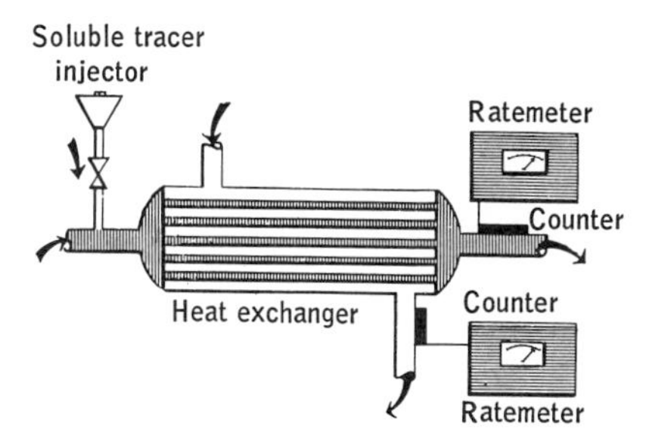

Fig. 19·20 The presence of leaks in a heat exchanger may be established by using the method indicated. If the soluble radioactive tracer is detected in the shell-side exit, there must be a leak. The relative size of the leak may be determined by comparing the readings of two counters. Leaks down to about 0.1 per cent can be detected easily. (*Tracerlabs, Inc.*)

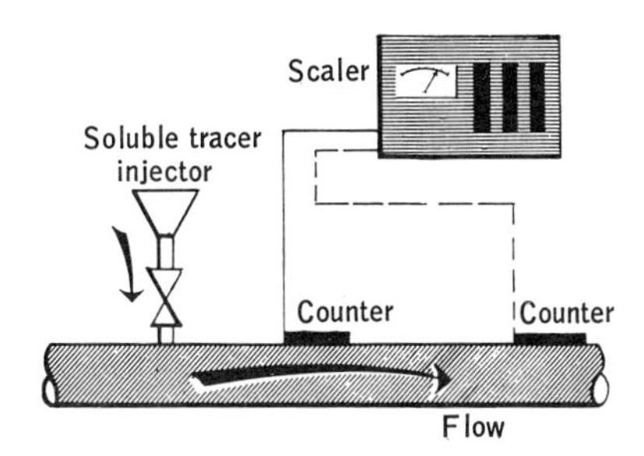

Fig. 19·21 If one knows the volume of liquid between two detectors, flow rates may be easily calculated from the time required for the tracer to travel between the detectors. (*Tracerlabs, Inc.*)

the time for the material to be moved from one point to another is determined. Or it might be that the material is injected at a constant rate. With the tracer injected at a constant rate, the degree of dilution of the tracer is an indication of the rate of flow; high rates of flow produce high degrees of dilution, and low rates produce a greater degree of concentration of the tracer.

To determine how a product is being processed, or how much is being lost in some process, or to determine the effectiveness of certain phases of the operation, the methods suggested by Figs. 19·22 to 19·24 should be considered.

Fig. 19·22 Evaporator performance may be evaluated by comparing the radioactivity of the outgoing products. If gasoline, in this instance, is radioactive, it will be a measure of the amount of tar which is being carried overhead. (*Tracerlabs, Inc.*)

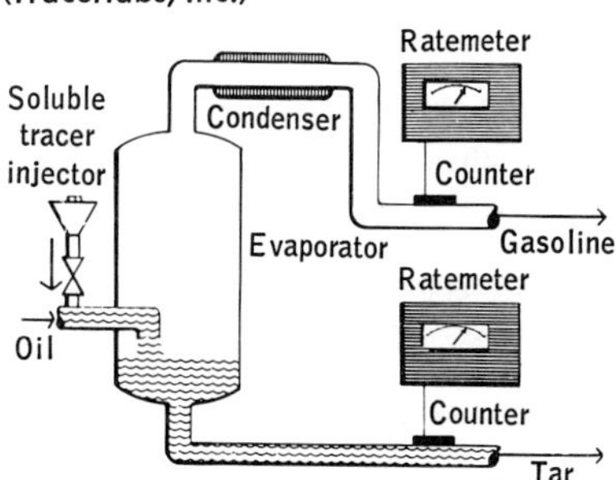

Fig. 19·23 The proper functioning of a bubble tray may be ascertained by noting whether or not the meters are consistent. (*Tracerlabs, Inc.*)

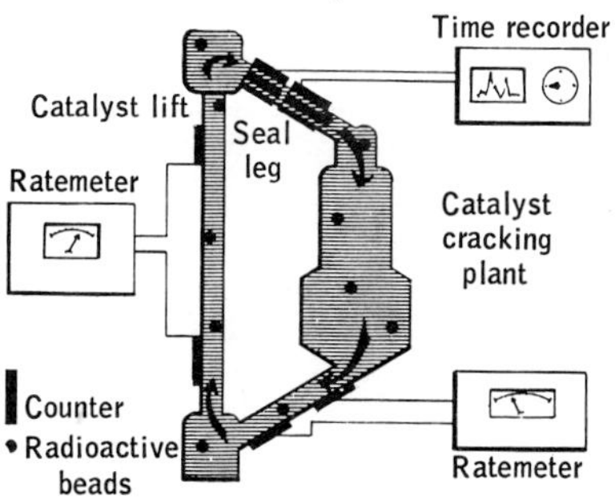

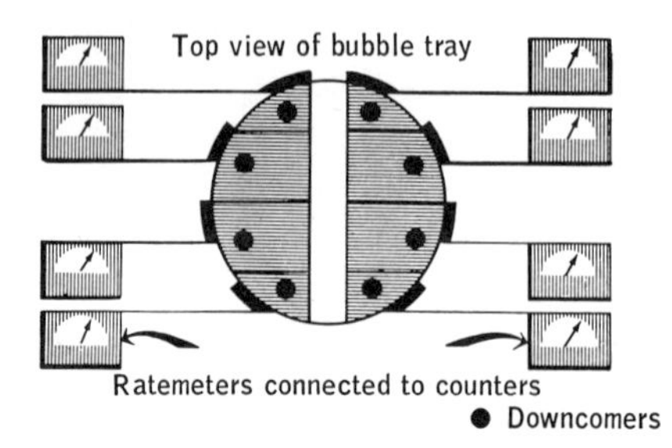

Fig. 19·24 The rate of flow of solids may also be determined by measuring the movement of radioactive beads or other materials which are added. In this example, the solid is the catalyst required for a catalytic cracking plant. (*Tracerlabs, Inc.*)

With continuous processing, there is always the possibility that some material or chemical from one batch may be carried over to some other portion of the work and alter the desired operating conditions. Although Fig. 19·25 shows a way to measure dye contamination, the same general method should be applicable whenever two liquids or even dry powders are carried along by the material being processed and are left to dilute subsequent materials.

SUMMARY

1. Radioisotopes produce emanations which can be used as measurements of thickness, density, liquid level, flow, and other phenomena. They produce different types of waves or energy as they spontaneously disintegrate. Each radioisotope has a characteristic rate of decay which ranges from periods of microseconds to thousands of years. The type and intensity of the emanation is also characteristic of the material.

2. For most industrial gaging work, the half-life should be several months or years in order to avoid frequent calibrations or loss in sensitivity. For certain types of work where the effect is to be merely transitory, then it may be desirable for the half-life to be of the order of hours. Short periods of time might be highly useful when the effects are used to gage contamination or position.

3. All measurements are of the inferential type. The only thing which is measured is the amount of absorption of the source radiation. For any given material, the thickness may be inferred from the absorption, *provided that the weight density of the material is constant.* Any variations in density will indicate apparent thickness deviations.

4. The isotope used depends upon the specific application in terms of the type of material and its composition. Some of the radiations are stopped by sheets of

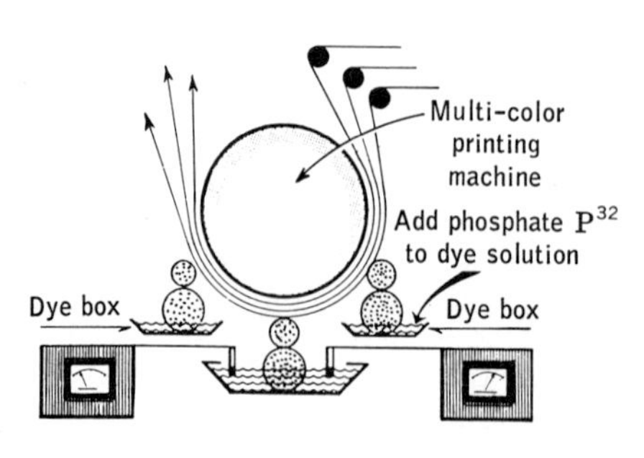

Fig. 19·25 Tracer instrumentation may be used to measure the amount of contamination present. In this instance, when the concentration of the offending color in the sensitive color reaches a preset soiling level, an alarm notifies operators to refill with fresh dye. Similar principles could be applied to many processes in which succeeding steps are contaminated by material which is carried along. (*Tracerlabs, Inc.*)

paper, while others can penetrate steel up to 1 in. thick in continuous strip-mill operation.

5. It is possible to detect thickness variations which exceed the mechanical capabilities of the associated equipment. Indications can be obtained for changes of 20 to 30 μin., but the mills are unable to roll closer than ten-thousandths.

6. All designs are basically alike in their requirement for ionization chambers, measurement of the potentials created by minute currents in resistors of very high values, and the measurement of these potentials by some type of vacuum tube voltmeter.

QUESTIONS

19·1 What is an isotope? How does a radioisotope differ from an isotope?

19·2 What are the most important industrial radioisotopes?

19·3 What determines the usefulness of a radioisotope?

19·4 If you were selecting a radioisotope for some specific gaging purpose, what are the characteristics which you would desire? Do any of the radioisotopes meet these requirements?

19·5 What types of radiation might be used industrially? Are all of them now used?

19·6 What are the main parts of an industrial ionization gage for use with radioisotopes? Describe completely the function of each part.

19·7 What does a radioisotope thickness gage really measure?

19·8 What are the conditions which must exist if one is to be correct in attributing meter deviations to thickness deviations?

19·9 What are the materials which have ten orbital electrons or fewer? Of what significance is this number in radiation gaging?

19·10 Describe how a backscatter gage differs from a conventional one. What are the conditions which would call for its use?

19·11 What is the function of each of the potentials and its associated potentiometer, if used, in Fig. 19·9?

19·12 Are there any precautions which must be observed with the use of resistors of the high values used in the ionization circuit? What might be termed the most critical component in such a circuit?

19·13 What are the conditions which would require a special electrometer tube to measure the potential on the grid? What are the objections to ordinary tubes? Industrial tubes?

19·14 Could you successfully substitute a transistor for the tube in Question 19·13? Why?

19·15 What conditions would have to be established if one were to use radiation absorption as a measure of specific gravity or density? Assume that both dry and liquid materials are to be gaged.

19·16 What advantages might the use of radioisotopes have over conventional means for determining corrosion rates? Might there be offsetting disadvantages?

19·17 If one were to reverse the polarity of battery *B*2, Fig. 19·9, describe what would happen to the output signal. Would the gage be operable?

19·18 From your knowledge of the circuit, Fig. 19·9, what must be true of battery *B*2?

19·19 If a single ionization tube were to extend the entire width of a 48-in. strip of steel, would you be justified in believing that the strip is being rolled to the correct dimension if the gage indicates that there is no deviation from the set point? If your answer is in the negative, what might you do to correct what you believe to be a fault?

19·20 What other gages that employ radioisotopes have you studied? Did the gage or gages differ in principle from the thickness gage which has been studied in this chapter?

19·21 What is the one property of radioisotopes which is of chief importance? Are there any substitutes for it?

19·22 How can radioisotopes be used to determine the rate of flow in chemical processing? In pipeline operations? What differences are there in the two types of measurement?

19·23 What is the function of the shielding which is used with radioisotope gages?

19·24 Are there any limitations on the use of radioisotope gages? Could you make a radioisotope gage and use it industrially if you wished to?

PROBLEMS

19·1 If battery *B*2, Fig. 19·9, is assumed to be 100 volts, what will be the resistance of the ionization tube if $2(10^{-9})$ amp flows in the ionization circuit? Assume that P_1 and P_3 do not add any potential to the circuit.

19·2 What will be the milligrams per square centimeter for steel if the steel is 0.030 in. thick?

19·3 What will be the milligrams per square centimeter for aluminum if the aluminum is 0.060 in. thick? 0.200 in.?

19·4 What will be the milligrams per square centimeter for lead if the lead is 0.100 in. thick? 0.500 in.?

19·5 Based on the milligrams per square centimeter in Probs. 19·2 to 19·4, what radioisotopes might be selected for gaging?

19·6 If you are using strontium 90 and the maximum allowable decrease in output is 10 per cent, when would you require a new source of radiation?

19·7 If a reduction of 50 per cent in the radiated energy is provided for in the design, how long could you use cobalt 60, cesium 137, cerium 144? Assume 100 per cent strength at the beginning.

20 Humidity Measurements

There are many industrial operations which must be carried on with known fixed amounts of moisture in the air if the process is to be satisfactory. In the case of textile mills, there must be the proper amount to ensure proper handling of the yarn. In the printing of multicolor illustrations, the paper must have the same dimensions and the same relative position each time that another color is printed, because variations in size or position will produce prints that are blurred or otherwise unsatisfactory. The presence of too much moisture will adversely affect fine-precision metal parts, while the absence of moisture may cause other materials to dry to such a degree that they are unsatisfactory. In the growing of bacteria and the manufacture of pharmaceuticals which depend upon bacteriological reactions, the proper amount of moisture may be one of the critical items.

The measurement of the amount of water present, the point at which condensation occurs, and other related measurements will require a variety of instruments employing quite different methods and principles. But first we should arrive at some definition of what we are measuring. There are two measures of humidity: one, *absolute humidity,* measures the amount of water vapor present in each cubic foot or other unit volume, while the other, *relative humidity,* compares the amount of water vapor present with what the air could hold at that temperature if it were holding its maximum amount. For most industrial applications, the per cent relative humidity is far more frequently employed than is the absolute value of the grains of water per cubic foot.

Relative humidity measurements are also of the inferred type. Some humidity measurements depend upon the temperature differences of two thermometers, while others rely upon the expansion or contraction of different materials. Some measurements depend upon the temperature at which the water vapor will condense, and still other measurements are based on the temperature at which certain salt solutions are in equilibrium.

PRINCIPLES, DEVICES, AND METHODS

Basic principles. It is a well-known fact that most natural and man-made organic fibers are hygroscopic in nature. As they absorb more moisture or lose moisture with changes in the humidity of their environment, they undergo a change in length. A very common example of hygroscopic material is the swelling of the planks of a boat to make the seams tight when the hull is wet and the wide cracks that develop during the time that the hull is out of water. Ropes on tents are known to undergo marked changes in length when they go from a dry to a wet condition. This behavior of many common materials is the basis for the construction of the simpler hygrometers for use when high degrees of accuracy are not required or when the temperature changes may be moderate.

The cooling effect of evaporation is well known to all of us. A breeze blowing over us when we emerge from a swim has a far greater cooling effect than the same breeze will have when we are dry. We also have experienced the cooling effect when alcohol, ether, or gasoline has been allowed to evaporate on the skin of the hand, for example. Evaporation is a heat-absorbing or cooling process, and evaporation will go on so long as the surrounding space is not saturated with the liquid. The drier the air, the faster do the droplets of water evaporate; the more humid the air, the slower is the rate of evaporation. Consequently, it is possible to use the decrease in temperature induced by the evaporation to evaluate the relative humidity. When the air is already saturated, there can be only a slow evaporation and a slight depression of the temperature. When the air is only slightly saturated there will be a rapid evaporation and a large depression below the ambient temperatures, Table 20·1.

When the determination of the dew point is used to evaluate humidity, we are employing a fact that we have probably observed many times. On cool days we may say that we can "see our breath." What we really mean is that the temperature is low enough for the water vapor in the breath to condense into droplets and thus become visible. When this occurs under normal natural conditions and we see this film of liquid forming on materials, we call it *dew*. The temperature at which the water vapor in the air forms into tiny droplets of water is called the *dew point*. At this temperature the air is holding all of the water vapor that it is possible for it to hold. If the temperature drops still further, some of the water vapor must drop out as droplets because the air can no longer hold it. The more water vapor there is in a cubic foot of air at some temperature, 60°F, for example, the sooner will the dew point be reached as the temperature is lowered. But if there is but a small amount of water vapor contained in the air, then it will be necessary to depress the temperature much more than in the preceding case if we are to reach the point at which the air can no longer hold the moisture. From studies which have been made, it is quite easy to determine the relative humidity at any temperature once the dew point has been established.

The last major measuring technique to be considered employs means which are also the newest. Dry salts, such as common table salt and others, are not very good conductors of electric currents. But if the salt is made damp, it will have some measure of conductivity. To a degree, the more liquid that the salt holds, the greater will be its conductivity. (We all know that a few grains of salt in water will make the solution an excellent conductor.) We are therefore interested in measuring either the amount of current that will flow under certain conditions or the temperature at which a condition of equilibrium can be established between the moisture in the atmosphere and the moisture of the salt solution.

Psychrometers. The general group of humidity instruments which are classed as *psychrometers* uses two temperature measurements and a chart to obtain the desired humidity values. In order to evaluate the evaporation rates, it is necessary to measure the temperature with a dry-bulb and a wet-bulb thermometer. The dry-bulb thermometer is just an ordinary thermometer in terms of installation and use, and it is so designated to differentiate it from its companion, which is kept wet in some manner in order that the temperature depression created by the evaporation of the water may be measured. Unless the wet bulb is always supplied with an adequate amount of water, it is impossible to know whether the reading of the wet bulb thermometer represents the maximum depression (and the minimum temperature) that could occur for the particular conditions prevailing. It is also necessary that the air surrounding the wet bulb be changed at a rate such that the air near the bulb will not become saturated with water vapor.

This last requirement led to the design of the sling psychrometer. As shown in Fig. 20·1, two mercury thermometers are mounted on a common support which can be rotated about a hand-held handle. One thermometer has its bulb covered with a wick which has been completely saturated with water. If the two thermometers are now rotated at a reasonable speed, the one with the wet wicking will be cooled by the evaporation of the water, while the other one will not be affected by the movement through the air. The dry-bulb thermometer and the air should, of course, be at the same temperature at the beginning of the test. From the difference in the two temperatures, Table 20·1 and Fig. 20·2, the relative humidity can be determined. In practice, the psychrometer is rotated for 15 to 20 sec and then the thermometers are read quickly, with the value of the wet-bulb thermometer being read first because it is the one which can change rapidly. The test should be repeated until two equal wet bulb readings are obtained. Precautions should be observed to

Table 20·1 Relative Humidity, or Per Cent of Saturation

(Difference between readings of wet and dry bulbs, °F)

Dry bulb minus wet bulb, °F:	1	2	3	4	5	6	7	8	9	10	11	12	13	14	15	16	17	18	19	20	21	22	23	24	25	26	27	28	29	30
Dry bulb, °F	Per cent of saturation																													
40	92	83	75	68	60	52	45	37	29	22	15	7	0																	
41	92	84	76	69	61	54	46	39	31	24	17	10	3																	
42	92	85	77	69	62	55	47	40	33	26	19	12	5																	
43	92	85	77	70	63	55	48	42	35	28	21	14	8	1																
44	93	85	78	71	63	56	49	43	36	30	23	16	10	4																
45	93	86	78	71	64	57	51	44	38	31	25	18	12	6																
46	93	86	79	72	65	58	52	45	39	32	26	20	14	8	2															
47	93	86	79	72	66	59	53	46	40	34	28	22	16	10	5															
48	93	86	79	73	66	60	54	47	41	35	29	23	18	12	7	1														
49	93	86	80	73	67	61	54	48	42	36	31	25	19	14	9	3														
50	93	87	80	74	67	61	55	49	43	38	32	27	21	16	10	5	0													
51	94	87	81	75	68	62	56	50	45	39	34	28	23	17	12	7	2													
52	94	87	81	75	69	63	57	51	46	40	35	29	24	19	14	9	4													
53	94	87	81	75	69	63	58	52	47	41	36	31	26	20	16	10	6	1												
54	94	88	82	76	70	64	59	53	48	42	37	32	27	22	17	12	8	3												
55	94	88	82	76	70	65	59	54	49	43	38	33	28	23	19	14	9	5	0											
56	94	88	82	76	71	65	60	55	50	44	39	34	30	25	20	16	11	7	2											
57	94	88	82	77	71	66	61	55	50	45	40	35	31	26	22	17	13	8	4	0										
58	94	88	83	77	72	66	61	56	51	46	41	37	32	27	23	18	14	10	6	1										
59	94	89	83	78	72	67	62	57	52	47	42	38	33	29	24	20	16	11	7	3										
60	94	89	83	78	73	68	63	58	53	48	43	39	34	30	26	21	17	13	9	5	1									
61	94	89	84	78	73	68	63	58	54	49	44	40	35	31	27	22	18	14	10	7	3									
62	94	89	84	79	74	69	64	59	54	50	45	41	36	32	28	24	20	16	12	8	4	1								
63	95	89	84	79	74	69	64	60	55	50	46	42	37	33	29	25	21	17	13	10	6	2								
64	95	90	84	79	74	70	65	60	56	51	47	43	38	34	30	26	22	18	15	11	7	4	0							

65	95	90	85	80	75	70	66	61	56	52	48	44	39	35	31	27	24	20	16	12	9	5	2							
66	95	90	85	80	75	71	66	61	57	53	48	44	40	36	32	29	25	21	17	14	10	7	3	0						
67	95	90	85	80	75	71	66	62	58	53	49	45	41	37	33	30	26	22	19	15	12	8	5	2						
68	95	90	85	80	76	71	67	62	58	54	50	46	42	38	34	31	27	23	20	16	13	10	6	3						
69	95	90	85	81	76	72	67	63	59	55	51	47	43	39	35	32	28	24	21	18	14	11	8	5	1					
70	95	90	86	81	77	72	68	64	59	55	51	48	44	40	36	33	29	25	22	19	15	12	9	6	3					
71	95	90	86	81	77	72	68	64	60	56	52	48	45	41	37	33	30	27	23	20	17	13	10	7	4	1				
72	95	91	86	82	77	73	69	65	61	57	53	49	45	42	38	34	31	28	24	21	18	15	12	9	6	3				
73	95	91	86	82	78	73	69	65	61	57	53	50	46	42	39	35	32	29	25	22	19	16	13	10	7	4	1			
74	95	91	86	82	78	74	69	65	61	58	54	50	47	43	39	36	33	29	26	23	20	17	14	11	8	5	3			
75	96	91	86	82	78	74	70	66	62	58	54	51	47	44	40	37	34	30	27	24	21	18	15	12	9	7	4	1		
76	96	91	87	82	78	74	70	66	62	59	55	51	48	44	41	38	34	31	28	25	22	19	16	13	11	8	5	3		
77	96	91	87	83	79	74	71	67	63	59	56	52	48	45	42	39	35	32	29	26	23	20	17	14	12	9	6	4		
78	96	91	87	83	79	75	71	67	63	60	56	53	49	46	43	39	36	33	30	27	24	21	18	16	13	10	8	5		
79	96	91	87	83	79	75	71	68	64	60	57	53	50	46	43	40	37	34	31	28	25	22	19	17	14	11	9	6		
80	96	91	87	83	79	75	72	68	64	61	57	54	50	47	44	41	38	35	32	29	26	23	20	18	15	12	10	7	5	3
82	96	92	88	84	80	76	72	69	65	61	58	55	51	48	45	42	39	36	33	30	28	25	22	20	17	14	12	10	7	5
84	96	92	88	84	80	76	73	69	66	62	59	56	52	49	46	43	40	37	35	32	29	26	24	21	19	16	14	12	9	7
86	96	92	88	84	81	77	73	70	66	63	60	57	53	50	47	44	42	39	36	33	31	28	26	23	21	18	16	14	11	9
88	96	92	88	85	81	77	74	70	67	64	61	57	54	51	48	46	43	40	37	35	32	30	27	25	22	20	18	15	13	11
90	96	92	89	85	81	78	74	71	68	65	61	58	55	52	49	47	44	41	39	36	34	31	29	26	24	22	19	17	15	13
92	96	92	89	85	82	78	75	72	68	65	62	59	56	53	50	48	45	42	40	37	35	32	30	28	25	23	21	19	17	15
94	96	93	89	85	82	79	75	72	69	66	63	60	57	54	51	49	46	43	41	38	36	33	31	29	27	24	22	20	18	16
96	96	93	89	86	82	79	76	73	69	66	63	61	58	55	52	50	47	44	42	39	37	35	32	30	28	26	24	22	20	18
98	96	93	89	86	83	79	76	73	70	67	64	61	58	56	53	50	48	45	43	40	38	36	34	32	29	27	25	23	21	19
100	96	93	89	86	83	80	77	73	70	68	65	62	59	56	54	51	49	46	44	41	39	37	35	33	30	28	26	24	22	21
102	96	93	90	86	83	80	77	74	71	68	65	62	60	57	55	52	49	47	45	42	40	38	36	34	32	30	28	26	24	22
104	97	93	90	87	83	80	77	74	71	69	66	63	60	58	55	53	50	48	46	43	41	39	37	35	33	31	29	27	25	23
106	97	93	90	87	84	81	78	75	72	69	66	64	61	58	56	53	51	49	46	44	42	40	38	36	34	32	30	28	26	24
108	97	93	90	87	84	81	78	75	72	70	67	64	62	59	57	54	52	49	47	45	43	41	39	37	35	33	31	29	27	25
110	97	93	90	87	84	81	78	75	73	70	67	65	62	60	57	55	52	50	48	46	44	42	40	38	36	34	32	30	28	26
112	97	94	90	87	84	81	79	76	73	70	68	65	63	60	58	55	53	51	49	47	44	42	40	38	36	35	33	31	29	27
114	97	94	91	88	85	82	79	76	74	71	68	66	63	61	58	56	54	52	49	47	45	43	41	39	37	35	34	32	30	28
116	97	94	91	88	85	82	79	76	74	71	69	66	64	61	59	57	54	52	50	48	46	44	42	40	38	36	34	33	31	29
118	97	94	91	88	85	82	79	77	74	72	69	67	64	62	59	57	55	53	51	49	47	45	43	41	39	37	35	34	32	30
120	97	94	91	88	85	82	80	77	74	72	69	67	65	62	60	58	55	53	51	49	47	45	43	41	40	38	36	34	33	31

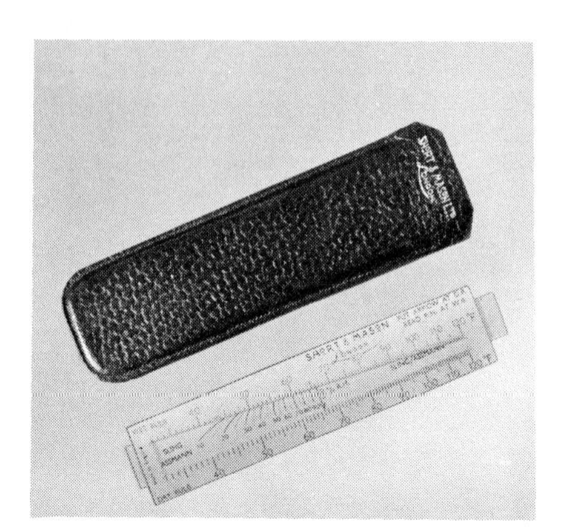

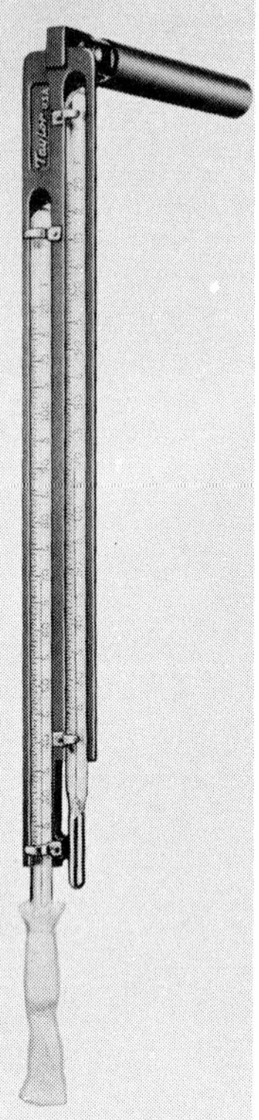

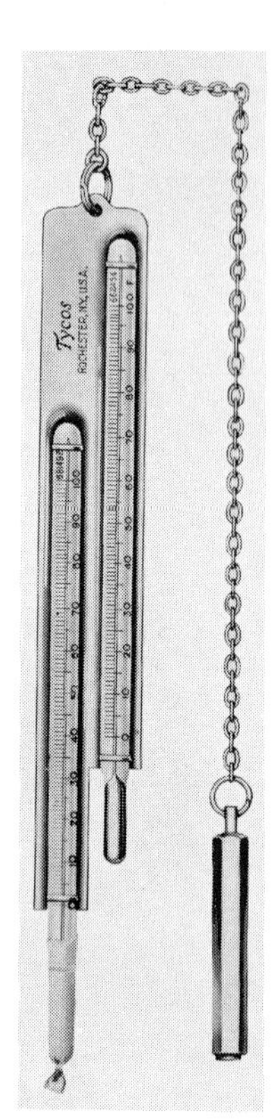

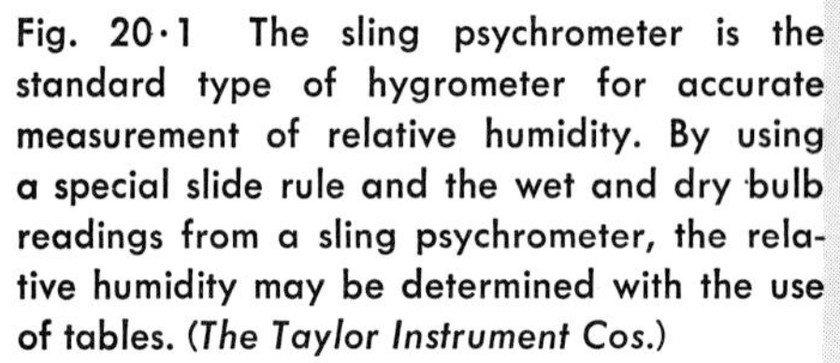

Fig. 20·1 The sling psychrometer is the standard type of hygrometer for accurate measurement of relative humidity. By using a special slide rule and the wet and dry bulb readings from a sling psychrometer, the relative humidity may be determined with the use of tables. (*The Taylor Instrument* Cos.)

make sure that the wet bulb has only clean water used on it, and the water should be close to the ambient temperature.

To use the chart shown in Fig. 20·2, assume a dry bulb reading of 80°F and a wet bulb reading of 66.5°F. We are required to find the relative humidity, the dew point, and the moisture content.

1. Relative humidity. Locate the point where 80°F dry bulb and 66.5°F wet bulb lines cross. This point is the 50 per cent relative humidity line.

2. Dew point. From the point of intersection of the wet and dry bulb curves, move left horizontally to the dew point temperature curve.

3. Moisture content. From the point of intersection of the wet and dry bulb curves, move horizontally to the right to the moisture content plot, in grains of water per pound of dry air.

For fixed installations, Fig. 20·3, essentially the same conditions must be secured if the results are to be valid. Unless there is this movement of air, the air near the wet bulb will approach a more saturated condition because of the evaporation of the water on the wick.

The water supply for the wick may be of a number of types. In certain designs the wick has been supplanted by a porous ceramic sleeve. There

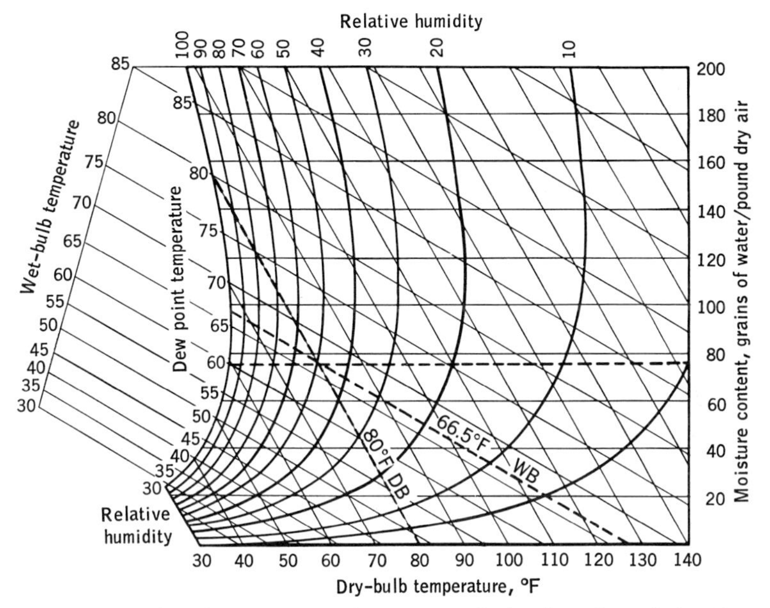

Fig. 20·2 The chart is a graphic means of showing the relationships among the various humidity characteristics of air. (*Minneapolis-Honeywell Regulator* Co.)

are certain definite conditions which should be maintained if satisfactory operation is to be obtained. Some of them are listed, in Installation and Operation Notes, later in this chapter.

Hygrometers. The change in length of fibers as their moisture content changes is employed in the usual design of hygrometer. Some hygrometers

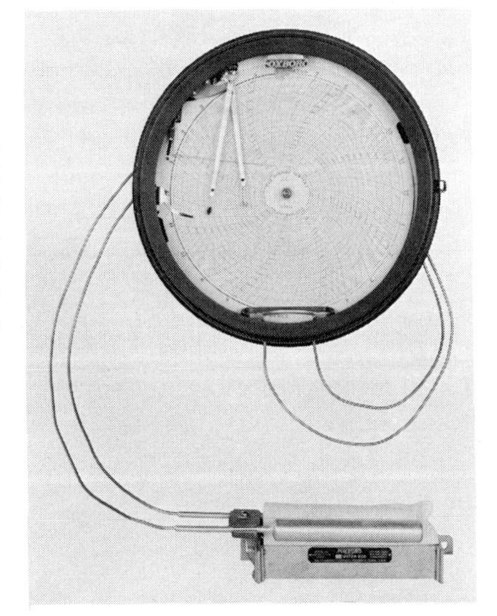

Fig. 20·3 Commercial installations which use the wet and dry bulb method for humidity determination require that the air move rapidly over the bulbs. This is in contrast to the sling psychrometer method by which the thermometers are moved through the air. (*The Foxboro Co.*)

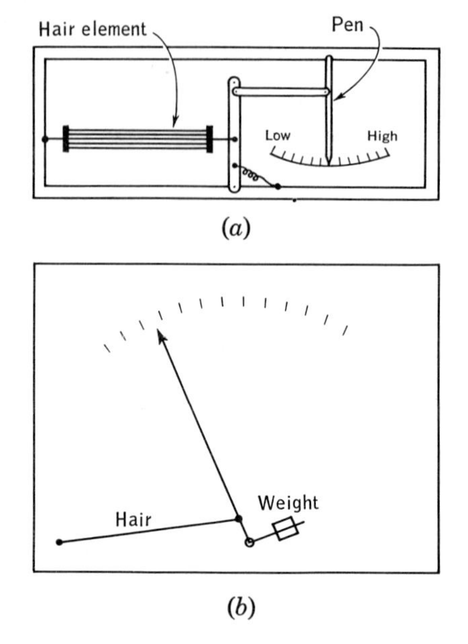

Fig. 20·4 Contraction and expansion of the hair element, in response to humidity conditions, positions the pointer. By suitable calibration, a direct reading of relative humidity is obtained.

employ only a small number of fibers; others utilize a membrane or skin; while still another design is based on a thin strip of wood that is cemented to a metal strip and wound into a helix. All the elements undergo changes in dimension as the relative humidity changes. A very simple representation of a hygrometer is shown in Fig. 20·4. The expansion and contraction of the fiber, either natural or man-made, creates a change in length which is reflected in the position of the pointer. A weight is shown acting in opposition to the pull of the fiber. A small spring might be employed instead.

Dew point determination. There are many ways of determining the dew point temperature. Some are very simple, Fig. 20·2, while others are quite complex and are designed for automatic operation. One simple way is to circulate cold water through a shiny cylinder on which the formation of a moisture film can be easily determined. The temperature of the cylinder when the film forms gives the dew point. Automatic means can be employed for detecting the film and for controlling the temperature of the medium which cools the cylinder, Fig. 20·5. Phototubes can be used to sense the reduced reflection when the film forms, and conventional temperature-sensing devices can record and control the temperature required to create moisture film. Such systems are objectionable in certain respects because of the requirements for cooling fluids and light measurements. To overcome the objections to cooling, one can use various gases and obtain the cooling effects by allowing the gases to expand. As the gases expand and cool, they change the temperature of the cylinder or other surface which is being monitored.

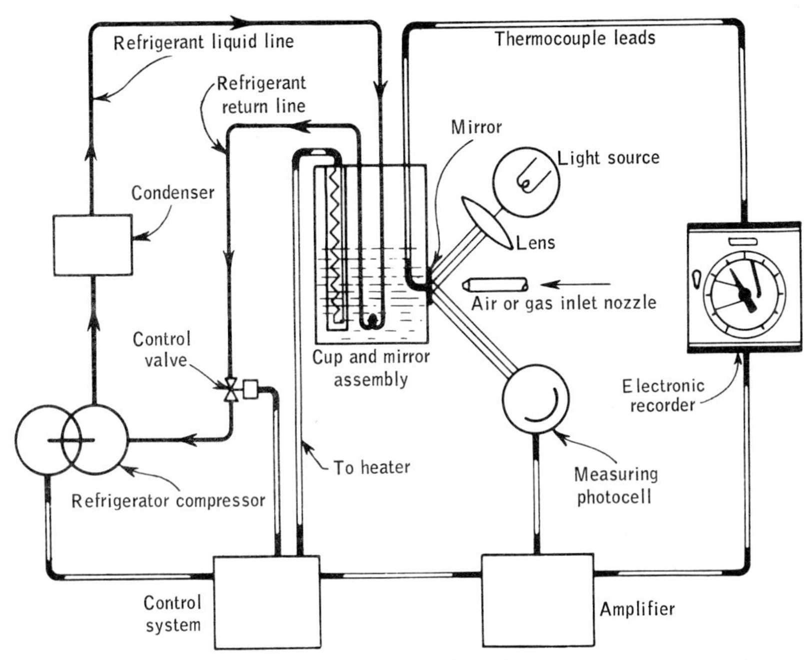

Fig. 20·5 When a polished mirror is cooled to the dew point, the formation of moisture on the surface greatly reduces the amount of light reflected. This large change in the amount of transmitted light can then be used to indicate the formation of dew. By alternately heating the mirror to dry it and then cooling it until dew forms, the process can be made continuous. (*Minneapolis-Honeywell Regulator* Co.)

A different design is found in the equipment which uses the salts of different metals, Fig. 20·6. The pure dry salt is a nonconductor of electricity, but as moisture is absorbed, the resistance decreases. If the supply voltage is adequate and if there is sufficient current capacity, the current can be of such a magnitude that it will heat the salt, which probably is supported on a fiber-glass fabric. The heating wire is covered by the salt. Then as the heat provided by the electric current dries out the salt, the conductivity of the salt decreases, and so will the heat input until a point at which there is equilibrium between the loss of moisture from the sample and its gain from the air is reached. By determining the temperature at which this equilibrium is reached, the relative and absolute humidities as well as the vapor pressure can be determined.

Determination of moisture content of fabrics and materials. In the weaving of cloth and in the processing of many materials, it is important that certain conditions of humidity be observed. If materials are made too dry, then not only may the product be unsatisfactory, but an unnecessary amount of heat may have been used to obtain the undesirably dry condition.

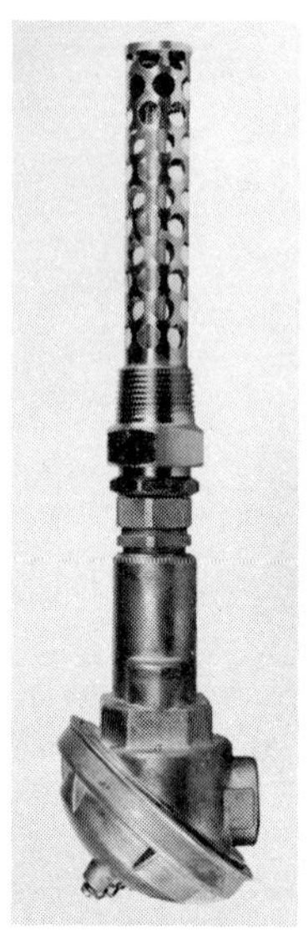

Fig. 20·6 Typical construction of a salt type of electrical conductivity humidity-measuring device. (*The Foxboro Co.*)

Many fabrics and other materials will have a high value of resistance when they are dry but a much lower value when they are wet or damp. The resistance of specified quantities of the material then may be a measure of the amount of moisture left in the product. The measurement may be made by passing the product or fabric between rollers, or two or more probes may make contact with the material.

One type of moisture-responsive device may be some hygroscopic material which may be so placed that it can respond quickly to changes in the ambient. Figure 20·7 gives an idea of how a moisture determina-

Fig. 20·7 Schematic diagram of a moisture meter used for industrial operations showing the relative positions of detector roll, work in process, and the grounded machine roll. (*The Brown Instrument Division, Minneapolis-Honeywell Regulator Co.*)

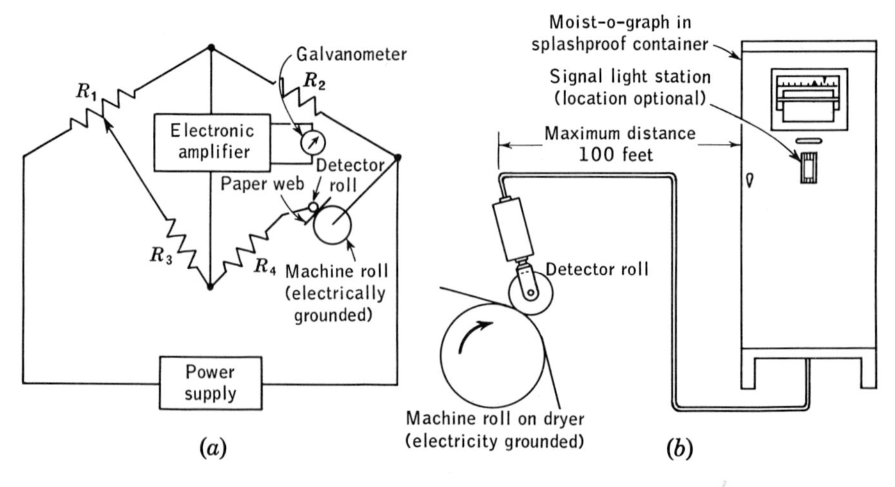

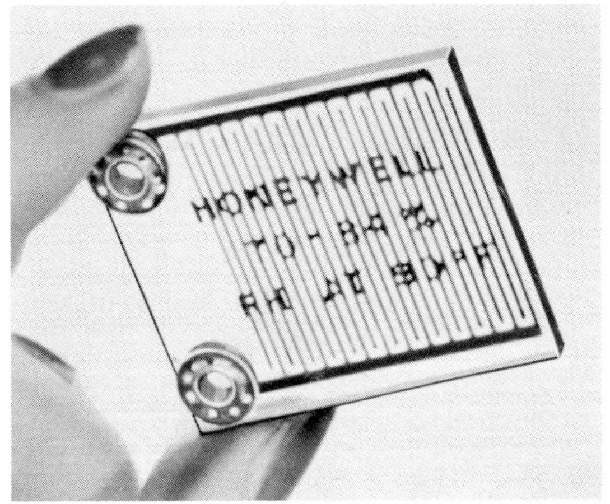

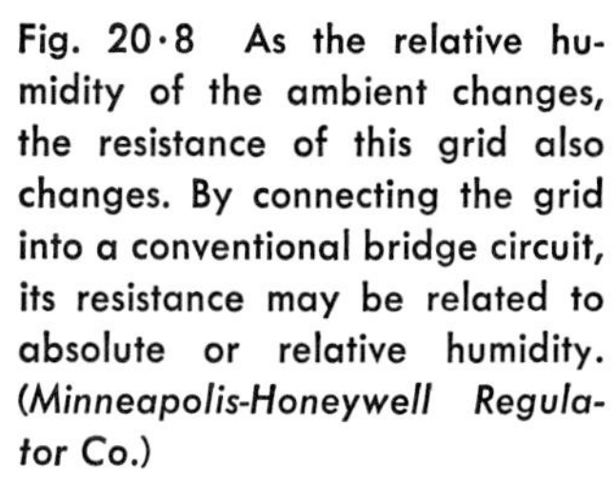
Fig. 20·8 As the relative humidity of the ambient changes, the resistance of this grid also changes. By connecting the grid into a conventional bridge circuit, its resistance may be related to absolute or relative humidity. *(Minneapolis-Honeywell Regulator Co.)*

tion might be made for paper, cloth, or sheet material. By exposing the grid type of sensing element, Fig. 20·8, to conditions involving the presence of moisture, the resistance may be correlated with humidity conditions.

Continuous water vapor recorder. A somewhat different approach to the measurement of water vapor in gases or air is illustrated by the sketches and schematics shown in Figs. 20·9 and 20·10. The instrument employs the fact that heat is given up when a vapor condenses and is taken up when a liquid is evaporated.

It will be noted in Fig. 20·9 that the sample from the plant is received under pressure and is then divided into two streams. One of the streams then proceeds through the valving and the heat exchanger, while the other one is dried in a regenerative dryer. The dried gas and the sample, as originally received from the line, are then allowed to pass through the heat exchanger for a 2-min period. During this time, the dry gas will absorb moisture from the desiccant in the heat exchanger. This will cool the desiccant. But at the same time, the wet sample is passing through the other half of the detector-desiccant cell, where the water vapor will condense. As it does so it will give up heat and thus raise the temperature of that part of the detector unit. With one half of the detector being heated and the other half being cooled, the two halves will develop a temperature difference. By arranging a group of thermocouples in series aiding, a usable thermopile output is available. The magnitude of the emf is then a measure of the water vapor.

At the end of the 2-min period, the valves are operated to interchange the two flows through the detector cell. This will also require that the thermopile output be switched, inasmuch as the two sections of the detector cell will reverse their roles.

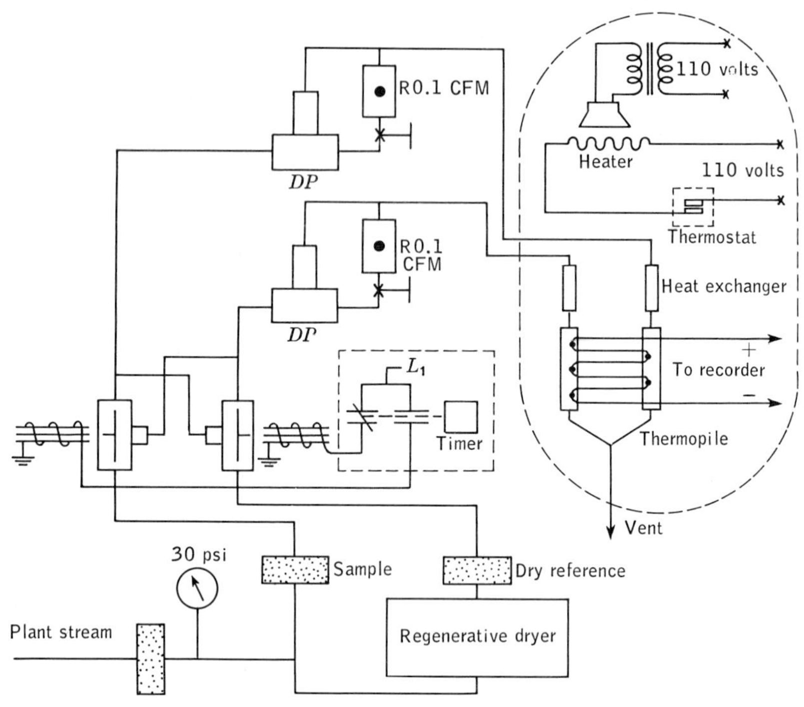

Fig. 20·9 A continuous water-vapor recorder which is based on the fact that a dry reference gas and the process gas will transfer unequal amounts of heat to the thermopile. This produces a temperature difference, which becomes a measure of the moisture content of the sample. (*Mine Safety Appliances Co.*)

INSTALLATION AND OPERATION NOTES

Sling psychrometer

1. Wet the wick only with *clean water,* which should be within a very few degrees of the ambient temperature.

2. Keep the dry bulb free of water or other liquids.

3. Swing the psychrometer in a smooth circular path for 15 to 20 sec. Avoid sudden or jerky motions.

4. If the period of rotation is too short, the temperature will not be depressed to its proper value. If the time is too great, the wick will dry and the bulb temperature will not remain at its minimum value.

5. Make outdoor observations in the shade. Observations made inside should not be too close to the operator's body or some source of heat.

6. Keep the wet bulb clean. *Use only clean water,* preferably distilled water. If the wick becomes dirty or contaminated, it will not conduct the water properly. Clean the wick by boiling it in clean water.

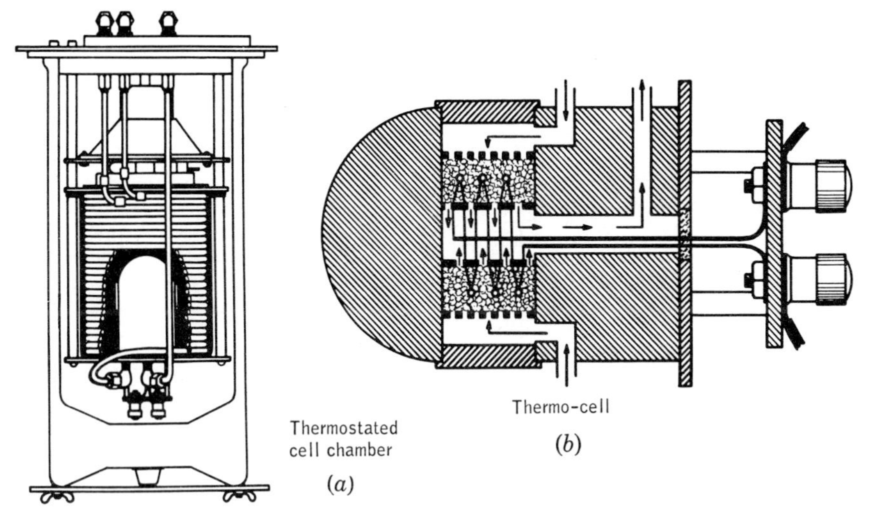

Fig. 20·10 A representation of the Thermo-Cell used in the continuous water-vapor recorder. By suitable valving, the dry reference gas and the sample process gas are made to pass first through one thermopile section and then through the other. One section is being warmed while the other is being cooled. (*Mine Safety Appliances Co.*)

Wet and dry bulb psychrometers

1. Supply only *clean water,* preferably distilled water.
2. Avoid any contamination of the wick by dirt or soluble salts, which will adversely affect the rate of evaporation and thus influence the temperature.
3. For accurate results there must be a rapid circulation of air over and around the bulbs, at least 15 fps.
4. For readings below the freezing point, remove the wick, because it will be of no use. Apply water in small quantities directly to the bulb. Evaporation will go on from the thin coat of ice which forms on the wet bulb.
5. If a porous sleeve is used, it must be handled carefully, because it is fragile. It must also be supplied with clean water. The sleeve should never be allowed to dry out, because a film will form which must be removed by cleaning.
6. Avoid installations near doors or other locations where the humidity is not representative of the values to be measured.
7. The best results will be obtained when the supply of water is about at the wet-bulb temperature. If water which is much above room temperature is supplied, its temperature may keep the wet bulb from attaining the value that it should have with the existing humidity conditions.
8. The instrument should be free from undue vibration.

9. Not only should the humidity being measured be representative of the area, but the temperature should also be representative. Avoid hot locations if they are not typical.

10. Replace wicks about once each week. Keep the replacement wick in water before using it; it will take some time for it to reach equilibrium conditions. Any attempt to use a new wick which has not been well soaked will introduce serious errors for some period of time.

11. Circulating air should strike the dry bulb first. Minimum air movement is about 15 fps.

12. The wet bulb must be thermally insulated so that it neither gains nor loses heat by conduction and radiation.

13. Normally, only linen or cotton wicks should be used. *Never use other materials for wicks.* The wick *must* cover the entire portion of the wet bulb.

14. The main section of the wick must be well immersed in water at all times.

15. Clean the thermometer bulbs. Dirt can create errors in heat transfer.

16. When using a lithium chloride humidity unit, alternating current must be used to avoid polarization.

17. Whereas the sling psychrometer and the wet and dry bulb hygrometers require a rapid movement of the air or gas in order to effect the maximum evaporation rate at the wet bulb, lithium chloride units should be used with relatively still air or gas. Where there is rapid movement of the medium, precautions must be taken to shield the sensing element from the rapid movement.

18. For hygrometers using the wet bulb principle, corrections should be made from suitable tables when the barometric pressures are more than about 1 in. different from the calibrating value. The lithium chloride type of device responds only to the pressure of the water vapor present and thus is not susceptible to atmospheric pressure changes.

19. Because the lithium chloride units require the addition of heat to bring the salt to a water vapor equilibrium point, these devices can be used only at temperatures above the ambient.

20. When humidity indicators rely upon the expansion or contraction of a material, time must be given for the material to respond to new conditions. Such devices often lag as much as an hour behind the humidity changes.

SUMMARY

1. There are two basic humidity measurements: the absolute humidity, which is a measure of the amount of water (in the form of water vapor) per unit weight or volume, and the relative humidity, which is a ratio of the amount of water actually present to the maximum amount of water which could be present at a particular temperature and pressure.

2. The humidity may be determined from the following: (*a*) Many materials are hygroscopic and change dimension with changes in the relative humidity. Dimensional changes then become a measure of the humidity. (*b*) Because the electrical resistance of certain substances and materials is dependent upon the amount of moisture present in the material, relationships between humidity and resistance may be established. (*c*) A liquid will evaporate until saturation of the atmosphere is reached, the rate of evaporation being much greater when the degree of saturation is low and greatly reduced as saturation is approached. Since evaporation is a cooling process, the reduction in temperature (below the ambient) caused by evaporation then becomes a measure of the relative humidity. (*d*) The temperature at which the air or gas becomes saturated (as indicated by the dew point) can be used to indicate the absolute and relative humidities at any higher temperatures.

3. Unless the cooling liquid is pure and the evaporating surfaces are clean, psychrometers will not give correct indications. Great care must be exercised to maintain the apparatus in a clean condition.

4. All instruments of the psychrometer class must have a proper flow of air or gas over the wet bulb if the evaporation is to continue at a rate which is representative of the humidity. However, instruments which employ the resistance of certain salts as a measure of the humidity must be shielded from high-speed movements of the air or gas.

5. The temperature of process materials from which water is being given off to the air, as in drying fabrics, and so on, may be close to that of a wet bulb thermometer exposed to the same ambient conditions. This reduction in temperature because of evaporation can seriously retard certain processes.

QUESTIONS

20·1 Of what importance are humidity measurements in industry? What are some of the industries which are primarily concerned with humidity determination?

20·2 What products depend upon correct humidity for quality control?

20·3 What are the principles upon which different commercial humidity measurements are made? Which lend themselves best to remote reading? To local reading? To fast response? To accuracy?

20·4 What are some of the important installation conditions which must be observed when making commercial humidity measurements?

20·5 To what extent might humidity determinations be erroneous even though the instruments are in correct operating condition and have the proper calibrations? What are the factors which should be considered as possible sources of error?

20·6 Why should the water supply for psychrometers be at the ambient temperature? What might happen if it is not?

20·7 Of what importance is the instruction that the air or gas should move first over the dry bulb and then over the wet bulb? Can you justify the instruction?

20·8 Could electrical temperature-sensing devices be used instead of "filled" mechanical systems? Are there any advantages? Any disadvantages?

20·9 If a film forms over the evaporating element, what would you expect of the indicated humidity in terms of high, low, or correct readings?

20·10 If materials other than linen or cotton are used for wicks, is there a possibility that the performance of the instrument will be altered? Why?

20·11 What might be the factors which limit the distance between detector roll and instrument to 100 ft, Fig. 20·7?

20·12 What is the purpose of alternating the flow through the thermopile, Figs. 20·9 and 20·10?

PROBLEMS

20·1 Using Table 20·1, for a dry bulb temperature of 70°F, what is the relative humidity when the difference in the wet and dry bulb thermometer readings is 5°? 10°? 20°? 25°?

20·2 Answer Prob. 20·1 when the wet bulb reading is 65°F, 60°F, 50°F, and 45°F.

20·3 Make a plot of relative humidity against difference in bulb temperatures for the dry bulb temperatures of 60, 80, and 100°F.

20·4 Check your plot in answer to Prob. 20·3 against the values given in Fig. 20·2.

20·5 At 10° intervals from 50° to 100°F, dry bulb, find the dew point for the indicated condition: the dry bulb temperature exceeds the wet bulb temperature by 5°F; 10°F.

Temp, °F	Rel humidity, %	Temp, °F	Rel humidity, %	Temp, °F	Rel humidity, %
0	30	40	28	80	16
10	20	50	41	90	16
20	21	60	32	100	12.7
30	35	70	27		

20·6 For the given temperatures and dew points, determine the relative humidities. Determine the dew points.

Temperature	0	10	20	30	40	50	50	60	70	80	90	100
Dew points	−24	−10	10	10	30	30	34	32	28	40	24	30

20·7 Check your answers to Prob. 20·5 against the values in Fig. 20·2 and Table 20·1.

20·8 Check your answers to Prob. 20·6 against the values in Fig. 20·2 and Table 20·1.

21 Specific-gravity and Viscosity Measurements

In industrial process work, it is often desirable or necessary to know the weight density, in pounds per cubic feet, or the relative weights of equal volumes of materials. Because of its common occurrence, water is generally used as the basis for comparison. Inasmuch as the weight per given volume is also dependent upon the temperature, it is necessary in many situations to specify the temperature at which the weights are to be determined. Water has its greatest density at 39.2°F. Some scales are referred to this, while others employ a different basis. Measurements involving the specific gravity of petroleum products are often referred to the weight of the material and water, both at 60°F.

It has been mentioned before that the term *density* as used by the scientist means the *mass per cubic foot or the slugs per cubic foot.* In the more common usage of the term, density is generally understood to mean the pounds per unit volume, which is generally the cubic foot. To avoid this confusion in terms, we shall use the term *weight density* when referring to the weight per cubic foot.

Weight density and specific-gravity measurements are most intimately related if the specific gravity is referred to the weight of an equal volume of water. The weight density tells us the number of pounds per cubic foot, while the specific gravity indicates the value of the fraction formed with the weight density of the material as the numerator and the weight density of water as the denominator. For example, the weight of a cubic foot of aluminum is 167 lb at room temperature, while that of water is 62.3 lb at the same temperature.

$$\text{Specific gravity} = \frac{167 \text{ lb}}{62.3 \text{ lb}} = 2.68 \qquad \text{dimensionless}$$

This value indicates that, volume for volume, aluminum is 2.68 times as heavy as water. It well might be that we are comparing the weight of some liquid with that of an equal volume of water; if we are considering liquids which are lighter than water, the ratio is less than 1.

The weight density of any material is equal to the product of the weight density of water and the specific gravity of the material.

$$\text{Weight density} = (62.3\ \text{lb/ft}^3)(\text{sp gr})$$

Thus, if the specific gravity is 0.5, the weight density of the material is

$$(62.3\ \text{lb/ft}^3)(0.5) = 31.15\ \text{lb/ft}^3$$

Using the same basic relationships, but changing the arrangement of the terms, if any two of the relationships—weight density, weight, or the specific gravity—are known, the other can be found easily.

Specific-gravity measurements. One of the simplest of the methods for determining specific gravity is to weigh a given volume of the material and then determine the weight of an equal volume of water. The ratio of the weight of the material to that of the water gives the desired ratio. The device used may be just a simple jar or tube which can be filled with each material and then weighed. The size of the container or its shape should be of no consequence. Thus, in practice, the determination should be a very easy one.

A very common method in practice is to use a float which must always displace its own weight of liquid. The higher or lower the float rests in the liquid is a measure of how much liquid has to be displaced to equal the weight of the float. The higher the float in the liquid, the higher the specific gravity; the lower the float, the lower the specific gravity. By using a calibrated scale on the float, Fig. 21·1, the specific gravity can be determined immediately. In a modification of this arrangement, shown in Fig. 21·2, a remote-reading electrical signal is desired. Changes in the specific gravity of the liquid change the amount of immersion of the float, and thus as the float moves up and down, the magnetic core for the variable inductance or differential transformer is also moved with respect to its reference position.

With any of the devices which are similar to Fig. 21·2, the height of the liquid must be maintained within close limits or the transmitter will be unable to distinguish between a change in specific gravity or a change in level if the float can also respond to level changes.

In Chap. 7 it was found that the level of a liquid could be measured by determining the pressure required to equal the pressure created by a column of the liquid. In that instance it was assumed that the specific gravity was known. In the present situation, if the height of the column of liquid and also the pressure necessary to balance the pressure created by the column are known, then the specific gravity can be found. For example, the weight of a column of liquid 10 ft high and 1 ft^2 in cross section equals the product of the volume and the weight density:

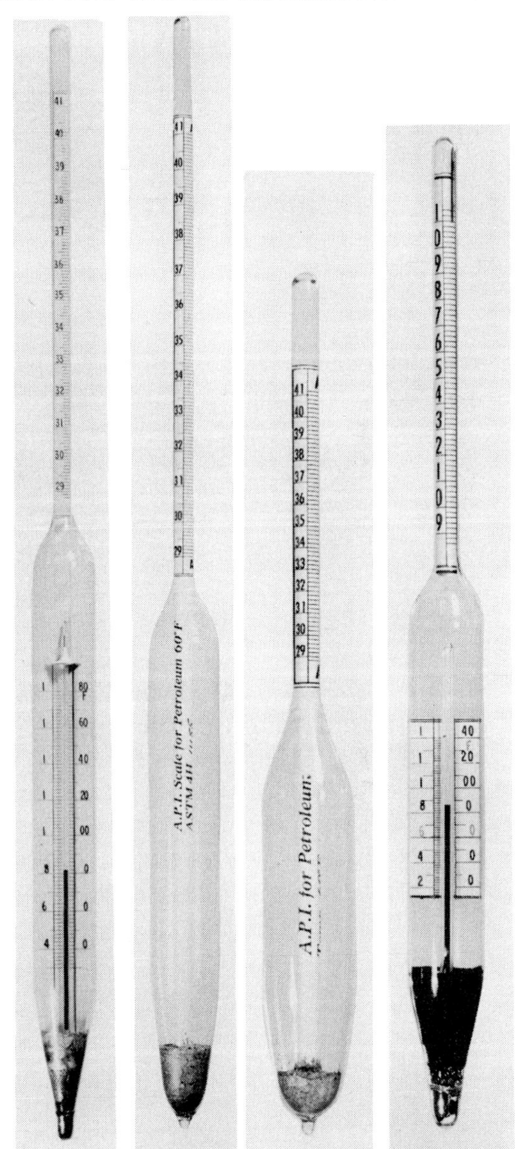

Fig. 21·1 Hydrometers have many different sizes, shapes, and scales. (*The Taylor Instrument Cos.*)

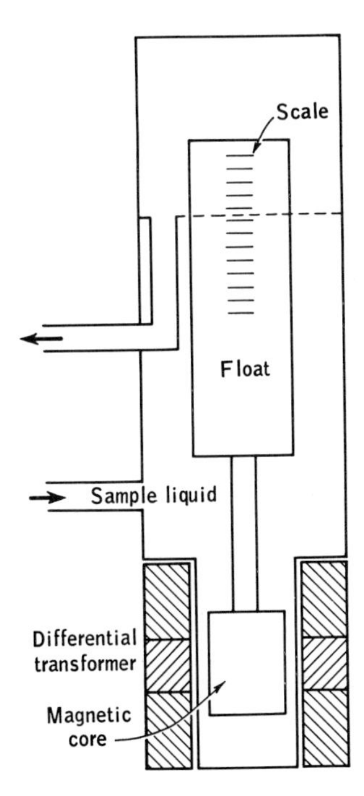

Fig. 21·2 A differential transformer whose core is positioned by a float provides an ideal way to convert specific-gravity values into electrical equivalents.

$$W = (10 \text{ ft})(1 \text{ ft}^2)(62.3 \text{ lb/ft}^3)(\text{sp gr}) \qquad \text{lb}$$

Then the pressure per square inch is equal to the total force divided by the area in square inches.

$$P = \frac{(10 \text{ ft})(1 \text{ ft}^2)(62.3 \text{ lb/ft}^3)(\text{sp gr})}{144 \text{ in.}^2/\text{ft}^2} = \text{lb/in.}^2$$

From these values, we find that

$$4.3(\text{sp gr}) = \text{pressure} \qquad \text{psi}$$

Then for a known height of 10 ft and a pressure of, let us assume, 4 psi, the specific gravity would be equal to

$$\frac{4 \text{ lb/in.}^2}{4.3 \text{ lb/in.}^2} = 0.93 \qquad \text{dimensionless}$$

In Fig. 21·3, the same technique is used except that in this case, one of the vessels contains liquid of known or desired specific gravity. For a fixed amount of pipe immersion a perfectly definite pressure will be established before any air can escape from the open end of the immersed pipe. This pressure is also transmitted to the **U**-tube manometer. In a similar vessel a similar pipe is immersed to the same depth. (The success of this measurement is entirely dependent upon both pipes having equal amounts of immersion in a vertical direction.) Unless the specific gravity of the process liquid is the same as that of the reference, it will

Fig. 21·3 U-tube manometers can be used for specific-gravity determinations. For equal lengths of pipe submergence, the pressures developed vary directly with the specific gravity of each fluid.

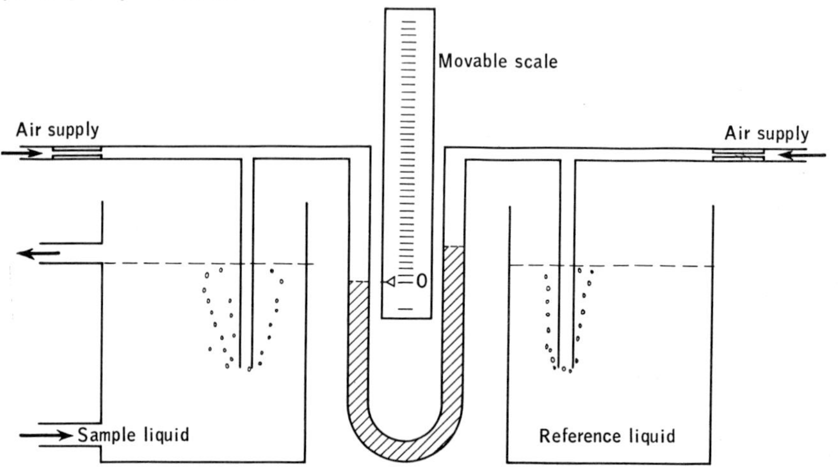

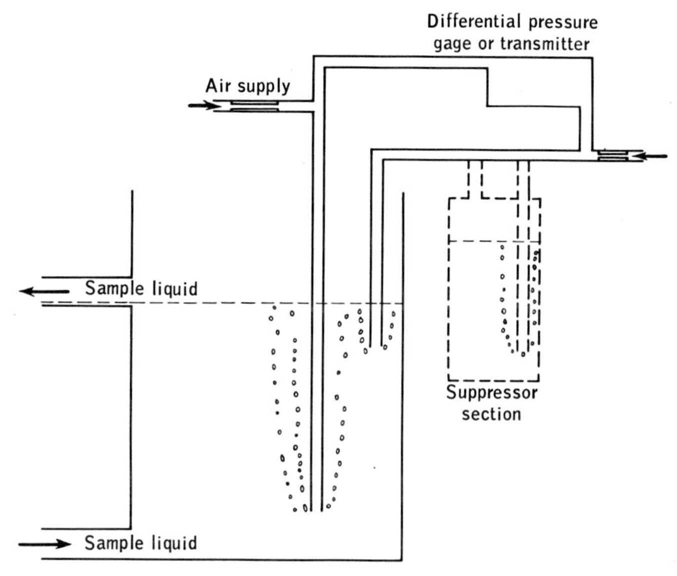

Fig. 21·4 No reference fluid is needed if two tubes with a fixed known distance between their open ends can be employed. The differential pressure then varies directly with the specific gravity.

be impossible to establish equal pressures by the two equal heads of liquid. Under these conditions, the manometer will indicate that one pressure is lower or higher than the other. The indicated differential pressure can then be used to infer the difference in the specific gravity. In place of the **U**-tube manometer which is shown, it would be equally feasible to install a differential pressure unit which could then indicate or transmit the difference between the standard and that being produced at the moment.

Occasionally it is even more desirable to use only a single container of liquid, as in Fig. 21·4. With such an installation the ends of the bubbler pipes must have a known difference in elevation. The differential pressures in such an installation are caused by the difference in both height and specific gravity which govern the weight density. It is sometimes advantageous to expand the readability of the instrument by suppressing part of the pressure which does not vary. The dotted portion of the figure indicates that which is often used for range-suppression purposes.

An entirely different method of determining specific gravity is shown in Fig. 21·5. With the buoyant body completely immersed, the buoyant force is dependent only upon the weight of the displaced liquid, which is in turn a function of the volume and the specific gravity. Inasmuch as the volume is constant, the force then varies directly with the specific

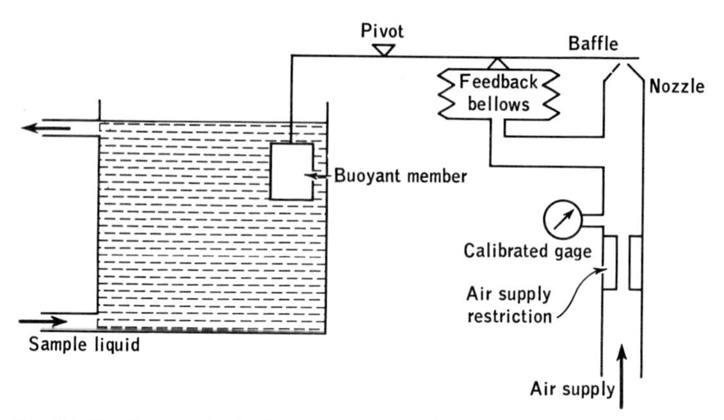

Fig. 21·5 One method of converting specific-gravity values to air signals for transmission or control purposes.

gravity. If we assume, for the moment, that there is an increase in the specific gravity of the liquid, then there will be a greater upward force on the left-hand end of the beam, which is assumed to be rigid. This will cause the right-hand end of the beam to move closer to the orifice. This then will create an increase in the pressure at the nozzle that will also be sensed by the bellows, which will try to expand. As the bellows tries to expand, it will force the baffle away from the orifice, thus causing a reduction in pressure in the pneumatic system. The bellows will move just enough to reestablish a new position of torque balance with a somewhat different pressure. By adding a weight on the left-hand end of the beam, Fig. 21·5, some of the buoyant force can be suppressed, and the force variations can be read on a calibrated gage of lower range. This should improve readability.

A somewhat different method for the determination of specific gravity is shown in Fig. 21·6. It is a well-known fact that materials of different specific gravities have different boiling points for any given pressure. In the figure shown, there is some reference liquid, such as water, and the main body of liquid in the evaporator. Both are subjected to the same pressure. With heat applied to both, each one will have a boiling point which is representative of its specific gravity, which in turn is a measure of dilution or purity. Any temperature differences in the boiling points then can be interpreted in terms of different degrees of concentration or specific gravity. Very often these differences in temperature are measured with resistance thermometers, but they can be made with any other system which will indicate what the temperature difference is. Two Bourdon tubes operating in the manner of a completely compensated thermometer with the two elements working in opposition to each other would perform this function.

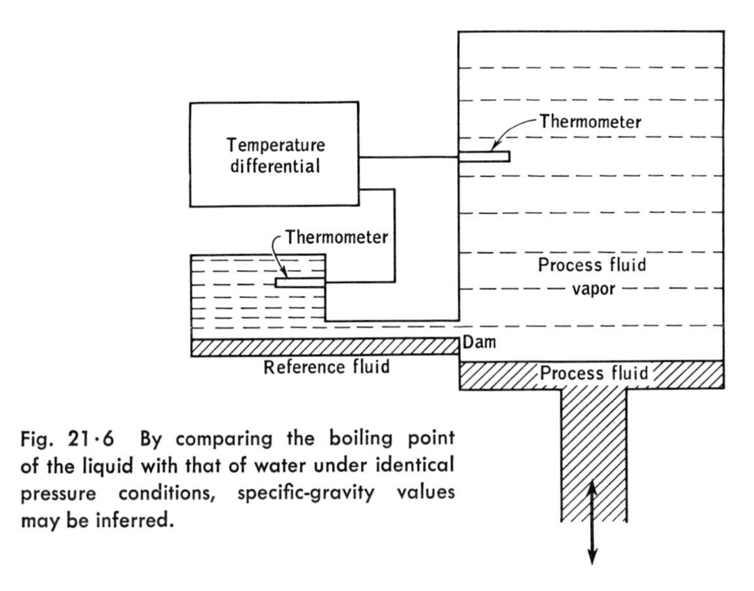

Fig. 21·6 By comparing the boiling point of the liquid with that of water under identical pressure conditions, specific-gravity values may be inferred.

There are many measuring problems in which there are changes in specific gravity and changes in weight density because of temperature changes. One means for minimizing temperature and specific-gravity changes is shown in Fig. 21·7. With such an arrangement, the level will vary with the weight density and the specific gravity, but the pressure in pounds per square inch at the time that the tank is emptied will always be the same. Thus, in one sense, we might say that the same "amount" of material is measured each time—the effects of dilution or volumetric changes with temperature have been eliminated.

In Fig. 21·4 it is indicated that there is a fixed level of operation because of the outflow. But the measurements could be made equally well with variability in height just so long as the shorter tube was always covered. The pressures created within each tube are a function of the depth of immersion. In Fig. 21·8 we have the essential elements of Fig. 21·4 with the exception that the pressure created within the longer tube will now be used for the additional purpose of controlling the liquid-level control valve which will permit the liquid to pass to storage. The differential pressure, which is a measure of the specific gravity, is now used to operate the density control valve which governs the amount of weak liquid which is permitted to enter the tank. Because there is no control of the amount of material which is to be dissolved or mixed with liquid, specific gravity or density control must be effected entirely by the amount of dilute liquid which is permitted to enter the dissolving tank.

When solids and liquids are to be mixed and specific gravity is to be maintained at some constant figure, the specific gravity can be controlled by an indirect method rather than by a direct measurement of the value.

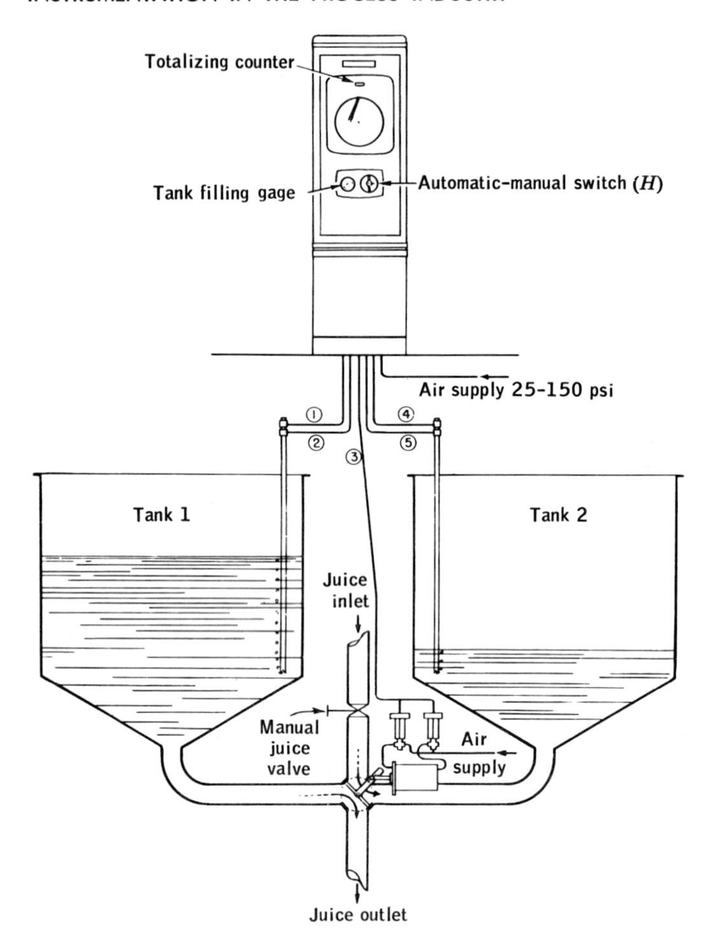

Fig. 21·7 If a known depth and specific gravity will create a certain hydraulic pressure, then variations in specific gravity will require variations in depth before the same pressure can be established. This ensures that each batch will contain the correct amount of ingredients, regardless of the specific gravity. (*The Foxboro Co.*)

If the materials to be mixed always have the same densities, then any specific gravity can be obtained by mixing the ingredients in a definite ratio. Generally, the control of solid flow is more difficult than is the control of liquid flow. For that reason, Fig. 21·9, the flow of solids is allowed to vary within certain limits imposed by the ingredients and the system. But the liquid with which the solid is to be mixed is controlled in such a manner as to ensure a fixed ratio between the two: a given number of gallons of liquid for each pound of solid material. In this way, the specific gravity is predetermined. If the specific gravity were

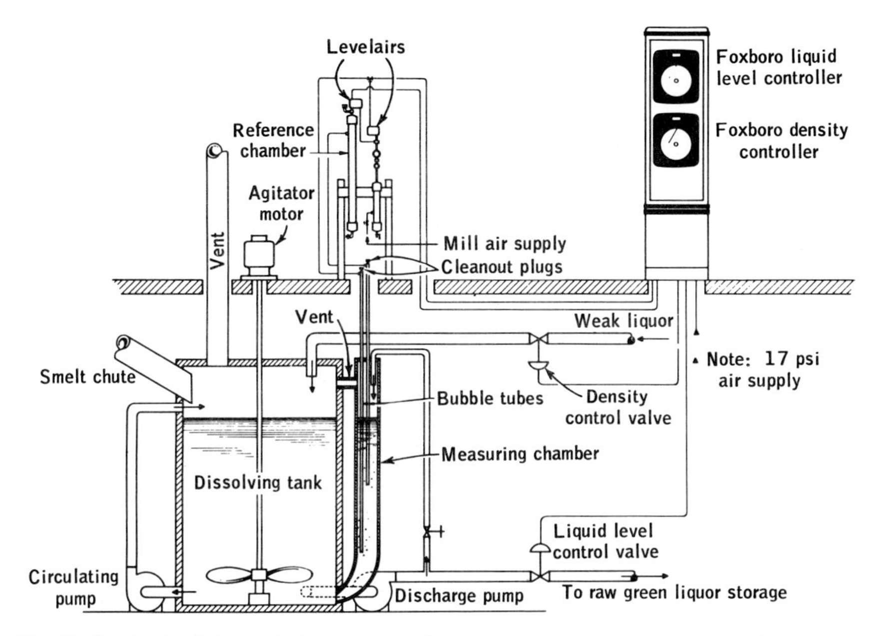

Fig. 21·8 A dissolving-tank density control system to obtain the correct specific gravity. (*The Foxboro Co.*)

Fig. 21·9 Instead of making a specific-gravity determination directly, the same results may be obtained by ensuring that the ratio of solids and liquids is some desired constant. (*The Foxboro Co.*)

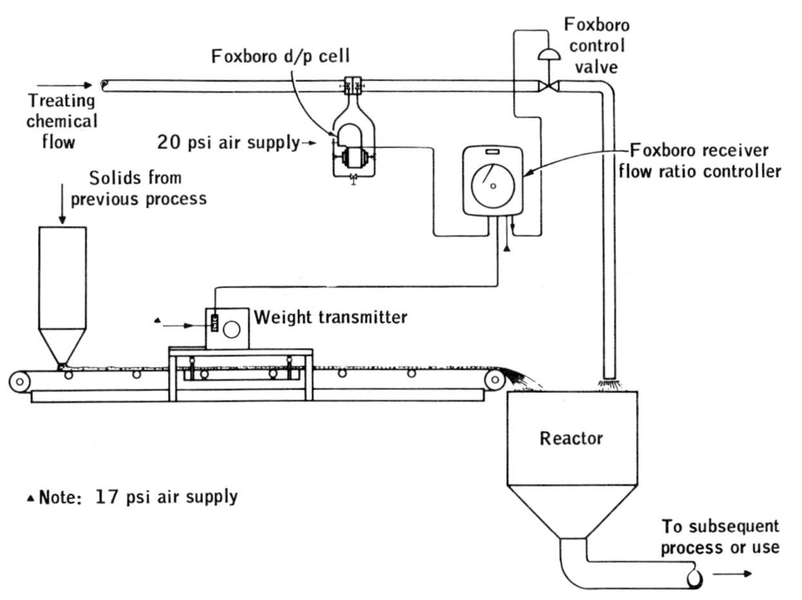

measured with one of the methods previously discussed, there would be two variables to control. That would require a system of far greater complexity.

Specific-gravity scales. The more common materials and liquids all have had their specific gravities determined entirely on the basis of the ratio of their weights to the weight of an equal volume of water. Several specific-gravity scales depart from this basis, and their values are not based on these simple ratios. One scale that is frequently used with petroleum products uses degrees API, which are the values established by the American Petroleum Institute. The values are listed in Table 21·1; they have been computed on the basis of

$$\text{Degrees API} = \frac{141.5}{\text{sp gr } 60°\text{F}/60°\text{F}} - 131.5$$

The denominator of the fraction indicates that the specific gravity of both the oil and the water should be determined at 60°F.

Another common scale is expressed in degrees Baumé, Tables 21·1 and 21·2. These values are obtained from:

$$\text{Degrees Baumé} = \frac{140}{\text{sp gr } 60°\text{F}/60°\text{F}} - 130$$

for liquids lighter than water

$$\text{Degrees Baumé} = 145 - \frac{145}{\text{sp gr } 60°\text{F}/60°\text{F}}$$

for liquids heavier than water

Still another scale, used where sugar solutions are employed, indicates the percentage of sucrose by weight. A fourth scale is expressed in degrees Twaddell, Table 21·2. The Twaddell degree is obtained from 200 (sp gr 60°F/60°F − 1). This expression is used to convert specific gravities of the heavier liquids (having specific gravities between 1.0 and 2.0) into 200 groups. Thus, each degree means a change of 0.005 in specific gravity.

Still another scale used for the commercial expression of specific gravity is the scale expressed in degrees Brix, the value of which is the percentage of sucrose by weight, Table 21·3.

Viscosity measurements. All of us have had experience with sticky, thick liquids which we call viscous fluids. We know that it takes far more force to move a round stick, for example, through such a liquid than it does to move it at an equal rate through water. The greater the force required, the higher the viscosity rating we are likely to give the fluid.

Table 21·1 Specific Gravity Conversion Tables, Liquids Lighter Than Water

$$\text{Degrees API} = \frac{141.5}{\text{sp gr } 60°\text{F}/60°\text{F}} - 131.5$$

$$\text{Degrees Baumé} = \frac{140}{\text{sp gr } 60°\text{F}/60°\text{F}} - 130$$

API reading		Deg API or Be	Baumé reading		API reading		Deg API or Be	Baumé reading	
lb/gal	Sp gr		Sp gr	lb/gal	lb/gal	Sp gr		Sp gr	lb/gal
5.086	0.6112	100	0.6087	5.066	6.316	0.7587	55	0.7568	6.300
5.109	0.6139	99	0.6114	5.088	6.350	0.7628	54	0.7609	6.334
5.131	0.6166	98	0.6140	5.110	6.385	0.7669	53	0.7650	6.369
5.154	0.6193	97	0.6167	5.132	6.420	0.7711	52	0.7692	6.404
5.176	0.6220	96	0.6195	5.155	6.455	0.7753	51	0.7735	6.440
5.199	0.6247	95	0.6222	5.178	6.490	0.7796	50	0.7778	6.476
5.222	0.6275	94	0.6250	5.201	6.526	0.7839	49	0.7821	6.511
5.246	0.6303	93	0.6278	5.225	6.563	0.7883	48	0.7865	6.548
5.269	0.6331	92	0.6306	5.248	6.600	0.7927	47	0.7910	6.586
5.293	0.6360	91	0.6335	5.272	6.637	0.7972	46	0.7955	6.623
5.316	0.6388	90	0.6364	5.296	6.675	0.8017	45	0.8000	6.661
5.341	0.6417	89	0.6393	5.320	6.713	0.8063	44	0.8046	6.699
5.365	0.6446	88	0.6422	5.345	6.752	0.8109	43	0.8092	6.738
5.390	0.6476	87	0.6452	5.370	6.790	0.8155	42	0.8140	6.777
5.415	0.6506	86	0.6482	5.395	6.830	0.8203	41	0.8187	6.817
5.440	0.6536	85	0.6512	5.420	6.870	0.8251	40	0.8235	6.857
5.465	0.6566	84	0.6542	5.445	6.910	0.8299	39	0.8284	6.898
5.491	0.6597	83	0.6573	5.471	6.951	0.8348	38	0.8333	6.939
5.516	0.6628	82	0.6604	5.497	6.993	0.8398	37	0.8383	6.980
5.542	0.6659	81	0.6635	5.522	7.034	0.8448	36	0.8434	7.022
5.568	0.6690	80	0.6667	5.549	7.076	0.8498	35	0.8485	7.065
5.595	0.6722	79	0.6699	5.576	7.119	0.8550	34	0.8537	7.108
5.622	0.6754	78	0.6731	5.602	7.163	0.8602	33	0.8589	7.152
5.649	0.6787	77	0.6763	5.629	7.206	0.8654	32	0.8642	7.196
5.676	0.6819	76	0.6796	5.657	7.251	0.8708	31	0.8696	7.241
5.703	0.6852	75	0.6829	5.685	7.296	0.8762	30	0.8750	7.286
5.731	0.6886	74	0.6863	5.712	7.341	0.8816	29	0.8805	7.332
5.759	0.6919	73	0.6897	5.741	7.387	0.8871	28	0.8861	7.378
5.788	0.6953	72	0.6931	5.769	7.434	0.8927	27	0.8917	7.425
5.817	0.6988	71	0.6965	5.798	7.481	0.8984	26	0.8974	7.473
5.845	0.7022	70	0.7000	5.827	7.529	0.9042	25	0.9032	7.522
5.874	0.7057	69	0.7035	5.856	7.578	0.9100	24	0.9091	7.570
5.904	0.7093	68	0.7071	5.886	7.627	0.9159	23	0.9150	7.620
5.934	0.7128	67	0.7107	5.916	7.676	0.9218	22	0.9211	7.670
5.964	0.7165	66	0.7143	5.946	7.727	0.9279	21	0.9272	7.721
5.994	0.7201	65	0.7179	5.976	7.778	0.9340	20	0.9333	7.772
6.025	0.7238	64	0.7216	6.007	7.830	0.9402	19	0.9396	7.825
6.056	0.7275	63	0.7254	6.038	7.882	0.9465	18	0.9459	7.877
6.087	0.7313	62	0.7292	6.070	7.935	0.9529	17	0.9524	7.931
6.119	0.7351	61	0.7330	6.102	7.989	0.9593	16	0.9589	7.986
6.151	0.7389	60	0.7368	6.134	8.044	0.9659	15	0.9655	8.041
6.184	0.7428	59	0.7407	6.166	8.099	0.9725	14	0.9722	8.096
6.216	0.7467	58	0.7447	6.199	8.155	0.9792	13	0.9790	8.153
6.249	0.7507	57	0.7487	6.233	8.212	0.9861	12	0.9859	8.211
6.283	0.7547	56	0.7527	6.266	8.270	0.9930	11	0.9929	8.269
					8.328	1.0000	10	1.0000	8.328

Table 21·2 Specific Gravity Conversion Table, Liquids Heavier Than Water

$$\text{Degrees Twaddell} = 200\,(\text{sp gr } 60°\text{F}/60°\text{F} - 1)$$

$$\text{Degrees Baumé} = 145 - \frac{145}{\text{sp gr } 60°\text{F}/60°\text{F}}$$

Twaddell reading		Deg Tw or Be	Baumé reading	
lb/gal	Sp gr		Sp gr	lb/gal
8.328	1.000	0	1.0000	8.328
8.370	1.005	1	1.0069	8.385
8.411	1.010	2	1.0140	8.445
8.453	1.015	3	1.0211	8.504
8.495	1.020	4	1.0284	8.565
8.536	1.025	5	1.0357	8.625
8.578	1.030	6	1.0432	8.688
8.619	1.035	7	1.0507	8.750
8.661	1.040	8	1.0584	8.814
8.702	1.045	9	1.0662	8.879
8.744	1.050	10	1.0741	8.945
8.786	1.055	11	1.0821	9.012
8.827	1.060	12	1.0902	9.079
8.869	1.065	13	1.0985	9.148
8.911	1.070	14	1.1069	9.218
8.952	1.075	15	1.1154	9.289
8.994	1.080	16	1.1240	9.361
9.035	1.085	17	1.1328	9.434
9.078	1.090	18	1.1417	9.508
9.118	1.095	19	1.1508	9.584
9.161	1.100	20	1.1600	9.660
9.203	1.105	21	1.1694	9.739
9.244	1.110	22	1.1789	9.818
9.286	1.115	23	1.1885	8.898
9.328	1.120	24	1.1983	9.979
9.369	1.125	25	1.2083	10.063
9.411	1.130	26	1.2185	10.148
9.452	1.135	27	1.2288	10.233
9.494	1.140	28	1.2393	10.321
9.535	1.145	29	1.2500	10.410
9.577	1.150	30	1.2609	10.501
9.619	1.155	31	1.2719	10.592
9.660	1.160	32	1.2832	10.686
9.702	1.165	33	1.2946	10.781
9.744	1.170	34	1.3063	10.879
9.785	1.175	35	1.3182	10.978
9.827	1.180	36	1.3303	11.079
9.868	1.185	37	1.3426	11.181
9.910	1.190	38	1.3551	11.285
9.951	1.195	39	1.3679	11.392
9.993	1.200	40	1.3810	11.501
10.035	1.205	41	1.3942	11.611
10.076	1.210	42	1.4078	11.724
10.118	1.215	43	1.4216	11.839
10.160	1.220	44	1.4356	11.956
10.201	1.225	45	1.4500	12.076

Twaddell reading		Deg Tw or Be	Baumé reading	
lb/gal	Sp gr		Sp gr	lb/gal
10.243	1.230	46	1.4646	12.197
10.284	1.235	47	1.4796	12.322
10.326	1.240	48	1.4948	12.449
10.367	1.245	49	1.5104	12.579
10.410	1.250	50	1.5263	12.711
10.452	1.255	51	1.5426	12.849
10.493	1.260	52	1.5591	12.984
10.535	1.265	53	1.5761	13.126
10.577	1.270	54	1.5934	12.270
10.618	1.275	55	1.6111	13.417
10.660	1.280	56	1.6292	13.568
10.701	1.285	57	1.6477	13.722
10.743	1.290	58	1.6667	13.880
10.784	1.295	59	1.6860	14.041
10.826	1.300	60	1.7059	14.207
10.868	1.305	61	1.7262	14.376
10.909	1.310	62	1.7470	14.549
10.951	1.315	63	1.7683	14.727
10.993	1.320	64	1.7901	14.908
11.034	1.325	65	1.8125	15.095
11.076	1.330	66	1.8354	15.285
11.117	1.335	67	1.8590	15.482
11.159	1.340	68	1.8831	15.683
11.200	1.345	69	1.9079	15.889
11.242	1.350	70	1.9333	16.101
11.451	1.375	75		
11.659	1.400	80		
11.867	1.425	85		
11.076	1.450	90		
11.284	1.475	95		
12.492	1.500	100		
12.908	1.550	110		
13.325	1.600	120		
13.741	1.650	130		
14.158	1.700	140		
14.574	1.750	150		
14.990	1.800	160		
15.407	1.850	170		
15.823	1.900	180		
16.240	1.950	190		
16.656	2.000	200		

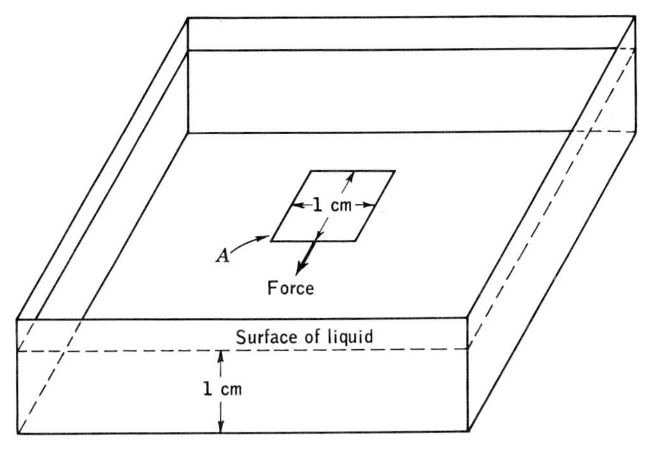

Fig. 21·10 One measure of the viscosity of a fluid is the force required to move an object through it at constant speed. By definition, the force, in dynes, required to move A at the rate of 1 cm/sec is numerically equal to the viscosity measured in poises.

The basic unit of viscosity is the poise. It is defined from the following, Fig. 21·10: If we have two plane surfaces, one of indefinite length and width while the other, A, is 1 cm square, separated from each other by a distance of 1 cm, and if a force of 1 dyne is required to move A at the rate of 1 cm/sec, then the viscosity, by definition, is 1 poise. If a force of 100 dynes is required, then the viscosity is 100 poises, and so on.

In measuring viscosity, we must recognize that there are two general classes of liquids: Newtonian and non-Newtonian. Newtonian fluids are those in which the shearing stress is directly proportional to the flow rate, Fig. 21·11*a*. Other fluids which do not maintain a fixed ratio of shear stress to flow rate are called non-Newtonian. It will be noted in Fig. 21·11*b* that not only does the force not bear a linear relation to the flow rate, but the value also depends upon whether or not the flow rate is increasing or decreasing.

Units of viscosity

1. The basic units of viscosity measurement are the *poise* and a unit one-hundredth as large called a *centipoise*. When using the cgs system, the unit is the poise and the viscosity is the *absolute viscosity*.

2. *Kinematic viscosity* is expressed in *stokes* or *centistokes*. The values are obtained by dividing the viscosity in poises by the density of the fluid in grams per cubic centimeter.

3. The *specific viscosity* is the ratio of the absolute viscosity of the liquid to that of a standard fluid, such as water. Measurements should be made with both fluids at the same temperature.

Table 21·3 Specific Gravity Conversion Table, Sugar Solutions at 20°C

Deg Brix or % sucrose by wgt	Sp gr at 20°/20°C	Deg Baumé	Deg Brix or % sucrose by wgt	Sp gr at 20°/20°C	Deg Baumé	Deg Brix or % sucrose by wgt	Sp gr at 20°/20°C	Deg Baumé
0	1.0000	0.00	25.0	1.1055	13.84	50.0	1.2317	27.28
0.5	1.0019	0.28	25.5	1.1078	14.11	50.5	1.2345	27.54
1.0	1.0039	0.56	26.0	1.1101	14.39	51.0	1.2373	27.81
1.5	1.0058	0.84	26.5	1.1125	14.66	51.5	1.2401	28.07
2.0	1.0078	1.12	27.0	1.1148	14.93	52.0	1.2429	28.33
2.5	1.0098	1.40	27.5	1.1171	15.20	52.5	1.2456	28.59
3.0	1.0117	1.68	28.0	1.1195	15.48	53.0	1.2484	28.86
3.5	1.0137	1.96	28.5	1.1219	15.75	53.5	1.2513	29.12
4.0	1.0157	2.24	29.0	1.1242	16.02	54.0	1.2541	29.38
4.5	1.0177	2.52	29.5	1.1266	16.29	54.5	1.2569	29.64
5.0	1.0197	2.79	30.0	1.1290	16.57	55.0	1.2598	29.90
5.5	1.0217	3.07	30.5	1.1314	16.84	55.5	1.2626	30.16
6.0	1.0237	3.35	31.0	1.1338	17.11	56.0	1.2655	30.42
6.5	1.0257	3.63	31.5	1.1362	17.38	56.5	1.2684	30.68
7.0	1.0277	3.91	32.0	1.1386	17.65	57.0	1.2712	30.94
7.5	1.0297	4.19	32.5	1.1410	17.92	57.5	1.2741	31.20
8.0	1.0318	4.46	33.0	1.1435	18.19	58.0	1.2770	31.46
8.5	1.0338	4.74	33.5	1.1459	18.46	58.5	1.2799	31.71
9.0	1.0359	5.02	34.0	1.1484	18.73	59.0	1.2829	31.97
9.5	1.0379	5.30	34.5	1.1508	19.00	59.5	1.2858	32.23
10.0	1.0400	5.57	35.0	1.1533	19.28	60.0	1.2887	32.49
10.5	1.0421	5.85	35.5	1.1558	19.55	60.5	1.2917	32.74
11.0	1.0441	6.13	36.0	1.1583	19.81	61.0	1.2946	33.00
11.5	1.0462	6.41	36.5	1.1608	20.08	61.5	1.2976	33.26
12.0	1.0483	6.68	37.0	1.1633	20.35	62.0	1.3006	33.51
12.5	1.0504	6.96	37.5	1.1658	20.62	62.5	1.3036	33.77
13.0	1.0525	7.24	38.0	1.1683	20.89	63.0	1.3066	34.02
13.5	1.0546	7.51	38.5	1.1709	21.16	63.5	1.3096	34.28
14.0	1.0568	7.79	39.0	1.1734	21.43	64.0	1.3126	34.53
14.5	1.0589	8.07	39.5	1.1760	21.70	64.5	1.3156	34.79
15.0	1.0610	8.34	40.0	1.1785	21.97	65.0	1.3187	35.04
15.5	1.0632	8.62	40.5	1.1811	22.23	65.5	1.3217	35.29
16.0	1.0653	8.89	41.0	1.1837	22.50	66.0	1.3248	35.55
16.5	1.0675	9.17	41.5	1.1863	22.77	66.5	1.3278	35.80
17.0	1.0697	9.45	42.0	1.1889	23.04	67.0	1.3309	36.05
17.5	1.0719	9.72	42.5	1.1915	23.30	67.5	1.3340	36.30
18.0	1.0740	10.00	43.0	1.1941	23.57	68.0	1.3371	36.55
18.5	1.0762	10.27	43.5	1.1967	23.84	68.5	1.3402	36.81
19.0	1.0784	10.55	44.0	1.1994	24.10	69.0	1.3433	37.06
19.5	1.0807	10.82	44.5	1.2020	24.37	69.5	1.3464	37.31
20.0	1.0829	11.10	45.0	1.2047	24.63	70.0	1.3496	37.56
20.5	1.0851	11.37	45.5	1.2073	24.90	70.5	1.3527	37.81
21.0	1.0873	11.65	46.0	1.2100	25.17	71.0	1.3559	38.06
21.5	1.0896	11.92	46.5	1.2127	25.43	71.5	1.3590	38.30
22.0	1.0918	12.20	47.0	1.2154	25.70	72.0	1.3622	38.55
22.5	1.0941	12.47	47.5	1.2181	25.96	72.5	1.3654	38.80
23.0	1.0964	12.74	48.0	1.2208	26.23	73.0	1.3686	39.05
23.5	1.0986	13.02	48.5	1.2235	26.49	73.5	1.3718	39.30
24.0	1.1009	13.29	49.0	1.2263	26.75	74.0	1.3750	39.54
24.5	1.1032	13.57	49.5	1.2290	27.02	74.5	1.3782	39.79

Table 21·3 (Continued)

Deg Brix or % sucrose by wgt	Sp gr at 20°/20°C	Deg Baumé	Deg Brix or % sucrose by wgt	Sp gr at 20°/20°C	Deg Baumé	Deg Brix or % sucrose by wgt	Sp gr at 20°/20°C	Deg Baumé
75.0	1.3814	40.03	85.0	1.4479	44.86	95.0	1.5181	49.49
75.5	1.3847	40.28	85.5	1.4514	45.09	95.5	1.5217	49.71
76.0	1.3879	40.53	86.0	1.4548	45.33	96.0	1.5254	49.94
76.5	1.3912	40.77	86.5	1.4582	45.57	96.5	1.5290	50.16
77.0	1.3944	41.01	87.0	1.4617	45.80	97.0	1.5326	50.39
77.5	1.3977	41.26	87.5	1.4652	46.03	97.5	1.5362	50.61
78.0	1.4010	41.50	88.0	1.4686	46.27	98.0	1.5399	50.84
78.5	1.4043	41.74	88.5	1.4721	46.50	98.5	1.5435	51.06
79.0	1.4076	41.99	89.0	1.4756	46.73	99.0	1.5472	51.28
79.5	1.4109	42.23	89.5	1.4791	46.97	99.5	1.5509	51.50
						100.0	1.5545	51.73
80.0	1.4142	42.47	90.0	1.4826	47.20			
80.5	1.4175	42.71	90.5	1.4861	47.43			
81.0	1.4209	42.95	91.0	1.4896	47.66			
81.5	1.4242	43.19	91.5	1.4932	47.89			
82.0	1.4276	43.43	92.0	1.4967	48.12			
82.5	1.4310	43.67	92.5	1.5003	48.35			
83.0	1.4343	43.91	93.0	1.5038	48.58			
83.5	1.4377	44.15	93.5	1.5074	48.81			
84.0	1.4411	44.38	94.0	1.5110	49.03			
84.5	1.4445	44.62	94.5	1.5145	49.26			

Fig. 21·11 (*a*) Newtonian fluids have a constant ratio of flow to force. (*b*) Non-Newtonian fluids do not have a constant ratio of flow to force. Moreover, their viscosity is controlled in part by the previous history. For a given flow rate, the force required will depend upon whether the rate is increasing or decreasing.

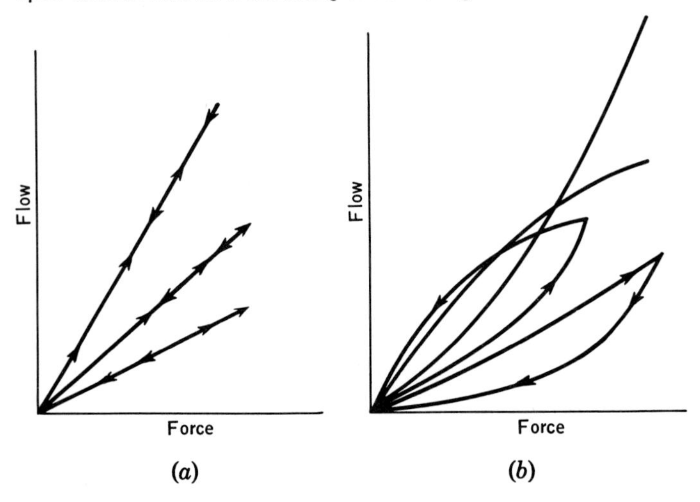

4. The *relative viscosity* is the ratio of the absolute viscosity of the liquid to that of water at 68°F. Because the viscosity of water at 68°F (or 20°C) is practically equal to 1, the absolute viscosity in centipoises also expresses the relative viscosity.

5. *Saybolt universal seconds* (SSU). These are the values obtained when using the Saybolt universal viscometer, which is often used for light fuel oil and lubricating oil determinations.

6. *Saybolt ferrol seconds* (SSF). These are units obtained when a capillary having a greater delivery rate is substituted for the standard tube in the Saybolt universal viscometer. Liquids of high viscosity, such as tars, heavy fuel oils, and road oils, are generally measured by this device.

By definition,

$$\text{Kinematic viscosity} = \frac{\text{absolute viscosity}}{\text{weight density, g per cc}} = \frac{\text{poises}}{\text{weight density, g per cc}}$$

In the English system, the absolute viscosity is expressed in pound-seconds per square foot.

$$\text{Kinematic viscosity in the English system} = \frac{g\mu e}{\text{weight density}}$$

To make the desired conversions, use the following:

$$\begin{aligned} 1 \text{ poise} &= 478.8 \text{ lb (force)-sec/ft}^2 \\ 1 \text{ centipoise} &= 4.78 \text{ lb (force)-sec/ft}^2 \\ 0.00209 \text{ poises} &= 1 \text{ lb (force)-sec/ft}^2 \\ 0.209 \text{ centipoises} &= 1 \text{ lb (force)-sec/ft}^2 \end{aligned}$$

There are many ways of making viscosity measurements, but there are relatively few that lend themselves to industrial practice where the measurement is to be used for purposes of control.

One method measures the pressure differential through a given length of pipe, with a given rate of flow, and from this the viscosity is inferred. For control purposes, if a given differential pressure is found to be satisfactory, then the system is designed to control at that differential, Fig. 21·12. Inasmuch as standard constant-volume pumps and differential pressure transmitters can be employed, this becomes an easy design or principle to follow. Thus, because a fixed rate of flow of a liquid having a definite viscosity should produce a given pressure drop, any change of pressure when the flow is constant is attributed to the viscosity. The necessary corrective measures then can be taken to return the viscosity to the desired value. In terms of purely arbitrary units, the output air pressure of the differential pressure transmitter is a measure of the viscosity.

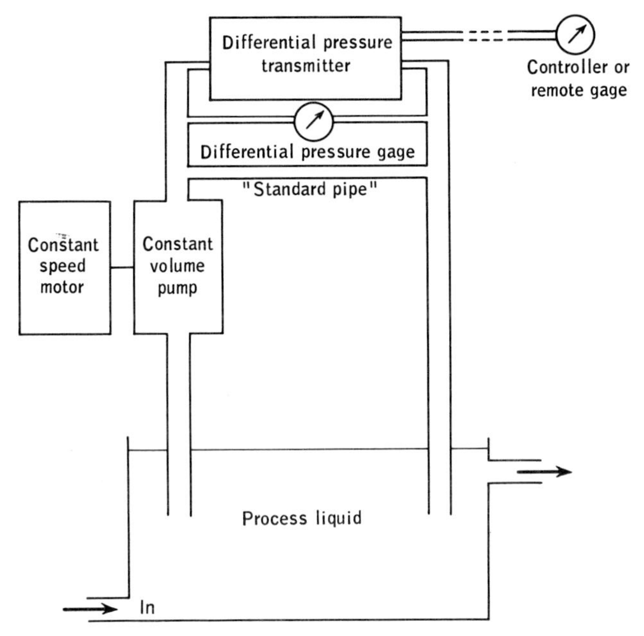

Fig. 21·12 The pressure required to produce a fixed flow rate of a liquid through a given pipe or tube may also be used as a measure of the viscosity of the liquid. This provides relative values rather than absolute ones.

A totally different approach to the measurement of viscosity is shown in Fig. 21·13. We are all aware that it requires more torque or effort to move an object in very thick, sticky fluids. Similarly, it will require more torque to rotate a drum or cylinder in a thick fluid than in a thin watery liquid. The total force or torque required depends upon the size of the cylinder and the viscosity of the liquid. Consequently, by varying the size of the cylinder, both in length and diameter, a given power input can be made to rotate a cylinder through liquids having a wide range of viscosities. If the cylinder is driven by a spring, the amount of twist in the spring will represent the torque which is required under stable operation conditions. The physical angular displacement of the two ends of the spring drive can be "read" directly, or the angular displacement may be used to rotate a potentiometer whose output voltage then is a measure of the viscosity. Or the positions of the two ends of the drive may alter the relative positions of two metal plates and thus alter their capacitance. Changes in capacitance can be determined with one of the many types of bridges, and the viscosity can be inferred from the capacitance. In general, a design employing capacitance techniques is likely to have less drag because of friction, and hence it should give responses to small viscosity changes.

A design based on a similar principle is sketched in Fig. 21·14*a* and *b*. In this instance, the tube which is rotated in the fluid is driven by the magnetic repulsion of two magnets: an electromagnet on an upper plate which is rotated by the driving motor and a permanent magnet which is mounted on the lower plate. For any given speed and viscosity, a definite torque must be applied to the cylinder. This is obtained by the mutual repulsion of the two fields. The current is so adjusted that the repulsion forces will furnish just the proper torque. Too much current will cause the gap between the magnets to become too great and cause an indicator circuit to be completed. Too little current will let the magnets come closer together and close another set of contacts. Adjusting the electromagnet's current will establish the proper operating point. The force required to overcome mechanical bearing forces can be determined easily and the meter can be returned to zero mechanically. Such a device eliminates many of the circuit complexities involved with changes in small capacitances.

A very recent approach to the problem of continuous viscosity measurement and control is found in the ultrasonic probe and computer de-

Fig. 21·13 A commercial viscometer, showing details of construction and the spindle which is immersed in the fluid. (*The Bristol Company.*)

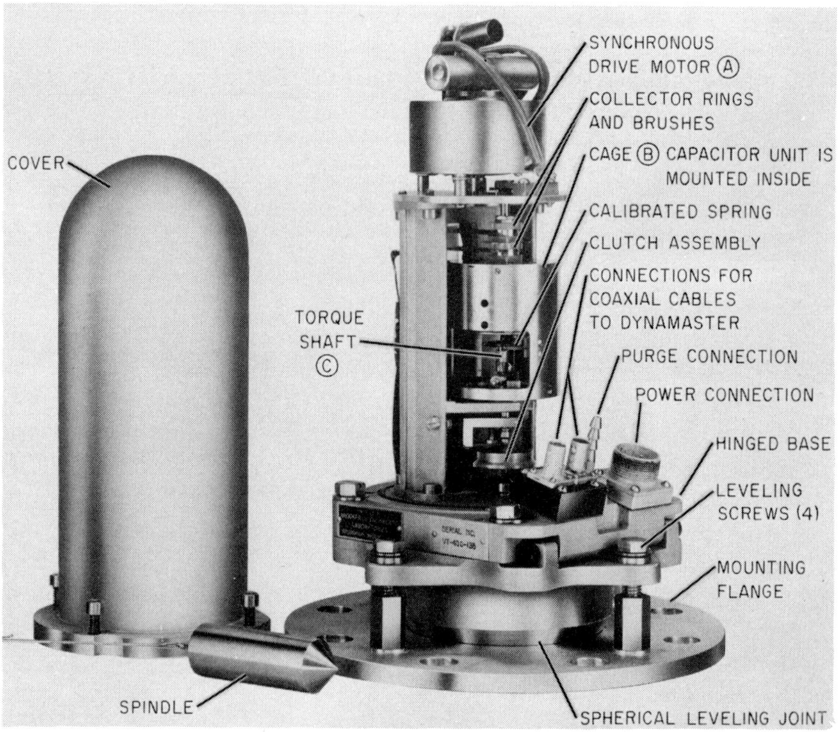

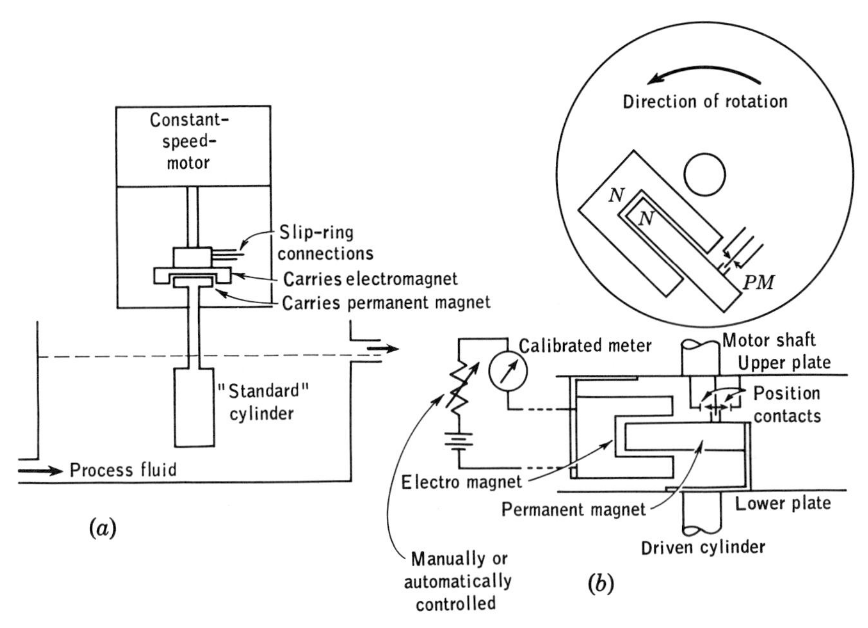

Fig. 21·14 (*a*) A sketch of a viscometer which uses the forces created by two magnets to drive the spindle or standard cylinder. The value of the current then becomes a measure of the viscosity. (*b*) With the correct value of current in the electromagnet, the force of repulsion between the magnets will be just equal to the viscous drag of the liquid.

signed by Bendix. In this design, an oscillator provides a frequency of about 28,000 cps. The active portion of the probe is magnetostrictive, so that as its length changes with the alternations of the current, the materials immediately adjacent to the probe are caused to slip with respect to adjacent particles. The power required to produce fixed movements then depends upon the viscosity of the liquid, which acts as a drag on the movement of the probe. The fact that the probe tip moves but a very minute amount does not alter the performance of the device. A computer is needed to convert the power required by the probe to a millivolt signal from which the viscosity can be determined. Figure 21·15*a* to *d* shows the probe and the more usual methods of installation.

Because viscosity may change so rapidly with temperature, it is often desirable to convert the viscosity to a value which would be observed at some base temperature quite different from the temperature at which the measurement is being made. The temperature may be determined with a resistance thermometer whose value can be fed into a temperature-compensating computer. The output of this second computer is then corrected for temperature.

In Chap. 11 it is stated that rotameters have very definite responses to the viscosity of the flowing fluid but that, through study and design,

Fig. 21·15 (*a*) Installation of probe in elbow or bend of large pipe ($3\frac{1}{2}$ in. or larger). (*b*) Up to 16 Visicon probes can be connected to a single unit. (*c*) The correct values of viscosity may be obtained through control of the temperature. (*d*) Automatic blending of ingredients may be used for control of viscosity. (*The Bristol Company.*)

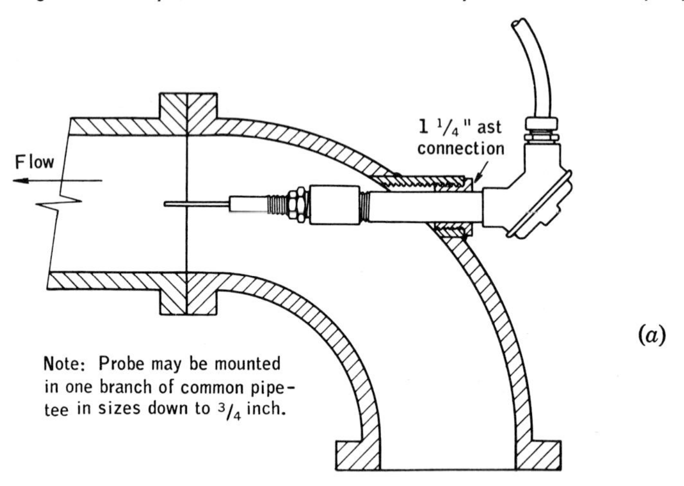

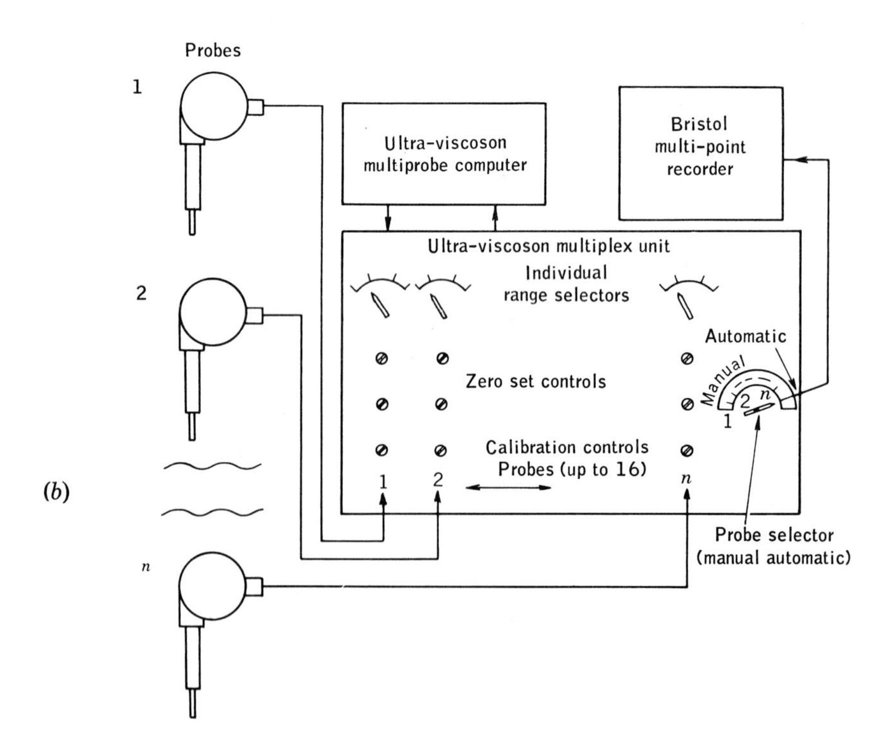

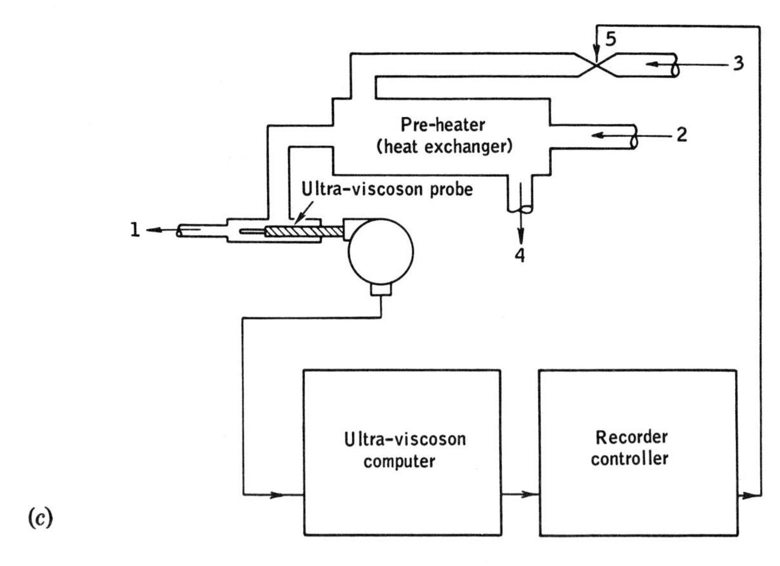

1. Output of constant desired viscosity
2. Input of uncontrolled viscosity
3. Heat exchange fluid input
4. Heat exchange fluid output
5. Control valve, pneumatic, or electric

For use in:

A- Fuel oil viscosity control
B- Extruder viscosity control
C- Evaporator viscosity control

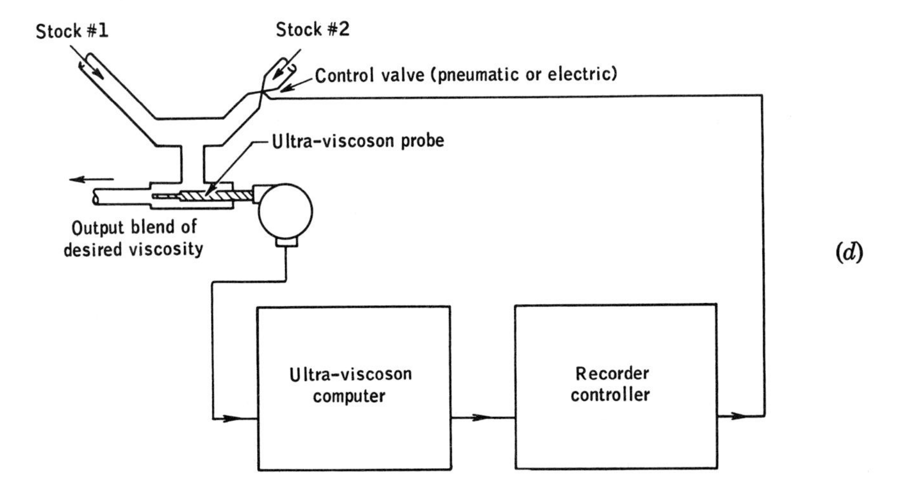

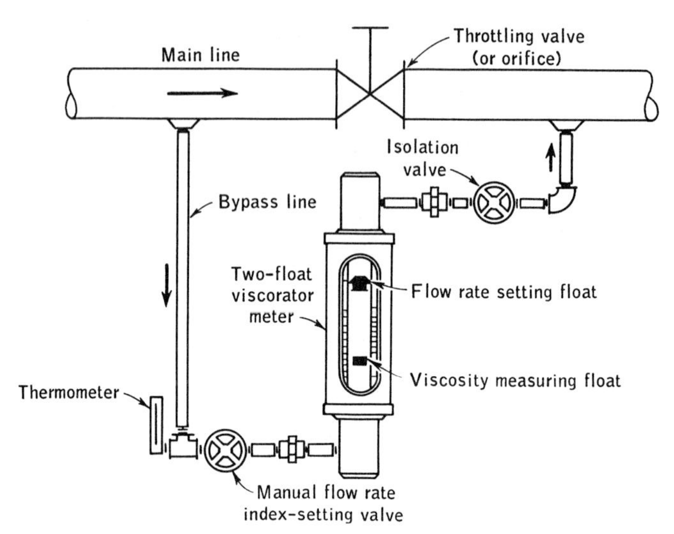

Fig. 21·16 **If the flow rate can be maintained constant by using a suitable pump, only the viscosity float is required to indicate the viscosity. (*The Fischer and Porter Company.*)**

it has been possible to make meters which are virtually unaffected by viscosity. For most purposes, the design which is unaffected by viscosity is the desirable one. But for the task at hand, advantage can be taken of viscosity-sensitive types of response. One such employment is indicated in Fig. 21·16. Inasmuch as there should be certain reference conditions, they are established by controlling the flow through the meter so that a known flow rate is established. A second float which is viscosity sensitive will then assume a position which is indicative of the viscosity at that particular temperature. When it is impracticable to install a throttling valve or orifice in the line, samples may be drawn from the line and caused to flow at a fixed rate through the rotameter. Since the rate is fixed, there is no need for a float to give this value. Consequently, only the viscosity-sensitive float is used.

A modification of the method which requires either manual control of the flow rate or the use of a constant-volume pump is to use a pressure regulator which will create a constant pressure differential across the terminals of the flowmeter. This will meet the requirement for a constant flow rate. The position of the viscosity-sensitive float then will be a measure of the viscosity, Fig. 21·17.

Still another way of measuring the viscosity of very viscous materials such as asphalt, clays, soaps, and waxes is to employ the well-known fact that such materials when in motion will try to carry other materials along with them. Thus, forces are set up which are a measure of the viscosity of the fluid at that particular temperature. Figure 21·18 shows

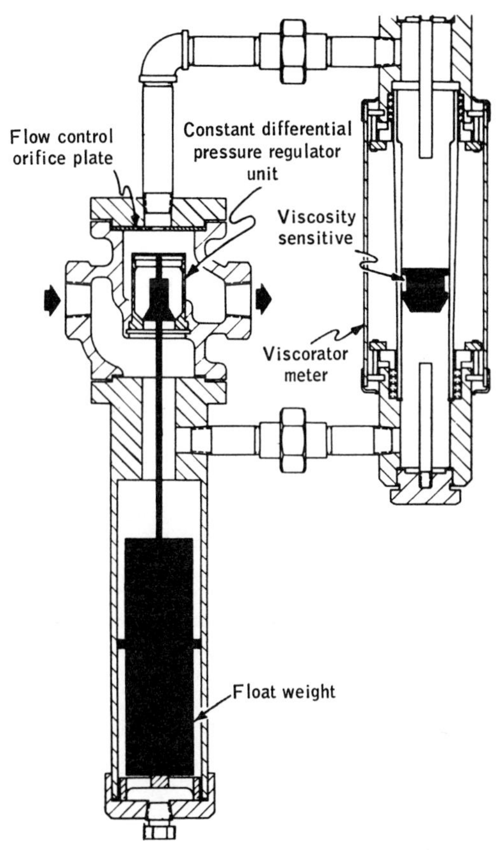

Fig. 21·17 A constant differential pressure, created by the differential pressure regulator, provides a flow rate which varies inversely with the viscosity. (*The Fischer and Porter Company.*)

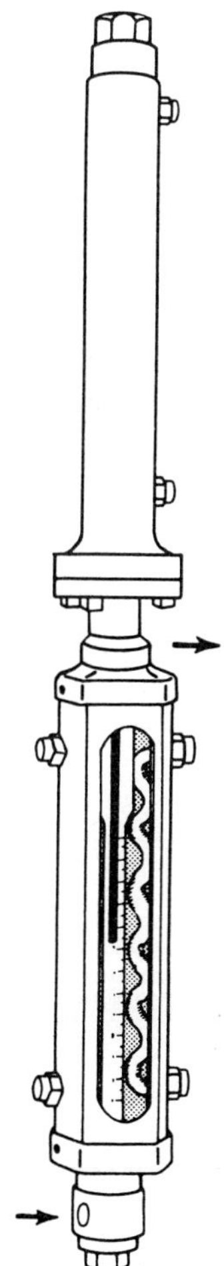

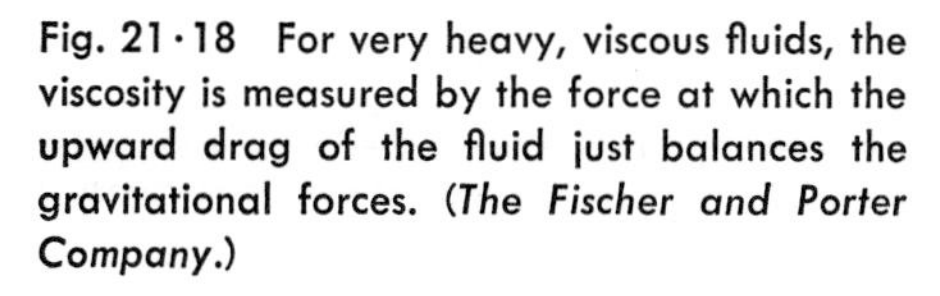

Fig. 21·18 For very heavy, viscous fluids, the viscosity is measured by the force at which the upward drag of the fluid just balances the gravitational forces. (*The Fischer and Porter Company.*)

a device in which a weighted glass tube is lifted by the upward movement of the fluid. But as the tube is lifted, it leaves the fluid and thus decreases the area which is in contact with the fluid. Soon a condition of balance is reached: the upward force of the flowing fluid is just equal to the downward force on the rod. Such a measurement obviously needs a constant flow rate in order to function correctly. Temperature measurement or control is a very important factor in the proper evaluation of the values measured.

When the scale is calibrated in terms of *kinedynamic units*, KDU, the instrument may be used with fluids which vary both in density and viscosity. However, variations in these two characteristics can still be read correctly on a scale employing the kinedynamic units. Figure 21·19 shows such a calibration and the nominal values of representative liquids.

Conversions into some of the more usual units can be made very readily:

$$\text{Centipoises} = \text{KDU} \times \sqrt{\frac{\rho_w(8.02) - \rho_w}{7.02}}$$

$$\text{Centistokes} = \text{KDU} \times \sqrt{\frac{8.02 - \rho_w}{7.02\rho_w}}$$

$$\rho_w = \text{mass density} = \text{grams/cm}^3$$

The numerical values 8.02 and 7.02 are to be used when the floats are

Fig. 21·19 The convenience of the kinedynamic unit is illustrated. Calibration points of liquids and gases at widely varying densities and viscosities, when expressed in KDU, fall on one smooth curve. (*The Fischer and Porter Company.*)

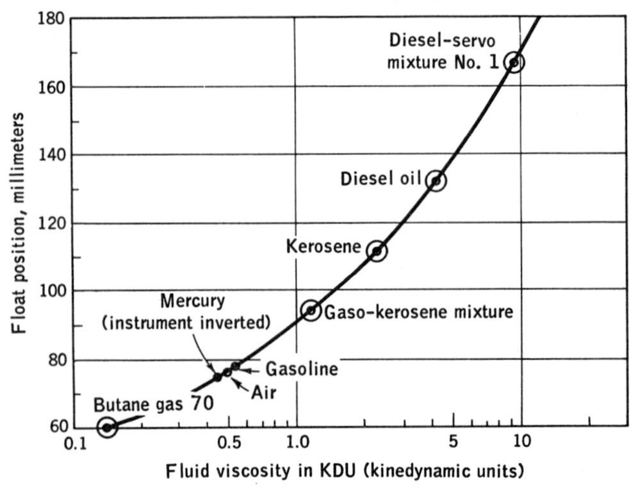

of no. 316 stainless steel having a specific gravity of 8.02 when referred to water. If other materials are used for the floats or if the floats have their specific gravity changed, then the proper values should be substituted in these equations.

SUMMARY

1. Many specific-gravity measurements are made by comparing the weight of the material with the weight of an equal volume of water. There are, however, a number of scales that do not use water as the standard.

2. The weight density of a material, in the English system, is expressed in pounds per cubic foot. It may be found by taking the product of the weight of a cubic foot of water and the specific gravity of the material.

3. As the specific gravity and the weight density of an object change, the weight of a given volume of liquid will vary linearly. A floating object, which will always displace its own weight, will then displace more or less fluid as the specific gravity changes. A suitable calibration can then convert amount of submergence into some suitable scale value. For a buoyant object which is kept submerged, the buoyant force will vary directly with the specific gravity and the weight density.

4. Because the weight of a column of liquid and the pressures which it will create vary directly and linearly with the specific gravity, the pressure created by any given fixed head of liquid may also be used for specific-gravity measurements.

5. Inasmuch as the boiling point of a liquid for given pressure conditions depends in part on the concentration or specific gravity of the liquid, the boiling point of a liquid may be used as a measure of the specific gravity.

6. Some of the more common commercial specific-gravity scales are

a. The one established by the American Petroleum Institute and commonly used when measuring petroleum products:

$$\text{Degrees API} = \frac{141.5}{\text{sp gr } 60°\text{F}/60°\text{F}} - 131.5$$

b. The Baumé scale, used for liquids both lighter and heavier than water:

$$\text{Degrees Baumé} = \frac{140}{\text{sp gr } 60°\text{F}/60°\text{F}} \qquad \text{liquids lighter than water}$$

$$\text{Degrees Baumé} = \frac{145}{\text{sp gr } 60°\text{F}/60°\text{F}} \qquad \text{liquids heavier than water}$$

c.

$$\text{Degrees Twaddell} = 200\left(\frac{\text{sp gr } 60°\text{F}}{\text{sp gr water } 60°\text{F}} - 1\right)$$

d. Degrees Brix are equal to the percentage of sucrose by weight in solution at 17.5°C.

7. Fluids which display a constant ratio of flow to force are called Newtonian, while those which do not possess this linearity—in fact, may vary widely, being different for an increase in flow than for a decrease—are called non-Newtonian.

8. The basic unit of viscosity is the poise. If two surfaces, one of indefinite length and width and the other 1 cm^2 in area, are separated by 1 cm and it takes a force of 1 dyne to move the 1 cm^2 surface parallel to the other at the rate of 1 cm/sec, then the liquid has a viscosity of 1 poise.

9. In making viscosity measurements, we may employ many of the relationships

previously studied: force balance, torque balance, and the fact that a constant rate of flow for materials of fixed viscosity requires a definite differential pressure. However, the use of an ultrasonic generator to power the probe and the use of magnetostriction devices are new applications.

QUESTIONS

21·1 What are the advantages of a design, such as shown in Fig. 21·1, which employs a free float to provide a measure of specific gravity? What factors could cause errors in indication? How could they be overcome?

21·2 What are the conditions which must be maintained if bubbler systems, Fig. 21·3, are to be used for accurate specific-gravity determinations?

21·3 For industrial use, would you consider the devices shown in Fig. 21·2 and Fig. 21·3 equally satisfactory for specific-gravity measurements? Which would be least affected by temperature? By changes in level or flow rates?

21·4 How does the force on the submerged float, Fig. 21·5, change with specific gravity? If the float is aluminum, would you consider this design subject to significant temperature errors? Could you suggest how to minimize the effect if it is appreciable?

21·5 Would you consider the design of Fig. 21·9 a more or less desirable way of controlling specific gravity than that of Fig. 21·8?

21·6 Is the equipment indicated in Fig. 21·9 measuring the specific gravity? What happens if the valve either loses its actuating air pressure or fails to respond in a normal manner?

21·7 What are the industrial needs which make viscosity determinations desirable?

21·8 Define simply and clearly what viscosity is and explain the units used.

21·9 What are the usual units for expressing viscosity?

21·10 Of what importance is viscosity measurement in any process with which you are familiar?

21·11 What are the differences between Newtonian and non-Newtonian fluids?

21·12 What physical characteristics of materials are used in viscosity determinations?

21·13 By means of a regulator, a constant pressure differential is provided across the rotameter, Fig. 21·17. How will the flow rate vary with viscosity?

21·14 What are the desirable characteristics of a viscometer? Which factors are the most important?

PROBLEMS

21·1 An empty container weighs 1 lb. When filled with liquid X, it weighs 4 lb, but when filled with water it weighs 5 lb. What is the specific gravity of X?

21·2 What is the weight density of liquid X, Prob. 15·1? The mass?

21·3 A submerged buoyant object displaces 20 oz of water. Its weight is 10 oz. What weight should be added to it so that it is just submerged in a liquid of specific gravity $\frac{3}{4}$? Specific gravity 1.5?

21·4 Without changing the weight or displacement of the buoyant object in Prob. 21·3, what specific gravity will it indicate if it is just submerged? If it is three-fourths out of the liquid?

22 pH Measurements

None of the measurements with which we have been concerned up to this point have involved any knowledge of the kind or type of material being processed. We have been measuring those factors which could affect the physical properties of the materials, but we have not been trying to measure the characteristics of the material after it had reacted to the pressure or temperature conditions. But a knowledge of the characteristics of the material in process is also desirable, and oftentimes necessary, if the desired end product is to be produced. Subsequent chapters will deal with other techniques for measuring the chemical constituents in terms of specific materials, or lack of them. In this portion of our study we are concerned mostly with whether a material has the properties of an acid or a base or is neutral in its reactions. For certain chemical reactions to proceed at their proper rate, or to the proper point, it may be quite necessary that acidic, basic, or neutral conditions be maintained within rather close limits.

An acid is a material which will dissociate or break down in a water solution and furnish hydrogen ions, which are electrically charged hydrogen atoms that carry positive charges. At the same time that the positively charged hydrogen ions are formed, a number of negatively charged particles (ions) will be formed. A *base* material is one which reacts with an acid. In general, water solutions of base materials will break down to form ions composed of one atom of oxygen and one atom of hydrogen which together carry one negative electrical charge. When acting as ions they are indicated as OH and are designated as *hydroxyl ions.* They are most commonly derived from water solutions of sodium and potassium hydroxide. The addition of an acid to a base will neutralize the base and produce a *salt.* The addition of a base to an acid will also neutralize the acid and in turn will also produce a salt.

Water has some of the characteristics of both acids and bases, since it will dissociate to form both the hydrogen and the hydroxyl ions in equal numbers.

PRINCIPLES, DEVICES, AND APPLICATIONS

Meaning of pH. From our definition of an acid as a material which in solution will provide hydrogen ions, we can say that the number of hydrogen ions or the concentration of hydrogen ions might be a measure of the acidity of the liquid, and conversely, the acidity of a liquid is also a measure of the concentration of hydrogen ions. In any given solution, the greater the acidity the smaller the concentration of hydroxyl ions; and the greater the concentration of hydroxyl ions the smaller the concentration of hydrogen ions. In any acidic solution at any one time, no matter how strongly acidic the liquid, there will also be a trace of hydroxyl ions, and by the same token a liquid which is strongly basic will also have a few hydrogen ions. In pure water we have equal numbers of each.

In setting up a rating scale to express numerically the value of the hydrogen ion concentration and thus the degree of acidity, we might assume that the greatest degree of acidity is found when there is 1 gram of hydrogen ions per unit volume. Then for lower degrees of acidity there would be a fractional part of a gram per unit volume, or 1 gram for a number of unit volumes. Table 22·1 lists, in column 1, concentrations in ions per unit volume. Instead of dealing with some of the larger numbers, it is easier to express them in logarithms which merely tell the power to which 10 would have to be raised to be equal to the same number. For example, $100 = 10^2$, $1{,}000 = 10^3$, $1{,}000{,}000 = 10^6$. When using 10 for the base number, the logarithm of 100 is 2; that for 1,000 is 3, and that for 1,000,000 is 6. The simplicity of using a small number for these comparisons will be evident after a little thought. Each change of a unit in the logarithm means that the quantity has changed by a factor of 10. If we did not use this scale, the last number in the column, Table 22·1, would always have to be written with 14 zeros. The negative sign in front of the exponent in the second column indicates that the value of the number is getting smaller and smaller, as will be evident from a study of column 1. In conventional use, the negative sign is omitted from the indicated value of the pH measurement. It is always to be understood that the value of 14 for pH means that there is an extremely small concentration of the hydrogen ion, while a value of 2 means that there is a very high concentration of hydrogen ions.

How to measure pH. Many of the earlier techniques for measuring the hydrogen ion concentration employed an *indicator* of some sort which when added to the sample would produce a certain color change if the hydrogen ion concentration were of a certain value. The resulting color could then be compared with a standard for an evaluation of the concentration. Such a method does not lend itself well to automatic deter-

Table 22·1 Concentration of Hydrogen Ions in Acid and Base Solutions

Concentration of hydrogen ions, grams/liter		pH	Concentration of hydroxyl ions, grams/liter Powers of 10	Nature of solution
Whole numbers	Powers of 10			
				STRONGLY ACID
1/1	$1/10^0 = 10^{-0}$	0	10^{-14}	Hydrochloric acid (10.0N) Hydrochloric acid (1.0N) Sulfuric acid (1.0N)
1/10	$1/10^1 = 10^{-1}$	1	10^{-13}	Hydrochloric acid (0.1N) Sulfuric acid (0.1N) Phosphoric acid (0.1N) Gastric (stomach) fluids Lime juice Hydrochloric acid (0.01N)
1/100	$1/10^2 = 10^{-2}$	2	10^{-12}	Lemon juice Alum (0.1N) Ginger ale
1/1,000	$1/10^3 = 10^{-3}$	3	10^{-11}	Vinegar (4% acetic acid) Grapefruit juice Fruit jellies Pickles Wines Orange juice
1/10,000	$1/10^4 = 10^{-4}$	4	10^{-10}	Very acid natural soil Tomatoes Beer
1/100,000	$1/10^5 = 10^{-5}$	5	10^{-9}	Hydrochloric acid (0.00001N) Boric acid (0.1N) Molasses Nickel plating bath Distilled water after standing in air
1/1,000,000	$1/10^6 = 10^{-6}$	6	10^{-8}	Corn Fresh milk

Table 22·1 (Continued)

Concentration of hydrogen ions, grams/liter		pH	Concentration of hydroxyl ions, grams/liter Powers of 10	Nature of solution
Whole numbers	Powers of 10			
				NEUTRAL
1/10,000,000	$1/10^7 = 10^{-7}$	7	10^{-7}	Theoretically pure water
				Human blood plasma Fresh eggs
etc.	$1/10^8 = 10^{-8}$	8	10^{-6}	Baking soda (1%) Sea water
	$1/10^9 = 10^{-9}$	9	10^{-5}	Borax solutions (0.1 to 10%) Very alkaline natural soil
	$1/10^{10} = 10^{-10}$	10	10^{-4}	Soap solutions Milk of magnesia Washing soda (1%)
	$1/10^{11} = 10^{-11}$	11	10^{-3}	Ammonium hydroxide (0.1*N*) Household ammonia (10%)
	$1/10^{12} = 10^{-12}$	12	10^{-2}	Copper cyanide plating bath Lime (saturated solution)
	$1/10^{13} = 10^{-13}$	13	10^{-1}	Sodium hydroxide (0.1*N*)
	$1/10^{14} = 10^{-14}$	14	10^{-0}	Sodium hydroxide (1.0*N*)
				STRONGLY BASIC

mination, nor can it be used with liquids which are normally colored or have such degrees of turbidity that color evaluations are difficult.

Because of these conditions, the determination of hydrogen ion concentration is more accurate and desirable if some value which is independent of color and turbidity can be employed. The method most favored today employs the measurement of the potential created by a special battery. The hydrogen ions react with a special glass electrode and produce a potential whose value can be measured.

In Chap. 19 it is noted that the current flow in the ionization cell is extremely small, being of the order of a billionth of an ampere. And the way that this extremely small current can be detected is by causing it to pass through a resistor of extremely high value. The potential developed under these conditions is then read by noting the response of a meter in the plate circuit of a special vacuum tube, the outstanding characteristic of which is the extremely small value of grid current. If the measuring circuit were to take no current to actuate it, then the measurement of extremely small currents would be possible. We have a similar situation here.

The resistance of the electrodes is of such a high value that robbing the circuit of more than a few electrons per second would adversely affect the performance of the device and the accuracy of the measurements. Before considering the basic circuit designs, we should consider the electrodes which provide the potential.

The electrodes. For many industrial applications two separate types of electrodes are used. One, the active cell, produces an emf which is proportional to the hydrogen ion concentration. A second cell is used for reference purposes. The reference cell or electrode is generally made with calomel, which is a compound of mercury, or with silver–silver chloride. For those situations in which mercury is not desired, such as in the manufacture of photographic emulsions and related products, other types of reference materials are employed. Sometimes these two sets of electrodes are combined in one unit; at other times both are separate units which may be used with a third component which is temperature responsive and is used to correct for temperature variations.

It should be stressed again that the performance of the potential methods of measuring hydrogen ion concentrations is based on the fact that when a piece of metal is placed in an ionic solution of the same metal, there will be a potential developed between the metal and the solution. The value of this emf is proportional to the concentration of ions. However, if we insert a piece of metal (of the same or some other material) into the solution in order to detect the value of the emf, then we completely upset the original conditions, because another potential will be developed at the new electrode. Therefore, we have to make our measurements in such a way as to eliminate any potentials introduced by the measuring system.

One way to accomplish this is shown in Fig. 22·1. This method requires the admission of hydrogen gas into the solutions in the presence of platinum electrodes. A definite potential is developed between the platinum electrode (which is coated with a finely divided platinum called platinum black) and the solution of known pH. A different potential is developed between the platinum electrode in the other cell and the solution in which it is immersed. Some means must now be used to connect the two cells without altering the emf's created at the two electrodes. This is done by

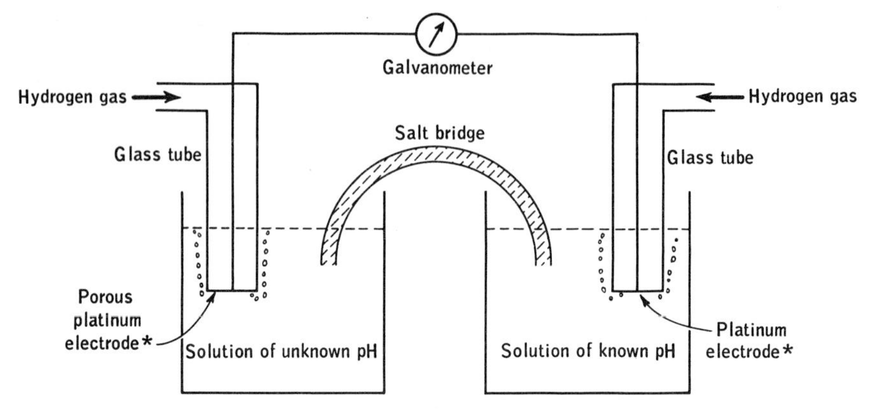

Fig. 22·1 When hydrogen gas is supplied to the two porous-platinum electrodes, a potential which is proportional to the pH concentration is developed in each cell. The salt bridge serves to connect the two cells in series opposition. The galvanometer then responds to the difference in potential.

using the conductivity of a salt solution which is more or less fixed in position and cannot mix freely with the solutions in which the electrodes are immersed. The performance of the parts here should be well understood, because the present commercial cells are duplicating the performance of these two units.

Electrically, the cells are connected in series opposition so that the resulting emf $= E = E_1 - E_2 = CT(\text{pH}_1 - \text{pH}_2)$. When E is expressed in millivolts and the temperature is expressed in the absolute temperature units of the Kelvin scale, the value of $C = 0.1983$. The change of pH with temperature is shown in Fig. 22·2.

Fig. 22·2 The change of potential with temperature is serious and cannot be ignored. Commercial instruments make the correction automatically, or provide for it.

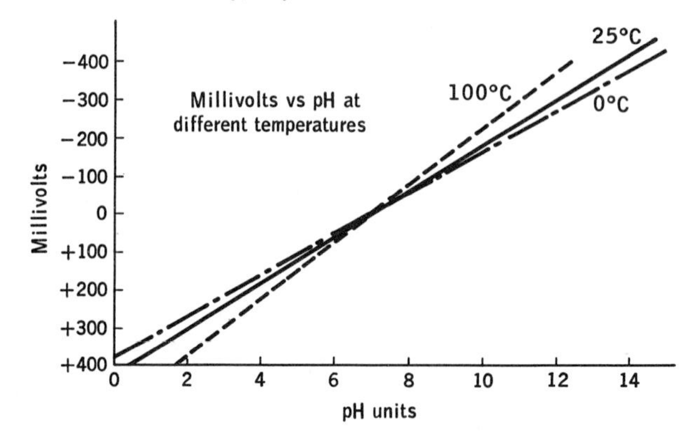

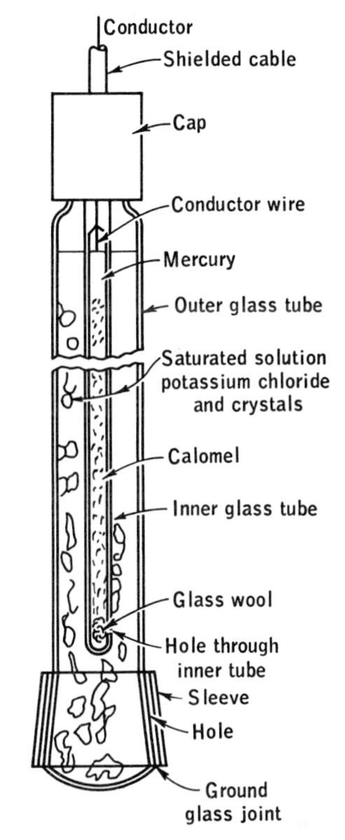

Fig. 22·3 Beckman type of reference electrode using calomel. (*The Bristol Company.*)

For a temperature of 25°C, or 298°K, the product of C and T is 59.12. Therefore, a change of 1 pH unit will change the emf output by 59.12 mv. Of very great importance from the standpoint of the design of the instrument, the above equation indicates that the emf is a linear function of pH. A major objection to such a device is the bother of providing for the generation of the hydrogen gas. A second objection is the undesirability of the salt bridge.

Present designs of reference half-cells simulate performance of the hydrogen units. Figures 22·3 and 22·4 indicate typical construction. A detailed explanation of Fig. 22·5 taken from "The Industrial pH Handbook" with permission of the Beckman Instruments, Inc., follows:

Beginning with the internal glass electrode, there is a potential existing, E_1, between the internal metallic electrode and the internal standard pH solution. The potential drop or resistance between the internal electrode and the inner surface of the glass membrane through the internal solution is represented by E_2. A potential exists, E_3, between the internal solution and the inner surface of the glass membrane. The relatively high resistance of the pH-sensitive glass membrane itself is represented as a potential drop, E_4. This high resistance, until recent years, was the chief factor in retarding widespread industrial pH development. The potential, E_5, between the outer surface of the glass electrode and the solution being measured is the desired potential that varies in accordance with change in hydrogen ion con-

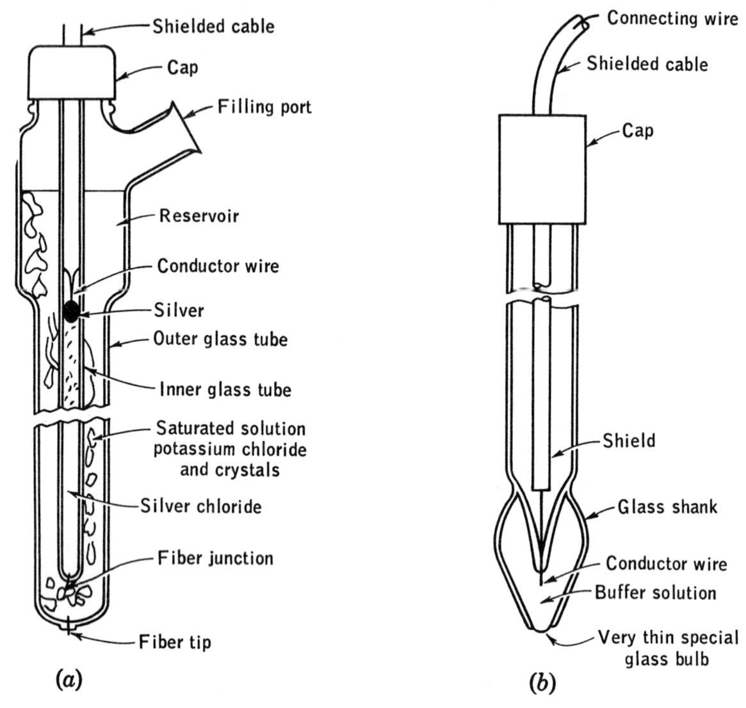

Fig. 22·4 (*a*) Beckman type of reference electrode, showing reservoir and fiber tip. (*b*) Beckman type of measuring electrode. (*The Bristol Company.*)

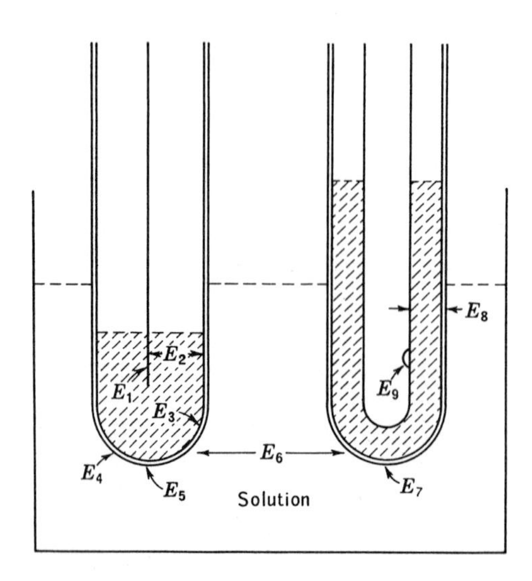

Fig. 22·5 There are nine different potentials established when an active and a reference cell are used for pH determination. Some of the potentials are quite small and may be ignored, but others are large and significant. (*Beckman, Scientific Instruments Division.*)

centration, or pH, of the test solution. E_6 represents the potential drop across the test solution. E_7 represents the liquid junction potential between the test solution and the solution within the reference electrode. The liquid junction is the opening or path of communication whereby the potassium chloride electrolyte maintains electrical contact with the test solution. E_8, the potential drop within the reference electrode, is due to the resistance of the electrolyte within the electrode. Finally, E_9 represents the potential developed by the reference electrode itself in its surroundings of saturated potassium chloride solution. The following list attempts to evaluate the importance of each of these factors.

E_1 Normally, this can be considered a constant potential as constant temperature. Extremely high temperatures can dissolve the silver chloride coating from the metal contacting the internal solution.

E_2 This is a negligible factor and normally can be ignored, with the present day types of internal filling solutions.

E_3 The stability of both the internal solution and the internal surface of the glass electrode membrane is sufficient to assume that this value is constant.

E_4 This will remain as a constant as long as the glass membrane does not undergo serious mechanical or chemical alterations.

E_5 This is a variable, depending upon the hydrogen ion concentration of the solution to be measured.

E_6 The potential drop across the test solution normally is negligible except in dealing with very low conductivity water or essentially nonaqueous solutions.

E_7 The liquid junction potential initially is insignificant, yet clogging or external pressure against the junction can seriously influence this element.

E_8 The filling solution has a negligible resistance as long as it does not become contaminated.

E_9 This solution can be considered constant if the filling solution is uncontaminated.

The potassium chloride solution performs the same function as the salt bridge in Fig. 22·1. The connection to the process fluid is made through either a fiber tip or by means of a connection through the liquid which wets the ground joint in Fig. 22·3. The resistances of these paths is of little or no significance with respect to the resistance of the glass electrode tip. The choice of the type of electrode depends upon the nature of the liquid. If the liquid is clear and not likely to form a scum or deposit, then the electrode with the small fiber at its tip will probably be preferred. If the liquid tends to coat the electrode, the greater amount of contact area will give the unit with the sleeve tip a definite advantage.

It will be recalled that there are two common designs of reference cells in use today, the silver chloride one (Fig. 22·4*a*) and the one based on the use of mercury compounds (Fig. 22·3). Figure 22·3 shows how the lead conductor makes contact with the cell through the medium of mercury which electrically connects the mercurous chloride electrode (calomel) with the outside circuit. A small hole in the lower part of the inner glass tube provides an electrical connection between the mercurous chloride and the saturated potassium chloride solution.

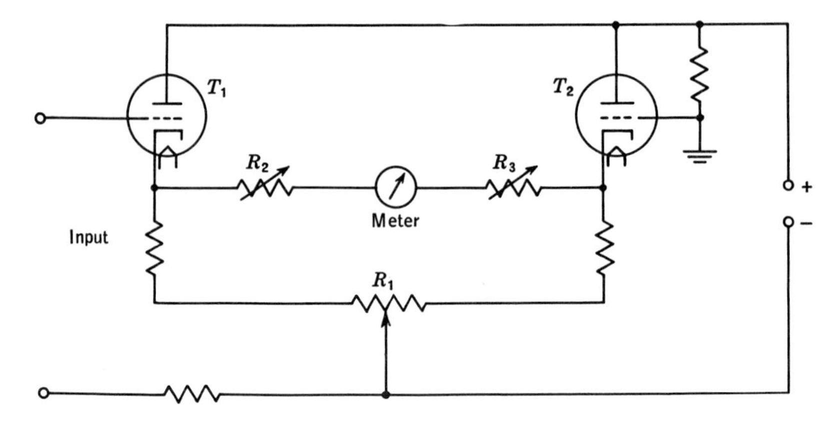

Fig. 22·6 Basic schematic of a high-impedance bridge used in pH measurements.

Principles of the pH meter. The use of the electrodes in solutions of various degrees of dissociation produces a potential which is to be used as a measure of the ionization, and hence of the concentration of hydrogen ions. As has been pointed out earlier in this chapter, the electrodes are made of glass, in part. They have resistance values which are in the neighborhood of hundreds of millions of ohms, which precludes the use of any measuring circuits which will draw current. There is one other problem involved when direct current flows: that of polarization of the electrodes because of the accumulation of gas. These conditions then establish the requirement that any measuring circuit be of such high resistance that for most practical purposes it could be said to be infinite. All operating conditions must be such as to minimize any leakage or pickup current which would invalidate the results.

One of the simplest of the measuring circuits is shown in Fig. 22·6. It will be seen that this is essentially a Wheatstone bridge of the type previously studied. There is also a very close resemblance to a vacuum tube voltmeter. The chief differences in this particular circuit which should be noted involve:

1. The use of special electrometer tubes whose grid currents are exceptionally small.

2. A correction for temperature inasmuch as the electrodes are temperature sensitive.

3. A regulated power supply, required because this is a deflection type of bridge from which high orders of accuracy are sought. Meters of this design have reproducibilities of the order of 0.025 pH units.

The extremely high resistance of the electrode circuit makes it impracticable to use the very simple, manually operated potentiometer such as is used for thermocouple measurements. To obtain a greater sensi-

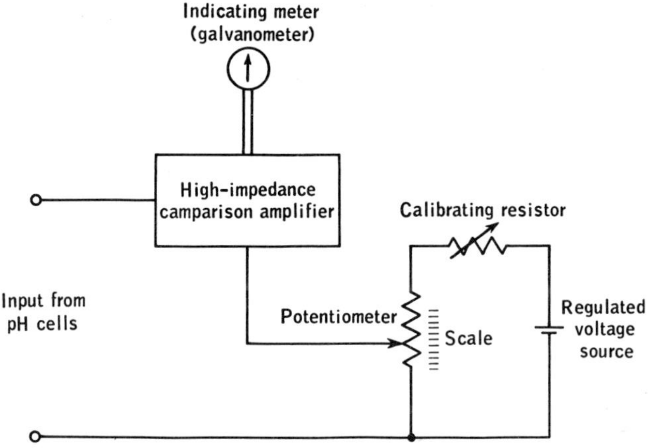

Fig. 22·7 By adjusting the potentiometer output so that it is equal to the output of the pH units, the impedance acts as though it were infinite, inasmuch as no current flows. The amplifier and the indicating galvanometer are used to facilitate the matching of the two potentials.

tivity and to ensure as nearly as possible that there will be no current flow, the main elements of the circuit shown in Fig. 22·7 are employed. This type of manually operated instrument can be replaced with instruments which employ automatic electronic potentiometric balancing or the conventional mechanical self-balancing potentiometer, Fig. 22·8. The main elements of Fig. 22·8 have been discussed in detail in Chap. 16. They are representative of the electronic self-balancing potentiometric devices in which particular attention has been given to isolating the high-resistance input circuit from the portions of the circuit which require appreciable current signals for their operation. *The high resistance of the*

Fig. 22·8 Simplified schematic diagram of a Leeds and Northrup Model 7664 pH indicator. (*The Instrument Society of America.*)

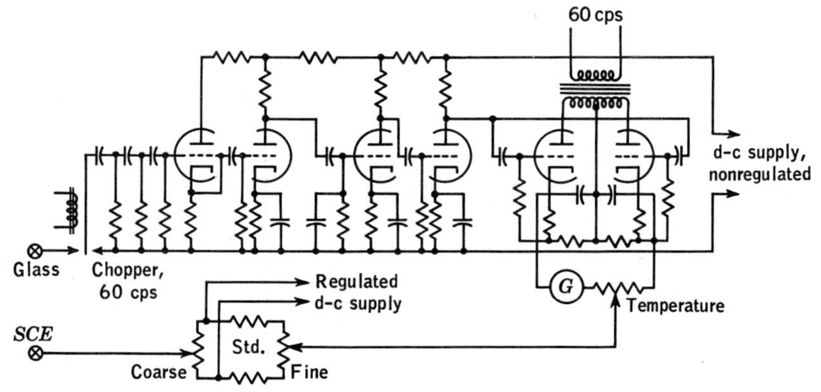

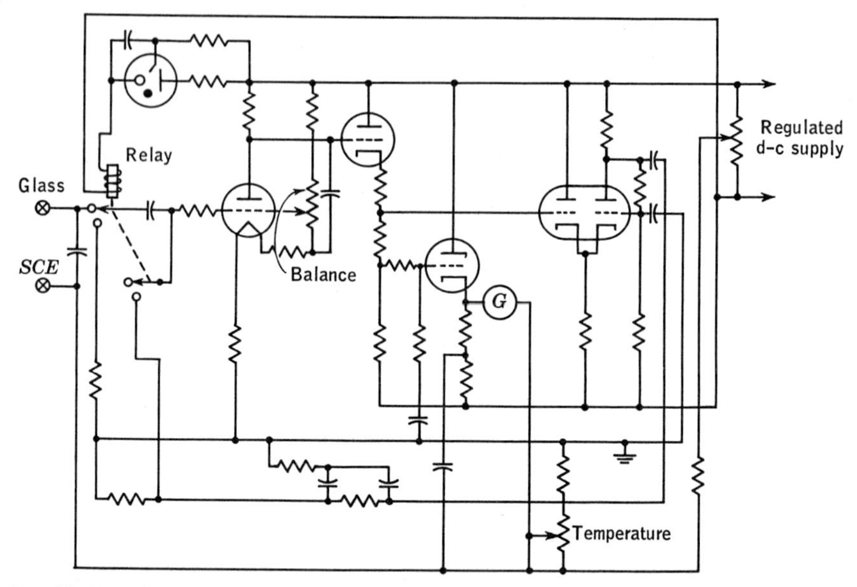

Fig. 22·9 **Schematic of the Beckman Zeromatic pH meter, simplified.** *(The Instrument Society of America.)*

input circuit must never be jeopardized when adding amplifiers or other circuit components.

For those situations which require the highest degrees of accuracy at all times, automatically self-balancing instruments may be used. One such circuit, shown in Fig. 22·9, employs a direct-current amplifier. To overcome one of the objections to direct-current amplifiers, which is drift, this amplifier employs continuous automatic correction. For approximately 0.015 sec each second the capacitor which is in series with the input grid of the first electrometer tube is switched to a zero correction amplifier which uses the dual triode on the right-hand side of the circuit. By taking such a short time for this periodic standardization, the instrument does not respond to the interruption of the input signal; so for all practical purposes there is no time lost for standardizing and one of the major objections to direct-current amplifiers is overcome.

Typical applications of pH measurement and control. When process liquids are being maintained at some specified value of pH concentration, the addition of dry materials of the proper type often is involved. Deviations of the controlled variable (pH) then cause corrective action to be taken to increase or decrease the amount of the neutralizer being added. It will soon be recognized that the methods shown in Fig. 22·10 are only schematic. Proper agitation and thorough mixing have to be provided if suitable control is to be obtained with a minimum of material.

Although controllers and controller action are not presented in detail here, it should be pointed out that a single type of controller may provide many different conditions as required. For example, in Fig. 22·10*a* the rate of adding the neutralizer is governed by the speed of the motor on the chemical feeder. However, in Fig. 22·10*b* a different sort of electrical signal is furnished to operate a shaker which governs the rate of adding chemicals. Figure 22·10*c* to *f* illustrates still other methods of controlling the desired variable. It will be noticed that in certain instances the liquid is pumped through the electrode assembly to ensure that the measured values are representative of the product being delivered. The system shown in Fig. 22·10*e* would be improved (in practice) by removing the liquid at the end where the sampling pump is working. Conceivably there could be quite a significant difference between the two ends if the mixing tank were of appreciable length. It must be remembered at all times that the measuring device must always be placed where it will provide representative readings of the product being delivered or where the best corrective action can be taken. Moreover, when control action is to be taken, any situation which will delay the response should be avoided.

Because of the extremely high resistance required for the grid of the electrometer tube, there is a definite restriction on the distance that the electrodes can be from the pH amplifier.

APPLICATION NOTES

1. Measuring electrodes when first put into service undergo a protracted change. Therefore, they must be checked frequently until they stabilize. The change is caused by the glass absorbing some of the chemical solution; also, the glass is being very slowly dissolved. Such effects will be minimized if the probe has been soaked in tap water or a dilute acid solution for at least 5 hr prior to installation.
2. Always install probes in such positions that inspection will be simple and convenient. Good practice indicates that a weekly inspection is in order even though the probe appears to be quite stable.
3. A good installation will ensure that the probe remains submerged even when out of service. If this is impracticable, the electrodes should be removed and stored so that they are immersed in the process liquid. If the measuring electrode is allowed to dry out by being exposed to the air for a long time, the potential which is developed will be unstable.
4. If the process liquid contaminates the probes by coating them with scum or sludge, the probes should be washed carefully with *tap water*. One manufacturer states: "Electrodes are destroyed faster by distilled water than by almost any other substance."
5. Certain proprietary materials which function as antiwetting agents may be employed to prevent the formation of solids or scum.
6. Under favorable conditions, reference electrodes should never have to be replaced unless they are broken. Electrodes should be stored with the tip down and never be permitted to dry out.

Fig. 22·10 By continuously monitoring the pH of the product, corrective action can be provided so that the desired value is always maintained. This is unlike installations which maintain a fixed ratio between two or more constituents. (*The Bristol Company.*)

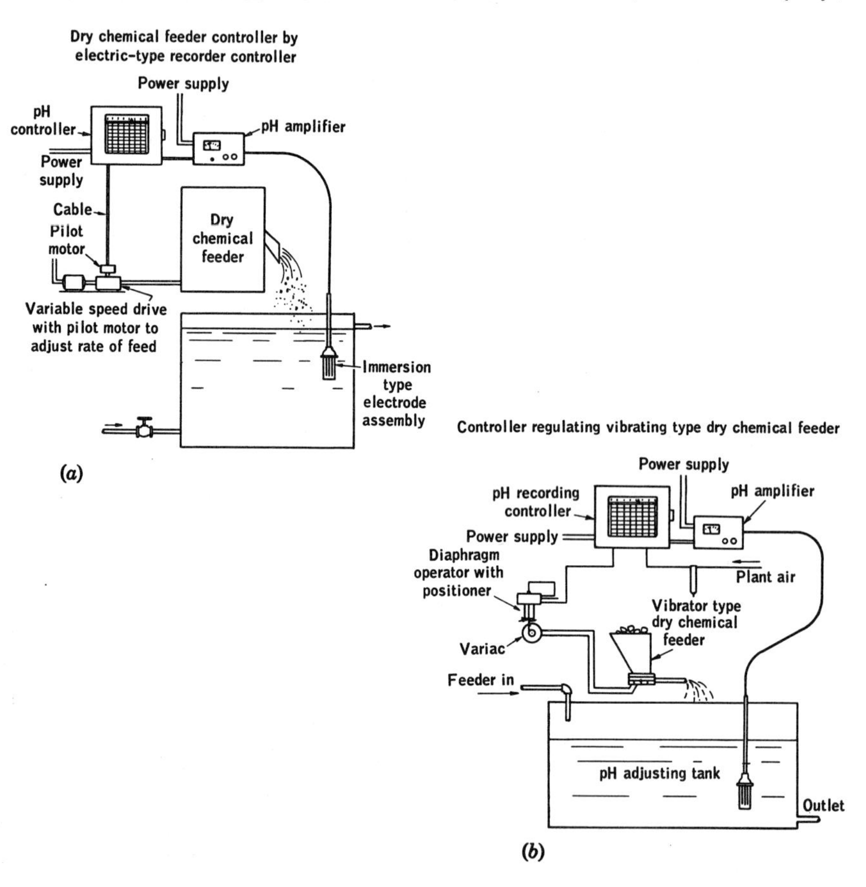

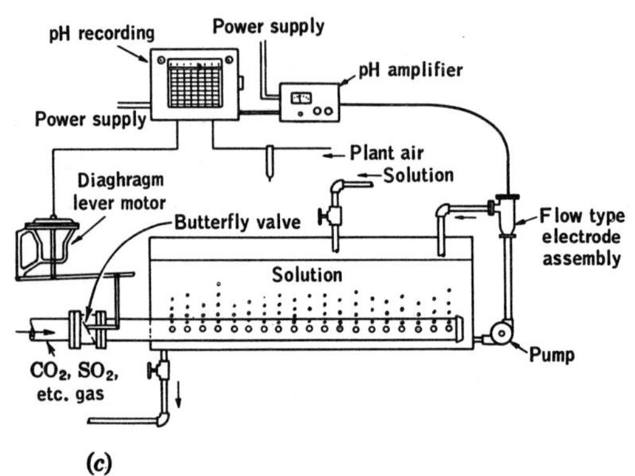

Means of adjusting pH using heavy suspensions such as lime slurry

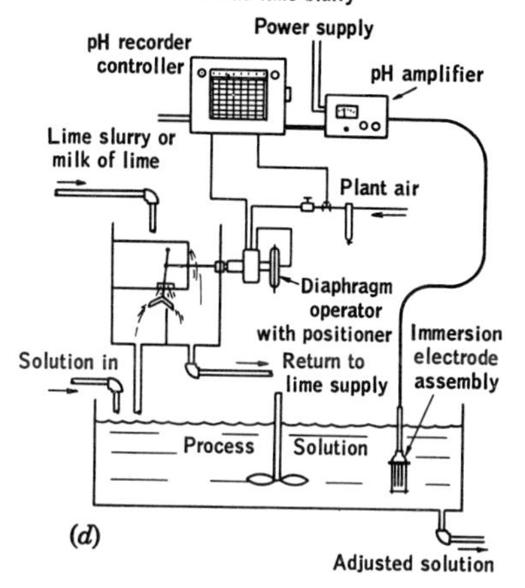

Control of pH of material under pressure using electric type control

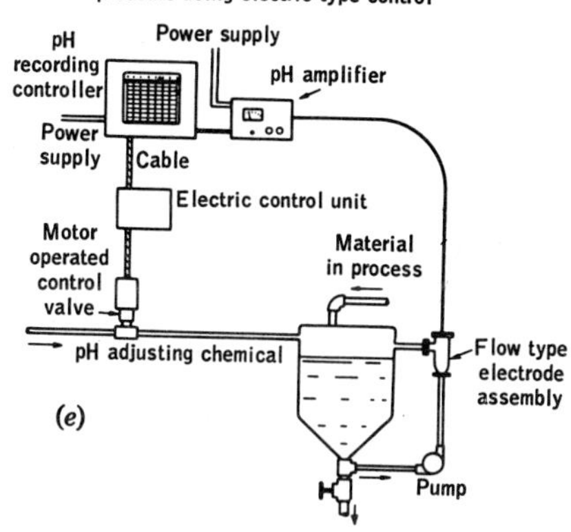

pH controller operating synchro valve to control pH

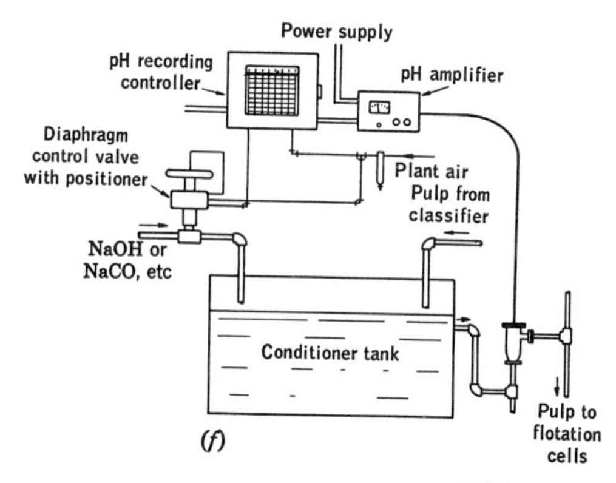

Remote recorder

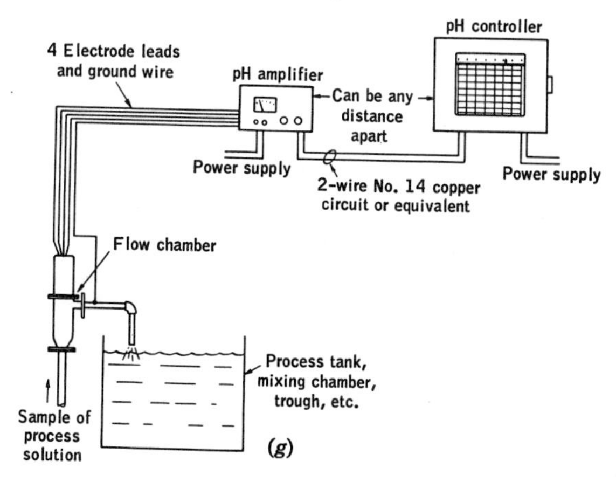

7. Handle probes carefully so that glass will not be scratched.

8. Always keep the top of the electrodes dry. Leakage paths will completely ruin the validity of the indicated pH values.

9. Probes are designed for specific pH and temperature conditions. To operate at values significantly different will impair the life of the units.

SUMMARY

1. Industrial processes are often of a chemical nature. They proceed at their best rate or use their materials most efficiently under certain conditions of acidity or alkalinity. It is therefore highly desirable in many instances to know the acidic or basic nature of materials which are being processed.

2. The most highly acidic material would have a concentration of 1 gram of hydrogen ions per liter. As a ratio, this becomes 1 gram/liter = 1, which may be written as $10/10 = 10^1/10^1 = 10^{1-1} = 10^0$. Inasmuch as it is customary to express this hydrogen ion concentration in terms of the power of 10, or the logarithm to the base 10, our highest degree of acidity will approach zero in numerical value.

Pure water has equal numbers of hydrogen and hydroxyl ions, there being 10^{-7} grams of each ion per liter. All acids contain some basic ions, and all bases contain acidic ions; the concentration of the ions of one kind varies inversely with the concentration of ions of the other kind. It will be noted in Table 22·1 that the sum of the exponents is always -14.

3. The usual manner of designating the concentration of the ions is to refer only to the hydrogen ions by using the reciprocal of the power of 10, which indicates the concentration. (The logarithm of the number to the base 10 would have the same value.) The use of powers of 10 to express the value makes it possible to use small numbers: a pH of 7 means the same thing as 1 gram of hydrogen ions in 10,000,000 liters, or 1 gram/10,000,000 liters. The system of notation is purely one of convenience.

4. The use of glass electrodes for determining the value of the emf created by the metal and the electrolyte is almost universal in industry today. Because the glass electrode is part of the electric circuit, its very high resistance demands that measuring techniques not require current when the reading is being made or taken. If any current is taken from the circuit, the portion of the emf used to force the current through the resistance will be subtracted from that originally available, and all values will be fictitiously low.

5. Great care must be used when handling the electrodes to prevent scratching or other damage.

6. Electrodes which are not in service must be given proper treatment to ensure that they do not dry out. If they do dry out, they cannot be restored to service.

7. The indicated value of pH is that of the material on the tip of the electrode. If the electrode becomes coated with some foreign material, the indicated pH will be that of the film and not that of the fluid.

8. The electrodes must be located where there is adequate agitation of the liquid and where the readings are representative of the process conditions.

9. When the measurements are being made for purposes of control, there should be the minimum possible dead time.

10. Because of the high resistances involved in the input circuits, adequate thought must be given to keeping the instrument dry and dust free.

QUESTIONS

22·1 What are the factors to be considered in installing pH-measuring electrodes?

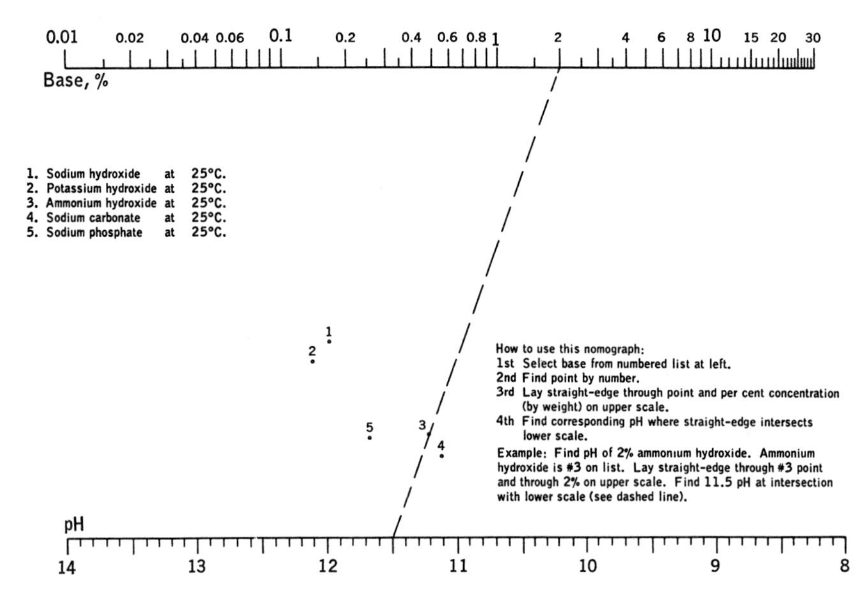

Fig. 22·11 Nomograph for pH versus concentration of bases. (*The Bristol Company.*)

22·2 What action goes on within a reference electrode?

22·3 What was the use of the salt bridge in early apparatus? What are the objections to its use?

22·4 What are the precautions which should be followed in storing electrodes which have been in service? Those which have never been in service?

22·5 Would you soak electrodes in distilled water rather than tap water, provided that both were equally available?

22·6 How often should electrodes in service be checked, assuming that they are functioning satisfactorily?

22·7 Why do electrodes which are first put in service require more attention and observation than older units which have been in service for a long time?

22·8 How can you reduce the stabilization time for new electrodes?

22·9 Why are pH measurements made? What is their significance? To what degree are they influenced by temperature?

22·10 Justify the use of the exponential value in describing pH conditions? What is the disadvantage of using ordinary means of notation?

22·11 Does a large pH number indicate a highly acidic or basic material?

22·12 What type of material has a pH of 7?

22·13 Can you measure the potential developed in pH measurements by using a high-resistance voltmeter of the D'Arsonval type? Why not, if your answer is in the negative? Justify your answers.

PROBLEMS

22·1 Find the pH value of a 0.1 per cent sodium hydroxide solution. If the percentage is raised to 1 per cent, what is the change in the pH value? Use Fig. 22·11.

22·2 Compare the pH values of the five materials listed on the nomograph, Fig. 22·11, for 1.5 per cent solutions. If this percentage is increased to 3, do all solutions undergo the same change in pH? Which has the greatest change? Which the least?

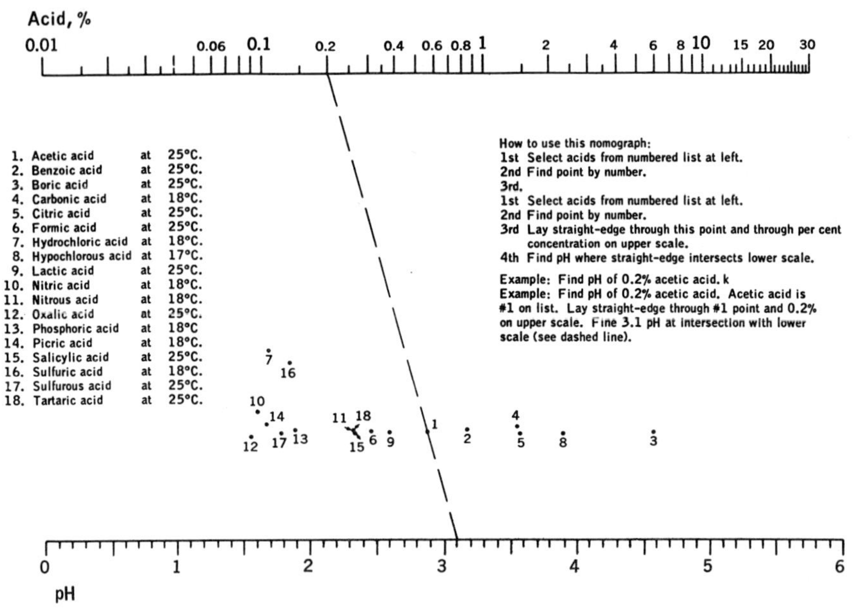

Fig. 22·12 Nomograph for pH versus concentration of acids. (*The Bristol Company.*)

22·3 Repeat Prob. 22·2 but use the first five materials listed on the nomograph, Fig. 22·12. What material has the highest pH at 1 per cent concentration? Which one has the least?

22·4 What common foods have the highest pH values? Which have the smallest? Which is most acid? Which is least acid?

23 Telemetering

The term *telemetering* may mean all sorts of things to different people. To a car manufacturer, the term may mean that you are able to read on the dash the level of gasoline in the tank—the indicating meter being electrical in nature and not restricted to position, in terms of a few feet more or less. To someone else, it may mean the ability to read an instrument when at a distance of 1,000 feet or more from the sensing element. This individual may be an instrument man, and he may expect the actuating medium to be compressed air or hydraulic pressure, in addition to some electrical device. To a man working for one of the major pipeline companies, telemetering may mean the transmission, either continuously or at specified intervals, of the rate of flow, the pressure, or any other variable that is desired, to the head office in New York or Chicago, hundreds or even thousands of miles away. Telemetering covers all these phases and many more. It is an indispensable part of the techniques which are being used to actuate computers and controllers which are designed to produce better products at lower costs.

The ways in which the signals are converted into a quantity suitable for transmission to some remote point (it may be only five or ten feet) range from the extremely simple to those which are most complex. Many of the methods and the required apparatus which will be discussed have probably been encountered in the study of the preceding chapters. They are grouped here in order to give a related picture of the mechanisms and the principles.

The following 14 groups may appear to be somewhat repetitious, but there is duplication in name only, not in the principle to be studied. With the exception of the hydraulic and the pneumatic principles which are considered last, the listing is more or less in the order of complexity.

1. Incremental-current devices; flow, pressure, level
2. Manometers, special mercury
3. Ratio telemetering
4. Wheatstone-bridge telemetering
5. Inductance bridge, or reactance, devices
6. Selsyn motor units
7. Pulse-type remote indicators
8. Time units for telemetering
9. Current variation through control of inductance or capacitance
10. Pulse type; frequency counters
11. Current transmitter types:
    Microsen
    Taylor
    Foxboro
12. Potential devices:
    Thermal
    Pressure, differential pressure, motion (differential transformer)
13. Pneumatic transmitters
14. Hydraulic telemetering

INCREMENTAL-CURRENT DESIGNS

One of the earliest methods for converting differential pressure signals into electrical equivalents employed a mercury manometer with a series of resistors in the low-pressure side, Fig. 23·1. These resistors were so

Fig. 23·1 Pressure-to-current converter. By using many resistance rods, each properly tailored to add the correct current increment when the liquid level reaches it, continuous indication is provided with relatively insignificant steps in the meter current.

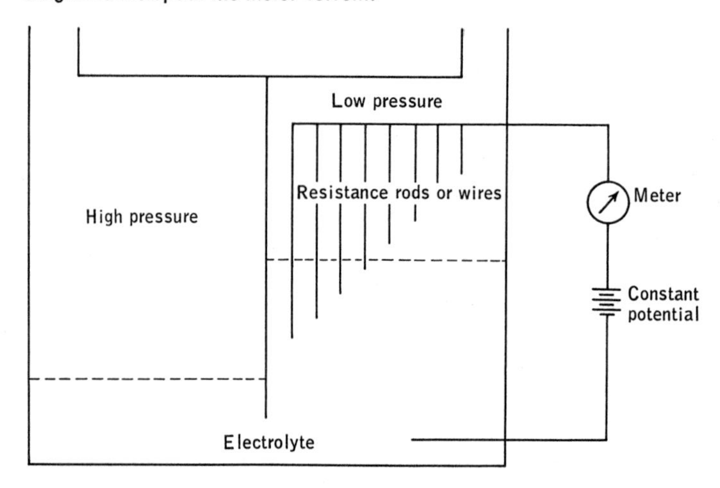

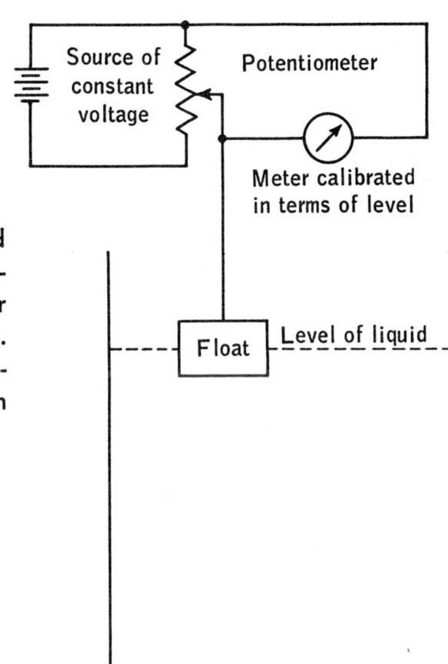

Fig. 23·2 Level-to-potential converter. Liquid levels can be converted into electrical indications if a float can position a potentiometer which has little mechanical drag or friction. To maintain an accurate calibration, the battery or other source of power must maintain a constant potential.

proportioned that as the mercury touched each one in turn, the additional current in the meter circuit would indicate the correct change in the level of the mercury, and hence the correct differential pressure. In effect, each resistor was tailor made so that there would be the correct change in the meter current. Such a design requires a constant-potential power supply. Any variation in the power supply will be reflected in the meter indication.

Despite the age of such a system, it is still perfectly good for the types of installations for which it was designed; any device or method which can displace or control the level of the electrolyte in one side of the device can make it work.

Figure 23·2 is also an incremental-current device, but it differs from the other design in that a potentiometer rheostat is used to furnish a potential which is proportional to the float position. This might well be the surface of the mercury in the preceding illustration.

Special mercury manometers. From our study of transformers we know that the current in the primary winding is governed by the characteristics of the magnetic circuit and by the amount of current taken by the secondary load. This current will vary directly with the secondary induced voltage and inversely with the secondary resistance or impedance. The method depicted in Fig. 23·3 employs the position of the mercury in two ways:

1. The higher the column of mercury, the greater will be the number of flux lines which will cut a conductor and consequently induce a voltage in it. (The mercury is the conductor.)

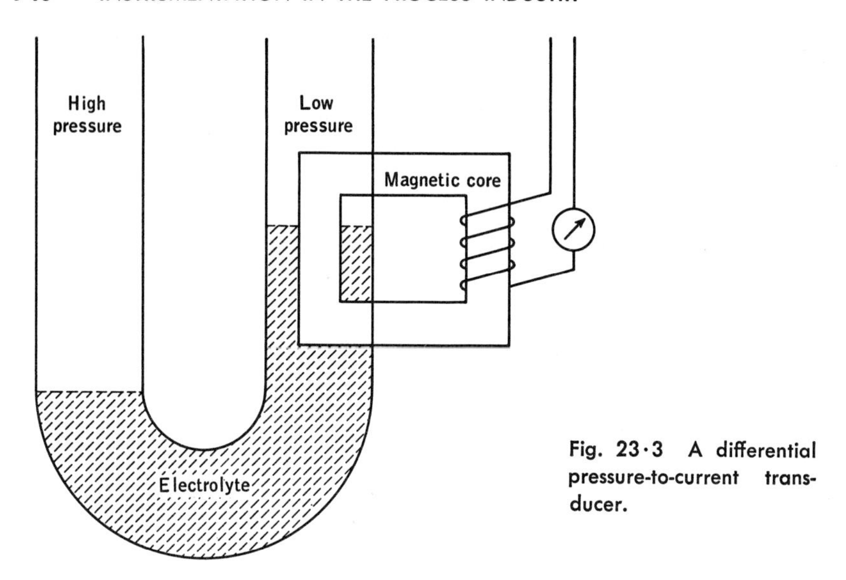

Fig. 23·3 A differential pressure-to-current transducer.

Fig. 23·4 By combining magnetic fields so as to obtain their vector sum, meters can be made relatively insensitive to voltage variations while maintaining their desired response to the variable being measured. Voltage variations will produce changes in the magnitude of the resulting vector field, but not in the direction in which it exists.

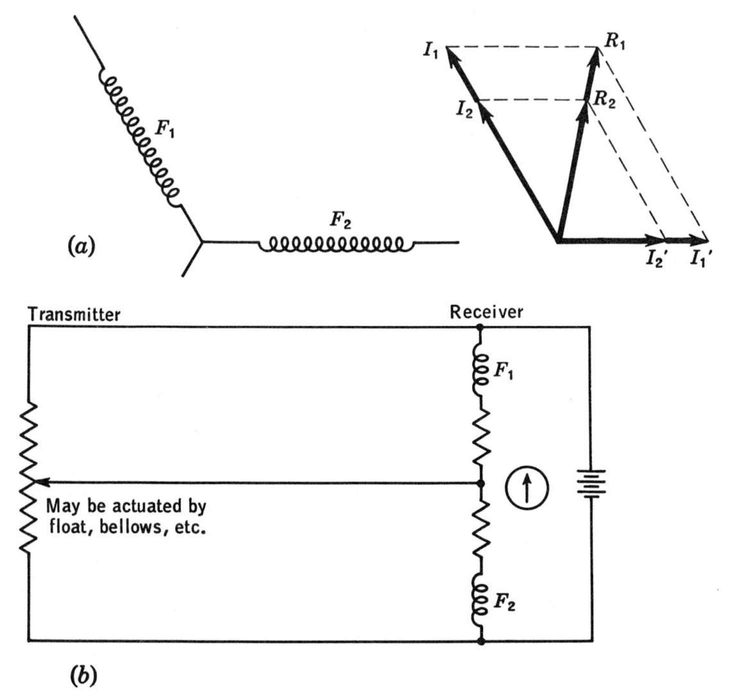

2. The greater the voltage induced in the secondary, the greater will be the current. Moreover, the resistance will be made smaller by having more paths in parallel through the mercury. All of this tends to change the primary current, which can be suitably calibrated for the differential pressure, or other phenomenon, that positions the mercury. As with both of the preceding designs, a constant voltage is required for accurate indications.

Ratio telemetering. If no regulated power supply is available, then the ratio type of meter is quite useful. In general, this design provides for two identical magnetic circuits which are connected in series across the power supply, Fig. 23·4*a*. The magnetic effects of the coils may be arranged to provide the vector relationships shown in Fig. 23·4*b*. If we assume that the magnetic effects of both coils will be changed by equal percentage amounts when the applied voltage changes, then the resultant vector will be changed in magnitude only—the direction will be unchanged with respect to the original position. It must be assumed that the currents do not reach such low values that they are unable to develop adequate positioning torques.

As the potentiometer rheostat is moved up or down by the actuating element, it will increase the current in one coil but decrease it in the other. Only at the halfway point will each coil develop the same magnetic effect.

Wheatstone-bridge telemetering. For those instances in which manual positioning of a potentiometer rheostat is adequate for obtaining remote indications, the circuit shown in Fig. 23·5 may be used. Two potentiometer rheostats, one at the transmitting end and one at the receiving end, and some suitable emf which depends upon the distance of transmission and the sensitivity of the meter, are used. If when the switch S is closed there is an indication by the galvanometer of current flow, the receiver potentiometer is repositioned until the galvanometer indicates no flow of current, and points A and B must be at the same electrical level. The distance of transmission is limited primarily by the resistance

Fig. 23·5 The basic elements of a Wheatstone bridge may be used for telemetering purposes.

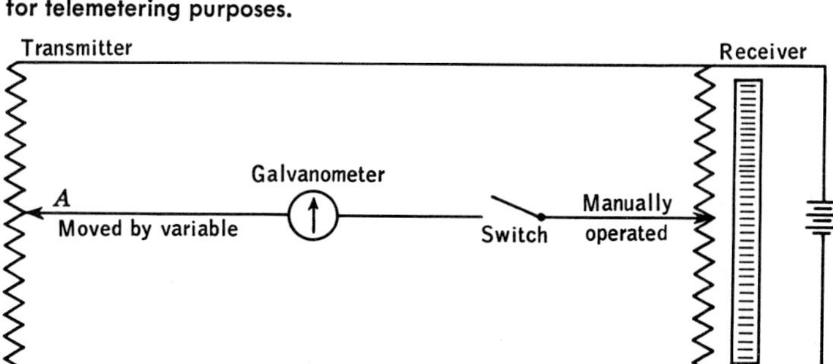

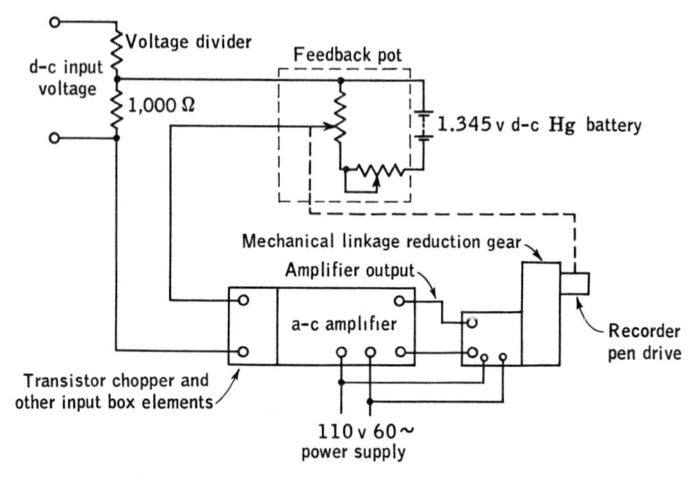

Fig. 23·6 Telemetering employs many principles to ensure that the received value is a true representation of the transmitted variable. In this instance a potentiometric device is used to eliminate consideration of the resistance of the line. (*The Hagan Corporation.*)

of the transmission line and pickup of electrical interference from power lines and other sources of interference.

Any one of a number of methods which have been discussed previously in connection with automatic potentiometers and bridges could be used to convert this bridge into a self-balancing device.

The direct-current type of potentiometric system is represented by Figs. 23·6 and 23·7. The performance of any such system depends upon the constancy of applied voltage at the transmitting end. Without an absolutely dependable source of potential the indicated value of the variable may be far from its true amount.

The alternating-current type of Wheatstone bridge has the same in-

Fig. 23·7 A simplified schematic of a potentiometric system for direct-current transmission. The accuracy of any system such as that represented depends upon the constancy of the source at the transmitter. Special types of batteries for this purpose maintain a fixed potential for a long period of time. (*The Hagan Corporation.*)

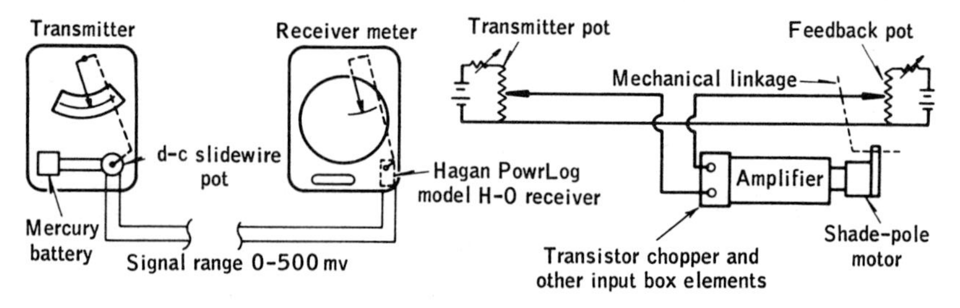

herent liabilities as does the direct-current model. However, the availability of an alternating-current error signal does simplify the problem of amplification. Whether one operates a bridge to reestablish conditions of potential equality, Figs. 23·8 and 23·9, or whether one converts the error signal into an analog current for transmission is of little importance, Fig. 23·10.

These basic differences have been encountered previously. They are included here to provide continuity to the section. We must, however, recall the various types of current transmitters:

1. Those which employ the change in plate current in an oscillator, Fig. 23·11, as its degree of oscillation is influenced by the degree of balance or position of certain of its mechanical elements.

2. Those which obtain balance of a bridge by controlling an auxiliary current through a portion of one bridge element.

It will be noted that the first type of converter will respond to mechanical inputs as well as electrical signals which can position a beam. The second type can be used to telemeter potentials, resistance, strain, or current.

ALTERNATING-CURRENT SYSTEMS

When alternating current is to be used for metering purposes, the change in magnetic reluctance as a movable core is positioned by the

Fig. 23·8 Schematic diagram for the remote measurement of resistance. By using three wires to the remote resistor, the effects of lead resistance can be nullified. This type of telemetering is quite common. (*The Hagan Corporation.*)

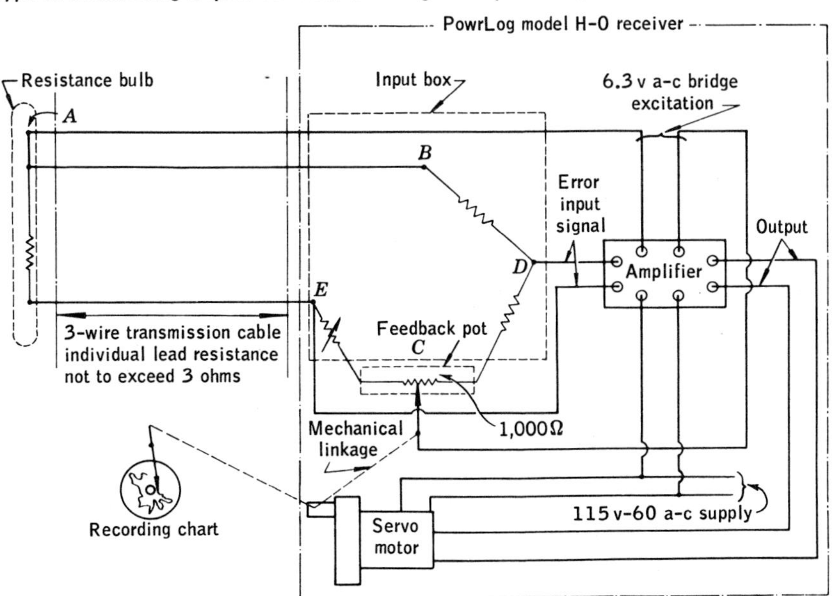

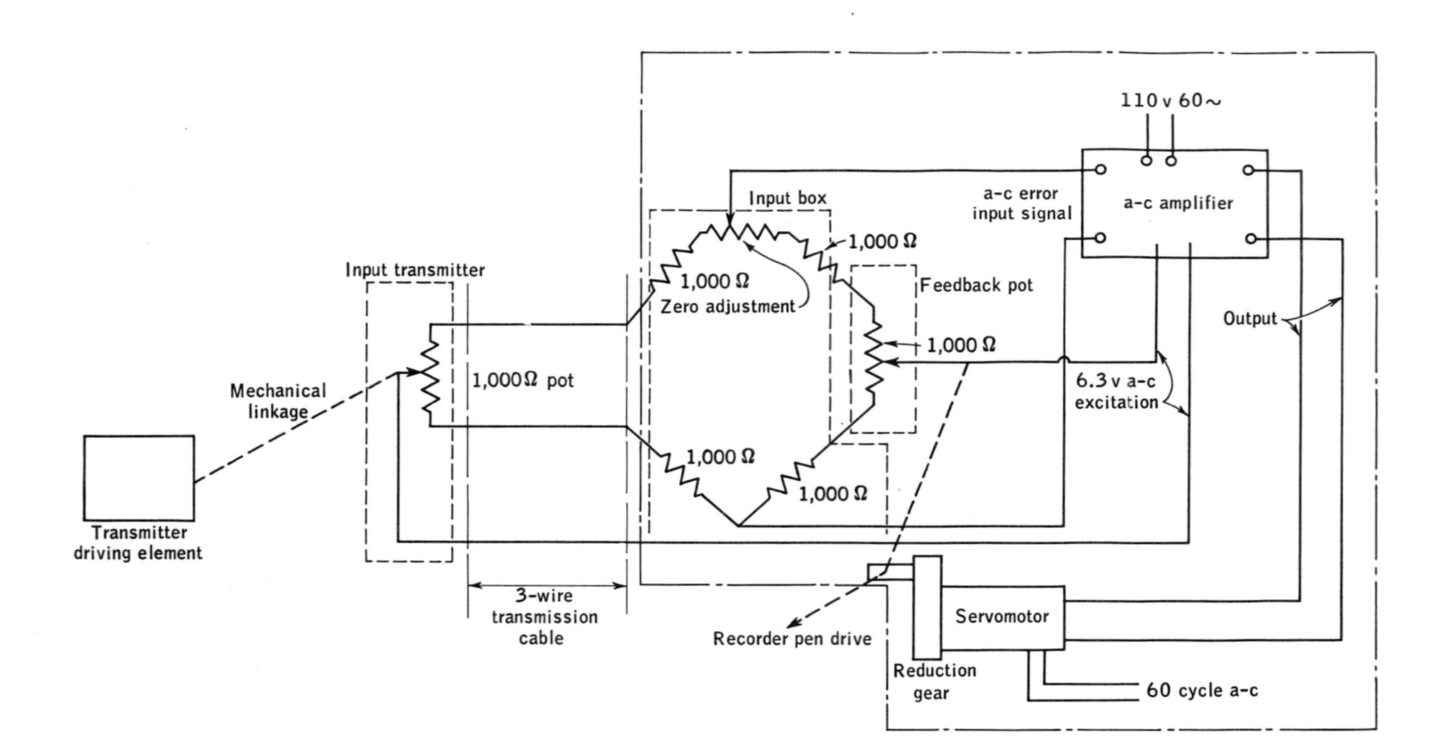

Fig. 23·9 Bridge type of circuit for alternating-current signal transmission. Both direct- and alternating-current bridges are commonly used to obtain remote values. This circuit employs a slight modification of the familiar Wheatstone bridge. (*The Hagan Corporation.*)

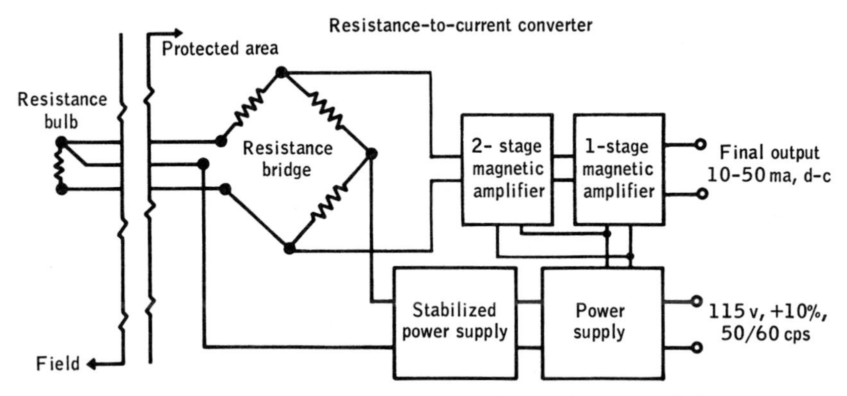

Fig. 23·10 A deflection type of Wheatstone bridge is the heart of this resistance-to-current converter. The current caused by any unbalance of the bridge is used to control the output of a magnetic amplifier. In this way the output current becomes an analogue of the bridge unbalance, and hence representative of the change in resistance which created the unbalance. When current is used as the telemetering agent, the distances can be significantly greater than they can be when potential is used as the transmitted value. (*The Foxboro Co.*)

variable will alter the potential applied to the two coils in exactly the same manner that positioning the potentiometer rheostat did in the preceding example. With the core at the mid-point of the coil, each coil will receive the same voltage. But any deviation from this position will increase the voltage of the one at the expense of the other.

Fig. 23·11 Displacement type of transmitter. This type of current transmitter employs conditions of torque balance in which the input signal—whether motion or electrical—must be balanced by a countertorque created by the transmitted current or some desired fraction of it. (*Manning, Maxwell and Moore.*)

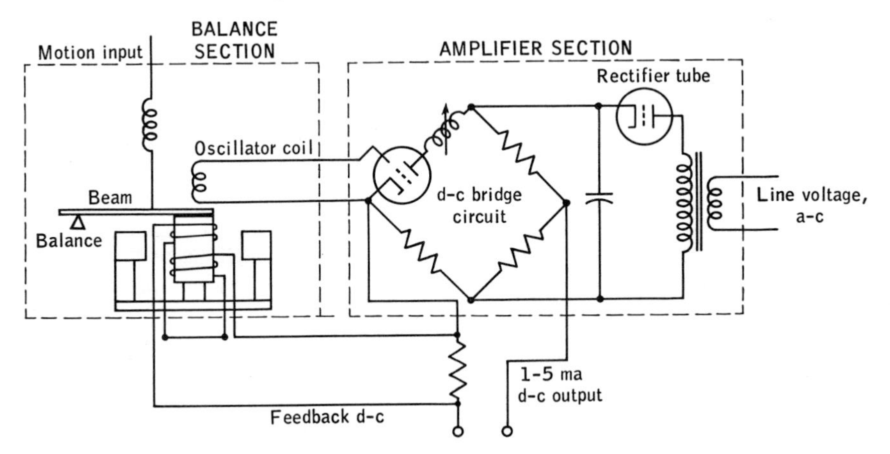

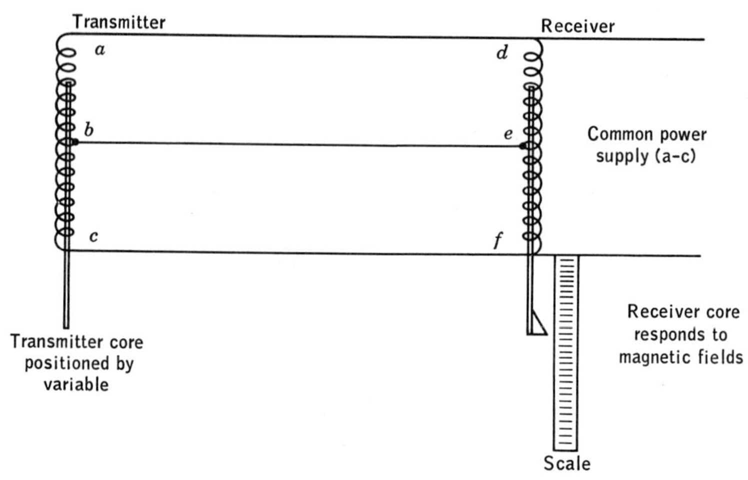

Fig. 23·12 The circulating current brought about by any lack of correspondence of position of the cores will try to position the receiver core.

An alternating-current version of such a bridge is shown in Fig. 23·12. The variable can position the core and create different voltages between points *ab* and *bc*. If the potential from *a* to *b* does not equal that from *d* to *e*, then current must flow because we have connected points which have unequal potentials. The same situation holds for the two other portions of the coils. Any current which results will try to reposition the receiver core and reestablish the condition of voltage equality for each pair of corresponding windings. By making the receiver core very light in weight, or by effectively counterbalancing it, the position of the receiver core will quite accurately indicate the position of the transmitter core. Means for obtaining the maximum performance of such a system will be found in Chap. 8.

Selsyn motor indicators. Where 360° rotation, or greater, is desired for the indicating units, then the selsyn system is frequently used. The term "selsyn" is made from the term "self-synchronous." The transmitter and the receiver, Fig. 23·13, are both energized from the same single-phase power supply. Unless this is done, or unless the relationships between the two power supplies are known and suitable phase shifting is done, the indications will be false. The rotating portion of each unit is a three-phase winding so positioned that each phase will have induced in it a voltage which depends upon the position of the rotor in the magnetic field created by the fixed stator. Thus, receiver and transmitter are generators (of alternating-current potentials, in this case) and they act just as do all generators when connected in parallel. If the potentials created by the transmitter are not the same as those of the receiver at any given

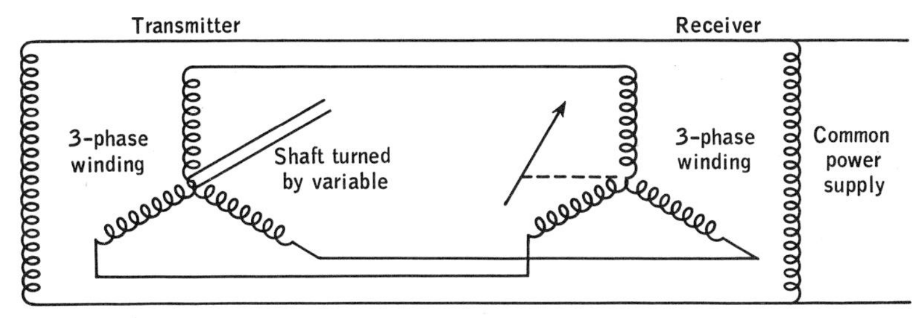

Fig. 23·13 Selsyn transmitters and receivers consist of a single-phase field and three-phase rotor (or the field may rotate and the three-phase element may be fixed). Both are energized from a common phase and frequency source. The accuracy or fidelity with which the receiver represents the positions of the transmitter is inversely proportional to the mechanical friction and the resistance of the three-phase circuits.

moment of time, then a current will circulate between transmitter and receiver. The effect of this current is to turn the receiver so that the circulating current approaches zero. Because the positioning of the synchro receiver depends upon this error current, every precaution is taken to reduce the friction of the receiver to a minimum so that even very small torques can position the rotor.

The transmitter is not free to move—it is positioned by the angular motion of a shaft or whatever the variable being measured may be. It is generally of greater electrical capacity than the receiver inasmuch as it may be desired to operate several receivers in parallel. There are a number of factors about such installations that should be kept in mind:

1. The accuracy of position of the receiver is dependent upon reducing all friction to a minimum.

2. In order that the circulating current may be maximum for any given angular error between transmitter and receiver, the resistance of the connecting wires must be kept to a minimum. This will be one of the factors in limiting distance between transmitter and receiver.

3. When there are several receivers in parallel, there can be interaction between them. For example, if one becomes jammed, it cannot come to agreement with the transmitter, so it will contribute a potential which will be added (vectorially) to that of the transmitter. This will cause all of the other receivers to display erroneous positions or values.

4. Because of the possibility that error may exist in the receiver units, this system is not used for critical installations where the indications must always be as free from error as it is possible to get them.

5. Within limits, the system is not responsive to voltage variations.

6. By using designs such that the current-carrying capacity is relatively high, it is possible to use the receiver to position valves or other

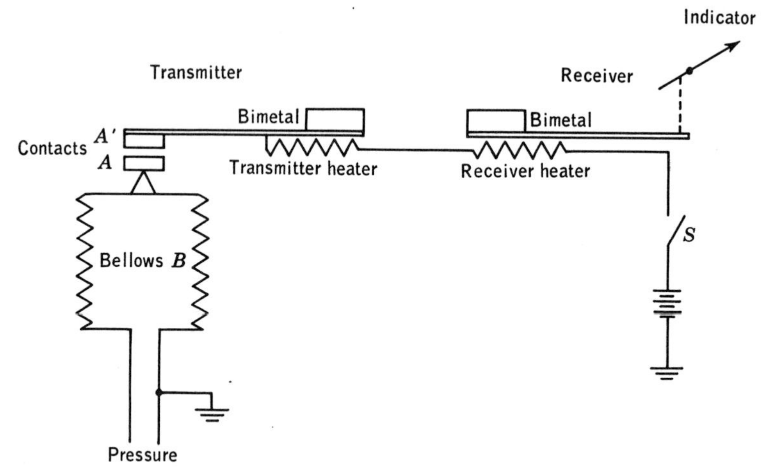

Fig. 23·14 Pulse-type telemetering equipment. Certain types of pulse-type telemetering equipment use very simple components which function under very unfavorable operating conditions. Such transmitters are independent of the normal voltage variations. Voltage variations influence the speed of response but not the accuracy.

devices, but the development of substantial torque always means that the transmitter and the receiver are not in exact agreement.

Pulse-type telemetering. This general classification covers a wide variety of designs, ranging from the extremely simple to the mechanically and electrically complex. One of the simplest is shown in Fig. 23·14. While a pressure measurement is indicated, the measurement can be any in which there is motion on the part of contact A. Starting from the indicated positive terminal of the battery, we find:

1. A resistance wire wound around a piece of bimetal strip which, when it deflects, will position the pointer or other indicating element.

2. A second resistance wire around a second piece of bimetallic strip. The circuit is terminated in contact A', which can complete the circuit when in contact with A.

Let us assume that the circuit has not been in use; switch S has been open. Whether or not there is still pressure on the system, contacts A and A' will be pushed together when the transmitter current is zero and the bimetal is cold. When the switch is closed, there will be a complete circuit through the two resistors. As current flows through them, the bimetallic strips will be heated. The receiver strip will move the pointer, while the transmitter strip will keep the contacts together until the strip has flexed enough to open the contact. As soon as the strip does open the circuit, the heating of the bimetallic strips will cease. As soon as the transmitter element has cooled slightly, it will again close contacts A

and A', thereby reestablishing the cycle. There are several characteristics of such a system that should be noted.

1. Once the system is in operation, the transmitter can determine the position of the primary element without putting any mechanical load of consequence on it. As soon as the circuit is complete, action is taken to lift the contact away. Consequently, the average force on the primary element, such as the bellows in this case, approaches zero.

2. The system is entirely responsive to current, not voltage. As the potential of the system changes, the speed of response will be altered but the accuracy of the indication will not be.

3. Because of the thermal storage system, the device cannot furnish instantaneous value, but it can furnish something close to the average. If one is interested in an average rather than all of the peak values, this provides a fairly simple way of obtaining it.

A close relative of the system just studied is that shown in Fig. 23·15. This one is far faster in its response because of the greatly reduced thermal mass. The transmitter consists of the bellows B, or some other positioning means, and contacts which complete the circuit through a length of Inconel-type wire. The circuit is completed through a meter.

Fig. 23·15 A pulse type of telemetering system which uses the change in length of a piece of wire. The small mass of a piece of wire makes it respond very quickly to changes in the amount of heat supplied to it. Any quantity or type of measurement can be made provided that the action will move the spring member so that contacts A and A′ will touch.

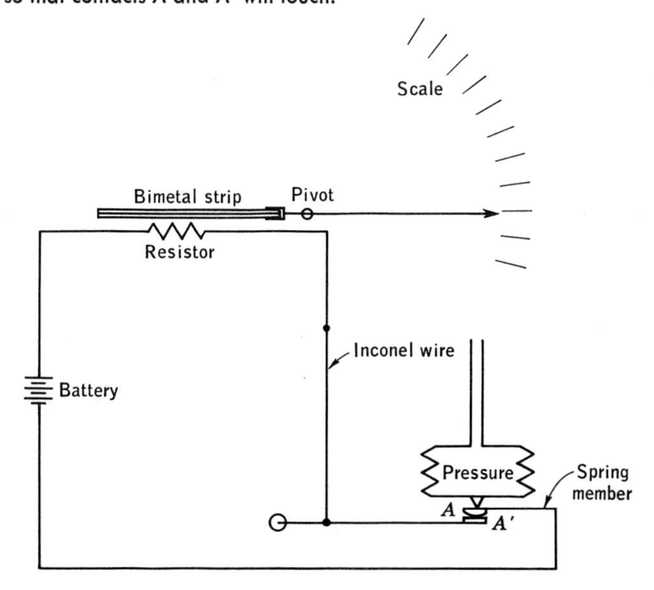

In operation, the heater wire pulls the contacts together, but the moment that this happens, the flow of current which results produces heating in the wire which will increase its length. The wire will continue to heat until the contacts are opened, but as soon as they open, the wire will cool and contract to bring the contacts together again. The wire heats again, and the cycle is repeated.

This device is also responsive to current only, but it has a far faster response than does the preceding design. It will be noted in operation that the ambient temperature has but little effect on the current-wire-temperature relationship. The air forms a sheath around the wire and in this manner effectively insulates it from the ambient. The wire should be small in diameter, of the order of 0.008 to 0.010 in., and have operating temperatures of 300 to 500°.

The circuit sketched in Fig. 23·16 is not of the pulse type but is included here because its basic principles are incorporated in certain instruments of the pulse type. This is one of the oldest circuits used for remote positioning of a pointer or indicator which is to move synchronously with the transmitter. Whether we move the battery contacts or whether we rotate the leads to the coils is of little importance as we consider the theory; in practice, rotating four connections might be more of a problem than positioning two.

In the position shown, coils A and A' receive full battery voltage, while coils B and B' receive no current because their leads are connected to

Fig. 23·16 Rotary motion or indication of the position of the transmitting variable controls the relative values of the currents in the receiver coils. The indicating element, carried by the rotatable permanent magnet, is positioned by the field created by the vector sum of the individual coil effects.

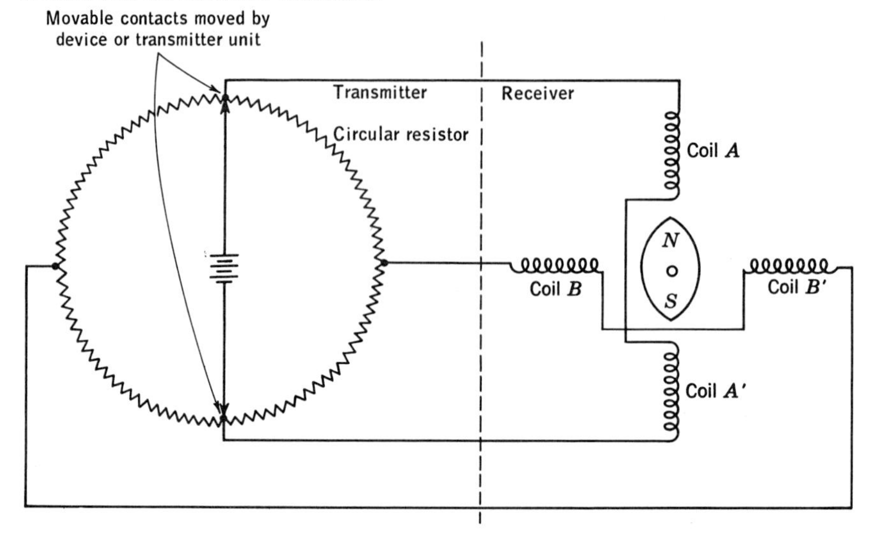

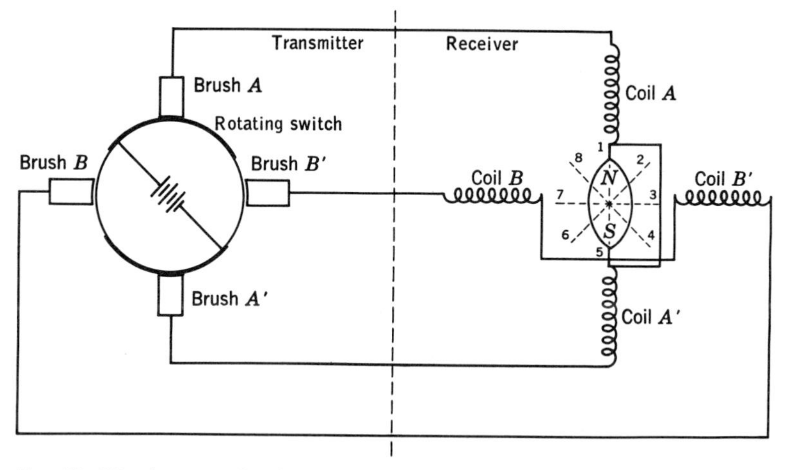

Fig. 23·17 A series of pulses can position a pointer. With the rotating switch in the position shown, coils A and A′ will be energized. This will orient the permanent magnet along the vertical axis. The magnet can be in any one of the eight positions which are indicated. It will not stop at any intermediate position unless the circuits are modified.

points at the same potential. This causes the permanent magnet in the receiver element to line up with the magnet field created by coils A and A'. But if the transmitter is rotated 45°, then the current in coils A and A' will be reduced while the current through coils B and B' will reach one-half (approximately) of their maximum value. With both sets of coils energized, the permanent magnet will rotate through an angle of 45°.

If the transmitter is rotated another 45° clockwise, coils B and B' will receive full current but coils A and A' will receive no current because they are now connected to points of equal potential. The indicating permanent magnet will now be in line with coils B and B'. If the transmitter is now rotated still more in a clockwise direction, the current in coils A and A' will be reversed. The vector sum of these two magnetic fields will now position the permanent magnet so that the north pole of the permanent magnet is between coil A' and coil B'. In this manner complete rotation of the permanent magnet can be obtained. Continuous rotation or indefinite positioning can be obtained with devices of this type.

The chief differences between Fig. 23·16 and Fig. 23·17 is that in the first instance the receiver is supplied current continuously while operating and the receiver magnet can assume any position—the position being determined by the polarity and the relative magnitudes of the coil currents. But in Fig. 23·17 no attempt is made to modulate or vary the magnitude of the current. Polarity and time of circuit completion are, however, controlled. Then the receiver moves in definite steps, keeping in exact

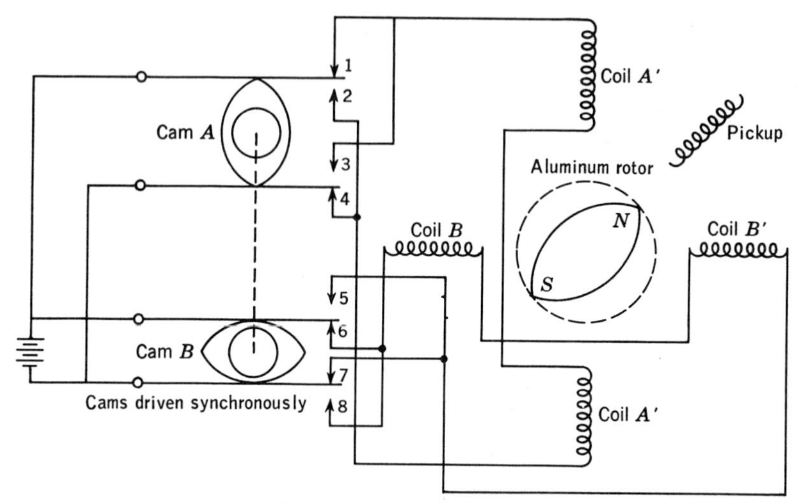

Fig. 23·18 A cam-actuated telemetering system. Four pairs of cam-actuated contacts may take the place of the rotary switch in Fig. 23·17. The action of the rotating element is the same in each design.

synchronism with the transmitter. Thus, with the circuit complete through coils A and A' only, the permanent magnet will assume position 1. But when the transmitter has rotated about 45°, current will also flow through brushes B and B', thus energizing coils B and B'. The permanent magnet in the receiver will now assume position 8 or 2. The direction of rotation of the transmitter will determine this. Further rotation of the transmitter will leave circuit BB' complete but will interrupt the flow of current to coils A and A'. In this manner, any one of eight definite positions will be assumed by the receiver, or continuous rotation will be obtained if the transmitter is moving fast enough.

Figure 23·18 suggests that the operation previously described may be obtained by using two cams to position contacts. Provided that both cams are driven synchronously, the action is identical with that previously described. It might also be noted that by using relays that are of a polarized type and have an off position when not energized, pulses coming over a telephone or radio link can produce the same type of performance. By using two frequencies which are mutually exclusive, the performance of cam A can be simulated. Similarly, two other frequencies can replace cam B. By sending one frequency or pulse, the action of cam A can be duplicated. By sending two frequencies, or pulses, the simultaneous closing of circuits to coils AA' and BB' can be obtained. Thus, a four-toned or four-frequency channel can be made to produce any one of eight definite positions or can provide continuous rotation.

If an aluminum or copper disk is positioned close to the permanent magnet, the rotation of the magnet will tend to make the disk follow it.

This is the same action that is used in the speedometer in your car. Restraining the rotation by means of a suitable spring provides a means of obtaining a measure of the transmitter rpm. At the same time, the rotation of the receiver magnet can create pulses within a pickup coil. The frequency is then a direct measure of the speed of the transmitter. Or it may be more desirable to measure the potential induced in the coil; the coil emf is also a direct measure of the speed of rotation of the permanent magnet and accordingly a measure of the speed of the transmitter.

Still another pulse-type system may be used to transmit rpm or any other variable which may be represented by a series of pulses, the number per minute or second being representative of the magnitude of the quantity being measured. In Fig. 23·19 the transmitter may be a single-pole, double-throw switch on a machine whose rpm is desired. (A wheel rotating with the movement of material under or over it can also generate a similar series of circuit closings.) If we assume that the contact completes the circuit from the battery through the d-c meter, an a-c meter, and the capacitor, we find that the charging current will cause a deflection of the meter pointer. For a fixed, constant value of battery potential a perfectly definite amount of electricity will be required to charge the capacitor. By keeping the capacitance and the resistance of the circuit within certain limits, time-constant values of a suitable value can be obtained. The time constant should be between one-tenth and one-fifth of the shortest pulse to be received, unless special calibration is permissible. When the contactor is moved to the opposite contact, the capacitor will be discharged and thus made ready for the next charging cycle.

It will be noted that the meter in the battery lead may be a milliammeter or microammeter, while the other meter must either be of the a-c

Fig. 23·19 Simple telemetering circuit for rpm. With a constant potential available from the battery, these meters become frequency meters. In this manner they indicate rpm.

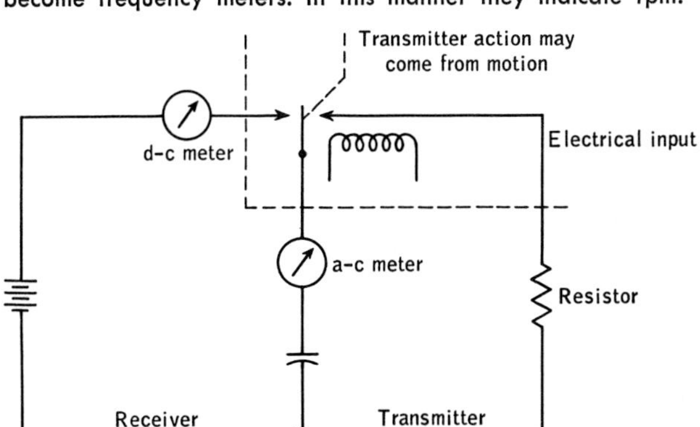

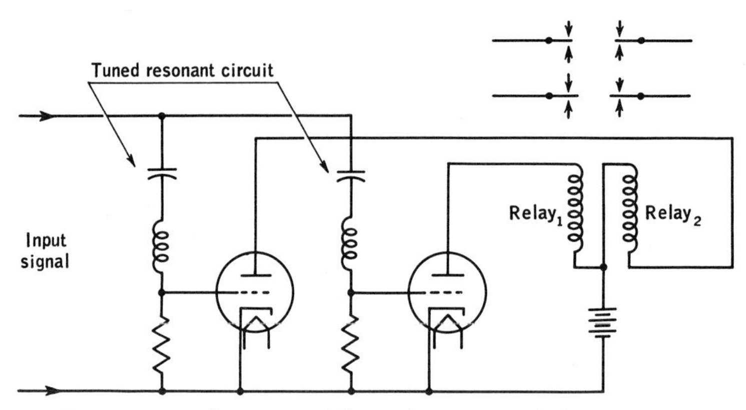

Fig. 23·20 By employing two different frequencies of alternating current instead of a direct-current battery, the range of a device such as shown in Fig. 23·19 can be greatly extended. Tuned resonant circuits control the tube current which passes through the relays.

type or be provided with a bridge type of rectifier. The meter in the capacitor lead will have twice the current to measure inasmuch as it measures both the discharge and charge currents. A thermal type of meter might well be substituted for either meter. Instead of the constant-potential battery which is indicated, a potentiometer circuit may be provided with some simple means for calibrating the potential applied to the charging circuit.

Remote actuation may be obtained by employing a relay to charge and discharge the capacitor, Fig. 23·20. By employing three or four meters (one for each digit to be transmitted) and by tuning each relay to a particular transmission frequency, digital transmission is easily available. The number of counts per minute or second for each frequency will govern the value indicated.

Quite unlike the two current-pulse devices just considered is another telemetering unit which uses pulses of a somewhat different nature, Fig. 23·21. The one under consideration is marketed by The Bristol Company. It uses but two wires; even radio links or telephone repeaters may be applied without interfering with the accuracy of the system. Basically there must be two synchronous electric motors which are running at the same speed because their power supplies are operating at the same frequency. The receiver then tries to duplicate the performance of the transmitter.

Some definite time interval such as 15 sec or 20 sec or some other suitable period, which in an extreme case might be as short as 2 sec, is selected. Thus the total time of the period is determined. This entire time interval is required for the transmitter unit to transmit the maximum signal. As

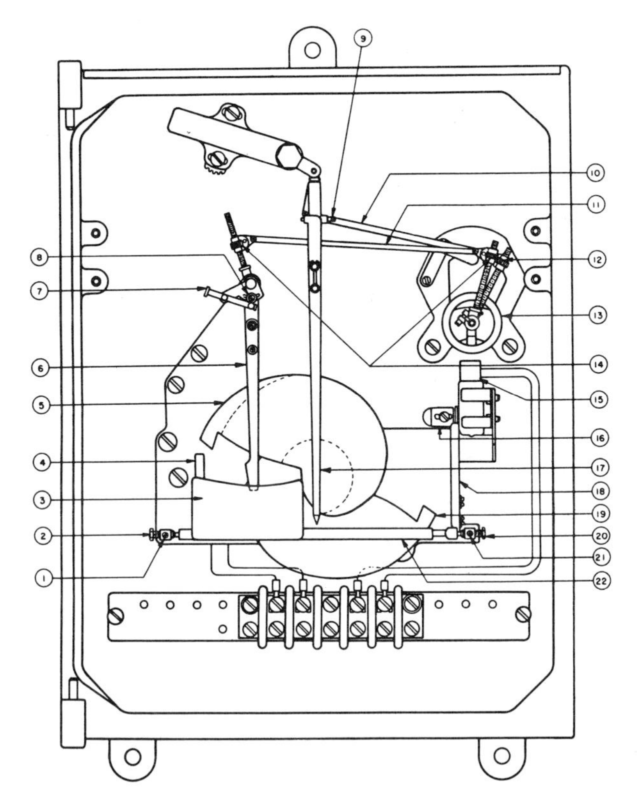

1. Locking screw
2. Bearing screw
3. Trip plate
4. Fixed arm
5. Cam - cut to any desired shape
6. Cam follower
7. Point adjustment for cam follower
8. Locking screw
9. Point adjustment for indicating pointer
10. Link
11. Link
12. Span adjustment for indicating pointer
13. Measuring element
14. Span adjustment for cam follower
15. Mercury switch
16. Permanent magnet
17. Indicating pointer
18. Lever with vane
19. Cam for balance
20. Bearing screw
21. Locking screw
22. Shaft

Fig. 23·21 Details of a typical Metameter transmitter. *(The Bristol Company.)*

some units are designed, it requires the entire interval to position the receiver if it is to register the maximum value. It therefore follows that inasmuch as the interval is a fixed number of seconds, then the transmitting time plus the nontransmitting time must always add up to the same value.

If we assume an interval of 15 sec and a value of 33 per cent of maximum is to be transmitted, then the following should happen: for 5 sec (or one-third of the period) the constantly rotating cam is in contact with the pen or indicator. This positions a magnetic switch so that the output circuit is closed during this interval.

During the period when the transmitter keeps its switch circuit closed, the solenoid located in the receiver is also actuated, and it causes a gear train to rotate the drive mechanism through one-third of its total travel, at which time it is released by the transmitter from further operation. A second gear train now causes the indicator to be driven in a direction from the maximum position toward the zero position. During this period, which is two-thirds of the cycle time, the actuating mechanism would have been driven back two-thirds of the total distance. This would leave it one-third of the way between the origin and the maximum value.

This ability to drive back is necessary when the position which is indicated exceeds the value of the transmitted variable. So long as the variable is increasing there is no need for the drive-back. But during periods of decreasing variable values it is the drive-back unit which actually positions the pen or indicator. The pen or indicator is thus positioned from two directions.

Such a system needs but two wires and the use of the same synchronous alternating-current frequency. There is no need for synchronizing the units inasmuch as they are controlled by the same pulses at the start of each operating cycle.

There are several aspects of pulse-type telemetering which are worthy of note, especially when extreme reliability of operation is concerned. If we use the system in which the length of the transmitted pulse is a measure of the variable, then there will be an off period for the time that remains before the next impulse starts. In certain cases, the off period is used by the receiver to transmit an acknowledgement to the transmitter. The length of the pulse sent by the receiver will vary inversely with the length of the transmitted pulse, Fig. 23·22. Any lack of coincidence will be evident at once, because the total length of time of the transmitted and received pulses should be a constant.

If improper pulses are returned to the transmitter, or if they are not received at all, then the transmitter knows that abnormal conditions exist and can send out the necessary alarm signals. Certain railway signaling systems use such a system to ensure that there is exact correspondence between transmitted values and those indicated by the receiving element.

This is not at all the sort of situation when several transmitters share a common channel, Fig. 23·23. If only a simple pulse system is used, then each transmitter functions in some predetermined sequence. But if a more complex system involving many frequencies is used, then all transmitters may be working at the same time inasmuch as it is possible to unscramble at the receivers the multiple frequencies, sending each one to its associated instrument.

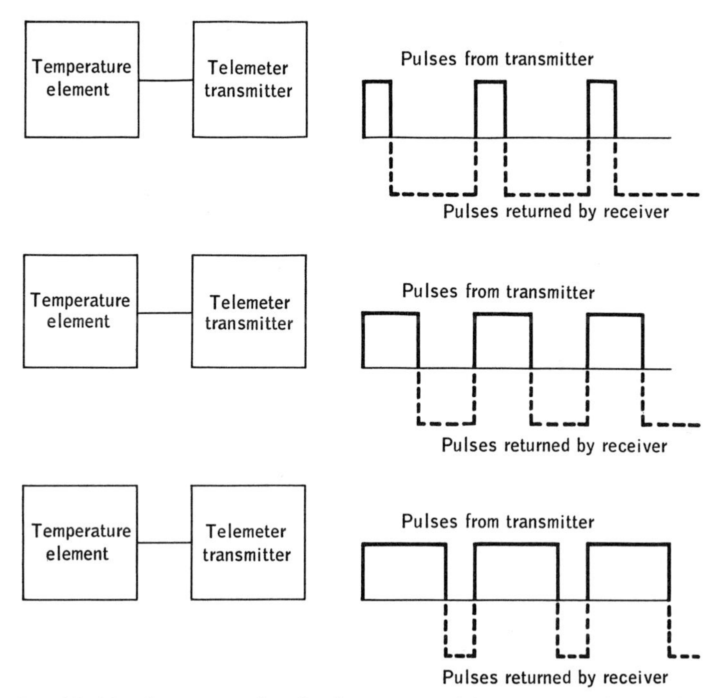

Fig. 23·22 One type of pulse-duration modulation. An acknowledgment pulse is often sent by the receiver. Its length is inversely proportional to the length of the transmitted pulse. Correct operation of both transmitter and receiver is thus continuously confirmed.

Fig. 23·23 Schematic diagram of an elementary multiplex system. (*The Bristol Company.*)

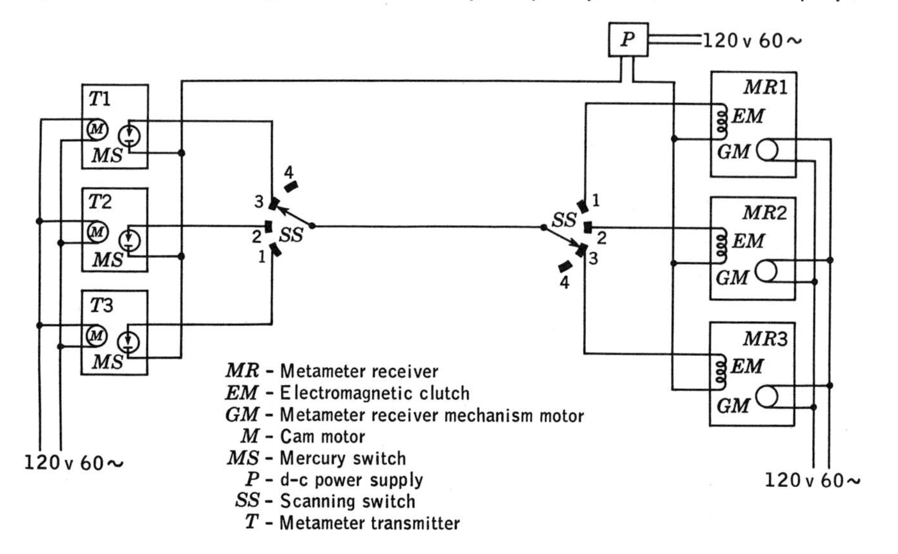

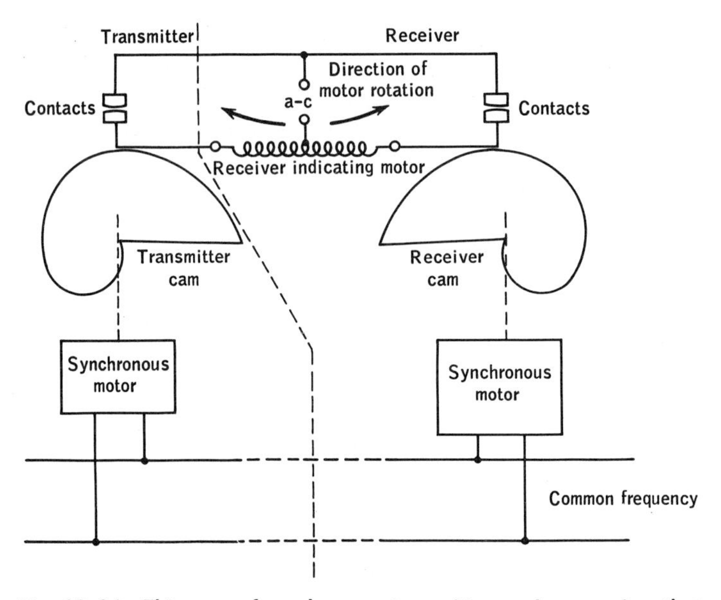

Fig. 23·24 This type of synchronous transmitter-receiver requires that both units be synchronized at all times. Any lack of synchronization will create erroneous indications.

Fig. 23·25 Pulse-duration modulation. The value of the variable is given by the length of the transmitted pulse.

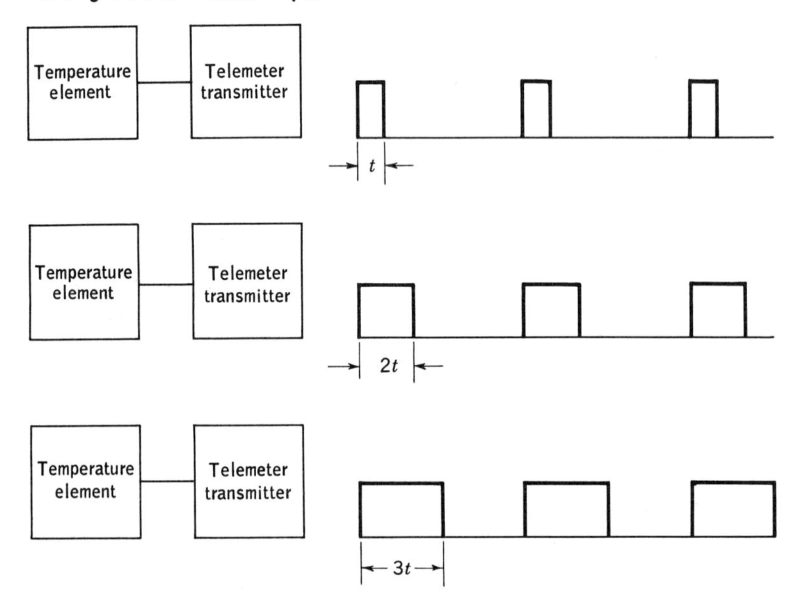

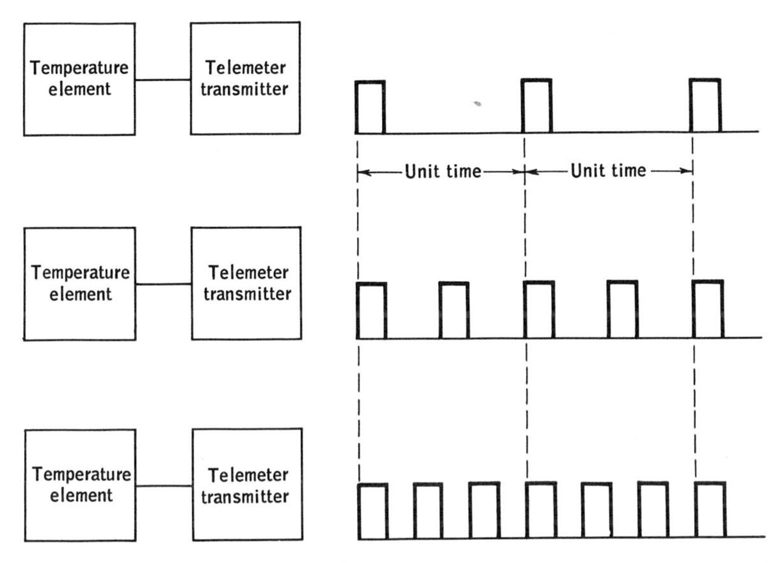

Fig. 23·26 Pulse-frequency modulation. The frequency of the transmitted pulse is a measure of the transmitted variable.

A different version of a remote telemetering system is shown in Fig. 23·24. The transmitter is essentially the same as the one previously considered. The heart-shaped cam and the position of the pointer which evaluates the variable determine the length of time that the transmitting circuit will be closed. The receiver contains a duplicate of the transmitter cam and indicator contactor arrangement. This system requires not only that the motors operate at the same speed but that each heart-shaped cam start from its zero position, for example, at precisely the same time.

The purpose of the two sets of contacts is to complete the drive circuits to the servomotoı which positions the receiver pen. If both sets of contacts are closed at the same moment, then the motor has equal torques trying to rotate it clockwise and counterclockwise. As a natural consequence of this conflict, the motor does not turn. But let one contact be closed when the other one is not, and the motor will then be energized to drive the pointer in the appropriate direction.

The transmission of discrete pulses when the number of pulses, Figs. 23·25 to 23·27, is the measure of the variable is simpler than the transmission of signals whose magnitude is a measure of the quantity, Fig. 23·28. The conversion of a magnitude into a number of pulses is represented by the sketch shown in Fig. 23·29. The direct-current input voltage is compared with the potential developed across the cathode resistor. Any variations between the values of the two are chopped and fed into a high-

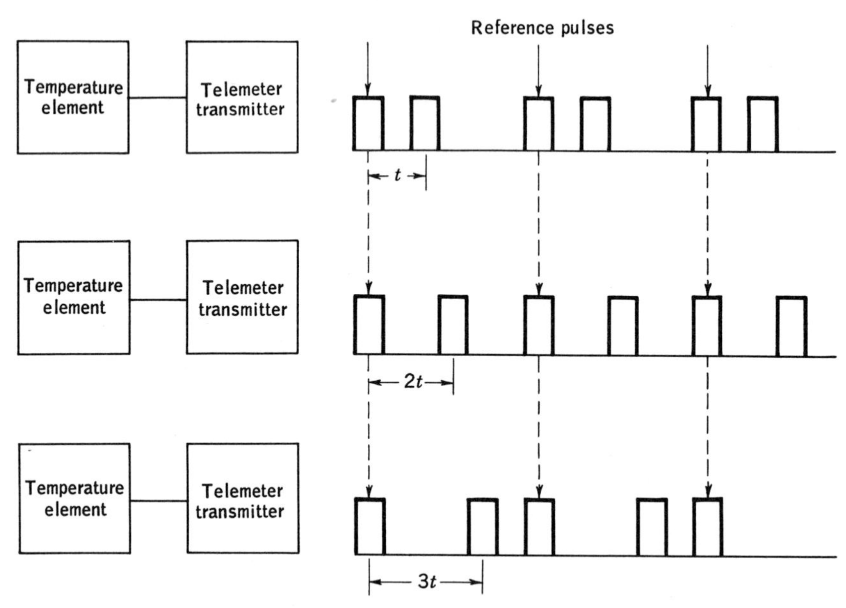

Fig. 23·27 Pulse-position modulation. The position of one pulse with respect to another may also be a measure of the variable.

Fig. 23·28 Amplitude-pulse modulation. Information can also be transmitted by varying the height or magnitude of the pulse.

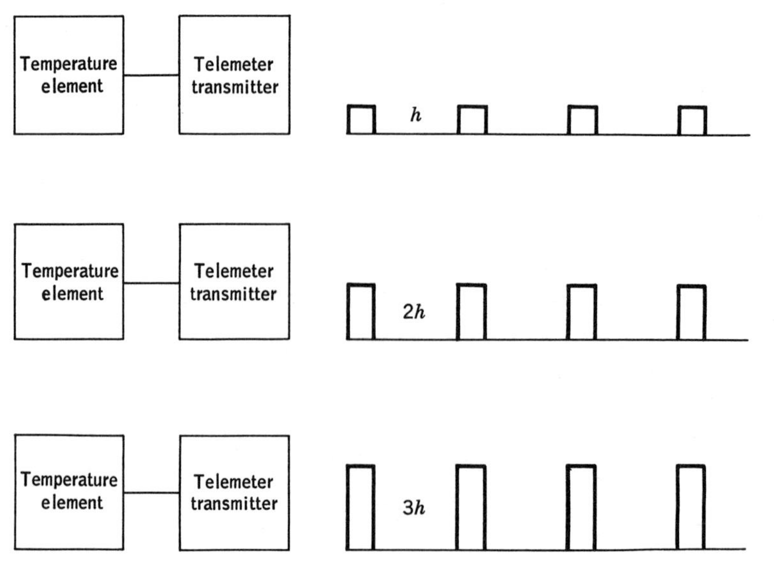

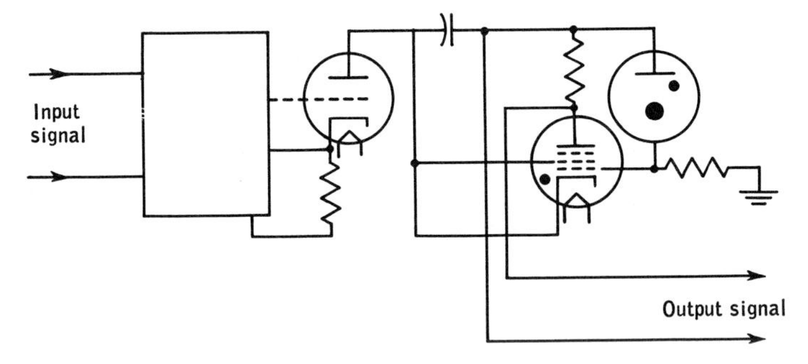

Fig. 23·29 This circuit can convert an input potential into a number of pulses of equal magnitude. The frequency with which the pulses are formed becomes a measure of the input signal. [*Douglas M. Considine (ed.), "Process Instrumentation and Control Handbook,"* McGraw-Hill Book Company, Inc., 1957.]

gain amplifier whose output is fed to the grid of the tube. As the input signal increases in value, the tube current must increase also. This charges a capacitor which is in series with the triode. But connected in parallel with the capacitor is a gas tube which will discharge when its voltage reaches some designed value. When it does discharge quickly, its current will flow through the grid resistor of the thyratron and cause it to conduct. The large current drawn by the thyratron will drop its plate voltage to the extinguishing point for the tube, and the tube will shut off. But this surge of current through the thyratron creates a voltage pulse across the plate resistor. This pulse is then sent to the receiver station. The frequency with which the pulses are sent depends upon the time required to charge the capacitor to the firing point of the gas tube: a small input signal will call for a small plate current in the triode and consequently a long time, relatively, to charge the capacitor. On the other hand, if the charging current through the tube is three times the first value, then the capacitor will charge in about one-third of the time. This then will produce three times as many pulses during any given time interval. The number of pulses per unit of time is then a measure of the variable.

When the information is to be telemetered over commercial telephone lines, there are definite upper and lower limits of frequency which are desirable. These requirements are based only on the charges for telephone lines and have nothing to do with the theory or practicability of operating at frequencies outside these limits. As a general rule, it may be said that the wider the frequency range the higher will be the charges.

The receiver which will convert the frequency pulses to significant signals is represented in Fig. 23·30. The input signal, which we will assume furnishes the indicated polarity to tubes 2 and 3, causes tube 2

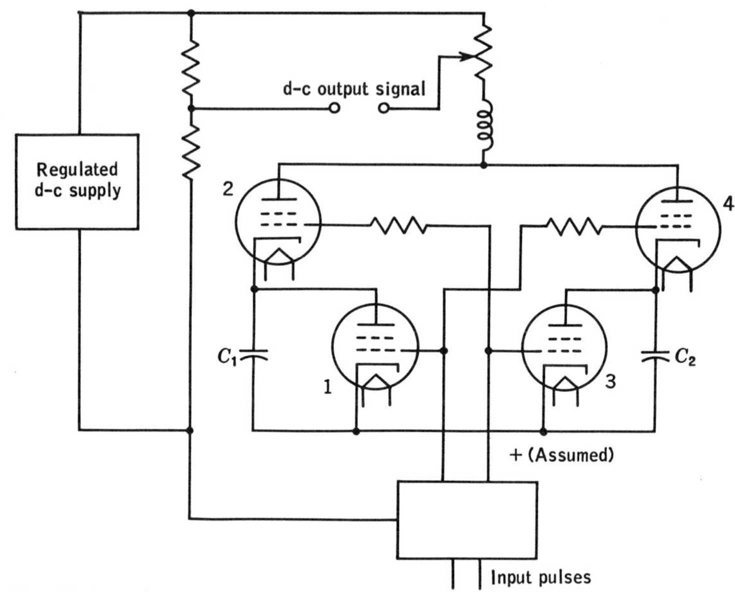

Fig. 23·30 This circuit indicates how it is possible to convert a series of pulses of fixed magnitude into a direct-current equivalent potential. The faster the input signals are received, the higher the output potential. [*Douglas M. Considine (ed.), "Process Instrumentation and Control Handbook,"* McGraw-Hill Book Company, Inc., 1957.]

to conduct and charge capacitor $C1$; tube 1 is negatively biased and will not discharge the capacitor. At the same time, tube 3 is conducting and discharging capacitor $C2$, which had previously been charged by tube 4.

Each time that tubes 2 and 4 conduct, they charge the capacitors to a value almost equal to the regulated value of the source. With a constant value of capacitance and a constant potential, there will be a definite charging current through the potentiometer rheostat P for each input pulse. The *average* value of the potential developed will then be a measure of the number of pulses received. By connecting in series the potentials developed across a number of these potentiometer rheostats, it should be feasible to find the sum of a number of quantities or to make corrections to obtain the values for standard conditions.

A somewhat different way to use the frequency method requires relative movement of the capacitor plates. Let us assume that the variable is pressure and that it will be measured by the deflection of the Bourdon tube (or bellows). The position of the tube tip will vary the value of the capacitance as it controls the separation of the plates. The value of the capacitance in turn can be made to control the frequency generated by the oscillator. In this manner, a definite frequency will be created for each pressure transmitted to the pressure element.

This variable frequency can then be transmitted over conventional telephone lines for any desired distance. It will be received on an automatic-frequency meter which displays or registers the pressure rather than the frequency. When many stations are telemetering to a common headquarters, the stations must also have characteristic identification. Such a system requires that a certain time be allowed for each conversion from frequency to variable value.

DIRECT-CURRENT SYSTEMS

For those situations in which telephone repeaters are not to be used and when the distance of transmission is not greater than about 20 miles, plus or minus a few miles, there are many who favor the use of a direct-current signal as the best way to transmit the value of the variable. Provided that the transmission lines are good, then the current at all points should be the same, and the same current might be used many times in the loop for purposes of indication or control. This eliminates the question of magnitude of voltages and also any problem concerned with counting pulses.

There are three systems which will be considered for this general class of remote indications. The simple schematic shown in Fig. 16·13 includes all the major components required for converting a thermocouple input in millivolts into a transmission equivalent of 1 to 5 ma, regardless of the magnitude of the millivolt input signal. The thermocouple input causes a current to flow in the coil which is supported by the movable beam, Fig. 16·13 modified—a coil replaces the mechanical input. The reaction of the magnetic field of the coil with that created by the permanent magnet will cause the coil to be driven upward (in the sketch). This will bring the beam closer to the oscillator coil in the grid circuit, thereby increasing the tube current and the bridge unbalance. The output current is divided between the feedback coil and a resistor. The value of the span resistor determines the amount of bridge unbalance current which is required to balance the torque created by the input thermocouple current.

It should be noted particularly that a similar circuit, Fig. 23·11, is not an electronic potentiometer, because there is no balancing of potentials. There is only the balancing of torques, so displacements can be converted into current analogs.

For certain operations, it is desirable to transmit a direct-current potential, particularly when the distance is relatively short. The differential pressure transmitter, Fig. 23·31, shows how the pressure can be converted into a potential which varies directly and linearly with the differences in the applied pressures. This potential is then converted into a direct-current signal for transmission.

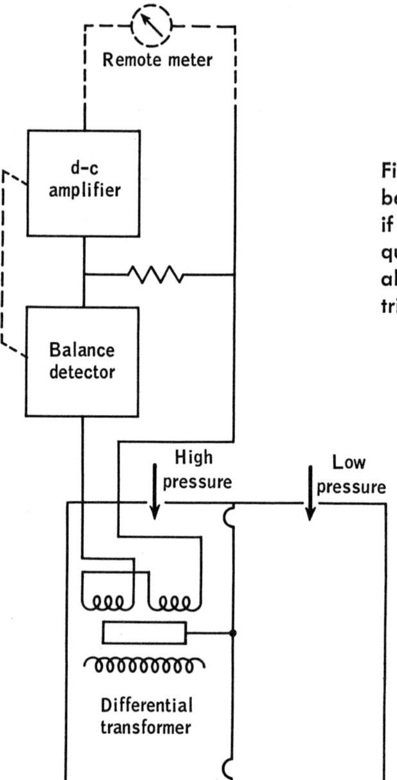

Fig. 23·31 Almost any variable can be converted into an analog current if remote registration of values is required. At the remote location it is also possible to convert from the electrical to the pressure equivalent.

With potential devices it is general to employ some means for measuring the variable which will not require the flow of current during the measuring period. Unless the value of the current is known and has been allowed for in the calibration, any current flow is likely to produce results which are too low.

If the variable is a current, particularly alternating current of the higher frequency values, it may be desirable to convert the current signal into a millivolt signal which can then be transmitted or handled by any one of the means previously discussed, Fig. 23·32. An electric-current signal then can be converted into a pneumatic equivalent, Fig. 23·33.

PNEUMATIC TELEMETERING

In most industrial plants, it is quite probable that the measured signals will be converted into pneumatic signals rather than electrical ones if it is feasible to use the pneumatic signal for transmission. In general, pneumatic transmission probably will be limited to distances under 1,000 ft. But within this distance the signal can be converted easily into pressure to actuate valves or other devices.

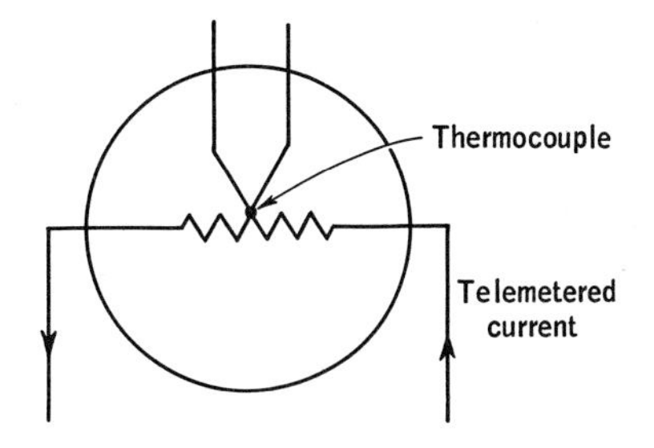

Fig. 23·32 High-frequency currents and complex waveforms can be converted into direct-current analogs when the heat which is created by the current is used to raise the temperature of the thermocouple junction.

The electrical signal may have a greater distance of transmission and speed, but to convert it into some useful function is still difficult. The pneumatic signal does possess the great advantage of being usable with an air-actuated valve motor, a very satisfactory situation.

The various pressure, temperature, level, and position converters to pneumatic signals have been considered in detail under the various types of measuring elements. They should be reviewed thoroughly for their principles of operation.

HYDRAULIC TELEMETERING

Hydraulic telemetering is in the same general class with pneumatic telemetering in that the distances involved are relatively short. Although very few converters of electrical or pneumatic signals to hydraulic signals have been considered, they employ only basic fundamental relationships. In general, the conversion from electrical values to hydraulic values will be more difficult than is the conversion to pneumatic ones

Fig. 23·33 Swartwout's power relay typifies electro-pneumatic transducers for operating pneumatic valves. At a remote point the transmitted current can be made to operate pneumatic valves or other devices. In this instance the outgoing air signal produces a condition of torque balance just equal to the torque produced by the electrical signal. Any change in the electrical signal immediately changes the torque and requires a change in the air output signal. (*Courtesy of Karp, Control Engineering.*)

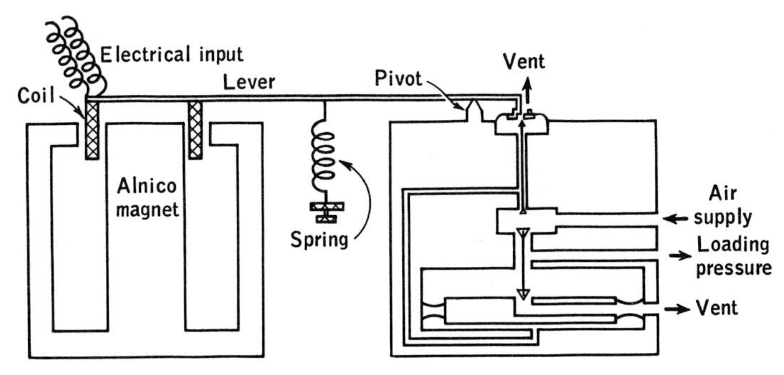

because of the greater forces required to position the hydraulic valves. Air is used with moderate pressures, and if a little escapes, it is not of terrific consequence. But if hydraulic fluids are to be retained under pressures of several hundreds of pounds per square inch, then the packing must be tight. Hydraulic actuators will therefore of necessity require appreciable forces for their actuation and control. The loss of hydraulic fluid is one problem that cannot be ignored; consequently, the systems must be tight.

PERFORMING MATHEMATICAL OPERATIONS

There is one phase of telemetering which is not concerned with remote transmission of data nearly so much as it is with the combining of results from two or more operations. Whether the amount of material stored in several places is to be totalized or whether the difference of two quantities is to be determined makes little difference, Figs. 23·34 to 23·36. These are operations which are not properly functions of the instruments which make the initial measurements. The results which are desired represent several operations, and the values of the variables therefore have to be transmitted to some central point for computational purposes. We find that this is one of the key problems in data logging and scanning: the measured variables generally must be evaluated and compared at some common point.

Fig. 23·34 By employing the correct circuit elements, many of the common mathematical steps can be performed.

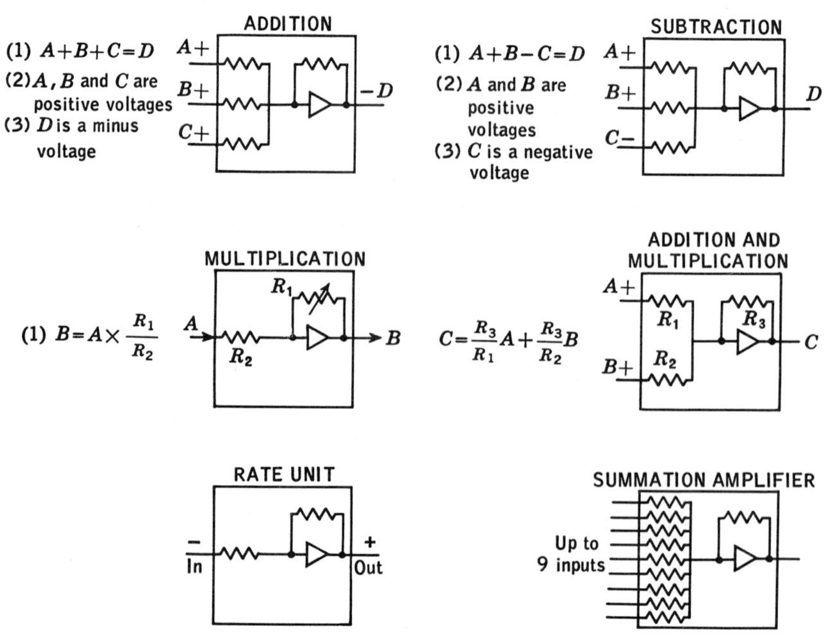

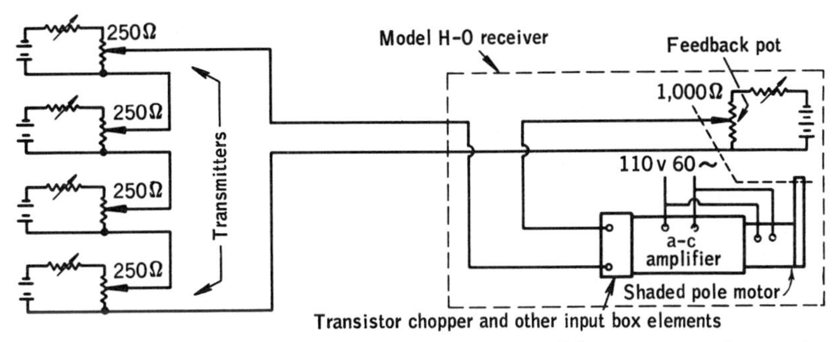

Fig. 23·35 Direct-current signal summation. Many different types of potentiometric circuits which have been studied may be used for getting the sum or difference of the quantities. (*The Hagan Corporation.*)

SUMMARY

1. The term "telemetering" is applied to any system in which the values of the variable being measured are indicated at a point remote from the sensing unit. According to this definition, a filled thermometer system with 10 ft of capillary to remove the meter from the point of reading would qualify. In general, however, the term suggests that some particular conversion or equipment has been inserted into the system in order to facilitate the distant reading.

2. With a greater and greater tendency to group instruments at some common point instead of having them scattered throughout a plant, there is a requirement that the instruments be able to transmit the appropriate values to the common center. The greater centralization of company operation and record keeping has

Fig. 23·36 A system of alternating-current signal summation. Alternating-current values may be added as readily as direct-current values if attention is paid to instantaneous polarity and phase angle. (*The Hagan Corporation.*)

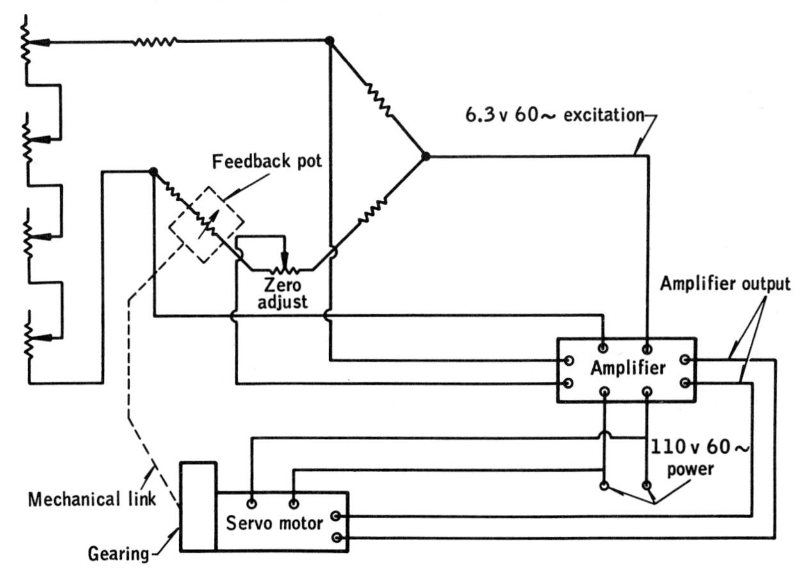

promoted the transmission of data to a central agency which may be hundreds of miles or more away from the point at which the measurement is to be made.

3. The greater use of data-logging systems and computers will require conversion of all variables into electrical equivalents and their transmission to a central agency. Greater dependence upon machine computations and control of operations will promote still further the requirements for telemetering in all its various phases.

4. The systems which are used for telemetering purposes vary in almost every conceivable respect. (*a*) They range from extreme simplicity to extreme complexity. (*b*) For high accuracy, they are more likely to employ pulse counts or frequency variation than magnitude counts. (*c*) Electrical systems predominate in terms of number of systems available. (*d*) Converting from pneumatic signals to electrical equivalents is relatively easy. The reverse transformation is also fairly standard. (*e*) Frequencies, the number of discrete pulses, or the duration of pulses can be transmitted long distances with less trouble than can other types of signals, particularly when there must be intervening amplifiers or relays.

QUESTIONS

23·1 What is meant by telemetering? Name several examples of telemetering.

23·2 What are some of the reasons for using telemetering apparatus?

23·3 In what industries is telemetering of particular significance? Where might it not be of any particular significance?

23·4 What is the basic difference between the methods of operation indicated by Figs. 12·1 and 12·2?

23·5 What would probably happen to the response of the ammeter if salt water were to be substituted for mercury in Fig. 23·3? What would probably be the response if distilled water were used?

23·6 What is the advantage of using the vector sum of two quantities to position a meter as compared with a magnitude response?

23·7 Are there any advantages in using a core to vary the voltage in a circuit as against using a potentiometer rheostat?

23·8 What are the limitations on the use of a system like the one shown in Fig. 23·5? Would you classify this as a bridge? If so, what type?

23·9 To what degree do bridge-type circuits of the balance type and potentiometers resemble each other in their test for balance?

23·10 What are the factors which would govern the errors in the receiver shown in Fig. 23·12?

23·11 What are the limitations in the use of the self-synchronous indicators?

23·12 If two selsyn units were connected with the transmitter connected to one phase of a three-phase power supply while the receiver was connected to another phase, what would you expect to happen in terms of accuracy of indication? What is the basic theory of the selsyn type of unit?

23·13 Suppose that a selsyn transmitter failed and there was no spare, but there was a spare receiver unit. Could you substitute the receiver with good results, if there was but one receiver? If there were several receivers?

23·14 How much force does the transmitter unit, Fig. 23·14, exert upon the sensing bellows *B*? What are your reasons for so believing?

23·15 Why should there be any difference in the response of the units represented in Figs. 23·14 and 23·15?

23·16 Of what significance is the off time in a device such as shown in Fig. 23·21? Is there anything to be gained by having the indicating element positioned from two directions?

Index